ELEMENTS

	IB	IIB	IIIA	IVA	VA	VIA	VIIA	
								Helium 4.003
			5 **B** Boron 10.81	6 **C** Carbon 12.01	7 **N** Nitrogen 14.01	8 **O** Oxygen 16.00	9 **F** Fluorine 19.00	10 **Ne** Neon 20.18
			13 **Al** Aluminum 26.98	14 **Si** Silicon 28.09	15 **P** Phosphorus 30.97	16 **S** Sulfur 32.06	17 **Cl** Chlorine 35.45	18 **Ar** Argon 39.95
28 **Ni** Nickel 58.71	29 **Cu** Copper 63.54	30 **Zn** Zinc 65.37	31 **Ga** Gallium 69.72	32 **Ge** Germanium 72.59	33 **As** Arsenic 74.92	34 **Se** Selenium 78.96	35 **Br** Bromine 79.91	36 **Kr** Krypton 83.80
46 **Pd** Palladium 106.4	47 **Ag** Silver 107.9	48 **Cd** Cadmium 112.4	49 **In** Indium 114.8	50 **Sn** Tin 118.7	51 **Sb** Antimony 121.8	52 **Te** Tellurium 127.6	53 **I** Iodine 126.9	54 **Xe** Xenon 131.3
78 **Pt** Platinum 195.1	79 **Au** Gold 197.0	80 **Hg** Mercury 200.6	81 **Tl** Thallium 204.4	82 **Pb** Lead 207.2	83 **Bi** Bismuth 209.0	84 **Po** Polonium (210)*	85 **At** Astatine (210)*	86 **Rn** Radon (222)*

63 **Eu** Europium 152.0	64 **Gd** Gadolinium 157.2	65 **Tb** Terbium 158.9	66 **Dy** Dysprosium 162.5	67 **Ho** Holmium 164.9	68 **Er** Erbium 167.3	69 **Tm** Thulium 168.9	70 **Yb** Ytterbium 173.0	71 **Lu** Lutetium 175.0
95 **Am** Americium (243)*	96 **Cm** Curium (247)*	97 **Bk** Berkelium (247)*	98 **Cf** Californium (251)*	99 **Es** Einsteinium (254)*	100 **Fm** Fermium (253)*	101 **Md** Mendelevium (256)*	102 **No** Nobelium (254)*	103 **Lr** Lawrencium (257)*

Fundamentals of Chemistry

Fundamentals of

FRED H. REDMORE

Chemistry Department
Highland Community College
Freeport, Illinois

Chemistry

Prentice-Hall, Inc., Englewood Cliffs, New Jersey 07632

Library of Congress Cataloging in Publication Data

Redmore, Fred H.
Fundamentals of chemistry.

Includes bibliographies and index.
1. Chemistry. I. Title.
QD31.2.R42 540 78-14294
ISBN 0-13-335158-0

FUNDAMENTALS OF CHEMISTRY
Fred H. Redmore

Printed in the United States of America

10 9 8 7 6 5 4 3 2

Editorial/production supervision by Karen J. Clemments
Cover/interior design by Mark A. Binn
Manufacturing buyer: Phil Galea

Cover photograph is entitled "Crystallizations"; photographed by Phillip Harrington; supplied by Peter Arnold.

PRENTICE-HALL INTERNATIONAL, INC., *London*
PRENTICE-HALL OF AUSTRALIA PTY. LIMITED, *Sydney*
PRENTICE-HALL OF CANADA, LTD., *Toronto*
PRENTICE-HALL OF INDIA PRIVATE LIMITED, *New Delhi*
PRENTICE-HALL OF JAPAN, INC., *Tokyo*
PRENTICE-HALL OF SOUTHEAST ASIA PTE. LTD., *Singapore*
WHITEHALL BOOKS LIMITED, *Wellington, New Zealand*

Contents

Foreword xv

Preface xvii

2 One
Introduction to the Study of Chemistry

Objectives

1 Science and the Scientific Method 2 Chemistry 3 Brief History of Chemistry Summary Exercises

10 Two
Measurement and Mathematical Operations

Objectives

1 Measurement of Matter 2 The Metric System and the International System of Units 3 Conversion Factors. The Factor Unit Method of Problem Solving 4 Temperature 5 Heat Energy—Specific Heat 6 Density and Specific Gravity Summary Exercises Problems

36 Three
Matter

Objectives

1 Classification of Matter—Atoms, Elements, Compounds, and Mixtures 2 Physical and Chemical Properties and Changes 3 Energy-Mass Relations 4 Relative Atomic Masses—Atomic Mass Units 5 Subatomic Particles 6 Basic Structure of Atoms. Atomic Number 7 Isotopes and the Atomic Weight Scale 8 Types of Compounds—Molecules, Formula Units, and Ions 9 Formula or Molecular Weight 10 The Mole 11 Introduction to the Periodic Table of the Elements: Metals and Nonmetals 12 Solutions Summary Exercises Problems

68 Four
The Structure of the Atom

Objectives

1 Historical Development of Theories of Atomic Structure 2 Arrangement of Electrons in Energy Levels 3 Arrangement of the Electrons in Sublevels 4 Orbitals 5 Occupancy of Sublevels and Orbitals 6 Shapes of Orbitals 7 Electronic Structure and the Periodic Table 8 Quantum Numbers Summary Exercises Problems

102 **Five**

The Periodic Table

Objectives

1 Development of the Periodic Table 2 Periods, Groups, and Families 3 Types of Elements 4 Valence Electrons 5 Differentiating Electron 6 Periodic Variations 7 Use of the Periodic Table in Predicting Properties 8 Oxidation Number (Oxidation State) Summary Exercises Problems

126 **Six**

Nomenclature of Inorganic Compounds

Objectives

1 Binary Compounds Containing Two Nonmetals 2 Binary Compounds Containing a Metal and a Nonmetal 3 Binary Acids 4 Ternary and Higher Compounds Summary Problems

142 **Seven**

Chemical Bonding

Objectives

1 Electrovalent or Ionic Bonding 2 Covalent Bonding 3 Lewis Structures (Electron Dot Formulas) 4 Resonance Forms of Molecules 5 Types of Covalent Bonds—Sigma (σ) and Pi (π) 6 Geometry of Molecules—The Valence Shell Electron Pair Repulsion Theory (VSEPR) 7 Hybrid Orbitals 8 Polarity of Molecules 9 Describing and Predicting the Bonding in Covalent Species 10 Molecular Orbital Theory Summary Exercises Problems

180 Eight

Chemical Equations

Objectives

1 Terms and Symbols Used in Equations 2 Writing and Balancing Simple Chemical Equations 3 Completing Chemical Equations. The Five Simple Types of Chemical Reactions 4 Combination Reactions 5 Decomposition Reactions 6 Replacement (Displacement) Reactions 7 Metathesis (Double Displacement) Reactions 8 Neutralization Reactions Summary Exercises Problems

198 Nine

Stoichiometry-Calculations Based on Formulas and Equations

Objectives

1 Percent Composition of Compounds 2 Relationships from the Formula of a Compound 3 Empirical (Simplest) and Molecular Formulas 4 Calculations Based on Balanced Equations Summary Exercises Problems

218 Ten

Thermochemistry

Objectives

1 Thermochemical Equations 2 Enthalpy (ΔH) 3 Hess's Law 4 Calorimetry 5 First Law of Thermodynamics 6 Second Law of Thermodynamics Summary Exercises Problems

236 Eleven
Gases

Objectives

1 General Properties of Gases and Comparison to Liquids and Solids 2 The Kinetic Theory 3 Pressure of Gases 4 Variables in Gases. Standard Temperature and Pressure (STP) 5 The Gas Laws—Relationships Among Volume, Temperature, and Pressure of Gases 6 Dalton's Law of Partial Pressures 7 Avogardo's Hypothesis and the Molar Volume of a Gas 8 Gas Law Relationships Involving Moles 9 Ideal Gas Law (Equation of State) 10 Graham's Law of Diffusion Summary Exercises Problems

268 Twelve
Liquids and Solids and Changes of State

Objectives

1 Attractive Forces in Liquids and Solids 2 The Liquid State 3 The Solid State 4 Heat Transformations in Changing from One State to Another 5 Phase Diagrams. Critical Temperature and Pressure 6 Water Summary Exercises Problems

292 Thirteen
Solutions and Colloids

Objectives

1 Solutions 2 Factors that Affect Solubility 3 Factors that Affect Rate of Solution 4 Saturated, Unsaturated, and Supersaturated Solutions 5 Concentration of Solutions 6 Solution Stoichiometry 7 Colligative Properties of Solutions 8 Osmosis and Osmotic Pressure 9 Colloids Summary Exercises Problems

324 Fourteen

Acids, Bases, and Ionic Equations

Objectives

1 Definitions and Properties of Acids and Bases 2 Equivalent Weight and Normality 3 Ionic Equations Summary Exercises Problems

348 Fifteen

Kinetics

Objectives

1 Reaction Rates and Rate Laws 2 Order of Reaction 3 Collision Theory of Reaction Rates 4 Factors That Control Reaction Rates 5 Reaction Mechanisms Summary Exercises Problems

360 Sixteen

Introduction to Chemical Equilibrium

Objectives

1 Reversible Reactions and Chemical Equilibrium 2 Equilibrium Constants 3 Factors that Cause A Shift in Equilibrium 4 Calculations Involving Equilibrium Constants Summary Exercises Problems

378 Seventeen

Equilibria in Solution I—Solubility, Ionization of Weak Acids and Bases, and pH

Objectives

1 Solubility Product Equilibria   2 Equilibria Involving Weak Acids and Bases   3 Ionization of Water   4 pH and pOH   5 Buffer Solutions
Summary   Exercises   Problems

410 Eighteen

Equilibria in Solution II—Polyprotic Acids and Hydrolysis of Salts

Objectives

1 Polyprotic Acids—Ionization of Hydrosulfuric Acid and Precipitation of Sulfides   2 Hydrolysis of Salts   Summary   Exercises   Problems

430 Nineteen

Balancing Oxidation-Reduction Equations

Objectives

1 Definition of Oxidation and Reduction. Oxidizing and Reducing Agents.
2 Balancing Oxidation-Reduction Equations By the Oxidation Number Method   3 Half-Reactions. Balancing Equations by the Ion-Election Method (Method of Half-Reactions).   Summary   Exercises   Problems

444 **Twenty**

Electrochemistry

Objectives

1 Electrolytic Cells 2 Quantitative Relationships in Electrolytic Cells 3 Voltaic or Galvanic Cells 4 Standard Electrode Potentials 5 The Nernst Equation 6 Equilibrium Constants From Redox Equations 7 Practical Cells Summary Exercises Problems

468 **Twenty One**

The Periodic Table-Periodic Variations

Objectives

1 Review of Electronic Structure and Related Topics 2 Periodic Variations in Acid–Base Strength of Compounds 3 Metals Summary Exercises Problems

482 **Twenty Two**

Properties of the Representative Elements

Objectives

1 Hydrogen 2 Group IA—The Alkali Metal Group: Lithium (Li), Sodium (Na), Potassium (K), Rubidium (Rb), Cesium (Cs), and Francium (Fr) 3 Group IIA—The Alkaline Earth Metals: Beryllium (Be), Magnesium (Mg), Calcium (Ca), Strontium (Sr), Barium (Ba), and Radium (Ra) 4 Group IIIA—Boron (B), Aluminum (Al), Gallium (Ga), Indium (In), and Thalium (Tl) 5 Group IVA—Carbon (C), Silicon (Si), Germanium (Ge), Tin (Sn), and Lead (Pb) 6 Group VA—Nitrogen (N), Phosphorus (P), Arsenic (As), Antimony (Sb), and Bismuth (Bi) 7 Group VIA—Oxygen (O), Sulfur (S), Selenium (Se), Tellurium (Te), and Polonium (Po) 8 Group VIIA—The Halogens: Fluorine (F), Chlorine (Cl), Bromine (Br), Iodine (I), and Astatine (At) Summary Exercises Problems

512 Twenty Three

Properties of the Transition Elements

Objectives

1 Introduction 2 Group IIIB—The Scandium Subgroup 3 Group IVB—The Titanium Subgroup: Ti, Zr, Hf 4 Group VB—The Vanadium Subgroup: V, Nb, Ta 5 Group VIB—The Chromium Subgroup: Cr, Mo, W 6 Group VIIB—The Manganese Subgroup: Mn, Tc, Re 7 Group VIII—The Iron Triad: Fe, Co, Ni 8 Group VIII—The Platinum Metals: Ru, Rh, Pd, Os, Ir, Pt 9 Group IB—The Copper Subgroup: Cu, Ag, Au 10 Group IIB—The Zinc Subgroup: Zn, Cd, Hg Summary Exercises

524 Twenty Four

Coordination Compounds

Objectives

1 Introduction 2 Geometry of Coordination Compounds 3 Experimental Determination of The Formula of a Complex 4 Theories of Bonding 5 Nomenclature Summary Exercises Problems

536 Twenty Five

Introduction to Nuclear Chemistry

Objectives

1 Nuclear Radiations 2 Nuclear Equations 3 Nuclear Stability and Natural Radioactivity 4 Half-Life 5 Artificial Radioactivity. Transmutation of Elements and Preparation of New Elements 6 Nuclear Fission 7 Nuclear Fusion 8 Detection of Radiation 9 Energy Changes in Nuclear Reactions 10 Applications of Nuclear Reactions and Radioactive Isotopes Summary Exercises Problems

558 Twenty Six

Introduction to Organic Chemistry

Objectives

1 General Principles 2 Hydrocarbons 3 Alkanes 4 Alkenes 5 Alkynes 6 Substituted Hydrocarbons 7 Aromatic Compounds 8 Alcohols and Phenols 9 Ethers 10 Carboxylic Acids and Their Derivatives 11 Aldehydes and Ketones 12 Amines Summary Exercises Problems

596 Twenty Seven

Introduction to Biochemistry

Objectives

1 Carbohydrates 2 Amino Acids and Proteins 3 Lipids—Fats and Oils 4 Enzymes 5 Nucleic Acids 6 Metabolism of Carbohydrates 7 Metabolism of Proteins 8 Metabolism of Lipids 9 Nutrition 10 Some Important Drugs and Their Action Summary Exercises

620 Appendices

I Exponents, Scientific Notation, Significant Digits, and Logarithms 620 II Some Conversion Factors 631 III SI Units 633 IV Thermodynamic Values—Enthalpy (ΔH), Free Energy (ΔG), and Entropy (S) 635 V Vapor Pressure of Water at Various Temperatures 637 VI Equilibrium Constants (at 20°C) 639 VII Logarithms of Numbers 644

647 Glossary

657 Answers to First Half of Problems

Index 689

Foreword

Those of you who are familiar with *Basic Chemistry* by Seese and Daub will notice some similarities between that book and this one. This is understandable since *Fundamentals of Chemistry* was developed out of an attempt to expand and upgrade *Basic Chemistry* into a general chemistry text. Originally, it was intended that *Fundamentals of Chemistry* would simply be a higher level version of *Basic Chemistry.* However, this was not feasible since the books are intended for different types of students; therefore, this current text is a completely independent and separate book.

Many of the positive features of *Basic Chemistry* are found in this text. For example, the text abounds with carefully worked-out problem examples, using the dimensional analysis (*factor-unit* or *factor-label*) method with careful attention paid to units and significant digits. In this text, as in *Basic Chemistry,* there are separate chapters on writing equations, balancing equations, and nomenclature of inorganic compounds. Too often in general chemistry texts these important topics are treated rather superficially or are relegated to the appendix, where many students never find them. Also, the reading level in these two books is approximately the same.

There are also notable differences between this text and *Basic Chemistry.* For example, the cartoon-type drawings, which have proven to be helpful and appropriate for an introductory chemistry text, are not used in this book. Also, *Fundamentals of Chemistry* includes more end-of-chapter problems, including many drill problems of varying degrees of difficulty. Finally, as would be expected, topics are treated in greater depth in *Fundamentals of Chemistry,* and more topics are included. However, it must be emphasized that this text stresses the fundamental principles of chemistry and does not put the emphasis on more advanced topics.

We hope that those of you who have found *Basic Chemistry* to be a suitable book for introductory chemistry will find this book equally useful for a general chemistry course. Those of you who have found *Basic Chemistry* to be somewhat elementary for your courses should find this book more suitable.

We have enjoyed working with Fred Redmore and hope that you share our enthusiasm for Redmore's *Fundamentals of Chemistry.*

WILLIAM S. SEESE
Casper College
Casper, Wyoming

GUIDO H. DAUB
University of New Mexico
Albuquerque, New Mexico

Preface

Since this book was written for beginning students in a general chemistry course, no previous background in chemistry is assumed for students taking this course. The emphasis (as suggested by the title) is on the fundamental principles of chemistry. Once the student has mastered these fundamentals, additional topics can be introduced by the instructor as time and ability of students allows. There are some topics included that may prove to be rather difficult for some classes. These have been indicated as being optional, and can be omitted without loss of continuity. However, all topics necessary for an understanding of the fundamental principles of chemistry have been included. Therefore, this should be suitable for a variety of classes in which students have varied abilities, interests, and goals.

The goal in writing this book has been to be as concise as possible, and yet to include all necessary topics in sufficient detail to explain them clearly. Many worked-out problem examples are included, which should prove helpful to the student. At the end of each chapter there are many problems, which are necessary for the student to learn the material thoroughly. It is not sufficient to watch the instructor solve problems in class, or to follow through the problem examples in the book. In order to learn the material, it is necessary to answer questions and solve problems on your own. These problems are arranged according to topics, and are generally of increasing difficulty within each group. Answers to the first half of all problems are included in the appendix. For example, where there are ten parts to a particular problem (a–j), the answer to the first five (a–e) are included; or when there are two similar problems (such as 19 and 20 in Chapter 2) the answer to the first one is given. This allows the student to work problems and to see immediately if the solution is correct, and also provides problems without answers that the instructor may want to assign to determine if the students are understanding the material.

Some other features of this book that should prove helpful to both students and instructors are the objectives, summaries, and glossary. At the beginning of each chapter is a list of general objectives for the chapter to give the student an overview of what will be covered. The instructor may wish to omit some of these that are not essential, and may want to add more. Each chapter ends with a summary of the material covered in the chapter, and a list of specific things that the student should be able to do upon completion of the chapter (with reference to specific problems that the student can do to determine if the material has been mastered). And finally a glossary of important terms is included in the appendix. All terms that are included in the glossary are printed in color in the text.

In all problem examples, careful attention has been paid to units and significant digits. The dimensional analysis (or factor-unit) method of problem solving has been used consistently throughout the book. In the calculations, atomic weights are generally rounded off to correspond to the significant digits in other parts of the problem. The student should learn when exact values are necessary, and when numbers can be rounded off. This is especially important in this age of readily available calculators for student use; often an answer is given on a calculator with several meaningless extra numbers included. Some SI units are used, and the SI system is explained in Chapter 2. However, for the most part, conventional metric units are used, such as the liter, calorie, mm Hg or torr, and so on.

When using this book, you, the student, should read the objectives and skim through the chapter before class. After class, read the chapter carefully, and make sure that each problem example is understood. If there are things in the book or from the lecture that are not clear, see the instructor or someone who can explain it. Do not wait to see if it gets cleared up later; it is essential that each topic is understood before going on to the next. The Exercises at the end of the chapter should help you determine if the material was understood; all terms should be defined or explained in your own words, not just by the memorization of a definition. And finally as many of the problems as necessary should be worked to check mastery of the material.

The Study Guide written to accompany this text should also prove helpful to the student. This gives an outline of all material covered in the text, and includes some additional exercises, as well as a sample quiz, with answers, for each chapter. There is also an Instruction Manual, a Solutions Manual, and a Laboratory Manual (written by Fred Redmore and C. H. Sorum) to accompany the text.

Many people have assisted in the writing of this text. The students who have used preliminary editions of the book at Highland Community College have made many helpful suggestions and comments. Several secretaries have helped with the typing of the manuscript through various stages; I am especially grateful to Betty Parker who many times cheerfully worked overtime to help meet deadlines. The staff at Prentice-Hall have been very helpful and patient with an inexperienced author. And my wife not only gave me encouragement, but also advice on proper grammar.

The following people have reviewed various parts of the text and provided many helpful suggestions: Dr. Wayne E. Dunbar of Winona State College, Professor Henry Heikkinen of the University of Maryland, Professor Floyd W. Kelly of Casper College, Professor Robert J. Munn of the University of Maryland, Dr. Malcolm Rundell of San Jacinto College, and Dr. Lawrence E. Wilkins.

Any suggestions or comments on this book, from students and instructors, would be welcome.

FRED H. REDMORE
Highland Community College
Freeport, Illinois 61032

Fundamentals of Chemistry

Objectives

1. To begin to understand what the study of chemistry involves and why it is important in your life.
2. To understand what science and the scientific method are and to be able to list the steps involved in the development of scientific theories and laws.
3. To learn what chemistry is and to be able to list the various branches into which the field of chemistry is generally divided.
4. To begin to understand how the field of chemistry has developed through the years by learning some of the major developments and by being able to show why these developments were important.

One

Introduction to the Study of Chemistry

Why study chemistry? This is a question asked by students for years. For many students the answer is that it is required for a degree in biology, agriculture, engineering, nursing, and so forth. But this is not a very satisfactory answer. Most students want to know why it is required for a degree in these various fields and why science and, more specifically, chemistry are important to them either in their chosen field or in their lives as educated and informed citizens.

An understanding of the basic principles of chemistry is important in such areas as biological systems or materials used in engineering, agriculture, and nursing. But even beyond that, the food you eat, the clothes you wear, the medicine you take, and most other aspects of your life are directly affected by the substances that you will study about in this course. If you will look for relationships between what you are studying and your life and try to keep in mind that there is a reason for what you are studying, you should find chemistry interesting. Of course, it must be remembered that some aspects of the study of chemistry may not come easily at first and will take time to master. Science and especially chemistry should help in developing *logical* reasoning, which is applicable to any field of study.

Another thing to keep in mind, which will make your study of chemistry much easier, is that it is essential for you to understand each topic before you go on to the next one. This should also help you to understand that even though there are topics that do not seem to have any direct application to anything that you will ever use, they are necessary for topics that will follow.

In this chapter we shall consider the definition of science and the scientific method; more specifically, we shall consider what chemistry is. Then we shall take a very brief look at the historical development of chemistry.

1-1 Science and the Scientific Method

Science is defined as knowledge that is obtained by study and experimentation; it is also concerned with the observation and classification of facts and the establishment of general theories and laws that can be verified by further experiments. The gathering of scientific knowledge is generally done by a process referred to as the *scientific method.* This consists of collecting facts and data by the observation of natural events. These facts are then correlated into a tentative theory called a *hypothesis.* This hypothesis is then tested to either show that it is correct or find where it is wrong. If after extensive experimentation the hypothesis is shown to be true, a *scientific law* is proposed. This scientific law is thus a statement of the behavior of nature.

Much of science is based on theories, such as the atomic theory of matter. Usually these theories are revised and refined as greater knowledge of nature is obtained or as more sophisticated methods of experimentation and mathematical calculation are devised.

Generally science is divided into the *natural* sciences and the *social* sciences. The natural sciences include the areas of physics, chemistry, astronomy, geology, and the biological sciences. Mathematics is used in the natural sciences to describe natural phenomena in precise terms. Very often scientific laws are expressed mathematically. The social sciences include the areas of history, political science, sociology, and economics.

1-2 Chemistry

Chemistry is that branch of science concerned with the study of matter (which, as we shall see in Chapter 3, is simply all the material the universe is composed of) and the changes matter undergoes. This is important not only to those of you who will be working in careers where a knowledge of chemistry is essential but also to everyone who is concerned about the world around him. For example, consider some of the topics that are of concern today—the environment, energy sources, food production, and medicine. These are the things about which chemistry is concerned.

In order to understand many of the topics mentioned here, however, it is first necessary to learn some fundamental principles. Therefore, in this text we shall begin with a study of the operations of measurement and the mathematics necessary to make precise, quantitative descriptions of the behavior of matter. Then we shall consider the basic properties, changes, and states of the matter that surround us.

We shall also consider some specific areas that are based on the fundamental principles studied earlier. This will include topics such as acids and bases, rates at which changes (reactions) take place, nuclear chemistry, and some important substances and their properties.

Branches of Chemistry

Chemistry is generally divided into five branches. There is much overlap among the branches, however, and the modern trend is to emphasize basic topics such as energy changes and chemical reactions rather than attempting to divide the study into these branches. In a general chemistry course, along with the fundamental topics that are basic to all branches of chemistry, you will at least touch on all of these branches. The five branches are as follows:

1. **Organic chemistry** is the study of the compounds of carbon. We shall consider this briefly in Chapter 26, as well as some organic compounds in discussing bonding and other topics.
2. **Inorganic chemistry** is the study of all elements and compounds other than organic compounds. This is what most of our study will concern.
3. **Physical chemistry** is the study of the theoretical aspects of the structure and changes of matter, such as how and why bonds are formed and energy changes take place. Some of the topics that we shall consider, at least in an introductory manner, are bonding, energy changes, and reaction rates.
4. **Analytical chemistry** is the study of what is present (qualitative analysis) and how much is present (quantitative analysis). In your laboratory work you will do some of each of these types of analysis. In this text these topics will be discussed in the section on chemical equilibrium.
5. **Biochemistry** is the study of the chemistry of biologically important elements and compounds. We shall consider some of these in Chapter 27.

1-3 Brief History of Chemistry

The beginning of chemistry could be considered as going back to prehistoric man, who used fire to do crude metallurgy and to make pottery and bricks. One of the first metals to be used was gold, probably because it often occurred as a free (uncombined) metal with a color that could easily be detected in the earth. Copper and silver were also used in very early times, around 4000 B.C. Glass and various dyes were first used around 1500 B.C.

The beginning of chemistry as a science could probably be set at about 400 B.C. when the theory was proposed that everything is composed of four elements: earth, air, fire, and water. Aristotle (384–322 B.C.) summarized some earlier theories and proposed that there were also four fundamental properties—hot, cold, wet, and dry. At the beginning of the Christian Era, in Egypt, there was a combination of the ancient arts of the Egyptians and the philosophical speculations of the Greeks to form the beginning of a body of chemical knowledge. The first books of chemistry were written in Egypt about 300 A.D. The term *chemistry* meant *the Egyptian art.*

During this time in China, *alchemy* emerged as a science concerned with the transmutation (changing) of one element into another. For the most part, its proponents hoped to change base metals, such as iron and copper, into gold. Alchemy probably had its beginnings in China about 140 B.C. and reached Europe, through Spain, during the twelfth century.

In Europe in the sixteenth and seventeenth centuries, chemistry was mainly concerned with the application to medicine. In fact, the first chemistry teachers were medical doctors. During this period the four-element theory was questioned, the study of gases was begun, and quantitative experiments were undertaken for the first time. The first textbook of chemistry was written in 1597. In the seventeenth century, Robert Boyle (1627–1691) did studies and found that when a metal is heated in air, the mass increases. He also discovered the relationship between volume and pressure of a gas and was the first to work with gases; up until then it was believed that a gas could not be contained in a vessel. In 1661 Boyle wrote the book *The Skeptical Chemist* in which he criticized the basic ideas of alchemy.

Early in the eighteenth century the Phlogiston Theory was proposed by George Ernst Stahl (1660–1734), based on earlier theories of Johann Becher (1635–1682). Phlogiston was described as a substance in combustible material that is given off when the material burns. This theory was probably the first great principle in chemistry. It persisted for about 100 years and was the center of much controversy during this time.

During the last half of the eighteenth century, much work was done with gases, especially by Joseph Black, Henry Cavendish, and Joseph Priestly in England and Carl Scheele in Sweden. Oxygen was probably discovered by Scheele between 1772 and 1774. His work was not published until 1777, however, and for this reason Priestly is usually credited with the discovery in 1774.

Joseph Priestly (1733–1804) was a Unitarian minister who did much scientific research, especially on gases such as air, carbon dioxide, and oxygen. Due to his unorthodox religious views and liberal political philosophy, about which he was very outspoken, he made many enemies and in 1794 was forced to leave England and come to America. He settled in Pennsylvania where he continued his scientific experimentations. Although his political views were very liberal, scientifically he was very conservative. Even after his discovery of oxygen he still believed in the Phlogiston Theory.

Antoine Laurant Lavoisier (1743–1794), a French chemist, is often called the father of modern chemistry. Even though he was not a great experimental chemist, he used others' results to propose many theories. He did do very careful quantitative experiments, however, to verify the work of others. His theory of combustion, in which he proposed that oxygen is necessary for combustion and that the substance being burned or the metal being heated combines with the oxygen, was largely responsible for the overthrow of the Phlogiston Theory. Among his other contributions he showed that water is composed of hydrogen and oxygen, proposed the theory of the indestructibility of matter, gave a clear definition of an element, proposed a system of chemical nomenclature, and showed that air contains oxygen.

Even though the theory of atoms was first proposed by the early Greek philosophers, it wasn't until the beginning of the nineteenth century that experimental evidence was used to show that matter must be composed of atoms. This was done by John Dalton (1766–1844), about 1803. This was followed by the development of the Periodic Law and the Periodic Table in the last half of the nineteenth century and development of theories of the structure of the atom and the formation of compounds in the twentieth century. Along with this there have been, and continues to be, discoveries made not only in the theoretical aspects of the nature of matter but also in the applications of these theories in the various branches of chemistry. This includes the development of new products, as well as a better understanding of living matter. Currently much research and development is being carried out in the areas of energy sources, medicine, and food and agriculture, to name just a few.

SUMMARY

1. *Science* is an organized system of knowledge that is based on experimentation.
2. The *scientific method* consists of the observation and classification of facts, based on experimental evidence. These observations are correlated into a tentative *theory,* called a *hypothesis;* this is often tested further and if found to be correct a *scientific law* is proposed, which gives a statement of the behavior of nature.
3. *Chemistry* is concerned with the study of *matter* and the changes that it undergoes. The study of chemistry is generally divided into the branches of *organic, inorganic, physical, analytical (qualitative* and *quantitative),* and *biochemistry.*
4. The history of chemistry can be classified approximately into the following areas:
   (a) Prehistoric man—used metals and made pottery and bricks.
   (b) 400 B.C.—beginning of chemistry as a science; four-element theory proposed.
   (c) First century—combination of the arts of the Egyptians and the Greeks; first book of chemistry written in Egypt; also the beginning of alchemy in China.
   (d) Twelfth century—alchemy reached Europe, through Spain.
   (e) Sixteenth and seventeenth centuries—chemistry applied to medicine; study of gases begun; quantitative experiments begun; first textbook of chemistry written (1597); Boyle studied gases and criticized the basic ideas of alchemy in his book *The Skeptical Chemist.*

(f) Eighteenth century—Phlogiston Theory proposed; oxygen discovered (1774); careful quantitative work by Lavoisier generally described as the beginning of modern chemistry.

(g) Nineteenth and twentieth centuries—atomic theory (Dalton), periodic law and theories of the structure of the atom led to modern theories of the nature of matter; further development and applications of these theories is currently taking place.

EXERCISES

1. Define or explain the following terms:
   (a) science
   (b) experimentation
   (c) scientific method
   (d) hypothesis
   (e) theory
   (f) scientific law
   (g) chemistry
   (h) organic chemistry
   (i) inorganic chemistry
   (j) analytical chemistry
   (k) physical chemistry
   (l) biochemistry
2. Distinguish between
   (a) hypothesis and theory
   (b) theory and law
   (c) inorganic and organic chemistry
   (d) qualitative and quantitative analysis
3. Explain, in terms of the scientific method, the three steps involved in the development of a scientific law.
4. List the major contributions to the field of chemistry made by each of the following:
   (a) Ancient Greeks
   (b) Aristotle
   (c) Egyptians
   (d) Robert Boyle
   (e) Stahl and Becher
   (f) Carl Scheele
   (g) Joseph Priestly
   (h) Antoine Lavoisier
   (i) John Dalton
   (j) Nineteenth-century chemists

REFERENCES

ASIMOV, ISAAC, *A Short History of Chemistry.* New York: Doubleday and Co., 1965.

CASSEBAUM, N., and J. A. SCHUFLE, "Scheele's Priority for the Discovery of Oxygen," *J. Chem. Educ.*, 1975, **52,** p. 442.

DALTON, JOHN, *A New System of Chemical Philosophy.* New York: The Citadel Press, 1964.

HALL, MARIE BOAS, "Robert Boyle," *Sci. Am.,* Aug. 1967, **217,** p. 96.

IHDE, AARON, *The Development of Modern Chemistry.* New York: Harper and Row, 1964.

PARTINGTON, J. R., *A Short History of Chemistry,* 3rd ed. New York: Harper and Brothers, 1957.

Objectives

1. To understand what the terms *mass* and *weight* represent and what the difference is between them.
2. To understand how mass (weight), length, and volume are measured and what the basic units are for each of these in the metric system.
3. To learn the prefixes used in the metric system and the International System of Units (SI).
4. To learn the basic units and some of the more common derived units used in the International System of Units (SI units).
5. To learn the common English-metric unit equivalents for mass, length, and volume (pounds to grams, inches to centimeters, and quarts to liters).
6. To learn to use *conversion factors* in problem solving.
7. To learn to do conversions within the metric system and between the English and the metric systems by using conversion factors (or what is often referred to as the *factor unit* or *dimensional analysis* method).
8. To learn to do conversions between the Fahrenheit (F), Celsius (C), and Kelvin (K) temperature scales.
9. To understand what the energy unit calorie represents and how this is related to the energy unit joule (J).
10. To understand what the specific heat of a substance represents and how it is measured experimentally and used in calculations.
11. To understand what density and specific gravity are and to learn to do calculations using these quantities.

Two

Measurement and Mathematical Operations

In this chapter we shall consider the units used to measure matter and some mathematical operations used in making and using measurements. We shall first consider the metric system and relationships between the English and the metric systems. Then we shall consider methods of doing conversions between the two systems and within the metric system. And finally we shall consider the measurement and use of temperature, heat, and density.

2-1 Measurement of matter

Mass and Weight

Matter is defined as anything that has mass and takes up space. *Mass* is the quantity of matter in a particular sample of matter. The mass of a body is constant and does not change, regardless of where it is measured. The *weight* of a body, however, is the gravitational force of attraction between the body's mass and the mass of the planet on which it is weighed. Thus, the weight of a body varies, depending on where it is being weighed, whereas the mass does not. For example, on the moon the gravitational force of attraction is considerably less than it is on the earth and, therefore, a body will weigh less on the moon than it does on the earth. The mass of the body would be the same on the moon, however, as it is on the earth.

The terms *mass* and *weight* are unfortunately often used interchangeably; however, you should recognize the difference in the terms.

The mass of a body can be measured on a *balance.* Several types of balances are shown in Fig. 2-1.

In the balance shown in Fig. 2-1(a) an object to be weighed is placed on the left-hand pan and weights are placed on the right-hand pan until the two balance. In the triple beam balance shown in Fig. 2-1(b) the same general principle applies, but weights are moved along the beams until they balance the object on the pan.

In the top loading balance shown in Fig. 2-1(c) and the single pan balance shown in Fig. 2-1(d) the weights are inside the balance and are added or removed by adjusting the knobs on the outside of the balance.

Thus, on a balance the mass of the object is obtained by comparing the object to weights of known mass. On a scale, a spring is stretched to give the weight of the object placed on the pan. This is calibrated to give a particular reading under certain conditions of gravitational attraction and will thus give a different reading on the moon than it would on the earth, as mentioned earlier. In chemistry, the balance is used exclusively and therefore the mass of the object is measured.

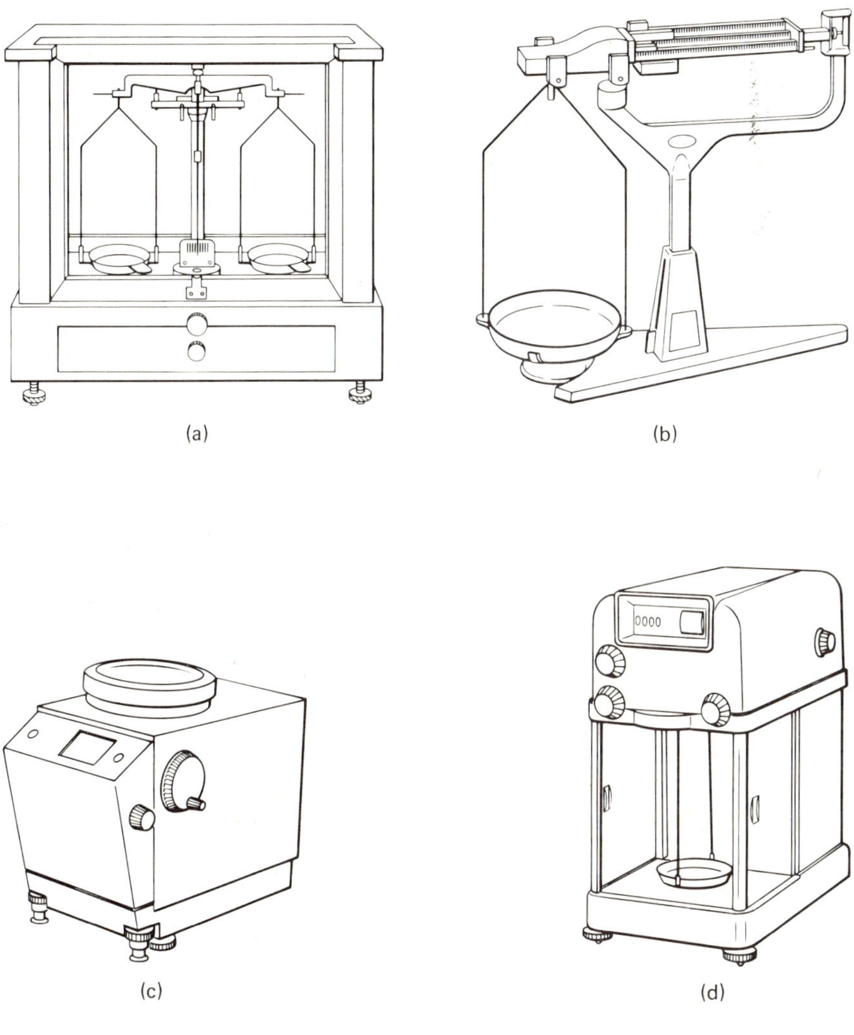

Figure 2-1
Balances. (a) Double pan analytical balance (b) Triple beam balance (c) Top loading balance (d) Single pan analytical balance

Length and Volume

Length is simply measured with a ruler, divided into inches or centimeters, or a meter stick.

Volume is measured in several ways. A liquid can be measured in a graduated cylinder, in a volumetric flask, or with a buret or pipet. Examples of these are shown in Fig. 2-2. To find the volume of a regular solid we can measure the dimensions of it and multiply them together. To find the volume of an irregular solid, we can place it in water and measure the amount of water that is displaced. Examples of these two methods of finding the volume of a solid are shown in Fig. 2-3.

2-2 The Metric System and the International System of Units

The metric system has as its basic units the gram (g) for mass (weight), the meter* (m) for length, the liter (l) for volume, and the second (s) for time. The units are related in multiples of 10, 100, 1000, and so on, like our monetary system. Table 2-1 lists the prefixes used in the metric system and gives an example of each. It is essential that you become familiar with these prefixes since you will use this system of measurement for all calculations.

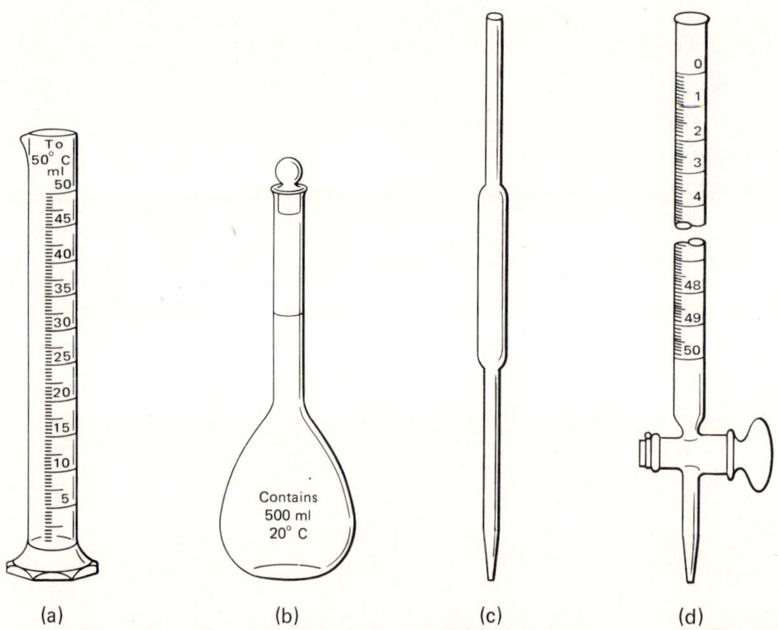

Figure 2-2
Apparatus used to measure volume. (a) Graduated cylinder (b) Volumetric flask (c) Pipet (d) Buret

*According to the International System of Units, the correct spelling is metre. In this text, however, we shall use the more familiar *meter.* You should recognize either spelling since both are currently used.

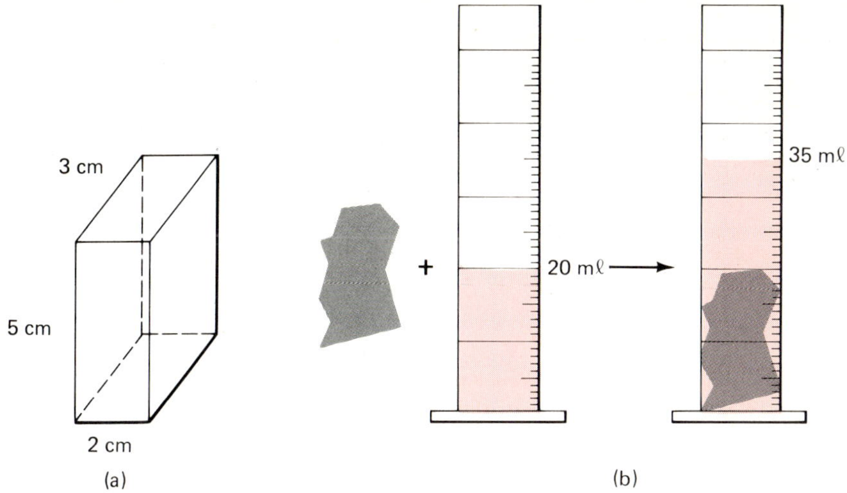

Figure 2-3
Methods of determining the volume of a regular and an irregular solid. (a) Volume of sol. = length × width × height or volume = 3 cm × 2 cm × 5 cm = 30 cm^3 = 30 ml (b) Volume of solid = (vol. of liquid + solid) − volume of liquid or volume = 35 ml − 20 ml = 15 ml = 15 cm^3

The unit **angstrom** (Å) is another unit of length that is often used to discuss size of atoms; **[1 Å = 10^{-8} cm = 10^{-10} m]**, or 1 m = 10^{10} Å. The unit nanometer (nm), however, which is a similar size unit (10 Å = 1 nm) is preferred and is often used in place of Å.

A cube 1 cm on a side will hold a volume of exactly 1 mℓ. Therefore, **[1 mℓ = 1 cm^3 (cc)]**. These units, since they are exactly the same, are often used interchangeably.

TABLE 2-1 *Prefixes used in the metric system*

PREFIX (ABBREVIATION)	VALUE	EXAMPLE
mega (M)	1,000,000 (10^6)	1 megagram (Mg) = 10^6 g
kilo (k)*	1,000 (10^3)	1 kilogram (kg) = 10^3 g
hecto (h)	100 (10^2)	1 hectoliter (hℓ) = 100 ℓ
deka (da)	10 (10^1)	1 decaliter (daℓ) = 10 ℓ
deci (d)*	1/10 (10^{-1})	1 decimeter (dm) = 10^{-1} m or 1 m = 10 dm
centi (c)*	1/100 (10^{-2})	1 centimeter (cm) = 10^{-2} m or 1 m = 10^2 cm
milli (m)*	1/1,000 (10^{-3})	1 millimeter (mm) = 10^{-3} m or 1 m = 10^3 mm
micro (μ)*	1/1,000,000 (10^{-6})	1 micrometer (μm) = 10^{-6} m or 1 m = 10^6 μm
nano (n)*	1/10^9 (10^{-9})	1 nanometer (nm) = 10^{-9} m or 1 m = 10^9 nm
pico (p)	1/10^{12} (10^{-12})	1 picometer (pm) = 10^{-12} m or 1 m = 10^{12} pm

*These are the commonly used ones, which should be memorized.

TABLE 2-2 *Basic SI units*

MEASUREMENT	UNIT (ABBREVIATION*)
mass	kilogram (kg)
length	meter (m)
time	second (s)
temperature	kelvin (K)*
amount	mole (mol)
electric current	ampere (A)*

*For units which are named after a person, such as kelvin and ampere, the abbreviation is capitalized, but not the name. This is true throughout the SI system.

SI Units

The International System of Units (SI units) are now being adopted throughout the world. This system is basically the same as the metric system, except that the standard unit of mass is the kilogram (kg) and the unit of volume is the cubic meter (m^3). This system also includes units for energy, force, pressure, and so forth. All of these are derived from the six basic units listed in Table 2-2.

Throughout this book we shall refer to the SI units since they will eventually be used exclusively in science. We shall also continue to use some of the more common and familiar metric units, however, such as the liter.

Some of the more common derived units, which we shall use later, are listed in Table 2-3. A complete list of these can be found on the back cover of this book.

TABLE 2-3 *Some derived units used in the SI system*

MEASUREMENT	UNIT
Volume	m^3
Density	kg/m^3 or $kg\ m^{-3}$*
Velocity	m/s or $m \cdot s^{-1}$
Force	newton (N)
	$N = kg \cdot m/s^2$
Energy	joule (J)
	$J = kg \cdot m^2/s^2$

*The notation $1/m^3$ is equivalent to m^{-3} (see Appendix I for a discussion of exponents). The dot between the kg and m simply means *times;* using $\times$ for this sometimes leads to confusion with an unknown quantity, for example. According to the SI system, the dot can be omitted since it could be confused with a decimal point. In this case a space is just left as $km\ m^{-3}$; a $-$ should never be used since this implies subtraction. Also, periods are never used with abbreviations, and the comma is not used to separate large numbers; here again, a space is left between each group of three numbers, such as 25 275 000.

Since the unit of volume is the m^3, the liter (ℓ) is defined as 1 dm^3, and therefore 1 mℓ = 1 cm^3 (cc). That is, a cube 1 cm on a side would hold a volume of exactly 10^{-3} ℓ or 1 mℓ.

Metric-English Equivalents

Table 2-4 lists a metric-English equivalent for each of the three units of measurement. You must learn these three equivalents to convert from one system to the other. There are also other equivalents that can be used but if you know these three and the prefixes used in the metric system (as listed in Table 2-1) you can do any conversions between the two systems.

TABLE 2-4 *Metric-English unit equivalents*

DIMENSION	ENGLISH UNIT	METRIC EQUIVALENT
Mass	1 pound (lb)	454 grams (g)
Volume	1.06 quarts (qt)	1 liter (ℓ)
Length	1 inch (in.)	2.54 centimeters (cm)
	1 mile (mi)	1.6 kilometers (km)
Time	1 second (s)	1 second (s)

2-3 Conversion Factors. The Factor Unit Method of Problem Solving

In making conversions within a given system or from one system to another, a very simple and logical method is to use what is generally called *conversion factors*. These are factors that change the units in a number without changing the value; therefore it is essential that the conversion factor always be equal to one.

For example, since 1 m = 100 cm, dividing both sides of this equation by 100 cm gives,

$$\frac{1\ \text{m}}{100\ \text{cm}} = 1$$

or, dividing both sides by 1 m gives

$$\frac{100\ \text{cm}}{1\ \text{m}} = 1$$

These are referred to as **conversion factors.** The first thing to remember about a conversion factor is that it must be equal to one. The next thing to keep in mind when using these to solve problems is that they must be used properly so that the units come out as desired. For example, using one of the conversion factors given here we can convert 2.5 cm to meters. To do this, we multiply the original number (2.5 cm) by the proper conversion factor to give the unit meter. That is,

$$2.5\ \text{cm} = ?\ \text{m}$$

$$2.5\ \cancel{\text{cm}} \times \frac{1\ \text{m}}{100\ \cancel{\text{cm}}} = 0.025\ \text{m}^*$$

Notice that we choose the conversion 1 m/100 cm since this gave the unit meter, which is what we wanted. If the factor 100 cm/1 m had been used, the units would have come out as cm^2/m, which would be a meaningless unit.

This was a very simple case but this same method can be used to solve any problem where we want to convert from a given unit to some new unit. We shall also see later that this same method can be used for many (in fact, most) of the calculations that are encountered in chemistry. In most cases, there is more than one step involved but the general method is still exactly the same.

In doing these calculations, and all future calculations, you should note that there are several steps that are followed. Some hints that should help you as you begin to solve problems now follow.

1. Read the problem over carefully to determine what is given, including the units, and what you are being asked for, including the units.
2. Whenever possible, try to condense the problem to a simple equality, as we did in the example above when we wrote 2.5 cm = ? m. As you become more proficient at this method of problem solving, you will probably not need to do this all the time but to start we shall follow this practice.
3. Determine what relationships will be needed. When you first start solving problems, you should write all of these down, along with the conversion factors. Later, when you become more proficient, this will not be necessary. In the example above, it could easily be seen that we needed the relationship 1 m = 100 cm.
4. Set up the conversion factors properly so that the units cancel to give the desired result. In the previous problem, we used 1 m/100 cm so that the centimeters would divide (or cancel) out.
5. Do the arithmetic and, if at all possible, try to decide if the answer sounds reasonable. That is, in the example above, you should realize that the answer, in meters, would be smaller than the original number, given in centimeters.
6. Check the units, by checking each conversion factor, if necessary.

*The answer here is expressed to two significant digits since the least accurate value here has two—the given value, 2.5 cm; the 1 m in the conversion factor is an exact value. See Appendix I for a discussion of significant digits.

Consider the following problem examples involving conversions within the metric system and between the English and metric systems.

PROBLEM EXAMPLE 2-1

Convert 5.8 cm to in.

SOLUTION:

The problem says

$$5.8 \text{ cm} = ?\text{ in.}$$

The relationship that we need to recall is

$$1 \text{ in.} = 2.54 \text{ cm}$$

The conversion factor will be either

$$\frac{1 \text{ in.}}{2.54 \text{ cm}} \quad \text{or} \quad \frac{2.54 \text{ cm}}{1 \text{ in.}}$$

Since we want the unit centimeters to divide out, we use 1 in./2.54 cm. Setting up the conversion, we find

$$5.8 \cancel{\text{cm}} \times \frac{1 \text{ in.}}{2.54 \cancel{\text{cm}}} = \frac{5.8}{2.54} \text{ in.} = \mathbf{2.3 \text{ in.}}$$

We chose the first conversion factor so that the centimeters would cancel out. We can see that this answer is reasonable since we know that an inch is longer than a centimeter and, therefore, the number (not the value) expressed in inches should be smaller than when expressed in centimeters. Also, the units come out to those desired in the problem. If the factor had been set up as 5.8 cm × 2.54 cm/1 in., the units would come out as cm^2/in., which obviously is incorrect.

PROBLEM EXAMPLE 2-2

Convert 55 mg to kilograms.

SOLUTION:

This problem says

$$55 \text{ mg} = ?\text{ kg}$$

The relationships that we need to recall are

$$1 \text{ kg} = 1000 \text{ g}$$
$$1 \text{ g} = 1000 \text{ mg}$$

Using these relationships and setting up the proper conversion factors gives

$$55\ \text{mg} \times \frac{1\ \text{g}}{1000\ \text{mg}} \times \frac{1\ \text{kg}}{1000\ \text{g}} = \frac{55}{(1000)(1000)}\ \text{kg} = \frac{55}{1 \times 10^6}\ \text{kg} = 55 \times 10^{-6}\ \text{kg}^* =$$

$$\mathbf{5.5 \times 10^{-5}\ kg}$$

This problem involved two steps, but in each case we had to choose the proper conversion factor to get the desired units. The answer sounds reasonable since we know that there should be a small number of kg in 55 mg.

PROBLEM EXAMPLE 2-3

Convert 105 yd to m.

SOLUTION:

$$105\ \text{yd} = ?\ \text{m}$$

Without writing all the relationships, we can see that we need to convert from yards to feet, feet to inches, inches to centimeters, centimeters to meters.

Therefore, we set up the problem as

$$105\ \text{yd} \times \frac{3\ \text{ft}}{1\ \text{yd}} \times \frac{12\ \text{in.}}{1\ \text{ft}} \times \frac{2.54\ \text{cm}}{1\ \text{in.}} \times \frac{1\ \text{m}}{100\ \text{cm}} = \frac{(105)(3)(12)(2.54)\ \text{m}}{(100)} = \mathbf{96.0\ m}$$

Since a yard and a meter are approximately the same, the answer in meters should be about the same as the original measurement in yards.

PROBLEM EXAMPLE 2-4

Convert 55 mi/hr to cm/s.

SOLUTION:

$$55\ \text{mi/hr} = ?\ \text{cm/s}$$

Here we need to convert hours to minutes and minutes to seconds. Also, we need to convert miles to kilometers, kilometers to meters, and meters to centimeters. Again, be careful to set up each conversion factor so that the units cancel out, and also make sure that each conversion factor is equal to one.

---

*Exponents, or powers of 10, are very useful in chemistry, where we often use very large or small numbers. A positive exponent means that the decimal point must be moved to the right and a negative exponent means that it is moved to the left. That is, a negative exponent indicates a small number (the larger the negative exponent, the smaller the number) and a positive exponent indicates a large number. Consider the following examples:

$$10 = 10^1; \quad 100 = 10^2; \quad \tfrac{1}{10} = 0.1 = 10^{-1}$$

$$\tfrac{1}{100} = 0.01 = 10^{-2}; \quad 200 = 2 \times 10^2; \quad 0.0032 = 3.2 \times 10^{-3}$$

See Appendix I for a complete discussion of exponents and scientific notation.

$$\frac{55\ \text{mi}}{\text{hr}} \times \frac{1.6\ \text{km}}{1\ \text{mi}} \times \frac{10^3\ \text{m}}{1\ \text{km}} \times \frac{10^2\ \text{cm}}{1\ \text{m}} \times \frac{1\ \text{hr}}{60\ \text{min}} \times \frac{1\ \text{min}}{60\ \text{s}} = \frac{(55)(1.6)(10^5)\ \text{cm}}{(60)(60)\ \text{s}} =$$

$$2.4 \times 10^3 \frac{\text{cm}}{\text{s}}$$

PROBLEM EXAMPLE 2-5

Convert 15.0 in.^{2} to m^2.

SOLUTION:

$$15.0\ \text{in.}^2 = ?\ \text{m}^2$$

Here we must convert square inches to square centimeters, and square centimeters to square meters. To do these conversions we use the relationships

$$(1\ \text{in.})^2 = (2.54\ \text{cm})^2 \quad \text{or} \quad 1^2\ \text{in.}^2 = (2.54)^2\ \text{cm}^2$$

and

$$(1\ \text{m})^2 = (100\ \text{cm})^2 \quad \text{or} \quad 1^2\ \text{m}^2 = (100)^2\ \text{cm}^2$$

Keep in mind with conversions of this type that *both the number and the unit must be squared* since what we are actually doing is taking

$$15.0\ \text{in.} \cdot \text{in.} \times \frac{2.54\ \text{cm}}{1\ \text{in.}} \times \frac{2.54\ \text{cm}^*}{1\ \text{in.}}$$

Of course, 1^2 is still 1, but we use 1^2 to emphasize this. Therefore, the conversion is

$$15.0\ \text{in.}^2 \times \frac{(2.54)^2\ \text{cm}^2}{1^2\ \text{in.}^2} \times \frac{1^2\ \text{m}^2}{(100)^2\ \text{cm}^2} = \frac{(15.0)(2.54)^2\ \text{m}^2}{(100)^2} = 9.68 \times 10^{-3}\ \text{m}^2$$

2-4 Temperature

There are three commonly used temperature scales. These are the *Fahrenheit* scale (°F), the *Celsius* scale (°C) (which is also sometimes referred to as the centigrade scale), and the *kelvin* scale (°K) (which is sometimes referred to as the *absolute* scale). In the SI system, the degree sign is not used, and thus the unit of temperature is simply K, as 273 K.

The Fahrenheit scale, named after the German physicist Gabriel Daniel Fahrenheit (1686–1736) is the scale with which you are probably most familiar. In scientific work, however, the Celsius and kelvin scales are used exclusively. On the Fahrenheit scale the freezing point of water is 32°, and the

*Another way of pointing this out is to recognize that 1 ft^2 does not equal 12 in.^{2} but rather (12 × 12) in.^{2} or 144 in.^{2}.

normal boiling point* of water is 212°. On the Celsius scale, named after the Swedish astronomer Anders Celsius (1701–1744) these points correspond to 0° and 100°, respectively.

In Fig. 2-4 these two scales are compared. Note that 32°F corresponds to 0°C, and 212°F corresponds to 100°C (as mentioned previously). Also note that at −40° the two scales turn out to be the same. On the Fahrenheit scale there are 180 divisions between the freezing point and the boiling point of water, and on the Celsius scale there are 100 divisions. Thus, 180 divisions on the Fahrenheit scale equal 100 divisions on the Celsius scale. This ratio is therefore, 100°C/180°F or 180°F/100°C, which can be reduced to 5°C/9°F or 9°F/5°C. This ratio is the same between any other two points. For example, between −40°F to 32°F and −40°C to 0°C the ratio is 72/40 = 9/5.

To convert a given temperature from degrees Celsius to degrees Fahrenheit or degrees Fahrenheit to degrees Celsius, we use this ratio, starting from some common point such as the freezing point of water or −40°. We shall derive a formula for the conversion using the common −40 point; an equally valid formula can be derived using the freezing point of water or any other common point. If we take F_x and C_x as the given temperatures, and F_s and C_s as the standard or reference point,

$$\frac{F_x - F_s}{C_x - C_s} = \frac{9}{5}.$$

Using −40 for F_s and C_s gives

$$\frac{F_x - (-40)}{C_x - (-40)} = \frac{9}{5}.$$

Solving for F_x gives

$$F_x + 40 = \frac{9}{5}(C_x + 40)$$

(2-1) $$F_x = \frac{9}{5}(C_x + 40) - 40 \quad \text{or} \quad \mathbf{°F = \frac{9}{5}(°C + 40) - 40}$$

And solving for C_x gives

(2-2) $$C_x = \frac{5}{9}(F_x + 40) - 40 \quad \text{or} \quad \mathbf{°C = \frac{5}{9}(°F + 40) - 40}$$

*We shall see in Chapter 12 that the boiling point of a liquid changes as the atmospheric pressure changes. Thus, the normal boiling point is the boiling point at one atmosphere pressure.

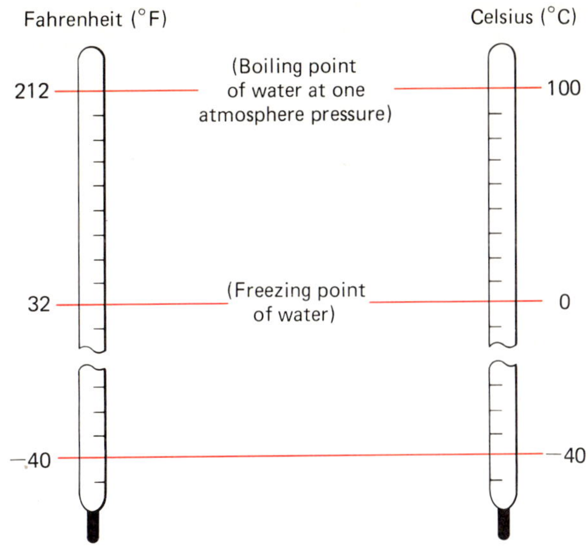

Figure 2-4
Comparison of the Fahrenheit and Celsius temperature scales

One way to remember these formulas is to think of three steps being involved: (1) Add 40. (2) Multiply by $\frac{5}{9}$ (°F to °C) or $\frac{9}{5}$ (°C to °F). (3) Subtract 40. Using the freezing point of water as F_s and C_s gives

$$°F = \frac{9}{5}°C + 32 \tag{2-3}$$

$$°C = \frac{5}{9}(°F - 32) \tag{2-4}$$

These are sometimes written as

$$°F = 1.8°C + 32$$

$$°C = \frac{°F - 32}{1.8}$$

Some students prefer the first form [Eq. (2-1) and (2-2)] since it is exactly the same except for the fraction; this fraction can be remembered by keeping in mind that the freezing point of water is higher on the Fahrenheit scale and, therefore, °F is a larger number than °C; thus to go from °C to °F, use $\frac{9}{5}$ and for °F to °C use $\frac{5}{9}$.

The kelvin scale, named after the British physicist and mathematician William Thompson (1824–1907), who was later titled Lord Kelvin, is set up with a theoretical lower limit of zero. This lower limit, often called *absolute zero,* corresponds to −273°C (or 273.15 to be more precise; in this text we shall use 273). Figure 2-5 compares these two scales. On the Celsius and

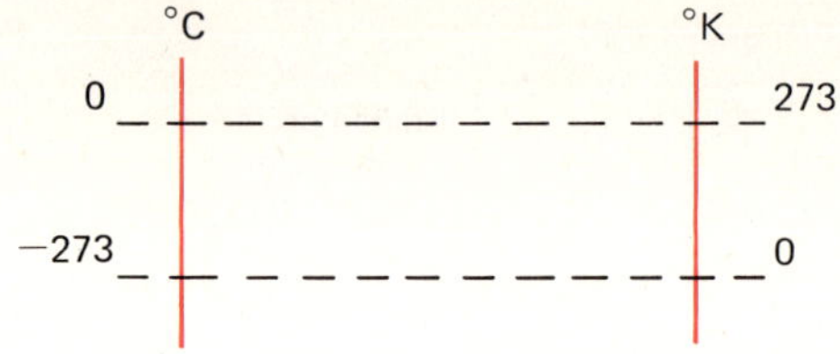

Figure 2-5
Comparison of the Celsius and kelvin temperature scales

kelvin scales the size of the degree is the same; just the zero point is different. Thus, to convert from °C to °K we need only to add 273.

$$°K = °C + 273 \quad \text{or} \quad °C = °K - 273 \tag{2-5}$$

Consider the following conversions between the various scales.

PROBLEM EXAMPLE 2-6

Convert 68°F (room temperature) to °C.

SOLUTION:

$$°C = \frac{5}{9}(°F + 40) - 40 = \frac{5}{9}(68 + 40) - 40 = \frac{5}{9}(108) - 40 = 60 - 40 = \mathbf{20°C}$$

Using the other formula gives

$$°C = \frac{5}{9}(°F - 32) = \frac{5}{9}(68 - 32) = \frac{5}{9}(36) = \mathbf{20°C}$$

PROBLEM EXAMPLE 2-7

Convert −25°C to °F.

SOLUTION:

$$°F = \frac{9}{5}(°C + 40) - 40 = \frac{9}{5}(-25 + 40) - 40 = \frac{9}{5}(15) - 40 = 27 - 40 = \mathbf{-13°F}$$

or

$$°F = \frac{9}{5}°C + 32 = \frac{9}{5}(-25) + 32 = -45 + 32 = \mathbf{-13°F}$$

(In these calculations, watch the signs on the numbers.)

PROBLEM EXAMPLE 2-8

Convert 25°C to °K.

SOLUTION:

To convert from °C to °K, we simply add 273. Thus,

$$°K = 25 + 273 = 298°K$$

PROBLEM EXAMPLE 2-9

Convert 140°F to °K.

SOLUTION:

To solve this we must first convert from °F to °C and then add 273.

$$°C = \frac{5}{9}(140 + 40) - 40 = 100 - 40 = 60°C$$

$$°K = 60 + 273 = 333°K$$

2-5 Heat Energy—Specific Heat

Heat energy is often measured in calories (cal) or kilocalories (kcal). A **calorie** is the amount of heat necessary to raise the temperature of 1 g of water 1°C (from 14.5° to 15.5°C to be precise). The calorie referred to in nutrition is actually a kilocalorie but it is often just designated as Calorie (Cal) and sometimes referred to as a *large calorie.* The sugar in a 10-oz soft drink, burned in the body, produces 110 kcal (Cal) or 110,000 cal of heat energy.* The average 21-year-old person needs between 2,000 and 3,000 kcal (Cal) per day.

The *specific heat* of a substance is the amount of heat required to produce a given change in temperature for a given mass of the substance. In the metric system this is usually expressed as the number of calories necessary to increase the temperature of one gram by one degree Celsius. Thus, in these units,

$$\text{specific heat} = \frac{\text{cal}}{(\text{g})(°\text{C})} \qquad (\text{or cal g}^{-1}\ °\text{C}^{-1})$$

In the International System of Units specific heat is the number of joules required to raise the temperature of 1.00 kg of a substance 1.00 K. Table 2-5 lists the specific heat of a few substances using both sets of units.

*A **joule** (J) is the unit of energy preferred by the SI system. It is the energy equal to a force of one newton (N) acting through a distance of one meter in the direction of the force. A newton is a unit of force in which a mass of one kilogram has an acceleration of one meter per second, per second. Therefore, the units of joules are $(\text{kg})(\text{m/s}^2)(\text{m}) = (\text{kg})(\text{m}^2/\text{s}^2)$. One calorie (cal) is equal to 4.184 joules (J), or to three significant digits, 1.00 cal = 4.18 J.

TABLE 2-5 *Specific heat of some substances*

SUBSTANCE	SPECIFIC HEAT (cal/g °C)	SPECIFIC HEAT (J/kg K)
Water (liquid)	1.00	4.18×10^3
Aluminum	0.217	9.07×10^2
Lead	0.031	1.30×10^2
Sugar	0.299	1.25×10^3
Silver	0.056	2.34×10^2
Sodium chloride (salt)	0.204	8.53×10^2

Consider the following calculations involving specific heat.

PROBLEM EXAMPLE 2-10

How many calories would be required to raise the temperature of 45.0 g of sodium chloride from 25.0° to 85.0°C?

SOLUTION:

From Table 2-5, the specific heat of sodium chloride is 0.204 cal/g °C or cal g^{-1} °C^{-1}.

$$\frac{0.204\ \text{cal}}{1\ \cancel{\text{g}} \times 1°\cancel{\text{C}}} \times 45.0\ \cancel{\text{g}} \times (85.0 - 25.0)°\cancel{\text{C}} = \mathbf{551\ cal}$$

PROBLEM EXAMPLE 2-11

If 8.5 g of metal, initially at 82.0°C, is placed in 45 g of water at 20.0°C and the final temperature of metal and water is 22.0°C, calculate the specific heat of the metal in cal/g°C.

SOLUTION:

The specific heat of the metal is found from the number of calories absorbed by the metal when it was heated, the number of grams of metal (8.5 g), and the temperature change of the metal (82.0°C − 22.0°C = 60.0°C). The number of calories given off by the metal is equal to the number of calories of heat energy absorbed by the water. This heat energy in calories can be calculated by using the specific heat of water (1.00 cal/g °C), the amount of water (45 g), and the temperature change of the water (22.0°C − 20.0°C = 2.0°C). Therefore, the number of calories is

$$\frac{1.00\ \text{cal}}{\cancel{\text{g}}\ °\cancel{\text{C}}} \times (45\ \cancel{\text{g}})(2.0\ °\cancel{\text{C}}) = 90\ \text{cal}$$

The specific heat is

$$\frac{90\ \text{cal}}{(8.5\ \text{g})(60.0°\text{C})} = \mathbf{0.18\ \frac{cal}{(g)(°C)}}$$

PROBLEM EXAMPLE 2-12

Calculate the number of joules of heat energy required to raise the temperature of 125 g of silver from 100° to 250°C.

SOLUTION:

From Table 2-5, the specific heat of silver is 2.34×10^2 J/kg K.

$$\frac{2.34 \times 10^2 \text{ J}}{1 \text{ kg K}} \times 125 \text{ g} \times \frac{1 \text{ kg}}{1000 \text{ g}} \times 150 \text{ K} = \mathbf{4.39 \times 10^3 \text{ J}}$$

The temperature change of 150°C is equal to a temperature change of 150K since the degree difference on both scales is the same, only the base is different.

2-6 Density and Specific Gravity

The **density** of a substance is defined as the mass of a substance occupying a unit volume.

$$\text{density} = \frac{\text{mass}}{\text{volume}}$$

Thus, the density expresses the mass for a given volume of the substance. Or, stated another way density gives the relationship between the mass and a given unit volume, for a particular substance.

We know that if we compare the same volume of various substances, some will be heavier than others. For example, a lead brick is much heavier than a piece of wood of the same size. Hence, we say that the lead is denser, or has a greater density than the wood.

In the metric system the density of liquids and solids is generally measured in grams per milliliter or grams per cubic centimeter. In SI units, density is expressed as kilogram per cubic meter. It is important that the units are included when expressing densities. For example, the density of water is 1.00 g/mℓ. It can also be expressed, however, as 62.4 lb/ft^3 or 8.35 lb/gal.

In expressing the density of a substance you will often see a number such as 20° or 4°, written as a superscript. This indicates the temperature (in °C) at which the density was measured. This is done because the density will vary somewhat with temperature since most substances expand when heated and hence the density decreases with increasing temperature. For most substances the density is reported at 20°C, which is usually considered to be room temperature. For water, however, it is often given at 4°C, which is the temperature at which water has a density of exactly 1.00 g/mℓ. We shall see in Chapter 12 that this is also the maximum density for water; at either higher or lower temperatures it is slightly less than 1.00 g/mℓ.

In calculating the density of a substance, the mass of a given sample is divided by the volume occupied by that sample.

Consider the following calculations involving density.

PROBLEM EXAMPLE 2-13

Calculate the density in grams per milliliter of a piece of metal that has a mass of 12 g and occupies a volume of 1.6 ml.

SOLUTION:

$$\frac{12\text{ g}}{1.6\text{ m}l} = \mathbf{7.5\ g/m}l$$

PROBLEM EXAMPLE 2-14

A cube of lead measures 3.00 cm on each edge and has a mass of 308 g. Calculate the density in g/cm^3.

SOLUTION:

The volume of the cube of lead is

$$3.00\text{ cm} \times 3.00\text{ cm} \times 3.00\text{ cm} = 27.0\text{ cm}^3$$

and the density is

$$\frac{308\text{ g}}{27.0\text{ cm}^3} = \mathbf{11.4\ g/cm^3}$$

In doing conversion problems involving density it is best to think of the density as giving the mass (weight) for a given unit volume. That is, if the density of a substance is 1.2 g/ml, this says 1.2 g = 1 ml and, therefore, we can use the conversion factor, for this particular substance, as

$$\frac{1.2\text{ g}}{1\text{ m}l} = 1 \quad \text{or} \quad \frac{1\text{ m}l}{1.2\text{ g}} = 1$$

Consider the following example.

PROBLEM EXAMPLE 2-15

The density of a liquid is 1.20 g/ml. Calculate (a) the mass in kilograms of 10.0 ml and (b) the volume in liters occupied by 10.0 g.

SOLUTION:

(a)

$$10.0\text{ m}l = ?\text{ kg}$$

$$10.0\,\cancel{\text{m}l} \times \frac{1.20\,\cancel{\text{g}}}{1\,\cancel{\text{m}l}} \times \frac{1\text{ kg}}{1000\,\cancel{\text{g}}} = \mathbf{0.0120\ kg}$$

(b)

$$10.0\text{ g} = ?\ l$$

$$10.0\,\cancel{\text{g}} \times \frac{1\,\cancel{\text{m}l}}{1.20\,\cancel{\text{g}}} \times \frac{1\ l}{1000\,\cancel{\text{m}l}} = \mathbf{8.33 \times 10^{-3}\ }l$$

The **specific gravity** of a substance gives the relative density of the substance as compared to some standard. In general, water at 4°C is taken as the standard, and thus specific gravity expresses the density of the substance as compared to water. This can be expressed as

$$\textbf{specific gravity} = \frac{\textbf{density of substance}}{\textbf{density of water at 4°C}}$$

Thus, if the specific gravity of a substance is 2, that substance has a density two times as great as the density of water at 4°C. And since the density of water at 4°C is 1.00 g/mℓ, the substance has a density of 2 g/mℓ.

Specific gravity has no units; it is just a ratio of one density to another. To find the density from the specific gravity it is merely necessary to multiply the specific gravity by the density of the reference substance, water. And since water has a density of 1.00 g/mℓ, density and specific gravity are numerically equal if the units grams per milliliter (or grams per cubic centimeter) are being used.

As with density, the temperature at which the densities of the substance and the reference (water) are measured is expressed as a superscript. For example, the specific gravity of benzene could be expressed as $0.88^{20°/4°}$. This indicates that the density of the benzene was measured at 20°C and the water was measured at 4°C. Table 2-6 lists the specific gravity of some substances.

Consider the following problem examples involving specific gravity.

PROBLEM EXAMPLE 2-16

The specific gravity of an organic liquid is 1.3. Calculate the number of liters in 84 g of the liquid.

SOLUTION:

The problem states

$$84 \text{ g} = ? \ \ell$$

TABLE 2-6 *Specific gravity of some substances*

SUBSTANCE	SPECIFIC GRAVITY
Water	1.00
Ether	0.708
Benzene	0.880
Acetic acid	1.05
Chloroform	1.49
Carbon tetrachloride	1.60
Sulfuric acid (conc)	1.83
Bromine	3.12

Since the specific gravity is 1.3, the density is 1.3 g/mℓ and, therefore,

$$1.3 \text{ g} = 1 \text{ m}\ell$$

Hence the answer is

$$84 \text{ g} \times \frac{1 \text{ m}\ell}{1.3 \text{ g}} \times \frac{1 \ \ell}{1000 \text{ m}\ell} = 0.065 \ \ell = 6.5 \times 10^{-2} \ \ell$$

PROBLEM EXAMPLE 2-17

If 2.0×10^4 mg of a substance occupy a volume of 15 mℓ, calculate the specific gravity of the substance.

SOLUTION:

Since we know that the density of water is 1.00 g/mℓ, we first want to find the density of the given substance in g/mℓ.

$$\frac{2.0 \times 10^4 \text{ mg}}{15 \text{ m}\ell} \times \frac{1 \text{ g}}{10^3 \text{ mg}} = 1.3 \frac{\text{g}}{\text{m}\ell}$$

Thus, the specific gravity is 1.3.

Notice that there are no units given with the specific gravity. This is because it was found by dividing 1.3 g/mℓ by 1.0 g/mℓ to give 1.3.

SUMMARY

1. The quantity of matter in a particular sample is referred to as the *mass.* The *weight* of a body is due to the gravitational force of attraction on the body. Therefore, the weight of an object can vary, but the mass cannot.
2. In making measurements by either the metric or the International System (SI) of Units the prefixes mega (M), kilo (k), deci (d), centi (c), milli (m), micro (μ), and nano (n) are used (among others) with the basic units. In the metric system the basic units of mass, length, volume, and time are the gram (g), meter (m), liter (ℓ), and second (s). The basic SI units are the kilogram (kg), meter (m), and second (s), among others. Other SI units are derived from these. For example, volume is expressed in cubic meters (m^3) and density in kilograms per cubic meter. There are a few units that are not SI units but are still used. These include the liter (1 ℓ = 1 dm^3; thus, 1 mℓ = 1 cm^3), calorie (1 cal = 4.184 J), and angstrom (1 Å = 10^{-10} m); however, the nanometer (1 nm = 10^{-9} m) is preferred to the Å.
3. To convert from any unit to a corresponding unit a *conversion factor* can be used.
4. In measuring temperature, the SI units use the kelvin scale (K), which is directly related to the Celsius scale by the fact that K = °C + 273.

To convert from °F to °C or °C to °F, the following equations can be used.

Summary

$$°F = \frac{9}{5}(°C + 40) - 40 \quad \text{or } °F = \frac{9}{5}°C + 32$$

5. Heat energy is usually measured in calories (cal) or kilocalories (kcal), where 1 cal is the amount of heat necessary to raise the temperature of 1 g of water 1°C. [The SI unit of energy is the joule (J); 1 cal = 4.18 J.]
6. The *specific heat* of a substance is the amount of heat necessary to raise the temperature of a given amount of the substance by a specified amount. In the metric system this is usually expressed in cal/g °C and in SI units as J/kg K.
7. The *density* of a substance gives the mass for a given unit volume and is usually expressed as grams per milliliter or grams per cubic centimeter for liquids and solids (or kilograms per cubic meter in SI units). *Specific gravity* is the ratio of the density of a substance to the density of water. Thus, a substance with a specific gravity of 2 has a density twice as great as water, or 2 g/mℓ. That is, specific gravity and density are numerically equal, *if* the density is expressed in grams per milliliter or grams per cubic centimeter. In doing calculations involving density it is best to think in terms of the number of grams that are equivalent to 1 mℓ. For example, a density of 2 g/mℓ says that 2 g = 1 mℓ.

Upon completion of this chapter you should be able to do the following things:*

1. Given any type of measurement (mass, length, or volume) in the metric system, convert to any equivalent unit in the metric system. (1, 2)
2. Given any measurement in the English system, convert it to any equivalent unit in the metric system, and vice-versa. (3, 4, 5)
3. Given a temperature measurement in degrees Fahrenheit, convert it to degrees Celsius and kelvin, and vice versa. (6, 7)
4. Given mass and temperature change of a sample of water in a calorimeter, calculate the number of calories transferred. (8)
5. Given any heat measurement in calories, convert to joules. (8)
6. Given calories or joules of heat energy, mass, and temperature change, calculate specific heat of a substance. (9)
7. Given the specific heat of a substance, either in cal/g °C or J/kg K units, and mass, heat energy, and/or temperature change, calculate the unknown mass, heat energy, or temperature change. (10)

*In any of these calculations, be able to use exponents, scientific notation, and significant digits properly. The numbers in parentheses refer to the problems that demonstrate the concept mentioned.

8. Given the mass and volume of a substance, calculate the density. (11)
9. Given the density (or mass and volume) of a substance in any units, calculate the specific gravity. (14)
10. Given the density or specific gravity of a substance and the mass or volume, calculate the unknown mass or volume. (12, 13)
11. Given any of the terms listed in Exercise 1, give a brief definition or explanation, in your own words.

EXERCISES

1. Define or explain the following terms:
   (a) mass
   (b) giga
   (c) mega
   (d) kilo
   (e) hecto
   (f) deka
   (g) deci
   (h) centi
   (i) milli
   (j) micro
   (k) nano
   (l) pico
   (m) angstrom
   (n) calorie
   (o) joule
   (p) liter
   (q) gram
   (r) meter
   (s) density
   (t) specific gravity
   (u) specific heat
   (v) conversion factor
   (w) dm^3
   (x) cm^3
   (y) m^3
   (z) m^{-2}
2. Distinguish between the following:
   (a) mass and weight
   (b) a balance and a scale
   (c) a calorie and temperature
   (d) calorie and joule
   (e) calorie and kilocalorie
   (f) the Celsius and Fahrenheit temperature scales
   (g) density and specific gravity
3. Compare a liter with a cubic meter, a cubic decimeter, and a cubic centimeter.
4. Devise a formula for temperature conversion (similar to Equations 2-1 through 2-4) using room temperature (20°C and 68°F) as the common point. Try your formula for a few values and see if it works.

PROBLEMS

Conversions

1. Carry out the following conversions within the metric system:
   (a) 2.3 m to cm
   (b) 0.120 *l* to m*l*
   (c) 8.5 kg to g
   (d) 1.85 m to μm
   (e) 0.25 dg to g
   (f) 25 cm to m
   (g) 345 m*l* to *l*
   (h) 85 g to kg
   (i) 0.150 μm to m
   (j) 38.0 g to dg

2. Carry out the following conversions within the metric system:
   (a) 1.80 km to mm (f) 0.15 mm to km
   (b) 0.50 cm to dm (g) 25 dm to cm
   (c) 32 m*l* to μ*l* (h) 7.5 μ*l* to m*l*
   (d) 255 nm to Å (i) 3.20 Å to nm
   (e) 2.8 *l* to cm^3 (j) 0.022 cm^3 to *l*
3. Carry out the following conversions between the English and the metric systems:
   (a) 8.20 in. to cm (d) 24.7 cm to in.
   (b) 2.55 lb to g (e) 58.5 g to lb
   (c) 2.85 qt to *l* (f) 4.25 *l* to qt
4. Carry out the following conversions between the English and the metric systems:
   (a) 24.8 lb to mg (f) 815 mg to lb
   (b) 25.0 m to ft (g) 4.5 ft to m
   (c) 8.5 qt to m*l* (h) 95 m*l* to qt
   (d) 4.50 km to yd (i) 45.0 yd to km
   (e) 2.80 tons to μg (j) 0.020 μg to tons
5. Carry out the following conversions:
   (a) 52.0 cm/s to ft/min (f) 525 ft/min to cm/s
   (b) 1.00 g/m*l* to lb/ft^3 (g) 80.0 lb/ft^3 to g/m*l*
   (c) 5.2 μ*l* to m^3 (h) 95 m^3 to μ*l*
   (d) 1.00 g/m*l* to kg/m^3 (i) 875 kg/m^3 to g/m*l*
   (e) 0.020 nm^2 to ft^2 (j) 8.5 ft^2 to nm^2

Temperature Conversion

6. Convert each of the following temperatures to °F and °K:
   (a) 40.0°C (d) 120°C
   (b) −80.0°C (e) −120°C
   (c) 500°C (f) 10°C
7. Convert each of the following temperatures to °C and °K:
   (a) 86.0°F (d) 450°F
   (b) −20.0°F (e) 210°F
   (c) −120°F (f) 12.0°F

Specific Heat (Use the specific heats given in Table 2-5)

8. Calculate the number of kilocalories and kilojoules required for each:
   (a) to raise the temperature of 100 g of liquid water from 12° to 82°C
   (b) to raise the temperature of 0.10 kg of water from 15° to 85°C
9. Calculate the specific heat of each:
   (a) Exactly 150.0 cal of heat will raise the temperature of 18.5 g of a liquid from 22.0° to 44.0°C.
   (b) A sample of metal with a mass of 25.0 g, initially at 75.0°C, is placed in 85.5 g of water at 22.0°C. The final temperature of the water and the metal is 27.5°C.
   (c) Exactly 80.0 cal of heat energy will raise the temperature of 10.0 g of an unknown metal from 20.0° to 60.0°C.

(d) A sample of metal with a mass of 48.2 g, initially at 90.5°C, is placed in 50.0 g of water at 25.0°C. The final temperature of the water and metal is 32.5°C.

10. How many calories of heat would be necessary to raise the temperature of
    (a) 85 mg of sodium chloride from 250° to 280°K?
    (b) 255 mg of aluminum from 185° to 280°K?

Density and Specific Gravity
(Use the specific gravity given in Table 2-6)

11. Calculate the density of each of the following in grams per cubic centimeter:
    (a) a substance with a mass of 365 g and a volume of 62 mℓ
    (b) a piece of metal measuring 1.0 cm by 0.10 dm by 25 mm having a mass of 5.0 g
    (c) a substance with a mass of 425 kg occupying a volume of 0.23 m^3
    (d) a piece of metal with a volume of 12.2 mℓ and a mass of 28.5 g
    (e) a piece of metal measuring 0.25 dm by 42 mm by 1.8 cm has a mass of 9.5 g
    (f) a substance with a mass of 83.5 kg occupying a volume of 0.015 m^3
12. Calculate the mass in grams of each of the following (use Table 2-6):
    (a) a 20.0-mℓ volume of ether
    (b) a 6.25-ℓ volume of carbon tetrachloride
    (c) a 2.2-qt volume of benzene
    (d) a 470-mℓ volume of benzene
    (e) a 0.100-ℓ volume of bromine
    (f) a 0.85-qt volume of ether
13. Calculate the volume, in milliliters, of each of the following (use Table 2-6):
    (a) 80.0 g of carbon tetrachloride (CCl_4)
    (b) 2.2 kg of benzene (C_6H_6)
    (c) 3.50 lb of chloroform ($CHCl_3$)
    (d) 3.5 g of acetic acid ($HC_2H_3O_2$)
    (e) 0.120 kg of ether
    (f) 25.2 lb of bromine
14. Calculate the specific gravity of each of the following:
    (a) a substance with a density of 3.75 g/mℓ
    (b) a substance with a density of 6.5×10^3 kg/m^3
    (c) a substance that has a volume of 3.25 ft^3 and a mass of 285 lb
    (d) a substance with a density of 6.8 g/mℓ
    (e) a substance with a density of 8.2×10^3 kg/m^3
    (f) a substance that has a volume of 5.2 ft^3 and a mass of 630 lb
15. Calculate the density in pounds per cubic foot of each of the following (use Table 2-6):
    (a) ether
    (b) bromine

General Problems

16. Calculate the volume in liters occupied by each of the following:
    (a) 0.012 mg of water

(b) 225 kg of water

17. Calculate the volume in gallons of each of the following (use Table 2-6):
    (a) 2.5 tons of sulfuric acid (H_2SO_4)
    (b) 0.025 ton of benzene (C_6H_6)
18. Calculate the density of each of the following in grams per cubic centimeter:
    (a) A piece of metal with a mass of 8.50×10^{-4} kg displaces 0.180 mℓ when placed in water.
    (b) A piece of metal with a mass of 2.3×10^{-4} kg displaces 8.0×10^{-2} mℓ of water.
19. If a car is traveling at a rate of 55 mi/hr, how many minutes would it take to travel 25 km?
20. If a car is traveling at a rate of 1.5 km/min, how many hours will it take to travel 52 mi?
21. Calculate the density of ether in kilograms per cubic meter.
22. Calculate the density of benzene in kg/m^3.
23. If the price of gold is \$150 per ounce, how much would 1.0 kg cost?
24. If platinum sells for \$5.50 per gram, what would the cost of 2.8 oz be?
25. The density of a liquid is 1.50 g/mℓ. Calculate the volume in gallons of 3.00 mg of the liquid.
26. Calculate the volume in liters of a sample of alcohol having a mass of 3.50 lb if the density of the alcohol is 6.6 lb/gal.
27. Calculate the specific heat, in cal/g °C and J/kg K, if 6.00 kcal of heat energy will raise the temperature of 2.00 lb of a metal from 25.0° to 52.0°F.
28. Calculate the specific heat, in cal/g °C and J/kg K, if 85.0 cal of heat energy will raise the temperature of 4.00×10^{-2} lb of a metal from 42.5° to 75.0°F.
29. If a wine contains 12.0 percent alcohol by volume, how many milliliters of alcohol are there in a gallon of the wine?
30. If a solution contains 25.0 percent water by volume, how many milliliters of water are there in a quart of the solution?
31. Body temperature is 98.6 °F. Calculate what this would be in °C and °K.
32. The boiling point of oxygen is −183 °C. Convert this to °F and °K.

REFERENCES

ASTON, ALLEN V., "Standards of Measurement," *Sci. Am.*, June 1968, **218,** p. 50.

BENSON, SIDNEY W., *Chemical Calculations,* 3rd ed. New York: John Wiley & Sons, Inc., 1971.

KENDIG, FRANK, "Coming of the Metric System," *Saturday Review,* Nov. 25, 1972, p. 40.

PAUL, MARTIN A., "International System of Units (SI)," *Chemistry,* Oct. 1972, **45,** no. 9, p. 14.

Objectives

1. To be able to define the term matter and to explain what is meant by states of matter.
2. To be able to explain what an atom, an element, and a compound are and how pure substances (elements and compounds) differ from mixtures.
3. To be able to tell what a symbol and a formula represent.
4. To be able to state what the law of definite proportions or constant composition is and what this tells about a compound.
5. To be able to differentiate between physical and chemical properties and to tell how a physical change differs from a chemical change.
6. To be able to tell what a chemical equation represents and what a balanced equation is.
7. To understand what the law of conservation of matter is and how this shows that a chemical equation must be balanced to be correct.
8. To understand the difference between potential and kinetic energy.
9. To understand what the law of conservation of energy is and how this is related to the law of conservation of matter.
10. To be able to explain what an atomic mass unit (amu) represents.
11. To learn what the relative mass and charge are for the electron, proton, and neutron.
12. To learn how the three subatomic particles are arranged in the atom.
13. To be able to tell what the atomic number and mass number of an atom represent, in terms of electrons, protons, and neutrons.
14. To be able to explain what isotopes are.
15. To understand what the atomic weight scale is and what the atomic weights, as given on the periodic table, represent.
16. To be able to define the term *ion* and explain how the charge is determined from the number of electrons gained or lost by the atom.
17. To understand what a molecule is and how a molecular (or covalent) compound differs from an ionic compound.
18. To understand what the formula, or molecular, weight of a compound represents and to be able to calculate the formula or molecular weight of compounds, using the atomic weight scale.
19. To be able to explain what a mole is.
20. To learn to do calculations involving moles, molecules (or formula units), and molecular or formula weight.
21. To understand what a solution is and to be able to identify the solute and solvent in any solution.
22. To be able to calculate percent (by mass) and molar concentrations and do calculations involving these quantities.

Three

Matter

In Chapter 2 we defined matter as anything that has mass and takes up space. Thus, matter is simply all the material of which the universe is composed. In this chapter we shall begin to consider the properties of matter and the classification of matter into various types. Many of the topics covered will just be introduced at this point and then considered in more detail later in the book. For example, we shall consider the basic structure of the atom and the types of compounds that atoms form. Then in Chapter 4 we shall consider atomic structure in detail and in Chapter 7 we shall study the bonds that hold atoms together in compounds.

Matter can exist in any one of three states—gas, liquid, or solid—depending on the temperature and pressure and the characteristics of the substance. For example, water in the solid form (ice) can be heated to form liquid water and heated further to form steam. A substance like sugar, however, can only exist in the solid state. Heating the sugar will cause it to decompose and not change the state as was the case with water.

3-1 Classification of Matter—Atoms, Elements, Compounds, and Mixtures

All matter is composed of tiny particles called **atoms.** Even though this idea had first been proposed by the ancient Greeks, it wasn't until about 1803 that the English chemist John Dalton (1766–1844) proposed his atomic theory based on experimental evidence. The main points of Dalton's theory were that all matter is composed of atoms, which are indivisible and preserve their identity in chemical changes. Also, he said that all atoms of a given element are identical in all properties and that when atoms combine to form molecules, they do so in the ratio of small whole numbers.

We now know there were some errors in Dalton's theory, namely, his statement that all atoms of a given element are identical in all properties. We shall see (Section 3-7) that all atoms of a given element are not identical in mass. Also, it is now known that atoms are composed of smaller subatomic particles (Section 3-5). Dalton's theory was generally correct, however, and was a major step in the development of the science of chemistry.

Thus, all matter is made up of atoms, either in the form of **pure substances** (elements or compounds) or mixtures. In this text the term *substance* will be understood to mean a pure substance—an element or compound. These atoms that make up all matter are extremely tiny. In general, they have diameters of between 1 and 5 Å (0.1–0.5 nm). If we were to place atoms of a diameter of 2.00 Å side by side, it would take over 100 million (127,000,000 to be exact) of them to occupy a 1-in. (2.54-cm) length.*

An **element** is a substance that is composed of only one kind of atom. There are presently 106 different elements known and therefore only 106 different kind of atoms. Of these 106, only 90 have been found to occur naturally. The remaining 16 have been produced synthetically (very small amounts of some of these may also exist in nature).

Each of the elements has a **symbol,** which is an abbreviation for the name. This symbol consists of the first letter of the name, which is capitalized, and in some cases one other letter, which is *not* capitalized. For example, the symbol for carbon is C, the symbol for cobalt is Co,† the symbol for chlorine is Cl, the symbol for chromium is Cr, and the symbol for copper is Cu. Some of these symbols come from the Latin name, such as Cu for copper from the Latin cuprum and Na for sodium from the Latin natrium. The following are the more common elements for which you should learn the symbols as soon as possible. The symbols and names for all the elements (along with other information that we shall consider later) are listed in a table called the **periodic table.** This is shown on the inside front cover of this text. We shall have much more to say about this table later, but for now you can use it to learn the names for the following elements. As you are doing this, compare the order in which they are listed here with the arrangement in the periodic table. The symbols that are underlined come from Latin names. These names are listed in Table 3-1.

H, Li, <u>Na</u>, <u>K</u>, Be, Mg, Ca, Sr, Ba, Ti, V, Cr, Mn, <u>Fe</u>, Co, Ni, Pt, <u>Cu</u>, <u>Ag</u>, <u>Au</u>, Zn, Cd, <u>Hg</u>, B, Al, C, Si, Sn, <u>Pb</u>, N, P, As, <u>Sb</u>, Bi, O, S, F, Cl, Br, I, He, Ne, Ar, U

An element may exist as individual atoms or as **molecules** made up of only one kind of atom. A molecule can be composed of a single atom, such as He, or two or more atoms held together by a force called a *covalent bond* (to be considered in Chapter 7), such as O_2 or H_2. For example, elements such as

*Starting with a 1-in. length we could calculate the number of atoms of 2.00 Å diameter as follows:

$$1.00 \text{ in.} \times \frac{2.54 \cancel{\text{cm}}}{1 \cancel{\text{in.}}} \times \frac{1 \cancel{\text{m}}}{10^2 \cancel{\text{cm}}} \times \frac{10^{10} \cancel{\text{Å}}}{1 \cancel{\text{m}}} \times \frac{1 \text{ atom}}{2.00 \cancel{\text{Å}}} = 1.27 \times 10^8 \text{ atoms}$$

†If the o had been capitalized here, the symbol would be CO, which is a compound composed of carbon (C), and oxygen (O). Therefore, it is important that you remember that only the first letter of a symbol is capitalized.

TABLE 3-1 *Latin names from which symbols are derived*

NAME	SYMBOL	LATIN NAME
Sodium	Na	Natrium
Potassium	K	Kalium
Iron	Fe	Ferrum
Copper	Cu	Cuprum
Silver	Ag	Argentum
Gold	Au	Aurum
Mercury	Hg	Hydrargyrum
Lead	Pb	Plumbum
Antimony	Sb	Stibium

aluminum (Al), iron (Fe), and helium (He) exist as individual atoms; however, several common elements exist as diatomic molecules—molecules composed of two atoms. These include hydrogen (H_2), oxygen (O_2), nitrogen (N_2), and chlorine (Cl_2).

A **compound** is a substance that is composed of two or more elements, chemically combined, in definite and constant proportions. The **formula** for a compound gives the elements (or atoms) that it is composed of and the relative proportions of these atoms. For example, the compound water is composed of two atoms of hydrogen and one atom of oxygen; therefore, the formula for water is H_2O. The **Law of Definite Proportions** or **Constant Composition** states that any pure compound always contains the same elements in exactly the same proportions by mass (weight).

A compound may consist of either molecules or ions (Section 3-8). In either case, however, the formula of a compound gives the relative number of atoms in a **formula unit.** Thus the formula unit can represent either a molecule or a group of ions. In this text we shall generally use the term *formula unit* for any type of compound.

Consider the following examples:

FORMULA OF COMPOUND (NAME)	NUMBER OF ATOMS OF EACH ELEMENT PRESENT IN A FORMULA UNIT
CO (carbon monoxide)	1 carbon, 1 oxygen
CO_2 (carbon dioxide)	1 carbon, 2 oxygen
H_2O_2 (hydrogen peroxide)	2 hydrogen, 2 oxygen
$Ca(OH)_2$ (calcium hydroxide)	1 calcium, 2 oxygen, 2 hydrogen*
$Mg(NO_3)_2$ (magnesium nitrate)	1 magnesium, 2 nitrogen, 6(2×3) oxygen
$(NH_4)_2SO_4$ (ammonium sulfate)	2 nitrogen, 8 hydrogen, 1 sulfur, 4 oxygen
$Al_2(C_2O_4)_3$ (aluminum oxalate)	2 aluminum, 6 carbon, 12 oxygen

*The parentheses in the formula denote the same thing as in mathematics; that is, each number inside the parentheses is multiplied by the number outside the parentheses.

A **mixture** is a combination of two or more pure substances in variable proportions. The mixture can be either *homogeneous* (the same throughout) or *heterogeneous*. For example, a *solution* is a homogeneous mixture whose composition can be varied, such as in a mixture of salt with water. Also, in a mixture the individual components retain their identity whereas in a compound the elements undergo a chemical change (Section 3-2) in combining to form the compound.

In a heterogeneous mixture the individual substances remain essentially the same as they were before being mixed. This is unlike a solution, such as salt in water, where one of the substances often changes in form. For example, when salt is mixed with water to form a solution, the individual salt crystals can no longer be seen, even though the salt and water both retain their identity. When salt and sand are mixed together, however, the two components can be seen as individual species.

One other difference between a mixture and a compound is that a mixture can be separated by ordinary physical means, whereas a compound cannot. This will be discussed further in Section 3-2, after considering the difference between physical and chemical changes.

The differences between mixtures and compounds can thus be summarized as follows:

COMPOUND	MIXTURE
Definite and constant composition.	Variable composition.
Elements that make up the compound lose their identity, due to chemical change taking place.	Elements or compounds that are mixed retain their identity since they are not chemically combined.
Cannot be broken down into simpler substances by ordinary physical means.	Can be separated by simple, physical operations.

Using the definitions considered in this section we can summarize the classification of matter according to the diagram shown in Fig. 3-1.

3-2 Physical and Chemical Properties and Changes

Physical properties are those properties that can be observed without changing the composition of the substance. These properties include color, odor, solubility, density, specific heat, melting point, and boiling point. If the change in a substance alters the form but not the composition of that substance, the change is then called a **physical change**.* For example, when solid water (ice) is heated, it first changes to liquid water and finally to steam. There is no change in the composition of the substance (H_2O), only in the

*A property distinguishes one substance from another; a change is a conversion from one to another.

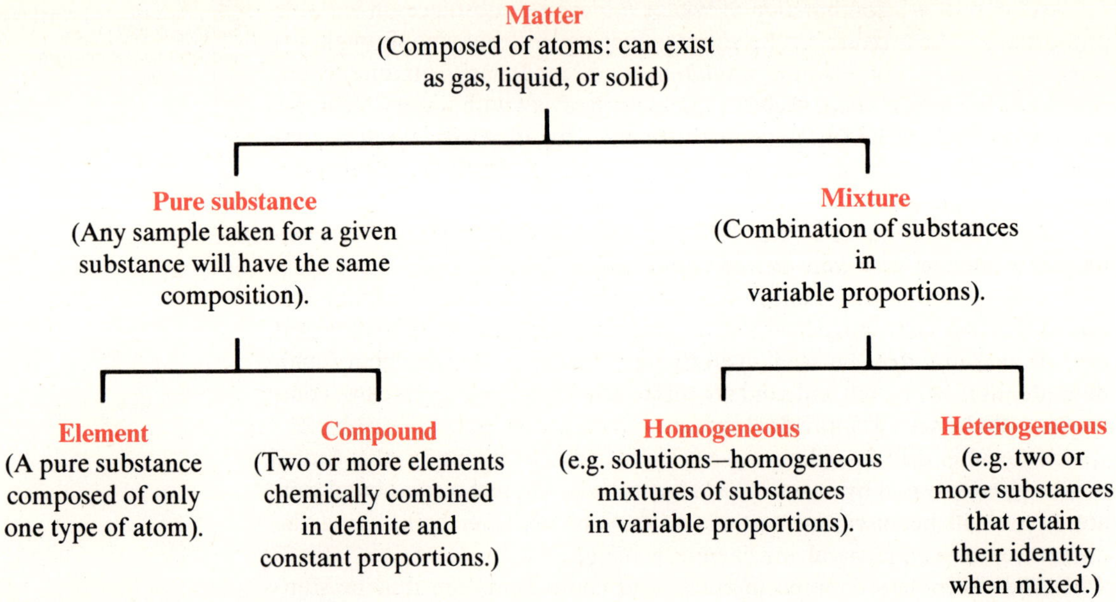

Figure 3-1
Classification of matter

form in which it exists. Thus, the changes taking place are simply physical changes. As another example, consider the change that takes place when salt dissolves in water. Even though the form of the salt changes, the composition does not. It is still sodium chloride (NaCl), but now it is dissolved in water.

A **chemical property** can only be observed when the substance undergoes a change in composition—that is, the substance undergoes a **chemical change.** For example, when a compound is formed between two or more elements, the elements undergo a chemical change to form a new substance (the compound). As an example of the difference between physical and chemical changes, let us consider again the separation of the components of a mixture, as opposed to the breakdown of a compound. If we have a mixture of salt and water, we can boil off the water and the salt will remain. If we pass an electric current through water, however, we can cause the water to decompose into the two elements hydrogen and oxygen. In the first case the mixture was simply separated to obtain the substances that made up the mixture (physical change). But in the second case, a chemical change took place in the water to form two completely different substances.

In the definition of a compound it was stated that the elements are chemically combined. This means that a chemical change has taken place in the elements forming the compound. One very striking example of this is the change that takes place when sodium metal (a very active metal that reacts violently with water) reacts with chlorine gas (a pale green, very poisonous, gas) to form the compound sodium chloride, which is ordinary table salt.

The chemical equation representing this change is

$$2\ Na + Cl_2 \longrightarrow 2\ NaCl$$

In a *chemical reaction* one or more elements or compounds react to form one or more different elements or compounds. No atoms disappear or are changed to different kinds of atoms, however; that is, a reaction is simply a process of rearrangement of atoms.

Thc **law of conscrvation of mass (matter)** states that, in a chemical reaction, mass (matter) is neither created nor destroyed.

A **chemical equation** shows what elements or compounds are reacting **(reactants)** and what new elements or compounds are being formed **(products).** In order to obey the law of conservation of mass, there must be exactly the same number of each atom on each side of a correct chemical equation. When we have exactly the same number of each atom on each side of the equation we say that the chemical equation is *balanced.* An equation is not correct until it is balanced. We shall have more to say about chemical equations and balancing equations later (in Chapters 8 and 19). For now, let us consider one example to point out what is meant by a balanced equation. When hydrogen gas (H_2) reacts with oxygen gas (O_2), the product formed is water (H_2O). Therefore the equation is

$$H_2 + O_2 \longrightarrow H_2O \qquad \text{(not balanced)}$$

This is not correct, however, since the equation is not balanced. If we left it this way, it would imply that an atom of oxygen is destroyed in the reaction. This, of course, cannot be the case. Therefore, in order to make the equation correct, we must write it as

$$2\ H_2 + O_2 \longrightarrow 2\ H_2O$$

Note that in balancing the equation we only change the coefficients before each substance (if there is no coefficient, it is understood to be one) and do not change the formulas for either the reactants or products. For example, it would not be correct to change the formula of water to H_2O_2, since this is an entirely different compound.

3-3 Energy–Mass Relations

All changes in nature, whether physical or chemical, are accompanied by changes in energy. Energy may be either potential or kinetic. **Potential energy** is the energy possessed by a substance by virtue of its position in space. **Kinetic energy** is energy possessed by a substance by virtue of its motion. A book on a desk has potential energy, but as it falls to the floor, its potential energy decreases and its kinetic energy increases.

There are several forms of energy, such as heat, mechanical, electrical, chemical, and light or radiant energy. Like matter, there is a Law of Conservation of Energy that states that energy cannot be created or destroyed. It can be changed from one form to another, however. For example, in starting a car, the battery produces chemical energy, which produces an electric current (electrical energy) which is transmitted to the spark plugs. These cause the compressed gas in the cylinders to ignite and produce the mechanical energy that causes the car to move.

It is now known that matter and energy are interconvertible. This was shown in 1905 by the German-American physicist Albert Einstein (1879–1955). This relationship between matter and energy is shown by the equation,

$$E = mc^2 \tag{3-1}$$

where E represents energy, m represents mass (or actually change in mass, often designated as Δm), and c represents the velocity of light (3.0×10^{10} cm/s). If m is expressed in grams and c in centimeters per second, E is expressed in an energy unit called an *erg* (g cm^2/s^2). This is a small unit of energy, which is related to calories and joules as follows:

$$1 \text{ erg} = 2.4 \times 10^{-8} \text{ cal} = 1 \times 10^{-7} \text{ J}$$

According to Eq. (3-1), if 1.0 g of mass were converted completely to energy, it would produce 2.2×10^{13} cal, calculated as follows:

$$(1.0 \text{ g})(3.0 \times 10^{10} \text{ cm/s})^2 = 9.0 \times 10^{20} \text{ g cm}^2/\text{s}^2 = 9.0 \times 10^{20} \text{ ergs}$$

$$9.0 \times 10^{20} \cancel{\text{ergs}} \times \frac{2.4 \times 10^{-8} \text{ cal}}{\cancel{\text{ergs}}} = 2.2 \times 10^{13} \text{ cal}$$

This amount of energy would raise the temperature of 2×10^8 l (about 220,000 tons) of water from 0° to 100°C.

In ordinary chemical reactions the energy changes are relatively small and, hence, the loss or gain in mass is quite small. For example, the maximum is about 5×10^5 cal, which corresponds to about 2×10^{-8} g. This quantity is too small to be detected on most balances.

As we shall see later, in nuclear changes (Section 25-9), the loss of mass converted to energy is quite significant and is great enough to detect, but we can conclude that the Law of Conservation of Mass is still valid for ordinary chemical reactions. To be even more correct, we must combine the Law of Conservation of Mass with the Law of Conservation of Energy and state that the total of the mass and the energy is neither created nor destroyed but may be transformed.

3-4 Relative Atomic Masses—Atomic Mass Units

An atom is a very tiny particle—much too small to be weighed on even the most sensitive balance. Therefore, a relative scale has been set up, comparing the masses of atoms. On this scale the lightest atom, hydrogen, has a relative mass of about 1, oxygen has a relative mass of about 16, and so forth. When this relative scale was first proposed, hydrogen was taken as the standard and everything was compared to it, but later oxygen was used as the standard, with a mass of 16 relative units. Finally it was agreed to use one isotope (Section 3-7) of carbon, called carbon-12. This then is assigned a value of exactly

12 and everything is compared to it. The relative units in this scale are called **atomic mass units** (amu). Thus, on this scale a hydrogen atom has a relative mass of almost 1 amu, oxygen almost 16 amu, and carbon-12 exactly 12 amu. Stated another way, an atom with a mass twice as great as a carbon-12 atom has a mass of 24 amu, and an atom with a mass one-half that of a carbon-12 atom has a mass of 6 amu. This is the basis of the atomic weight scale, to be discussed in Section 3-7.

3-5 Subatomic Particles

It is now known that, unlike Dalton's statement that atoms are indivisible, the atom is composed of three smaller subatomic particles. These are called **electrons, protons,** and **neutrons.**

The electron (e^-) is a particle with a relative charge of -1 and a very small mass (0.00055 amu). This mass is so small that for all practical purposes it is considered to be negligible.

The proton (p) is a particle having a relative unit positive charge ($+1$)* and a mass of just over 1 amu (1.0073).

The neutron (n) is a neutral (uncharged) particle with a mass just over one amu (1.0087). Since the mass of both the proton and neutron are so close to one they are generally both rounded off to one amu for most purposes.

Table 3-2 summarizes the data for these three subatomic particles.

TABLE 3-2 *Properties of subatomic particles*

PARTICLE (ABBREVIATION)	APPROXIMATE MASS (AMU)	RELATIVE CHARGE
Electron (e^-)	Negligible	-1
Proton (p)	1	$+1$
Neutron (n)	1	0

3-6 Basic Structure of Atoms. Atomic Number

We now want to consider how the three subatomic particles are arranged in an atom. The following are the fundamental facts about atomic structure.

1. The protons and neutrons are found in the center of the atom, called the **nucleus.** The electrons are found outside the nucleus in *shells* or *energy levels.*

*The charge on an electron and a proton is the same but opposite in sign. The actual charge on each of these species is 1.602×10^{-19} coul. The coulomb is a unit used for measuring electrical charge but as you can see, the value of the charge of an electron or proton in coulombs is quite awkward to handle. Since atomic particles which are charged have charges which are the same or integral multiples of the charge of an electron, the relative charge of an electron may be chosen as -1, and hence the relative charge on a proton is $+1$.

2. The nucleus is only a very small part of the atom, in terms of size, but it contains most of the mass of the atom. Therefore, the nucleus has a very high density (1×10^{14} g/cm^3). One cubic centimeter of nuclear matter would have a mass of 1.1×10^8 tons. Also, since each proton has a relative charge of +1, the relative charge on the nucleus is positive and equal to the number of protons.

3. The number of protons in the nucleus is defined as the **atomic number.** Thus, all atoms of the same element have the same number of protons.

4. An atom is electrically neutral. Therefore, there are the same number of electrons outside the nucleus as there are protons in the nucleus. In forming compounds, atoms sometimes gain or lose electrons to form charged particles called *ions* (we shall have more to say about this in Section 3-8).

5. The shells or energy levels of electrons are dispersed at a relatively great distance from the nucleus. The nucleus has a diameter of about 10^{-5} Å (10^{-6} nm), whereas the diameter of the entire atom is in the range of 1 to 5 Å (0.1 to 0.5 nm). Therefore, these electrons are dispersed at distances that extend up to 100,000 times the diameter of the nucleus. This means that if the nucleus of an atom were the size of a baseball, the electrons would be dispersed at a distance up to 3 miles away.

6. The number of protons in an atom plus the number of neutrons is called the **mass number,** which is essentially equal to the atomic mass in amu, since the mass of the electron is negligible. Hence, the number of neutrons present in a given atom is equal to the mass number minus the number of protons or

$$\text{number of neutrons} = \text{mass number} - \text{atomic number}$$

The general notation that is used to represent the mass number and atomic number of a given type of atom is

$$^{A}_{Z}\text{E}$$

where E = the symbol of the element
A = the mass number (in amu)
Z = the atomic number

Recall that the atomic number is equal to the number of protons found in the nucleus. It is also equal to the number of electrons found outside the nucleus in a neutral atom. The mass number is equal to the sum of the protons and neutrons in the nucleus.

In the periodic table the elements are arranged in order of increasing atomic numbers. These numbers are given above the symbol for each element, in the periodic table on the inside front cover of this book. For example, hydrogen (H) is atomic number 1, helium (He) is number 2, and so on.

Now let us consider the general arrangement of the subatomic particles in the atoms of some elements.

1. $^{1}_{1}H$

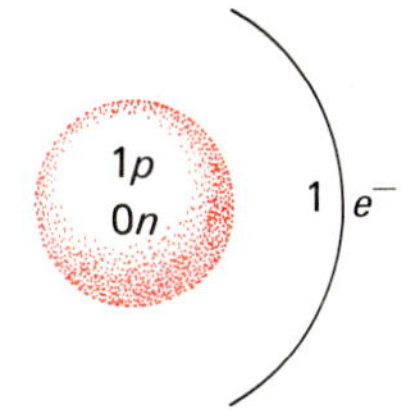

   Atomic number = 1 = number of protons in nucleus = number of electrons
   Mass number = 1 = number of neutrons + protons; hence, number of neutrons = 1 − 1 = 0

2. ^{14}C

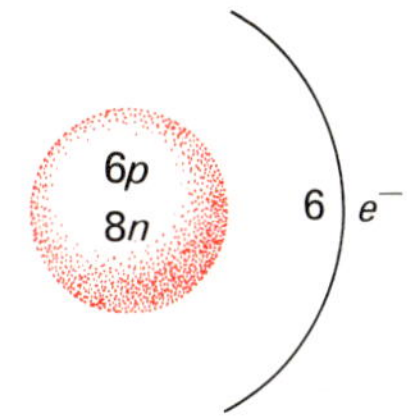

   Atomic number (from periodic table) = 6; therefore, there are 6 protons and 6 electrons
   Mass number = 14; therefore, number of neutrons = 14 − 6 = 8

Since the nucleus has a positive charge and the electron a negative charge, it seems as though the electrons should be drawn into the nucleus. We shall see in Chapter 4 that the electrons remain in various energy levels around the nucleus and are always found in a particular level (or shell). We shall also consider at that time how the electrons are arranged in these energy levels and what effect this arrangement has on the properties of the atom.

3-7 Isotopes and the Atomic Weight Scale

In the periodic table (see inside front cover of this text) you will note that along with the symbol, name, and atomic number for each element, there is also another number given. This is called the relative atomic mass or **atomic weight.** Since these are relative masses, hereafter in this text we shall refer to them as the *atomic weight.*

You will notice that many of these are not close to whole numbers. This is not what we would expect from what has been said previously (Section 3-5) about the relative mass of electrons, protons, and neutrons. Since the mass of a proton and a neutron are each almost one and the mass of an electron is negligible, we would expect the relative mass of an atom to be close to a whole number—certainly not halfway between as is the case with chlorine. The reason that many of these atomic weights are not close to whole numbers is that they represent the average mass of a particular type of atom, since all atoms of the same element do not necessarily have the same mass. There are atoms of the same element that have different masses. Atoms having different atomic masses, or mass numbers, but the same atomic number are called **isotopes.** Thus, isotopes are atoms of the same element and hence have the same number of protons and the same number of electrons, but different numbers of neutrons. Isotopes of the same element have the same chemical properties but slightly different physical properties. For example, there are two isotopes of chlorine, called chlorine-35 (^{35}Cl) and chlorine-37 (^{37}Cl), and three isotopes of hydrogen, ^{1}H, ^{2}H, and ^{3}H. These isotopes have the following numbers of protons, neutrons, and electrons:

ISOTOPE	ELECTRONS	PROTONS	NEUTRONS
${}^{35}_{17}Cl$	17	17	18
${}^{37}_{17}Cl$	17	17	20
${}^{1}H$ (hydrogen)	1	1	0
${}^{2}H$ (deuterium)	1	1	1
${}^{3}H$ (tritium)	1	1	2

The atomic weight, in atomic mass units, for the elements, such as 12.01 for carbon, is an average mass based on the abundance of the isotopes of that element in nature. The atomic weight for the element is found by multiplying the exact atomic weight of each isotope by its percent abundance in nature and then taking the sum of the values obtained. Keep in mind that this represents the *relative, average* mass for a particular element.

For example, any naturally occurring sample of chlorine consists of about 75.5 percent chlorine-35 and 24.5 percent chlorine-37. This means that for each 1000 atoms of chlorine, 755 are chlorine-35 and 245 are chlorine-37. Therefore, the average mass of chlorine atoms is found by taking the sum of the masses of the chlorine-35 atoms and the chlorine-37 atoms, and dividing by 1000, as follows,

$$\begin{array}{lr} 755 \text{ atoms} \times 35 = & 26{,}425 \\ 245 \text{ atoms} \times 37 = & 9{,}061 \\ \hline 1000 \text{ atoms} \qquad = & 35{,}490 \end{array}$$

Therefore, the average mass is about 35.5.

By very accurate measurements, the exact mass of a particular isotope is found, and this is used to calculate the exact atomic weight.

Consider the following problem example:

PROBLEM EXAMPLE 3-1

Calculate the atomic mass to the proper number of significant digits for carbon, given the following data:

ISOTOPE	EXACT ATOMIC MASS (AMU)	% ABUNDANCE IN NATURE
C-12	12.00000*	98.89
C-13	13.00335	1.110

*Even if we consider the exact masses of the protons, the neutrons, and the electrons, the total mass does not equal the mass found for a particular atom of an isotope; it is greater. For example, carbon-12 has an exact atomic mass of 12.00000 amu. Calculations based on the number of protons,

SOLUTION:

$$12.00000 \text{ amu } (0.9889) + 13.00335 \text{ amu } (0.0110) = \mathbf{12.01 \text{ amu}}$$

Chlorine is found in nature as two isotopes: Cl-35 (34.969 amu) and Cl-37 (36.966 amu). These are found in the ratio of 75.53% Cl-35 and 24.47% Cl-37. Therefore, the average, relative mass of chlorine is calculated as

$$(34.969 \text{ amu})(0.7553) + (36.966 \text{ amu})(0.2447) = 35.46 \text{ amu}$$

Based on the average mass, the atomic weight of chlorine is found to be 35.46, or about 35.5. But we would never find an atom of chlorine that would have a relative mass of 35.5 amu; it would have a relative mass of about 35.0 or 37.0 amu, depending on the isotope with which we were working. However, the average relative mass of any naturally occurring sample of chlorine is about 35.5 amu. Thus, we say that the atomic weight of chlorine is 35.5. The same reasoning applies to all the other elements and their atomic weights; these are given on the periodic table.

Atomic weights are therefore relative weights with the isotope carbon-12 (often written as ^{12}C) taken as a standard. For example, one atom of He with an atomic weight of 4 is about one-third as heavy as one atom of ^{12}C. The atomic weight as given in the periodic table is a weighted average of all isotopes found in any naturally occurring sample of the element. That is, even though all atoms of a given element do not have exactly the same mass, any naturally occurring sample will contain the same percentages of the naturally occurring isotopes. Therefore, *the atomic weight gives the average relative mass of a given type of atom.*

3-8 Types of Compounds—Molecules, Formula Units, and Ions

A compound may consist of either **ions** (ionic compounds) or **molecules** (molecular or covalent compounds). An ion is an atom that has lost or gained one or more electrons (in general, up to a maximum of three) and thus has a charge. An atom that has lost electrons has a positive charge since the electron is negative, and it is called a **cation.** An atom that has gained electrons has a negative charge, and it is called an **anion.** Since each electron has a −1

(*Table footnote continued*)
neutrons, and electrons, and their respective masses in atomic mass units, give the following for carbon-12:

$$\begin{aligned} 6 \text{ protons} \times \frac{1.0073 \text{ amu}}{1 \text{ proton}} &= 6.0438 \text{ amu} \\ 6 \text{ neutrons} \times \frac{1.0087 \text{ amu}}{1 \text{ neutron}} &= 6.0522 \text{ amu} \\ 6 \text{ electrons} \times \frac{0.00055 \text{ amu}}{1 \text{ electron}} &= \underline{0.0033 \text{ amu}} \\ & 12.0993 \text{ amu} \end{aligned}$$

This difference in mass (in this case, 0.0993 amu) is considered to be converted to energy, according to Einstein's equation [Eq. (3-1)] and its liberation is required to hold the positive protons and neutrons together in the nucleus (binding energy). This will be discussed in more detail in Chapter 25.

charge, an atom that has lost one electron has a +1 charge; if two electrons are lost, the charge is +2, and so forth. Likewise, when an atom gains one electron, it attains a −1 charge; if two electrons are gained, the charge is −2. In forming ionic compounds, the electrons are transferred from one atom to another. Consider the following examples:

SYMBOL	ELECTRONS LOST OR GAINED	TOTAL NUMBER OF PROTONS	TOTAL NUMBER OF ELECTRONS	CHARGE
Na^+	1 electron lost	11	10	+1
S^{2-}	2 electrons gained	16	18	−2

(Notice in each of these examples that only electrons are gained or lost—the number of protons remains the same.)

Since the formula of a compound gives the relative number of atoms of each element present (Section 3-1), this represents what is called a *formula unit.* Thus, a formula unit of helium consists of a He atom, a formula unit of hydrogen gas consists of an H_2 molecule, a formula unit of sodium chloride (NaCl) consists of one sodium ion (Na^+) and one chloride ion (Cl^-), and so forth.

In a molecular, or covalent, compound, electrons are shared by the atoms making up the compound. In this case, the formula unit is a molecule, as opposed to the ions that make up an ionic compound. But in either case, the formula for the compound gives the number of atoms of each kind that make up a formula unit.

3-9 Formula or Molecular Weight

The knowledge of the atomic weight scale, as found in the periodic table, can now be combined with the information obtained from the formula of a compound. From this we can calculate the **formula weight** (or **molecular weight**) of a compound. This is done by considering the number of atoms in a formula unit of the compound and the atomic weights of the elements. The term *formula weight* is the more general term since it can be used for atoms, ions, and compounds that exist as molecules or compounds that exist as formula units. The term *molecular weight,* however, is very often used for any type of compound. That is, it is used for compounds that do not actually exist as molecules as well as those that do. In this text we shall use formula weight and molecular weight interchangeably, with the more general formula weight preferred.*

*These formula weights are sometimes expressed in grams and called *gram-atomic weight* or *gram-molecular weight* or *gram-formula weight.* We shall see in Section (3-10), however, that there is a more convenient method of expressing these, in terms of the mole concept.

Consider the following examples of determining the formula weight of some compounds, using the atomic weights as found in the periodic table.

PROBLEM EXAMPLE 3-2

Calculate the formula weight of each of the following:
(a) H_2O (b) $Ca(OH)_2$ (c) $Mg(NO_3)_2$

SOLUTION:

(a) Since there are 2 atoms of hydrogen and 1 atom of oxygen in a formula unit of water, the formula (molecular) weight is

$$2 \text{ atoms H} \times \frac{1.0 \text{ amu}}{1 \text{ atom H}} = 2.0 \text{ amu}$$

$$1 \text{ atom O} \times \frac{16.0 \text{ amu}}{1 \text{ atom O}} = 16.0 \text{ amu}$$

$$\text{formula weight} = 2.0 \text{ amu} + 16.0 \text{ amu} = \mathbf{18.0 \text{ amu}}$$

This can be simplified as follows:

$$\begin{aligned} H &= 2 \times 1.0 = 1.0 \\ O &= 1 \times 16.0 = \underline{16.0} \\ \text{formula weight} &= \mathbf{18.0} \end{aligned}$$

(b) Here we have 1 atom of calcium (Ca), 2 atoms of oxygen (O), and 2 atoms of hydrogen (H). Therefore, the formula (molecular) weight is

$$\begin{aligned} Ca &= 1 \times 40.1 = 40.1 \\ O &= 2 \times 16.0 = 32.0 \\ H &= 2 \times 1.0 = \underline{2.0} \\ \text{formula weight} &= \mathbf{74.1} \quad \text{or} \quad \mathbf{74.1 \text{ amu}} \end{aligned}$$

Keep in mind that these are all relative numbers and can therefore be expressed as a pure number; however, in all cases we can call them atomic mass units.

(c) Here we have 1 atom of magnesium (Mg), 2 atoms of nitrogen (N), and 6 (3 × 2) atoms of oxygen (O). Therefore, the formula (molecular) weight is

$$\begin{aligned} Mg &= 1 \times 24.3 = 24.3 \\ N &= 2 \times 14.0 = 28.0 \\ O &= 6 \times 16.0 = \underline{96.0} \\ \text{formula weight} &= \mathbf{148.3} \quad \text{or} \quad \mathbf{148.3 \text{ amu}} \end{aligned}$$

3-10 The Mole

A **mole (mol)** is a number of particles, just as a dozen is; in fact, the mole is sometimes referred to as the "chemists dozen." We shall consider the exact value to the number of particles in a mole a little later; for now let us just say that it is a very large number. Using the letter N to stand for this number of particles, we can thus say that, just as there are 12 donuts in a dozen donuts, there are N atoms in a mole of atoms. Also, just as there are 12 of anything in a dozen, there are N of any particles in a mole. For example, a mole of sodium atoms contains N atoms, a mole of sodium ions (Na^+) contains N ions, a mole of water molecules contains N molecules, and a mole of electrons contains N electrons.

Since it was stated previously that all sodium atoms have the same average mass, a mole of sodium atoms will always have the same mass. Likewise, a mole of water molecules will always have the same mass, and so forth.

Just as the relative masses of atoms are based on carbon-12 as a standard so is the mole. That is, the number N is the number of atoms that there are in a certain, constant amount of carbon-12. This amount is taken as the atomic weight of carbon-12 (12.0000) expressed in grams. That is, the number N is equal to the number of atoms that there are in exactly 12 grams of carbon-12.

If we now combine this idea with what we know about the atomic weight scale, we can determine the mass of a mole of any particles such as atoms, ions, molecules, and so on. That is, since a hydrogen atom is approximately one-twelfth as heavy as a carbon-12 atom and since N atoms of carbon-12 have a mass of 12 g, we can see that N atoms of hydrogen have a mass of approximately 1 g and N molecules of H_2 have a mass of approximately 2 g.

Therefore, we can say that 1 mole of ^{12}C has a mass of 12.0 g, by definition. From this the following relationships are obtained:

$$1 \text{ mole H atoms} = 1.0 \text{ g}$$
$$1 \text{ mole } H_2 \text{ molecules} = 2.0 \text{ g}$$
$$1 \text{ mole } H_2O \text{ molecules} = 18.0 \text{ g}$$

Although a mole represents a number of particles, it can also be expressed as the mass in grams that contains that number of particles. It must be kept in mind that the mole is defined in terms of grams; that is, the number of atoms that there are in exactly 12 g of carbon-12. To emphasize this, the term *gram-mole* is sometimes used, but this is not necessary.*

*Occasionally the terms pound-mole, ton-mole, and so on are used. These would represent the formula weight expressed in pounds or tons, respectively. This is not necessary, however, if the concept of the mole is used properly; therefore, we shall not consider these terms in this text. If these are encountered, you would simply convert the given units to grams and use the proper definition of the mole (or gram-mole).

Now that we have defined the mole, we want to consider the actual value of N. It is found experimentally, by diffraction of X rays and other methods, that the number of atoms in exactly 12 g of carbon-12 is 6.02×10^{23} (to three significant digits). This number is known as **Avogadro's number,** named in honor of the Italian physicist and chemist Amedeo Avogadro (1776–1856). Thus, 1 mole of formula units contains 6.02×10^{23} of these units, whether they are atoms, molecules, groups of ions, or whatever. The mass of 6.02×10^{23} formula units (atoms, molecules, ions, and so on) is thus the formula weight expressed in grams.

The mass of 1 mole of a substance is sometimes referred to as the *molar mass.* That is, the molar mass is the mass of any substance that contains Avogadro's number of units, where the units can be atoms, molecules, formula units, individual ions, and so on.

The concept of the mole also gives us another way of expressing atomic, molecular, or formula weights as grams per mole. This is often useful in doing calculations, as we shall see in some of the following problem examples. For example, since the formula weight of water is 18.0 amu, we can say that 1 mole of water has a mass of 18.0 g. This can also be expressed as 18.0 g H_2O/mol.

In this text we shall generally use the term *formula weight* for elements and compounds, even if they are known to exist in the form of molecules. This is an important point to consider in doing calculations, and, if done carefully, much confusion can be avoided. For example, the formula weight of hydrogen atoms (H) is 1.0, but the formula weight of hydrogen molecules (H_2) is 2.0; thus, 1 mol of H atoms = 1.0 g and 1 mol of H_2 = 2.0 g. Likewise, the formula weight of NaCl is 58.5, and thus 1 mol of NaCl is 58.5 g.

We shall now consider several types of calculations that can be done using the mole concept. It must be kept in mind when doing calculations of this type that the mole represents an amount of a substance. This amount can be expressed as a number of particles (atoms, molecules, ions, groups of ions, and so on) or as a mass in grams. Also, in discussing moles, we often find it useful to use the prefix milli- for 1/1000 mole; this is generally written as the abbreviation *mmol.*

PROBLEM EXAMPLE 3-3

Calculate the number of moles of NaOH in 85.0 g of NaOH.

SOLUTION:

The formula weight of NaOH is 40.0; therefore, there are 40.0 g of NaOH in 1 mole. Thus, the number of moles in 85.0 g of NaOH is calculated as

$$85.0 \text{ g } \cancel{\text{NaOH}} \times \frac{1 \text{ mol NaOH}}{40.0 \text{ g } \cancel{\text{NaOH}}} = \mathbf{2.12 \text{ mol NaOH}}$$

PROBLEM EXAMPLE 3-4

Calculate the mass in grams of 0.720 mol of $Ca_3(PO_4)_2$.

SOLUTION:

The formula weight of $Ca_3(PO_4)_2$ is $(3 \times 40.1) + (2 \times 31.0) + (8 \times 16.0) = 310$. Therefore, 1 mol $Ca_3(PO_4)_2$ = 310 g, and the mass of 0.720 mol is calculated as

$$0.720 \cancel{\text{mol } Ca_3(PO_4)_2} \times \frac{310 \text{ g } Ca_3(PO_4)_2}{1 \cancel{\text{mol } Ca_3(PO_4)_2}} = \mathbf{223 \text{ g } Ca_3(PO_4)_2}$$

PROBLEM EXAMPLE 3-5

Calculate the number of molecules in 24.5 g of CO_2.

SOLUTION:

In this calculation we must first calculate the number of moles, using the formula weight (44.0 g CO_2/mol) and then use the relationship between molecules and moles to find the number of molecules.

$$24.5 \text{ g } \cancel{CO_2} \times \frac{1 \text{ mol } \cancel{CO_2}}{44.0 \text{ g } \cancel{CO_2}} \times \frac{6.02 \times 10^{23} \text{ molecules } CO_2}{1 \text{ mol } CO_2} = \mathbf{3.35 \times 10^{23} \text{ molecules } CO_2}$$

PROBLEM EXAMPLE 3-6

Calculate the mass in grams of one atom of carbon-12.

SOLUTION:

In this case, we must use both the number of atoms in a mole and the mass (in grams) of a mole. The problem says

$$1 \text{ atom } ^{12}C = ?\text{ g}$$

and therefore we proceed as follows:

$$1 \text{ atom } \cancel{^{12}C} \times \frac{1 \cancel{\text{mol } ^{12}C}}{6.02 \times 10^{23} \cancel{\text{atom } ^{12}C}} \times \frac{12 \text{ g}}{1 \cancel{\text{mol } ^{12}C}} = \mathbf{2.0 \times 10^{-23} \text{ g}}$$

PROBLEM EXAMPLE 3-7

Calculate the volume in cubic centimeters occupied by 3.50×10^{26} atoms of aluminum if the density of Al is 2.70 g/cm^3.

SOLUTION:

We must first convert from atoms to moles (using Avogadro's number), then from moles to grams (using the atomic weight), and finally from grams to cubic centimeters (using the density).

$$3.50 \times 10^{26} \text{ atoms Al} \times \frac{1 \text{ mol Al}}{6.02 \times 10^{23} \text{ atoms Al}} \times \frac{27.0 \text{ g Al}}{1 \text{ mol Al}} \times \frac{1 \text{ cm}^3}{2.70 \text{ g Al}} = \mathbf{5.81 \times 10^3 \text{ cm}^3}$$

PROBLEM EXAMPLE 3-8

If 25.00 g of a compound contain 0.4200 mol, what is the formula weight of the compound?

SOLUTION:

Since formula weight can be expressed in terms of grams per mole, we divide the number of grams by the number of moles:

$$\frac{25.00 \text{ g}}{0.4200 \text{ mol}} = \mathbf{59.52 \text{ g/mol}}, \text{ or } \mathbf{59.52 \text{ amu}}$$

PROBLEM EXAMPLE 3-9

If 4.20×10^{20} atoms of an element have a mass of 55.2 mg, what is the atomic weight of the element?

SOLUTION:

This is similar to the last example except that more steps are involved. Again, we want grams/mole, but we are starting with milligrams/atom.

$$\frac{55.2 \text{ mg}}{4.20 \times 10^{20} \text{ atoms}} \times \frac{1 \text{ g}}{10^3 \text{ mg}} \times \frac{6.02 \times 10^{23} \text{ atoms}}{1 \text{ mol}} = \mathbf{79.1 \frac{g}{mol}}$$

PROBLEM EXAMPLE 3-10

Calculate the mass, in kilograms, of 3.25 mmol of H_2O

SOLUTION:

Recall that one mole contains 10^3 mmol, just like milligrams, milliliters, etc.

$$3.25 \text{ mmol } H_2O \times \frac{1 \text{ mol } H_2O}{10^3 \text{ mmol } H_2O} \times \frac{18.0 \text{ g } H_2O}{1 \text{ mol } H_2O} \times \frac{1 \text{ kg } H_2O}{10^3 \text{ g } H_2O} = \mathbf{5.85 \times 10^{-5} \text{ kg } H_2O}$$

PROBLEM EXAMPLE 3-11

Calculate the number of hydrogen atoms in 0.25 mg of H_2

SOLUTION:

There are several steps involved here. First, we must convert from milligrams to grams (since the mole is defined in terms of grams); then go from moles to molecules; and finally from molecules to atoms.

$$0.25\ \text{mg H}_2 \times \frac{1\ \text{g H}_2}{10^3\ \text{mg H}_2} \times \frac{1\ \text{mol H}_2}{2.0\ \text{g H}_2} \times \frac{6.02 \times 10^{23}\ \text{molecules H}_2}{1\ \text{mol H}_2} \times \frac{2\ \text{atoms H}}{1\ \text{molecule H}_2} = \mathbf{1.5 \times 10^{20}\ atoms\ H}$$

3-11 Introduction to the Periodic Table of the Elements: Metals and Nonmetals

The elements are arranged according to atomic numbers in a table called the *periodic table* of the elements.* This lists the elements with their symbols, atomic numbers, atomic weights, and so forth. The elements are divided into *metals* and *nonmetals,* and the differentiating factors are the physical and chemical properties of the elements.

Slightly to the right of the center of the periodic table is a dark stairstep line that in general separates the metals from the nonmetals. To the left of this line are the metals and to the right are the nonmetals. Such elements as sodium, copper, iron, gold, lead, and platinum are metals. Elements such as carbon, nitrogen, oxygen, sulfur, chlorine, helium, and neon are nonmetals. You will notice from the periodic table that the majority of the elements are metals.

The last column of elements in the periodic table is a special group, called the *noble gases.* Since they are relatively unreactive, they occur in nature in the free state. For example, helium (He) is found in gas wells, and small amounts of both helium and argon (Ar) are found in the air. Originally, it was thought that these elements would not react with anything and for this reason they were called the inert gases. It is now known, however, that they can be made to react but not very readily; hence, they are now called the noble gases.

The general physical characteristics of metals are high luster, electrical and thermal conductivity, hardness, high density, and high melting point. Of course, there are exceptions to these general properties, such as sodium, a soft metal which is less dense than water, and mercury, which is a liquid at room temperature.

*We shall consider the periodic table in detail in Chapter 5 and again in Chapters 21 through 23.

The chemical characteristics of metals will be considered later. For now, we might note that they do not combine with each other. Most of them combine with nonmetals, however, and for this reason they usually occur in nature combined with such nonmetals as oxygen or sulfur.

The general properties of the nonmetals are for the most part opposite the properties of the metals. For example, they are poor conductors, generally are soft (if they exist as solids at ordinary temperatures), and often have low melting and boiling points; thus several exist as gases at ordinary temperatures, such as hydrogen, nitrogen, oxygen, fluorine, and chlorine. As with the metals there is considerable variation in some of these properties such as melting and boiling point.

One general chemical characteristic of the nonmetals is that they combine with metals, and also some of them combine with each other, such as in compounds of nitrogen and oxygen, hydrogen and sulfur, or carbon and chlorine.

3-12 Solutions

A **solution** is a homogeneous mixture, made up of a **solute** dissolved in a **solvent.** For example, in a water (aqueous) solution of sodium chloride, the sodium chloride is the solute and the water is the solvent. Most of the solutions that you will work with in the laboratory will be aqueous (water) solutions. In Chapter 13 solutions will be considered in detail, and at that time we shall consider solutions in all three states of matter. For now we want to consider briefly some of the properties of aqueous (water) solutions and a couple of methods that are used to express *concentration* of these solutions.

When a solid that is made up of ions (an ionic compound) is dissolved in a liquid, the solution formed has the ability to conduct an electric current, as shown in Fig. 3-2. These substances are thus known as **electrolytes,** whereas substances that do not ionize are called **nonelectrolytes.**

If the solution shown in Fig. 3-2 is an electrolyte, the light bulb will light when the electrodes are placed in the solution. If the solution is a nonelectrolyte, the bulb will not light.

The *concentration* of a solution refers to the amount of solute in a given amount of solvent or solution. There are several methods used to express concentration of solutions. We shall consider percent (by mass) and molar concentration at this time.

Percent concentration (by mass) is simply calculated by dividing the mass of solute by the total mass of solution and multiplying times 100 (recall that percent means *fraction of a hundred* or *parts per 100*).

This is expressed as

$$\text{percent (\%)} = \frac{\text{mass of solute}}{\text{mass of solution}} \times 100$$

Consider the following problem examples.

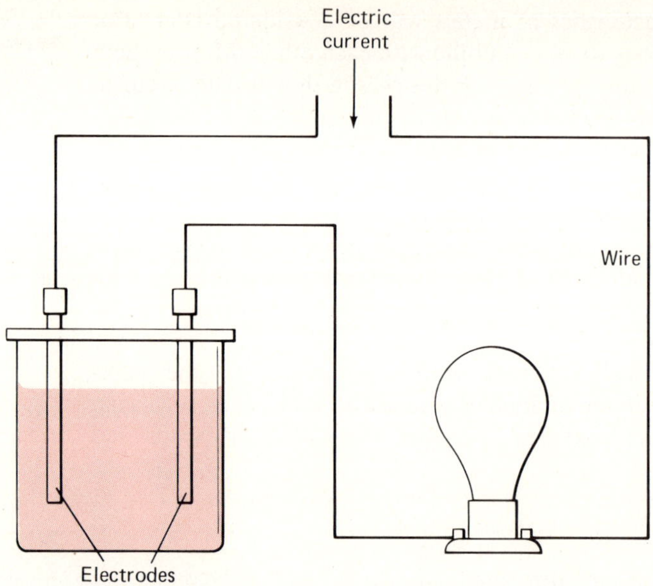

Figure 3-2
Apparatus used to determine if a solution contains an electrolyte or nonelectrolyte

PROBLEM EXAMPLE 3-12

Calculate the percent concentration of sodium chloride (NaCl) in a solution that contains 2.8 g of NaCl in 75 g of solution.

SOLUTION:

$$\% \text{ NaCl} = \frac{2.8}{75} \times 100 = \mathbf{3.7\%}$$

PROBLEM EXAMPLE 3-13

Calculate the percent concentration of NaCl in a solution made by adding 8.50 g of NaCl to 24.0 g of water.

SOLUTION:

First we must find the total mass of the solution, by adding the 8.5 g of solute to the 24.0 g of water, to give

$$8.50 + 24.0 = 32.5 \text{ g solution}$$

Then, the percent concentration is calculated as

$$\% \text{ NaCl} = \frac{8.50 \text{ g}}{32.5 \text{ g}} \times 100 = \mathbf{26.2\%}$$

Molar (M) *concentration* expresses the moles of solute per liter of solution. That is,

$$\textbf{molar } (M) = \frac{\textbf{moles of solute}}{\textbf{liters of solution}}$$

This is a very useful method of expressing concentration of solution and the most commonly used method in the laboratory.

Consider the following calculations involving molar concentration.

PROBLEM EXAMPLE 3-14

Calculate the molar (M) concentration of a solution that contains 3.25 mol of sodium chloride in 250 mℓ of solution.

SOLUTION:

We shall use the volume as 0.250 ℓ (that is, 250 mℓ = 0.25 ℓ), to give

$$\text{M} = \frac{3.25 \text{ mol solute}}{0.250\ \ell \text{ solution}} = \mathbf{13.0}\ \boldsymbol{M}$$

PROBLEM EXAMPLE 3-15

Calculate the molar concentration of sodium hydroxide (NaOH) in a solution that contains 24.0 g of NaOH in 375 mℓ of solution.

SOLUTION:

In this case we must first calculate the number of moles of NaOH present as

$$24.0 \text{ g } \cancel{\text{NaOH}} \times \frac{1 \text{ mol NaOH}}{40.0 \text{ g } \cancel{\text{NaOH}}} = 0.600 \text{ mol NaOH}$$

Then calculate the molar concentration by dividing the 0.600 mol of NaOH by the 0.375 ℓ (375 mℓ) of solution, to give,

$$\text{M} = \frac{0.600 \text{ mol NaOH}}{0.375\ \ell \text{ soln}} = \mathbf{1.60\ M\ NaOH}$$

An alternate solution would be to start with 24.0 g/375 mℓ and convert to mol/ℓ, as follows:

$$\frac{24.0 \text{ g } \cancel{\text{NaOH}}}{375 \cancel{\text{ m}\ell \text{ soln}}} \times \frac{1000 \cancel{\text{ m}\ell \text{ soln}}}{1\ \ell \text{ soln}} \times \frac{1 \text{ mol NaOH}}{40.0 \text{ g } \cancel{\text{NaOH}}} = \mathbf{1.60}\ \boldsymbol{M}\ \textbf{NaOH}$$

PROBLEM EXAMPLE 3-16

Calculate the number of grams of NaOH necessary to prepare 2.5 ℓ of 0.15 M solution.

SOLUTION:

The problem says, 2.5 l soln = ? g NaOH, given that it is a 0.15 M solution (which contains 0.15 mole NaOH in each liter of solution). Therefore, it is calculated as

$$2.5\,\cancel{l} \times \frac{0.15\ \cancel{\text{mol NaOH}}}{\cancel{l}} \times \frac{40\ \text{g NaOH}}{1\ \cancel{\text{mol NaOH}}} = \mathbf{15\ g\ NaOH}$$

SUMMARY

1. Matter is composed on tiny particles called *atoms.* Atoms of a given *element* have identical chemical properties. Atoms of different elements have different properties. There are presently 106 different kinds of atoms (elements) known. Each of these is represented by a *symbol.*
2. A *compound* is formed by the chemical combination of atoms in definite and constant proportions. The *formula* of a compound tells what atoms are present and in what proportions. A *mixture* is a combination of substances in variable proportions. This can be *homogeneous* (as in a solution) or *heterogeneous.*
3. A *pure substance* (substance) thus has a symbol (for an element) or a formula (for a compound) whereas a mixture does not.
4. A *physical property* of a substance is a property that can be observed without changing the composition, and thus a *physical change* is simply a change in form; that is, it does not change the composition.
5. A *chemical property* can only be observed when a substance undergoes a change in composition. That is, in a *chemical change* a substance is changed into one or more new substances.
6. According to the Law of Conservation of Matter (Mass), in a *chemical reaction* no atom disappears or is changed into an atom of another element. Therefore, in a *chemical equation* there must be exactly the same number of each atom on each side of the equation. That is, the equation must be *balanced.*
7. All changes (physical and chemical) are accompanied by a change in *energy.* It has been shown that small amounts of matter (mass) can be converted to energy, and thus the laws of conservation of matter and energy are often combined into one Law of Conservation of Matter-Energy.
8. Since individual atoms are too small to be weighed, the masses of atoms are assigned relative values, compared to each other, taking carbon-12 as the standard. Thus, on this scale the average hydrogen atom has a relative mass of about 1, oxygen about 16, and so on. These numbers are sometimes expressed as *atomic mass units* (amu).

9. In an atom the *protons* (mass of almost 1 amu, charge of +1) and *neutrons* (mass of almost 1 amu, no charge) are found in the *nucleus* (center of the atom). The *electrons* (negligible mass, charge of −1) are found outside the nucleus in various energy levels. The number of protons in the nucleus (which is equal to the number of electrons outside the nucleus for a neutral atom) corresponds to the *atomic number* of the atom. The nucleus makes up most of the mass of the atom but only a tiny fraction of the size.
10. The *mass number* of an atom corresponds to the sum of the protons and neutrons in the atom. The general notation $^{A}_{Z}E$ gives the symbol for the element (E), the atomic number (Z), and the mass number (A). The number of neutrons in the atom is equal to $A - Z$. For example, the symbol $^{18}_{8}O$ means that in this atom of oxygen (O) there are 8 protons, 8 electrons, and $18 - 8 = 10$ neutrons.
11. *Isotopes* are atoms with the same atomic number but different mass numbers, due to different numbers of neutrons. That is, they are atoms of the same element, but they have different relative masses. All the elements occur naturally as a constant percentage of a small number of isotopes. Therefore, by using the relative masses of different isotopes and percent abundance of each one in any naturally occurring sample of that element, the average relative atomic mass of that element can be calculated. This gives the *atomic weight scale,* which lists the average, relative atomic mass for each of the elements.
12. A compound may consist of *molecules* (in which electrons are shared between atoms) or *ions* (in which electrons are transferred from one atom to another). In either case, the *formula* of the compound gives the relative number of atoms present in a *formula unit* of the compound (the formula unit for an element is simply represented by the symbol for that element). The formula, or molecular, weight of a substance is calculated by considering the number of atoms in a formula unit of the compound and the atomic weight of the elements.
13. A *mole* is the amount of a substance that contains the same number of individual entities as there are atoms in 12.0 g of carbon-12. This number, known as *Avogadro's number,* has been found experimentally to be 6.02×10^{23}. The individual entities that make up a mole can be atoms, molecules, ions, or any specified group of ions such as a formula unit of an ionic compound. Using the atomic weight scale it can be seen that 1 mole of any substance is the atomic, molecular, or formula weight *expressed in grams* (where the general term formula weight can be used for substances that exist as individual atoms, molecules, ions, or groups of ions). Therefore, a mole can be thought of (and used in calculations) as a number of particles or as the mass of a substance that contains that number of particles.
14. The elements are arranged in the *periodic table* according to increasing atomic number; this table also includes the symbol for the element and the atomic weight. The elements are divided into *metals*

and *nonmetals,* according to their physical and chemical properties. In the table, the nonmetals are on the right side and the upper portion generally (see the dividing line on the periodic table).

15. A *solution* is a homogeneous mixture made up of a *solute* dissolved in a *solvent.* The *concentration* of the solution expresses the amount of solute in a given amount of solution. This can be expressed (among other methods) as percent by mass or volume

$$\left(\frac{\text{amount of solute}}{\text{amount of solution}} \times 100\right)$$

or as molar (M) concentration (moles of solute per liter of solution).

Upon completion of this chapter, you should be able to do the following things (using a periodic table when necessary).

1. Given the formula of any compound, tell how many atoms of each element are present in a formula unit. (1)
2. Given the mass number and the name, symbol, or atomic number for the element, tell how many electrons, protons, and neutrons there are in that particular isotope. (2)
3. Given the symbol and charge on a particular ion, tell how many protons and electrons there are in that ion. (3)
4. Given the exact atomic weight of each isotope and the percent of each isotope present in nature, for a given element, calculate the atomic weight. (4)
5. Given the formula for a compound, calculate the formula weight of the compound. (5)
6. Given the formula for a compound, calculate the mass in grams of a given number of moles of the compound. (6)
7. Given the formula and mass of a substance, calculate the number of moles present. (7)
8. Given the number of moles of a substance and the formula, calculate the number of atoms or molecules present; or, given the number of atoms or molecules, calculate the mass or volume (if the density is known). (8, 9, 10, 11)
9. Given the mass (or the volume and density) of a substance and the number of moles, atoms, or molecules, calculate the atomic or molecular weight. (12)
10. Given the formula of a substance, calculate the number of moles of a particular element in a given amount of the substance. (13)
11. Given the mass of solute and the amount of solution, calculate the percent or molar (M) concentration. (14, 15)

12. Given the concentration (percent or molar), calculate the amount of solute necessary to prepare a given amount of a solution.
13. Given any of the terms listed in Exercise 1, give a brief definition or explanation in your own words.

EXERCISES

1. Define the following terms:
   - (a) matter
   - (b) atom
   - (c) element
   - (d) symbol
   - (e) compound
   - (f) formula
   - (g) molecule
   - (h) ion
   - (i) anion
   - (j) cation
   - (k) formula unit
   - (l) pure substance
   - (m) mixture
   - (n) electron
   - (o) proton
   - (p) neutron
   - (q) atomic number
   - (r) mass number
   - (s) nucleus
   - (t) isotopes
   - (u) atomic mass unit
   - (v) atomic weight scale
   - (w) formula (molecular) weight
   - (x) mole
   - (y) states of matter
   - (z) chemical equation
   - (aa) Law of Conservation of Matter
   - (bb) solution
   - (cc) solute
   - (dd) solvent
   - (ee) percent (by mass) concentration
   - (ff) molar concentration
2. Distinguish between each of the following, and give an example of each case.
   - (a) pure substance and mixture
   - (b) element and compound
   - (c) physical property and chemical property
   - (d) physical change and chemical change
   - (e) potential and kinetic energy
   - (f) anion and cation
   - (g) solute and solvent
3. List two differences between isotopes of the same element.
4. Boron, with an atomic weight of 10.8, consists of two isotopes: boron-10 and boron-11. Which one of these isotopes is predominate in any naturally occurring sample? Explain your answer.

PROBLEMS

Symbols and Formulas

1. Tell how many atoms of each element are present in a formula unit of each of the following: (Write the name of the element and the number of atoms.)
   - (a) H_2O
   - (b) CH_4
   - (c) NH_3
   - (d) $Ca(OH)_2$
   - (e) $Mg(NO_3)_2$
   - (f) $MgCl_2$
   - (g) C_6H_6
   - (h) Na_2SO_4
   - (i) $Fe_3(PO_4)_2$
   - (j) $Cu(BrO_4)_2$

Composition of Atoms

2. Tell how many protons, how many electrons, and how many neutrons there are in each of the following atoms:
(a) ^{12}C (f) ^{4}He
(b) ^{80}Br (g) ^{18}O
(c) ^{2}H (h) ^{31}P
(d) ^{214}Bi (i) ^{214}Pb
(e) ^{206}Pb (j) ^{238}U

Ions

3. Tell how many protons and how many electrons there are in each of the following ions:
(a) Na^+ (f) Fe^{3+}
(b) Ca^{2+} (g) S^{2-}
(c) Br^- (h) K^+
(d) Ni^{3+} (i) Ba^{2+}
(e) O^{2-} (j) Cl^-

Atomic and Formula (Molecular) Weights

4. Calculate the atomic weight to the proper number of significant digits for each of the following from the given data:

(a)

Isotope	Exact Atomic Mass (amu)	% Abundance in Nature
Ga-69	68.94	60.40
Ga-71	70.92	39.60

(b)

Isotope	Exact Atomic Mass (amu)	% Abundance in Nature
N-14	14.00	99.63
N-15	15.00	0.3700

5. Calculate the formula (molecular) weight of each of the following:
(a) $CaCl_2$ (f) $MgBr_2$
(b) NH_3 (g) CH_4
(c) $Ba(OH)_2$ (h) $Al(OH)_3$
(d) $Fe_3(SO_4)_2$ (i) $Ca_3(PO_4)_2$
(e) $KClO_4$ (j) $NaClO_3$

Moles

6. Calculate the mass in grams of 1 mole of each of the following:
(a) $CaCrO_4$ (c) $Na_2C_2O_4$
(b) $CaSO_4$ (d) $C_6H_{12}O_6$

7. Calculate the number of moles in each of the following:
(a) 4.80 g Al (f) 8.00 g of NaCl
(b) 3.00 lb NH_4Cl (g) 2.50 oz of $Ca(OH)_2$
(c) 2.00×10^{24} atoms of Na (h) 8.30×10^{20} molecules H_2O
(d) 0.050 mg H_2O (i) 0.020 kg H_2O
(e) 2.0 μg Mg (j) 3.5 mg Ca

8. Calculate the number of atoms in each of the following:

(a) 0.500 mol C
(b) 3.5 mol H_2
(c) 0.025 g He
(d) 2.2 kg Al
(e) 0.50 lb NH_3
(f) 0.85 mol Al
(g) 4.2 mol O_2
(h) 12.5 g Na
(i) 0.020 mg Fe
(j) 4.8 oz CH_4

9. Calculate the number of molecules in each of the following:

(a) 10.5 g O_2
(b) 2.5 lb NH_3
(c) 0.020 mg SO_2
(d) 2.5 l of water
(e) 8.00 mol CH_4
(f) 20.0 g H_2
(g) 8.52 lb CH_4
(h) 1.25 gal H_2O
(i) 0.015 mg CO_2
(j) 3.50 mol H_2

10. Calculate the mass, in grams, of each of the following:

(a) 1.8 mol CO_2
(b) 12.5 mmol NaCl
(c) 2.25×10^{22} molecules H_2O
(d) one atom nitrogen
(e) 0.25 mol SO_3
(f) 8.25 mmol $CaCl_2$
(g) 4.02×10^{24} molecules NH_3
(h) one atom aluminum

11. Calculate the volume, in cubic centimeters, occupied by each of the following:

(a) one mol sulfur (density = 2.1 g/cc)
(b) 25 mmol gold (density = 19 g/cc)
(c) 8.5×10^{24} molecules CCl_4 (density = 1.6 g/ml)
(d) 1 atom lead (density = 11.3 g/cc)
(e) 1.2 kg NaCl (density = 2.2 g/cc)
(f) 8.5 mol NaCl (density = 2.2 g/cc)
(g) 0.022 mmol lead (density = 11.3 g/cc)
(h) 1 atom gold (density = 19 g/cc)
(i) 6.2×10^{22} molecules CCl_4 (density = 1.6 g/ml)
(j) 85 mg sulfur (density = 2.1 g/cc)

12. Calculate the molecular (formula) weight of each of the following substances:

(a) 98.5 g contains 2.3 mol
(b) 2.0×10^5 mg contains 3.5 mol
(c) 42 g contains 8.5×10^{22} molecules
(d) 2.8 kg contains 1.2×10^{24} molecules
(e) 2.0 mg contains 1.0×10^{18} molecules
(f) 8.5×10^4 mg contains 1.8 mol
(g) 92.0 g contains 1.2 mol
(h) 55 g contains 2.2×10^{22} molecules
(i) 3.5 kg contains 1.0×10^{24} molecules
(j) 4.2 mg contains 2.2×10^{18} molecules

13. Calculate the number of moles of oxygen atoms in each of the following:

(a) 8.20 mol O_2
(b) 10.5 g O_2
(c) 24.0 mmol Al_2O_3
(d) 2.52 g $Ca(NO_3)_2$
(e) 0.87 lb $Ca(OH)_2$
(f) 0.53 mol O_2
(g) 0.020 mg O_2
(h) 15 mmol $Ca(OH)_2$
(i) 2.5 tons $Mg(OH)_2$
(j) 2.2 kg $Mg_3(PO_4)_2$

Solutions

14. Calculate the percent of solute in each of the following solutions:
    (a) 8.60 g of NaCl in 95.0 g of solution
    (b) 25.0 g of K_2CO_3 in 100 ml of water
    (c) 0.150 kg of KOH in 850 g of solution
    (d) 300 mg of $CaCl_2$ in 2.80 ml of water
    (e) 1.8 mg of NaOH in 8.8 μl of water
    (f) 12.5 g of KOH in 135 g of solution
    (g) 32.2 g of NaCl in 150 ml of water
    (h) 425 mg of NaOH in 0.120 kg of solution
    (i) 0.035 kg of $Ca(OH)_2$ in 875 ml of water
    (j) 750 mg of $CaCl_2$ in 0.0115 μl of water
15. Calculate the molar (M) concentration of each of the following solutions:
    (a) 1.75 moles of H_2SO_4 in 0.375 l of solution
    (b) 24 g of NaOH in 1.5 l of solution
    (c) 18.5 g of $CaCO_3$ in 235 ml of solution
    (d) 4.2 mg $Ca(OH)_2$ in 2.5 ml solution
    (e) 8.5 lb $CaCl_2$ in 1.2 gal solution
    (f) 4.5 moles of H_3PO_4 in 0.45 l of solution
    (g) 35 g of $Ca(OH)_2$ in 0.82 l of solution
    (h) 12.3 g of $CaCl_2$ in 350 ml of solution
    (i) 87 mg $CaCO_3$ in 50.0 ml solution
    (j) 5.2 lb NaOH in 3.5 gal solution
16. Calculate the number of grams of solute necessary to make each of the following solutions:
    (a) 350 ml 1.8 M NaOH solution
    (b) 125 ml 6.25 M $Ca(NO_3)_2$ solution
    (c) 240 g 3.5% NaCl solution
    (d) 225 ml 12.0% $CaCl_2$ solution (density = 1.08 g/ml)
    (e) 1.25 l 8.25% $NaNO_3$ solution (specific gravity = 1.05)
    (f) 75 ml 6.0 M NaCl solution
    (g) 475 ml of 1.85 M $Mg(OH)_2$ solution
    (h) 35 g 12% $NaNO_3$ solution
    (i) 875 ml 45.0% $Ca(NO_3)_2$ solution (density = 1.15 g/ml)
    (j) 2.75 l 60.0% $CaCl_2$ solution (specific gravity = 1.18)

General problems

17. Carry out the following conversions:
    (a) 125 g H_2O to millimoles
    (b) 25 mmol Cu to ounces
    (c) 8.2×10^{25} molecules CH_4 to pounds
    (d) 0.20 kg Al to millimoles
    (e) 2.5 l CCl_4 to molecules (specific gravity = 1.6)
    (f) 1.8 mmol water to liters
    (g) 18.5 g NaCl to millimoles
    (h) 2.0 mmol H_2O to pounds
    (i) 1.2×10^{20} molecules CO_2 to ounces
    (j) 8.50 kg CCl_4 to millimoles
    (k) 24.8 mmole CCl_4 to liters (specific gravity = 1.6)
    (l) 0.24 l water to millimoles

18. Calculate the number of moles in each of the following:
   (a) Hydrochloric acid (HCl) in 52.8 g of 35.0% HCl solution
   (b) H_3PO_4 in 425 ml of 70.0% H_3PO_4 solution (the specific gravity of 70.0% H_3PO_4 is 1.53)
   (c) Sulfuric acid (H_2SO_4) in 35 ml of 60% sulfuric acid solution (the specific gravity of 60% H_2SO_4 is 1.5)
   (d) Sulfuric acid (H_2SO_4) in 56 g of 48% sulfuric acid solution
   (e) Phosphoric acid (H_3PO_4) in 360 ml of 1.5 M H_3PO_4 solution
   (f) Hydrochloric acid (HCl) in 225 ml of 15.5% HCl solution (the specific gravity of 15.5% HCl is 1.12)
19. Look up the following temperatures and convert them to °F and K:
   (a) melting point of mercury
   (b) boiling point of ethyl alcohol
   (c) boiling point of chloroform
   (d) melting point of sodium chloride
20. Calculate the volume, in liters, of each of the following:
   (a) 8.5×10^{22} molecules of a substance with a specific gravity of 1.37 and a molecular weight of 125 amu
   (b) 4.2×10^{21} molecules of a substance with a specific gravity of 1.22 and molecular weight of 185 amu
21. Calculate the number of calories necessary to carry out each of the following temperature changes:
   (a) 4.2×10^{24} molecules of water from 0° to 20°C
   (b) 8.2×10^{23} atoms of zinc from 25° to 55°C (specific heat of Zn is 0.092 cal/g °C)
   (c) 2.1×10^{25} atoms of tin from 10° to 85°C (specific heat of Sn is 0.054 cal/g °C)
   (d) 5.0×10^{26} molecules of water from 20.0° to 100.0°C

References

REFERENCES

DINGA, GUSTAV P., "The Elements and the Derivation of Their Name and Symbols," *Chemistry,* Feb. 1968, **41,** p. 20.

FEINBERG, GERALD, "Ordinary Matter," *Sci. Am.,* May 1967, **216,** p. 126.

GUGGENHEIM, E. A., "The Mole and Related Quantities," *J. Chem. Educ.,* 1961, **38,** p. 86.

KIEFFER, WILLIAM F., *The Mole Concept in Chemistry.* New York: Reinhold Publishing Corporation, 1963.

SEABORG, GLENN T., "Prospects for Further Considerable Extension of the Periodic Table," *J. Chem. Educ.,* Oct. 1969, **46,** p. 626.

WHITAKER, ROBERT D., "An Historical Note on the Conservation of Mass," *J. Chem. Educ.,* Oct. 1975, **52,** p. 658.

Objectives

1. To understand the major aspects of the historical development of theories of atomic structure.
2. To understand what is meant by the term *energy level* and to be able to tell what the maximum number of electrons is for each energy level.
3. To understand what is meant by a sublevel and to be able to tell into what sublevels each energy level is divided.
4. To be able to tell the maximum number of electrons for each type of sublevel.
5. To understand what an orbital and a set of orbitals represent and to be able to tell how many orbitals there are in a set for each sublevel.
6. To understand the relationship between the number of orbitals and the maximum number of electrons in each sublevel.
7. To be able to give the order in which electrons occupy sublevels (the aufbau order) and using this to be able to write the electron configuration for any atom.
8. To learn Hund's rule and to be able to use it, along with the aufbau order, to show the orbital diagram for any atom and to tell how many unpaired electrons there are in any atom.
9. To learn the shapes for *s*, *p*, and *d* orbitals.
10. To understand how the periodic table is set up, following the aufbau order, and to be able to set up a periodic table with all atomic numbers.
11. To understand how the allowed solutions to the wave equation give the possible sublevels for a given energy level and the number of orbitals in a given sublevel.
12. Applying the spin quantum number and the Pauli Exclusion Principle, to show the maximum number of electrons for a given orbital, sublevel, and energy level.

Four

The Structure of the Atom

In order to understand the arrangement of the periodic table, and to be able to use it, we must consider the arrangement of the electrons in atoms. Even though the periodic table was set up before anything was known about the structure of the atom, we shall find it useful to consider the correlation between the electronic arrangement of atoms and the organization of the periodic table.

In this chapter we shall first consider the structure of atoms and the arrangement of the periodic table. Then in Chapter 5 we shall study the characteristics of the periodic table and some periodic variations within the table.

4-1 Historical Development of Theories of Atomic Structure

Following Dalton's atomic theory in 1803, a series of developments led to the modern theory of atomic structure. About 1830, Michael Faraday, an English chemist, showed that there is a relationship between electricity and matter. This was the first suggestion that perhaps the atom was not the fundamental particle of all matter but was made up of smaller particles. About 1880, experiments were carried out by Sir William Crookes that showed that if high voltage was passed through a gas discharge tube, some sort of invisible rays moved from the negative to the positive electrode. Since the negative electrode is referred to as the *cathode,* these rays were called *cathode rays.* The gas discharge tube, or cathode-ray tube, is often called the Crookes tube in honor of its inventor. A diagram of this is shown in Fig. 4-1. It was later shown that the cathode rays were actually negatively charged particles. About 6 years after this, in 1886, Eugene Goldstein found that when the cathode rays are formed in the gas discharge tube, positive particles remain. These were later shown to be positive ions, left after the electrons were lost. All of these discoveries, along with the discovery of X rays by Roentgen in 1895 and radioactivity by Becquerel in 1896, showed that the atom is not invisible, as Dalton

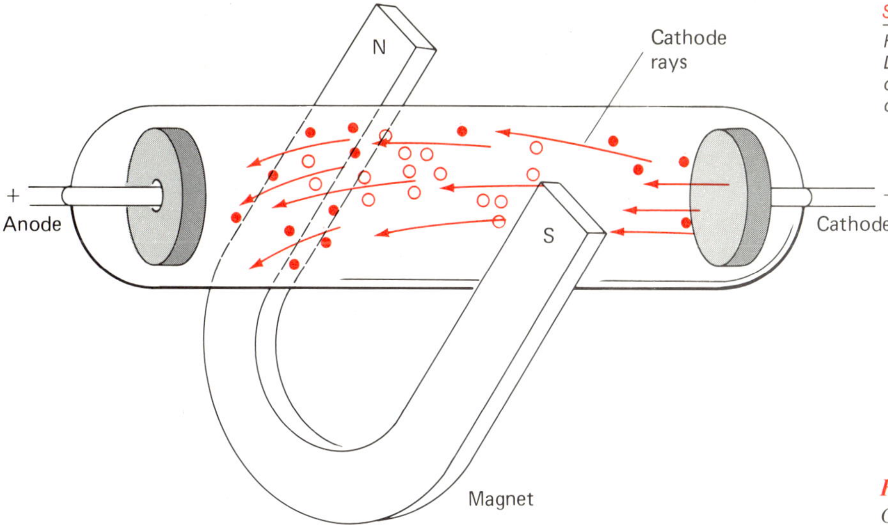

Figure 4-1
Gas discharge tube

thought, but is composed of smaller particles. Some of these particles were found to have negative and some positive charges. It was also known at this time that the atomic number of an element is equal to the number of electrons in an atom of that element. And it was found that the number of protons is equal to the number of electrons. Since the neutron was not discovered until several years later, in about 1930, however, there was some difficulty in explaining what made up the rest of the mass of the atom. It was generally considered at that point that there were more protons and electrons bound together, which made up the difference in mass.

In 1897, J. J. Thomson determined the charge to mass ratio on cathode rays and found that it was always the same for any gas. He also found that the charge to mass ratio was different for the positive ions formed from different gases. He therefore concluded that electrons could be removed from gas particles, leaving a positively charged particle that was different for different gases. In 1908, Robert Milliken determined the charge on the electron and proton with his oil drop experiment. A sketch of the apparatus used for this is shown in Fig. 4-2. Oil droplets were sprayed between the charge plates. These would fall due to the force of gravity, but Milliken showed that if they could be given a charge, such as with an X ray, they would rise toward the positive plate. The rate of this rise was measured and from this rate the amount of charge on the oil droplet was calculated. He found that this was always some small whole number multiple of -1.60×10^{-19} coul, which he concluded to be the charge on one electron. The charge on various positive particles was measured the same way, and it was found that the smallest one obtained, from the hydrogen atom, had a positive charge of 1.60×10^{-19} coul.

The next step was to determine how these various particles are arranged in the atom. One of the first theories was that of J. J. Thompson, in which he

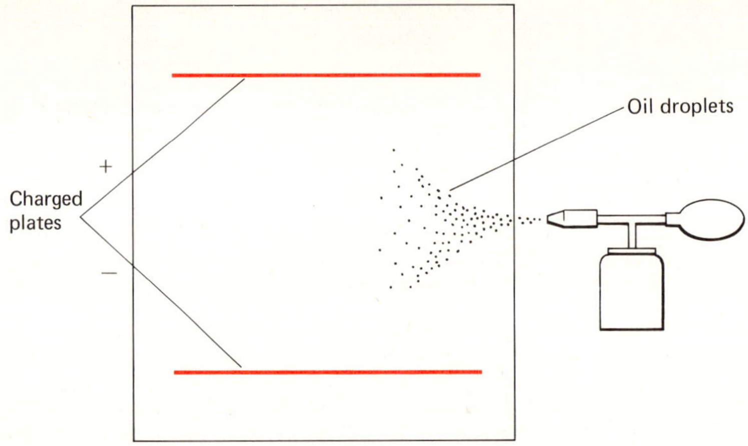

Figure 4-2
Apparatus for oil drop experiment

proposed that the atom is a mass of positive charge with electrons scattered throughout it as far apart as possible. This is sometimes referred to as the "raisin pudding" theory, in which the mass of positive charge is the pudding and the electrons are the raisins. In 1911, Ernest Rutherford carried out experiments, however, in which he bombarded a thin sheet of gold foil with tiny positively charged particles called *alpha particles.* He found that most of the alpha particles passed right through the gold foil, as would be expected from the Thompson theory. A few were deflected, however, and this led Rutherford to the conclusion that the positive part of the atom must be concentrated in a small part of the atom. Therefore, he proposed the planetary, or nuclear, model of the atom. In this model the positive part (protons) is in the center of the atom called the *nucleus,* and the electrons are considered to be spinning around the nucleus as the planets do around the sun.

There were also difficulties, however, with this model of the atom. First of all, according to classical physics, a charged particle moving in a curved path around the nucleus should radiate energy, and thus it would be expected that the electron would eventually spiral into the nucleus.

Also, it was found that the electrons did not give off a continuous spectrum of light when they were allowed to fall back into their original ground state after having been in an "excited state" (when energy was put into the atom). Only light of a definite wavelength was given off for a particular element. This can be shown visibly for some elements that give off light at a wavelength that we see as a particular color. For example, when compounds of sodium are heated, they give off light of a yellow-orange color and potassium compounds give a purple color.

In 1913, Niels Bohr proposed a model of the atom that would account for these experimental results. He proposed that the electrons in an atom could only be in certain orbits, or energy levels, around the nucleus; that is, his theory was that the energy of the electrons is *quantized* and can only be

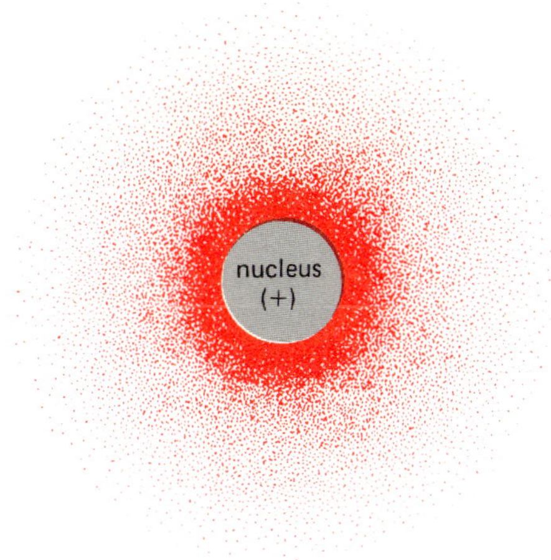

Figure 4-3
An electron cloud in an atom (not drawn to scale—the nucleus is a very *tiny part of the atom)*

lost or gained in discrete amounts. This can be compared to a ladder on which you can climb from one step to the next, but not part way in between.

Refinements of the Bohr theory, in the 1920's and 1930's, led to the modern theory of atomic structure, which is based on *quantum mechanics* (or wave mechanics). This is a mathematical description of the laws of motion that apply to very small particles, like electrons, which have properties of both particles and waves. This theory provides a mathematical calculation of the energy of each electron in an atom. It shows that the electrons occupy various *energy levels* around the nucleus. Each energy level consists of one or more *sublevels,* and each sublevel consists of a set of one or more *orbitals.* An orbital represents a region in space with a specific shape, where a pair of electrons is most likely to be found. It is also found from quantum mechanics that there is a limit to the number of electrons that can occupy any given energy level.*

The energy levels described in this theory represent an *electron cloud* of negative charge spread out over a given region in space. This electron cloud represents a probability of finding a given electron in an atom. The greatest probability is in that region where the electron cloud is heaviest, or where the electron density is greatest. Each of the orbitals that will be described in this chapter, and again in Chapter 7, should therefore be thought of as an electron cloud with a certain shape.

Thus, these electron clouds represent regions of greatest electron density; that is, they represent regions in space where a given electron, or set of electrons, is most likely to be found at any given time. One way of describing this is to imagine that we could take very rapid pictures of the electron as it

*We shall see later in this chapter that this is due to the following facts:
(1) There is a limit to the number of electrons that can occupy any orbital.
(2) There is a limit to the number of orbitals in any sublevel.
(3) There is a limited number of sublevels in each energy level.

moves at random outside the nucleus. If we took many pictures, with the electron represented by a dot in each picture, and put them together, we would obtain the electron cloud diagram that we shall be using here to describe the structure of the atom. Figure 4-3 gives a representation of how we might picture these electron clouds.

4-2 Arrangement of Electrons in Energy Levels

The electrons in the atom exist in **energy levels** or shells, which increase in energy as they increase in distance from the nucleus. That is, the nearer the electron is to the nucleus, the less energy the electron has; the further away it is, the more energy it has. These energy levels are designated by whole numbers or sometimes by capital letters as 1 or *K*, 2 or *L*, 3 or *M*, 4 or *N*, 5 or *O*, 6 or *P*, 7 or *Q*, respectively. In the modern theory of atomic structure these energy levels represent areas of greatest electron density probability. A particular energy level represents the region in space where the electrons occupying that level will be found about 90% of the time; that is, there is a 90% probability of finding the electrons in this region in space at any given time.

It is found that the maximum number of electrons in each energy level is equal to $2n^2$ where n is the number of the energy level. Therefore, in the first energy level ($n = 1$) the maximum is $2 \times 1^2 = 2$; in the second energy level the maximum is $2 \times 2^2 = 8$; in the third it is $2 \times 3^2 = 18$; and so forth. Table 4-1 lists the maximum number of electrons in each energy level for the first six energy levels. For the first 18 elements the electrons are thus arranged in energy levels as shown in Fig. 4-4. Keep in mind that in these drawings the electrons are being represented as existing most of the time at some general distance from the nucleus. This is not intended to represent orbits in which electrons are spinning around the nucleus, as in the Bohr theory. Recall that the atomic number is equal to the number of protons in the nucleus, which is equal to the number of electrons.

These energy levels, which represent an electron cloud of given size, correspond to the orbits in the Bohr theory. As an electron absorbs energy it goes from one energy level to another, and not partway between. Thus, we say that the energy of the electron is *quantized.*

Figure 4-4
Arrangement of electrons in energy levels for the first 18 elements

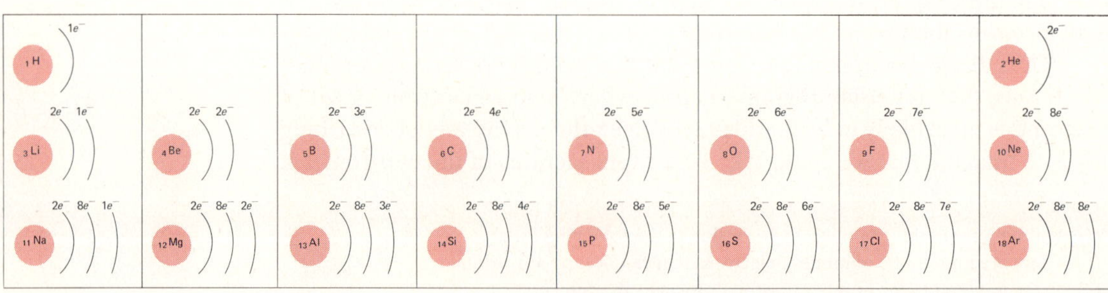

TABLE 4-1 *Maximum number of electrons for each energy level*

	ENERGY LEVEL (LETTER DESIGNATION)	MAXIMUM NUMBER OF ELECTRONS
Increasing Energy ↓	1 (*K*)	$2(1)^2 = 2$
	2 (*L*)	$2(2)^2 = 8$
	3 (*M*)	$2(3)^2 = 18$
	4 (*N*)	$2(4)^2 = 32$
	5 (*O*)	$2(5)^2 = 50$
	6 (*P*)	$2(6)^2 = 72$

4-3 Arrangement of the Electrons in Sublevels

The energy levels are further divided into **sublevels,** or subshells. These sublevels are designated by letters: *s* for the first possible sublevel, *p* for the second sublevel, *d* for the third, *f* for the fourth, *g* for the fifth, and from here on they simply go in alphabetical order. *The number of sublevels that each energy level is divided into is equal to the number of the energy level.* That is, the first energy level only has one sublevel, called the 1*s*; the second energy level has two sublevels, called 2*s* and 2*p*; the third has three, called 3*s*, 3*p*, and 3*d*; and so forth. The sublevels for the first five energy levels are listed in Table 4-2.

Each of these sublevels also has a maximum number of electrons that it can hold. That is, any *s* sublevel can hold a maximum of 2 electrons; any *p* sublevel can hold a maximum of 6 electrons; any *d* sublevel can hold a maximum of 10 electrons; and any *f* sublevel can hold a maximum of 14 electrons. Following this pattern they go up as 18 for *g*, 22 for *h*, and so forth. These are summarized in Table 4-3.

TABLE 4-2 *Sublevels for the first five energy levels*

ENERGY LEVEL	NUMBER OF SUBLEVELS	DESIGNATION OF SUBLEVELS
1	1	1*s*
2	2	2*s*, 2*p*
3	3	3*s*, 3*p*, 3*d*
4	4	4*s*, 4*p*, 4*d*, 4*f*
5	5	5*s*, 5*p*, 5*d*, 5*f*, 5*g*

TABLE 4-3 *Maximum number of electrons in each sublevel (or subshell)*

SUBLEVEL (OR SUBSHELL)	MAXIMUM NUMBER OF ELECTRONS
s	2
p	6
d	10
f	14
g	18

4-4 Orbitals

Each sublevel consists of a set of **orbitals.** Each of these orbitals represents *an electron cloud of a given shape.* The set of orbitals that make up a particular sublevel are all of equal energy. For example, an *s* sublevel consists of a set of *s* orbitals, a *p* sublevel consists of a set of *p* orbitals, and so forth. The number of orbitals in each set is as follows:

$$s = 1, \quad p = 3, \quad d = 5, \quad f = 7, \quad g = 9, \quad \text{and so on}$$

Each of these orbitals can hold a maximum of two electrons. Therefore, an orbital can be thought of as a region in space, with a particular shape, in which two electrons can be found. Combining this maximum of two electrons per orbital with the number of orbitals in each set, we can see that the maximum number of electrons in each sublevel is as follows (compare this to Table 4-3):

1. Two for the *s* orbital (1 orbital $\times$ 2 electrons $=$ 2) in any given set.
2. Six for the *p* orbitals (3 orbitals $\times$ 2 electrons $=$ 6) in any given set.
3. Ten for the *d* orbitals (5 orbitals $\times$ 2 electrons $=$ 10) in any given set. And so forth.

All this information can be summarized as in Table 4-4 for the first three energy levels. In this table, the following information is given: (1) energy level, (2) the sublevels within that energy level, (3) the number of orbitals in each particular sublevel (that is, the number of orbitals in each set), (4) the maximum number of electrons in each sublevel (considering two electrons for each orbital), and (5) the maximum number of electrons in that energy level. Compare the information given here with that in Tables 4-1 through 4-3. Note that the energy level determines the number of sublevels, and the number of orbitals in each sublevel determines the maximum number of electrons in that sublevel.

TABLE 4-4 *Summary of energy levels, sublevels, orbitals, and maximum number of electrons for the first three energy levels*

PRINCIPAL ENERGY LEVEL	SUBLEVEL	NUMBER OF ORBITALS IN SUBLEVEL	MAXIMUM NUMBER OF ELECTRONS IN SUBLEVEL (2/ORBITAL)	MAXIMUM NUMBER OF ELECTRONS IN ENERGY LEVEL
1	*s*	1	2	2
	s	1	2	
2	*p*	3	6	8
	s	1	2	
3	*p*	3	6	18
	d	5	10	

PROBLEM EXAMPLE 4-1

For the fourth energy level show (a) what sublevels are available, (b) the maximum number of electrons in each sublevel, and (c) the maximum number of electrons in the energy level.

SOLUTION:

(a) The sublevels of the fourth energy level are 4*s*, 4*p*, 4*d*, and 4*f*. (b) The maximum number of electrons for each sublevel is

$$4s = 2, \quad 4p = 6, \quad 4d = 10, \quad 4f = 14$$

(c) The maximum number of electrons in the energy level is $2 + 6 + 10 + 14 = 32$. Note that this is exactly the same result we would find by taking $2n^2$, where $n = 4$. This shows in more detail, however, why the maximum number is 32.

The two electrons in a given orbital are said to have opposite spins; that is, apparently each electron is spinning about its own axis. This is known from experimental evidence that shows that in an atom with one unpaired electron there is a slight attraction toward a magnetic field. If there are two unpaired electrons, the attraction toward the magnetic field is approximately twice as strong, and so forth. Since it is known that a spinning particle will create a magnetic field, it is assumed that each electron is spinning about its own axis. When two electrons are in the same orbital, the spins cancel each other out so it is assumed that they must be in opposite directions. Therefore, when an orbital is filled with its maximum of two electrons, we say that the electrons are *paired.* When an orbital has only one electron in it, we say that the electron is *unpaired.*

PROBLEM EXAMPLE 4-2

Explain the maximum number of electrons in the fourth energy level, in terms of sublevels, orbitals, and electron spins.

SOLUTION:

The fourth energy level consists of 4*s*, 4*p*, 4*d*, and 4*f* orbitals. The set of 4*s* consists of a single orbital, the 4*p* contains three orbitals, the 4*d* contains five orbitals, and the 4*f* contains seven orbitals. With two electrons in each orbital (with opposite spins), the maximum number of electrons in the fourth energy level is

$$
\begin{aligned}
4s &= 1 \text{ orbital} \times 2 \text{ electrons} = 2 \\
4p &= 3 \text{ orbitals} \times 2 \text{ electrons} = 6 \\
4d &= 5 \text{ orbitals} \times 2 \text{ electrons} = 10 \\
4f &= 7 \text{ orbitals} \times 2 \text{ electrons} = 14 \\
\hline
\text{Total} &= 32
\end{aligned}
$$

4-5 Occupancy of Sublevels and Orbitals

Within each energy level, the *s* sublevel is of lower energy than the *p* sublevel, which is of lower energy than the *d* sublevel, and so on. It must be kept in mind, however, that the orbitals within a given set are all of equal energy. For example, all the orbitals that make up the 2*p* sublevel are of equal energy. Often these orbitals are represented by a box, or series of boxes, which can be arranged to show how they increase in energy, such as

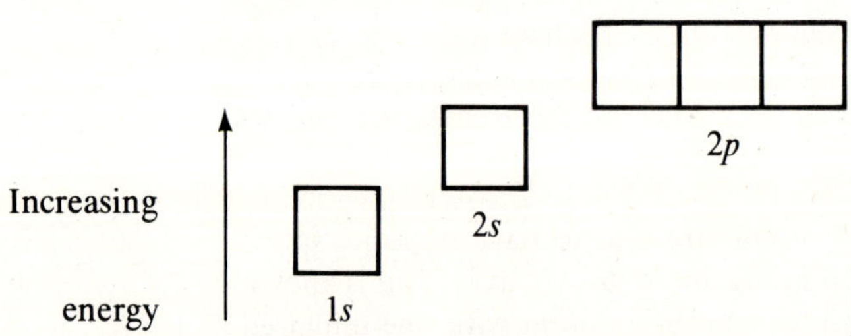

Also, we sometimes find it useful to differentiate between the three orbitals that make up a *p* sublevel, by referring to them as p_x, p_y, and p_z (the reason for this notation will be explained in Section 4-6), such as

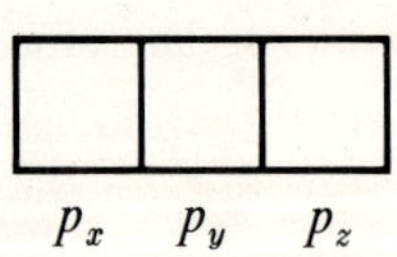

For the first 18 elements the electrons occupy orbitals starting with the lowest energy level and filling each to capacity before the next is started. That is, the 1*s* orbital fills with two electrons, then the 2*s* fills with two electrons, then the 2*p* fills with six, the 3*s* with two, and finally the 3*p* with six. This is usually written as $1s^2$, $2s^2$, $2p^6$, and so on. This notation is sometimes designated as the nl^x notation where n represents the energy level; l, the sublevel; and x, the number of electrons. Using this notation, the electron configuration for element 18, argon, would be $1s^2, 2s^2, 2p^6, 3s^2, 3p^6$.

Beyond the 3*p* sublevel there is an energy overlap, and we find that in occupying sublevels beyond this electrons fill the 4*s* sublevel before the 3*d*. The same is true with the 5*s* and 4*d*. This is shown for the first four energy levels, in Fig. 4-5. With higher energy levels, there is even more overlap of this type.

It turns out that the order in which electrons occupy sublevels (or orbitals) is as follows:

1*s*, 2*s*, 2*p*, 3*s*, 3*p*, 4*s*, 3*d*, 4*p*, 5*s*, 4*d*, 5*p*, 6*s*, 4*f*,* 5*d*, 6*p*, 7*s*, 5*f*,* 6*d*, 7*p*, 8*s*

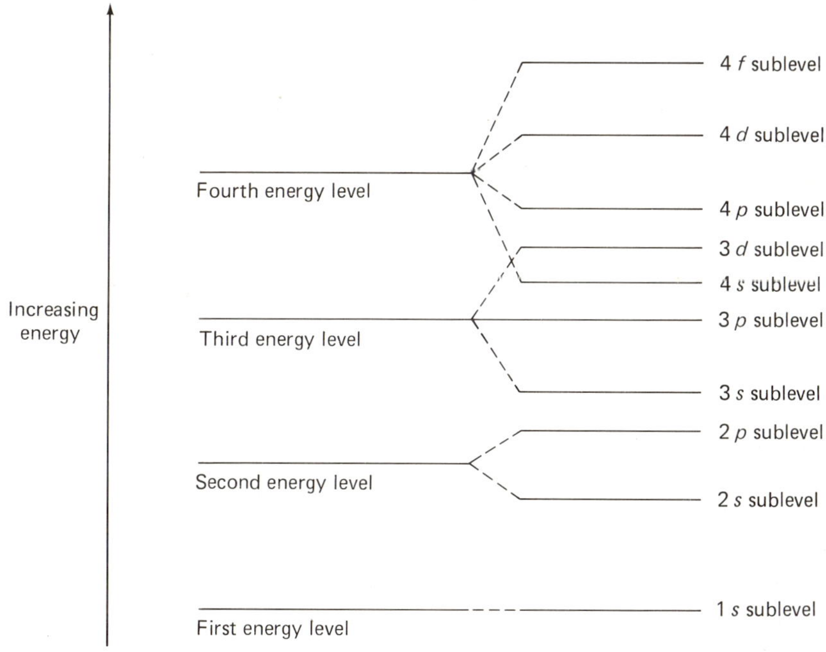

Figure 4-5
Energy diagram of the sublevels in the first four energy levels

*It is found that in occupying sublevels one electron goes into the 5*d*, then the 4*f* fills, and then the rest of the 5*d*. The same thing happens with the 5*f* and 6*d*.

This order of occupancy of sublevels is called the **aufbau order,** which means *building up* in German. As we shall see in Section 4-7, the periodic table is arranged following this order.

In any atom, electrons occupy the lowest energy sublevels that are available. When one sublevel is filled to capacity, the next lowest one begins to fill. This gives the arrangement of the electrons in an atom in the lowest energy condition, called the *ground state.*

Hund's Rule

For any set of orbitals of equal energy, such as the set of $2p$ orbitals, it is found that there is one electron in each orbital before there is any pairing. This is known as **Hund's Rule.** It apparently takes less energy for an electron to occupy an orbital by itself than to pair up with another electron in an orbital of equal energy.

If we represent the sublevels (sets of orbitals) as boxes, or sets of boxes, and the electrons with arrows, we can show the orbital diagram for any atom. This is done in Fig. 4-6 for the first 10 elements. By using arrows to represent electrons, we can indicate a direction for the spin of the electron and also indicate which ones are paired and which are unpaired. A diagram of this type shows how the electrons are arranged not only in energy levels and sublevels but also in orbitals within a sublevel. It also shows how many unpaired electrons there are in each case. For example, from Fig. 4-6, we can see that boron has one unpaired electron, carbon has two, nitrogen has three, oxygen has two, fluorine has one, and neon has zero.

PROBLEM EXAMPLE 4-3

Tell how many unpaired electrons there are in a silicon (Si) atom.

SOLUTION:

Using the aufbau order we can find that the electron configuration for silicon (atomic number 14) is $1s^2\ 2s^2\ 2p^6\ 3s^2\ 3p^2$. Since all orbitals through the $3s$ sublevel are filled, the only ones we need to be concerned about are the $3p$ orbitals. Using boxes to represent these we can see that the electrons would be arranged as

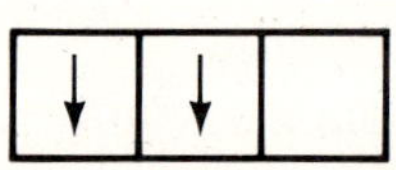

and thus there are two unpaired electrons in an atom of silicon.

Atom (with atomic number)	Orbital diagram				
$_1H$	↓				
$_2He$	↓↑				
$_3Li$	↓↑	↓			
$_4Be$	↓↑	↓↑			
$_5B$	↓↑	↓↑	↓		
$_6C$	↓↑	↓↑	↓	↓	
$_7N$	↓↑	↓↑	↓	↓	↓
$_8O$	↓↑	↓↑	↓↑	↓	↓
$_9F$	↓↑	↓↑	↓↑	↓↑	↓
$_{10}Ne$	↓↑	↓↑	↓↑	↓↑	↓↑
	1*s*	2*s*	2*p*		

Figure 4-6
Orbital diagram for first ten elements

4-6 Shapes of Orbitals

Each type of orbital (*s, p, d, f,* and so on) has a unique shape; that is, the region in space that represents the electron cloud for that particular pair of electrons has a particular shape. These shapes are as follows:

1. Spherical shape for *s*.
2. Dumbbell shape for *p* (with the dumbbell arranged along either the *x* axis, the *y* axis, or the *z* axis).
3. Four-lobed electron cloud for *d* orbitals.
4. Complex shapes for all orbitals beyond this.

These shapes are represented in Fig. 4-7. Keep in mind that these represent three-dimensional diagrams. Note that in the *p* orbitals there is

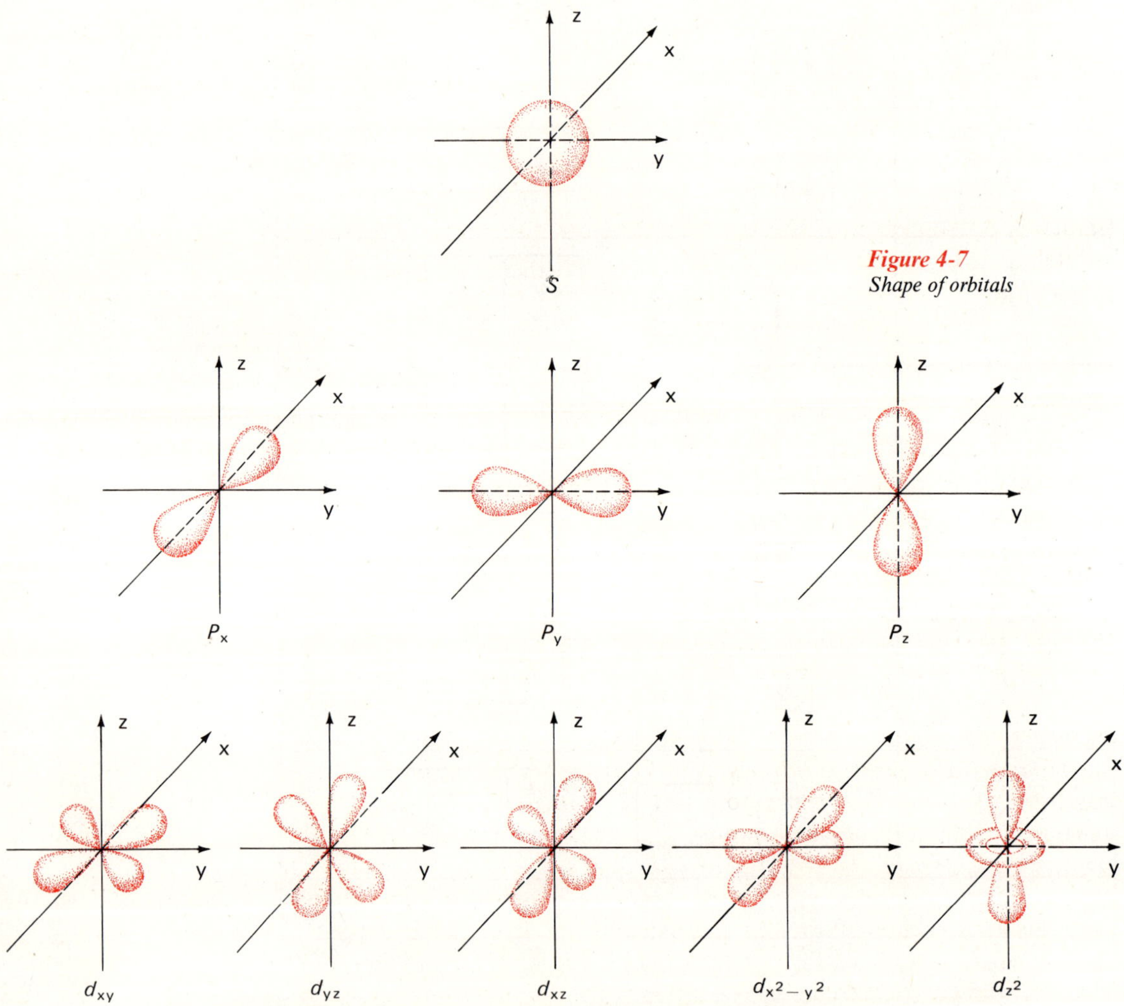

Figure 4-7
Shape of orbitals

an area of zero probability density, often referred to as a *nodal plane,* along the axis that contains the nucleus. However the electron clouds that make up the two parts of the dumbbell represent the region in space where that particular pair of electrons will be found 90 percent of the time. Do not think of these as having one electron in a cloud above the nucleus and one electron in the cloud below the nucleus; this is not the case. The shape of these orbitals, especially for the *s* and *p*, is very important in the discussion of formation of chemical bonds, which will be considered in Chapter 7.

The three different *p* orbitals are sometimes represented as p_x, p_y, and p_z. For example, the three boxes which represent a set of 2*p* orbitals (2*p* sublevel) in Figure 4-6 could be labeled as $2p_x$, $2p_y$, and $2p_z$. Likewise the *d* orbitals are given a similar notation to indicate which plane the five electron clouds occupy. The names given to these are d_{xy}, d_{xz}, d_{yz}, $d_{x^2-y^2}$ and d_{z^2}. For our purposes, however, we do not need to be concerned about the representation of specific *d* orbitals, as we shall be with the *p* orbitals.

A set of *p* orbitals all taken together gives roughtly a spherical shape, and the same thing is true for a set of *d* orbitals. To understand this you must keep in mind that the dumbbell-shaped cloud for a particular *p* orbital does not have a definite surface area boundary but that the electron density probability simply becomes less and less as you move out away from the electron cloud represented by this dumbbell-shaped drawing. The axes that we use are not part of the atom; they are simply a notation that we use in discussing orbitals and electron clouds in our model of atomic structure.

4-7 Electronic Structure and the Periodic Table

The electronic structure of the elements varies in a regular way with increasing atomic number. Therefore, we can see that there is a direct relationship between the electronic structure of an atom and the properties of that atom.

In setting up the periodic table the *aufbau* order is followed, with the maximum number of electrons going into each sublevel before the next one starts to fill. Each time a new *s* orbital, and hence a new energy level, starts to fill, a new row across the table, called a *period,* is started. For the first 20 elements this gives the arrangement as shown in Fig. 4-8.

After element 20 (calcium) a set of *d* orbitals (the 3*d* sublevel) fills, and thus the table is split after element 20 in order to fit in this series of 10 elements. Also, after element 57 (lanthanum, La), there is another break in the table as the series of 4*f* orbitals (the 4*f* sublevel) fills. Since this sublevel can hold a maximum of 14 electrons, this series contains 14 elements. The table is not split at this point, however, since that would make it awkward to work with. This gives the common long form of the periodic table, as shown on the inside front cover of this text. Figure 4-9 shows the periodic table with the electron configuration for the elements.

It can be seen from Fig. 4-9 that there is considerable variation in the electron configuration of the series in which the *f* orbitals are filling. This

A	IIA	IIIA	IV A	V A	VI A	VII A	VIII A
1 H $1s^1$							2 He $1s^2$
3 Li $1s^2$ $2s^1$	4 Be $1s^2$ $2s^2$	5 B $1s^2$ $2s^2, 2p^1$	6 C $1s^2$ $2s^2, 2p^2$	7 N $1s^2$ $2s^2, 2p^3$	8 O $1s^2$ $2s^2, 2p^4$	9 F $1s^2$ $2s^2, 2p^5$	10 Ne $2s^2$ $2s^2, 2p^6$
11 Na $1s^2$ $2s^2, 2p^6$ $3s^1$	12 Mg $1s^2$ $2s^2, 2p^6$ $3s^2$	13 Al $1s^2$ $2s^2, 2p^6$ $3s^2, 3p^1$	14 Si $1s^2$ $2s^2, 2p^6$ $3s^2, 3p^2$	15 P $1s^2$ $2s^2, 2p^6$ $3s^2, 3p^3$	16 S $1s^2$ $2s^2, 2p^6$ $3s^2, 3p^4$	17 Cl $1s^2$ $2s^2, 2p^6$ $3s^2, 3p^5$	18 Ar $1s^2$ $2s^2, 2p^6$ $3s^2, 3p^6$
19 K $1s^2$ $2s^2, 2p^6$ $3s^2, 3p^6$ $4s^1$	20 Ca $1s^2$ $2s^2, 2p^6$ $3s^2, 3p^6$ $4s^2$						

Figure 4-8
Electron configuration for first twenty elements

is due to the fact that the sublevels are very close together at this point. There are two other common variations to this order of filling with which you should be familiar. These occur at atomic number 24 (chromium, Cr) and 29 (copper, Cu). In the case of chromium, instead of having a $4s^2\ 3d^4$, it is actually $4s^1\ 3d^5$. There is added stability in two sets of half-filled orbitals (recall that from Hund's Rule this half-filled set of *d* orbitals would have one electron in each orbital, and hence it has a symmetrical distribution around the nucleus). With copper, number 29, instead of having the $4s^2\ 3d^9$ configuration it is actually $4s^1\ 3d^{10}$. In this case there is added stability by having a half-filled set of *s* orbitals and a completely filled set of *d* orbitals.

We can now use the periodic table to follow through the aufbau order and to determine the electron configuration for any atom. Note that in the periodic table the first two columns represent atoms in which electrons are entering *s* orbitals. The last six columns represent atoms in which electrons are entering *p* orbitals. The 10 columns in the center of the table represent atoms in which electrons are entering *d* orbitals. Also note that going across any period the *s* and *p* orbitals being filled are in the outermost (highest) energy level, whereas the *d* orbitals are of one energy level lower. For the *f* orbitals they are two energy levels lower than the highest energy level. For example, going across the sixth period (row) in the table, electrons are entering 6*s*, 4*f*, 5*d*, and 6*p* orbitals (with one going into the 5*d* before the 4*f* fills, as noted earlier).

In order to set up a periodic table the aufbau order must be known. A memory aid for this is shown in Fig. 4-10. What is done here is to list each *possible* set of orbitals in each energy level, such as 1*s*; then 2*s*, 2*p*; then 3*s*, 3*p*, 3*d*; and so forth. Then diagonal lines are drawn through this, and these lines follow the aufbau order of occupancy of orbitals. You should now be able to set up a periodic table, with all atomic numbers, and from this determine the electron configuration for any element.

Now that we see how the periodic table is set up, we can get the electron configuration for any atom. Also, we can tell from the periodic table how many electrons are in the outermost energy level, what sublevels they are in, and using Hund's Rule we can tell how many unpaired electrons there are. Consider the following examples.

PROBLEM EXAMPLE 4-4

Using the periodic table, give the complete electron configuration for bromine.

SOLUTION:

Since bromine (Br) is element number 35 there are a total of 35 electrons in a bromine atom. Following the periodic table these would be arranged as follows:

$$1s^2\ 2s^2\ 2p^6\ 3s^2\ 3p^6\ 4s^2\ 3d^{10}\ 4p^5 \quad \text{or} \quad 1s^2\ 2s^2\ 2p^6\ 3s^2\ 3p^6\ 3d^{10}\ 4s^2\ 4p^5$$

(Of course, this same thing could be done by following the aufbau order. Once we see how the periodic table is set up, however, it is easier to find the electron configuration from that.)

PROBLEM EXAMPLE 4-5

For an atom of tin (Sn), tell how many electrons there are in the outermost energy level and what orbitals they are in.

SOLUTION:

From the periodic table we see that there are four electrons in the outermost level, with the configuration $5s^2\ 5p^2$. From Hund's Rule we also know that there are two unpaired electrons (in two of the 5*p* orbitals).

	IA	IIA	IIIB	IVB	VB	VIB	VIIB	VIII	VIII	VIII	IB	IIB	IIIA	IVA	VA	VIA	VIIA	VIIIA
1	1 H $1s^1$																	2 He $1s^2$
2	3 Li $1s^2$ $2s^1$	4 Be $1s^2$ $2s^2$											5 B $1s^2$ $2s^2$ $2p^1$	6 C $1s^2$ $2s^2$ $2p^2$	7 N $1s^2$ $2s^2$ $2p^3$	8 O $1s^2$ $2s^2$ $2p^4$	9 F $1s^2$ $2s^2$ $2p^5$	10 Ne $1s^2$ $2s^2$ $2p^6$
3	11 Na (Ne) $3s^1$	12 Mg (Ne) $3s^2$											13 Al (Ne) $3s^2$ $3p^1$	14 Si (Ne) $3s^2$ $3p^2$	15 P (Ne) $3s^2$ $3p^3$	16 S (Ne) $3s^2$ $3p^4$	17 Cl (Ne) $3s^2$ $3p^5$	18 Ar (Ne) $3s^2$ $3p^6$
4	19 K (Ar) $4s^1$	20 Ca (Ar) $4s^2$	21 Sc (Ar) $3d^1$ $4s^2$	22 Ti (Ar) $3d^2$ $4s^2$	23 V (Ar) $3d^3$ $4s^2$	24 Cr (Ar) $3d^5$ $4s^1$	25 Mn (Ar) $3d^5$ $4s^2$	26 Fe (Ar) $3d^6$ $4s^2$	27 Co (Ar) $3d^7$ $4s^2$	28 Ni (Ar) $3d^8$ $4s^2$	29 Cu (Ar) $3d^{10}$ $4s^1$	30 Zn (Ar) $3d^{10}$ $4s^2$	31 Ga (Ar) $(3d^{10})$ $4s^2$ $4p^1$	32 Ge (Ar) $(3d^{10})$ $4s^2$ $4p^2$	33 As (Ar) $(3d^{10})$ $4s^2$ $4p^3$	34 Se (Ar) $(3d^{10})$ $4s^2$ $4p^4$	35 Br (Ar) $(3d^{10})$ $4s^2$ $4p^5$	36 Kr (Ar) $(3d^{10})$ $4s^2$ $4p^6$
5	37 Rb (Kr) $5s^1$	38 Sr (Kr) $5s^2$	39 Y (Kr) $4d^1$ $5s^2$	40 Zr (Kr) $4d^2$ $5s^2$	41 Nb (Kr) $4d^3$ $5s^2$	42 Mo (Kr) $4d^5$ $5s^1$	43 Tc (Kr) $4d^5$ $5s^2$	44 Ru (Kr) $4d^6$ $5s^2$	45 Rh (Kr) $4d^7$ $5s^2$	46 Pd (Kr) $4d^8$ $5s^2$	47 Ag (Kr) $4d^{10}$ $5s^1$	48 Cd (Kr) $4d^{10}$ $5s^2$	49 In (Kr) $(4d^{10})$ $5s^2$ $5p^1$	50 Sn (Kr) $(4d^{10})$ $5s^2$ $5p^2$	51 Sb (Kr) $(4d^{10})$ $5s^2$ $5p^3$	52 Te (Kr) $(4d^{10})$ $5s^2$ $5p^4$	53 I (Kr) $(4d^{10})$ $5s^2$ $5p^5$	54 Xe (Kr) $(4d^{10})$ $5s^2$ $5p^6$

6	55 Cs (Xe) $6s^1$	56 Ba (Xe) $6s^2$	57 La * (Xe) $5d^1$ $6s^2$	72 Hf (Xe) $(4f^{14})$ $5d^2$ $6s^2$	73 Ta (Xe) $(4f^{14})$ $5d^3$ $6s^2$	74 W (Xe) $(4f^{14})$ $5d^4$ $6s^2$	75 Re (Xe) $(4f^{14})$ $5d^5$ $6s^2$	76 Os (Xe) $(4f^{14})$ $5d^6$ $6s^2$	77 Ir (Xe) $(4f^{14})$ $5d^7$ $6s^2$	78 Pt (Xe) $(4f^{14})$ $5d^9$ $6s^1$	79 Au (Xe) $(4f^{14})$ $5d^{10}$ $6s^2$	80 Hg (Xe) $(4f^{14})$ $5d^{10}$ $6s^1$	81 Tl (Xe) $(4f^{14})$ $(5d^{10})$ $6s^2$ $6p^2$	82 Pb (Xe) $(4f^{14})$ $(5d^{10})$ $6s^2$ $6p^1$	83 Bi (Xe) $(4f^{14})$ $(5d^{10})$ $6s^2$ $6p^3$	84 Po (Xe) $(4f^{14})$ $(5d^{10})$ $6s^2$ $6p^4$	85 At (Xe) $(4f^{14})$ $(5d^{10})$ $6s^2$ $6p^5$	86 Rn (Xe) $(4f^{14})$ $(5d^{10})$ $6s^2$ $6p^6$
7	87 Fr (Rn) $7s^1$	88 Ra (Rn) $7s^2$	89 Ac ** (Rn) $6d^1$ $7s^2$	104	105	106												

58 * Ce (Xe) $4f^2$ $5d^0$ $6s^2$	59 Pr (Xe) $4f^3$ $5d^0$ $6s^2$	60 Nd (Xe) $4f^4$ $5d^0$ $6s^2$	61 Pm (Xe) $4f^5$ $5d^0$ $6s^2$	62 Sm (Xe) $4f^6$ $5d^0$ $6s^2$	63 Eu (Xe) $4f^7$ $5d^0$ $6s^2$	64 Gd (Xe) $4f^7$ $5d^1$ $6s^2$	65 Tb (Xe) $4f^9$ $5d^0$ $6s^2$	66 Dy (Xe) $4f^{10}$ $5d^0$ $6s^2$	67 Ho (Xe) $4f^{11}$ $5d^0$ $6s^2$	68 Er (Xe) $4f^{12}$ $5d^0$ $6s^2$	69 Tm (Xe) $4f^{13}$ $5d^0$ $6s^2$	70 Yb (Xe) $4f^{14}$ $5d^0$ $6s^2$	71 Lu (Xe) $4f^{14}$ $5d^1$ $6s^2$
90 ** Th (Rn) $5f^0$ $6d^2$ $7s^2$	91 Pa (Rn) $5f^2$ $6d^1$ $7s^2$	92 U (Rn) $5f^3$ $6d^1$ $7s^2$	93 Np (Rn) $5f^4$ $6d^1$ $7s^2$	94 Pu (Rn) $5f^6$ $6d^0$ $7s^2$	95 Am (Rn) $5f^7$ $6d^1$ $7s^2$	96 Cm (Rn) $5f^7$ $6d^1$ $7s^2$	97 Bk (Rn) $5f^9$ $6d^0$ $7s^2$	98 Cf (Rn) $5f^{10}$ $6d^0$ $7s^2$	99 Es (Rn) $5f^{11}$ $6d^0$ $7s^2$	100 Fm (Rn) $5f^{12}$ $6d^0$ $7s^2$	101 Md (Rn) $5f^{13}$ $6d^0$ $7s^2$	102 No (Rn) $5f^{14}$ $6d^0$ $7s^2$	103 Lw (Rn) $5f^{14}$ $6d^1$ $7s^2$

Figure 4-9
Electronic configuration of the elements

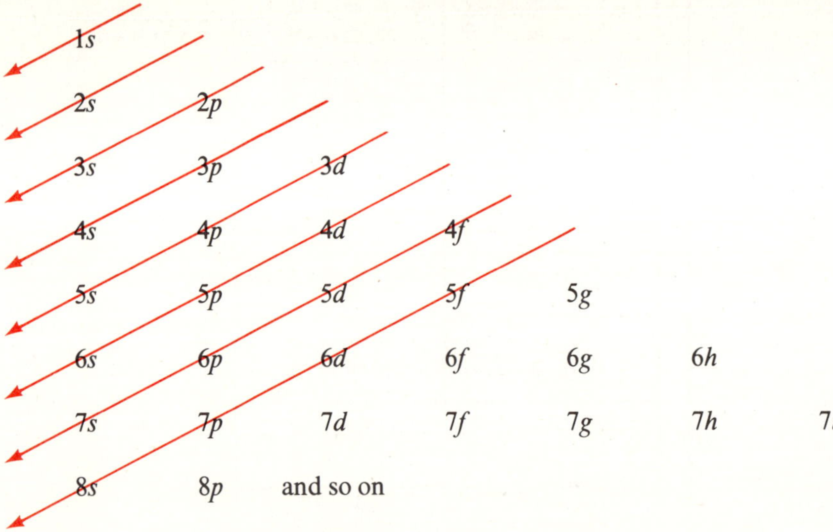

Figure 4-10
Order of filling the sublevels. Write down the energy levels with their sublevels and then draw diagonal lines which follow the order of filling

4-8 Quantum Numbers*

The modern theory of atomic structure that we are using here is based on the mathematical calculation of the probability of finding a given electron in a given region in space around the nucleus. We have considered the arrangement of these electrons in energy levels, sublevels, and orbitals. But so far we have not explained where all of these numbers come from: for example, why there is a maximum of 2 electrons in the first energy level, 8 in the second, 18 in the third, and so on; why the first energy level only contains an *s* sublevel, whereas the second contains an *s* and a *p* sublevel, and so on; and, finally, why there is only one *s* orbital in a set, three *p* orbitals in a set, five *d* orbitals in a set, and so on.

All of these values come from allowed solutions to the wave equation. This equation, called the *Schrödinger wave equation,* is a complex equation that has many different possible solutions. Each of the allowed solutions give a set of three quantum numbers. The first of these is the principal quantum number (n), which is related to the energy level. The next one is the secondary quantum number (l), which is related to the sublevel within a given energy level. And the third is the magnetic quantum number (m_l), which is related to the orientation in space of a particular orbital.

A fourth quantum number, the spin quantum number (m_s) is added to these to account for the fact that there can be two electrons, with opposite spins, in any orbital.

The Principal Quantum Number (n)

The **principal quantum number,** n, can have any positive integral value from one on; that is, there is one set of solutions to the wave equation for each posi-

*This section is optional and may be omitted without loss of continuity.

tive integral value of n. This specifies the energy level, and is related to the size of the electron cloud or the general distance from the nucleus for a particular group of electrons. For each allowed energy, there is a positive integral value of n.

The Secondary Quantum Number (l)

The allowed values for l, the **secondary quantum number,** depend on the value of n; l can have any integral value from 0 to $(n - 1)$. Therefore, if $n = 1$, l can only be 0; if $n = 2$, l can be 0 or 1; if $n = 3$, l can be 0, 1, or 2; and so forth. *The value of l describes a particular sublevel* and determines the shape of the electron probability distribution. As mentioned in Section 4-3, these sublevels are designated by letters as s for $l = 0$, p for $l = 1$, d for $l = 2$, f for $l = 3$, g for $l = 4$, and so forth.* The shape of each of these electron probability distributions was discussed in Section 4-6 and shown in Fig. 4-7.

Looking at the possible values for n and l, it can be seen that when $n = 1$, l can only be 0. Thus in the first energy level, there is only one sublevel, called $1s$ $(n = 1, l = 0)$. In the second energy level $(n = 2)$, there are two sublevels: $2s$ $(n = 2, l = 0)$ and $2p$ $(n = 2, l = 1)$. In the third energy level $(n = 3)$, there are three sublevels: $3s$ $(n = 3, l = 0)$, $3p$ $(n = 3, l = 1)$, and $3d$ $(n = 3, l = 2)$. Table 4-5 summarizes all the possible values for sublevels for the first four energy levels. Compare the information shown here with that given in Table 4-2.

PROBLEM EXAMPLE 4-6

Explain, in terms of quantum numbers, what sublevels there are in the sixth energy level.

SOLUTION:

Since $n = 6$, l can have the values 0, 1, 2, 3, 4, or 5 and therefore there are s, p, d, f, g, and h sublevels in this energy level (designated as $6s$, $6p$, $6d$, $6f$, $6g$, and $6h$).

The Magnetic Quantum Number (m_l)

The allowed values for this third quantum number (**magnetic quantum number**) depend on the value of l. This can have any integral number from $+l$ to $-l$, including zero; that is, for each allowed solution to the wave equation, there is a particular value of n (positive integral number). For each value of n there are certain allowed values for l (integral values from 0 to $n - 1$). Finally, for each particular allowed value of l, there are a set of allowed values of m_l (integral values from $+l$ to $-l$, including zero).

*This is simply a short way of designating a particular value of the quantum number. For example, it is simpler to refer to the $2s$ orbital than to call it the orbital where $n = 2$ and $l = 0$.

TABLE 4-5 *Sublevels for the first four energy levels*

VALUE OF *n* (ENERGY LEVEL)	POSSIBLE VALUES OF *l* (0 to *n* − 1)	SUBLEVEL	NOTATION FOR SUBLEVEL
1	0	*s*	1*s*
2	0	*s*	2*s*
	1	*p*	2*p*
3	0	*s*	3*s*
	1	*p*	3*p*
	2	*d*	3*d*
4	0	*s*	4*s*
	1	*p*	4*p*
	2	*d*	4*d*
	3	*f*	4*f*

This quantum number is related to the possible orientation in space with respect to an applied magnetic field.* This is therefore related to the possible orbitals within a given sublevel. Each sublevel (designated by the value of l) can have $2l + 1$ orientations with respect to an applied magnetic field or we can say there are $2l + 1$ orbitals in a given sublevel. For example, for $l = 0$ (the *s* sublevel), m_l can only be 0; therefore, there is only one orientation in space or a single orbital in any given set of *s* orbitals. When $l = 1$, m_l can be $+1$, 0, or -1; therefore, there are three orientations in space or three *p* orbitals in a set. When $l = 2$, m_l can be $+2$, $+1$, 0, -1, or -2; therefore, there are five orientations in space or five *d* orbitals in a set.

Unlike the *n* and *l* quantum numbers, the actual value of m_l is not so important as the number of values that are possible. If there is only one possible value of m_l (when $l = 0$), there is only one orbital in that sublevel (*s* sublevel). If there are three possible values of m_l (when $l = 1$), there are three orbitals in that sublevel (*p* sublevel, or set of *p* orbitals).

PROBLEM EXAMPLE 4-7

Explain, in terms of quantum numbers, the number of orbitals in any set of *d* orbitals (that is, in any *d* sublevel).

SOLUTION:

For the *d* sublevel, $l = 2$. Therefore, m_l can have the values $+2$, $+1$, 0, -1, or -2. Since there are five possible values of m_l, there are five orbitals in the set.

*In the presence of a magnetic field, along a particular axis, some of the orbitals in a particular set are of higher energy than the others. This magnetic field can be caused by another atom or group of atoms. For our purposes, we shall simply think of this as a name for the third quantum number. Also, keep in mind that in the atom itself, not in the presence of a magnetic field, all the orbitals of a given set are of equal energy.

The Spin Quantum Number (m_s)

This fourth quantum number is added to account for the spin of the electrons —the **spin quantum number.** It does not come from the allowed solutions to the wave equation. Since it is found that any given orbital can hold a maximum of two electrons with opposite spins, there must be two possible values of m_s for each value of m_l (that is, for each particular orbital). These values for m_s can either be $+\frac{1}{2}$ or $-\frac{1}{2}$; that is, for each value of m_l, there are two possible values of m_s, or two allowed energies within each orbital.

The Pauli Exclusion Principle

The **Pauli Exclusion Principle** relates the four quantum numbers to the maximum possible number of electrons in each orbital, sublevel, and energy level. This principle states that *no two electrons in any given atom can have exactly the same set of four quantum numbers.* This means that once we have specified the first three quantum numbers, which describe a particular orbital, and then add the spin quantum numbers of $+\frac{1}{2}$ or $-\frac{1}{2}$, we have two different sets of quantum numbers for one particular orbital. Since there are only two values for m_s, there cannot be more than two sets of quantum numbers for this particular orbital. Note that this Pauli Exclusion Principle is simply another way of stating the fact that within any given orbital there can be two electrons with opposite spins. But as with everything else that we have been considering in this section, this principle gives a sound, mathematical basis for stating this fact and is not just an empirical rule.

All this information concerning allowed solutions to the wave equation, possible quantum numbers, and maximum number of electrons, can be summarized as in Table 4-6. Combining this information with the aufbau order, and Hund's Rule if necessary, electron configuration can be written for any element, as done in Section 4-4.

A set of allowed solutions to the wave equation gives a set of three quantum numbers which specify a particular energy level, sublevel, and orbital; that is, the n quantum number specifies a particular energy level, the l quantum number specifies a particular sublevel within that energy level, and the m_l specifies a particular orbital within a sublevel. Therefore, a set of three quantum numbers describes a particular orbital. For example, the quantum numbers $n = 1$, $l = 0$, $m_l = 0$ specify the $1s$ orbital. The set of quantum numbers $n = 2$, $l = 1$, $m_l = -1$ describes one of the $2p$ orbitals.

If we assign a particular value of m_l to a particular orbital, we can show that for any atom each electron has its own set of quantum numbers. For example, using the p orbitals ($l = 1$), we can equate them as follows:

$$m_l = +1 = p_x \text{ orbital}$$
$$m_l = 0 = p_y \text{ orbital}$$
$$m_l = -1 = p_z \text{ orbital}$$

We can then use this to write the sets of quantum numbers for all the electrons in any atom. In doing this we shall use the value for m_s of $+\frac{1}{2}$ first and then $-\frac{1}{2}$.

TABLE 4-6 *Energy levels, sublevels, orbitals, and maximum number of electrons for the first four energy levels*

VALUE OF n= ENERGY LEVEL	VALUE OF l (AND TYPE OF SUBLEVEL)	NUMBER OF VALUES OF m_l (AND ALLOWED VALUES)	NUMBER OF ENERGY STATES OR MAXIMUM NUMBER OF ELECTRONS	MAXIMUM NUMBER OF ELECTRONS IN ENERGY LEVEL
1	0 (s)	1 (0)	2 (opposite spins)	2
2	0 (s)	1 (0)	2 (two in each orbital)	8
	1 (p)	3 (+1, 0, −1)	6	
3	0 (s)	1 (0)	2	18
	1 (p)	3 (+1, 0, −1)	6	
	2 (d)	5 (+2, +1, 0, −1, −2)	10	
4	0 (s)	1 (0)	2	32
	1 (p)	3 (+1, 0, −1)	6	
	2 (d)	5 (+2, +1, 0, −1, −2)	10	
	3 (f)	7 (+3, +2, +1, 0, −1, −2, −3)	14	

For example, the 10 sets of quantum numbers for the 10 electrons in a neon (Ne) atom would be written as follows:

n	l	m_l	m_s	
1	0	0	$+\frac{1}{2}$	1s sublevel
1	0	0	$-\frac{1}{2}$	
2	0	0	$+\frac{1}{2}$	2s sublevel
2	0	0	$-\frac{1}{2}$	
2	1	+1	$+\frac{1}{2}$	2p sublevel
2	1	0	$+\frac{1}{2}$	
2	1	−1	$+\frac{1}{2}$	
2	1	+1	$-\frac{1}{2}$	
2	1	0	$-\frac{1}{2}$	
2	1	−1	$-\frac{1}{2}$	

In writing sets of quantum numbers of this type, you must keep in mind that the 1s is lower in energy than the 2s, and the 2s is lower than the 2p. Within the set of 2p orbitals, however, they are all of equal energy. Writing these in this manner also helps to emphasize Hund's Rule.

Consider the following problem examples involving quantum numbers.

PROBLEM EXAMPLE 4-8

Tell how many electrons in a given atom can have the following sets of quantum numbers:

(a) $n = 2$
(b) $n = 2, l = 1$
(c) $n = 2, l = 1, m_l = -1$
(d) $n = 2, l = 1, m_l = -1, m_s = +\frac{1}{2}$

SOLUTION:

(a) Since this just represents the second energy level, there can be eight electrons; that is, there is one *s* orbital (2*s*) and three *p* orbitals (2*p*), each with two electrons.
(b) Since this represents the 2*p* orbitals, there can be six electrons with this set of quantum numbers.
(c) Since this represents *one* of the 2*p* orbitals, there can only be two electrons; that is, the two possible sets of quantum numbers are

$$n = 2, \quad l = 1, \quad m_l = -1, \quad m_s = +\tfrac{1}{2}$$

and

$$n = 2, \quad l = 1, \quad m_l = -1, \quad m_s = -\tfrac{1}{2}$$

(d) Since this is a set of four quantum numbers, only one electron in a given atom can have this set.

PROBLEM EXAMPLE 4-9

Tell what is wrong with each of the following sets of quantum numbers:

(a) $2, 2, 0, +\frac{1}{2}$
(b) $2, 0, 1, +\frac{1}{2}$
(c) $2, 0, 0, 0$
(d) $-1, 0, 0, +\frac{1}{2}$
(e) $2, 1, +\frac{1}{2}, -\frac{1}{2}$

SOLUTION:

(a) When $n = 2$, l can only be 0 or 1, *not* 2.
(b) When $l = 0$, m_l can only be 0, not 1.
(c) m_s can only be $+\frac{1}{2}$ or $-\frac{1}{2}$, not 0.
(d) n can only be positive, integral values, not negative.
(e) m_l can only be integral values between $+l$ and $-l$; therefore it cannot be $+\frac{1}{2}$.

PROBLEM EXAMPLE 4-10

Tell what is wrong with each of the following electron configurations.

(a) $1s^2, 2s^4$
(b) $1s^2, 2s^2, 2p^6, 2d^{10}$
(c) $1s^2, 2s^2, 2p^6, 3s^2, 3p^6, 3d^{10}$

SOLUTION:

(a) Any s orbital has a maximum of two electrons. Therefore, it cannot have $2s^4$.

(b) There are no d orbitals in the second energy level; that is, when $n = 2$, l can only be 0 or 1.

(c) According to the aufbau order, the $4s$ orbital fills after the $3p$, not the $3d$.

SUMMARY

1. After Dalton's atomic theory it was shown that the atom is made up of the smaller particles that we now know to be protons, neutrons, and electrons. The arrangement of these particles was an important step in the history of the development of chemistry. The arrangement of the electrons in an atom is especially useful since this determines many of the properties of an element.
2. According to the modern theory of atomic structure, electrons are arranged in *energy levels* around the nucleus. Each of these energy levels is divided into one or more *sublevels,* and each sublevel consists of a set of *orbitals.* It is found that the maximum number of electrons in any energy level is $2n^2$, where n is the number of the energy level. Also the number of sublevels corresponds to the number of the energy level. These sublevels are designated as s, p, d, f, and so on. An s sublevel consists of a set of s orbitals (with one orbital in each set); a p sublevel consists of a set of three p orbitals; a d sublevel consists of a set of five d orbitals; an f sublevel consists of a set of seven orbitals; and so on. Since each orbital can hold a maximum of two electrons (with opposite spins), it can be seen that the maximum number of electrons in any s sublevel is 2 (1×2); in any p sublevel, it is 6 (3×2); in any d sublevel, it is 10 (5×2); and in any f sublevel, it is 14 (7×2).
3. Electrons enter energy levels according to the lowest available sublevel; the sublevels increase in energy according to the *aufbau order.* Also, in filling a particular sublevel, electrons occupy orbitals of equal energy one at a time before there is any pairing (*Hund's Rule*).
4. The periodic table is arranged according to increasing atomic numbers, following the aufbau order, with a new period starting each time a new energy level begins to fill. Therefore we can set up a periodic table following this order and then use this table to determine the electron configuration for any atom, as well as determining the number of unpaired electrons by applying Hund's Rule.
5. The allowed solutions to the wave equation give a set of three quantum numbers that describe a particular allowed energy. The *principal quantum number (n)* describes the energy level; the *secondary quantum number (l)* describes a particular sublevel within that energy level [the

possible values of l depend on the value of n and determine the type of sublevel(*s*) in a given energy level]; the *magnetic quantum number* (m_l) describes a particular orbital within a sublevel (the value of m_l depends on l and, therefore, the possible values of m_l determine the number of orbitals in a particular type of sublevel). The fourth quantum number, called the *spin quantum number* (m_s), is added to account for the two electrons with opposite spins in any given orbital.

Summary

6. The quantum number n can have any integral value from 1 on; l can be any integral value from 0 to $(n - 1)$; m_l can be any integral value from $+l$ to $-l$, including 0; m_s can be $+\frac{1}{2}$ or $-\frac{1}{2}$.
7. The sublevels are given letter designations, as *s* for $l = 0$, *p* for $l = 1$, *d* for $l = 2$, *f* for $l = 3$, and so on (going in alphabetical order from here on). The energy level and sublevel are thus designated as 1*s*, 2*s*, 2*p*, and so on.
8. The *Pauli Exclusion Principle* states that in a given atom no two electrons can have exactly the same set of quantum numbers. This means that there can be no more than two electrons in any given orbital and that these two shall have opposite spins (different values for the spin, m_s, quantum number). This tells us that any set of *s* orbitals can hold a maximum of 2 electrons; any set of *p* orbitals, a maximum of 6; for *d* orbitals, 10; for *f*, 14; and so on.
9. All this shows why there is a definite limit to the number of electrons in any given energy level, sublevel, and orbital. These limits are determined by considering the sublevels that are available in each level: *s* for the first level (1*s*), *s* and *p* for the second (2*s* and 2*p*), and so forth, as well as the maximum number of electrons (two) that can be found in any orbital.

Upon completion of this chapter, you should be able to do the following things (assume a periodic table is available):

1. Given a value for an energy level, tell (a) what the maximum number of electrons is in that energy level; (b) how many sublevels there are in the energy level; and (c) what the names would be for these sublevels (such as 1*s* or 2*s*, 2*p*, and so on). (1, 2)
2. Given a particular sublevel, tell (a) what the maximum number of electrons is for that sublevel and (b) how many orbitals make up the sublevel. (3, 4)
3. Given a value for any energy level, explain the maximum number of electrons in that energy level, in terms of sublevels, orbitals, and electron spins. (5)
4. Given a particular type of orbital (*s*, *p*, or *d*), sketch the shape of that orbital.
5. Given the aufbau order, be able to set up a periodic table, with all atomic numbers.

6. Given the name, symbol, or atomic number for any atom, (a) write the complete electron configuration for that atom, using the nl^x notation; (b) tell how many unpaired electrons there are in the atom; (c) draw an orbital diagram for the atom; (d) write the electron configuration for the electrons in the outermost energy level or levels for the atom. (6, 7, 8, 9, 10)
7. Given the electron configuration for the outermost electrons for any element, identify the element. (11)
8. Given a particular quantum number (n, l, m_l, or m_s), tell what it represents and what values it can have.
9. Given a numerical value for a particular energy level, explain in terms of quantum numbers what sublevels there are in that energy level. (12)
10. Given any type of sublevel (*s*, *p*, *d*, *f*, and so on), tell how many orbitals there are in a set, and explain in terms of quantum numbers. (13)
11. Given any set of n and l quantum numbers, tell what type of orbital is described by that set of numbers (e.g., $n = 1, l = 0$ describes the 1*s* orbital). (14)
12. Given the value for an energy level, explain the maximum number of electrons in that energy level in terms of the four quantum numbers. (15)
13. Tell the maximum number of electrons that can have the particular set of quantum numbers, given (a) the value of n; (b) the value of n and l; (c) the value of n, l, and m_l; (d) the value of n, l, m_l, and m_s. (16)
14. Given a set of four quantum numbers in which one of the values does not correspond to an allowed solution to the wave equation, tell which value is incorrect and why. (17)
15. Given any incorrect electron configuration, tell what is wrong with it, and why it is wrong. (18)
16. Given any of the first 10 elements in the periodic table, write correct sets of quantum numbers for each electron in an atom of the element. (19)
17. Given any of the terms listed in Exercise 1, give a brief definition or explanation, in your own words.

EXERCISES

1. Define or explain the following terms:
   (a) energy level
   (b) orbital
   (c) sublevel
   (d) aufbau order
   (e) Hund's Rule
   (f) principal quantum number
   (g) secondary quantum number
   (h) magnetic quantum number
   (i) spin quantum number
   (j) Pauli Exclusion Principle
2. How does the quantum mechanical theory or model of atomic structure differ from the Bohr theory, and how are they similar?
3. How does a *p* orbital differ from an *s* orbital?
4. How does a *d* orbital differ from a *p* orbital?
5. How does Hund's Rule account for the fact that oxygen has two unpaired electrons?
6. Distinguish between
   (a) subshell and orbital
   (b) principal quantum number and secondary quantum number
   (c) secondary quantum numbers and magnetic quantum number
7. Explain why the maximum number of electrons in the fifth energy level is 50 (2×5^2), in terms of sublevels, orbitals, and electron spins.
8. Set up a periodic table, following the aufbau order, putting in all atomic numbers.

PROBLEMS

Electron structure

1. Tell what the maximum number of electrons is for each of the following energy levels:
   (a) 4
   (b) 7
   (c) 5
   (d) 8
2. Tell what sublevels (*s*, *p*, *d*, and so on) there are in each of the following levels:
   (a) 1
   (b) 4
   (c) 6
   (d) 2
   (e) 3
   (f) 7
3. Tell how many orbitals there are in each of the following sets:
   (a) *d*
   (b) *s*
   (c) *g*
   (d) *i*
   (e) *p*
   (f) *f*
   (g) *h*
   (h) *j*
4. What is the maximum number of electrons in each sublevel in Problem 3 (for example, what is the maximum number of electrons in any *d* sublevel)?

5. Make a table, like Table 4-4, for each of the following:
   (a) fourth energy level
   (b) fifth energy level
6. Give the complete electron configuration for each of the following, using the nl^x notation:
   (a) phosphorus
   (b) iron
   (c) arsenic
   (d) tin
   (e) mercury
   (f) sulfur
   (g) bromine
   (h) nickel
   (i) iodine
   (j) lead
7. Tell how many unpaired electrons there are in each of the following:
   (a) aluminum
   (b) chlorine
   (c) lead
   (d) manganese
   (e) nickel
   (f) silicon
   (g) argon
   (h) tin
   (i) cobalt
   (j) iron

Periodic table

8. Give the electron configuration (using the nl^x notation) for the electrons in the outermost energy level for each of the following:
   (a) Cl
   (b) S
   (c) Ca
   (d) Sn
   (e) I
   (f) P
   (g) Br
   (h) As
   (i) Ba
   (j) Pb
9. Give the electron configuration for the electrons in the outer *d* and *s* sublevels for the following elements:
   (a) Ti
   (b) Fe
   (c) Co
   (d) Hg
   (e) Cu
   (f) V
   (g) Ni
   (h) Mn
   (i) Zn
   (j) Cr
10. Show the orbital diagram (like Fig. 4-6) for the electrons in the outermost energy level(s) for each of the following:
   (a) Ba
   (b) P
   (c) Ni
   (d) Br
   (e) Cr
   (f) N
   (g) Ca
   (h) Fe
   (i) Cl
   (j) Cu
11. Give the atomic number and symbol for the elements with the following electron configurations for the outermost electrons:
   (a) $5s^2$
   (b) $3s^2\ 3p^4$
   (c) $4s^2\ 3d^6$
   (d) $4s^1\ 3d^{10}$
   (e) $5s^2\ 5p^6$
   (f) $3s^1$
   (g) $4s^1$
   (h) $3s^2\ 3p^5$
   (i) $4s^2\ 3d^8$
   (j) $4s^1\ 3d^5$

Quantum numbers

12. Tell what sublevels (s, p, d, and so on) there are in each of the following energy levels, and explain each answer in terms of quantum numbers:
(a) 2 (d) 4
(b) 5 (e) 6
(c) 8 (f) 9

13. Tell how many orbitals there are in each of the following sets, and explain each answer in terms of quantum numbers:
(a) s (e) p
(b) d (f) f
(c) g (g) h
(d) i (h) j

14. Tell what type of orbital (such as 1s, 3d, and so on) is described by each of the following sets of quantum numbers:
(a) $n = 1, l = 0$ (f) $n = 2, l = 0$
(b) $n = 3, l = 2$ (g) $n = 3, l = 1$
(c) $n = 6, l = 1$ (h) $n = 6, l = 3$
(d) $n = 5, l = 3$ (i) $n = 5, l = 2$
(e) $n = 7, l = 5$ (j) $n = 7, l = 6$

15. Complete a table, like Table 4-6, for the following energy levels:
(a) $n = 5$ (b) $n = 6$

16. Tell how many electrons there can be, in a given atom, with the following sets of quantum numbers, and explain each answer:
(a) $n = 3$ (f) $n = 2$
(b) $n = 2, l = 1$ (g) $n = 3, l = 1$
(c) $n = 4, l = 2, m_l = 0$ (h) $n = 3, l = 2, m_l = 1$
(d) $n = 3, l = 2, m_l = 1, m_s = +\frac{1}{2}$ (i) $n = 4, l = 1, m_l = 0, m_s = -\frac{1}{2}$
(c) $n = 7$ (j) $n = 8$

17. Tell what is wrong with each of the following sets of quantum numbers:
(a) $2, 2, 0, +\frac{1}{2}$ (f) $3, 4, 2, -\frac{1}{2}$
(b) $3, 1, 2, -\frac{1}{2}$ (g) $3, 2, 1, 0$
(c) $2, 0, 0, 0$ (h) $4, 2, 3, +\frac{1}{2}$
(d) $3, \frac{1}{2}, 0, -\frac{1}{2}$ (i) $2, -1, 0, -\frac{1}{2}$
(e) $-3, 2, 0, +\frac{1}{2}$ (j) $3, 1, +\frac{1}{2}, -\frac{1}{2}$

18. Tell what is wrong with each of the following electron configurations:
(a) $1s^2\,2s^2\,2p^8$ (e) $1s^2\,2s^2\,2p^6\,3s^2\,3p^6\,3d^{10}$
(b) $1s^2\,2s^2\,2p^6\,2d^{10}$ (f) $1s^2\,2s^2\,2p^6\,3s^2\,3p^8$
(c) $1s^2\,2s^2\,2p^4\,3s^2$ (g) $1s^2\,1p^6\,2s^2\,2p^6$
(d) $1s^2\,2s^2\,2p^6\,3s^2\,3p^6\,4s^2\,3d^{10}\,4p^6\,4d^{10}$ (h) $1s^2\,2s^2\,2p^6\,2d^3$

19. Write a set of quantum numbers for each electron in each of the following atoms:
(a) He (c) Li
(b) Be (d) B

General problems

20. If the spin quantum number could be $+\frac{1}{2}$, $-\frac{1}{2}$ or 0, what would be the maximum number of electrons in each of the following?
    (a) third energy level
    (b) a set of p orbitals
    (c) the $2s$ orbital
    (d) fourth energy level
    (e) the $3d$ orbitals
    (f) any s orbital
21. Which of the following are possible and which are impossible according to quantum mechanics (explain each answer)?
    (a) $2d$
    (b) $4f$
    (c) $8s$
    (d) $7g$
    (e) $5h$
    (f) $7h$
    (g) $6g$
    (h) $3f$
    (i) $9s$
    (j) $7j$
22. Explain what is wrong with each of the following statements:
    (a) When $n = 4$, l can be 1, 2, 3, or 4.
    (b) When $l = 2$, m_l can be 2, 3, 4, 5, or 6.
    (c) Orbitals where $l = 6$ would be called m orbitals.
    (d) The fifth energy level contains four sublevels.
    (e) The quantum numbers $n = 3$, $l = 2$ describe one of the $3p$ orbitals.
    (f) When $l = 3$, m_l can only be $-2, -1, 0, +1$, or $+2$.
    (g) When $n = 2$, l can be $-2, -1, 0, +1$, or $+2$.
    (h) Orbitals where $l = 7$ would be called g orbitals.
    (i) The sixth energy level contains seven sublevels.
    (j) The quantum numbers $n = 4$, $l = 3$, $m_l = 0$ describe one of the $4d$ orbitals.
23. Give the electron configuration for the outermost electrons in each of the following undiscovered elements:
    (a) 108
    (b) 116
    (c) 120
    (d) 114
    (e) 110
    (f) 119
24. Tell how many unpaired electrons there are in an atom of each of the elements in Problem 23.
25. Give the atomic number for the element that would have each of the following electron configurations:
    (a) $5f^{14}\,6d^5\,7s^2$
    (b) $8s^1$
    (c) $7p^4$
    (d) $7s^2\,7p^5$
    (e) $5f^{14}\,6d^7\,7s^2$
    (f) $7p^2$

REFERENCES

References

GAMOW, GEORGE, "The Principle of Uncertainty," *Sci. Am.*, Jan. 1958, **198,** p. 19.

LAMBERT, FRANK, "Atomic Orbitals From Wave Patterns," *Chemistry,* Feb. 1968.

RYSCHKEWITSCH, GEORGE E., *Chemical Bonding and the Geometry of Molecules.* New York: Reinhold Publishing Corporation, 1963 (Chapters 1 and 2).

SISLER, HARRY H., *Electronic Structure, Properties and the Periodic Law.* New York: Reinhold Publishing Corporation, 1963.

Objectives

1. To understand how the periodic table is set up, according to atomic numbers and the electronic structure of the atoms, and to be able to set up a periodic table, with all atomic numbers in the correct positions.
2. To be able to give the electron configuration for any atom, using the periodic table.
3. To understand what the terms *period, group,* and *family* refer to in relation to the periodic table and to be able to tell what period and group any element is in.
4. To understand the basis of dividing the elements into representative elements, noble gases, transition metals, and the inner transition series and to be able to classify any element as one of these types.
5. To understand what is meant by the term *valence electrons* and to be able to tell from the position in the periodic table how many valence electrons there are for any representative element, as well as the number of unpaired electrons.
6. To understand what is meant by the term *differentiating electron* and to be able to tell from the periodic table what sublevel the differentiating electron is in, for any element.
7. To understand and be able to predict and explain periodic variations in metallic properties, atomic size, ionic size, ionization energy, electron affinity, and electronegativity.
8. To learn to use the periodic table to predict what monatomic ions various elements might form, to write names and formulas by comparison to known names and formulas, and to determine what ions or atoms are isoelectronic with each other.
9. To be able to predict minimum and maximum oxidation states for nonmetals and all oxidation states for representative metals.

Five

The Periodic Table

The relationship of the periodic table to the electronic structure of the atoms was discussed in Chapter 4. Also, it was shown how the periodic table is arranged, following the aufbau order, and how it is possible to determine the number of unpaired electrons in any atom by applying Hund's Rule. In this chapter we shall build on this knowledge of the periodic table and learn to use it to explain and predict various properties and variations of these properties within the table.

5-1 Development of the Periodic Table

Early in the nineteenth century, more and more elements were being discovered and their properties determined. It was noticed that certain groups of elements had similar properties. This led chemists of that time to try to classify into groups the elements that were similar. In 1869, Lothar Meyer (1830–1895), a German chemist, devised a classification of the elements into a table that accounted for the periodic variations in properties. His table included 56 elements.

In the same year, a Russian chemist by the name of Dmitri Mendeleev (1834–1907) devised a similar table. His table went further than Meyers, however, in that Mendeleev left gaps and predicted that new elements would be discovered to fill them. He also predicted the properties of these undiscovered new elements, based on their position in his table. Within Mendeleev's lifetime, some of these elements were discovered and found to have properties quite similar to those he predicted.

One error in both of these early **periodic tables** was the fact that the elements were arranged in order of increasing atomic weights. This caused several elements to be misplaced in the table. When H. G. Moseley (1887–1915) determined the charge on the nucleus, he concluded that the elements should be arranged by increasing atomic number. This corrected the earlier discrepancies of the periodic table.

This classification of the elements by atomic number is the basis of the **periodic law,** which states that some of the physical and many of the chemical properties of the elements are periodic functions of their atomic numbers. We now know that these periodic variations are due to the electron configuration of the elements. In the remainder of this chapter, we shall consider some uses of the periodic table and expand on this in several places throughout the remainder of the book. If necessary, you should review the material in Chapter 4 concerning the arrangement of the periodic table in relation to the electronic configuration of the elements.

5-2 Periods, Groups, and Families

The horizontal rows in the periodic table are called **periods,** and the vertical columns are called **groups.** You will note that these groups are divided into A and B subgroups. The A subgroups, due to their similarities within a group, are often called **families.** Some of these families are referred to by special names, such as the alkali metals for Group IA, alkaline earth metals for Group IIA, and halogens for Group VIIA. The other A subgroups are sometimes classified according to the first member of the subgroup or family. Thus the IIIA elements are sometimes referred to as the boron family, the IVA elements as the carbon family, the VA elements as the nitrogen family, and the VIA elements as the oxygen family.

The last group, called Group zero (or sometimes Group VIIIA), contains a series of elements with a very stable configuration. These elements have very little tendency to react with anything else. For this reason, they were originally called the inert gases (since they are all gases at ordinary temperature). It is now known that they will react under the proper conditions, although the conditions are rather drastic. For this reason, they are now called the **noble gases,** rather than the inert gases. The term *noble gas* implies that these elements tend to exist by themselves, in what is often called the free state. They do not have much tendency to react with other elements.

5-3 Types of Elements

As well as being classified as metals or nonmetals, as discussed in Section 3-11, the **elements** are also divided into the following four types: representative elements, noble gases, transition elements, and inner transition elements.

Figure 5-1 shows the areas in the periodic table where these four types of elements are found.

Representative Elements

The **representative elements** are the A subgroups—IA through VIIA. In these elements, the outer energy level is incomplete and the electrons are occupying *s* or *p* orbitals in going from one representative element to the next. Therefore, the electron configuration for these will be from ns^1 to np^5. Now that you have seen how the periodic table is built up according to the

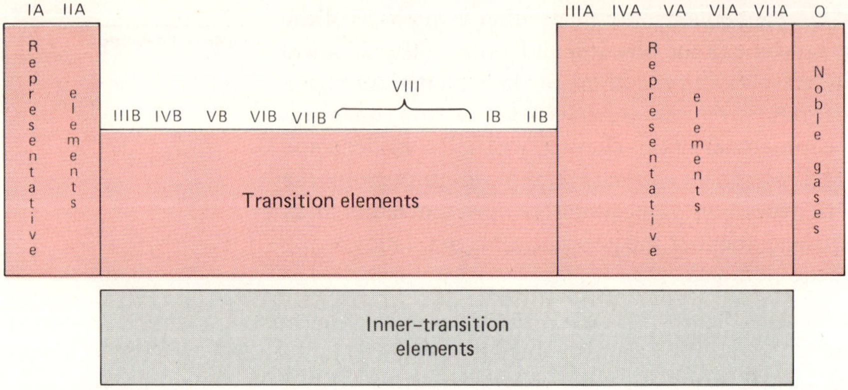

Figure 5-1
Types of elements in the periodic table

arrangement of the electrons in energy levels and sublevels, you should be able to determine quickly the electron configuration for the outermost energy levels by looking at the position in the periodic table. For example, sodium (Na), atomic number 11, being in the third period and the IA subgroup would have one electron in the $3s$ orbital, with all lower orbitals being completely filled. Nitrogen (N), atomic number 7, being in the second period and Group VA would have two electrons in the $2s$ orbital and three in the $2p$, or for the outermost electrons we would say the configuration is $2s^2\ 2p^3$.

Noble Gases

As mentioned earlier, the **noble gases** are the last group in the table. Each element in this group has a completely filled set of s and p orbitals. Thus, the electron configuration for the outermost electrons is $s^2\ p^6$ in each case (with the exception of helium). This is apparently a very stable electron configuration. Not only do these elements have little tendency to react or combine with anything else, but also we shall see later that in many compounds the atoms attain this noble gas electron configuration. The electron configuration for the noble gas krypton (Kr), being in the fourth period, would be $4s^2\ 4p^6$ for the electrons in the outermost energy level.

Transition Elements

The **transition elements** are the series in which a set of d orbitals is being filled. In general, the outermost energy level here will have an ns^2 configuration (with the exception as noted earlier for Groups VIB and IB, see Section 4-7). There is also an $(n-1)d^1$ through $(n-1)d^{10}$ configuration along with this. Therefore, in these elements there are two incomplete energy levels, which are similar in energy, and thus there is more variation in the properties of these elements than there is with the representative elements. For example, the electron configuration for nickel (Ni), atomic number 28, being in the fourth period would be $3d^8\ 4s^2$. Keep in mind that for the transition elements the electrons in the *d* orbitals are always one energy level lower than the electrons in the *s* orbitals.

Inner Transition Elements

The **inner transition series** are the two series of elements from 58–71 and 90–103, in which a set of *f* orbitals is being filled. These are also sometimes called the long transition series. In general, these elements have three incomplete energy levels since one electron enters a *d* orbital before the set of *f* orbitals begin filling. For example, uranium (U), element number 93, has a $5f^3\ 6d^1\ 7s^2$ configuration. As noted earlier, at these higher energy levels the sublevels are very close together in energy. Therefore, there are many exceptions to the expected order of filling of orbitals going across the inner transition series, as can be seen in Fig. 4-9.

Consider the following examples:

1. Chlorine (Cl) is a representative element.
2. Manganese (Mn) is a transition element.
3. Magnesium (Mg) is a representative element.
4. Argon (Ar) is a noble gas.
5. Uranium (U) is an inner transition element.

5-4 Valence Electrons

The electrons in the outer energy level or levels are the ones that will be used in forming compounds and are thus often referred to as **valence electrons.** For the representative elements the number of valence electrons corresponds to the group number. For example, sodium has one valence electron since it is in Group IA, magnesium (Group IIA) has two, aluminum (IIIA) has three, silicon has four, phosphorus has five, sulfur has six, and chlorine has seven.

For the transition elements the *s* electrons (electrons in *s* orbitals) in the outermost energy level are used in forming compounds as well as some of the electrons in *d* orbitals. For the first half of the transition series, all the *d* electrons are valence electrons, but beyond this only some of the *d* electrons are used, the maximum number being six. Therefore, we can say that scandium (Sc) in Group IIIB has three valence electrons, titanium (Ti) has four, vanadium (V) has five, chromium (Cr) has six, and manganese (Mn) has seven.

For the representative elements, it is often convenient to show the valence electrons in the atom by using what is called the *kernel notation,* or an *electron dot notation.* In this notation, the symbol for an element is used to represent the nucleus and all inner electrons, and a dot is put in for each of the valence electrons. We shall find this especially useful when

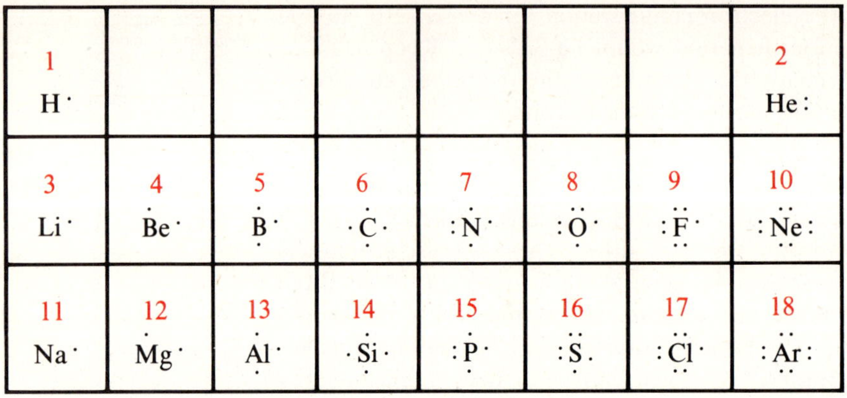

Figure 5-2
Electron dot notation for first eighteen elements

we discuss the bonding of atoms to form compounds. Figure 5-2 gives the electron-dot notation for the first 18 elements.

Consider the following example.

PROBLEM EXAMPLE 5-1

Using the periodic table, tell (i) how many valence electrons there are, (ii) what the electron configuration is for the valence electrons, and (iii) how many unpaired electrons there are for each of the following:

(a) N (b) Ca (c) Sn

SOLUTION:

(a) Since this is in Group VA, there are five valence electrons, with the configuration $2s^2\ 2p^3$ (since it is in the second period). Therefore, there are three unpaired electrons (one in each of the $2p$ orbitals–recall Hund's Rule).
(b) Ca, in Group IIA, has two valence electrons, with the configuration $4s^2$. The two electrons are paired.
(c) Sn has four valence electrons–$5s^2\ 5p^2$ (the $4d$ electrons are not valence electrons). The 2 in the $5p$ orbitals would be unpaired.

5-5 Differentiating Electron

The **differentiating electron** in an atom is the electron that makes that atom different from the one before it in the periodic table. This can also be stated as the last electron added in building up the periodic table following the aufbau order. In general, for the representative elements the differentiating electron is in an s or p orbital; for the transition elements the differentiating electron is in a d orbital (with some exceptions such as following Groups VIB and IB as noted earlier); and for the inner transition elements the differentiating electron is in an f orbital, with several exceptions as noted earlier. Keep in mind that going across a period in the table the differentiating electron is in the outermost energy level for the representative elements, one energy level lower for the transition elements, and two energy levels lower for the inner transition elements. For example, going across the fourth period the differentiating electron would be first in the $4s$ orbital, then in the $3d$, and finally in the $4p$. Likewise, going across the sixth period it would be in the $6s$, $5d$, $4f$, and finally $6p$.

Consider the following examples:

1. P—the differentiating electron is in the $3p$ sublevel (that is, one of the $3p$ orbitals).
2. Fe—the differentiating electron is in the $3d$ sublevel.
3. Zn—the differentiating electron is in the $4s$ sublevel (recall that Cu is $4s^1\ 3d^{10}$).

5-6 Periodic Variations

We now want to consider some of the properties of atoms and look at the variations in these properties within the periodic table.

Metallic Properties

In Chapter 3, when the periodic table was introduced, we considered in general which elements in the table are classified as metals and which are classified as nonmetals. Within any period in the table, the elements become less metallic, or more nonmetallic, going across the period from left to right.

Within a group or family, metallic properties increase in going down the group. This is especially noticeable in the IIIA, IVA, and VA subgroups, where in each case the elements vary from nonmetals to metals, going down the group.

Atomic Size

Going across any period in the table, there is a small but rather general decrease in size of the atomic radius. This is due to the fact that going across a period the valence electrons are all in the same energy level, or the same general distance from the nucleus, but in proceeding from one element to the next the nuclear charge increases by one. Even though there is also an increase in the number of electrons, each electron is attracted toward the nucleus; thus the greater the nuclear charge, the greater the overall pull on the electrons toward the nucleus.

Going down any group in the table there is a rather large increase in atomic size. In this case, even though there is increased nuclear charge, there is also one more energy level of electrons. Since the size of an atom depends on where the electrons are distributed outside the nucleus, this increase in

3	4	5	6	7	8	9	10
Li	Be	B	C	N	O	F	Ne
1.23	0.90	0.82	0.77	0.75	0.73	0.72	0.71
11	12	13	14	15	16	17	18
Na	Mg	Al	Si	P	S	Cl	Ar
1.54	1.36	1.18	1.11	1.06	1.02	0.99	0.98
19	20	31	32	33	34	35	36
K	Ca	Ga	Ge	As	Se	Br	Kr
2.03	1.74	1.26	1.22	1.20	1.16	1.14	1.12

Figure 5-3
Size of some representative elements in angstroms (Å)

the number of energy levels causes the increase in the atomic radius. Figure 5-3 shows the atomic radii of some of the representative elements.

Consider the following example.

PROBLEM EXAMPLE 5-2

Using the periodic table, arrange the following elements in order of decreasing size: Br, Sr, Sn, I, Cs, Ba.

SOLUTION:

Going across the table from left to right, and up the groups, the order is Cs, Ba, Sr, Sn, I, Br.

Ionic Size

In Chapter 3 it was pointed out that an ion is an atom that has gained or lost electrons. Since an electron has a negative charge, when an atom loses one electron, it attains a plus one charge. For example, when a sodium atom loses one electron, it becomes a Na^{+} ion. Likewise, when two electrons are lost from an atom, it becomes an ion with a plus two charge, such as in the case of Ca^{2+}. If an atom gains one electron, it becomes an ion with a minus one charge, such as Cl^{-}; if it gains two electrons, it becomes an ion with a minus two charge, such as S^{2-}. It is found that a simple monatomic ion never attains a charge greater than three, and even three is quite rare. We shall see later, in Chapter 7, that ions are often formed, especially for the representative elements, in such a way that the ions attain a noble gas electron configuration.

When a metal atom loses electrons, there is usually a relatively large decrease in size; that is, the radius of the ion formed is considerably smaller than the radius of the atom from which it was formed. This is especially true when all the valence electrons are lost, as in the case of the IA and IIA metals. When this happens, the ion formed has one less energy level than the original atom had and thus the size is smaller. For the nonmetals, when electrons are gained to form negative ions, there is a rather large increase in the size. This is due to the repulsion of the electrons for one another. For example, when a chlorine atom gains an electron to form a chloride ion, there is an increase in size; the radius of the chloride ion is greater than the radius of the chlorine atom.

Figure 5-4 gives the ionic radius for some ions of representative elements. Comparison of this with Fig. 5-3 points out the difference between atomic radius and ionic radius.

Consider the following example.

PROBLEM EXAMPLE 5-3

In each of the following sets, tell which would be the smallest:
(a) Cl, Cl^-
(b) Na, Na^+
(c) S^{2-}, Cl^-, Br^-
(d) Na, Na^+, Mg, Mg^{2+}, Ca, Ca^{2+}

SOLUTION:

(a) Cl (The Cl^- ion would have an extra electron and thus be larger.)
(b) Na^+ (This has lost its one valence electron and is thus smaller.)
(c) Cl^- (S^{2-} has gained two electrons and would thus be larger; Br^- has one more energy level of electrons than Cl^-.)
(d) Mg^{2+} (Mg^{2+} is smaller than either Na or Ca; all the ions have lost all valence electrons.)

Ionization Energy

Ionization energy is the amount of energy required to remove an electron from an atom or ion. The first ionization energy for a particular atom is therefore the amount of energy required to remove one electron from that atom; the second ionization energy is the amount of energy required to remove the second electron from the ion formed; and so forth. The second ionization energy is always higher than the first since an electron is being removed from a positive ion, and the third is likewise higher than the second. Figure 5-5 lists some ionization energies. Note that in this table once a noble gas electron configuration is attained, such as in the case of Na^+, Be^{2+}, and Mg^{2+}, the next ionization energy becomes very high. This explains why once a noble gas electron configuration is attained, no more electrons can be removed from an atom by a simple chemical reaction.

It can be seen from Fig. 5-5 that if we compare the first ionization energies for the atoms going across a period, there is a rather general increase. As with the comparison of atomic size, this is due to the fact that the electron being removed in each case is at the same energy level but there is a greater nuclear charge holding that electron. You will note in this table that there are two places where there is a decrease in going across a period; that is, in going

3 Li^+ 0.60	4 Be^{2+} 0.31							
11 Na^+ 0.95	12 Mg^{2+} 0.65							
19 K^+ 1.33	20 Ca^{2+} 0.99	21 Sc^{3+} 0.81	22 Ti^{3+} —	23 V^{3+} 0.74	24 Cr^{3+} 0.69	25 Mn^{2+} 0.80	26 Fe^{2+} 0.76	27 Co^{2+} 0.75

from Group IIA to IIIA and from VA to VIA, the ionization energy decreases. This can be explained from the fact that Group IIA has a complete set of electrons in an *s* orbital and, therefore, the differentiating electron in IIIA is the first one in a set of *p* orbitals. Thus, it is a little easier to remove that one electron, which leaves a rather stable s^2 configuration, than it would be to remove that electron if it were not for the s^2 configuration. Likewise, Group VA has a half-filled set of *p* orbitals, which you will recall is a symmetrical distribution of electrons around the nucleus. Therefore, in VIA the first electron is removed somewhat more easily than it would be if it were not for the remaining half-filled set of *p* orbitals.

Going down a group in the table, the first ionization energy decreases from one atom to the next. This again is due to the fact that in each case the valence electrons are in a higher energy level or at a greater distance from the nucleus than they were in the preceding atom. Therefore, it is easier to remove these electrons.

Consider the following example.

PROBLEM EXAMPLE 5-4

In each set, choose the one with the highest ionization energy:

(a) Na, Al, Cl, Br

(b) Na, Na^+, Mg, Mg^+

SOLUTION:

(a) Cl—since this has the smallest atomic size (that is, valence electrons closer to the nucleus), the ionization energy is highest.

(b) Na^+—this has a noble gas electron configuration and, thus, has the highest ionization energy.

			5 B^{3+} 0.20	6 C^{4+} 0.44	7 N^{3-} 1.71	8 O^{2-} 1.40	9 F^{-} 1.36
			13 Al^{3+} 0.50	14 Si^{4+} 0.41	15 P^{3-} 2.12	16 S^{2-} 1.84	17 Cl^{-} 1.81
28 Ni^{2+} 0.72	29 Cu^{2+} 0.69	30 Zn^{2+} 0.74	31 Ga^{3+} 1.13	32 Ge^{4+} 0.53	33 As^{3-} 2.22	34 Se^{2-} 1.98	35 Br^{-} 1.95

Figure 5-4
Ionic radii (in Å) for some common ions

Electron Affinity

The amount of energy released when an atom gains an electron is called the **electron affinity.** Figure 5-6 gives the electron affinities for some nonmetals. It can be seen from this table that going across a period, such as from nitrogen to oxygen to fluorine, the electron affinity increases. As with atomic size and ionization energy, this is due to the fact that the valence electrons

IA	IIA	IIIA	IVA	VA	VIA	VIIA	VIIIA
1 H 313							2 He 567 1250
3 Li 124 1744	4 Be 215 420 3550	5 B 191 580 875	6 C 260 560 1100	7 N 335 680 1100	8 O 314 811 1270	9 F 402 805 1450	10 Ne 497 950 1500
11 Na 119 1090	12 Mg 176 347 1850	13 Al 138 435 656					

Figure 5-5
Ionization energies of some elements (in kcal/mole; listed as: first
second
third)

N	O	F
almost zero	35	85
	S	Cl
	53	87
		Br
		82

Figure 5-6
Electron affinities for some nonmetals (in kcal/mole)

are in the same energy level but the nuclear charge becomes greater. Therefore, with the valence electrons somewhat closer to the nucleus and the nuclear charge greater, more energy is released when an electron is added to the valence shell. We shall see in Chapter 7 that when the nonmetals form monatomic negative ions, a noble gas electron configuration is always attained. Going down a group of nonmetals, such as the halogens (Group VIIA), the electron affinity decreases. This is due to the fact that going down the group the valence electrons are further from the nucleus and, therefore, not as much energy is released when an electron is accepted into the valence shell. Note in Fig. 5-6 that fluorine (F) is an exception to the general rule. This is due to the small size of the fluorine atom and, hence, the somewhat lower tendency to accept an electron from what would be expected. The small size causes more repulsion among the valence electrons.

H 2.1								
Li 1.0	Be 1.5							
Na 0.9	Mg 1.2							
K 0.8	Ca 1.0	Sc 1.3	Ti 1.5	V 1.6	Cr 1.6	Mn 1.5	Fe 1.8	

Electronegativity

The **electronegativity** of an atom is defined as the general tendency of that atom to attract electrons toward itself in a compound. This is determined from the ionization energies and the electron affinities. It is not an energy measurement, however, but simply a general tendency for atoms to attract electrons. There are several different electronegativity scales that have been set up, but the one most commonly used is the table arranged by Linus Pauling, which is referred to as the Pauling scale. Figure 5-7 lists the electronegativity of the elements according to the Pauling scale. Note from this table, as you would expect from what has been previously said about atomic size, ionization energy, and electron affinity, that there is an increase in electronegativity going across a period from left to right and a decrease going down a group.

This concept of electronegativity is very useful in writing names and formulas for compounds and predicting the type of bonding in compounds and polarity of bonds and molecules.

Consider the following example.

PROBLEM EXAMPLE 5-5

Arrange the following elements in order of increasing electronegativity: Ba, Br, I, Sn, Sr.

SOLUTION:

Looking at position in the periodic table, including both the period and group that each one is in, gives the following order:
Ba, Sr, Sn, I, Br.

				B 2.0	C 2.5	N 3.0	O 3.5	F 4.0
				Al 1.5	Si 1.8	P 2.1	S 2.5	Cl 3.0
Co 1.8	Ni 1.8	Cu 1.9	Zn 1.6	Ga 1.6	Ge 1.8	As 2.0	Se 2.4	Br 2.8

Figure 5-7
Electronegativities of some common elements (Pauling scale)

5-7 Use of the Periodic Table in Predicting Properties

We have seen in this chapter that there are several periodic variations that we can predict by using the periodic table. We shall see as we proceed further that there are many other correlations that can be made by using the periodic table. To complete our introduction to the use of the table, let us consider a few other types of correlations and predictions that can be made.

Monatomic Ions

From what has been said previously about metals versus nonmetals, and the concepts of ionization energy, electron affinity, and electronegativity, we can begin to predict what ions various elements might form. In doing this, we shall also keep in mind that a noble gas electron configuration is a very stable condition, and very often atoms tend to attain this configuration in forming compounds. The ions that we are considering here are simple monatomic (one atom) ions. In forming these ions, only one or two electrons are gained or lost, in general. There are a few cases where +3 (such as Al^{3+}) and −3 ions (such as N^{3-}) are known to exist, but these are quite rare. There are no cases known where the charge is greater than three.

In looking at the periodic table, therefore, we would predict that the elements in Group IA could lose their one valence electron to form +1 ions, such as Na^{+}. In Group IIA we would predict that +2 ions would form, such as Ca^{2+}. In each of these groups the lower members more readily lose their valence electrons and are considered to be the more active metals. For example, calcium, strontium, and barium are much more active metals than beryllium.

The atoms that form monatomic anions (negative ions) are the nonmetals on the far right-hand side of the periodic table, not including the noble gases. In this case, the atoms are gaining electrons, and this is always done in such a way that a noble gas electron configuration is obtained. Thus, the VIIA elements, called the *halogens,* form −1 ions, such as Cl^{-}. The first two elements in Group VIA can form −2 ions, such as S^{2-}. And occasionally nitrogen, in Group VA, forms a −3 ion, N^{3-}. With these elements, the ones toward the top of each group have the greatest tendency to form ions. This is what we would expect from what has been said previously about electron affinity and electronegativity. These are sometimes referred to as the more active nonmetals. For example, chlorine is considered to be a more active nonmetal than bromine.

Isoelectronic Species

Substances, such as atoms, ions, or molecules, that have the same number of electrons are said to be *isoelectronic.* We shall have more to say about molecules of this type later; for now, let us consider some atoms and ions that are isoelectronic.

We have seen that atoms of the Group IA elements, such as sodium, very readily lose their one valence electron to form +1 ions, such as Na^+. When the sodium atom reacts in this way, the resulting Na^+ ion has the same electron configuration as a Ne atom. Therefore, we say that Ne and Na^+ are isoelectronic. By considering their position in the periodic table, we can see that there are several other ions that are also isoelectronic with the Ne atom. These are Mg^{2+}, Al^{3+}, F^-, O^{2-}, and N^{3-}. Each of these ions has a $1s^2\,2s^2\,2p^6$ configuration, exactly the same as neon.

PROBLEM EXAMPLE 5-6

Write the formulas for four ions that are isoelectronic with argon (Ar).

SOLUTION:

From their position in the periodic table we can see that the following ions would all have a $3s^2\,3p^6$ configuration: S^{2-}, Cl^-, K^+, and Ca^{2+}.

Names and Formulas

In Chapter 6 we shall consider in detail the writing of correct names and formulas of inorganic compounds. Before we do this, however, we can see how the periodic table can be used to determine the name or formula of a compound by comparing it with another compound for which we do know the correct name and formula. For example, given the fact that the compound calcium chlorate has the formula $Ca(ClO_3)_2$, we can predict the names and formulas for some related substances. In doing this, we shall use the relationship, from the periodic table, of the various elements included. First we shall consider some names and then some formulas; keep in mind that in all cases we are comparing the compounds to $Ca(ClO_3)_2$.

$Mg(ClO_3)_2$ = magnesium chlorate

$Ca(BrO_3)_2$ = calcium bromate

$Ba(BrO_3)_2$ = barium bromate

strontium chlorate = $Sr(ClO_3)_2$

calcium iodate = $Ca(IO_3)_2$

magnesium iodate = $Mg(IO_3)_2$

5-8 Oxidation Number (Oxidation State)

The valence of an atom describes the combining power of that atom in a compound. Another related concept, which is even more useful, is the concept **oxidation number** or **oxidation state.** This is a positive or negative value that not only describes the combining capacity of an atom but also gives an indication of how the electrons are arranged in the compound. The terms *oxidation number* and *oxidation state* are often used interchangeably, although there is a slight difference in their correct usage. Oxidation state is generally used to describe what the oxidation number is in a particular compound. For example, in the compound $FeCl_2$, the oxidation number of iron is $+2$; therefore, in this compound, iron is in the $+2$ oxidation state.

For ionic compounds the oxidation number corresponds to the actual charge on each ion. For covalent (molecular) compounds, oxidation numbers are assigned in such a way that the more electronegative part has a negative oxidation number and the more positive (less electronegative) part has a positive oxidation number. These oxidation numbers are assigned according to the electronic structure and the electronegativity for several key elements. Oxidation numbers for other elements are derived from these, using the fact that the sum of all the oxidation numbers in a compound must be zero. For a polyatomic ion (an ion consisting of more than one atom) the sum of all the oxidation numbers is equal to the charge on the ion.

Oxidation number is, therefore, sometimes defined as an apparent charge on an atom in a compound. In some cases, it is an actual charge, when electrons are transferred from one atom to another to form ions. But in many other cases it just gives an indication of which atom in the compound has a greater attraction for electrons and how many electrons are shifted toward or away from a particular atom. In using the concept of oxidation number, we are always comparing the various atoms in a compound. Thus, we say that any element in the free state has an oxidation number of zero.

The following are some general rules for assigning or determining oxidation numbers (states):

1. The oxidation number of an element in the *free* or *uncombined* state is always *zero.*

2. The algebraic sum of the oxidation numbers of all the atoms in the formula for a compound is *zero.*

3. The oxidation number of an ion is the same as the charge on the ion.

4. The sum of the oxidation numbers of the atoms in a polyatomic ion must equal the charge on the ion.

5. Some common oxidation numbers:
   (a) Group IA elements are always $+1$.
   (b) Group IIA elements are always $+2$.
   (c) Hydrogen is usually $+1$, except in hydrides (ionic compounds with IA or IIA elements); then it is -1.
   (d) Oxygen is usually -2, except in peroxides, such as H_2O_2 and Na_2O_2; then it is -1 (in superoxides, such as KO_2, it is $-\frac{1}{2}$).
   (e) Group VIIA elements are -1 when in binary compounds with more electropositive (less electronegative) elements.
   (f) Sulfur in binary compounds with more electropositive elements is -2.
   (g) Nitrogen in binary compounds with more electropositive elements is -3.

(h) Al (Group IIIA) is +3; Zn (Group IIB) is +2; Ag (Group IB) is +1.
(i) Sn and Pb are +2 or +4; Cu and Hg are +1 or +2; Fe is +2 or +3.
(j) All other metals, like these last five, have more than one oxidation state or are said to have variable oxidation numbers. These five are the most common ones and must be learned.

Using these general rules, we can determine the oxidation state of most elements in common compounds. If these rules do not cover all elements, use your knowledge of the periodic table to try to determine what an unknown element is. For example, since sulfur (S) is −2 in binary compounds with more electropositive elements, selenium (Se) is probably the same.

Consider the following examples:

1. $KMnO_4$: K is +1; O is −2; therefore, Mn must be +7 (+1 + 7 − 8 = 0).
2. SO_4^{2-}: O is −2; therefore, S must be +6 (−8 + 6 = −2).
3. $FeCl_2$: Cl is −1; therefore, Fe is +2 in this compound.
4. PbO_2: O is −2; therefore, Pb must be +4 in this compound.
5. CrSe: Se (like S) is probably −2 here; therefore, Cr must be +2.

Periodic Variations in Oxidation States

In general, the group number not only tells how many valence electrons there are in an atom but also gives the maximum oxidation state. There are some exceptions to this, just as there are with the order of filling of orbitals. For example, oxidation states higher than seven are very rare and, therefore, the transition elements in Group VIII do not generally exhibit an oxidation state of eight. For most of these elements, especially in the first row, the oxidation states are +2 or +3 (as we have seen for iron). For some of the heavier elements in this group, however, there are compounds that exhibit an oxidation state of eight. Another common exception is copper, in Group IB. The maximum, and most common, oxidation state of this element is +2.

In predicting periodic variations in oxidation states, we must consider three different types of elements: nonmetals, representative metals, and transition metals.

The *nonmetals* generally have several different oxidation states, with the exception of oxygen and fluorine.* The minimum oxidation state is equal to the group number minus eight, such as −1 for the VIIA elements (7 − 8),

*Due to the fact that these are the two most electronegative elements known, they would not be expected to form many different oxidation states. In fact, since fluorine is the most electronegative element and only needs one more electron to attain a noble gas electron configuration, it is always −1 in compounds.

-2 for the VIA nonmetals $(6 - 8)$, and so forth. The maximum oxidation state is equal to the Group number.

The *representative metals* in Groups IA and IIA, and aluminum, as we have seen, only exhibit a single oxidation state. For the metals following transition series there are two possible oxidation states for each. These correspond to the case where the *p* electrons (electrons in *p* orbitals) are used or the *s* and *p* electrons are used. For this reason, tin and lead, with s^2p^2 configurations, can be either $+2$ or $+4$.

For the *transition metals* there are generally two or more oxidation states possible. For the first half of the transition elements (and some of the elements in Group VIII) there are usually several oxidation states possible. This is due to the fact that for these elements the electrons in *d* orbitals are valence electrons. Table 5-1 lists the possible oxidation states for the first row transition elements.

Consider the following examples of predicting oxidation states.

PROBLEM EXAMPLE 5-7

Predict the minimum and maximum oxidation state for each of the following:
(a) P (b) Se (c) Br

SOLUTION:

From the position in the periodic table, the minimum and maximum for each is
(a) -3 $(5 - 8)$, $+5$ (Group VA)
(b) -2 $(6 - 8)$, $+6$ (Group VIA)
(c) -1 $(7 - 8)$, $+7$ (Group VIIA)

PROBLEM EXAMPLE 5-8

Predict the possible oxidation states for each of the following:
(a) Rb (b) In

SOLUTION:

(a) Since Rb is in Group IA, it can only be $+1$.
(b) Since In is in Group IIIA and has a $5s^2\ 5p^1$ electron configuration, it could be $+1$ or $+3$.

TABLE 5-1 *Oxidation states for first row transition elements*

IIIB	IVB	VB	VIB	VIIB	VIII			IB	IIB
Se	Ti	V	Cr	Mn	Fe	Co	Ni	Cu	Zn
3	3, 4	2, 3, 4, 5	2, 3, 6	2, 3, 4, 6, 7	2, 3	2, 3	2, 3	1, 2	2

SUMMARY

1. The *periodic law* states that some physical and many chemical properties of the elements vary periodically with the atomic numbers since the electron structure of the elements varies in a regular way with increasing atomic number.
2. Since the periodic table is arranged following the aufbau order of occupancy of orbitals, the electron configuration for any atom can be found directly from the *periodic table.* In general, the number of the *period* (horizontal row) tells what energy level the outer electrons are in, and the *group number* indicates how many electrons there are in the outermost energy level(s). This is especially useful for the representative elements, in which the group number tells how many *valence electrons* there are. The A *subgroups* are sometimes referred to as *families.*
3. The elements in the periodic table are often classified as *representative* (IA–VIIA), *noble gases* (VIIIA or 0), *transition metals* (B subgroups), and *inner transition elements.* These can be determined from their position in the table or by considering the number of incomplete energy levels there are or by what type of orbital the *differentiating* electron is in (*s* or *p* for representative, *d* for transition, and *f* for inner transition).
4. The valence electrons are often represented by *electron dot notation* in which the symbol for an element is used to represent the nucleus and all inner energy levels of electrons, and a dot represents one valence electron.
5. Going across a period and up a group in the periodic table, *atomic size* decreases and *ionization energy, electron affinity,* and *electronegativity* all increase. This explains the variation in metallic properties of elements within the table. These variations are due to the nuclear charge on the atom and the energy level where the valence electrons are found.
6. We can also use the periodic table to predict what monatomic ions could be formed from the elements, what ions and atoms are *isoelectronic* with each other, and what the names and formulas are for compounds by comparison with known compounds. Also, we can determine *oxidation states* for many elements, in compounds, using the periodic table to determine some common oxidation numbers.

Upon completion of this chapter you should be able to do the following things:

1. Set up a periodic table with all atomic numbers included in the proper order.
2. Given a periodic table with atomic numbers and symbols, tell (a) if a given element is a metal or nonmetal; (b) what period and/or group

any element is in; (c) if a given element is a representative element, a noble gas, a transition element, or an inner transition element; (d) how many valence electrons there are in a given element, and how many unpaired electrons there are; and (e) what sublevel (or type of orbital) the differentiating electron is in for a given element. (1 through 9)

3. Define, or explain, the terms and predict periodic variations in these properties, given the terms (a) atomic size, (b) ionic size, (c) ionization energy, (d) electron affinity, and (e) electronegativity. (10 through 13)
4. Given any atom or ion, write the formula for four atoms or ions that are isoelectronic with the atom or ion. (14)
5. Given any representative element, predict the minimum and maximum oxidation states, and for metals, predict all possible oxidation states. (15, 16)
6. Given any formula for a compound, tell the oxidation state of each element in the compound. (17)
7. Given any of the terms in Exercise 1, give an explanation or definition in your own words.

EXERCISES

1. Define the following terms:
   (a) periodic law
   (b) periodic table
   (c) period
   (d) group
   (e) family
   (f) differentiating electron
   (g) valence electrons
   (h) ionization energy
   (i) electron affinity
   (j) electronegativity
   (k) noble gas
   (l) transition elements
   (m) representative elements
   (n) inner transition elements
   (o) oxidation number (state)
2. Explain the variation in each of the following properties, going across a period (left to right) and down a group:
   (a) atomic size
   (b) ionization energy
   (c) electronegativity
3. Distinguish between
   (a) ionization energy and electron affinity
   (b) ionization energy and electronegativity
   (c) representative elements and noble gases
   (d) transition elements and inner transition elements
   (e) representative elements and transition elements

PROBLEMS

Arrangement of the periodic table

1. Tell what group in the periodic table each of the following is in:
   (a) potassium (b) phosphorus (c) tin (d) magnesium (e) chlorine (f) calcium (g) sodium (h) sulfur (i) bromine (j) lead
2. Tell what period each of the following is in.
   (a) carbon (b) lead (c) silver (d) iron (e) mercury (f) neon (g) manganese (h) oxygen (i) zinc (j) tin
3. Classify each of the following as representative, transition, noble gas, or inner transition:
   (a) magnesium (b) copper (c) neon (d) uranium (e) manganese (f) phosphorus (g) iron (h) gold (i) helium (j) mercury

Electron configuration

4. Tell how many valence electrons there are in each of the following:
   (a) bromine (b) nitrogen (c) aluminum (d) sulfur (e) lead (f) carbon (g) chlorine (h) arsenic (i) potassium (j) tin
5. Give the electron configuration for the valence electrons for each of the following:
   (a) iodine (b) oxygen (c) arsenic (d) calcium (e) tin (f) sulfur (g) aluminum (h) bromine (i) phosphorus (j) lead
6. Write electron dot notation for each of the following:
   (a) potassium (b) sulfur (c) silicon (d) chlorine (e) hydrogen (f) sodium (g) oxygen (h) bismuth (i) nitrogen (j) iodine
7. Give the electron configuration for the electrons in the outer *d* and *s* sublevels for the following transition elements:
   (a) titanium (b) iron (c) cobalt (d) mercury (e) copper (f) vanadium (g) nickel (h) manganese (i) zinc (j) chromium
8. Tell what type of orbital (such as 2*s*, 3*p*, and so on) the differentiating electron is in for each of the following:
   (a) oxygen (b) potassium (c) aluminum (d) tin

(e) nickel
(f) sodium
(g) sulfur
(h) calcium
(i) lead
(j) iron

9. Using the periodic table, tell how many unpaired electrons there are in each of the following:
(a) fluorine
(b) phosphorus
(c) silicon
(d) bromine
(e) chromium
(f) aluminum
(g) arsenic
(h) chlorine
(i) calcium
(j) copper

Periodic variations

10. In each group, choose the one with the highest ionization energy:
(a) Na, P, Cl, Al
(b) Na, K, Rb, Li, Cs
(c) N, Cl, O, C, F, Br
(d) Ca, Mg, Be, Sr, Ba
(e) C, F, N, O
(f) P, Na, S, Mg, Cl, I, Br

11. Explain the difference in ionization energy between each of the following:
(a) Ca and Zn
(b) Na^+ and Mg^+
(c) Mg and Mg^+
(d) K and Cu
(e) K^+ and Ca^+
(f) Ca^+ and Ca^{2+}

12. Arrange the following in order of increasing atomic sizes:
(a) Cs, Mg, K, Na, Rb
(b) F, O, Ne, S, As
(c) Ca, Al, Sr, Mg, Ba
(d) P, F, N, O, As

13. Arrange the following in order of increasing electronegativity:
(a) O, Na, N, Rb, F, Mg
(b) As, F, P, Sb, O
(c) Mg, Na, P, Al, Cl
(d) S, F, Se, O

Isoelectronic species

14. Write the formula (symbol and charge) for four atoms or ions that are isoelectronic with each of the following:
(a) Br^-
(b) S^{2-}
(c) Xe
(d) Mg^{2+}
(e) Cl^-
(f) Kr

Oxidation states

15. List the oxidation states for each of the following:
(a) Ca
(b) Pb
(c) Al
(d) Fe
(e) Hg
(f) K
(g) Sn
(h) Zn
(i) Cu
(j) Ba

16. List the minimum and maximum oxidation state for each of the following:
(a) Se
(b) P
(c) Cl
(d) As
(e) S
(f) I

17. Find the oxidation state of the underlined element in each of the following:
(a) $\underline{Cr}Cl_3$
(b) $H_2\underline{S}O_3$
(c) $\underline{Mn}_2O_7$
(d) $Ca(\underline{Br}O_3)_2$
(e) $\underline{Cr}O_4{}^{2-}$
(f) $\underline{Mn}Br_2$
(g) $H_3\underline{P}O_4$
(h) $\underline{Cr}_2O_5$
(i) $Al(\underline{Cl}O_4)_3$
(j) $\underline{Mn}O_4{}^-$

General problems

18. Explain the difference in electronic configuration between
    (a) K and Sc^{2+}
    (b) Ca and Ti^{2+}
19. Group the following electronic configurations of elements together according to those you would expect to show similar chemical properties:
    (a) $1s^2$, $2s^2\,2p^6$, $4s^2\,3d^{10}$, $5s^2\,4d^{10}$, $5s^2\,5p^4$
    (b) $2s^2$, $2s^2\,2p^4$, $2s^2\,2p^5$, $4s^2\,4p^4$, $5s^2$
    (c) $1s^2$, $2s^2\,2p^6$, $4s^2\,4p^3$, $6s^2\,6p^3$, $3s^2\,3p^6$
    (d) $2s^1$, $3s^2\,3p^3$, $3s^2\,3p^4$, $5s^2\,5p^3$, $4s^1$
20. Using the periodic table, determine the chemical symbols for the electronic configurations in Problem 19 and then check your answers by referring to the periodic table to see if you placed the elements of similar chemical properties in the same group.
21. Using the periodic table, indicate which one of the following pairs of elements is the most metallic:
    (a) phosphorus and arsenic
    (b) sulfur and tellurium
    (c) magnesium and phosphorus
    (d) silicon and aluminum
    (e) germanium and lead
    (f) calcium and arsenic
22. Given the following formulas of compounds, sodium sulfate (Na_2SO_4), magnesium phosphate [$Mg_3(PO_4)_2$], and aluminum oxide (Al_2O_3), and using the periodic table, write the formulas for the following compounds:
    (a) cesium sulfate
    (b) magnesium arsenate
    (c) sodium selenate
    (d) gallium oxide
    (e) barium phosphate
    (f) strontium phosphate
    (g) aluminum sulfide
    (h) indium oxide
    (i) sodium tellurate
    (j) potassium sulfate
23. For each of the following new or undiscovered elements, tell which group it should be in and what the electron configuration should be for the outermost electrons:
    (a) 106
    (b) 114
    (c) 119
    (d) 120
    (e) 107
    (f) 116
24. Predict what oxidation states each of the following elements could exhibit:
    (a) 115
    (b) 120
    (c) 114
    (d) 119

REFERENCES

HYDE, J. F., "A Newly Arranged Periodic Chart," *Chemistry,* Sept. 1976, **49,** p. 15.

KNORRE, YELENA, "Expanding Mendeleev's Table," *Chemistry,* July-Aug. 1973, **26,** p. 27.

SANDERSON, R. T., *Chemical Periodicity.* New York: Reinhold Publishing Corp. 1960.

SANDERSON, R. T., "Ionization Energy and Atomic Structure," *Chemistry,* May 1973, **46,** p. 12.

SZABADVARY, F., "A Mysterious Table," *Chemistry,* Sept. 1974, **47,** p. 11.

Objectives

To learn to write correct names and formulas for the following types of compounds:

1. Binary
   - (a) Two nonmetals
   - (b) Metal + nonmetal
     - (1) Metal with constant oxidation number
     - (2) Metal with variable oxidation numbers (using both the classical and Stock system)
   - (c) Binary acids
   - (d) Compounds named like binary compounds (ammonium, cyanide, and hydroxide compounds)
2. Bases (hydroxides)
3. Oxyacids and their salts (including acid salts)
4. Mixed (double) salts

Six

Nomenclature of Inorganic Compounds

Before considering the formation of compounds from atoms, we shall learn to write correct names and formulas for inorganic compounds. To do this, we shall use our knowledge of the electronic structure of atoms and the position of elements in the periodic table. We shall start with binary compounds (those consisting of only two elements) and then proceed to more complex ternary and higher compounds. The systematic nomenclature will be used; however, in some cases, common names are used almost exclusively and these shall also be discussed.

The systematic nomenclature, which will be used in this chapter, was developed by an organization of chemists called the International Union of Pure and Applied Chemistry (IUPAC). These rules were set up by this group and are periodically revised and updated.

The names and formulas of inorganic compounds are written in such a way that every compound can be named from its formula, and each formula has a name peculiar to that formula. The more positive portion is written and named first. This could be a metal, a positive polyatomic ion, a hydrogen ion, or simply the less electronegative portion of the compound. The more negative portion is written and named last. This can be an anion or simply the more electronegative atom.

6-1 Binary Compounds Containing Two Nonmetals

All **binary compounds** take an *-ide* ending on the last element named. As mentioned previously, the element with the highest electronegativity is written and named last.

If two nonmetals combine to form only one compound, the compound is simply named by writing the name of the second element with an -ide ending. Compounds of this type contain hydrogen with another nonmetal. For example, HBr is hydrogen bromide and H_2S is hydrogen sulfide. For

TABLE 6-1 *Greek prefixes*

GREEK PREFIX	NUMBER	GREEK PREFIX	NUMBER
Mono-	1	Hexa-	6
Di-	2	Hepta-	7
Tri-	3	Octa-	8
Tetra-	4	Nona- (or ennea-)	9
Penta-	5	Deca-	10

other binary compounds of this type, where more than one binary compound can be formed from two elements, the Greek prefixes, as listed in Table 6-1, are used with each element. The mono prefix is not used with the first element. For example, CO_2 is carbon dioxide, and CO is carbon monoxide. It is also acceptable to use Roman numerals here to indicate the oxidation state of the more positive elements; however, the Greek prefixes are preferred.

Some binary compounds of two nonmetals have common names, such as *water* for H_2O and *ammonia* for NH_3 (ammonia is an exception to the general rule that the most positive part is written first).

6-2 Binary Compounds Containing a Metal and a Nonmetal

The metal is always written first, in both the name and the formula. As with all other binary compounds, the nonmetal takes an -ide ending. There are two types that we must consider: metals with a fixed (only one) oxidation number and those with variable (more than one) oxidation numbers.

Metals with Fixed Oxidation Numbers

The metals with fixed (constant) oxidation numbers are the IA and IIA metals, plus Al, Zn, and Ag.* All other metals have variable oxidation numbers. The names of compounds in which the metal has a fixed oxidation number are simply written as in the following examples (note that no prefixes are used):

FORMULA	NAME
NaCl	Sodium chloride
Na_2S	Sodium sulfide
AgBr	Silver bromide
Al_2O_3	Aluminum oxide
CaH_2	Calcium hydride
Mg_3N_2	Magnesium nitride

*Compounds of silver are known in which the Ag is +2 or +3, but +1 is by far the most common.

In writing the formulas of compounds you must know the oxidation numbers, as stated in the general rules given in Section 5-8. Consider the following examples:

NAME	FORMULA
Potassium chloride	KCl (K is +1; Cl is −1)
Calcium bromide	$CaBr_2$ (Ca is +2; Br is −1)
Barium oxide	BaO (Ba is +2; O is −2)
Aluminum sulfide	Al_2S_3 (Al is +3; S is −2)

Metals with Variable Oxidation Numbers

Except for IA metals, IIA metals, aluminum, zinc, and silver, all metals have variable oxidation numbers. In a binary compound of a metal of this type with a nonmetal, the oxidation number of the metal must be indicated in the name. There are two methods of doing this that we shall consider here.

In the *classical* system, an *-ic* ending is used on the name of the metal for the highest oxidation state and an *-ous* ending is used for the lowest. Also, the Latin name is used for iron (ferric and ferrous), copper (cupric and cuprous), tin (stannic and stannous), and lead (plumbic or plumbous). There are several disadvantages to this system. For example, the name doesn't give any indication of the actual oxidation state, only if it is the higher or lower. Also, for metals with more than two possible oxidation states, it cannot be used. Unfortunately, this system is still used and thus you must become familiar with it.

The newer *Stock system,* or *Roman numeral system,* is more accurate. In this system, the actual oxidation number of the metal is indicated by a Roman numeral in parentheses immediately following the name of the metal. Consider the following examples of the names of some common cations, using both the Roman numeral system and the classical -ic, -ous system.

Cu^+ = cuprous or copper(I)
Cu^{2+} = cupric or copper(II)
Fe^{2+} = ferrous or iron(II)
Fe^{3+} = ferric or iron(III)

As you may have noticed in the preceding examples, the -ous, -ic system can become confusing. For example, in copper the -ous represents the +1 oxidation number, whereas in iron it is +2. Also, Cu^{2+} is cupric, but Fe^{2+} is ferrous. Consider the following examples.

In writing formulas from names, recall that the oxidation number of each element is used to determine how many of each atom there must be in the formula.

FORMULA	NAME
$CuCl_2$	Copper(II) chloride or cupric chloride
FeO	Iron(II) oxide or ferrous oxide
SnF_4	Tin(IV) fluoride or stannic fluoride
PbS	Lead(II) sulfide or plumbous sulfide
HgO	Mercury(II) oxide or mercuric oxide

NAME	FORMULA
Iron(III) oxide	Fe_2O_3
Lead(IV) oxide	PbO_2
Cuprous chloride	$CuCl$
Stannous fluoride	SnF_2

A third method that can be used for compounds of this type is to use the Greek prefixes, as discussed in Section 6-1. This is not recommended but is sometimes used. For example, PbO_2 could be named as lead dioxide.

Compounds Named Like Binary Compounds

There are a few other compounds that take an -ide ending, like binary compounds. These include the following:

1. Ammonium (NH_4^+) compounds—in these the NH_4^+ ion acts like a simple metal ion; for example $(NH_4)_2S$ is ammonium sulfide.
2. Cyanides (CN^-)—here the CN^- ion acts like a simple nonmetal anion; for example, $Fe(CN)_2$ is iron(II) cyanide or ferrous cyanide.
3. Hydroxides (OH^-)—these compose a special type of compound, which will be discussed in Section 6-4; the hydroxide has the -ide ending of binary compounds.

6-3 Binary Acids

Substances that produce H^+ ion in solution are **acids.*** A binary compound composed of hydrogen with a more electronegative element can act as a **binary acid,** when in water solution. For example, when the gas hydrogen chloride (HCl) is placed in water solution, it forms hydrochloric acid. For acids of this type, the prefix hydro- is added, and the -ide ending is replaced by -ic acid. Consider the following examples:

*In Chapter 14, we shall define acids and bases in more exact terms.

COMPOUND	NAME	NAME OF ACID FORMED IN WATER SOLUTION
HBr	Hydrogen bromide	*Hydro*bromi*c* acid
H_2S	Hydrogen sulfide	*Hydro*sulfur*ic* acid
HCN	Hydrogen cyanide	*Hydro*cyan*ic* acid

6-4 Ternary and Higher Compounds

Hydroxides

These are inorganic compounds that contain the —OH group. They are commonly called **bases, alkalies,** or just **hydroxides.** To write the correct formulas for these compounds it must be remembered that the —OH group has a −1 charge. Consider the following examples:

NAME	FORMULA
Sodium hydroxide	NaOH
Calcium hydroxide	$Ca(OH)_2$
Iron(III) hydroxide	$Fe(OH)_3$
Stannic hydroxide	$Sn(OH)_4$

(Recall that for metals with variable oxidation numbers we must indicate the oxidation number in the name.)

Ternary Acids and Their Salts

Acids that contain more than two elements always contain oxygen, along with the hydrogen, and are thus often referred to as **oxyacids.** The names and formulas of the more common ones are listed in Table 6-2 and must be memorized. A *salt* is a compound formed when one or more of the hydrogen ions of an acid is replaced by a cation (metal or positive polyatomic ion) or when one or more of the hydroxide ions of a base is replaced by an anion (nonmetal or negative polyatomic ion). The binary compounds of metal cations with nonmetal anions and the ternary compounds of metal cations or ammonium ions with negative polyatomic ions are examples of salts. Therefore, the names and formulas of salts, other than the binary ones already considered, are directly related to the names and formulas of the oxyacids. Once you have learned the names and formulas of the oxyacids, you can determine from them the oxyanions and salts, using the following general rules.

1. Since the hydrogen ion (H^+) has a +1 charge, the charge on the oxyanion is equal to the number of hydrogen ions that can be produced by a molecule of the acid.

TABLE 6-2 *Common oxyacids, oxyanions, and an example of a salt of each*

ACID		OXYANION		EXAMPLE OF SALT	
Formula	Name	Formula	Name	Formula	Name
H_2SO_3	Sulfurous acid	SO_3^{2-}	Sulfite	Na_2SO_3	Sodium sulfite
H_2SO_4	Sulfuric acid	SO_4^{2-}	Sulfate	$FeSO_4$	Iron(II) sulfate
HNO_2	Nitrous acid	NO_2^-	Nitrite	$Ca(NO_2)_2$	Calcium nitrite
HNO_3	Nitric acid	NO_3^-	Nitrate	NH_4NO_3	Ammonium nitrate
H_2CO_3	Carbonic acid	CO_3^{2-}	Carbonate	$Al_2(CO_3)_3$	Aluminum carbonate
H_3PO_3	Phosphorous acid	PO_3^{3-}	Phosphite	$Mg_3(PO_3)_2$	Magnesium phosphite
H_3PO_4	Phosphoric acid	PO_4^{3-}	Phosphate	$Sn_3(PO_4)_2$	Tin(II) phosphate or stannous phosphate
$HClO$	Hypochlorous acid*	ClO^-	Hypochlorite	$Ca(ClO)_2$	Calcium hypochlorite
$HClO_2$	Chlorous acid*	ClO_2^-	Chlorite	$NaClO_2$	Sodium chlorite
$HClO_3$	Chloric acid*	ClO_3^-	Chlorate	$Al(ClO_3)_3$	Aluminum chlorate
$HClO_4$	Perchloric acid*	ClO_4^-	Perchlorate	$KClO_4$	Potassium perchlorate

*Other halogens form compounds analogous to these last four examples.

2. An -ous acid always forms an -ite oxyanion or salt; an -ic acid always forms an -ate ion or salt.

Applying these rules, we can see that since sulfuric acid is H_2SO_4, the sulfate ion is SO_4^{2-} or since HNO_2 is nitrous acid, NO_2^- is the nitrite ion and so on.

Notice also, from Table 6-2, that where there are two compounds with the same central atom, the one in the lower oxidation state is the -ous acid and the higher is the -ic acid. For more than two (e.g., $HClO$, $HClO_2$, $HClO_3$, $HClO_4$) the general rules given in Table 6-3 hold for many compounds (but not all). These rules are always followed if there are four acids of this type with different oxidation states, such as the series listed in Table 6-2. (One common exception to these rules is H_2CO_3—carbon*ic* acid.)

TABLE 6-3 *General rules for determining names of oxyacids and oxyanions from oxidation states*

OXIDATION STATE OF CENTRAL ATOM	NAME OF ACID		NAME OF OXYANION	
1 or 2	Hypo-	-ous	Hypo-	-ite
3 or 4		-ous		-ite
5 or 6		-ic		-ate
7 or 8	Per-	-ic	Per-	-ate

Acid Salts

Acid salts are salts of polyprotic acids (more than one replaceable H^+ ion) in which one or more hydrogen atoms remain in the salt. For example, if only one of the H^+ ions in H_2SO_4 is replaced, an HSO_4^- ion remains. In naming these, *hydrogen* is added to the name of the oxyanion. For example, HSO_4^- is named hydrogen sulfate.

Salts of this type are sometimes named using the prefix *bi* or *acid* in place of *hydrogen.* For example, HCO_3^- is often called *bicarbonate.* The use of hydrogen is preferred, however. For example, $NaHCO_3$ is sodium hydrogen carbonate or sodium bicarbonate or sodium acid carbonate; $Ca(HSO_4)_2$ is calcium hydrogen sulfate or calcium bisulfate or calcium acid sulfate.

In the case of acids like H_3PO_4 where there are more than two H^+ ions, *dihydrogen* is used for the ion that contains two hydrogens; that is, $H_2PO_4^-$ is called *dihydrogen phosphate.* For example, Na_2HPO_4 is sodium hydrogen phosphate (sometimes called disodium hydrogen phosphate), and NaH_2PO_4 is sodium dihydrogen phosphate.

Consider the following examples of various types of salts:

FORMULA	NAME
$CaSO_4$	Calcium sulfate (it is not necessary to indicate the oxidation state for calcium)
$Fe_2(CO_3)_3$	Iron(III) carbonate or ferric carbonate (iron must be +3 in this case since carbonate is −2)
$Ca(BrO_4)_2$	Calcium perbromate (since perchlorate is ClO_4^-, perbromate is BrO_4^-; we need two of the BrO_4^- groups since Ca is +2; it is *not* necessary to indicate that there are two perbromate groups—just like $CaCl_2$ is calcium chloride and *not* dichloride)
$Pb_3(PO_4)_2$	Lead(II) phosphate (the II refers to the oxidation state of Pb; since the charge on the phosphate is −3, because phosphoric acid is H_3PO_4, there must be three lead ions for each two phosphate ions)
$Ba(H_2PO_4)_2$	Barium dihydrogen phosphate (the *di*hydrogen phosphate refers to the hydrogen atoms in the group, not to the fact that there are two of these groups—for example, NaH_2PO_4 is sodium *di*hydrogen phosphate)
Calcium chlorite	$Ca(ClO_2)_2$ (since Ca is +2 and the chlorite ion is −1, there must be two chlorites for each Ca)
Stannic iodate	$Sn(IO_3)_4$ (tin, in this case, is in the highest oxidation state, +4; since chlorate is ClO_3^-, iodate is IO_3^-)
Magnesium hydrogen sulfate	$Mg(HSO_4)_2$ (since sulfuric acid is H_2SO_4, hydrogen sulfate is HSO_4^-; therefore, since Mg is +2 (group IIA), there must be two of the hydrogen sulfate groups for each Mg^{2+} ion)

*Mixed Salts**

If a salt contains two cations, the one with the lower charge is usually written and named first; otherwise, the nomenclature is just the same as for other salts. For example, $KCaPO_4$ is potassium calcium phosphate; $Na_2Ca(SO_4)_2$ is disodium calcium sulfate. Unlike simple salts of Na and Ca, the number of each atom must be indicated here by Greek prefixes, again omitting the mono. Occasionally, the same thing is done with salts that contain O^{2-} or OH^- along with another anion (negative ion). For example, BiOCl is bismuth oxide chloride, or bismuth oxychloride; $Co(OH)(NO_3)$ is cobalt(II) hydroxide nitrate, or cobalt(II) hydroxynitrate. Many of these so-called *mixed salts* are actually complex ions (or coordination compounds), which will be discussed in Chapter 24. In many cases, with these as well as other complicated inorganic compounds, it is actually necessary to know something about the structure and properties of the compound before a correct name can be written. This is the subject of more advanced inorganic chemistry, however, and need not concern us at this point.

Using the Periodic Table to Determine Names and Formulas

From what you have learned here, and from your knowledge of the periodic table and atomic structure, you should be able to predict names and formulas correctly for other less common substances. Consider the following examples:

FORMULA	NAME
$KMnO_4$	Potassium permanganate (see Table 6-3)
Na_2SeO_4	Sodium selenate (since sulfate is $SO_4{}^{2-}$, $SeO_4{}^{2-}$ must be selenate)
$Al_2(CrO_4)_3$	Aluminum chromate (Cr is +6 here; therefore it is chromate)

NAME	FORMULA
Barium phosphide	Ba_3P_2 (since nitride is N^{3-}, phosphide must be P^{3-}; Ba, in Group IIA, is +2)
Nickel(II) arsenate	$Ni_3(AsO_4)_2$ (since phosphate is $PO_4{}^{3-}$, arsenate must be $AsO_4{}^{3-}$; from the name we know that Ni is +2)
Cesium silicate	Cs_2SiO_3 (since carbonate is $CO_3{}^{2-}$, silicate must be $SiO_3{}^{2-}$; cesium, in Group IA, is +1)

*The portion of Section 6-4 dealing with mixed salts is optional. It may be omitted without loss of continuity.

SUMMARY

In both names and formulas, the most positive part is written first; also, the total positive oxidation state must equal the total negative. Learn the rules for determining oxidation states; remember which metals have constant oxidation states (IA, IIA, Al, Zn, Ag)—all others are variable, and the oxidation number for these must be indicated in the name (by -ic, -ous ending or Roman numeral) for all types of compounds.

1. Binary compounds—take -ide ending.
   (a) Two nonmetals—for those that form more than one compound, a Greek prefix is used to indicate the number of each atom (mono is omitted for the most positive part).
   (b) Metal and nonmetal—for metals with variable oxidation states, use
      (1) -ic ending on metal for higher oxidation state and -ous for lower (with Latin stem for Fe, Cu, Sn, and Pb in both cases).
      (2) Roman numeral to indicate oxidation state of metal.
      (3) Can use Greek prefixes—not recommended.
   (c) Compounds named like binary—compounds of NH_4^+, OH^-, CN^-.
2. Bases (OH^-)—named as hydroxides.
3. Acids (H^+ compounds)
   (a) Binary—*hydro-* prefix and *-ic acid* ending.
   (b) Ternary (oxyacids)—name is determined from the oxidation state of the central atom (you must learn the ones listed in Table 6-2).
4. Salts of oxyacids—named for acid they are derived from; -ic acid forms -ate salt, -ous acid forms -ite salt. For acid salts, *hydrogen* is added to the name of the oxyanion (can also use bi- or acid; hydrogen is preferred); for two hydrogen (as in $H_2PO_4^-$), use *dihydrogen.*

Upon completion of this chapter, you should be able to do the following: For each of the following types of compounds (assume that a periodic table, with symbols and atomic numbers, is available), given the name, write the correct formula; given the formula, write the correct name.

---

1. Binary compounds composed of two nonmetals. (1, 2)
2. Binary compounds composed of a metal and a nonmetal, in which the metal has only one possible oxidation state (number). (3, 4)
3. Binary compounds composed of a metal and a nonmetal in which the metal has more than one possible oxidation state (number), including both the classical and the Stock (Roman numeral) system. (3, 4)
4. Compounds named like binary compounds, including ammonium compounds, cyanides, and hydroxides. (5, 6)
5. Binary acids. (7, 8)

6. Bases (or hydroxides), including those containing metals that have more than one possible oxidation state. (9, 10)
7. Monatomic and polyatomic ions. (11, 12)
8. Ternary acids (oxyacids) and compounds formed from oxyacids, including those containing metals that have more than one possible oxidation state. (13, 14)
9. Acid salts. (15, 16)
10. Mixed (or double) salts. (17, 18)
11. Compounds that contain elements for which you must use your knowledge of the periodic table to figure out the name or formula since they contain elements or polyatomic ions for which you have not learned any general rules for determining oxidation states or formulas. (19, 20)

PROBLEMS

Binary compounds of two nonmetals

1. Name the following:
   - (a) SO_2
   - (b) N_2O_3
   - (c) PBr_5
   - (d) SO_3
   - (e) P_2O_5
   - (f) PCl_3
2. Write the correct formula for each of the following:
   - (a) phosphorus tribromide
   - (b) diphosphorus pentoxide
   - (c) sulfur trioxide
   - (d) dinitrogen trioxide
   - (e) sulfur dioxide
   - (f) phosphorus pentachloride

Binary compounds of metal and nonmetal

3. Name the following (for those metals with variable oxidation states, give both the classical and Stock system names):
   - (a) $MgCl_2$
   - (b) $SnCl_4$
   - (c) $HgBr_2$
   - (d) PbS_2
   - (e) FeO
   - (f) $CaBr_2$
   - (g) $PbBr_4$
   - (h) $AlCl_3$
   - (i) SnO_2
   - (j) HgO
4. Write the correct formula for each of the following:
   - (a) magnesium bromide
   - (b) aluminum oxide
   - (c) ferrous bromide
   - (d) lead(II) sulfide
   - (e) stannic oxide
   - (f) potassium sulfide
   - (g) iron(III) oxide
   - (h) zinc bromide
   - (i) stannous chloride
   - (j) ferric oxide

Compounds named like binary compounds

5. Name the following:
   - (a) NH_4Cl
   - (b) $Ca(CN)_2$
   - (c) $NaCN$
   - (d) $(NH_4)_2S$

6. Write the correct formula for each of the following:
(a) magnesium cyanide
(b) ammonium sulfide
(c) ammonium bromide
(d) ferrous cyanide

Binary acids

7. Name the following as acids:
(a) HCl
(b) H_2S
(c) HBr
(d) HCN

why do you need 2

8. Write the correct formula for each of the following:
(a) hydrobromic acid
(b) hydrocyanic acid
(c) hydroiodic acid
(d) hydrosulfuric acid

Bases

9. Name the following (for those metals with variable oxidation states, give both the classical and Stock system names):
(a) $Mg(OH)_2$
(b) $Fe(OH)_3$
(c) $Zn(OH)_2$
(d) $Sn(OH)_4$
(e) $Ca(OH)_2$
(f) $Fe(OH)_2$
(g) $Al(OH)_3$
(h) $Pb(OH)_4$

10. Write the correct formula for each of the following:
(a) barium hydroxide
(b) aluminum hydroxide
(c) mercury(I) hydroxide
(d) lead(II) hydroxide
(e) lithium hydroxide
(f) zinc hydroxide
(g) chromium(III) hydroxide
(h) stannous hydroxide

Ions

11. Write the formula for each of the following ions:
(a) ammonium
(b) carbonate
(c) chlorite
(d) tin(II)
(e) phosphite
(f) bromate
(g) periodate
(h) nitrate
(i) sulfite
(j) hypochlorite
(k) nitrite
(l) ferric
(m) barium
(n) stannous
(o) cyanide
(p) cuprous
(q) copper(II)
(r) nitrite
(s) hypobromite
(t) iodate

12. Name each of the following ions:
(a) S^{2-}
(b) SO_3^{2-}
(c) PO_4^{3-}
(d) Pb^{2+}
(e) ClO_4^-
(f) Cl^-
(g) ClO_3^-
(h) BrO^-
(i) IO_2^-
(j) CrO_4^{2-}
(k) NO_2^-
(l) CO_3^{2-}
(m) SO_4^{2-}
(n) NH_4^+
(o) Br^-
(p) PO_3^{3-}
(q) Sn^{2+}
(r) IO^-
(s) BrO_4^-
(t) MnO_4^-

Acids and salts

13. Name the following (for metals with variable oxidation states, give both the classical and the Stock system name):
   (a) H_2SO_3 (in water)
   (b) $Ca_3(PO_4)_2$
   (c) H_3PO_3 (in water)
   (d) $Mg(ClO_3)_2$
   (e) NH_4NO_2
   (f) HBrO (in water)
   (g) $Pb(SO_4)_2$
   (h) $Sn(IO_3)_2$
   (i) HNO_2 (in water)
   (j) $Al_2(CO_3)_3$
   (k) $FeCO_3$
   (l) $HClO_2$ (in water)
   (m) $FeSO_4$
   (n) NH_4NO_3
   (o) $HClO_3$ (in water)
   (p) $Sn_3(PO_4)_2$
   (q) $Ca(ClO_2)_2$
   (r) $HBrO_3$ (in water)
   (s) $Sn(SO_3)_2$
   (t) $Pb(IO_2)_2$
14. Write the correct formula for each of the following:
   (a) calcium nitrate
   (b) iron(III) sulfate
   (c) sulfurous acid
   (d) mercuric phosphite
   (e) magnesium chlorate
   (f) plumbic carbonate
   (g) ammonium phosphate
   (h) tin(II) sulfite
   (i) ferrous iodite
   (j) perbromic acid
   (k) sodium sulfate
   (l) hypochlorous acid
   (m) magnesium nitrite
   (n) cuprous sulfate
   (o) ammonium chlorate
   (p) chromium(III) carbonate
   (q) zinc phosphate
   (r) lead(IV) sulfite
   (s) iodic acid
   (t) stannous bromate

Acid salts

15. Name the following:
   (a) $Al(HSO_4)_3$
   (b) KH_2PO_4
   (c) $Cu(HCO_3)_2$
   (d) $CaHPO_3$
   (e) NH_4HSO_3
   (f) $Ca(HCO_3)_2$
   (g) $Fe(HSO_4)_2$
   (h) NaH_2PO_4
   (i) $Zn(HSO_3)_2$
   (j) $(NH_4)_2HPO_3$
16. Write the correct formula for each of the following:
   (a) sodium hydrogen sulfite
   (b) ferrous hydrogen carbonate
   (c) calcium dihydrogen phosphate
   (d) potassium hydrogen phosphate
   (e) ammonium hydrogen phosphite
   (f) potassium hydrogen sulfate
   (g) lead(II) hydrogen carbonate
   (h) aluminum hydrogen phosphate
   (i) magnesium dihydrogen phosphate
   (j) ammonium hydrogen sulfite

Mixed (double) salts

17. Name the following:
   (a) $NaCaPO_4$
   (b) $K_2Ba(SO_4)_2$
   (c) $KCa(NO_3)_3$
   (d) $Na_2Mg(SO_4)_2$
18. Write the correct formula for each of the following:
   (a) sodium aluminum sulfate
   (b) disodium calcium nitrate
   (c) potassium calcium phosphate
   (d) disodium barium sulfate

Compounds for which knowledge of the periodic table must be used

19. Name the following:
    - (a) K_2SeO_4
    - (b) $Al_2(SiO_3)_3$
    - (c) $Ca_3(AsO_4)_2$
    - (d) Rb_3P
20. Write the correct formula for each of the following:
    - (a) silver arsenate
    - (b) cesium phosphide
    - (c) sodium silicate
    - (d) radium selenite

General

21. Name the following:
    - (a) HNO_2 in water
    - (b) $CaBr_2$
    - (c) N_2O_4
    - (d) H_3PO_4 in water
    - (e) $Cr(OH)_3$
    - (f) $Cd(NO_3)_2$
    - (g) Fe_2O_3
    - (h) KOH
    - (i) $Hg(NO_2)_2$
    - (j) $HBrO_4$ in water
    - (k) KH_2PO_4
    - (l) NH_4I
    - (m) $Mg(CN)_2$
    - (n) $FeSO_4$
    - (o) SnS_2
    - (p) Ag_3PO_4
    - (q) Mg_3N_2
    - (r) $NaKSO_4$
    - (s) $BaCrO_4$
    - (t) $NaHSO_3$
    - (u) $LiBrO_4$
    - (v) NH_4HCO_3
    - (w) $SrSiO_3$
    - (x) $Co_3(AsO_4)_2$
    - (y) $KMnO_4$
    - (z) SnS_2
    - (aa) P_2O_5
    - (bb) H_2SO_3 in water
    - (cc) $CuCO_3$
    - (dd) $HBrO$ in water
    - (ee) $MgBr_2$
    - (ff) $NaCN$
    - (gg) $Mn(OH)_2$
    - (hh) NH_4CN
    - (ii) $PbSO_4$
    - (jj) $AgNO_2$
    - (kk) $NaClO_2$
    - (ll) PCl_3
    - (mm) $Ca(H_2PO_3)_2$
    - (nn) $Ba(OH)_2$
    - (oo) $Zn(MnO_4)_2$
    - (pp) $LiHCO_3$
    - (qq) $BaCl_2$
    - (rr) $Fe(BrO_4)_3$
    - (ss) $BaSeO_4$
    - (tt) $Cu(HSO_4)_2$
    - (uu) $NH_4Al(SO_4)_2$
    - (vv) $Cr_2(SiO_3)_3$
    - (ww) $Ca(MnO_4)_2$
    - (xx) $(NH_4)_3N$
22. Write the correct formula for each of the following:
    - (a) silver phosphate
    - (b) calcium nitrate
    - (c) stannic iodide
    - (d) zinc hydrogen carbonate
    - (e) sulfur dioxide
    - (f) potassium bisulfite
    - (g) lithium sulfite
    - (h) aluminum sulfate
    - (i) ammonium nitride
    - (j) lead(II) sulfide
    - (k) bromic acid
    - (l) potassium calcium phosphate
    - (m) ferrous iodate
    - (n) zinc cyanide
    - (o) phosphorus trichloride
    - (p) mercuric hydroxide
    - (q) iron(II) chloride
    - (r) ammonium carbonate
    - (s) plumbic silicate
    - (t) cupric nitrite
    - (u) stannous carbonate

(v) ferrous bromide
(w) magnesium arsenate
(x) aluminum hypobromite
(y) barium dihydrogen phosphate
(z) ammonium sulfate
(aa) sodium chlorite
(bb) strontium sulfite
(cc) hydrobromic acid
(dd) cadmium carbonate
(ee) ferrous nitrite
(ff) cupric phosphate
(gg) iron(III) nitrate
(hh) mercuric chloride
(ii) dinitrogen tetroxide
(jj) barium carbonate
(kk) sodium potassium sulfate
(ll) lithium iodide
(mm) stannous fluoride
(nn) lead(II) phosphate
(oo) aluminum cyanide
(pp) sulfur trioxide
(qq) iron(III) perbromate
(rr) magnesium sulfite
(ss) strontium selenate
(tt) potassium sodium sulfate
(uu) barium hydrogen carbonate
(vv) potassium dihydrogen phosphate
(ww) calcium arsenate
(xx) calcium nitride

REFERENCES

WALKER, RUTH A., and HELEN JOHNSTON, *The Language of Chemistry.* Englewood Cliffs, N.J.: Prentice-Hall, Inc., 1967.

Objectives

1. To understand what an ionic bond is and how it is formed and to be able to predict which compounds are ionic and which are covalent.
2. To learn to predict what anions and cations might be formed, using the periodic table.
3. To understand the energy changes involved in the formation of ionic compounds.
4. To understand what a covalent bond is and how it is formed.
5. To understand what a coordinate covalent bond is and to be able to show how this type of bond is formed for a particular compound.
6. To understand what multiple bonds are and what the maximum number of bonds between two atoms is.
7. To understand what polarity is and to be able to predict what bonds would be polar and how great the polarity would be for various bonds (that is, to choose the most polar bond from a group).
8. To be able to draw correct Lewis structures (electron dot formulas) for molecules, including not only those that exhibit an *octet* of electrons but also with less than or more than eight.
9. To understand what resonance forms are and to be able to tell what molecules exhibit resonance.
10. To be able to explain and predict molecular geometry in terms of the VSEPR theory.
11. To understand what sigma (σ) and pi (π) bonds are, how they are formed, and how they differ from each other, and to be able to tell from a Lewis structure how many sigma (σ) bonds and how many pi (π) bonds there are in a compound.
12. To be able to explain resonance forms in terms of delocalized pi bonds.
13. To understand what a hybrid orbital is and to be able to use the concept of hybridization in explaining and predicting molecular geometry.
14. To learn how the electron pairs are arranged in the following types of hybrid orbitals: sp, sp^2, sp^3, dsp^3, d^2sp^3.
15. To be able to explain and predict polarity of molecules, using molecular geometry and electronegativity.
16. To understand the basic principles of the molecular orbital theory of bonding and to be able to find the bond order and number of unpaired electrons for simple diatomic molecules, given the order of energy levels for molecular orbitals.

Seven

Chemical Bonding

In previous chapters we have considered (1) what matter is and how it is measured; (2) what the structure of the atom is that makes up matter and how this is related to the arrangement of the elements in the periodic table; and (3) how to write correct names and formulas for the compounds formed when atoms combine. Now we shall consider how atoms combine to form compounds. To do this we shall look at the two most important types of chemical bonds—electrovalent (or ionic), in which electrons are transferred from one atom to another, and covalent, in which a pair of electrons is shared between two atoms. Later, in Chapter 12, we shall also consider metallic bonding, hydrogen bonding, and van der Waals forces of attraction. These last three types of bonding are important in considering states of matter and changes of state, but they are not necessary for a consideration of the chemical bond that forms compounds.

You will recall from Chapter 3 that when a compound is formed, a chemical change takes place. The compound formed is different from the atoms that made it up. For example, when sodium metal combines with chlorine gas, the sodium chloride formed is different from the individual elements. This is true for both ionic and covalent compounds.

Also, in forming a compound from the elements, energy is always released or absorbed. This will be considered in Chapter 10.

7-1 Electrovalent or Ionic Bonding

An electrovalent or ionic bond is formed when one or more electrons are transferred from one atom to another. This forms a positive ion and a negative ion, which are then held together by an electrostatic attraction. Due to the nature of this type of bonding, the ions formed do not form a molecule consisting of one or more positive ions and one or more negative ions. Instead each positive ion is surrounded by negative ions and each negative ion is surrounded by positive ions. Thus the formula for an ionic compound, such as sodium chloride (NaCl), does not indicate that one sodium ion combines with one chloride ion but only that the ratio of sodium to chloride ions is one Na^+ to one Cl^-. Therefore, when speaking of ionic compounds the term *formula unit* is normally used, and the relative mass of this should be referred to as the *formula weight*. In Chapter 3 it was pointed out that the term *molecular weight* is often used for any type of compound.

The way that the positive and negative ions are arranged in an ionic compound depends mainly on the size and charge of the ions. In Chapter 12 in the discussion of solids, we shall consider some crystal structures for various types of ionic compounds.

The electrostatic attraction between the individual ions is a rather weak interaction. In the crystal structure formed, however, each ion is surrounded by several of the oppositely charged ions. This makes the bonding force in ionic compounds quite strong. We shall see later, in discussing changes of state, that ionic compounds have relatively high melting and boiling points due to this strong interaction. Figure 7-1 shows the arrangement of Na^+ and Cl^- ions in a crystal of NaCl.

The formula of an ionic compound simply depends on the charge on the ions that make it up, as considered in Chapter 6. For example, since the sodium ion has a +1 charge and the chloride has a −1 charge, the formula for sodium chloride is NaCl. With calcium chloride, since the calcium ion has a +2 charge, the formula is $CaCl_2$.

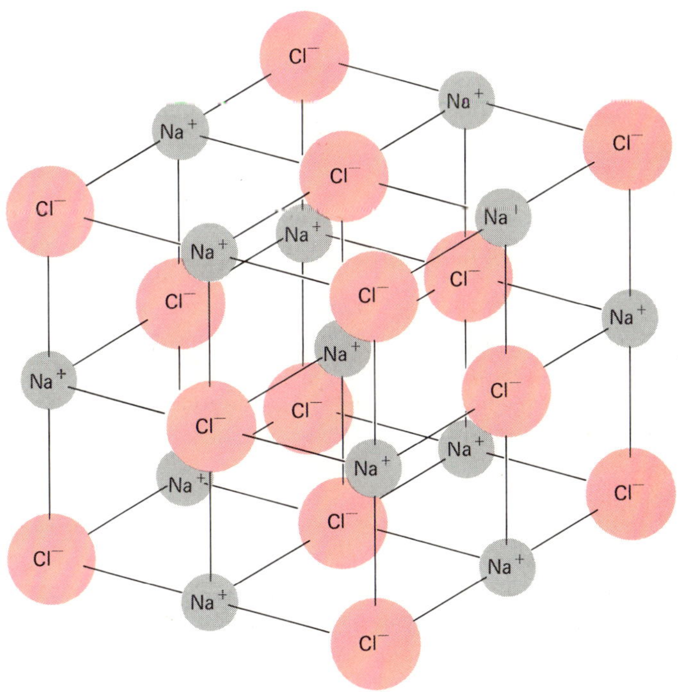

Figure 7-1
Crystal of sodium chloride, NaCl

Formation of Ions

In the discussion of oxidation states in Section 5-8, it was pointed out that in general the maximum oxidation state corresponds to the group number. Also, for the nonmetals, the minimum oxidation state corresponds to the group number minus eight. In terms of both possible oxidation states and formation of ions, the metals can be divided into three groups:

1. Group IA, IIA, and aluminum.
2. Post transition metals.
3. Transition metals.

Of course there is a direct correlation between oxidation states and formation of monatomic ions. The ion formed in all cases is simply one of the possible oxidation states for the element being considered. It must be kept in mind, however, that a simple monatomic ion never has a charge greater than 3. Thus, only certain of the oxidation states can be actual ions. Also, you should recall, for monatomic ions, that metals only form positive ions and nonmetals only form negative ions.

In forming monatomic ions the nonmetals always gain enough electrons to attain a noble gas electron configuration. Thus, they will always be in their lowest possible oxidation state.* Therefore, these atoms will always gain one, two, or possibly three electrons in forming monatomic ions, such as Cl^-, S^{2-}, or N^{3-}.

In forming monatomic ions, the metals can lose one, two, or possibly three electrons to form cations with a charge of $+1$, $+2$, or $+3$; however, a charge of either $+3$ or -3 is quite rare. Compounds in which the oxidation number is three or greater are almost always covalent.

In considering either oxidation states or ions formed, for the metals the only generalization that can be made is that electrons in *s* or *p* subshells are always used in groups; that is, all the electrons in a particular *s* or *p* sublevel are used, never part of them. For example, the IIA metals never lose just one electron to form a $+1$ ion; they always lose both of the electrons in the *s* orbital to give $+2$.

For the IA and IIA metals and aluminum, the ions formed always have a noble gas configuration; that is, these metals lose all their valence electrons in forming cations. For metals following the transition metals, either just the *p* electrons (electrons in *p* orbitals) or the *p* and the *s* are lost. Thus, bismuth (Bi) can form a Bi^{3+} ion; tin or lead can form $+2$ ions (the $+4$ oxidation state would be covalent); and thallium (Tl) can form Tl^+ or Tl^{3+} ions.

For the transition metals, the situation is more complex when considering oxidation states (see Section 5-8). In the formation of ions, however,

*We are only considering simple monatomic ions here. Polyatomic ions, such as nitrate, and sulfate, are groups of nonmetals covalently bonded. The bonding within these groups will be considered later with covalent bonds.

the usual ones are +2, where the two electrons in the *s* orbital are used, or +3, where the *s* electrons and one *d* electron are used. You must keep in mind when considering these that the *s* electrons are always in a higher energy level than the *d* (such as 4*s* and 3*d*) and are therefore lost first, even though it is not the order that they fill according to the aufbau principle. Also, with the transition metals, it is possible to use just one of the *d* electrons. This accounts for the variety of oxidation states, and possible ions, with these elements.

To summarize what has been said about the formation of monatomic ions, keep in mind the following facts:

1. The limit to the charge on the monatomic ion is 3.
2. Metals only form positive ions (cations) and nonmetals only form negative ions (anions).
3. Nonmetals always form anions with a noble gas electron configuration.
4. Electrons in *s* and *p* orbitals are always used in groups—that is, all electrons in a given *s* or *p* sublevel are used.
5. For transition metals, the *ns* electrons are always lost before the $(n - 1)d$.
6. Transition metals can lose one *d* electron, in addition to the *s* electrons.

Consider the following example.

PROBLEM EXAMPLE 7-1

Using the periodic table, explain, in terms of electron configurations, what monatomic ions each of the following could form.

(a) sulfur
(b) magnesium
(c) lead
(d) iron

SOLUTION:

(a) Since sulfur is a nonmetal, the only possible ion is S^{2-}, in which the atom attains a noble gas electron configuration.
(b) The Mg atom has an s^2 configuration; thus, it can only lose these two valence electrons, to form Mg^{2+}. It would never lose just one of the *s* electrons.
(c) Lead has an s^2p^2 configuration; thus it can only form a Pb^{2+} ion. Lead can also exist in a +4 oxidation state, where all four valence electrons are used, but this would not be ionic.
(d) Since Fe has a $3d^6\,4s^2$ configuration, it can form either an Fe^{2+} ion, which would be $3d^6$, or an Fe^{3+} ion, which would be $3d^5$. It would never lose only one of the *s* electrons and would never lose more than three electrons.

Energy Changes*

In considering the energy changes involved in the formation of ionic compounds, it is important to consider not only the ionization energy of the metal and the electron affinity of the nonmetal (Section 5-6) but also the following properties:

1. Energy of sublimation, which is the amount of energy necessary to get the solid metal into the gaseous state.
2. Dissociation energy, which is the amount of energy necessary to dissociate the nonmetal molecules into the individual atoms.†
3. Lattice energy, which is the energy released when the positive and negative ions form a crystal.

Since all the energy transferred in these processes is in the form of heat, these first two are often referred to as heat of sublimation (S), and heat of dissociation (D). The lattice energy is abbreviated as U, the ionization energy as I, the electron affinity as A, and the heat (energy) of formation as Q. Combining all of these thermodynamic properties to calculate the energy difference between the individual elements and the ionic compound formed gives what is often referred to as the Born-Haber cycle. This shows that the heat of formation of an ionic compound is equal to the amount of energy released when the nonmetal atom gains an electron and the crystal lattice is formed, minus the amount necessary to vaporize the metal, remove an electron from it, and dissociate the gaseous nonmetal. This can be stated as

$$Q = A + U - S - I - \tfrac{1}{2}D \qquad (7\text{-}1)$$

or, considering this as a cycle,

$$A + U - S - I - \tfrac{1}{2}D - Q = 0 \qquad (7\text{-}2)$$

The reason that $\frac{1}{2}D$ is used is that we are only using one atom, from the reaction

$$X_2 + D \longrightarrow 2X \qquad (\text{where } D = \text{energy})$$

or

$$\frac{1}{2}X_2 + \frac{1}{2}D \longrightarrow X$$

This can be diagrammed as in Fig. 7-2. In this diagram, Q represents the energy it would take to break MX up into M and $\frac{1}{2}X_2$. Positive values here indicate energy required and negative values represent energy released.

*The portion of Section 7-1 dealing with energy changes is optional. It can be omitted without loss of continuity.

†All the nonmetals, except the noble gases, exist in the gaseous state diatomic (or higher) molecules. These must be dissociated into individual atoms before the anion can form.

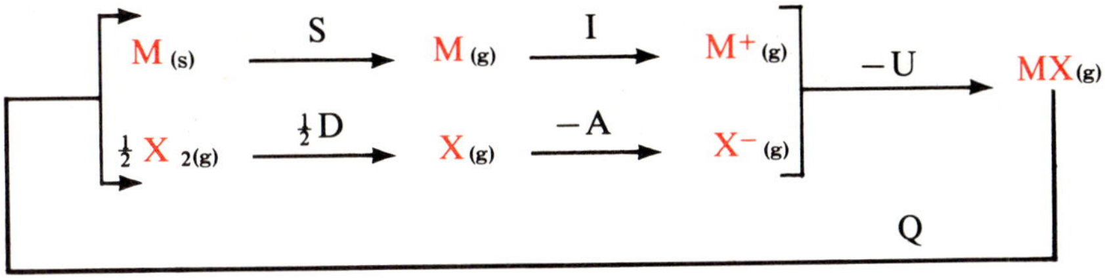

Figure 7-2
Born-Haber cycle (the negative signs refer to energy released)

7-2 Covalent Bonding

In Chapter 3 a covalent or molecular compound was defined as a compound in which covalent bonds are formed between the atoms. In forming the **covalent bond,** a pair of electrons are shared between the atoms, in a new type of orbital. Since the covalent bond is formed from a pair of electrons, it is sometimes referred to as an *electron pair bond.* We shall see in discussing covalent bonds that it is also important to consider (1) the orbital that the pair of electrons occupies; (2) how the electron pairs are arranged; (3) if there are **multiple** (two or three) **bonds** between two atoms; (4) if there are nonbonding pairs, or lone pairs, of electrons; and (5) if the bond formed is polar or nonpolar.

In many covalent compounds one electron comes from each atom, in forming the bond. There are some cases, however, where one atom furnishes both electrons. A bond of this type is often referred to as a **coordinate covalent bond.** Once this bond is formed however, it is no different than any other covalent bond.

If two pairs of electrons are shared between two atoms, the bond is called a *double bond;* if three pairs are shared, it is a *triple bond.* The **bond length,** which is the distance between the two nuclei making up the bond, is shorter for a double bond than for a single and shorter for a triple bond than for a double. Likewise, the **bond strength,** which is the amount of energy necessary to break the bond, is greatest for a triple bond and least for a single bond. The maximum number of bonds that can form between two atoms is three.

In forming these covalent bonds, the new orbital formed is sometimes referred to as a **molecular orbital.** To avoid any confusion with the molecular orbital theory of bonding (which we shall consider briefly in Section 7-10), however, it is best to think of these simply in terms of overlap of atomic orbitals. In considering the formation of the covalent bond by the overlap of orbitals, there are three things that should be remembered:

1. The new orbital is formed by the overlap of atomic orbitals of different atoms.
2. Electrons occupy the new orbitals in pairs.
3. Only valence electrons are used to form the bond. Inner electrons are bound too strongly to their own atom to be used in bond formation.

The hydrogen molecule, H_2, is a simple example of a covalent substance. Each hydrogen atom has only one electron in the 1*s* orbital. By joining together, the two atoms can each have a share of two electrons in their valence shell. When the covalent bond is formed between the two hydrogen atoms, the pair of electrons is shared in the new molecular orbital that is formed. This is shown in Fig. 7-3, which first shows an electron dot notation to represent a pair of electrons shared between the two atoms; then the overlap of the 1*s* orbitals is represented, to show in more detail how the electrons are arranged. Finally, an electron cloud diagram is used to represent the overlap of the electron clouds on the individual atoms to form a new type of electron cloud that surrounds both nuclei.

Polarity of Covalent Bonds

When the two atoms being joined together by a covalent bond have different electronegativities, the electrons will be attracted more toward the atom with the higher electronegativity. This means that the electron cloud is not equally distributed among the atoms. Thus one atom has a somewhat negative character and the other one, a somewhat positive. This causes the bond to be a **polar bond;** that is, one part is partially positive and the other part is partially negative. The greater the difference in electronegativity between the two

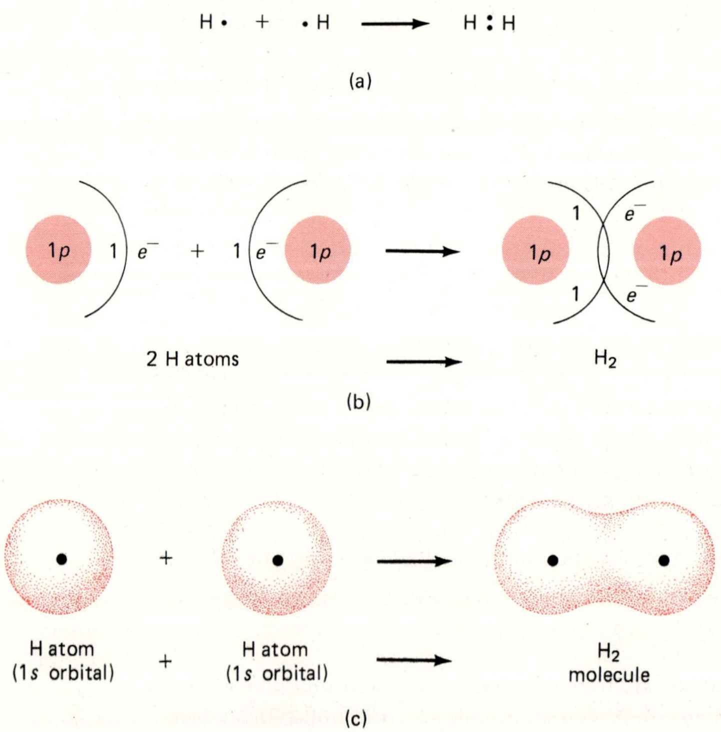

Figure 7-3
The formation of hydrogen, H_2, from two hydrogen atoms showing (a) the electron dot notation b) the energy levels involved, and (c) the overlap of atomic orbitals to form a molecular orbital

atoms, the more polar the bond is. An electrovalent, or ionic, bond can be thought of as an extreme case of a polar covalent bond, where one atom has an electronegativity so much greater than the other that the electron is completely removed from the one atom and transferred to the other.

Consider the following problem example.

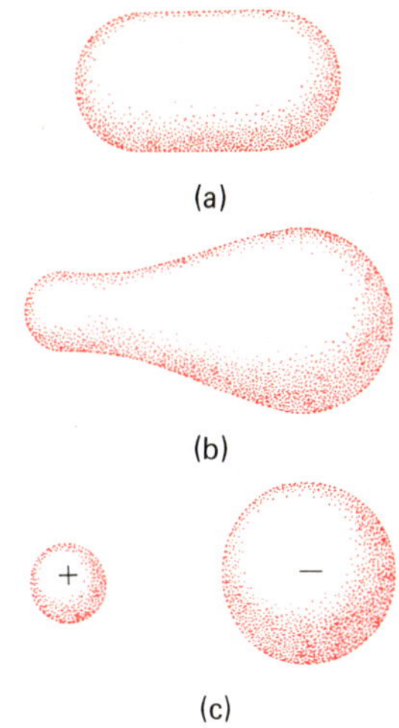

Figure 7-4
Electron cloud representation of (a) a covalent (nonpolar) bond with equal sharing of electrons, (b) a covalent (polar) bond with unequal sharing of electrons, and (c) an ionic (electrovalent) bond

PROBLEM EXAMPLE 7-2

Using the periodic table, arrange the following bonds in order of decreasing polarity and indicate with (+) and (−) the more positive and more negative part: O—O, N—O, C—O, Be—O.

SOLUTION:

Due to the differences in electronegativity between the atoms, the order of decreasing polarity would be

$$\overset{(+)(-)}{\text{Be—O}} \quad \overset{(+)(-)}{\text{C—O}} \quad \overset{(+)(-)}{\text{N—O}} \quad \underset{\text{(nonpolar)}}{\text{O—O}}$$

We shall see later (Section 7-8) that the polarity of a molecule depends not only on the polarity of the bonds but also on the shape of the molecule. Figure 7-4 represents a nonpolar covalent bond, a polar covalent bond, and an ionic or electrovalent bond.

7-3 Lewis Structures (Electron Dot Formulas)

In many compounds there are eight electrons, or four pairs, around each atom, just as there are in a noble gas (hydrogen, of course, only has two, like the noble gas He). Therefore, the formation of a *complete octet,* or a noble gas electron configuration, seems to be a very strong driving force in the formation of molecules.

In representing a molecule, for purposes of describing the bonding, the first step is to show how the electron pairs are arranged within the molecule. This is done by using what is called a **Lewis structure** or an *electron dot formula.* In drawing these the symbol for each element is used to represent the nucleus and all inner electrons. The valence electrons are then shown as dots around the symbol. All the valence electrons in the molecule are arranged in such a way that each atom has eight electrons around it, except hydrogen, which only has two. In some cases this may require double or triple bonds between some atoms. We shall see later that there are molecules in which some atoms have more or less than eight electrons around them, but to begin we shall just consider those that have eight. In drawing Lewis structures the following general rules may be helpful:

1. Arrange the atoms in the most probable arrangement, from the formula for the compound. For example, in CO_2 the C is in the center

with the two oxygen atoms around it; in SO_3 the sulfur is the central atom; and so forth. In some cases experimental evidence is needed to determine correctly which is the central atom, but in the vast majority of cases logic and common sense will tell you how to arrange the atoms. There are a few special cases that you will learn as you continue your study of chemistry. For now the most important one is oxyacids; in these compounds all replaceable hydrogens are connected to oxygen atoms. For example, $HClO_2$ is HOClO.

2. Count up and use the total number of valence electrons that are available—no more and no less.
3. Put one pair of electrons between each two atoms and arrange the remaining electrons in pairs around the atoms in such a way as to obtain eight around each atom except H, which only has two.
4. Rearrange the electron pairs if necessary by putting in double or triple bonds, in order to place eight electrons around each atom, except H.

Consider the following examples.

1. H_2O

   The most probable arrangement is H O H. Putting in two electrons between each two atoms gives

   $$\mathrm{H:O:H}$$

   Placing the remaining four valence electrons (from the total 6 + 1 + 1 = 8) around the oxygen gives

   $$\mathrm{H:\underset{\cdot\cdot}{\overset{\cdot\cdot}{O}}:H}$$

   This gives the oxygen atom eight electrons around it; each H, two; and all eight valence electrons are accounted for. Thus, it must be the correct Lewis structure.

2. NH_3

   The most probable arrangement is

   $$\begin{matrix}\mathrm{H} & \mathrm{N} & \mathrm{H}\\ & \mathrm{H} & \end{matrix}$$

   Putting one pair of electrons between each two atoms gives

   $$\begin{matrix}\mathrm{H:} & \underset{\cdot\cdot}{\mathrm{N}} & \mathrm{:H}\\ & \mathrm{H} & \end{matrix}$$

   There are two more valence electrons (from a total of 5 + 1 + 1 + 1 = 8), and N needs two more. Therefore, the correct structure is

   $$\begin{matrix}\mathrm{H:} & \underset{\cdot\cdot}{\overset{\cdot\cdot}{\mathrm{N}}} & \mathrm{:H}\\ & \mathrm{H} & \end{matrix}$$

3. SO_2

The most probable arrangement with two electrons between each two atoms is

O:S:O

There are 14 valence electrons remaining, from the total of 18 (6 + 6 + 6), which can be placed as

This only gives 6 electrons around S, however, and since all valence electrons have been used, we cannot add 2 more. Therefore, the only way to get 8 around the S and still leave 8 around each O is to put one double bond in; that is, we shall put one of the lone pairs (non-bonding pairs) from one of the oxygens between the S and the O, to form a double bond. This gives

4. N_2

The only arrangement here is NN. Putting two electrons between the two atoms and the remaining eight around the nitrogen gives

Since this only gives six electrons around each N, we rearrange one of the pairs to give a double bond:

This still leaves one N with only six electrons, however. Therefore, we rearrange another pair (from the N on the left) to give a triple bond:

This gives eight electrons around each N, using the 10 valence electrons that are available.

5. NO_3^-

Since this is an ion, with a -1 charge, the total number of valence electrons is $5 + (3 \times 6) + 1 = 5 + 18 + 1 = 24$. Using the most probable arrangement and putting in two electrons between each two atoms and the remaining electrons around the atoms gives

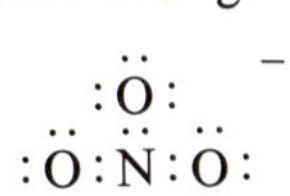

Since N only has six electrons around it however, we must rearrange to give one double bond as

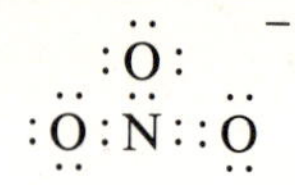

When drawing Lewis structures, always check your final result to make sure that you have all the valence electrons included (and no more) and eight electrons around each atom, except hydrogen. Also, keep in mind that this does *not* represent the shape of the molecule; therefore, it does not matter how the atoms are arranged around the central atom. For example, H_2O can be shown as H—O—H, H—O (with H below the O), H (with O—H below the H), and so on.

Molecules Without an Octet of Electrons

There are a few molecules for which no good Lewis structure can be drawn. These are odd electron molecules, such as NO and NO_2. These are quite rare, however, and can only be explained adequately using another theory of bonding, such as the molecular orbital theory.

There are also molecules known for which a Lewis structure can be drawn but which are known from experimental evidence to have less than eight electrons around an atom other than hydrogen. For example, the compound BF_3 is known to have only single bonds in it. Therefore, the only possible Lewis structure for this compound is

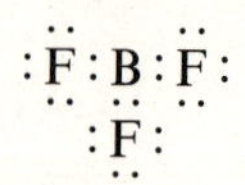

These compounds are also quite rare, being limited for the most part to atoms in Group IIIA, especially boron.

Another type of molecule that does not have an octet of electrons around each atom is the case where there are more than eight electrons around an atom; especially the central atom in the molecule. These cases are much more common than the two cases mentioned previously. The only restriction on the formation of compounds of this type is that there must be *d* orbitals available in the valence shell in order to form molecules of this type. Therefore, atoms of elements in the first and second period of the periodic table cannot form molecules of this type.

In drawing Lewis structures for molecules of this type, the same general rules are followed, except that if necessary more than 8 electrons are placed around the central atom. For example, in the molecule PCl_5, the only logical arrangement is to put the phosphorus in the center with the five chlorines around it. Putting in a pair of electrons for each bond thus gives 10 electrons around the P. The remaining 30 electrons are arranged around the Cl atoms, as in the previous examples, to give

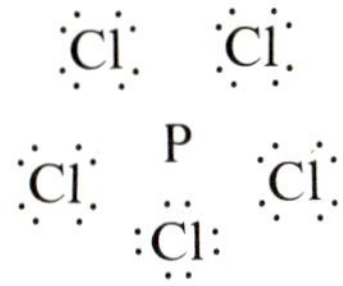

As another example, consider the molecule XeF_4. The most probable arrangement is

F

F Xe F

F

Putting in 2 electrons for each bond and 6 more around each F only accounts for 32 of the 36 valence electrons [$8 + (4 \times 7) = 36$]. Therefore, there must be two lone pairs around the Xe, to give

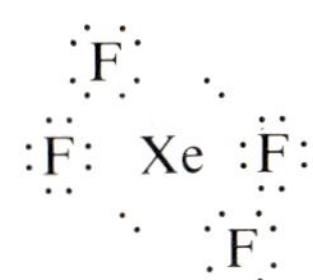

These cannot form double bonds since that would put 10 electrons around a fluorine atom; since fluorine has no *d* orbitals in the valence shell, this is not possible.

7-4 Resonance Forms of Molecules

For some compounds, two or more equivalent Lewis structures can be drawn. When this is the case, all possible forms are shown with a double-headed arrow (⟷) between them. This indicates that no one of them is the actual structure but that it is a combination of all of these. These are called **resonance forms.** For example, the SO_2 molecule would have two resonance forms:

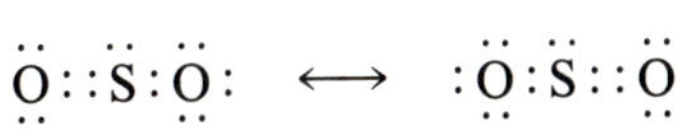

These are sometimes simplified and shown with a structural formula, in which a dash (—) is used to represent a pair of electrons. Using this notation, the resonance forms of SO_2 would be shown as

$$O{=}S{-}O \longleftrightarrow O{-}S{=}O$$

The NO_3^- ion would have three resonance forms:

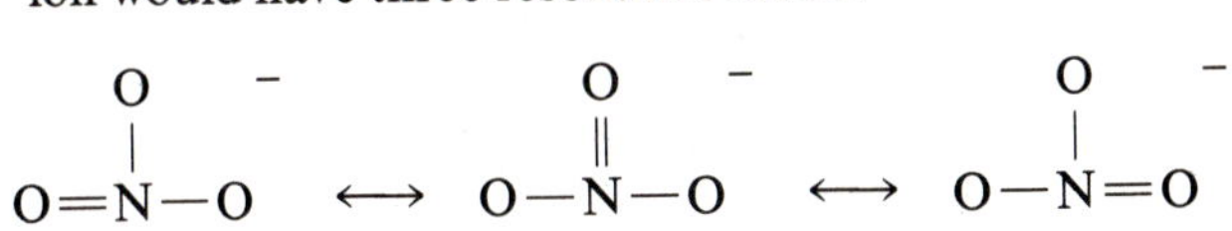

We shall see later (Section 7-5) that there is another way of describing these resonance forms. The main thing to keep in mind at this point is that these represent ways of drawing the Lewis structure for one molecule; it does not mean that the molecule changes from one form to another. That is, the resonance forms only show different positions where the multiple bonds can be. And finally it should be noted that resonance allows us to account for the fact that there are two identical S—O bonds in the SO_2 molecule, and three identical N—O in the NO_3^- ion; any one of the Lewis structures cannot show this.

7-5 Types of Covalent Bonds—Sigma (σ) and Pi (π)

There are two types of covalent bonds that we shall consider. These are called **sigma** (σ) and **pi** (π) **bonds.** A sigma bond is the first bond formed between two atoms in a covalent compound. It is formed by the direct, or head-on, overlap of orbitals and is thus the strongest bond and the one that determines the shape of the molecule. The new orbital formed is an electron cloud surrounding the two nuclei. Figure 7-5 shows a representation of a sigma bond formed by the overlap of two *s* orbitals and also one formed by the overlap of two *p* orbitals. When two electrons occupy this new orbital, a σ covalent bond has been formed.

A pi bond is formed after the σ bond; that is, it is the second (or third) bond formed between two atoms. This is formed by the sideways overlap of *p* orbitals (or sometimes *p* and *d*). This is a weaker bond than a sigma bond since there is a nodel plane (zero electron probability density) along the internuclear axis, as shown in Fig. 7-5.

There is also a third type of bond, called a delta (δ) bond, which is formed by the overlap of *d* orbitals. This will not be needed for our discussions in this book.

In Fig. 7-3 the formation of the H_2 molecule was used as an example of a σ bond formed between two *s* orbitals. Let us now look at two other molecules, O_2 and N_2, in terms of σ and π-bond formation.

From the Lewis structure for O_2 we can see that there are two bonds between the two oxygen atoms:*

$$\ddot{\underset{\cdot\cdot}{O}}::\ddot{\underset{\cdot\cdot}{O}}$$

Therefore there is one sigma bond and one pi bond. These can be thought of as being formed by overlap of *p* orbitals on the oxygen atoms. The first bond formed would be the sigma bond, formed by the direct overlap of two *p* orbitals, such as the p_x. Then the pi bond could be formed by the sideways overlap of p_y orbitals.

*It is known experimentally that there are actually two unpaired electrons in an O_2 molecule. This can only be explained adequately in terms of the molecular orbital theory (Section 7-10). For present purposes, however, we are only concerned with the double bond formed between the two oxygen atoms.

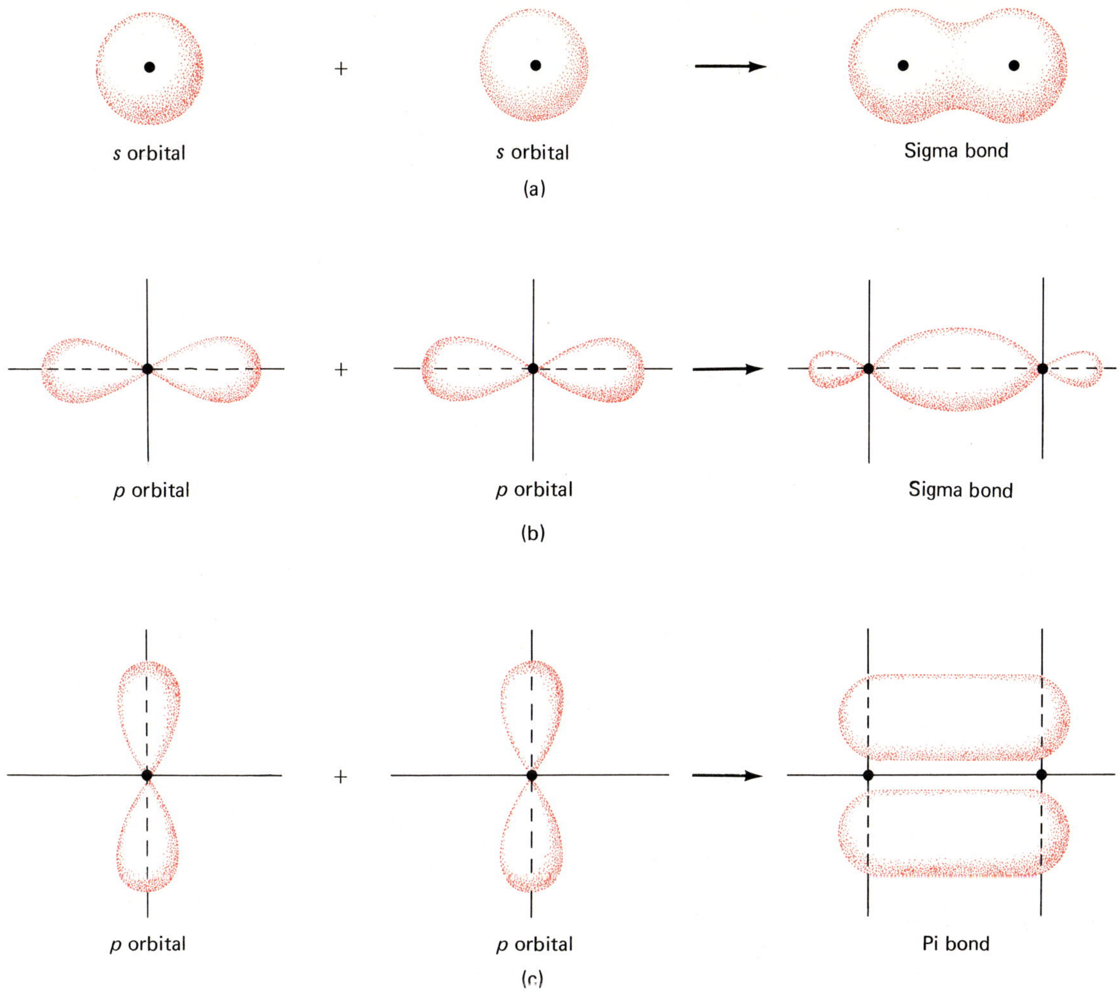

Figure 7-5
Formation of sigma and pi bonds (dots represent the nucleus of individual atoms): (a) sigma bond formed from s *orbitals; (b) sigma bond formed from* p *orbitals; (c) pi bond formed from sideways overlap of* p *orbitals*

The Lewis structure for N_2 shows that there must be a triple bond between the two atoms:

$$\ddot{N}\vdots\vdots\ddot{N}$$

Therefore, there is one σ bond and two π bonds. These can be thought of as being formed from the three *p* orbitals on the nitrogen atoms, with the lone pair in each case being in an *s* orbital. Figure 7-6 shows the formation of the σ bond by the direct overlap of the p_x orbitals and the π bonds by the overlap of the p_y and p_z orbitals. Keep in mind that, as with atomic orbitals,

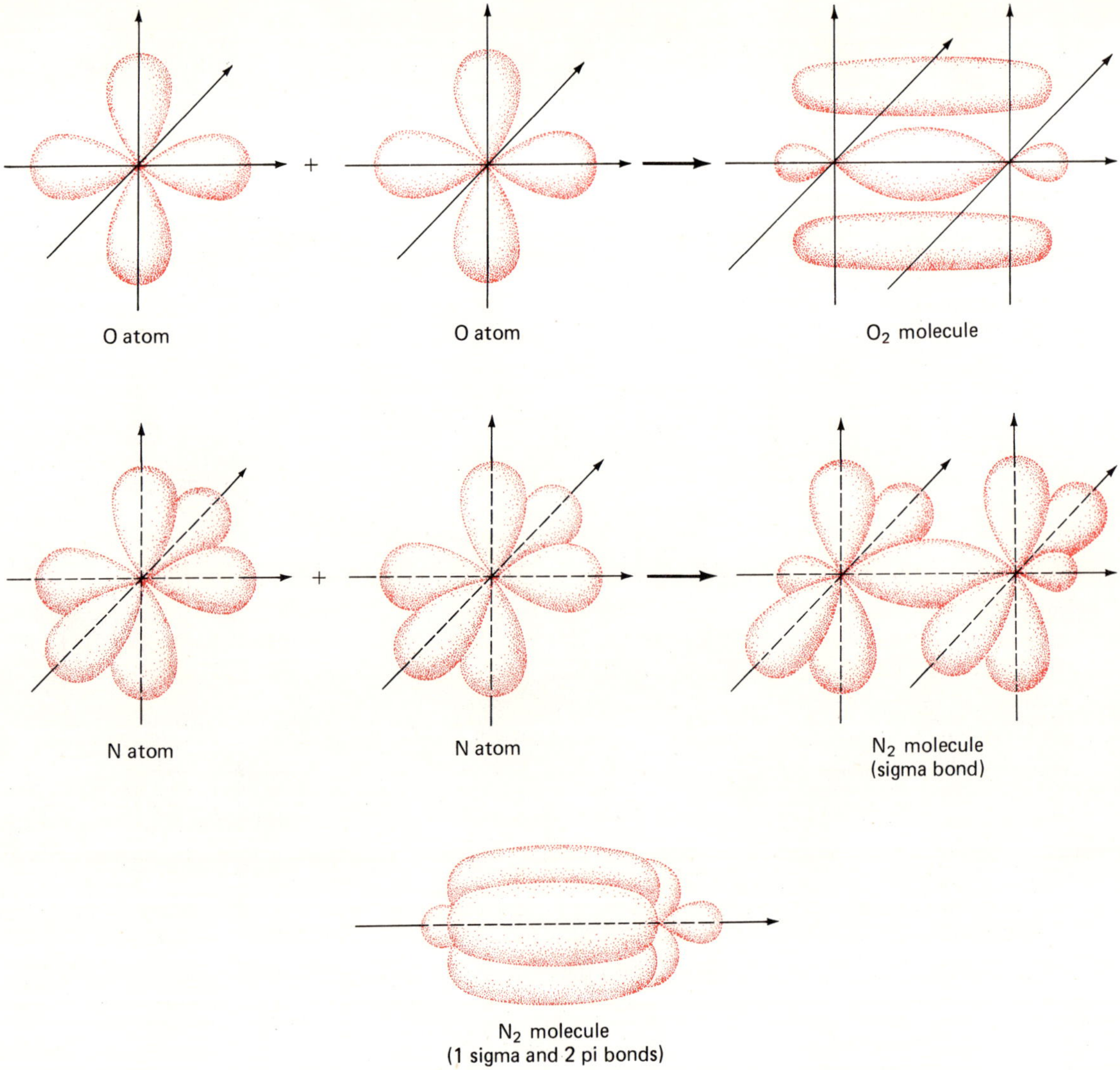

Figure 7-6
Formation of sigma and pi bonds in oxygen, O_2, molecule and nitrogen, N_2, molecule

these clouds do not represent solid boundaries. Therefore, the overall electron cloud around the N_2 molecule is cylindrical, just as the combination of all three *p* orbitals in a set gives a spherical electron cloud.

Another way of describing resonance forms is in terms of a **delocalized pi bond.** For example, in the SO_2 molecule two resonance forms were shown, with the double bond in different positions, and it was stated that the actual structure was somewhere between these two. If we now consider this in terms of the sigma and pi bonds formed, these two forms would be shown as in Fig. 7-7(a). The pi bond is not actually on either one side or the other, however, but is delocalized among all three of the *p* orbitals rather than being localized between any two of them. This is shown in Fig. 7-7(b). In Fig.

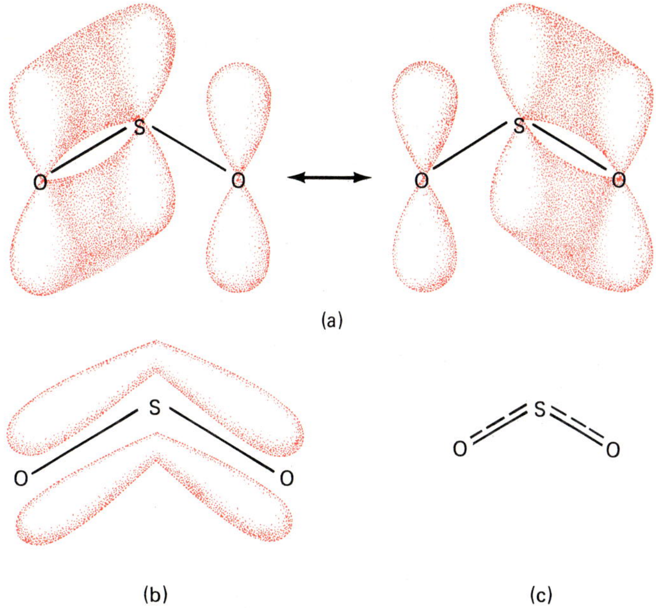

Figure 7-7
Representation of (a) pi bonds in resonance forms, (b) delocalized pi bond, and (c) structural formula showing a sigma and ½ pi bond between two atoms

7-7(c) the SO_2 molecule is represented as having one sigma bond between the sulfur and each oxygen and one-half of a bond. Experimentally, it is found that the bond length is equivalent to one and one-half bonds.

7-6 Geometry of Molecules—the Valence Shell Electron Pair Repulsion Theory (VSEPR)

One method used to explain and predict the geometry of molecules is to consider the repulsion of electron pairs around the central atom. Since electrons all have a negative charge, the electron clouds will repel each other and they separate as far apart as possible. Keep in mind two things about this. First, a double or triple bond acts like a single group since it represents pairs of electrons between two atoms. Also, the nonbonding or lone pairs around the central atom must also be considered since they also consist of electron clouds.

Since we are considering the repulsion of the electrons in the valence shell, this is often called the *v*alence *s*hell *e*lectron *p*air *r*epulsion *t*heory, or VSEPR Theory.

In using this theory or model to describe molecular geometry, first draw a Lewis structure. From that, count the number of pairs, or groups of pairs for multiple bonds, that will determine the arrangement of the electron pairs. These are often called **stereoactive pairs.** For example, by referring back to the Lewis structures drawn in Section 7-3 if necessary, it can be seen that water and ammonia each have four stereoactive pairs, SO_2 and NO_3^- each have three, and CO_2 (O::C::O) has two. Two factors must be kept in mind here: First, we are considering stereoactive pairs around the central atom; second, we count bonding and lone pairs, but we only count a double or triple bond as one.

Another way of determining stereoactive pairs, which gives exactly the same results, is to count the sigma bonding and nonbonding pairs around the central atom.

From the number of stereoactive pairs around the central atom, we can predict how these pairs will arrange themselves, to get as far apart as possible. Then we consider which pairs actually form bonds and predict the bond angles in the molecule. If there are nonbonding pairs of electrons present, the bond angles are less than they would be if all pairs were bonding. This is due to the fact that a lone (nonbonding) pair is "spread out" more than a bonding pair, which is held between the two nuclei. The arrangement of these lone pairs can only be inferred since there is no way of measuring the angles between an atom and a lone pair of electrons.

There are five types of arrangements that we must consider around the central atom. These are shown in Tables 7-1 and 7-2. Table 7-1 includes the cases where there are 8 electrons around the central atom, and 7-2 shows those that have more than 8. Note in the examples in Table 7-1 that there are four pairs of electrons around the central atom, but in some cases one or two of these pairs are π bonding pairs, and hence, are not stereoactive. Also, in both tables it should be noted that the first column gives the theoretical arrangement of electron pairs around the central atom for a particular type. However, if there are lone (nonbonding) pairs, this does not give the shape of the molecule. For example, with three stereoactive pairs around the central atom, if all three are sigma bonding pairs, the angles in the molecule are 120°. However, if there is one lone pair and two bonding pairs, the molecule just has a V shape, with angles less than 120°. Figure 7-8 shows some molecules, using the information given in Tables 7-1 and 7-2.

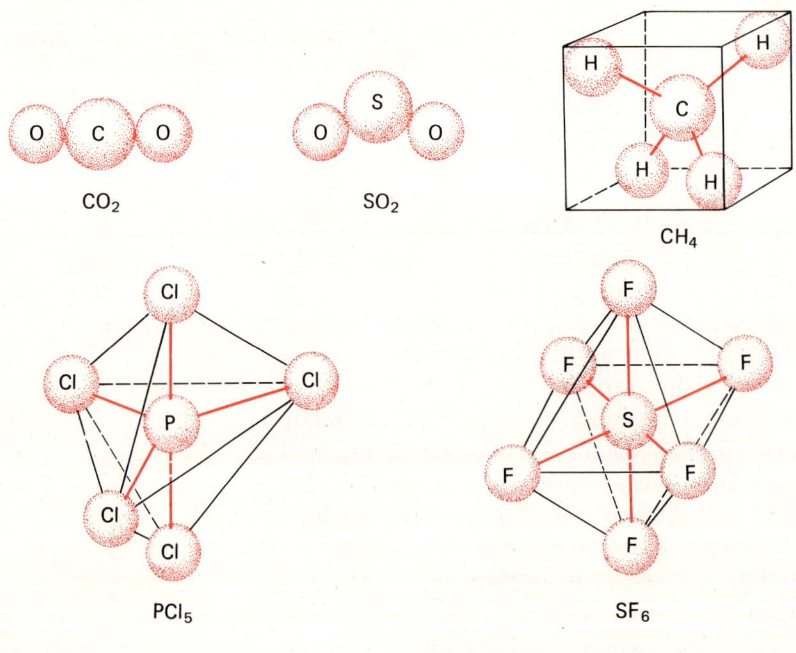

Figure 7-8
Geometry of some molecules

Geometry of molecules for those with an octet of electrons

NUMBER OF STEREOACTIVE ELECTRON PAIRS AND ARRANGEMENT OF ELECTRON PAIRS AROUND CENTRAL ATOM	SIGMA BONDING PAIRS	NONBONDING PAIRS	MOLECULAR STRUCTURE		EXAMPLES
2-linear—(180°)	2	0	Linear (angles of 180°)	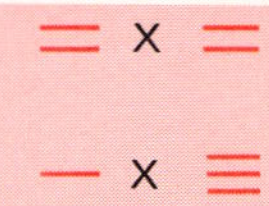	CO_2 HCN
3-plane triangle—(120°)	3	0	Plane triangle (angles of 120°)	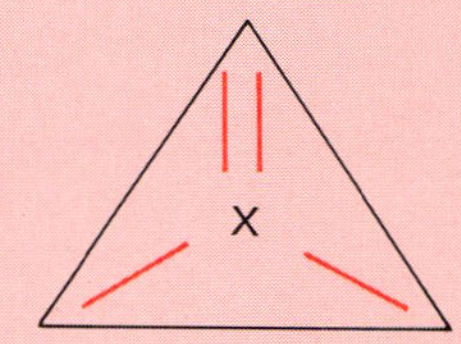	SO_3 NO_3^-
	2	1	V-shape (angles less than 120°)	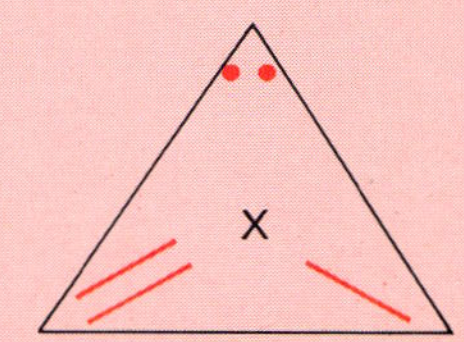	SO_2 NO_2^-
4-tetrahedral—(109°28′)	4	0	Tetrahedral (angles of 109°)	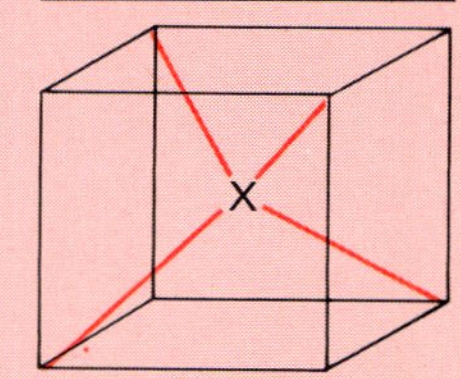	CH_4 NH_4^+
	3	1	Trigonal pyramid (angles less than 109°)	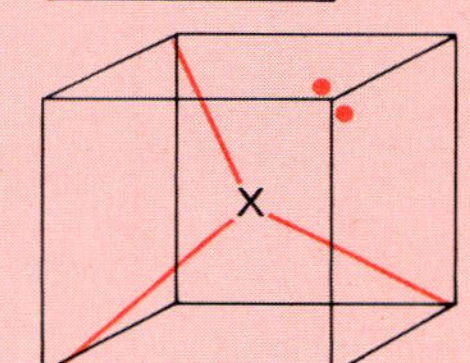	NH_3 SO_3^{2-}
	2	2	V-shaped (angles less than 109°)	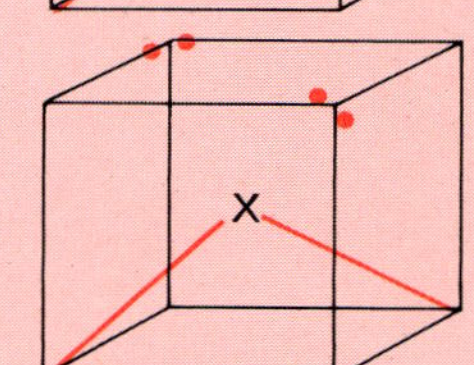	H_2O ClO_2^-

TABLE 7-2 *Geometry of molecules for those with more than eight electrons around the central atom*

NUMBER OF STEREOACTIVE ELECTRON PAIRS AND ARRANGEMENT OF ELECTRON PAIRS AROUND CENTRAL ATOM	SIGMA BONDING PAIRS	NONBONDING PAIRS	MOLECULAR STRUCTURE		EXAMPLES
5-trigonal bipyramid—(90°, 120°)	5	0	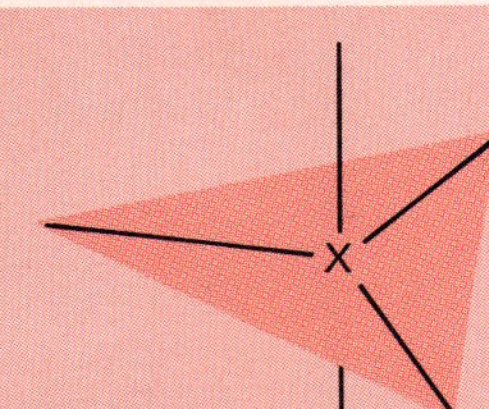	Trigonal bipyramid (angles = 90° and 120°)	PCl_5
	4	1	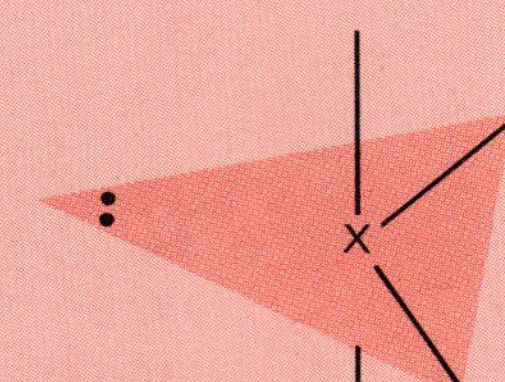	Irregular or teeter totter (angles less than 120° and 90°)	SF_4
	3	2	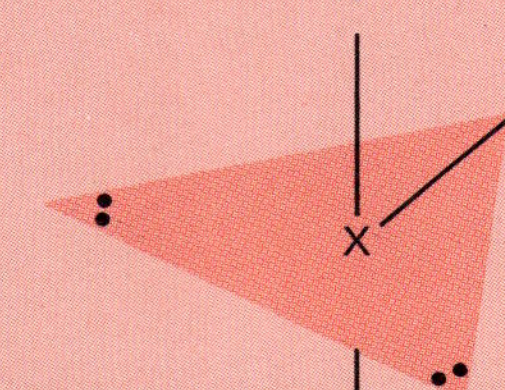	T-shaped (angles less than 90°)	ClF_3

6-octahedral—(90°)	2	3	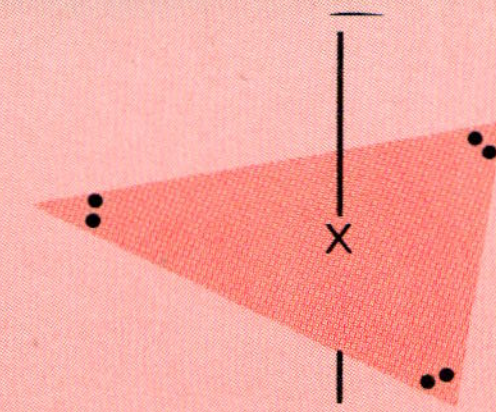	Linear (angles = 180°)	XeF_2
	6	0	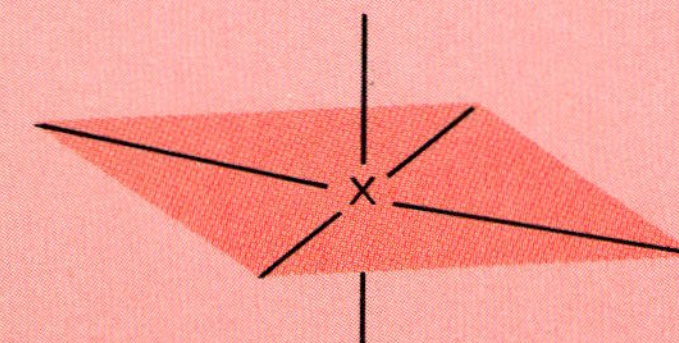	Octahedral (angles = 90°)	SF_6
	5	1	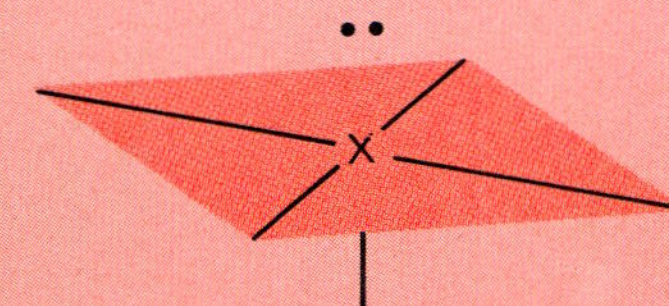	Square pyramid (angles less than 90°)	BrF_5
	4	2	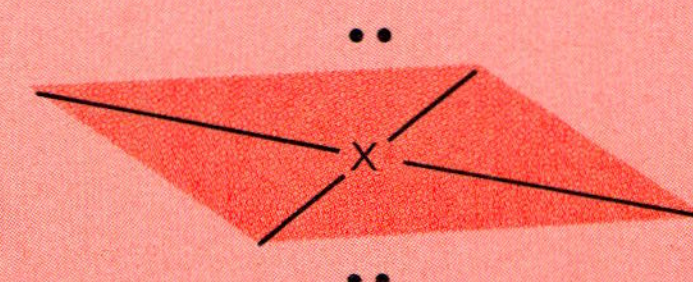	Square plane (angles = 90°)	BrF_4

7-7 Hybrid Orbitals

For many compounds there is no way to explain the bonding just in terms of *s*, *p*, and *d* orbitals. For example, there is no way to describe adequately the bonding in CH_4 which contains four identical C—H bonds. If one of the bonds was formed using an *s* orbital, and the other three using *p* orbitals, they would not be identical. Therefore, the concept of *hybridization* is often used as a method of explaining and predicting the bonding in compounds. This is simply a different way of thinking of the atomic orbitals that are used to form covalent bonds. It appears as though orbitals on the same atom that are similar in energy can combine with each other to form some new type of orbital.

Keep in mind that the concept of hybridization is just one method of explaining and predicting bonding and is most accurately used to explain the bonding from known geometry of the molecule. It is also useful, however, for predicting molecular geometry. There are some cases where it appears obvious that there is no hybridization. For example, where angles are known to be 90°, it appears as though just *p* orbitals are used.

The concept of hybridization is especially useful for carbon (organic) compounds. Since carbon has a $2s^2\ 2p^2$ configuration for the valence electrons, it appears as though one of the *s* electrons is probably promoted up to the empty *p* orbital before bonding takes place. This gives carbon four half-filled orbitals that can form covalent bonds. The orbital diagram for this is

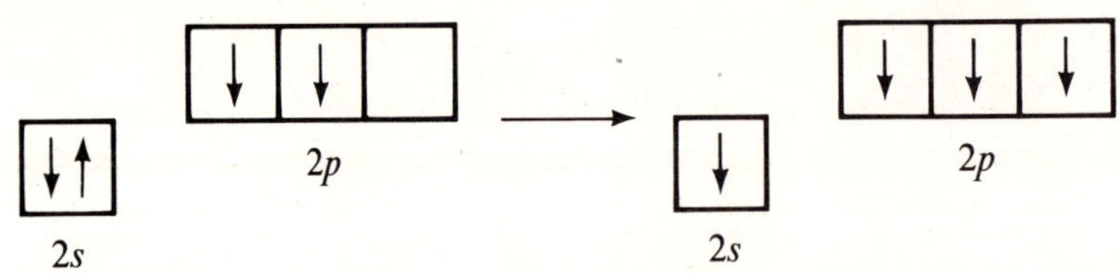

If these four orbitals are used to form bonds, however, the bonds would not all be equivalent as they are known to be in a compound like methane (CH_4). Therefore, it appears as though some new type of orbitals must be formed before bonding takes place. If all four of the orbitals, one *s* plus three *p*, combine, four orbitals of a new type are formed. These are called sp^3 **hybrid orbitals** since each one consists of one part *s* character and three parts *p* character. This can be shown, just considering energy differences between orbitals, as

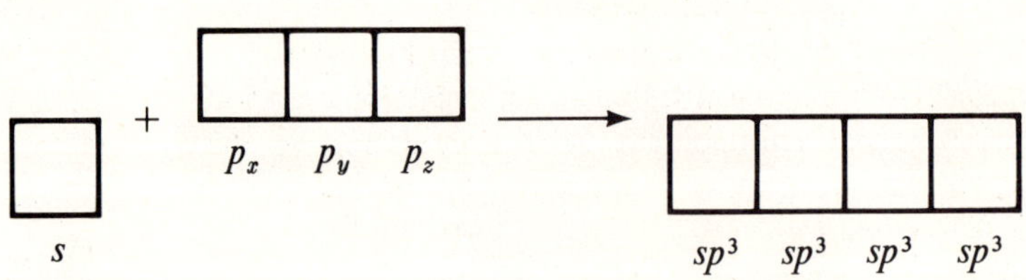

The arrangement of these orbitals is shown in Fig. 7-9(c). These form a tetrahedral arrangement with the four orbitals as far apart as possible.

Similarily, for a compound like $H_2C{=}CH_2$, where there are three equivalent sigma bonds and one pi bond around each carbon, there must be some new type of orbital formed. This would be formed from the *s* and two of the *p* orbitals, leaving one *p* orbital to form the pi bond. Since these are formed by the combinations of one *s* and two *p* orbitals, they are called sp^2 hybrid orbitals. Using the boxes again to represent orbitals, this would be shown as

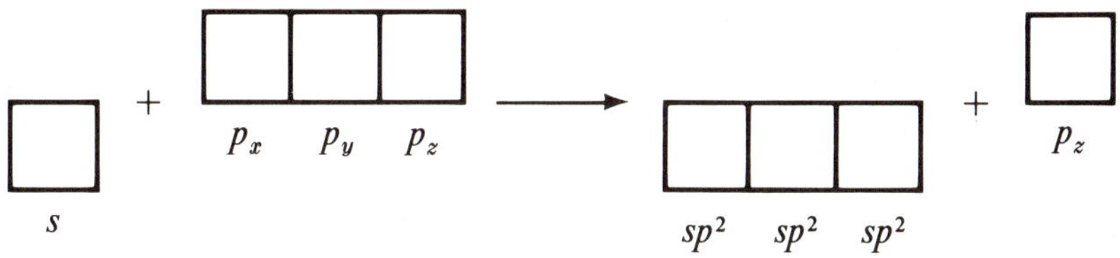

These orbitals are arranged in two dimensions, such as in the *xy* plane. Therefore, the three orbitals arrange themselves in a plane triangle arrangement, with angles of 120°, as shown in Fig. 7-9(b).

For a compound like $HC{\equiv}CH$ where there are two equivalent sigma bonds and two pi bonds around each carbon, the hybrid orbitals are formed from one *s* and one *p* orbital, leaving two *p* orbitals for pi-bond formation. These are thus called *sp* hybrid orbitals. This would be shown as

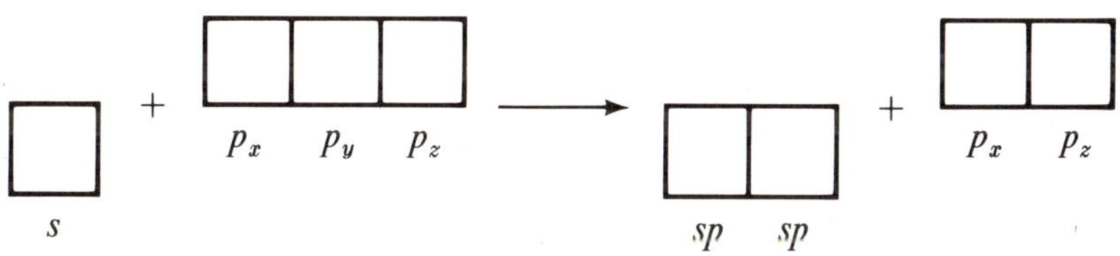

Since they are arranged along one axis, such as the *x* axis, the arrangement for the two orbitals as far apart as possible would be linear, as shown in Fig. 7-9(a).

Similar descriptions can be used to describe the formation of hybrid orbitals from *d*, *s*, and *p* orbitals. For five equivalent orbitals around the central atom, the *s* orbital, all three *p* orbitals, and one *d* orbital combine to form five equivalent hybrid orbitals, called sp^3d (or dsp^3). With the orbitals as far apart as possible, three are at the corner of a triangle, one above and one below the plane of the triangle, as shown in Fig. 7-10(a).

For six equivalent orbitals around the central atom, the *s* orbital, all three *p* orbitals, and two *d* orbitals combine to form six equivalent hybrid orbitals, called d^2sp^3 (or sp^3d^2). The arrangement of these six orbitals, as far apart as possible, gives an octahedral arrangement, as shown in Fig. 7-10(b).

These can be summarized as follows:

1. *sp*—combination of one *s* orbital and one *p* orbital to form two *sp* hybrid orbitals. These are in a linear arrangement [see Fig. 7-9(a)].

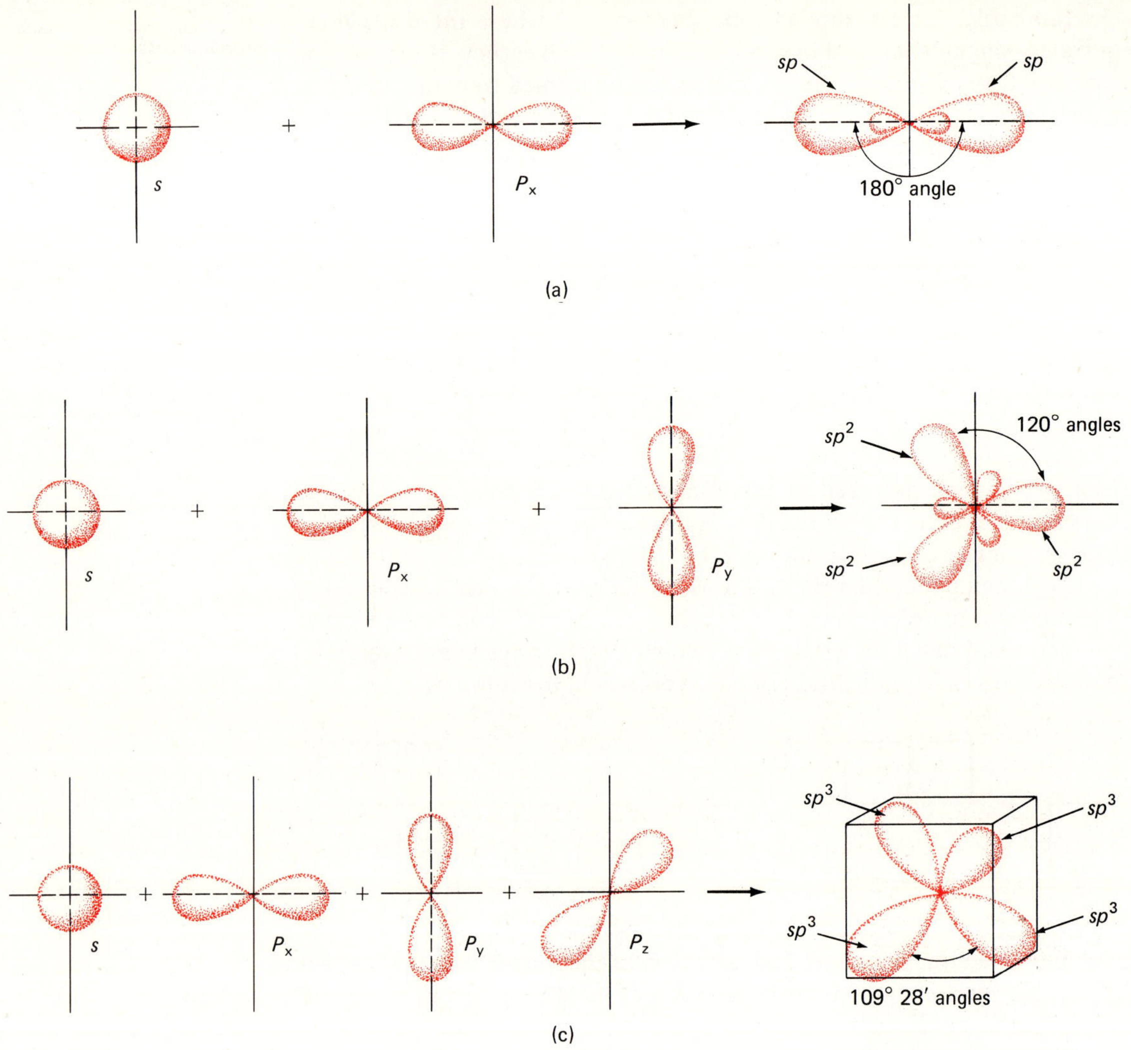

Figure 7-9
Arrangement of electron pairs: (a) two sp *hybrid orbitals with two* p *orbitals,* p_y *and* p_z*, left to form pi bonds; (b) three* sp^2 *hybrid orbitals with one* p *orbital,* p_z*, left to form pi bonds; (c) four* sp^3 *hybrid orbitals with no* p *orbitals left to form pi bonds*

2. sp^2—combination of one *s* and two *p* orbitals to form three sp^2 hybrid orbitals. These are in a plane triangle arrangement, with angles of 120° [see Fig. 7-9(b)].

3. sp^3—combination of one *s* and three *p* orbitals to form four sp^3 hybrid orbitals. These are arranged in the shape of a tetrahedron with angles of 109°28′ [see Fig. 7-9(c)].

4. dsp^3—combination of one *d*, one *s*, and three *p* orbitals to form five dsp^3 hybrid orbitals. These are arranged in the form of a trigonal bipyramid, with angles of 120° and 90° [see Fig. 7-10(a)].

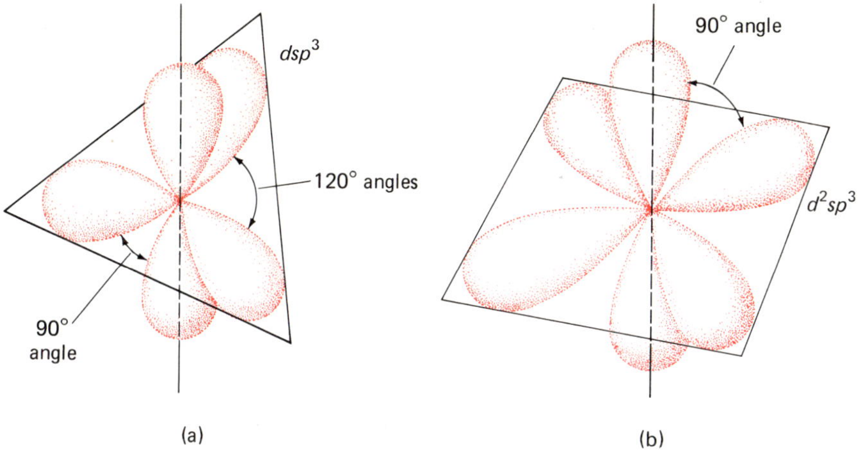

Figure 7-10 *Arrangement of electron pairs: (a) five* dsp³ *hybrid orbitals; (b) six* d²sp³ *hybrid orbitals*

5. d^2sp^3—combination of two d, one s, and three p orbitals to form six d^2sp^3 hybrid orbitals. These are arranged in an octahedral shape with all angles equal to 90° [see Figure 7-10(b)].

There are several things which should be noted about these hybrid orbitals:

1. Only orbitals of similar energy combine: for example, $2s$ and $2p$; or $3s$, $3p$, and $3d$; or $3d$, $4s$, and $4p$.
2. The number of orbitals that combine equal the number of hybrid orbitals formed. For example, one s and two p orbitals form three sp^2 hybrid orbitals.
3. Once an orbital is used for hybridization, it is no longer available in the pure form. For example, in sp^3 hybridization, there are no p orbitals left to form pi bonds; in sp^2, there is one p orbital left to form a pi bond; and in sp, there are two p orbitals left to form pi bonds.

In many cases, the concept of hybridization can also be used to describe and predict molecular geometry. For example, since the bond angles in H_2O are about 105°, it appears as though there are probably sp^3 hybrid orbitals formed, with two of them used to form sigma bonds and the other two containing the nonbonding pairs. The distortion from the predicted 109°28′ for the tetrahedral arrangement of sp^3 hybrid orbitals can be explained, as done earlier, by the fact that the lone pairs are spread out more than bonding pairs. This causes them to repel the bonding pairs more than the bonding pairs repel each other and this makes the angles less than 109°. There are other cases, such as AsH_3, where the angles are very close to 90° and thus it appears that there is no hybridization.

Applying the concept of hybridization to the discussion of molecular geometry (Section 7-6), we could simply say that for two stereoactive pairs (sigma bonding and nonbonding) *sp* hybrid orbitals could be used; for three pairs it could be sp^2 hybridization; for four it could be sp^3 hybridization; for five it could be dsp^3; and for six it could be d^2sp^3.

It should be pointed out here that this gives the same result as the VSEPR Theory discussed earlier. It is simply a way of explaining what types of orbitals are occupied by electron pairs around the central atom. These hybrid orbitals are still as far apart as possible, just as the electron pairs discussed earlier were. The reason for this is that an orbital is a region in space occupied by a pair of electrons. Hence, there is no disagreement between the VSEPR Theory and the concept of hybridization. They are simply two models used to describe the same principle.

7-8 Polarity of Molecules

The polarity of a molecule refers to whether the electron clouds that make up the various bonds are equally distributed or not. If they are not, one part of the molecule will be more positive and the other part will be more negative. That is, we say that the molecule is a *dipole* or that it is a **polar molecule,** if one part is more positive and another part more negative. This is similar to the poles of a magnet, often referred to as the north and south poles. On the other hand, if the electron clouds are equally distributed, we say that the molecule is *nonpolar.*

The polarity of a molecule depends on two things: the electronegativity of the atoms and the shape of the molecule. If there is no difference in the electronegativity of the atoms that make up the molecule, it will be a **nonpolar molecule.** On the other hand, if there is a difference in electronegativity, the polarity depends on the shape of the molecule. For example, in the CO_2 molecule there is a difference in electronegativity between carbon and oxygen. Since the molecule is linear ($O{=}C{=}O$), however, the electron cloud will not be shifted one way or the other, and hence it is nonpolar. On the other hand, in the water molecule, where there is not only a difference in electronegativity but also a bent shape (with angles less than 109°), the molecule is polar. This has a great effect on the properties of water, as we shall see in Chapter 12.

All organic compounds contain carbon and most also contain hydrogen. Since there is only a slight difference in electronegativity between carbon and hydrogen, most of these compounds are nonpolar. We shall see later (in Chapters 12 and 26) that this also has an effect on the properties of organic compounds.

7-9 Describing and Predicting the Bonding in Covalent Species

We can use all the information considered up to this point either to describe the bonding in a known covalent species or to predict what it would be for an unknown substance. This includes not only molecules but also polyatomic ions.

In doing this we need to consider the arrangement of electron pairs around the central atom, the possible presence of sigma or pi bonds and lone or nonbonding pairs, the bond angles and shape of the molecule, and finally the possibility of resonance forms. From all this and through the use of electronegativities of the atoms involved, we can also predict whether the molecule will be polar or nonpolar.

No matter whether we use the VSEPR method or the concept of hybridization to describe molecular geometry, the following steps should be followed:

1. Draw a Lewis structure to determine the arrangement of electron pairs around the central atom.
2. Count the number of stereoactive pairs around the central atom. This can be done either by counting the sigma bonding and nonbonding pairs, or by considering the bonding pairs (counting multiple bonds as one) and nonbonding pairs (using the VSEPR theory).
3. From this, determine the arrangement of *electron pairs* around the central atom—that is, if they are in a linear arrangement, plane triangle, tetrahedral, and so on.
4. Determine the shape of the molecule from the bonding pairs; that is, the information in step 3 does not give the shape of the molecule but only the arrangement of the electron pairs. The geometry of the molecule is determined by those electron pairs that form sigma bonds and those that are lone (nonbonding) pairs. For example, in the water molecule, the electron pairs would be predicted to be in a tetrahedral arrangement around the central atom (oxygen). However, the molecule is just a V shape, with angles less than 109° (due to the tetrahedral arrangement of electron pairs and the lone pairs of electrons).
5. For molecules that contain pi bonds, determine whether there are resonance forms possible (or delocalized pi bonds).

Consider the following example.

PROBLEM EXAMPLE 7-3

Describe the bonding in the SO_2 molecule, as completely as possible.

SOLUTION:

In the *Lewis structure* there must be one lone pair of electrons around the S,

$$\underset{\cdot\cdot}{\overset{\cdot\cdot}{O}}::\overset{\cdot\cdot}{S}:\underset{\cdot\cdot}{\overset{\cdot\cdot}{O}}:$$

and one (but only one) double bond.

Therefore, there are two σ bonds, one π bond, and one lone pair around the central atom. This π bond could be described in terms of two resonance

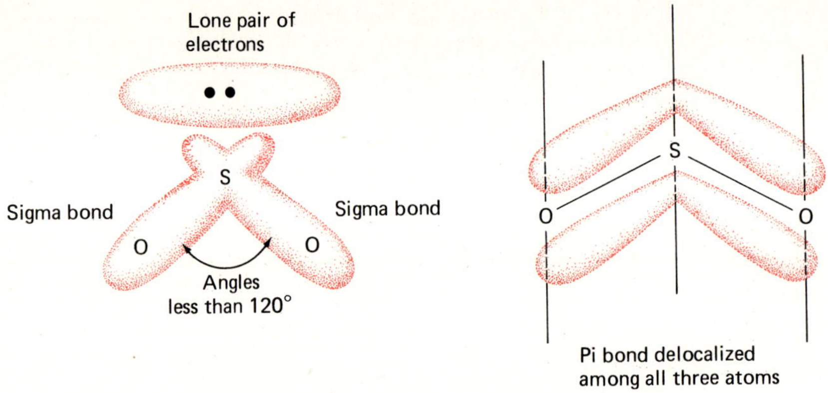

Figure 7-11
Bonding in the SO_2 molecule

forms

$$O{=}\ddot{S}{-}O \quad \longleftrightarrow \quad O{-}\ddot{S}{=}O$$

or as being delocalized among all three atoms rather than being localized between the S and one of the oxygens, as shown in Fig. 7-11.

The arrangement of electron pairs around the central atom would be in a plane triangle arrangement since there are *three stereoactive pairs* (two bonding and one nonbonding)—this could be thought of as occupying *sp^2 hybrid orbitals.* Since only two of these pairs form bonds, the shape of the molecule would be a V shape. *The angles would be less than 120°*, due to the lone pair of electrons. Keep in mind that the σ bonding pairs and lone pair are in the sp^2 hybrid orbitals; thus, the angles are about 120°. Due to the lone pair of electrons, however, the angle is less than 120°, as shown in Fig. 7-11.

Due to the difference in electronegativity and the shape of the molecule, it would be polar.

7-10 Molecular Orbital Theory*

Another method of explaining covalent bonding is to consider energy levels of molecular orbitals, similar to those considered for atomic orbitals. This is actually more accurate since it is again based on mathematical calculations of energy, but does not give as good a picture as the simpler theories discussed earlier.

In this method of representation, there is one molecular orbital formed for each atomic orbital that is used. These orbitals are associated with the molecule as a whole and are represented as energy levels within the molecule. Electrons then are put into these energy levels, or molecular orbitals, following a new type of aufbau order.

*This section is optional and can be omitted without loss of continuity.

When two atomic orbitals overlap to form molecular orbitals, there is one molecular orbital formed for each atomic orbital that is used. Thus when two atomic orbitals combine, there are two new molecular orbitals formed. One of these is of lower energy than the original atomic orbitals, and the other is of higher energy. The one of lower energy is called a *bonding molecular orbital* (BMO), and the one of higher energy is called an *antibonding molecular orbital* (ABMO).* These molecular orbitals can be either sigma or pi and are designated according to the type of atomic orbitals that form them, as σ 1*s*, σ 2*s*, $\sigma\ 2p_x$, $\pi\ 2p_y$, and so on for the bonding molecular orbitals. The antibonding molecular orbitals are designated as σ^* 1*s*, σ^* 2*s*, $\sigma^*\ 2p_x$, $\pi^*\ 2p_y$, and so on. Figure 7-12 shows the formation of the σ 1*s* and σ^* 1*s* molecular orbitals from two 1*s* orbitals on different atoms.

Electrons occupy these orbitals in pairs, just as they do atomic orbitals. Also, Hund's Rule still applies for molecular orbitals of equal energy, as we shall see in some examples later.

The **bond order,** which is equivalent to the number of bonds formed between two atoms, is equal to the number of electron pairs in bonding molecular orbitals (BMO's) minus the number of electron pairs in antibonding molecular orbitals (ABMO's). This can also be found by taking the total number of electrons and dividing by two, to obtain electron pairs. Using this method, we can find the bond order as follows:

$$\text{bond order} = \frac{\text{no. of electrons in BMO's} - \text{no. of electrons in ABMO's}}{2}$$

The number of unpaired electrons in a molecule can also be found by considering how the electrons are arranged in the molecular orbitals.

For example, using the σ 1*s* and σ^* 1*s* molecular orbitals, the bonding in the H_2 molecule would be determined as follows:

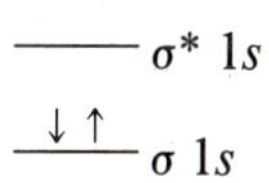

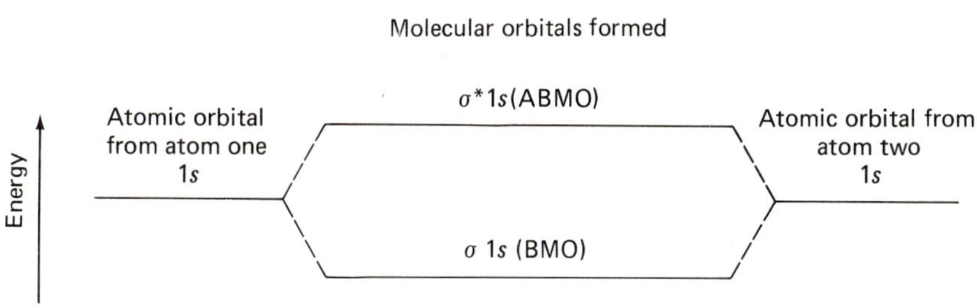

Figure 7-12 *Formation of molecular orbitals from 1s atomic orbitals*

*In some molecules there are three atomic orbitals that combine. In this case, one is a BMO, one an ABMO, and the third one is at the same energy as the original atomic orbital and is called a *nonbonding molecular orbital.*

Since there are two electrons in the σ 1s (BMO) and zero electrons in the σ * 1s (ABMO), the bond order is

$$\frac{2-0}{2}=1$$

Also, there are no unpaired electrons in this molecule.

For the He_2 molecule it would be

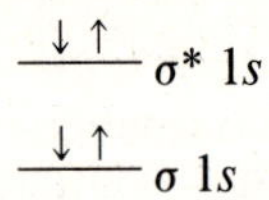

Thus, there are two electrons in the σ 1s (BMO) and two electrons in the σ* 1s (ABMO), and the bond order is

$$\frac{2-2}{2}=0.$$

Therefore, we would say that we would not expect the He_2 molecule to exist, which is the case.

For more complex molecules the order of energy levels must be calculated. This gives the order as shown in Fig. 7-13 for most diatomic molecules of second row elements.

Using this order, we can explain the bonding as N_2 and O_2 as follows:

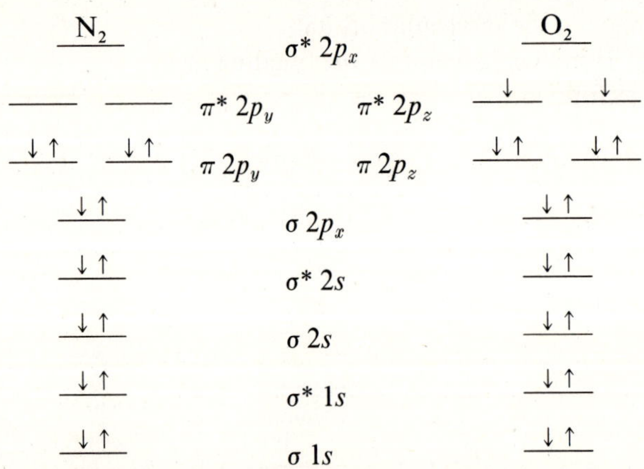

Notice here that all electrons are included, not just the valence electrons. Also, note that Hund's Rule is followed, and thus O_2 has two unpaired electrons (which is known to be the case from experimental evidence). The bond order of each of these is calculated as follows:

$$N_2=\frac{10-4}{2}=3; \qquad O_2=\frac{10-6}{2}=2$$

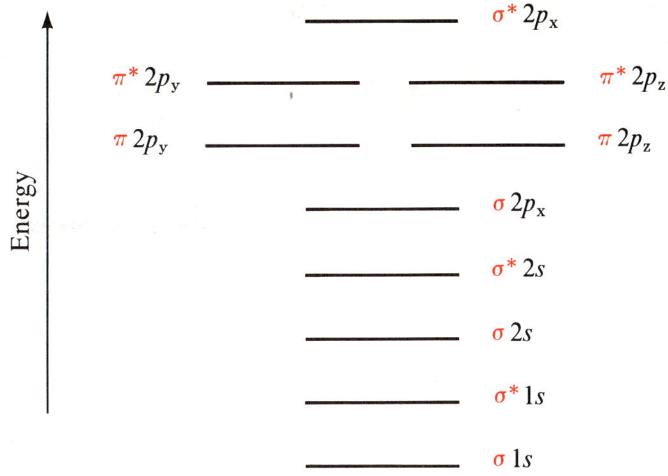

Figure 7-13
Order of molecular orbitals

SUMMARY

1. An *ionic bond* is formed when electrons are transferred from one atom to another. The bond is an electrostatic attraction between oppositely charged ions.
2. *Monatomic anions* always gain enough electrons to attain a noble gas electron configuration.
3. *Monatomic cations* are one of the following types:
   (a) IA or IIA metal or aluminum—lose all valence electrons to form a noble gas electron configuration.
   (b) Post transition metal—form ions in which all *p* electrons, or all *p* and all *s* electrons, in the valence shell are lost.
   (c) Transition metals—in which outermost *s* electrons are lost first (and all are lost, not just one); then one of the *d* electrons can be lost (in the formation of any ion, no more than three electrons are ever gained or lost).
4. A *covalent bond* is formed when a pair of electrons are shared between two atoms (very often in such a way that in the resulting molecule each atom has the equivalent of a noble gas electron configuration). If both electrons come from one atom, the bond is called a *coordinate covalent bond.*
5. If there are two pairs of electrons between two atoms, it is called a *double bond,* and if there are three pairs, it is a *triple bond.*
6. In a *polar covalent bond,* the electron cloud is shifted more toward one of the atoms; in a *nonpolar covalent bond,* the electron cloud is equally distributed.
7. The shared pair of electrons is in a new type of orbital, often referred to as a *molecular orbital,* which is formed by the overlap of atomic orbitals in the original atoms.

8. A *Lewis structure* (electron dot formula) shows how the electron pairs are arranged in the molecule—it does not represent the shape of the molecule. In drawing these, there are eight electrons (four pairs) around each atom (except H) if possible; there *can* be more than eight if there are *d* orbitals available (that is not for the second period elements). There are also cases where there are less than eight electrons around some atoms.
9. With some molecules there are two or more equally correct Lewis structures that can be drawn. These are called *resonance forms.*
10. The shape (geometry) of molecules can be explained and predicted by considering the repulsion of electron pairs (or groups of pairs for multiple bonds), according to the *Valence Shell Electron Pair Repulsion* (VSEPR) Theory. Two pairs (or groups of pairs) give a linear shape; three pairs, a plane triangle; four pairs, a tetrahedral arrangement; five pairs, a trigonal bipyramid; and six pairs, an octahedral arrangement.
11. When there are nonbonding pairs of electrons present, they force the bonding pairs closer together and cause distortion of the predicted bond angles (180°, 120°, 109°, 90° and 120°, and 90°, respectively, for the five types listed in item 10).
12. The first bond formed between two atoms is called a *sigma (σ) bond.* It is the strongest bond and the one that determines the shape of the molecule.
13. If there is a second or third bond formed, they are *pi (π) bonds* and are formed by the sideways overlap of *p* orbitals in the two adjacent atoms. These bonds are not so strong as the sigma bonds and do not affect the shape of the molecule.
14. The orbitals used to form covalent bonds are often thought of as being *hybrid orbitals;* that is, they are apparently different from the *s*, *p*, and *d* orbitals ordinarily considered to exist in atoms. There are five types of hybrid orbitals that can be used to explain the five arrangements of electron pairs listed in items 10 and 11. These are sp, sp^2, sp^3, dsp^3, and d^2sp^3.
15. The bonding in covalent compounds can also be explained in terms of the *molecular orbital theory.* In this theory the molecular orbitals that are formed when the atomic orbitals overlap to form the bond are arranged in order of increasing energy. Then electrons are placed in these orbitals, in the same way as is done for atomic orbitals in the atom. From this can be determined the *bond order* and the number of unpaired electrons in the molecule.

Summary

Upon completion of this chapter, you should be able to do the following (in all cases, assume that a periodic table is available with symbols and atomic numbers):

1. Predict what monatomic ions any element could form. (2)
2. Given a group of covalent bonds,
   (a) Predict which would be polar and which nonpolar. (3)
   (b) Arrange in order of increasing polarity. (4)
3. Given a formula or name for a compound or polyatomic ion, show the correct Lewis structure (electron dot formula). (5)
4. Given the Lewis structure of a compound, tell
   (a) How many sigma bonds there are around the central atom. (8)
   (b) How many pi bonds there are around the central atom, and how they are formed. (8)
   (c) How many lone pairs of electrons there are around the central atom. (8)
   (d) How many resonance forms there would be and how the pi bonds could be described as being delocalized in cases where there are resonance forms. (6)
   (e) The approximate bond angles. (7)
   (f) What type of hybrid orbital could be formed on the central atom. (9)
   (g) If the molecule would be polar or nonpolar. (10)
5. Given the name or formula for any covalent compound or polyatomic ion, describe the bonding as completely as possible, starting with a Lewis structure and including all the things listed in item 4. (11)
6. Given the order in which electrons fill molecular orbitals (according to the molecular orbital theory) and the formula for a diatomic molecule, tell what the bond order would be in the molecule and how many unpaired electrons there would be. (12)
7. Given any of the terms listed in Exercise 1, define or explain, in your own words.

EXERCISES

1. Define or explain the following terms
   (a) ionic bond
   (b) cation
   (c) anion
   (d) covalent bond
   (e) molecular orbital
   (f) coordinate covalent bond
   (g) multiple bond
   (h) bond length
   (i) bond strength
   (j) polar bond
   (k) Lewis structure
   (l) sigma (σ) bond
   (m) pi (π) bond
   (n) nodal plane
   (o) resonance forms
   (p) delocalized π bond
   (q) stereoactive pairs of electrons
   (r) hybrid orbital
   (s) polar molecule
   (t) nonpolar molecule
   (u) bond order
2. Distinguish between
   (a) ionic (electrovalent) bond and covalent bond
   (b) single bond, double bond, and triple bond
   (c) sigma bond and pi bond
   (d) atomic orbital and molecular orbital
   (e) polar bond and nonpolar bond
   (f) polar molecule and nonpolar molecule

PROBLEMS

Ionic bonding

1. Give the electron configuration for the outer electrons for each of the following ions:

(a) K^+	(i) Sn^{2+}
(b) Ca^{2+}	(j) Mn^{2+}
(c) Cl^-	(k) Ba^{2+}
(d) S^{2-}	(l) Br^-
(e) Zn^{2+}	(m) Na^+
(f) Cu^+	(n) O^{2-}
(g) Fe^{3+}	(o) Ag^+
(h) V^{2+}	(p) Fe^{2+}

(q) Cd^{2+}
(r) Ti^{2+}
(s) Pb^{2+}
(t) Cu^{2+}

2. Predict what monatomic ions each of the following could form:
(a) Na
(b) Al
(c) Br
(d) N
(e) Ba
(f) Ti
(g) Fe
(h) Sn
(i) Ca
(j) K
(k) S
(l) Cl
(m) V
(n) Hg
(o) Pb
(p) Ni

Polarity of bonds

3. Place a (+) above the atom or atoms that are relatively positive and a (−) above the atom or atoms that are relatively negative in the following covalent bonded molecules.
(a) HF
(b) H_2O
(c) ICl
(d) PCl_3
(e) OF_2
(f) HI
(g) BrCl
(h) $SiCl_4$
(i) NH_3
(j) Cl_2O

4. Arrange each of the following groups in order of increasing polarity:
(a) N—F, Mg—O, N—O, N—N
(b) C—O, N—O, O—O, C—F

Lewis structure

5. Draw a Lewis structure for each of the following:
(a) H_2O
(b) SO_2
(c) SO_3
(d) HOCl
(e) NH_3
(f) C_2H_6
(g) C_2H_4
(h) C_2H_2
(i) HCN
(j) NO_3^-
(k) SO_4^{2-}
(l) CO_3^{2-}
(m) sulfite ion
(n) phosphate ion
(o) BH_3
(p) PCl_5
(q) SeF_4
(r) SF_6
(s) IF_5
(t) XeF_4
(u) H_2S
(v) CO_2
(w) PCl_3
(x) CCl_4
(y) CS_2
(z) N_2
(aa) PH_3
(bb) CH_3Cl
(cc) C_2H_3Cl
(dd) CN^-
(ee) ClO_2^-
(ff) phosphite ion
(gg) nitrite ion
(hh) sulfate ion
(ii) BF_3
(jj) AsF_5
(kk) SF_4
(ll) BrF_5
(mm) KrF_4
(nn) SeF_6

Resonance forms

6. Tell which parts of Problem 5 would have resonance forms and how many there are for each.

Geometry of molecules

7. Using the VSEPR Theory, predict the bond angles in each part of Problem 5.

Sigma (σ) and pi (π) bonds, and lone pairs of electrons

8. Tell how many sigma bonds and how many pi bonds there are for each part of Problem 5; also, tell how many lone (nonbonding) pairs of electrons there are around the central atom.

Hybrid orbitals

9. Tell what type of hybrid orbital could possibly be formed for each part of Problem 5.

Polarity

10. Tell if each part of Problem 5 is polar or nonpolar.

Covalent bonding—general

11. Describe, as completely as possible, the bonding in each of the following (including types of bonds, bond angles, resonance forms, type of hybridization possible, and draw a sketch), and tell if each molecule is polar or nonpolar.
    (a) CH_4
    (b) SiO_2
    (c) SeO_2
    (d) ClF_3
    (e) IF_5
    (f) AsH_3
    (g) HOBr
    (h) TeO_2
    (i) XeF_2
    (j) ClF_5

Molecular orbital theory

12. For each of the following molecules, give the bond order and the number of unpaired electrons, using the order of molecular orbitals as given in Section 7-10.
    (a) H_2
    (b) He_2
    (c) N_2
    (d) CO
    (e) NO
    (f) F_2
    (g) Ne_2
    (h) O_2
    (i) BeO
    (j) OF

REFERENCES

COMPANION, AUDREY L., *Chemical Bonding.* New York: McGraw-Hill Book Co., 1964, Chapter 4.

DAHL, PETER, "The Valence Shell Electron Pair Repulsion Theory," *Chemistry,* March 1973, **46,** p. 17.

PAULING, LINUS, *The Chemical Bond.* New York: Cornell University Press, 1967.

RYSCHKEWITSCH, GEORGE E., *Chemical Bonding and the Geometry of Molecules.* New York: Reinhold Publishing Corp., 1963 (Chapters 3 and 4).

SANDERSON, R. T., "The Nature of 'Ionic' Solids," *J. Chem. Educ.,* 1967, **44,** p. 516.

SEBERA, DONALD K., *Electronic Structure and Chemical Bonding.* New York: Blaisdell Publishing Co., 1964.

Objectives

1. To understand what a balanced chemical equation is and what information it gives, including the meaning of various symbols used in equations.
2. To learn to balance simple equations.
3. To understand what a combination reaction is and what general types of substances undergo this kind of reaction and to be able to write balanced equations for the more common examples of combination reactions, such as a metal plus a nonmetal, a metal oxide or a nonmetal oxide plus water, and a metal oxide plus a nonmetal oxide.
4. To understand what a decomposition reaction is and what general types of substances undergo this kind of reaction and to be able to write balanced equations for some common examples of this, such as heating chlorates, hydrates, carbonates, hydrogen carbonates (bicarbonates), and certain oxides.
5. To understand what a replacement reaction is and to be able to write balanced equations for reactions of this type using an activity (electromotive) series for metals or from a general knowledge of which are the more active metals or nonmetals from their position in the periodic table.
6. To understand what a metathesis (double replacement) reaction is and what type of products are formed and to be able to write balanced equations for metathesis reactions in which a precipitate is formed, using the general solubility rules, and to write ionic equations for these reactions.
7. To understand what a neutralization reaction is and what types of substances undergo this kind of reaction and to be able to write balanced equations for neutralization reactions involving acid and base, nonmetal oxide and base, metal oxide and acid, or nonmetal oxide and metal oxide.

Eight

Chemical Equations

In Chapter 3 the concept of chemical equations was introduced. You should recall that an equation is a short way of expressing a chemical change, in terms of symbols and formulas. Also, you should recall that according to the Law of Conservation of Mass, matter cannot be created nor destroyed in a chemical reaction. Therefore, the equation must be balanced to be correct; that is, there must be exactly the same number of each atom on each side of the equation.

In this chapter we shall consider the balancing of simple chemical equations and the use of various terms and symbols used in writing equations. Then we shall consider predicting the products and writing equations. Since it would be impossible to learn all the possible reactions that can take place, we shall consider five types of reactions. By learning these five general types of reactions and what substances undergo these reactions, you can predict the products and write equations for many chemical reactions.

Equations may be written as molecular or as ionic equations. Most of the ones considered in this chapter will be written in molecular form. We shall introduce the idea of writing ionic equations, however, and do some of these involving the formation of precipitates. In Chapter 14 we shall consider ionic equations in detail.

8-1 Terms and Symbols Used in Equations

In writing equations, various terms and symbols are used. The *reactants* are always written on the left and the *products* on the right, separated by a single arrow ($\longrightarrow$), a double ($\rightleftharpoons$), or an equals sign ($=$). Often the physical state of the reactants and products is indicated by the following notation:

1. (g) for gas, or in some cases if the gas is a product, an arrow pointing upward is used, such as, $CO_{(g)}$, or $CO\uparrow$.

2. (l) for liquid, such as $H_2O_{(l)}$.
3. (s) for solid, or if the solid is a product, sometimes an arrow pointing down after the formula for that product, or the product can be underlined, such as $FeS_{(s)}$, $FeS\downarrow$, or $\underline{FeS}$.
4. (aq) for a substance in aqueous (water) solution, such as $NaCl_{(aq)}$.

A Δ in an equation represents heat. Either this can be written over the arrow to indicate that heat is necessary to cause the reaction to take place or in some cases it is shown as a product in the reaction. We shall consider the heat and energy changes in reactions in Chapter 10. A **catalyst** is indicated above or below the arrow also. A catalyst is a substance that alters the speed of the reaction without being consumed in the reaction. In general, a catalyst is used to speed up a reaction, but there are also some catalysts that are used to slow down reactions. These are sometimes called *negative catalysts* or *inhibitors.* Catalysts are important in many types of reactions. For example, many industrial processes depend on finding the correct catalyst to produce a desired product at a reasonable rate. Also, enzymes in the body catalyze many biological reactions that take place (these will be considered in Chapter 27).

In writing equations, often the symbols are used to indicate the state of a substance or to show that a gas or a solid is produced. There are also times when writing equations however, that we do not include all this information. In the equations written in this chapter, the states will all be included. Not only should you become familiar with the symbols that are used but also you should begin to become familiar with the state of various substances at ordinary conditions. Many of these you are already familiar with, and others you will learn more about as you proceed further in your study of chemistry. For example, you are probably aware that most metals in the free state are solids at ordinary temperature and that substances like oxygen, hydrogen, and nitrogen are gases. You know that at ordinary temperatures water is a liquid but at elevated temperatures it is a gas. In Section 8-7 you will begin to learn what common substances are soluble in water and would therefore be in aqueous solution and what substances are insoluble and therefore would exist in the solid state. Your experience in the laboratory will be helpful in learning the solubilities in water.

8-2 Writing and Balancing Simple Chemical Equations

The equations that we shall consider in this chapter will be balanced by inspection, that is, simply by putting coefficients before any of the formulas until there are exactly the same number of each atom on both sides of the equation. In Chapter 19, we shall consider more complex oxidation-reduction equations and learn some methods for balancing reactions of this type.

Some general guidelines for balancing these simple equations are given below. There is no set of rules, however, that can be learned to do this.

We are simply putting in whatever numbers are needed to obtain the same number of atoms on each side of the equation. The first and most important rule to remember is that the formulas are never changed in balancing an equation. The following general guidelines should be helpful when learning to balance these equations.

1. Write the correct formulas for the reactants and the products, with the reactants on the left and the products on the right separated by $\longrightarrow$, $\rightleftharpoons$, or $=$. Each reactant and each product is separated from each other by a + sign. *Once the correct formula is written, it must not be changed during the subsequent balancing operation.*

2. Start with the more complex parts—that is, the compounds that have several elements in them. In some cases, it is just a matter of balancing all atoms other than hydrogen and oxygen first.

3. Balance hydrogen and oxygen by putting in water, if necessary, after all other elements have been balanced.

4. Leave any elements in the free state until last since changing the coefficients on these only changes that one kind of atom. For example, when a 2 is placed before H_2O, it doubles the number of H atoms and O atoms, but when a 2 is placed before Al, it only changes the number of Al atoms.

5. For reactions involving polyatomic ions, balance the ion as a group. For example, with SO_4^{2-}, balance as sulfate ions, not as S atoms and O atoms.

6. Generally, if fractions appear in an equation, we multiply everything by the smallest number that will eliminate the fraction. It is not essential to get rid of fractions, however; it is just more simple in most cases. Also, make sure at the end that all coefficients are in the lowest possible ratio; if they are not, simplify to make them so.

Consider the following examples.

1. $$Al_{(s)} + Cl_{2(g)} \longrightarrow AlCl_{3(s)} \qquad \text{(unbalanced)}$$

To balance the Cl, we put a 3 before the Cl_2 and a 2 before the $AlCl_3$. Then we put a 2 before the Al to balance it, to give

$$2\,Al_{(s)} + 3\,Cl_{2(g)} \longrightarrow 2\,AlCl_{3(s)}$$

2. $$C_2H_{6(g)} + O_{2(g)} \xrightarrow{\Delta} CO_{2(g)} + H_2O_{(g)} \qquad \text{(unbalanced)}$$

We start with the C, by putting a 2 before the CO_2. Next, we balance H by putting 3 before H_2O. This gives

$$C_2H_{6(g)} + O_{2(g)} \xrightarrow{\Delta} 2\,CO_{2(g)} + 3\,H_2O_{(g)} \qquad \text{(unbalanced)}$$

To balance the oxygen, we need 7 atoms on the left, so we would put in $\frac{7}{2}O_2$, to give

$$C_2H_{6(g)} + \tfrac{7}{2}\,O_{2(g)} \xrightarrow{\Delta} 2\,CO_{2(g)} + 3\,H_2O_{(g)}$$

This is balanced but contains a fraction. To get rid of the fraction, we can multiply through by 2 to give

$$2\,C_2H_{6(g)} + 7\,O_{2(g)} \xrightarrow{\Delta} 4\,CO_{2(g)} + 6\,H_2O_{(g)}$$

3. $$Ca(OH)_{2(aq)} + H_3PO_{4(aq)} \longrightarrow Ca_3(PO_4)_{2(s)} + H_2O_{(l)}$$

(unbalanced)

Starting with the calcium phosphate (the most complex part) and considering the phosphate ion as a group, we place a 2 before the H_3PO_4 to balance the phosphate groups and a 3 before the $Ca(OH)_2$ to balance the Ca. This gives

$$3\,Ca(OH)_{2(aq)} + 2\,H_3PO_{4(aq)} \longrightarrow Ca_3(PO_4)_{2(s)} + H_2O_{(l)}$$

(unbalanced)

This just leaves the hydrogen and oxygen. Considering the H, there are 12 atoms on the left and, therefore, we need to place a 6 before the H_2O. This also balances the O atoms, with 6 other than those in the phosphate, or a total of 14 altogether. (We could have used oxygen in the last step in place of hydrogen and obtained the same result.) This gives

$$3\,Ca(OH)_{2(aq)} + 2\,H_3PO_{4(aq)} \longrightarrow Ca_3(PO_4)_{2(s)} + 6\,H_2O_{(l)}$$

(balanced)

To check this we can add up all the Ca, O, H, and P atoms on each side of the equation. This gives

LEFT	RIGHT
Ca = 3 × 1 = 3	Ca = 1 × 3 = 3
O = (3 × 2) + (2 × 4) = 14	P = 2 × 1 = 2
H = (3 × 2) + (2 × 3) = 12	O = (4 × 2) + (6 × 1) = 14
P = 2 × 1 = 2	H = 6 × 2 = 12

Therefore, the equation is balanced, and all numbers are whole numbers and are in the smallest possible ratio.

If an equation is given with both the reactants and products indicated, it is only necessary to write all the correct formulas and balance the equation. Consider the following example.

A sulfuric acid solution is mixed with an aqueous solution of sodium hydroxide to form water and an aqueous solution of sodium sulfate.

$$H_2SO_{4(aq)} + NaOH_{(aq)} \longrightarrow H_2O + Na_2SO_{4(aq)} \quad \text{(unbalanced)}$$

Once the correct formulas have all been written, they must not be changed in the balancing operation. To balance this we first consider the sodium atoms and put a 2 before the NaOH on the left side. This balances everything except the hydrogen and the oxygen; counting the hydrogen on the left side, we see there are four atoms of hydrogen, and therefore we must put a 2 on the right side before the H_2O. This gives

$$HSO_{4(aq)} + 2\,NaOH_{(aq)} \longrightarrow 2\,H_2O_{(l)} + Na_2SO_{4(aq)}$$

We now count up the oxygen atoms to make sure they are balanced. There are 6 on the left side (4 + 2) and 6 on the right (2 + 4). Therefore, all the atoms are balanced, the coefficients are all whole numbers and are in the lowest possible ratio, and thus the equation is balanced.

8-3 Completing Chemical Equations. The Five Simple Types of Chemical Reactions

In order to begin to be able to predict the products in a reaction, we shall consider five types of reactions and we shall look at some general cases. The five types that we shall consider here are the following:

1. Combination reactions.
2. Decomposition reactions.
3. Replacement (displacement) reactions.
4. Metathesis (double replacement) reactions.
5. Neutralization reactions (these are actually included in types 1 and 4 but are sufficiently important to be included as a special type).

Later, we shall study other general types of reactions, such as oxidation-reduction, and learn some specific reactions for various elements and compounds.

When writing equations, follow these steps.

1. Write the correct formulas for the reactants, if necessary.
2. Decide what type of reaction it is and thus what the products are.
3. Write the correct formula(s) for the product(s).
4. Balance the equation.

8-4 Combination Reactions

In a **combination reaction,** two or more substances (either elements or compounds) react to form one product. There are several general types of combination reactions that we will consider here.

1. **Metal + nonmetal $\longrightarrow$ binary compound** (such as oxide, sulfide or halide)

For the IA and IIA metals, as well as aluminum, zinc, and silver, we can always predict the product that will form. For other metals, which have variable oxidation states, the final product depends on the reaction conditions, and generally more information is needed to predict the product. Consider the following examples.

$$4\,Al_{(s)} + 3\,O_{2(g)} \longrightarrow 2\,Al_2O_{3(s)}$$

$$Mg_{(s)} + S_{(s)} \xrightarrow{\Delta} MgS_{(s)}$$

$$2\,Na_{(s)} + Cl_{2(g)} \longrightarrow 2\,NaCl_{(s)}$$

In writing these equations, we have included the state for all the reactants and products. You may need to consult a handbook or other sources to determine what the state of some of these substances will be; however, you should begin to become familiar with some of the properties of the more common types of substances.

2. **Nonmetal + oxygen $\longrightarrow$ nonmetal oxide**

In general, there are a variety of products that can be formed in reactions of this type. For example when carbon combines with oxygen, if there is a limited amount of oxygen present, the product is carbon monoxide (CO); if there is an excess of oxygen, the product is carbon dioxide (CO_2), according to the following equations:

$$2\,C_{(s)} + \underset{\text{(limited)}}{O_{2(g)}} \xrightarrow{\Delta} 2\,CO_{(g)}$$

$$C + \underset{\text{(excess)}}{O_{2(g)}} \xrightarrow{\Delta} CO_{2(g)}$$

The formula of the product can only be determined by a knowledge of the oxides of the nonmetal, such as CO and CO_2 in this case.

3. **Metal oxide + water $\longrightarrow$ metal hydroxide (base)**

For example, calcium oxide (CaO) combines with water to form calcium hydroxide, according to the following equation:

$$CaO_{(s)} + H_2O_{(l)} \longrightarrow Ca(OH)_{2(aq)}$$

The formula of the hydroxide formed is determined from a knowledge of the oxidation number of the metal and the charge on the hydroxide ion (OH^-). If the metal can have more than one oxidation state, the oxidation state in the hydroxide is always the same as it was in the oxide. For example, iron(III) oxide forms iron(III) hydroxide. Due to the formation of a metal hydroxide, or base, from a metal oxide in water, the metal oxide is sometimes called a *basic oxide,* or **basic anhydride.**

4. **Nonmetal oxide + water ⟶ oxyacid**

When a nonmetal oxide combines with water, the product formed is an oxyacid. For this reason, nonmetal oxides are often called *acid oxides,* or **acid anhydrides.** To determine what acid will be formed, there are two things that are used:

(a) Knowledge of the **oxyacids** that a particular nonmetal can form, such as sulfuric acid and sulfurous acid for sulfur.
(b) Oxidation number of the central atom, which will always be the same in the oxyacid as it is in the oxide.

Consider the following example:

$$SO_{2(g)} + H_2O_{(l)} \longrightarrow H_2SO_{3(aq)}$$

To determine what acid will be formed here, we consider the two common oxyacids H_2SO_3 and H_2SO_4. Then we consider the oxidation number of sulfur; since it is +4 in SO_2, we know that the acid formed is H_2SO_3, where sulfur also has an oxidation number of +4. This type of combination reaction explains why the various nonmetal oxides that are often found as air pollutants are harmful. For example, the sulfur dioxide that is produced in many industrial processes, and in the burning of coal, will mix with water to form sulfurous acid. Also, the SO_2 can be oxidized in the air to form SO_3. When this combines with water, an even more harmful substance, sulfuric acid, is produced:

$$SO_{3(g)} + H_2O_{(l)} \longrightarrow H_2SO_{4(aq)}$$

5. **Metal oxide + nonmetal oxide ⟶ salt**

Reactions of this type will be considered in Section 8-8 with neutralization reactions. Since a metal oxide is a basic anhydride, and a nonmetal oxide is an acid anhydride, a combination of these to form a salt is actually a type of acid-base reaction. One way of predicting the product formed in reactions of this type is to consider what base or hydroxide the metal oxide would form if it were in water and what acid the nonmetal oxide would form if it were in water. Then by deciding what salt the given acid and base would form, we can determine the product. For example

$$CaO_{(s)} + SO_{2(g)} \longrightarrow CaSO_{3(s)}$$

Since the CaO is the anhydride of $Ca(OH)_2$ and SO_2 is the anhydride of H_2SO_3, the salt formed is $CaSO_3$.

8-5 Decomposition Reactions

In a **decomposition reaction,** one compound decomposes to form two or more new substances. The products formed can be either elements or compounds. Often heat is necessary to cause these reactions to take place. In general, these are more specific reactions, for various types of compounds. We shall consider a few of the more common ones.

1. **Hydrates,** when heated, decompose to yield water and the anhydrous salt. A hydrate is a salt that contains one or more molecules of water for each formula unit of salt, built right into the crystal structure. When these substances are heated, the water is driven off. For example, barium chloride dihydrate decomposes according to the following equation,

$$BaCl_2 \cdot 2\,H_2O_{(s)} \xrightarrow{\Delta} BaCl_{2(s)} + 2\,H_2O_{(g)}$$

2. **Chlorates,** when heated, decompose to form chlorides plus oxygen gas. For example, potassium chlorate decomposes according to the following equation:

$$2\,KClO_{3(s)} \xrightarrow{\Delta} 2\,KCl_{(s)} + 3O_{2(g)}$$

This reaction is often used to prepare small amounts of oxygen in the laboratory.

3. A few **metal oxides** decompose when heated to form the free metal and oxygen gas. For example, mercuric oxide decomposes according to the following equation:

$$2\,HgO_{(s)} \xrightarrow{\Delta} 2\,Hg_{(l)} + O_{2(g)}$$

4. Most **carbonates,** when heated, decompose to form oxides and CO_2 For example, calcium carbonate (limestone), when heated, forms calcium oxide (lime) and CO_2, according to the following equation:

$$CaCO_{3(s)} \xrightarrow{\Delta} CaO_{(s)} + CO_{2(g)}$$

The Group IA carbonates do not decompose readily.

5. Most **hydrogen carbonates (bicarbonates),** when heated, decompose to form an oxide plus water and CO_2. For example, calcium hydrogen carbonate decomposes according to the following equation:

$$Ca(HCO_3)_{2(s)} \xrightarrow{\Delta} CaO_{(s)} + 2\,CO_{2(g)} + H_2O_{(g)}$$

When hydrogen carbonates of the Group IA metals are heated, they form a carbonate plus water and CO_2 (recall that the IA metal carbonates are stable toward heat, whereas other carbonates are not). For example, sodium bicarbonate decomposes according to the following equation:

$$2\,NaHCO_{3(s)} \xrightarrow{\Delta} Na_2CO_{3(s)} + CO_{2(g)} + H_2O_{(g)}$$

6. **Water** can be broken down into hydrogen gas and oxygen gas by a direct electric current, according to the following equation:

$$2\,H_2O_{(g)} \xrightarrow[\text{electric current}]{\text{direct}} 2\,H_{2(g)} + O_{2(g)}$$

8-6 Replacement (Displacement) Reactions

The most common type of **replacement** (or displacement) **reaction** is one in which a metal replaces another metal ion from a solution. In order to do this, the free metal must be more active than the metal that is in solution. The metals can be arranged in a series called the **electromotive series** or *activity series.* The arrangement of this will be considered in Chapter 20. For now we will just use this activity series to predict the products of replacement reactions. The activity series is given on the inside back cover of this text and is also listed on the margin of this page. Notice that hydrogen is also included here, even though it is not a metal. In using this series, any free metal higher on the list will displace another metal that is lower from solution. Also, any metal above hydrogen on the activity series will displace hydrogen gas from an acid. Consider the following examples:

Li
K
Ba
Ca
Na
Mg
Al
Zn
Fe
Cd
Ni
Sn
Pb
(H)
Cu
Hg
Ag
Au

1. $Fe_{(s)} + CuSO_{4(aq)} \longrightarrow FeSO_{4(aq)} + Cu_{(s)}$

   Since iron is higher on the activity series, it will displace Cu from solution. For metals that can exist in more than one oxidation state, the lower one is often formed, but it depends on the conditions. Hence, the new salt formed in this reaction is iron(II) sulfate.

2. $Cu_{(s)} + FeSO_{4(aq)} \longrightarrow$ no reaction (Cu is less active than Fe)
3. $Zn_{(s)} + 2\,HCl_{(aq)} \longrightarrow H_{2(g)} + ZnCl_{2(aq)}$ (Zn is above hydrogen)
4. A very active metal, such as sodium, will displace hydrogen gas from water. In reactions of this type, only one of the hydrogen atoms is displaced, due to the slight amount of H^+ formed from the H—OH molecule (this will be discussed in Chapter 17, after weak acids and bases have been considered).

$$2\,Na_{(s)} + 2\,H_2O_{(l)} \longrightarrow H_{2(g)} + 2\,NaOH_{(aq)}$$

There are also replacement reactions involving nonmetals. For example, for the halogens, the series would be fluorine-chlorine-bromine-iodine; that is, free chlorine will displace bromine from a compound, and bromine will displace iodine from a compound. Notice here that the activity of nonmetals is related to their ability to gain electrons and form negative ions, whereas the activity of metals is related to their tendency to lose electrons to form positive ions. In writing replacement reactions involving the halogens, keep in mind that in the free state all the halogens exist as diatomic molecules. Consider the following examples:

1. $Cl_{2(g)} + 2\,NaBr_{(aq)} \longrightarrow 2\,NaCl_{(aq)} + Br_{2(aq)}$
2. $I_{2(s)} + 2\,NaBr_{(aq)} \longrightarrow$ no reaction

8-7 Metathesis (Double Displacement) Reactions

In a **metathesis** (or double displacement) **reaction,** two compounds react to form two new compounds. In general, these involve ions, and the positive ions exchange partners with the negative ions to form the two new compounds. In many of these reactions an insoluble substance is formed. Therefore, to predict the products formed here, you must be familiar with some common solubility rules. The general solubility rules listed on the inside of the back cover of this book are useful in predicting many reactions. These can be summarized as follows:

1. Soluble–compounds of the IA metals; ammonium (NH_4^+) compounds; nitrates; acetates; most chlorides except AgCl, $PbCl_2$, and Hg_2Cl_2; and most sulfates, except Ca^{2+}, Sr^{2+}, Ba^{2+}, and Pb^{2+} sulfates.
2. Insoluble–oxides, hydroxides, carbonates, and phosphates, except for those of the IA metals and ammonium; and sulfides, except those of IA and IIA elements and ammonium.

Consider the following examples.

1. $AgNO_{3(aq)} + HCl_{(aq)} \longrightarrow AgCl_{(s)} + HNO_{3(aq)}$
2. $Na_2SO_{4(aq)} + Pb(NO_3)_{2(aq)} \longrightarrow PbSO_{4(s)} + 2\ NaNO_{3(aq)}$
3. $Ba(NO_3)_{2(aq)} + 2\ NaOH_{(aq)} \longrightarrow Ba(OH)_{2(s)} + 2\ NaNO_{3(aq)}$

   Note that in each of these cases one of the products formed is an insoluble substance, as determined from the solubility rules.

Another type of metathesis reaction involves the formation of nonionized substances, such as water or carbon dioxide. For example, any carbonate, either in the solid state or in aqueous solution, will react with acid to form water and carbon dioxide gas, plus a salt. Consider the following example:

$$CaCO_{3(s)} + 2\ HCl_{(aq)} \longrightarrow CaCl_{2(aq)} + H_2O_{(l)} + CO_{2(g)}$$

All the neutralization reactions involving acids and bases, which we shall consider in Section 8-8, are actually metathesis reactions in which the nonionized substance water is formed, along with a salt.

It is often useful to write metathesis reactions as **ionic equations.** In doing this, salts in solution are written in the form of ions, and solid substances or nonionized substances are written in molecular form. Also weak acids and bases are written in molecular form; these will be considered along with a detailed study of ionic equations in Chapter 14, after considering acids and bases. Consider the following example showing the reaction between an aqueous solution of silver nitrate and an aqueous solution of sodium chloride written as an ionic equation.

$$Ag^{+} + NO_3^{-} + Na^{+} + Cl^{-} \longrightarrow AgCl_{(s)} + Na^{+} + NO_3^{-}$$

Notice here that both the silver nitrate and sodium chloride are written in the form of the ions, as they actually exist in the aqueous solution. The silver chloride, however, since it is insoluble (see the solubility rules), is written in "molecular" form or as formula units. Finally, the other product, sodium nitrate, is written in the form of ions since that is how it actually exists in aqueous solution. Notice also in this equation that we did not indicate aqueous solution for all the ions present. This is understood from the ionic equation; that is, the silver nitrate and sodium chloride, for example, would not be in the form of individual ions unless they were in aqeuous solution.

8-8 Neutralization Reactions

In a **neutralization reaction** an acid or acid oxide (nonmetal oxide) reacts with a base or basic oxide (metal oxide). A **salt** is formed in the reaction, and for most of these, water is also formed. The only case where water is not formed would be in a simple combination of a nonmetal oxide with a metal oxide; for all other neutralization reactions, the formation of water is the driving force behind the reaction. That is, if the nonionized substance water were not formed, many of these would simply consist of a mixture of various ions, with no actual reaction taking place. Therefore, it can be seen that most of these reactions are simply a special case of metathesis reaction involving the formation of water; however, these are sufficiently important to consider as a special class.

As noted in Section 8-4, combination reactions involving a metal oxide and a nonmetal oxide can also be considered as neutralization reactions. The reason for this is that a metal oxide is a basic anhydride and a nonmetal oxide is an acid anhydride.

Consider the following examples of various types of neutralization reactions.

1. **Acid + base $\longrightarrow$ water + salt**

$$HCl_{(aq)} + NaOH_{(aq)} \longrightarrow H_2O_{(l)} + NaCl_{(aq)}$$

The hydrogen ion from the acid combines with the hydroxide ion from the base to form water. This leaves sodium ion and chloride ion in the solution, which therefore is an aqueous solution of sodium chloride. Writing this as an ionic equation, it would be

$$H^{+} + Cl^{-} + Na^{+} + OH^{-} \longrightarrow H_2O_{(l)} + Na^{+} + Cl^{-}$$

The solubility rules must also be kept in mind when indicating the state in these equations. For example, when phosphoric acid reacts with calcium hydroxide, the salt that is formed (calcium phosphate) is a solid rather than an aqueous solution, as shown by the following equation:

$$2\,H_3PO_{4(aq)} + 3\,Ca(OH)_{2(aq)} \longrightarrow 6\,H_2O_{(l)} + Ca_3(PO_4)_{2(s)}$$

2. **Metal oxide (basic anhydride) + acid ⟶ water + salt**

$$CaO_{(s)} + 2\,HCl_{(aq)} \longrightarrow H_2O_{(l)} + CaCl_{2(aq)}$$

3. **Nonmetal oxide (acid anhydride) + base ⟶ water + salt**

$$SO_{2(g)} + 2\,NaOH_{(aq)} \longrightarrow H_2O_{(l)} + Na_2SO_{3(aq)}$$

Since SO_2 is the anhydride of sulfurous acid (H_2SO_3), the salt formed is sodium sulfite (Na_2SO_3).

4. **Basic oxide (metal oxide) + acid oxide (nonmetal oxide) ⟶ salt**

$$MgO_{(s)} + SO_{3(g)} \longrightarrow MgSO_{4(s)}$$

The SO_3 will form the SO_4^{2-} polyatomic ion, since SO_3 is the anhydride of H_2SO_4. Therefore, the salt formed is magnesium sulfate. This type of reaction is actually a combination reaction because only one substance is formed. It is also a neutralization, however, in that a basic oxide reacts with an acid oxide to form a salt, but no water is formed in this case.

5. **Ammonia + acid ⟶ ammonium salt**

$$NH_{3(g)} + HCl_{(g)} \longrightarrow NH_4Cl_{(s)}$$

Since ammonia (NH_3) is a base, the ammonia gas reacts with the hydrogen chloride gas to form the ammonium salt, ammonium chloride. The thin film you often see on the reagent bottles or on the windows of the laboratory is partly due to the formation of ammonium chloride, as the preceding equation shows.

SUMMARY

1. To determine what the products will be in a chemical reaction, it is often helpful to think in terms of types of reactions. But this is just the beginning since a knowledge of the chemical properties of the particular substances being considered is also necessary. The greater your knowledge of chemistry, the better your ability to predict products of chemical reactions.
2. In writing chemical equations, first it must be determined what the products will be, then the correct formula(s) must be written for all

reactants and products, and finally the equation must be balanced; also the state of each substance can be indicated by (g), (l), or (s).

3. The following types of reactions have been considered:
   (a) Combination–two elements to form a compound, or an oxide plus water (metal oxide $\longrightarrow$ hydroxide; nonmetal oxide $\longrightarrow$ acid, with the same oxidation state for the central atom).
   (b) Decomposition–upon heating, hydrates give off water, some oxides yield oxygen gas, most carbonates (except IA) form oxide plus CO_2, and hydrogen carbonates form CO_2, H_2O, and oxide (or carbonate, for IA compounds).
   (c) Replacement–metals and hydrogen, predicted according to the activity of metals or nonmetals from their position in the periodic table, or from the electromotive (activity) series of metals (and hydrogen).
   (d) Metathesis–formation of an insoluble substance (determined by the general solubility rules) or a nonionized substance such as CO_2 or H_2O (also, formation of weak acids and bases, or complex ions–to be considered later).
   (e) Neutralization–acid plus base or acid plus metal oxide or base plus nonmetal oxide, all yield a salt and water; a metal oxide (basic anhydride) plus a nonmetal oxide (acid anhydride) can also form a salt.

Upon completion of this chapter, you should be able to do the following things:

1. Given the reactants and products (either by name or formula), write a balanced equation to represent the reaction. (1, 2)
2. Given the name or formula for any of the following reactants, write a complete balanced equation to represent the reaction:
   a) Combination of metal and nonmetal (especially for metals with constant oxidation numbers), nonmetal with oxygen, a metal oxide with water, a nonmetal oxide with water. (3)
   b) Decomposition, upon heating, of chlorates, hydrates, carbonates, hydrogen carbonates, and some oxides (such as HgO). (4)
   c) Replacement of a metal by a more active metal in aqueous solution (given the electromotive series or from position in the periodic table), hydrogen from an acid by an active metal, hydrogen from water by a very active metal (Na, K, Ca); a nonmetal by a more active nonmetal (for the halogens). (5)
   d) Methathesis (double displacement) reaction involving the formation of an insoluble substance (by being familiar with the general solubility rules). (6)
   e) A carbonate with an acid. (6)

f) An acid plus a base. (7)
g) A basic oxide (metal oxide) plus an acid. (7)
h) An acid oxide (nonmetal oxide) plus a base. (7)
i) A metal oxide plus a nonmetal oxide. (7)
j) Ammonia plus an acid. (7)

3. Given any of the terms listed in Exercise 1, give a definition or explanation in your own words.

EXERCISES

1. Define or explain the following terms:
   (a) chemical equation
   (b) balanced equation
   (c) electromotive series (activity series)
   (d) catalyst
   (e) acid anhydride
   (f) basic anhydride
   (g) ionic equation
   (h) metathesis reaction
   (i) decomposition reaction
   (j) combination reaction
   (k) neutralization reaction
   (l) replacement reaction
   (m) salt
2. Distinguish between
   (a) a catalyst and an inhibitor
   (b) activity of metal and activity of nonmetal
   (c) acid anhydride and basic anhydride

PROBLEMS

Balancing equations

1. Balance each of the following equations by inspection:
   (a) $CaCl_{2(aq)} + (NH_4)_2CO_{3(aq)} \longrightarrow CaCO_{3(s)} + NH_4Cl_{(aq)}$
   (b) $KClO_{3(s)} \longrightarrow KCl_{(s)} + O_{2(g)}$
   (c) $Fe(OH)_{3(s)} + H_2SO_{4(aq)} \longrightarrow Fe_2(SO_4)_{3(aq)} + H_2O_{(l)}$
   (d) $K_{(s)} + H_2O_{(l)} \longrightarrow KOH_{(aq)} + H_{2(g)}$
   (e) $Mg_{(s)} + N_{2(g)} \longrightarrow Mg_3N_{2(s)}$
   (f) $Ca_{(s)} + O_{2(g)} \longrightarrow CaO_{(s)}$
   (g) $AgNO_{3(aq)} + BaCl_{2(aq)} \longrightarrow AgCl_{(s)} + Ba(NO_3)_{2(aq)}$
   (h) $MgCl_{2(aq)} + Na_3PO_{4(aq)} \longrightarrow Mg_3(PO_4)_{2(s)} + NaCl_{(aq)}$
   (i) $CaCO_{3(s)} + H_3PO_{4(aq)} \longrightarrow Ca_3(PO_4)_{2(aq)} + CO_{2(g)} + H_2O_{(l)}$
   (j) $Al_{(s)} + H_2SO_{4(aq)} \longrightarrow Al_2(SO_4)_{3(aq)} + H_{2(g)}$

2. Change the following word equations into chemical equations and balance by inspection (indicate the state of all reactants and products):
   (a) calcium chloride solution + lead(II) nitrate solution ⟶ solid lead (II) chloride + calcium nitrate solution
   (b) ferric oxide + hydrochloric acid ⟶ ferric chloride solution + water
   (c) sodium hydrogen carbonate solution + phosphoric acid ⟶ sodium phosphate solution + carbon dioxide gas + water
   (d) mercury + oxygen ⟶ mercury(II) oxide
   (e) barium nitrate solution + sulfuric acid ⟶ barium sulfate solid + nitric acid
   (f) solid sodium carbonate + sulfuric acid ⟶ sodium sulfate solution + carbon dioxide gas + water
   (g) aluminum sulfate solution + sodium hydroxide solution ⟶ aluminum hydroxide precipitate + sodium sulfate
   (h) iron + chlorine ⟶ iron(III) chloride solution
   (i) potassium hydroxide solution + sulfuric acid ⟶ potassium hydrogen sulfate solution + water
   (j) tin(II) oxide + hydrochloric acid ⟶ tin(II) chloride solution + water

Combination reactions

3. Complete and balance the following equations:
   (a) $Ca_{(s)} + O_{2(g)}$
   (b) $C_{(s)} +$ excess $O_{2(g)}$
   (c) $CaO_{(s)} + H_2O_{(l)}$
   (d) $SO_{2(g)} + H_2O_{(l)}$
   (e) $Al_{(s)} + O_{2(g)}$
   (f) $Al_{(s)} + N_{2(g)}$
   (g) $Al_2O_{3(s)} + H_2O_{(l)}$
   (h) $P_4O_{10(s)} + H_2O_{(l)}$
   (i) $Al_{(s)} + Cl_{2(g)}$
   (j) $Li_2O_{(s)} + H_2O_{(l)}$
   (k) $SO_{3(g)} + H_2O_{(l)}$
   (l) $C_{(s)} +$ limited $O_{2(g)}$
   (m) $Mg_{(s)} + S_{(g)}$
   (n) $Na_2O_{(s)} + H_2O_{(l)}$
   (o) $N_2O_{3(g)} + H_2O_{(l)}$
   (p) $Mg_{(s)} + N_{2(g)}$

Decomposition reactions

4. Complete and balance the following equations:
   (a) $HgO_{(s)}$ + heat
   (b) $KClO_{3(s)}$ + heat
   (c) $MgCO_{3(s)}$ + heat
   (d) $Na_2CO_3 \cdot H_2O_{(s)}$ + heat
   (e) $Ca(HCO_3)_{2(s)}$ + heat
   (f) $SrCO_{3(s)}$ + heat
   (g) $CaSO_4 \cdot 2\,H_2O$ + heat
   (h) $Ba(HCO_3)_{2(s)}$ + heat
   (i) $NaHCO_3$ + heat
   (j) KNO_3 + heat

Replacement reactions

5. Complete and balance the following equations:
   (a) $Pb_{(s)} + HCl_{(aq)}$
   (b) $Ca_{(s)} + H_2O_{(l)}$
   (c) $Fe_{(s)} + CuCl_{2(aq)}$
   (d) $Cl_{2(g)} + NaBr_{(aq)}$
   (e) $Al_{(s)} + HCl_{(aq)}$
   (f) $Al_{(s)} + SnCl_{2(aq)}$
   (g) $Hg_{(s)} + SnCl_{2(aq)}$
   (h) $Zn_{(s)} + NiCl_{2(aq)}$
   (i) $Na_{(s)} + H_2O_{(l)}$
   (j) $Cu_{(s)} + NaCl_{(aq)}$
   (k) $Au_{(s)} + NiSO_{4(aq)}$
   (l) $Br_{2(aq)} + NaCl_{(aq)}$
   (m) $Cu_{(s)} + FeCl_{2(aq)}$
   (n) $Br_{2(l)} + NaI_{(aq)}$

Metathesis reactions

6. Complete and balance the following equations. Indicate any precipitate by (s) and any gas by (g).
   (a) $Pb(NO_2)_{2(aq)} + HCl_{(aq)}$
   (b) $Pb(NO_3)_{2(aq)} + H_2S_{(g)}$
   (c) $Bi(NO_3)_{3(aq)} + NaOH_{(aq)}$
   (d) $MnSO_{4(aq)} + (NH_4)_2S_{(aq)}$
   (e) $CaCO_{3(s)} + HCl_{(aq)}$
   (f) $ZnCO_{3(s)} + H_3PO_{4(aq)}$

(g) $BaCO_{3(s)} + HNO_{3(aq)}$
(h) $FeSO_{4(aq)} + (NH_4)_2S_{(aq)}$
(i) $AgNO_{3(aq)} + H_2S_{(g)}$
(j) $Bi(NO_3)_{2(aq)} + H_2S_{(g)}$
(k) $FeCO_{3(s)} + H_2SO_{4(aq)}$
(l) $Ba(NO_3)_{2(aq)} + NaOH_{(aq)}$

Neutralization reactions

7. Complete and balance the following equations. Indicate any precipitate by (s).
(a) $Zn(OH)_{2(s)} + H_2SO_{4(aq)}$
(b) $BaO_{(s)} + H_2SO_{4(aq)}$
(c) $CO_{2(g)} + Ca(OH)_{2(aq)}$
(d) $Al(OH)_{3(s)} + HCl_{(aq)}$
(e) $KOH_{(aq)} + CO_{2(g)}$
(f) $BaO_{(s)} + HCl_{(aq)}$
(g) $CaO_{(s)} + CO_{2(g)}$
(h) $NH_{3(g)} + HBr_{(g)}$
(i) $Fe(OH)_{3(s)} + H_3PO_{4(aq)}$
(j) $N_2O_{3(g)} + NaOH_{(aq)}$
(k) $Na_2O_{(s)} + HNO_{2(aq)}$
(l) $SO_{2(g)} + NaOH_{(aq)}$
(m) $Pb(OH)_{2(s)} + HNO_{3(aq)}$
(n) $MgO_{(s)} + HCl_{(aq)}$
(o) $BaO_{(s)} + SO_{3(g)}$
(p) $NH_{3(g)} + HCl_{(g)}$

General equations

8. Write a complete balanced equation for each reaction; indicate the state of each product.
(a) $BaO_{(s)} + H_2O_{(l)}$
(b) $MgCO_{3(s)}$ + heat
(c) $Ca_{(s)} + ZnCl_{2(aq)}$
(d) $NaCl_{(aq)} + Br_{2(g)}$
(e) $Mg_{(s)} + Br_{2(l)}$
(f) $P_2O_{5(s)} + H_2O_{(l)}$
(g) $Ca_{(s)} + HCl_{(aq)}$
(h) $ZnO_{(s)} + HClO_{3(aq)}$
(i) $H_2SO_{4(aq)} + BaCl_{2(aq)}$
(j) $KHCO_{3(s)}$ + heat
(k) $Sn_{(s)} + CaCl_{2(aq)}$
(l) $CO_{2(g)} + Al_2O_{3(s)}$
(m) $BaCl_2 \cdot 2\,H_2O_{(s)}$ + heat
(n) $SO_{3(g)} + Fe(OH)_{3(aq)}$
(o) mercuric oxide + heat
(p) copper metal + aluminum chloride solution
(q) sodium bicarbonate + heat
(r) aluminum + sulfuric acid
(s) calcium bicarbonate + heat
(t) sodium sulfide solution + zinc chloride solution
(u) calcium hydroxide + hydrochloric acid
(v) aluminum + ferrous sulfate solution
(w) magnesium oxide + nitrous acid
(x) dinitrogen trioxide + water
(y) sulfur dioxide + sodium hydroxide solution
(z) $CaO_{(s)} + H_2O_{(l)}$
(aa) $BaCO_{3(s)}$ + heat
(bb) $Mg_{(s)} + AlCl_{3(aq)}$
(cc) $NaI_{(aq)} + Br_{2(l)}$
(dd) $Ba_{(s)} + Cl_{2(g)}$
(ee) $N_2O_{3(g)} + H_2O_{(l)}$
(ff) $Zn_{(s)} + HCl_{(aq)}$
(gg) $CaO_{(s)} + HClO_{(aq)}$
(hh) $CaCl_{2(aq)} + H_2SO_{4(aq)}$
(ii) $Mg(HCO_3)_{2(s)}$ + heat
(jj) $Zn_{(s)} + CaCl_{2(aq)}$
(kk) $SO_{2(g)} + Al_2O_{3(s)}$
(ll) $MgSO_4 \cdot 7\,H_2O_{(s)}$ + heat
(mm) $NaOH_{(aq)} + CO_{2(g)}$
(nn) potassium chlorate + heat
(oo) potassium hydrogen carbonate + heat
(pp) aluminum + chloric acid
(qq) ammonium sulfide solution + aluminum bromide solution
(rr) ferric hydroxide solution + sulfuric acid
(ss) zinc + cuprous chloride solution
(tt) aluminum oxide + sulfurous acid
(uu) diphosphorus pentoxide + water
(vv) sulfur trioxide + magnesium hydroxide
(ww) barium carbonate + heat
(xx) barium chloride solution + sodium carbonate solution

REFERENCES

Benson, Sidney W., *Chemical Calculations,* 3rd ed. New York: John Wiley & Sons, Inc., 1971 (Appendix III).

Walker, Ruth A., and Helen Johnston, *The Language of Chemistry.* Englewood Cliffs, N.J.: Prentice-Hall, Inc., 1967.

Objectives

1. To understand what relationships can be obtained from the formula of a compound, such as percent and other mass relationships, and to be able to do calculations involving these relationships.
2. To learn to calculate empirical (and molecular) formulas of compounds from percent composition or weight relationships (and molecular weight if the molecular formula is to be calculated).
3. To be able to do calculations involving mass or weight quantities from balanced equations, using mole relationships.
4. To be able to calculate the limiting reagent, the amount of excess reagent, and the theoretical yield in a chemical reaction.
5. To be able to calculate the percent yield in a reaction if the actual yield is known.

Nine

Stoichiometry-Calculations Based on Formulas and Equations

In previous chapters we have considered elements, compounds, and chemical equations. In this chapter we shall again consider elements, compounds, and chemical equations but in regard to quantitative calculations. In doing this we shall be using the concepts of formula (molecular) weight and moles, as well as the factor-unit method of calculations; therefore, this material should be reviewed at this point, if necessary.

The term **stoichiometry** refers to the quantitative relationship of chemical combinations. Thus, it refers to chemically equivalent quantities, as determined either from the formula of a compound or from a balanced chemical equation.

First we shall use atomic and formula (molecular) weights to calculate equivalent quantities from chemical formulas and to calculate simplest (empirical) and molecular formulas. Then we shall use the coefficients in balanced chemical equations to do stoichiometric calculations involving mass (weight) relationships from the balanced equation.

9-1 Percent Composition of Compounds

Percent is defined as parts per hundred, or a fraction of one hundred. For example, if a compound contains 15 g of oxygen in 100 g of compound, the substance contains 15% oxygen by mass. In general, percent is taken to mean by mass unless some other units are specified, such as volume percent.

Any type of mass unit can be used to calculate percentage; that is, to find the percent composition of a compound we can use either relative units such as the atomic mass unit (amu) or other mass units such as grams. For example, to calculate the percent of hydrogen and oxygen in water, we can simply use the relative masses of the elements and the total relative mass, called the *formula weight,* or we can consider the mass of 1 mole and use the common mass unit grams. In either case, the percent composition of water would be calculated as follows:

The formula (molecular) weight is 18.0, from

$$2.0 \text{ amu H } (1.0 \times 2) + 16.0 \text{ amu O } (16.0 \times 1) = 18.0 \text{ amu}$$

Therefore, the percent of H and O are

$$\% \text{ H} = \frac{2.0 \text{ amu}}{18.0 \text{ amu}} \times 100 = 11\% \text{ H}$$

$$\% \text{ O} = \frac{16.0 \text{ amu}}{18.0 \text{ amu}} \times 100 = 89\% \text{ O}$$

Or, using 1 mole of H_2O gives

$$1 \text{ mole } H_2O = 18.0 \text{ g } H_2O$$

and it contains 2.0 g H and 16.0 g O. Therefore, the percent of H and O are:

$$\% \text{ H} = \frac{2.0 \text{ g}}{18.0 \text{ g}} \times 100 = 11\% \text{ H}$$

$$\% \text{ O} = \frac{16.0 \text{ g}}{18.0 \text{ g}} \times 100 = 89\% \text{ O}$$

Consider the following problem examples.

PROBLEM EXAMPLE 9-1

Calculate the percent composition of calcium nitrate.

SOLUTION:

The formula for calcium nitrate is $Ca(NO_3)_2$, and the formula weight is calculated as

$$\begin{aligned} Ca &= 1 \times 40 = 40 \\ N &= 2 \times 14 = 28 \\ O &= 6 \times 16 = 96 \\ \hline \text{formula weight} &= 164 \end{aligned}$$

The percent of each element in the compound is calculated by dividing the mass of each element by the total mass and multiplying by 100, to give,

$$\% \text{ Ca} = \frac{40}{164} \times 100 = 24\%$$

$$\% \text{ N} = \frac{28}{164} \times 100 = 17\%$$

$$\% \text{ O} = \frac{96}{164} \times 100 = 59\%$$

PROBLEM EXAMPLE 9-2

It is found that 8.25 g of a metal combine with oxygen to form 17.80 g of metal oxide. Calculate the percent of metal and oxygen in the compound.

SOLUTION:

Since the compound contains only the two elements, metal and oxygen, the mass of oxygen is 17.80 g − 8.25 g = 9.55 g.
Therefore, the percent of metal (M) and oxygen (O) is calculated as

$$\% \text{ M} = \frac{8.25}{17.80} \times 100 = \mathbf{46.3\%}$$

$$\% \text{ O} = \frac{9.55}{17.80} \times 100 = \mathbf{53.7\%}$$

9-2 Relationships from the Formula of a Compound

There are several relationships that can be obtained from the formula of a compound. For example, the formula gives us the relationship in terms of atoms, such as two atoms of hydrogen to one atom of oxygen in one molecule of water. We can also combine this with the relative masses of atoms, using the atomic weight scale, to obtain relationships from the formula in terms of either atomic mass units or grams (using the concept of the mole). In general, we find the mole concept to be very useful in quantitative applications of principles; thus we shall use this in our consideration of the relationships that can be obtained from the formula of a compound. For example, considering the compound water again, we can see from the formula that 1 mole of water molecules contains 2 moles of hydrogen atoms and 1 mole of oxygen atoms. Thus, we can say that for this compound 1 mole of water is equivalent to 2 moles of hydrogen atoms or that 1 mole of water is equivalent to 1 mole of oxygen atoms or that 2 moles of hydrogen atoms are equivalent to 1 mole of oxygen atoms. Likewise, we can also find the relationship in terms of grams as follows.

$$1 \text{ mol } H_2O = 2 \text{ mol H} = 1 \text{ mol O}$$

or

$$1 \text{ mol } H_2O = 18.0 \text{ g } H_2O = 2.0 \text{ g H} = 16.0 \text{ g O}$$

That is, for this compound, 2 moles of H are chemically equivalent to 1 mole of O or 16.0 g of oxygen are chemically equivalent to 2.0 g of hydrogen, and so forth. Consider the following examples of calculations based on formulas.

PROBLEM EXAMPLE 9-3

How many grams of sulfur will combine with 15.0 g of Zn to form ZnS?

SOLUTION:

This problem can be approached in two ways:

(a) A method that points out the relationship from the formula is to consider the relationship between moles of S atoms and moles of Zn atoms in ZnS (a ratio of 1 to 1 in this case).

$$15.0\ \text{g Zn} \times \frac{1\ \text{mol Zn}}{65.4\ \text{g Zn}} \times \frac{1\ \text{mol S}}{1\ \text{mol Zn}} \times \frac{32.1\ \text{g S}}{1\ \text{mol S}} = 7.36\ \text{g S}$$

(b) The most direct method is just to consider the relationship between grams of Zn and grams of S in ZnS; that is, 1 mole of ZnS contains 32.1 g of S and 65.4 g of Zn and therefore 32.1 g of S is equivalent to 65.4 g of Zn in this compound.

$$15.0\ \text{g Zn} \times \frac{32.1\ \text{S}}{65.4\ \text{g Zn}} = 7.36\ \text{g S}$$

PROBLEM EXAMPLE 9-4

How many grams of oxygen are there in 2.5 moles of $Ca(NO_3)_2$?

SOLUTION:

$$2.5\ \text{mol}\ Ca(NO_3)_2 \times \frac{6\ \text{mol O atoms}}{1\ \text{mol}\ Ca(NO_3)_2} \times \frac{16\ \text{g O atoms}}{1\ \text{mol O atoms}} = 240\ \text{g O atoms}$$

or

$$2.5\ \text{mol}\ Ca(NO_3)_2 \times \frac{164\ \text{g}\ Ca(NO_3)_2}{1\ \text{mol}\ Ca(NO_3)_2} \times \frac{96\ \text{g O atoms}}{164\ \text{g}\ Ca(NO_3)_2} = 240\ \text{g O atoms}$$

9-3 Empirical (Simplest) and Molecular Formulas

The **empirical formula** of a compound gives the smallest whole number ratio of atoms that make up the compound. This can be determined from the percent composition of the compound or from the experimentally determined mass relationships of elements that make up the compound.

The **molecular formula** of the compound is the true formula and shows the actual number of atoms of each element present in one molecule of the compound. For example, the empirical formula for the compound hydrogen peroxide is HO; however, the actual molecular formula is found to be H_2O_2.

Both the empirical and molecular formulas for a compound are obtained from experimental evidence. For example, when a new compound is prepared, it is analyzed to find what elements are present, and in what ratios. This can be done either in simple mass ratios or as percent composition. This procedure is often carried out by instrumental methods, which are usually quite rapid. Once the percent composition is determined experimentally, the empirical formula can be calculated. Then the molecular weight is determined, also by experimental methods, and this is used along with the empirical formula to determine the correct molecular formula for the compound. For the majority of inorganic compounds, the molecular formula and the empirical formula turn out to be the same. For organic compounds, however, this is very often not the case.

To calculate the empirical formula of a compound, the following steps are carried out:

1. If not given directly, find the relative number of grams of each element present. That is, if the composition is given as a percentage, consider a total of 100 grams and determine from that what the relative number of grams of each element would be. For example, water is found to be about 11% hydrogen and 89% oxygen. Therefore, we could use the relative masses of 11 g of hydrogen to 89 g of oxygen in each 100 g of compound.

2. Convert each of these quantities in grams to moles of atoms, using the atomic weight of each element.

3. Convert the ratio of numbers obtained to the smallest possible whole numbers. One way of doing this is to divide through by the smallest value and then to multiply by whatever number is necessary to obtain whole numbers. Keep in mind that in this third step we are simply converting the ratio calculated in the first two steps to small whole numbers. Therefore, each operation must be done exactly the same to each number involved or the ratio will no longer be correct.

To calculate the molecular formula from the empirical formula, the molecular weight must be known. This is done by adding all the relative masses of the atoms in an empirical formula and dividing this total into the molecular weight. This tells how many of the *empirical units* are present in one molecule. For example, the empirical formula for hydrogen peroxide is HO and the molecular weight is known to be 34. The relative mass of one empirical formula is 17 ($1 + 16 = 17$), and $34 \div 17$ is 2. Therefore, there are two of these empirical units in one molecule, and thus the molecular formula is H_2O_2.

Consider the following examples of calculations of empirical and molecular formulas.

PROBLEM EXAMPLE 9-5

Calculate the empirical formula for a compound that contains 26.6% potassium, 35.4% chromium, and 38.1% oxygen. (The error of 0.1% here emphasizes the experimental aspect.)

SOLUTION:

Since we are given percent composition here, we can consider how many grams of each element there would be in 100 g of compound. Each of these is converted to moles of atoms in order to find the ratio of atoms in the compound.

$$26.6\ \cancel{g\,K} \times \frac{1\text{ mol K atoms}}{39.1\ \cancel{g\,K}} = 0.680\text{ mol K atoms}$$

$$35.4\ \cancel{g\,Cr} \times \frac{1\text{ mol Cr atoms}}{52.0\ \cancel{g\,Cr}} = 0.681\text{ mol Cr atoms}$$

$$38.1\ \cancel{g\,O} \times \frac{1\text{ mol O atoms}}{16.0\ \cancel{g\,O}} = 2.38\text{ mol O atoms}$$

This gives the ratio of atoms but not in small whole numbers. We now need to do whatever mathematical operations are necessary to get these to small whole numbers. First, divide each one by the smallest value, as follows:

For K,

$$\frac{0.680}{0.680} = 1\text{ K}$$

For Cr,

$$\frac{0.681}{0.680} = 1\text{ Cr}$$

For O,

$$\frac{2.38}{0.680} = 3.5\text{ O}$$

Then we multiply each by two to eliminate the fraction:

$$1\text{ K} \times 2 = 2\text{ K}$$

$$1\text{ Cr} \times 2 = 2\text{ Cr}$$

$$3.5\text{ O} \times 2 = 7\text{ O}$$

Thus, the empirical formula is **$K_2Cr_2O_7$.**

PROBLEM EXAMPLE 9-6

A compound is found to contain 27.3 g of carbon, 4.55 g of hydrogen, and 36.4 g of oxygen. The molecular weight is found to be 90.0. Calculate the molecular formula.

SOLUTION:

First we calculate the empirical formula, as in the previous problem.

$$27.3\ \cancel{g\,C} \times \frac{1\text{ mol C}}{12.0\ \cancel{g\,C}} = 2.28\text{ mol C}$$

$$4.55\ \cancel{g\,H} \times \frac{1\text{ mol H}}{1.00\ \cancel{g\,H}} = 4.55\text{ mol H}$$

$$36.4\ \cancel{g\,O} \times \frac{1\text{ mol O}}{16.0\ \cancel{g\,O}} = 2.28\text{ mol O}$$

Dividing each by the smallest value (2.28) gives

For C,

$$\frac{2.28}{2.28} = 1\ \text{C}$$

For H,

$$\frac{4.55}{2.28} = 2\ \text{H}$$

For O,

$$\frac{2.28}{2.28} = 1\ \text{O}$$

Therefore, the empirical formula is CH_2O.
The mass of one of these "empirical formula units" is

$$\begin{aligned} C &= 1 \times 12.0 = 12.0 \\ H &= 2 \times 1.00 = 2.0 \\ O &= 1 \times 16.0 = 16.0 \\ \hline & \qquad\qquad\quad 30.0 \end{aligned}$$

If we divide this into the molecular weight, we find that there are three of these formula units in a molecule ($90.0 \div 30.0 = 3$). Therefore, the molecular formula is **$C_3H_6O_3$.**

9-4 Calculations Based on Balanced Equations

In Chapter 8 we considered the process of writing and balancing chemical equations. We are now going to use these balanced equations to calculate the amount of material produced or required in a given reaction. You will recall that a balanced chemical equation can be interpreted in terms of atoms and molecules, or in terms of moles. We shall find this to be a very useful concept and will thus do all calculations in terms of moles. For example, the reaction describing the combination of hydrogen gas and oxygen gas, which we have considered several times, is written as follows:

$$2\ H_{2(g)} + O_{2(g)} \longrightarrow 2\ H_2O_{(l)}$$

This is read as 2 moles of hydrogen gas combine with 1 mole of oxygen gas to form 2 moles of liquid water.

There are many practical applications to the types of calculation that we will be doing here. For example, if a particular process is to be carried out commercially to prepare a substance, the reactions are first carried out in the laboratory to find the best conditions under which the reactions will take place. Then the amount of product that is theoretically obtainable from a given amount of reactants is calculated. This **theoretical yield** is compared with the actual yield obtained in the laboratory to find what is called the **percent yield.** If the process is to be carried out commercially, the next step is for the chemical engineer to work out a small-scale operation of the process, called a *pilot plant,* and find the best conditions under which to carry out the process. We shall be concerned with the calculations involved in finding the theoretical yield for a given reaction.

In doing these calculations, we shall use the mole method. The central part of any calculation of this type is to determine the relative relationships between moles of reactants and products from the balanced equation. For example, the equation representing the burning of methane (CH_4) in oxygen to form carbon dioxide and water is written as follows:

$$CH_{4(g)} + 2\,O_{2(g)} \longrightarrow CO_{2(g)} + 2\,H_2O_{(g)}$$

This balanced equation gives us the relationships between moles of methane and moles of oxygen, moles of methane and moles of carbon dioxide, moles of carbon dioxide and moles of water, and so on. The following relationships can be obtained from this balanced equation:

$$1\text{ mol }CH_4 = 2\text{ mol }O_2$$
$$1\text{ mol }CH_4 = 1\text{ mol }CO_2$$
$$1\text{ mol }CO_2 = 2\text{ mol }H_2O$$

and so on.

In this chapter we shall only consider mass relationships (or weight-weight calculations) from balanced equations. Later we shall consider this again in terms of gases (Chapter 11), and solutions (Chapter 13).

In general, there are four steps in calculations of this sort, although in some problems some of these steps are not necessary.

1. Write a complete, *balanced* equation.
2. Convert from the given units to moles.
3. Convert from moles of the given quantity to moles of the desired quantity—from the balanced equation.
4. Convert from moles of the new quantity to the desired units, using formula weight, density, Avogadro's number, and so on.

Consider the following examples of calculations from balanced equations.

PROBLEM EXAMPLE 9-7

Calculate the number of moles of NaOH that are necessary to produce 7.5 moles of Na_2SO_4 from the reaction

$$2\,NaOH + H_2SO_4 \longrightarrow Na_2SO_4 + 2\,H_2O$$

SOLUTION:

Since the given quantity is in moles and the desired quantity is also in moles, there is just one step involved—the conversion from moles of NaOH to moles of Na_2SO_4, using the relationships from the balanced equation.

$$7.5\ \cancel{\text{mol Na}_2\text{SO}_4} \times \frac{2\ \text{mol NaOH}}{1\ \cancel{\text{mol Na}_2\text{SO}_4}} = \mathbf{15\ mol\ NaOH}$$

PROBLEM EXAMPLE 9-8

Calculate the number of grams of magnesium chloride that could be obtained from 8.50 g of hydrochloric acid when the latter is reacted with an excess of magnesium oxide.

SOLUTION:

This problem involves all four steps: writing the balanced equation, converting from grams of HCl to moles, converting from moles of HCl to moles of $MgCl_2$ (using the balanced equation), and converting from moles of $MgCl_2$ to grams. The balanced equation is

$$\text{MgO} + 2\ \text{HCl} \longrightarrow \text{H}_2\text{O} + \text{MgCl}_2$$

The problem says

$$8.50\ \text{g HCl} = ?\ \text{g MgCl}_2$$

The calculation is done as follows:

$$8.50\ \cancel{\text{g HCl}} \times \frac{1\ \cancel{\text{mol HCl}}}{36.5\ \cancel{\text{g HCl}}} \times \frac{1\ \cancel{\text{mol MgCl}_2}}{2\ \cancel{\text{mol HCl}}} \times \frac{95.3\ \text{g MgCl}_2}{1\ \cancel{\text{mol MgCl}_2}} = \mathbf{11.1\ g\ MgCl_2}$$

Limiting Reagent

In general, when a chemical reaction is carried out, one of the reagents will be used in excess of the amount needed. The reagent that is not present in excess is the one that will determine how much product can be obtained and is thus referred to as the **limiting reagent.** For example, in the reaction of hydrogen gas and oxygen gas to form water, we have seen that 2 moles of hydrogen will combine with 1 mole of oxygen to form 2 moles of water. If there is only 1 mole of oxygen present, and an excess of hydrogen, only 2 moles of water can be obtained. No matter how much hydrogen is present, the oxygen will limit the amount of water that can be produced. Thus, the oxygen is the limiting reagent.

The calculations involved here are exactly the same as those done previously. There are usually several calculations, however, that must be done in each problem. In general, the following steps can be used in doing calculations of this type:

1. Calculate the number of moles of product that could be obtained for each reagent given.

2. The reagent that gives the least number of moles of product is the limiting reagent and is the one that will determine the theoretical yield in the reaction; that is, no matter how much of the excess reagent is present, no more product can be obtained than that calculated from the limiting reagent.

3. Next, the moles of theoretical yield are converted to any other desired units, such as grams.

4. To find the amount of excess reagent present, if desired, we first calculate the amount of the excess reagent that will be used to produce the theoretical yield. The difference between this amount and the amount present to start is the amount of excess. The amount of excess reagent that is used can be calculated either from the theoretical amount of product obtainable, as calculated in Step 2, or from the amount of limiting reagent present. In either case, the mole relationships from the balanced equation are used to do the calculations.

Consider the following problem example.

PROBLEM EXAMPLE 9-9

A 50.0 g sample of calcium carbonate is reacted with 35.0 g of phosphoric acid. Calculate:
(a) The number of grams of calcium phosphate that could be produced.
(b) The number of grams of excess reagent that will remain.

SOLUTION:

The balanced equation is

$$3\ CaCO_3 + 2\ H_3PO_4 \longrightarrow Ca_3(PO_4)_2 + 3\ CO_2 + 3\ H_2O$$

(a) To calculate the theoretical yield, we first calculate the number of moles of $Ca_3(PO_4)_2$ that could be produced by each reagent, *if* it were all used up.

$$50.0\ \text{g}\ CaCO_3 \times \frac{1\ \text{mol}\ CaCO_3}{100\ \text{g}\ CaCO_3} \times \frac{1\ \text{mol}\ Ca_3(PO_4)_2}{3\ \text{mol}\ CaCO_3} = 0.167\ \text{mol}\ Ca_3(PO_4)_2$$

$$35.0\ \text{g}\ H_3PO_4 \times \frac{1\ \text{mol}\ H_3PO_4}{98.0\ \text{g}\ H_3PO_4} \times \frac{1\ \text{mol}\ Ca_3(PO_4)_2}{2\ \text{mol}\ H_3PO_4} = 0.179\ \text{mol}\ Ca_3(PO_4)_2$$

This tells us that $CaCO_3$ is the limiting reagent, and the maximum amount of product that we can obtain is 0.167 mole of $Ca_3(PO_4)_2$. To find the number of grams that can be produced, we convert 0.167 mole of $Ca_3(PO_4)_2$ to grams:

$$0.167\ \text{mol}\ Ca_3(PO_4)_2 \times \frac{310\ \text{g}\ Ca_3(PO_4)_2}{1\ \text{mol}\ Ca_3(PO_4)_2} = \mathbf{51.8\ g\ Ca_3(PO_4)_2}$$

This is the theoretical yield.

(b) To find the grams of excess reagent, we must first find how much of the reagent that is in excess (H_3PO_4) is used. Then we subtract this from the amount of H_3PO_4 present at the beginning, to find how much excess remains. We can either start with the theoretical yield [0.167 mol $Ca_3(PO_4)_2$] and calculate how much H_3PO_4 it would take to produce this or start with the limiting reagent (50.0 g $CaCO_3$) and find how much H_3PO_4 will react with that. Either way, we find the same result. We shall start with the 0.167 mole of $Ca_3(PO_4)_2$.

$$0.167 \text{ mol } Ca_3(PO_4)_2 \times \frac{2 \text{ mol } H_3PO_4}{1 \text{ mol } Ca_3(PO_4)_2} \times \frac{98.0 \text{ g } H_3PO_4}{1 \text{ mol } H_3PO_4} = 32.7 \text{ g } H_3PO_4$$

This is the amount of H_3PO_4 used; therefore, the amount remaining is 35.0 g − 32.7 g = **2.30 g** in excess.

Percent Yield

All the calculations that have been done up to this point are theoretical yields. This is the maximum amount of product that can be obtained in a particular reaction. In general, however, when these reactions are carried out in the laboratory, we do not actually obtain the theoretical amount of product. This is especially true in organic reactions, where there are often side reactions taking place and where some of the products are lost in the process of isolation and purification. The amount of product that is actually obtained is called the *actual yield.* The percent yield is the percent of the theoretical yield that is actually obtained, calculated as follows:

$$\text{percent yield} = \frac{\text{actual yield}}{\text{theoretical yield}} \times 100$$

Consider the following examples.

PROBLEM EXAMPLE 9-10

If 35.4 g $Ca_3(PO_4)$ is actually obtained in Problem Example 9-9, what is the percent yield?

SOLUTION:

Actual yield = 35.4 g and theoretical yield = 51.8 g (from Problem Example 9-9); therefore,

$$\text{percent yield} = \frac{35.4 \text{ g}}{51.8 \text{ g}} \times 100 = \mathbf{68.3\%}$$

PROBLEM EXAMPLE 9-11

If 15.0 g of MgO is treated with 18.5 g of H_3PO_4 and 17.6 g of $Mg_3(PO_4)_2$ is obtained, calculate the percent yield and the grams of excess reagent.

SOLUTION:

The balanced equation is

$$3\ \mathrm{MgO} + 2\ \mathrm{H_3PO_4} \longrightarrow \mathrm{Mg_3(PO_4)_2} + 3\ \mathrm{H_2O}$$

(a) Find the limiting reagent:

$$15.0\ \mathrm{g\ MgO} \times \frac{1\ \mathrm{mol\ MgO}}{40.3\ \mathrm{g\ MgO}} \times \frac{1\ \mathrm{mol\ Mg_3(PO_4)_2}}{3\ \mathrm{mol\ MgO}} = 0.124\ \mathrm{mol\ Mg_3(PO_4)_2}$$

$$18.5\ \mathrm{g\ H_3PO_4} \times \frac{1\ \mathrm{mol\ H_3PO_4}}{98.0\ \mathrm{g\ H_3PO_4}} \times \frac{1\ \mathrm{mol\ Mg_3(PO_4)_2}}{2\ \mathrm{mol\ H_3PO_4}} = 0.0944\ \mathrm{mol\ Mg_3(PO_4)_2}$$

Therefore, H_3PO_4 is the limiting reagent.

(b) Calculate the theoretical yield, using the limiting reagent.

$$0.0944\ \mathrm{mol\ Mg_3(PO_4)_2} \times \frac{262.9\ \mathrm{g\ Mg_3(PO_4)_2}}{1\ \mathrm{mol\ Mg_3(PO_4)_2}} = 24.8\ \mathrm{g\ Mg_3(PO_4)_2}$$

(c) Calculate the amount of excess reagent.
There are several ways that this can be done—but any of these are equivalent; that is, we can find how many grams of MgO (the reagent present in excess) will react with 18.5 g H_3PO_4 (the limiting reagent) or we can calculate how many grams of MgO would be used in forming 0.0944 mole of $Mg_3(PO_4)_2$. In either case, after finding the amount used, we find the amount of excess.
Using the 0.0944 mole $Mg_3(PO_4)_2$ gives

$$0.0944\ \mathrm{mole\ Mg_3(PO_4)_2} \times \frac{3\ \mathrm{moles\ MgO}}{1\ \mathrm{mole\ Mg_3(PO_4)_2}} \times \frac{40.3\ \mathrm{g\ MgO}}{1\ \mathrm{mole\ MgO}} = 11.4\ \mathrm{g\ MgO}$$

If we have 15.0 g MgO to start and use 11.4 g, the amount of excess is (15.0 − 11.4) = **3.6 g MgO** in excess.

(d) Calculate the percent yield.
To do this we use the theoretical yield calculated in Step (b) and the actual yield given in the problem.

$$\text{percent yield} = \frac{\text{actual yield}}{\text{theoretical yield}} = \frac{17.6\ \mathrm{g}}{24.8\ \mathrm{g}} \times 100 = \mathbf{71.0\%}$$

SUMMARY

1. The *percent* of each element in a compound is found by dividing the mass of each element by the molecular or formula weight and multiplying by 100.
2. From the formula of a compound it is possible to do quantitative calculations involving *stoichiometric (chemically equivalent) quantities.* These are based either on the relationship between moles of each atom present or simply from the weight relationships (which are calculated from the formula and the atomic weights of the elements present).
3. From the mass relationships of the elements in a compound, or the percent composition, the *empirical (simplest) formula* can be calcu-

lated. This is done by converting from mass relationships to mole relationships (using atomic weights) and then converting the ratio to the smallest possible whole number ratio mathematically.

4. If the molecular weight of the compound is known, the *molecular formula* can be calculated from the empirical formula by finding the relative mass of the empirical formula (formula weight of the empirical formula) and dividing that into the molecular weight, to find how many of these empirical units there are in a formula unit (or molecule). If this turns out to be one, which it does for most inorganic compounds, it simply means that the empirical formula and molecular formula are the same.
5. To do stoichiometric calculations from a balanced equation the units given are first converted to moles (if necessary), then moles of the given quantity are converted to moles of the desired quantity (using the mole relationships from the balanced equation), and finally moles of the desired quantity are converted to any other units (if necessary).
6. If the amounts of two or more reagents are given, the first step is to calculate the *limiting reagent,* by finding which reagent would give the least amount of product (as outlined in item 5 above). This amount of product is the *theoretical yield.* Along with this calculation, the amount of excess reagent can also be calculated, again using the procedure as outlined in item 5 above.
7. The *percent yield* can be calculated if the actual (experimentally determined) yield is known. This is done by dividing the actual yield by the theoretical yield and multiplying by 100.

Upon completion of this chapter, you should be able to do the following things (assume a periodic table with atomic weights is available).

1. Given the formula of a compound, calculate the percent composition or percent of a given element. (1, 2)
2. Given the amounts of the different elements in a compound, calculate the percent of any one element in the compound. (3)
3. Given the formula of a compound, do stoichiometric calculations based on the mass relationships from the formula. (4, 5)
4. Given the mass relationships in a compound (either in terms of percent or mass units), calculate (a) the empirical formula of the compound and (b) if the molecular weight is given, calculate the molecular formula. (6, 7, 8)
5. Given a chemical equation (or the reactants for a given equation), do stoichiometric calculations from the balanced equation and express the answer in any units requested. (9, 10)
6. Given the amounts of more than one reactant in a calculation of the type listed in item 5, calculate the limiting reagent, the theoretical yield (in any units called for), and the amount of excess reagent (in any units called for). (11, 12, 13)

7. Given the information listed in items 5 or 6 along with the actual yield from a reaction, calculate the percent yield. (14, 15)
8. Given any of the terms listed in Exercise 1, give a brief definition or explanation in your own words.

EXERCISES

1. Define or explain the following terms:
   (a) percent composition
   (b) stoichiometry
   (c) empirical formula
   (d) molecular formula
   (e) limiting reagent
   (f) theoretical yield
   (g) excess reagent
   (h) percent yield
2. Distinguish between
   (a) empirical and molecular formula
   (b) theoretical and actual yield

PROBLEMS

Percent composition

1. Calculate the percent composition of each of the following:
   (a) NaOH
   (b) $Mg(OH)_2$
   (c) $Ca(NO_3)_2$
   (d) KOH
   (e) $Ba(NO_3)_2$
   (f) $(NH_4)_2S$
2. Calculate the percent of oxygen in each of the following:
   (a) sulfuric acid
   (b) barium hydrogen carbonate
   (c) ammonium phosphate
   (d) stannous nitrate
   (e) iron(II) chlorate
   (f) magnesium hydroxide
   (g) nitric acid
   (h) calcium hydrogen sulfate
   (i) lead(II) phosphate
   (j) ammonium bicarbonate
3. Calculate the percent of metal in each of the following:
   (a) 0.45 g of metal combines with 0.32 g of oxygen
   (b) 0.85 g of metal sulfide contains 240 mg of sulfur
   (c) a 7.50-g sample of impure iron ore contains 6.00 g of FeO
   (d) 2.5 g of metal combine with 1.8 g of sulfur
   (e) 0.42 g of metal oxide contains 120 mg of oxygen
   (f) a 3.50-g sample of lye contains 2.75 g of sodium hydroxide

Relationships from the formula of a compound

4. Calculate the number of
   (a) grams of calcium in 8.50 g of calcium phosphate
   (b) grams of silver in 1.25 moles of silver chloride
   (c) grams of zinc nitrate containing 0.35 g of zinc
   (d) milligrams of tin in 0.180 g of tin(II) oxide
   (e) grams of iron in 4.20 mmol of ferrous phosphate
   (f) grams of aluminum in 0.75 g of aluminum sulfate

(g) grams of potassium in 6.3 moles of potassium phosphate
(h) grams of calcium bromide containing 1.4 g of calcium
(i) grams of tin in 250 mg of stannic carbonate
(j) milligrams of lead in 1.50 mmol of lead(II) oxide

5. Calculate the number of grams of oxygen in each of the following:
(a) 5.2 moles of sodium sulfite
(b) 0.48 g of zinc carbonate
(c) 1.75 lb of $Ca(ClO_4)_2$
(d) 12.5 mmol of calcium phosphite
(e) 3.5×10^{-2} μg of ferrous sulfate
(f) 8.25 g of sodium carbonate
(g) 1.5 moles of ammonium chlorate
(h) 0.45 lb of Mn_2O_3
(i) 25 mmol of chromium(III) oxide
(j) 8.3×10^{-3} μg of stannic hydroxide

Empirical and molecular formula

6. Determine the empirical formula of each of the following:
(a) A compound that contains 25.9% iron and 74.1% bromine.
(b) A compound that is 44.9% potassium, 18.4% sulfur, and 36.7% oxygen.
(c) A compound that contains 43.66% manganese and 56.34% chlorine.
(d) A compound composed of 4.64 g of magnesium and 1.78 g of nitrogen.
(e) A compound composed of 2.63 mg of calcium, 4.22 mg of sulfur, and 3.15 mg of oxygen.
(f) A compound that contains 62.6% tin and 37.4% chlorine.
(g) A compound that contains 33.4% calcium, 26.7% sulfur, and 39.9% oxygen.
(h) A compound that contains 32.8% chromium and 67.2% chlorine.
(i) A compound that contains 2.55 g of aluminum, 8.78 g of phosphorus, and 4.53 g of oxygen.
(j) A compound composed of 2.02 mg of magnesium, 5.33 mg of sulfur, and 2.66 mg of oxygen.

7. Calculate the molecular formula of each of the following compounds:
(a) empirical formula = CH
molecular weight = 39.0
(b) empirical formula = HgCl
molecular weight = 472.2
(c) empirical formula = C_2H_5
molecular weight = 58.0
(d) empirical formula = CH
molecular weight = 65.0

8. Calculate the molecular formula of each of the following compounds:
(a) A compound consists of 32.3% sodium, 45.1% sulfur, and 22.5% oxygen; the molecular weight is 142.2.
(b) A compound contains 6.44 g of nitrogen and 14.72 g of oxygen and has a molecular weight of 92.0.
(c) A compound contains 82.8% carbon and 17.2% hydrogen and has a molecular weight of 58.0.
(d) A compound consists of 85.7% carbon and 14.3% hydrogen and has a molecular weight of 84.0.

(e) A compound consists of 44.8% potassium, 36.8% sulfur, and 18.3% oxygen; the molecular weight is 174.4.

(f) A compound contains 34.5 g of carbon and 5.76 g of hydrogen and has a molecular weight of 56.0

Calculations from balanced equations

9. For the reaction

$$4\,FeS + 7\,O_2 \longrightarrow 2\,Fe_2O_3 + 4\,SO_2$$

(a) calculate the number of moles of Fe_2O_3 that could be produced from 7.20 moles of FeS.
(b) calculate the number of grams of SO_2 that could be produced from 3.25 moles of oxygen.
(c) calculate the number of grams of oxygen that would react with 0.125 g of FeS.
(d) if 18.5 g of Fe_2O_3 are produced in a reaction, how many moles of SO_2 would be produced?
(e) calculate the number of milligrams of FeS that would be necessary to produce 185 lb of SO_2.
(f) calculate the number of moles of O_2 that would react with 4.3 moles of FeS.
(g) calculate the number of grams of oxygen necessary to produce 0.180 mole of Fe_2O_3.
(h) calculate the number of grams of FeS that would react with 8.35 g of oxygen.
(i) if 2.38 moles of SO_2 are produced in a reaction, how many grams of Fe_2O_3 would be produced?
(j) calculate the number of pounds of oxygen that would be needed to produce 0.150 mg of SO_2.

10. For the reaction

$$2\,C_2H_6 + 7\,O_2 \longrightarrow 4\,CO_2 + 6\,H_2O$$

(a) how many moles of C_2H_6 would be needed to produce 1.85 g of CO_2?
(b) how many grams of oxygen would be needed to produce 9.25 g of CO_2?
(c) how many millimoles of CO_2 could be produced from 6.8 lb of C_2H_6?
(d) how many milligrams of CO_2 would be produced in a reaction that forms 7.25×10^3 lb of water?
(e) how many liters of liquid water could be produced from 2.5×10^{-2} kg of oxygen?
(f) how many moles of oxygen will react with 4.72 moles of C_2H_6?
(g) how many grams of CO_2 could be produced from 42.5 g of C_2H_6?
(h) how many pounds of CO_2 could be produced from 4.5 moles of C_2H_6?
(i) how many kilograms of C_2H_6 would react with 0.24 mmol of oxygen?
(j) how many milligrams of C_2H_6 would be needed to produce 3.5 *l* of liquid water?

Limiting reagent and excess reagent

11. For each of the following, calculate the number of grams of $Ca_3(PO_4)_2$ that could be produced, given the following equation:

$$3\ CaCO_3 + 2\ H_3PO_4 \longrightarrow Ca_3(PO_4)_2 + 3\ CO_2 + 3\ H_2O$$

(a) from 12.5 g of $CaCO_3$ and 17.3 g of H_3PO_4
(b) from 3.25 moles of $CaCO_3$ and 475 g of H_3PO_4
(c) from 3.25 mmol of H_3PO_4 and 52.8 mg of $CaCO_3$
(d) from 1.85 moles of $CaCO_3$ and 325 g of H_3PO_4
(e) from 0.275 g of $CaCO_3$ and 0.425 g of H_3PO_4
(f) from 4.25 mmol of $CaCO_3$ and 125 mg of H_3PO_4

12. Calculate the number of grams of excess reagent present in each part of Problem 11.

13. For the reaction

$$2\ Al + 3\ H_2SO_4 \longrightarrow Al_2(SO_4)_3 + 3H_2$$

if 2.25 g of Al is added to 4.36 g of H_2SO_4, calculate
(a) how many grams of $Al_2(SO_4)_3$ would be formed.
(b) how many moles of hydrogen would be formed.
(c) how many grams of excess reagent will remain after the reaction is complete.
(d) how many moles of $Al_2(SO_4)_3$ would be formed.
(e) how many grams of hydrogen would be formed.
(f) how many moles of excess reagent will remain after the reaction is complete.

Percent yield

14. Calculate the percent yield in each of the following:
(a) The theoretical yield in a reaction is 8.50 g and the actual yield is 7.35 g.
(b) 12.5 g of $CaCO_3$ (limestone) is heated and 4.20 g of CaO (lime) is obtained.
(c) 7.50 g of nitrogen is reacted with 1.30 g of hydrogen and 5.75 g of ammonia is produced.
(d) The theoretical yield in a reaction is 49.3 g and the actual yield is 37.2 g.
(e) 3.28 g of $BaCl_2 \cdot 2\ H_2O$ is heated and 2.16 g of $BaCl_2$ is obtained.
(f) 4.30 moles of nitrogen are reacted with 9.20 moles of hydrogen and 98.0 g of ammonia are produced.

15. For the reaction

$$CaO + 2\ HCl \longrightarrow H_2O + CaCl_2$$

(a) if 1.23 g of CaO is reacted with excess hydrochloric acid and 1.85 g of $CaCl_2$ is formed, what is the percent yield?
(b) if 30.2 g of CaO is added to 34.5 g of HCl and 6.35 g of water is formed, what is the percent yield?

(c) if 25.4 g of CaO is reacted with excess hydrochloric acid and 34.8 g of $CaCl_2$ is formed, what is the percent yield?

(d) if 3.65 g of CaO is added to 4.02 g of HCl and 5.85 g of $CaCl_2$ is formed, what is the percent yield?

General problems

16. If calcium carbonate is reacted with nitric acid, calculate
    (a) how many moles of calcium nitrate could be formed from 15.2 g of calcium carbonate.
    (b) how many milligrams of CO_2 could be produced from 3.85 g of calcium carbonate.
    (c) the percent yield if 1.15 g of calcium carbonate is used and 1.23 g of calcium nitrate is formed.
    (d) the number of grams of excess reagent remaining if 12.2 g of calcium carbonate is combined with 16.3 g of HNO_3.
    (e) how many grams of calcium carbonate would react with 36.2 m*l* of 15.0% nitric acid (density = 1.12 g/m*l*).
    (f) how many grams of nitric acid would react with 3.26 moles of calcium carbonate.
    (g) how many kilograms of $CaCO_3$ would be needed to produce 285 g of calcium nitrate.
    (h) the percent yield if 24.2 g of calcium carbonate is reacted with excess acid and 18.5 g of calcium nitrate is produced.
    (i) the number of grams of excess reagent remaining if 3.75 g of calcium carbonate is combined with 4.63 g of HNO_3.
    (j) how many milliliters of 26.0% nitric acid would react with 16.5 g of calcium carbonate (density of nitric acid = 1.15 g/m*l*)

17. Calculate the number of grams of oxygen that will react with each of the following:
    (a) 14.2 g of calcium to form CaO
    (b) 1.75 g of aluminum to form aluminum oxide
    (c) 0.13 lb of iron to form iron(III) oxide
    (d) 4.34 moles of tin to form stannous oxide
    (e) 2.82 g of barium to form BaO
    (f) 24.2 g of silver to form silver oxide
    (g) 1.12 lb of lead to form lead(II) oxide
    (h) 0.54 mole of iron to form ferric oxide

18. Calculate the molecular formula of each of the following compounds:
    (a) A bromide of mercury that contains 71.5% mercury has a molecular weight of 562.
    (b) A chloride of mercury that contains 15.0% chlorine has a molecular weight of 472.2.

REFERENCES:

BENSON, SIDNEY W., *Chemical Calculations,* 3rd ed. New York: John Wiley & Sons, Inc., 1971.

PETERS, EDWARD I., *Problem Solving for Chemistry,* 2nd ed. Philadelphia: W. B. Saunders Co., 1977.

Objectives

1. To understand what a thermochemical equation is and what the terms *exothermic* and *endothermic* mean.
2. To be able to do stoichiometric calculations involving heat transfer from thermochemical equations.
3. To understand what enthalpy (ΔH) represents and how this is related to total energy change (ΔE).
4. To understand what ΔH of reaction (ΔH_r) and ΔH of formation (ΔH_f) represent.
5. To understand the relationship between ΔH_r and ΔH_f of products and reactants and to be able to do calculations involving this relationship.
6. To understand what a state function is and to be able to explain why ΔH is a state function.
7. To learn what Hess's Law is and to be able to do calculations involving this law.
8. To understand how calorimetric measurements are made and the basic principles of a calorimeter.
9. To learn the basic principles of the First Law of Thermodynamics and to understand how this is related to the principles of thermochemistry.
10. To understand what entropy is.
11. To learn the basic principles of the Second Law of Thermodynamics and to understand how the free energy change can be calculated and used to predict spontaneity of reaction.

Ten

Thermochemistry

Along with the weight (mass) relationships from balanced chemical equations, which were considered in Chapter 9, energy relationships are also important. In this chapter we shall consider the amount of heat either evolved or absorbed by a chemical reaction.

There are also other forms of energy that are sometimes involved in chemical reactions, such as light, mechanical energy, and electrical energy. In this chapter, however, we shall be concerned only with energy in the form of heat.

Also, the energy change in a chemical reaction is related to the equilibrium conditions. Therefore, by studying the energy changes it is possible to predict whether a reaction will take place or not and to determine the conditions under which the reaction will proceed best.

10-1 Thermochemical Equations

A **thermochemical equation** is a chemical equation that includes the amount of energy, in the form of heat, that is transferred in the reaction. This heat, usually expressed in kilocalories (kcal) or kilojoules (kJ), can be considered as a product or a reactant in the balanced equation. A reaction that gives off heat is called an **exothermic** reaction and one that absorbs heat is called an **endothermic** reaction. The amount of heat either absorbed or released in a reaction is called the *heat flow,* designated as Q.

For any thermodynamic system, *if heat is given off by the system, the heat flow is given a negative sign; if heat is absorbed by the system, it is positive.* For example, in an exothermic reaction, the energy of the surroundings is increased due to the heat given off and thus the energy of the system (the chemical reaction) has decreased; therefore, Q is negative. For an endothermic reaction, the system is increasing in heat energy and thus Q is positive. This is shown in Fig. 10-1.

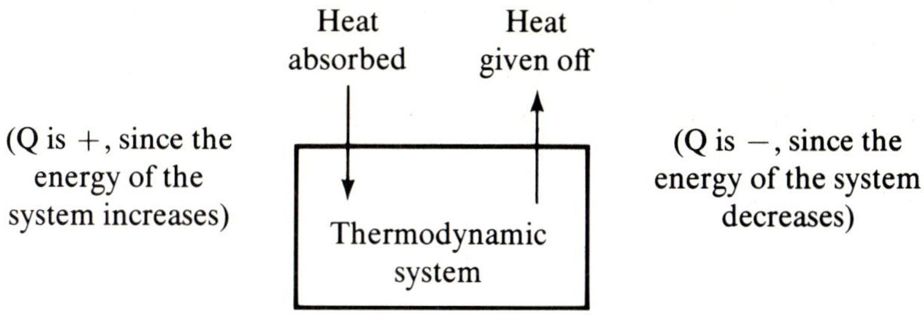

Figure 10-1 Convention for designating heat flow in a thermodynamic system

Thermochemical equations are like any other chemical equations in that the heat flow is simply part of the total equation, like the other products and reactants. Therefore, the amount of heat is handled in calculations stoichiometrically just like the other reactants and products from the balanced equation. Also, thermochemical equations can be handled like algebraic equations. For example, from the equation

$$2\,H_{2(g)} + O_{2(g)} \longrightarrow 2\,H_2O_{(l)} + 136 \text{ kcal}$$

there are several relationships that can be obtained, such as the following:

1. Stoichiometric relationships between moles of products or reactants and amount of heat, such as,

   $$2 \text{ moles } H_2 = 136 \text{ kcal} \quad \text{or} \quad 2 \text{ moles } H_2O = 136 \text{ kcal}$$

2. For the reverse reaction, 136 kcal of heat are *absorbed* (an endothermic reaction).
3. If the total equation is divided by 2, the following equation is obtained:

   $$H_{2(g)} + \tfrac{1}{2}\,O_{2(g)} \longrightarrow H_2O_{(l)} + 68 \text{ kcal}$$

Consider the following problem example.

PROBLEM EXAMPLE 10-1

(a) Calculate the amount of heat produced from the combustion of 10.0 g of C_2H_6, which burns in oxygen according to the following equation:

$$2\,C_2H_{6(g)} + 7\,O_{2(g)} \longrightarrow 4\,CO_{2(g)} + 6\,H_2O_{(l)} + 750 \text{ kcal} \quad (3135 \text{ kJ})$$

(b) What is the heat flow (Q)?
(c) Is this reaction exothermic or endothermic?
(d) Calculate Q for the reaction

$$C_2H_6 + \tfrac{7}{2}\,O_2 \longrightarrow 2\,CO_2 + 3\,H_2O$$

SOLUTION:

(a) In equations of this type it is understood that moles are being considered, not atoms and molecules; that is, *2 moles* of C_2H_6 will produce 750 kcal of heat.

$$10.0\ \text{g } C_2H_6 \times \frac{1\ \text{mol } C_2H_6}{30.0\ \text{g } C_2H_6} \times \frac{750\ \text{kcal (3135 kJ)}}{2\ \text{mol } C_2H_6} = \mathbf{125\ kcal} \quad (522\ \text{kJ})$$

(b) Thus, 125 kcal (522 kJ) of heat are given off by the system. Therefore,

$$Q = \mathbf{-125\ kcal} \quad (-522\ \text{kJ})$$

(c) Therefore, the reaction is **exothermic**—heat is given off (and the sign on the heat flow is negative).

(d) Since the original equation has been divided by 2, Q for this equation is $-\frac{750}{2}$ kcal = **−375 kcal.**

10-2 Enthalpy (ΔH)

The heat change, or change in heat content, in a chemical reaction is often referred to as the change in **enthalpy.** This is designated by the notation ΔH, where Δ indicates *change of* (and is read "delta" H). It is not possible to measure the actual heat content of any particular substance. The best we can do is measure the apparent changes in heat content when reactants are transformed into products. This change in heat content is thus what is commonly referred to as ΔH, the change in enthalpy of the system. For example, in the reaction

$$2\ H_{2(g)} + O_{2(g)} \longrightarrow 2\ H_2O_{(l)} + 136\ \text{kcal}$$

the difference in heat content between 2 moles of H_2 gas plus 1 mole of O_2 gas and 2 moles of H_2O liquid is 136 kcal. Therefore, ΔH for this equation is -136 kcal since heat is given off by the system (exothermic reaction).

ΔH is a **state function.** This means that it is not dependent on the path by which the reaction takes place; that is, if the reaction takes place in one step or in several successive steps, the overall change in enthalpy is the same. Another example of a state function is potential energy. A rock on a cliff has a certain amount of potential energy, no matter what path it takes getting down. On the other hand, the *distance* between two points is not a state function since this does depend on the path followed in going from one point to the other. This is shown in Fig. 10-2.

ΔE versus ΔH

The energy change in a chemical reaction is designated by ΔE. If a reaction is carried out at constant volume, this would be the measured amount of energy released in a particular exothermic reaction. We generally carry out

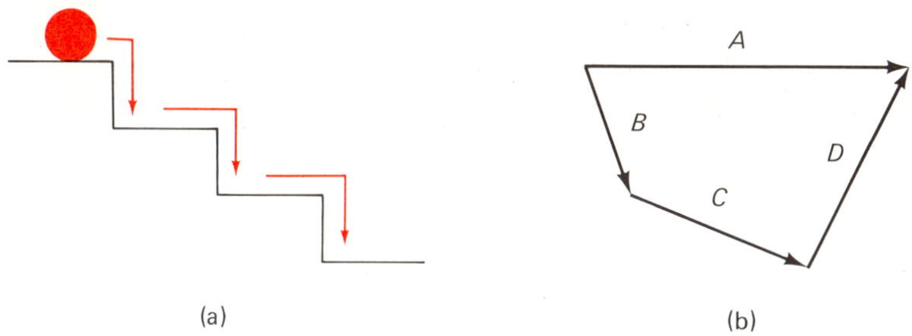

Figure 10-2 *Potential energy versus distance between two points. (a) Total potential energy is equal to the sum of the three steps (a state function) (b) Distance* A *does* not *equal* B + C + D *(not a state function)*

reactions at constant pressure, however, and not at constant volume.* Therefore, some of the energy produced actually is doing work against the outside force, such as the pressure of the atmosphere. Thus it is only the excess energy that is given off as heat. As an analogy, we could compare ΔE and ΔH to salary versus take-home pay. ΔE is the total amount (salary) and ΔH is the amount actually available (take-home pay). We shall have more to say about this in Sections 10-6 and 10-7.

Heat of Reaction and Heat of Formation

From what was stated above, we can say that the heat flow (Q) in a reaction at constant pressure is equal to the change in enthalpy for the reaction. This is called the *heat of reaction,* or ΔH of reaction (ΔH_r).

The *heat of formation* of a compound (ΔH_f) is defined as the ΔH_r for the formation of 1 mole of the compound from the elements, in the standard state. The *standard state* is defined as the most stable form of an element, such as $O_{2(g)}$ for oxygen. The heat content of a free element, in the standard state, is taken as zero.

For example, to find the heat of formation (ΔH_f) of CO_2, we can measure the heat of reaction (ΔH_r) for the reaction:

$$C_{(s)} + O_{2(g)} \longrightarrow CO_{2(g)}$$

At 25°C and 1 atm (atmosphere) pressure, this is found experimentally to be -94.1 kcal (-393 kJ); that is, 94.1 kcal of heat are given off in this reaction. Therefore, ΔH_f of $CO_{2(g)}$ is -94.1 kcal/mol. Table 10-1 gives the heat of formation of several compounds.

Consider the following problem example.

PROBLEM EXAMPLE 10-2

Calculate ΔH_f of carbon monoxide from the following reaction:

$$2\,C_{(s)} + O_{2(g)} \longrightarrow 2\,CO_{(g)} + 52.8 \text{ kcal} \quad (221 \text{ kJ})$$

*We shall see in Chapter 11 that if the number of moles of gas present in a system is increased, the pressure increases, and vice versa. Therefore, in a closed system (constant volume) the pressure often changes when a reaction is carried out.

TABLE 10-1 *Heat of formation of some common substances, at 25°C and 1 atm pressure*

COMPOUND	ΔH kcal/mol	ΔH kJ/mol
$H_2O_{(g)}$	−57.8	−242
$H_2O_{(l)}$	−68.3	−285
$NO_{(g)}$	21.5	89.9
$NO_{2(g)}$	8.0	33.4
$NH_{3(g)}$	−11.0	−46.0
$NH_4NO_{3(s)}$	−87.3	−365
$HNO_{3(aq)}$	−41.4	−173
$SO_{2(g)}$	−71.0	−297
$SO_{3(g)}$	−94.5	−395
$H_2SO_{4(aq)}$	−217	−907
$CO_{(g)}$	−26.4	−110
$CO_{2(g)}$	−94.1	−393
$CH_3OH_{(l)}$	−57.0	−238
$C_{12}H_{22}O_{11(s)}$	−530	−2215
$C_2H_{2(g)}$	54.2	227
$C_2H_{4(g)}$	12.5	52.2
$C_2H_{6(g)}$	−20.2	−84.4

SOLUTION:

The ΔH_r for the reaction as written is −52.8 kcal (−221 kJ). This is not the ΔH_f of CO, however, since the reaction as written involves the formation of 2 moles of CO. Therefore, to calculate ΔH_f, we must divide by 2 through the equation to give

$$C_{(s)} + \tfrac{1}{2}O_{2(g)} \longrightarrow CO_{(g)} + 26.4 \text{ kcal} \quad (110 \text{ kJ})$$

Thus, ΔH_f of CO = **−26.4 kcal/mol** (−110 kJ/mol).

Total Energy Change in a Reaction

In Chapter 3 it was pointed out that energy cannot be created nor destroyed. Therefore, the total amount of energy change in a reaction, which is ΔH_f at constant pressure, is equal to the difference between the ΔH_f of products and ΔH_f of reactants; that is, the total heat content of the products minus the total heat content of the reactants is equal to the heat of reaction. This can be stated as

$$\Delta H_r = \Sigma\, \Delta H_f \text{ products} - \Sigma\, \Delta H_f \text{ reactants} \qquad (10\text{-}1)$$

In this equation, the symbol Σ means *the sum of.*

Using the relationship given in Eq. (10-1), the heat of reaction can be calculated from the heat of formation of the products and reactants. Likewise, if the heat of formation of all but one of the products or reactants is known, the remaining heat of formation can be calculated if ΔH_r is known. Thus, from a table of ΔH_f values of some common compounds, many other ΔH_f values can be calculated by measuring the ΔH_r for reactions involving these compounds. Consider the following problem examples.

PROBLEM EXAMPLE 10-3

Calculate ΔH_r for the following reaction:

$$2\,CH_3OH_{(l)} + 3\,O_{2(g)} \longrightarrow 2\,CO_{2(g)} + 4\,H_2O_{(l)}$$

SOLUTION:

Using the ΔH_f values from Table 10-1, Eq. (10-1) gives

$$\Delta H_r = 2\,(\Delta H_f\,CO_2) + 4\,(\Delta H_f\,H_2O) - [2\,(\Delta H_f\,CH_3OH)]$$

$$= 2\,\cancel{mol}\left(-94.1\,\frac{kcal}{\cancel{mol}}\right) + 4\,\cancel{mol}\left(-68.3\,\frac{kcal}{\cancel{mol}}\right) - \left[2\,\cancel{mol}\left(-57.0\,\frac{kcal}{\cancel{mol}}\right)\right]$$

$$= -188.2\text{ kcal} - 273.2\text{ kcal} + 114.0\text{ kcal} = \mathbf{-347.4\ kcal}$$

Note that the coefficients from the balanced equation are used. Also, recall that ΔH_f of free elements in the standard state is taken as zero.

PROBLEM EXAMPLE 10-4

Calculate ΔH_f for C_4H_{10} from the following reaction:

$$2\,C_4H_{10(g)} + 13\,O_{2(g)} \longrightarrow 8\,CO_{2(g)} + 10\,H_2O_{(g)} + 1271\text{ kcal}$$

SOLUTION:

$$\Delta H_r = 8\,(\Delta H_f\,CO_2) + 10\,(\Delta H_f\,H_2O) - [2\,(\Delta H_f\,C_4H_{10})]$$

$$-1271\text{ kcal} = 8\text{ mol}\left(-94.1\,\frac{kcal}{mol}\right) + 10\text{ mol}\left(-57.8\,\frac{kcal}{mol}\right) - [2\text{ mol}\,(\Delta H_f\,C_4H_{10})]$$

$$-1271\text{ kcal} = 8(-94.1)\text{ kcal} + 10(-57.8)\text{ kcal} - 2\text{ mol}\,(\Delta H_f\,C_4H_{10})$$

$$-1271\text{ kcal} + 752.8\text{ kcal} + 578\text{ kcal} = -2\text{ mol}\,(\Delta H_f\,C_4H_{10})$$

$$2\text{ mol}\,(\Delta H_f\,C_4H_{10}) = (1271 - 752.8 - 578)\text{ kcal}$$

$$2\text{ mol}\,(\Delta H_f\,C_4H_{10}) = -59.8\text{ kcal}$$

$$\Delta H_f\,C_4H_{10} = \frac{-59.8\text{ kcal}}{2\text{ mol}} = \mathbf{-29.9\ kcal/mol}$$

10-3 Hess's Law

Since ΔH is a state function, the ΔH of reaction for the combination of two or more reactions is equal to the sum of the ΔH_r values for each individual reaction. This is known as **Hess's Law** and simply tells us that if we add two or more equations to obtain an overall equation, the ΔH_r is the sum of all the parts. For example, from the two reactions

$$Sn_{(s)} + Cl_{2(g)} \longrightarrow SnCl_{2(s)} + 84 \text{ kcal}$$
$$SnCl_{2(s)} + Cl_{2(g)} \longrightarrow SnCl_{4(s)} + 47 \text{ kcal}$$

we can find ΔH_f of $SnCl_4$ by adding the two reactions to give

$$Sn_{(s)} + SnCl_{2(s)} + 2\ Cl_{2(g)} \longrightarrow SnCl_{2(s)} + SnCl_{4(s)} + (84 + 47) \text{ kcal}$$

or

$$Sn_{(s)} + 2\ Cl_{2(g)} \longrightarrow SnCl_{4(s)} + 131 \text{ kcal}$$

Therefore, ΔH_f of $SnCl_4$ is -131 kcal (-548 kJ).

Bond Energies

The bond energy in a compound is the amount of energy necessary to break the bonds in the compound and form atoms. This can be calculated by using Hess's Law and some known ΔH_r values. Consider the following example.

PROBLEM EXAMPLE 10-5

Given the following equations, calculate the bond energy in ClF:

$$F_{2(g)} + Cl_{2(g)} \longrightarrow 2\ ClF_{(g)} + 27 \text{ kcal}$$
$$F_{2(g)} \longrightarrow 2\ F_{(g)} - 37 \text{ kcal}$$
$$Cl_{2(g)} \longrightarrow 2\ Cl_{(g)} - 58 \text{ kcal}$$

SOLUTION:

To find the bond energy of a compound, we must write the equation to represent the process of breaking the bond to form individual atoms. In this case,

$$ClF_{(g)} \longrightarrow Cl_{(g)} + F_{(g)}$$

To solve for this equation, we shall need to use the equations given in the problem and apply Hess's Law. Therefore, writing the first equation in the reverse order and then adding the three equations together gives

$$
\begin{array}{rcl}
2\ ClF_{(g)} & \longrightarrow & Cl_{2(g)} + F_{2(g)} - 27 \text{ kcal} \\
F_{2(g)} & \longrightarrow & 2\ F_{(g)} - 37 \text{ kcal} \\
Cl_{2(g)} & \longrightarrow & 2\ Cl_{(g)} - 58 \text{ kcal} \\
\hline
2\ ClF_{(g)} + F_{2(g)} + Cl_{2(g)} & \longrightarrow & Cl_{2(g)} + F_{2(g)} + 2\ F_{(g)} + 2\ Cl_{(g)} \\
 & & - (27 + 37 + 58) \text{ kcal}
\end{array}
$$

or

$$2\,ClF_{(g)} \longrightarrow 2\,F_{(g)} + 2\,Cl_{(g)} - 122\text{ kcal}$$

Dividing everything by 2 gives the desired equation:

$$ClF_{(g)} \longrightarrow F_{(g)} + Cl_{(g)} - 61\text{ kcal}$$

Therefore, it takes 61 kcal to break 1 mole of ClF into the individual atoms. Thus, the bond energy is **61 kcal/mol** (255 kJ/mol).

10-4 Calorimetry

Heats of reaction are measured in a *calorimeter.* There are several types of calorimeter. One of the simplest is a styrofoam cup that can be used to make approximate measurements of things such as specific heat of a metal or heat of neutralization for the reaction of an acid with a base. More complex, and hence more accurate, calorimeters consist of a well-insulated container with a reaction vessel of some type immersed in water inside the container. An example is shown in Fig. 10-3. An accurately measured amount of water is placed in the calorimeter and the temperature is measured. Then a reaction takes place in the reaction vessel, and the change in the temperature of the water is measured. Using the specific heat of water as 1.00 cal/g °C, the number of calories of heat produced, or absorbed, can be measured. A correction must be made, however, for the substances other than water in the calorimeter; that is, the container, the stirring apparatus, and the thermometer will all absorb some heat. To make this correction, called *calibrating the instrument,* a reaction for which ΔH_r is accurately known is carried out in the apparatus. The difference between the amount of heat absorbed by the water (measured by the rise in temperature) and the total ΔH_r is the correction factor; it is called the *water equivalent of the calorimeter.*

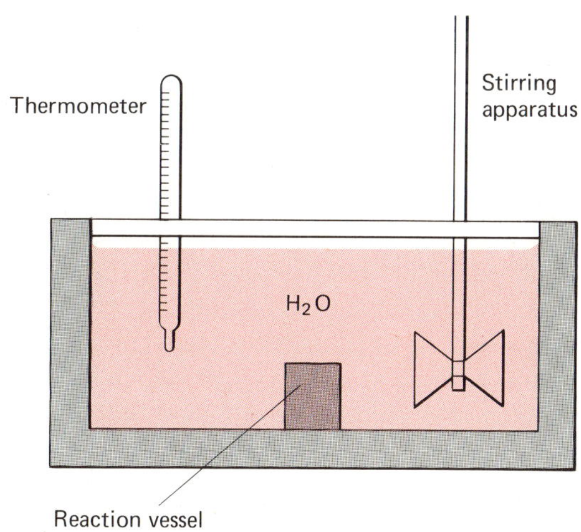

Figure 10-3
A typical calorimeter

10-5 First Law of Thermodynamics

The **First Law of Thermodynamics** is simply the Law of Conservation of Energy, which states that energy cannot be created nor destroyed. As stated in Chapter 3, it is now known that small amounts of matter can be converted to energy. Therefore, the laws of conservation of matter and energy are often combined to state that matter-energy cannot be created nor destroyed. Another way of saying this is by the statement that the total amount of matter-energy in the universe is constant. The heat produced in an exothermic reaction is due to a very small amount of matter being converted to energy.

This First Law of Thermodynamics accounts for all the known facts concerning thermochemical equations. For example, since energy must be conserved, the sum of all the heats of reaction must be equal to the overall heat of reaction, as stated in Hess's Law. Also, this accounts for the fact that the ΔH_r is equal to the sum of the ΔH_f of products minus reactants.

The difference between ΔE and ΔH is also a consequence of the first law; that is, at constant volume, no work is being done and thus ΔE is measured by the total amount of heat energy released in the reaction. At constant pressure, however, some of the energy released is used to do work against the atmosphere; therefore, what is measured is the change in heat content (enthalpy), ΔH. Another way of expressing this relationship between the total energy change and the enthalpy change, or heat flow (Q), is the following:

$$\Delta E = Q - W \qquad (10\text{-}2)$$

where Q is the heat flow and W is the work done against the surroundings. Since heat flow measures the amount of energy absorbed (that is, if Q is positive, the reaction is endothermic), the total energy change (ΔE) is measured by the heat flow (Q) minus the work done (W). For example, if 200 kcal of heat energy is absorbed by a system and 50 kcal of work is done on the surroundings, the change in energy is

$$\Delta E = 200 \text{ kcal} - 50 \text{ kcal} = 150 \text{ kcal}$$

Thus, the energy of the system has increased by 150 kcal.

10-6 Second Law of Thermodynamics*

If ΔH is negative (that is, heat is released or the reaction is exothermic), there is a good chance that the reaction will go spontaneously. This is not always true, however, and there are reactions for which ΔH is positive (endothermic) that go spontaneously. Thus, there must be another factor involved. This other factor is the randomness or disorder of the system, which is called the **entropy** (S or ΔS). It was stated in Sections 10-2 and 10-5 that at constant pressure the total energy change is equal to the change in enthalpy (ΔH) minus

*This section is optional and can be omitted without loss of continuity.

TABLE 10-2 *Entropy (ΔS) values of some common substances*

SUBSTANCE	ΔS	
	kcal/mol K	kJ/mol K
$H_2O_{(g)}$	0.0451	0.189
$H_2O_{(l)}$	0.0167	0.067
$H_{2(g)}$	0.0312	0.130
$Cl_{2(g)}$	0.0533	0.221
$N_{2(g)}$	0.0458	0.189
$O_{2(g)}$	0.0490	0.204
$NH_{3(g)}$	0.0460	0.192
$CO_{(g)}$	0.0473	0.197
$CO_{2(g)}$	0.0511	0.213
$SO_{2(g)}$	0.0594	0.247
$C_2H_{2(g)}$	0.0480	0.200
$C_2H_{4(g)}$	0.0525	0.217
$C_2H_{6(g)}$	0.0549	0.225
$CH_3OH_{(g)}$	0.0568	0.234
$CH_3OH_{(l)}$	0.0303	0.125

the work done against the atmosphere.* Likewise, at constant temperature the total amount of energy available to do work, called the **free energy (ΔG),** is equal to ΔH minus the change in entropy multiplied times the absolute temperature. This can be expressed as

(10-4) $$\Delta G = \Delta H - T\Delta S$$

A reaction will only occur spontaneously if ΔG is negative. Recall that a negative sign means energy is given off by the system. Therefore, to determine if a reaction will take place spontaneously, we can calculate ΔG from tables of ΔH and ΔS (see Table 10-2) values and apply the following criteria:

1. If ΔG is ($-$), the reaction is spontaneous as written.
2. If ΔG is ($+$), the reverse reaction is spontaneous.
3. If ΔG is zero, the reaction is at equilibrium (that is, the rate at which the forward reaction is taking place is equal to the rate at which the reverse reaction is taking place).

Consider the following problem examples.

*This work done against the atmosphere can be expressed as the pressure (P) times the change in volume (ΔV), or $P\Delta V$. Therefore, we can say that

(10-3) $$\Delta E = \Delta H - P\Delta V$$

PROBLEM EXAMPLE 10-6

Calculate (a) ΔH, (b) ΔS, and (c) ΔG for the following reaction, at 25°C, and tell if the reaction would go spontaneously:

$$C_2H_{4(g)} + 3\,O_{2(g)} \longrightarrow 2\,CO_{2(g)} + 2\,H_2O_{(g)}$$

SOLUTION:

(a) $\Delta H_r = 2\,(\Delta H_f\,CO_{2(g)}) + 2\,(\Delta H_f\,H_2O_{(g)}) - \Delta H_f\,C_2H_4$

$$= 2\,\cancel{mol}\left(-94.1\,\frac{kcal}{\cancel{mol}}\right) + 2\,\cancel{mol}\left(-57.8\,\frac{kcal}{\cancel{mol}}\right) - 1\,\cancel{mol}\left(12.5\,\frac{kcal}{\cancel{mol}}\right)$$

$$= \mathbf{-316\ kcal}$$

(b) $\Delta S = 2\,(\Delta S\,CO_{2(g)}) + 2\,(\Delta S\,H_2O_{(g)}) - [(\Delta S\,C_2H_{4(g)}) + 3\,(\Delta S\,O_{2(g)})]$

$$= 2\,\cancel{mol}\left(0.0511\,\frac{kcal}{\cancel{mol}}\right) + 2\,\cancel{mol}\left(0.0451\,\frac{kcal}{\cancel{mol}}\right)$$

$$- \left[1\,\cancel{mol}\left(0.0525\,\frac{kcal}{\cancel{mol}\,K}\right) + 3\,\cancel{mol}\left(0.0490\,\frac{kcal}{\cancel{mol}\,K}\right)\right]$$

$$= \mathbf{-0.0071\,\frac{kcal}{K}}$$

(c) $\Delta G = \Delta H - T\Delta S$

$$= -316\ kcal - (298\ K)\left(-0.0071\,\frac{kcal}{K}\right)$$

$$= \mathbf{-314\ kcal}$$

Therefore, the reaction would go spontaneously.

PROBLEM EXAMPLE 10-7

For the reaction, $H_2O_{(l)} \longrightarrow H_2O_{(g)}$, calculate ΔG at (a) 20°C, (b) 100°C, and (c) 150°C.

SOLUTION:

$$\Delta H_r = \Delta H_f\,H_2O_{(g)} - \Delta H_f\,H_2O_{(l)}$$

$$= 1\,\cancel{mol}\left(-57.8\,\frac{kcal}{\cancel{mol}}\right) - 1\,\cancel{mol}\left(-68.3\,\frac{kcal}{\cancel{mol}}\right)$$

$$= 10.5\ kcal$$

$$\Delta S = 1\,\cancel{mol}\left(0.0451\,\frac{kcal}{\cancel{mol}\,K}\right) - 1\,\cancel{mol}\left(0.0167\,\frac{kcal}{\cancel{mol}\,K}\right)$$

$$= 0.0284\,\frac{kcal}{K}$$

(a) At 20°C (293 K),

$$\Delta G = 10.5 \text{ kcal} - 293 \text{ K}\left(0.0284 \frac{\text{kcal}}{\text{K}}\right)$$

$$= \mathbf{2.2 \text{ kcal}}$$

(b) At 100°C (373 K),

$$\Delta G = 10.5 \text{ kcal} - 373 \text{ K}\left(0.0284 \frac{\text{kcal}}{\text{K}}\right)$$

$$= \mathbf{0}$$

(c) At 150°C (423 K),

$$\Delta G = 10.5 \text{ kcal} - 423 \text{ K}\left(0.0284 \frac{\text{kcal}}{\text{K}}\right)$$

$$= \mathbf{-1.5 \text{ kcal}}$$

Therefore, at 150°C the reaction takes place spontaneously; at 100°C (the boiling point of water at 1 atm pressure) the reaction is at equilibrium; and at 20°C the reverse reaction takes place spontaneously.

Thus, the Second Law of Thermodynamics tells the following things:

1. Some energy absorbed is always going to increase the entropy of the system except at absolute zero. (At 0 K the $T\Delta S$ term would become zero.)
2. The entropy of the universe is increasing (the First Law of Thermodynamics can be stated *the energy of the universe is constant*).
3. Any spontaneous process is a change to a more disordered or more probable state.
4. At constant temperature and pressure a change can occur spontaneously only if there is a decrease in free energy.

SUMMARY

1. A *thermochemical equation* is a balanced equation in which the heat transferred is included in the equation. In an *exothermic* reaction, heat is given off and in an *endothermic* reaction heat is absorbed.
2. The *enthalpy* (or change in enthalpy, ΔH) of a reaction is a measure of the change in heat content, or the amount of heat transferred in the reaction as written. Since the enthalpy does not depend on the path of the reaction, it is a *state function*. For an exothermic reaction, ΔH is negative; for an endothermic reaction, it is positive.

3. The heat of formation (ΔH_f) of a compound is the ΔH of reaction (ΔH_r) for the reaction in which 1 mole of the compound is formed from the elements in the standard state.
4. From a table of ΔH_f values, ΔH_r can be calculated, for a particular reaction, using the relationship

$$\Delta H_r = \Sigma\, \Delta H_f \text{ products} - \Sigma\, \Delta H_f \text{ reactants}$$

5. *Hess's Law* states that the ΔH_r for the combination of two or more reactions is equal to the sum of the ΔH_r values for each individual reaction.
6. The *bond energy* in a compound is the amount of energy necessary to break the bonds in the compound. It can be calculated using known ΔH values and Hess's Law.
7. A *calorimeter* is a device in which heat flow is measured, using water as a standard. Thus, by measuring the change in temperature of an accurately measured amount of water, the amount of heat given off or absorbed can be calculated.
8. The *First Law of Thermodynamics* states that energy (or matter-energy) cannot be created nor destroyed and is thus often described as the Law of Conservation of Energy.
9. The *entropy (S)* of a system is a measure of the randomness or disorder of the system. For any spontaneous process there is always an increase in entropy. Thus the total amount of energy available, called the *free energy (G)*, is related to the total amount of energy (H) by the factor $\Delta G = \Delta H - T\Delta S$.
10. The *Second Law of Thermodynamics* states that the entropy of the universe is increasing, and that any spontaneous change is always to a more disordered state.

Upon completion of this chapter, you should be able to do the following things:

1. Given a balanced thermochemical equation, (a) determine if it is endothermic or exothermic, (b) tell what the ΔH is for the reaction, (c) tell what ΔH would be for any related reaction (that is, the reverse reaction, the reaction all divided through by any number, etc.), and (d) do stoichiometric calculations involving heat flow from the equation. (1, 2)
2. Given a table of ΔH_f values, calculate ΔH_r for any balanced equation. (3)
3. Given ΔH_r and ΔH_f values for all but one compound in a balanced equation, calculate the unknown ΔH_f. (4)
4. Given two or more thermochemical equations, write an overall thermochemical equation using Hess's Law. (5)

5. Given a table of ΔH and ΔS values, calculate ΔG for a reaction and tell if the reaction would go spontaneously. (6)
6. Given any of the terms listed in Exercise 1, give a brief definition or explanation in your own words.

EXERCISES

1. Define or explain the following terms:
   (a) thermochemical equation
   (b) heat flow
   (c) endothermic reaction
   (d) exothermic reaction
   (e) enthalpy
   (f) state function
   (g) ΔH of reaction (ΔH_r)
   (h) ΔH of formation (ΔH_f)
   (i) standard state
   (j) Hess's Law
   (k) First Law of Thermodynamics
   (l) Second Law of Thermodynamics
   (m) entropy
   (n) free energy
2. Explain the convention used to designate the sign on Q or ΔH for an exothermic and an endothermic reaction.
3. What is the relationship between ΔE and ΔH?
4. Distinguish between ΔH of reaction (ΔH_r) and ΔH of formation (ΔH_f).
5. Explain why ΔH_f of the elements in the standard state is taken as zero.

PROBLEMS

Stoichiometry

1. For the reaction:

$$2\ CH_3OH_{(l)} + 3\ O_{2(g)} \longrightarrow 2\ CO_{2(g)} + 4\ H_2O_{(l)} + 347\ \text{kcal}$$

   (a) is the reaction endothermic or exothermic?
   (b) what is ΔH_r?
   (c) calculate the amount of heat in kilocalories and kilojoules released when 25.4 g of CH_3OH reacts.
   (d) how much heat, in kilocalories and kilojoules, would be required in order to cause 185 g of CO_2 to react with water to form CH_3OH and O_2?
   (e) calculate ΔH_r for the reaction

$$2\ CO_{2(g)} + 4\ H_2O_{(l)} \longrightarrow 2\ CH_3OH_{(l)} + 3\ O_{2(g)}$$

   (f) calculate ΔH_r for

$$CH_3OH_{(l)} + \tfrac{3}{2}\ O_{2(g)} \longrightarrow CO_{2(g)} + 2\ H_2O_{(l)}$$

2. For the reaction:

$$C_2H_{4(g)} + 3\ O_{2(g)} \longrightarrow 2\ CO_{2(g)} + 2\ H_2O_{(g)} + 316\ \text{kcal}$$

(a) is the reaction endothermic or exothermic?
(b) what is ΔH_r?
(c) calculate the amount of heat in kilocalories a kilojoules released when 64.5 g of C_2H_4 reacts.
(d) how much heat, in kilocalories and kilojoules, would be required in order to cause 125 g of CO_2 to react with water to form C_2H_4 and O_2?
(e) calculate ΔH_r for the reaction

$$2\,CO_{2(g)} + 2\,H_2O_{(g)} \longrightarrow C_2H_{4(g)} + 3\,O_{2(g)}$$

(f) calculate ΔH_r for the reaction:

$$\tfrac{1}{2}\,C_2H_{4(g)} + \tfrac{3}{2}\,O_{2(g)} \longrightarrow CO_{2(g)} + H_2O_{(g)}$$

ΔH_r and ΔH_f

3. Using Table 10-1, calculate ΔH_r for each of the following reactions. Tell if each is exothermic or endothermic.
   (a) $2\,NO_{(g)} + O_{2(g)} \longrightarrow 2\,NO_{2(g)}$
   (b) $2\,C_2H_{6(g)} + 7\,O_{2(g)} \longrightarrow 4\,CO_{2(g)} + 6\,H_2O_{(g)}$
   (c) $N_{2(g)} + 3\,H_{2(g)} \longrightarrow 2\,NH_{3(g)}$
   (d) $2\,CH_3OH_{(l)} + 3\,O_{2(g)} \longrightarrow 2\,CO_{2(g)} + 4\,H_2O_{(g)}$
   (e) $2\,SO_{2(g)} + O_{2(g)} \longrightarrow 2\,SO_{3(g)}$
   (f) $C_2H_{4(g)} + 3\,O_{2(g)} \longrightarrow 2\,CO_{2(g)} + 2\,H_2O_{(g)}$
   (g) $2\,CO_{(g)} + O_{2(g)} \longrightarrow 2\,CO_{2(g)}$
   (h) $N_{2(g)} + O_{2(g)} \longrightarrow 2\,NO_{(g)}$
4. Calculate ΔH_f for each of the following (using Table 10-1).
   (a) $C_2H_{2(g)}$ from the reaction $2\,C_2H_{2(g)} + 5\,O_{2(g)} \longrightarrow 4\,CO_{2(g)} + 2\,H_2O_{(g)} + 600$ kcal
   (b) $CH_{4(g)}$ from the reaction $CH_{4(g)} + 2\,O_{2(g)} \longrightarrow CO_{2(g)} + 2\,H_2O_{(g)} + 192$ kcal
5. Given the following reactions,
   (a) $S + O_2 \longrightarrow SO_2 + 71.0$ kcal
   $2\,SO_2 + O_2 \longrightarrow 2\,SO_3 + 47.0$ kcal
   calculate ΔH_r for $2\,S + 3\,O_2 \longrightarrow 2\,SO_3$
   (b) $2\,C + O_2 \longrightarrow 2\,CO + 52.8$ kcal
   $2\,CO + O_2 \longrightarrow 2\,CO_2 + 125.4$ kcal
   calculate ΔH_r for $C + O_2 \longrightarrow CO_2$

ΔS and ΔG

6. For each of the following reactions, calculate ΔH, ΔS, and ΔG, at 25°C, and tell if the reaction will go spontaneously (use Tables 10-1 and 10-2).
   (a) $N_{2(g)} + 3\,H_{2(g)} \longrightarrow 2\,NH_{3(g)}$
   (b) $2\,CO_{(g)} + O_{2(g)} \longrightarrow 2\,CO_{2(g)}$
   (c) $2\,C_2H_{2(g)} + 5\,O_{2(g)} \longrightarrow 4\,CO_{2(g)} + 2\,H_2O_{(g)}$
   (d) $2\,H_{2(g)} + O_{2(g)} \longrightarrow 2\,H_2O_{(g)}$
   (e) $2\,NH_{3(g)} \longrightarrow N_{2(g)} + 3\,H_{2(g)}$
   (f) $2\,CH_3OH_{(l)} + 3\,O_{2(g)} \longrightarrow 2\,CO_{2(g)} + 4\,H_2O_{(g)}$

General problems

7. How much heat will be given off when 18.5 g of C_2H_4 is burned in oxygen to form $CO_{2(g)}$ and $H_2O_{(g)}$?
8. How much heat will be given off when 0.760 g of C_2H_6 is burned in oxygen to form $CO_{2(g)}$ and $H_2O_{(g)}$?
9. Tell if each of the following are exothermic or endothermic:
   (a) $N_{2(g)} + 3\,H_{2(g)} \longrightarrow 2\,NH_{3(g)}$
   (b) $N_{2(g)} + 2\,O_{2(g)} \longrightarrow 2\,NO_{2(g)}$
   (c) $2\,H_{2(g)} + O_{2(g)} \longrightarrow 2\,H_2O_{(l)}$
   (d) $2\,CO_{(g)} + O_{2(g)} \longrightarrow 2\,CO_{2(g)}$

REFERENCES

DAVIES, WILLIAM G., *Introduction to Chemical Thermodynamics: A Non-Calculus Approach.* Philadelphia: W. B. Saunders Co., 1972.

PIMENTEL, GEORGE C., and RICHARD D. SPRATLEY, *Understanding Chemical Thermodynamics.* San Francisco, California: Holden-Day, Inc., 1970.

PORTER, GEORGE, "The Laws of Disorder," *Chemistry,* May-Sept. 1968.

ROGERS, DONALD R., "An Informal History of the First Law of Thermodynamics," *Chemistry,* Dec. 1976, **49,** p. 11.

SPENCER, J. N., DAVID GORDON, and H. D. SCHREIBER, "Entropy and Chemical Reactions," *Chemistry,* Dec. 1974, **47,** p. 12.

Objectives

1. To become familiar with the basic characteristics of gases and the differences among gases, liquids and solids.
2. To understand the postulates of the kinetic theory as applied to gases and to be able to explain how these account for the properties of gases.
3. To understand what the following units, used to measure the pressure of gases, represent: pounds per square inch (psi), millimeters (or centimeters or inches) of Hg, torr, atmospheres, and pascals (Pa); and to be able to convert from any of these to any other.
4. To learn what the units standard temperature and pressure (STP) represent.
5. To understand and be able to explain, in terms of kinetic theory, the relationship between volume and pressure, volume and temperature, temperature and pressure, pressure and number of moles, volume and number of moles, and temperature and number of moles, for gases.
6. To be able to do calculations involving relationships between volume and pressure, volume and temperature, and temperature and pressure, or any combination of these factors; mixtures of gases, using Dalton's Law of Partial Pressures; the molar volume of a gas; the relationships among volume, pressure, or temperature and number of moles; the Ideal Gas Law or equation of state for an ideal gas (including the calculation of the value of *R* with the desired units); and Graham's Law of Diffusion.

Eleven

Gases

In Chapter 3, it was stated that there are three physical states of matter: solid, liquid, and gas. In this chapter and the next, we shall consider these physical states in relation to the properties of each state and the changes each one may undergo. We shall first consider gases since they are the simplest of the three physical states and because more is known of the gaseous state than of the other two states. We shall begin with a brief discussion of the relationship among gases, liquids, and solids, in order to understand the gas state better.

11-1 General Properties of Gases and Comparison to Liquids and Solids

In the gas state, the particles (molecules*) are very widely separated and are moving very rapidly, in contrast to the solid state in which the particles are held in a definite arrangement and move very little. The liquid state is an intermediate case, in which the particles are touching and are attracted to each other but still move quite freely. In Chapter 12 we shall consider the forces of attraction between particles in the liquid and solid state. For now, just keep in mind that the particles are attracted to each other and are held together in the liquid and solid states but are not in the gas state. Figure 11-1 shows this difference in properties in the three states of matter.

Due to these differences in the three states of matter, we observe the following properties of gases, which we shall compare to liquids and solids.

1. Gases have *no shape or volume.* They will fill any container they are placed in. Liquids have a definite volume but no definite shape, as you will notice when a beaker of water or other liquid is spilled in the

*We can now define a molecule more precisely as the smallest particle of a substance that exists in the gas phase.

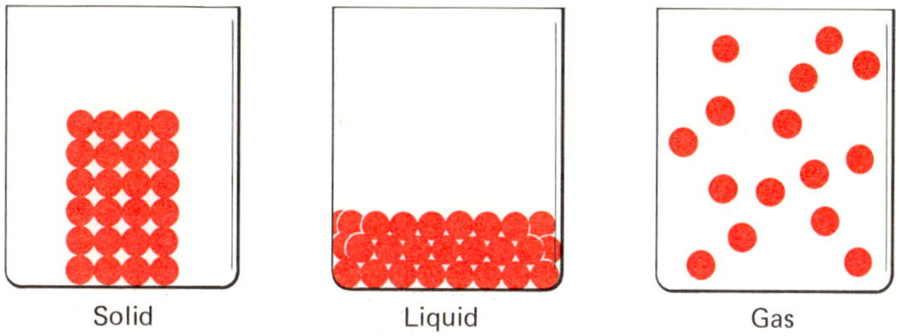

Figure 11-1
Comparison of solid, liquid and gas states of matter

laboratory. Solids have both a definite shape and volume, like an ice cube.

2. Gases *diffuse* very readily, and any two gases will *mix* completely when combined. Liquids diffuse much more slowly than gases but more rapidly than solids, which hardly diffuse at all. Two liquids will mix if they are similar (we shall also have more to say about this in Chapter 12).

3. Gases can be *compressed* very readily, due to the large amount of space between the molecules. In fact, when a gas is compressed, the molecules are simply forced a little closer together and have less space between them. On the other hand, liquids and solids can only be compressed under extremely high pressures. This is due to the fact that the particles are touching in a liquid or solid. Thus, there is no space between them and they cannot very readily be forced closer together.

4. Gases have much lower *densities* than liquids or solids. Again, this is due to the larger amount of space between the molecules in a gas. Due to the low densities of gases, they are generally measured in grams per liter rather than grams per milliliter or grams per cubic centimeter. Also, as we shall consider in more detail in Section 11-7, the density of a gas varies greatly with changes in temperature and pressure. However, the density of a liquid or solid changes very little with changes in temperature or pressure.

Table 11-1 lists the general properties of gases, liquids, and solids and shows how the three states of matter differ.

In going from the gas to the liquid to the solid state, the particles become more ordered and restricted in their movement. We shall see in Chapter 12 that there is a direct relationship between the forces of attraction between particles and the physical properties, such as melting point and boiling point; that is, it takes energy to convert from the solid to the liquid to the gas phase. The amount of energy depends on how tightly the particles are held together in the solid and liquid phases.

As a comparison between the gas and the liquid state, if 1 mℓ of water is changed to steam at 100°C and 1 atm pressure, it will occupy a volume of about 1.7 ℓ.

TABLE 11-1 *Comparison of gases, liquids, and solids*

PROPERTIES	GAS	LIQUID	SOLID
Particle arrangement	Widely separated	Not widely separated (touch)	Packed close together
Movement	Very freely (increases with temperature)	Freely, but slower than gases	Very little
Diffusion	Easily	Slower than gases	Very slowly
Shape and volume	Fits container; no volume of its own	Fits container; has fixed volume	Both fixed
Compressibility	Easily	Difficult (only under extremely high pressures)	Extremely difficult
Density	Low	High	High
Mixability	Readily	Slower than gases	Extremely slow

11-2 The Kinetic Theory

The kinetic theory explains many of the properties of matter. It is especially concerned with the relationship between heat and motion. According to kinetic theory, the temperature of a substance is a measurement of the average kinetic energy of the molecules that make up the substance. In applying the kinetic theory to gases, the following assumptions are made.

1. Gases are composed of molecules, which are in very rapid motion. The amount of space between the individual molecules is so great that we can say that the volume occupied by the molecules themselves is negligible. This explains why gases have low densities, are easily compressed, and diffuse very readily, and also why they will fill any container they are placed in.

2. The average kinetic energy of the molecules is proportional to the kelvin (absolute) temperature. Also, at a given temperature, all gases have the same average kinetic energy. It must be kept in mind, however, that we are considering *average* kinetic energy; in any sample of gas at a given temperature, some of the molecules will have higher energy and some will be lower.

3. The molecules are so far apart and are moving so rapidly that the forces of attraction between them are negligible.

All these assumptions are only absolutely true for **ideal gases.** At ordinary temperatures and pressures, however, real gases behave as ideal gases. It is only under extremely high pressures, or at very low temperatures, that there is significant deviation from ideal behavior. At very high pressures the molecules are forced close together. Thus, the volume of the molecules becomes significant, and the forces of attraction between molecules also must be taken into account. At low temperatures the motion of the molecules is slowed down to the point when attractive forces between molecules may become significant. There are correction factors that can be used to account for deviation from ideal behavior. For our purposes in this text, however, we shall only be considering gases that behave as ideal gases and shall not need to be concerned with these correction factors.

11-3 Pressure of Gases

As the molecules of a gas strike the sides of their container, they cause a *pressure* on the container. Pressure is defined as force per unit area and can be measured in units such as pounds per square inch.

Gases in the atmosphere (mostly nitrogen, oxygen, and small amounts of argon in pure air) also exert a pressure. Standard atmospheric pressure, called 1 **atmosphere** (atm), is taken as 14.7 pounds per square inch (psi). This is the pressure exerted by the atmosphere at sea level under "normal" weather conditions; that is, this value is used as a standard, and called 1 atm.

One method of measuring the pressure of the atmosphere is with a *barometer.* The mercury barometer was first devised by Evangelista Torricelli (1608–1647), an Italian mathematician and physicist. A barometer consists of a sealed glass tube, about a meter long, that is filled with mercury. This is inverted over a container of mercury, which is open to the atmosphere, as shown in Fig. 11-2. The level of mercury in the tube will fall to the point where the pressure of the atmosphere is just great enough to support the column of mercury. At 1 atm (14.7 psi) there is a column of mercury 76 cm (760 mm) high. Thus, 1 atm of pressure can also be expressed as 76 cm of mercury (76 cm Hg) or 760 mm of mercury (760 mm Hg).

Atmospheric pressure can also be expressed as inches of mercury (in. Hg), where 1 atm = 29.9 in. Hg. This is simply calculated by converting

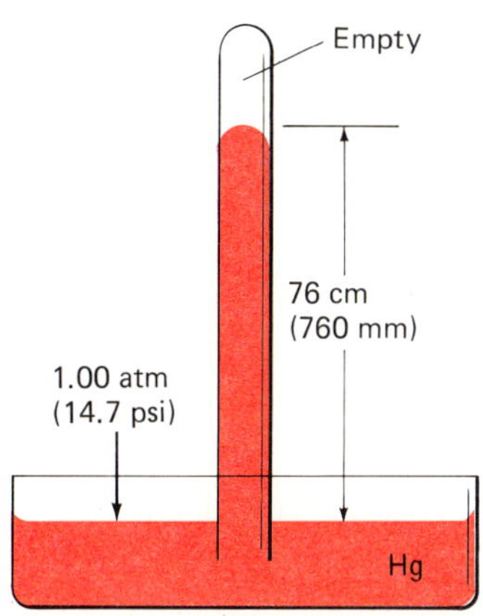

Figure 11-2
Mercury barometer

centimeters (or millimeters) to inches. A newer unit used to express pressure is the **torr,** named in honor of Torricelli. One torr is equivalent to one mm Hg; thus,

$$1 \text{ atm} = 760 \text{ mm Hg} = 760 \text{ torr}$$

The unit of pressure that is used in the SI units is the pascal (Pa), where

$$1 \text{ atm} = 1.01 \times 10^5 \text{ Pa}$$

This unit is named after the French scientist Blaise Pascal (1623–1662). One pascal is the pressure caused by a force of one newton (N) acting on an area of one square meter (N/m^2). A newton is the amount of force that gives a mass of one kilogram an acceleration of one meter per second squared ($kg\ m/s^2$). Therefore, the units of pascals are $kg/m\ s^2$. These units that can be used to express pressure can be summarized as follows:

$$1 \text{ atm} = 14.7 \text{ psi} = 76 \text{ cm Hg} = 760 \text{ mm Hg} = 760 \text{ torr} = 1.01 \times 10^5 \text{ Pa}$$

In this book we shall use the units atmospheres (atm) and torr almost exclusively.

11-4 Variables in Gases. Standard Temperature and Pressure [STP]

There are four variables that must be taken into account when considering the behavior of gases. These are volume, temperature, pressure, and amount of gas.

The volume of a gas is simply the space occupied by the sample. Due to the low densities of gases, volume is usually expressed in liters but any volume units can be used.

In making quantitative calculations involving gases, the temperature must be expressed on the kelvin (absolute) temperature scale. You will recall (Section 2-4) that this scale is set up such that at the zero point all molecular motion will stop, with $0 \text{ K} = -273°\text{C}$. Since temperature is a measure of the average kinetic energy of a substance, the molecular motion of a gas is directly proportional to the temperature, using this absolute scale; that is, as the temperature in degrees kelvin is doubled, the average kinetic energy is twice as great.

The pressure of a gas is the force exerted when the molecules strike the sides of the container. As noted in Section 11-3, there are several units that can be used to express pressure, with atmospheres, millimeters of Hg, or torr being the most common. In this text we shall use either atmospheres or torr almost exclusively.

When considering the amount of gas in a sample, the main factor is the number of particles (molecules) present. Therefore, amount of gas is expressed in moles and abbreviated (n).

Standard temperature and pressure (STP) is taken as 0°C (273 K) and 1 atm (760 torr).

11-5 The Gas Laws—Relationships Among Volume, Temperature, and Pressure of Gases

We now need to consider the relationships among volume, temperature, and pressure of gases. In order to do this we shall consider the relationship between two variables when the third one is constant. For example, we shall first consider how pressure changes as the volume of a gas is changed, at constant temperature. Then we shall consider how volume and temperature are related, at constant pressure. Finally, we shall consider the relationship between temperature and pressure, at constant volume. In all cases, we shall be considering the variation for a given amount of gas. Later (Section 11-8) we shall consider how volume, pressure, and temperature are related to the amount of gas present, measured in moles.

Relationship between Volume and Pressure, at Constant Temperature (Boyle's Law)

It is found experimentally that at constant temperature, as the volume of a gas is increased, the pressure decreases. Likewise, if the pressure is going to be increased (at constant temperature), the volume must be decreased; or, if the pressure is to be decreased, the volume must increase. Thus, we say that at constant temperature, for a given amount of gas, *volume is inversely proportional to pressure.* This means that if the volume is doubled, the pressure will only be half as great; if the pressure is three times as great, the volume must be one-third of what it was originally.

This can be expressed as

$$V \propto \frac{1}{P} \quad \text{(at constant temperature)}$$

By introducing a proportionality constant, we can write this as the equation

$$V = k\frac{1}{P} \quad \text{or} \quad PV = k$$

This tells us that the pressure times the volume is constant, at a constant temperature. Thus, as stated earlier, if the pressure is doubled (taken times 2) the volume would have to be one-half, in order for P times V to remain constant.

This relationship was first shown in 1660 by Robert Boyle and hence is known as **Boyle's Law.** This relationship can be explained in terms of the kinetic theory of gases. If the temperature is constant, the molecules have some constant, average kinetic energy. If the volume is decreased, such as to half of what it was originally, the pressure will be greater (twice as great in this case) since the molecules will strike the sides of the container more frequently. Figure 11-3 demonstrates Boyle's Law.

In doing calculations involving changes in pressure or volume, at constant temperature, we simply consider what effect a particular change in volume (or pressure) will have on the initial pressure (or volume) and multiply the original condition by the proper factor.

Consider the following problem examples.

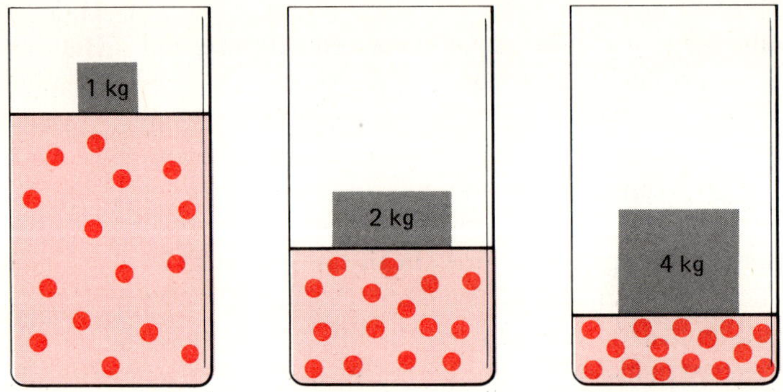

Figure 11-3 Demonstration of Boyle's law (temperature is constant)

PROBLEM EXAMPLE 11-1

A sample of gas occupies a volume of 225 mℓ at a pressure of 720 torr and a temperature of 20°C. Calculate the new pressure if the volume is increased to 350 mℓ, at constant temperature.

SOLUTION:

In working these problems, arrange the data in an orderly form, such as

$$P_1 = 720\text{ torr} \qquad\qquad P_2 = ?$$

$$V_1 = 225\text{ m}\ell \xrightarrow[\text{pressure decreases}]{\text{volume increases}} V_2 = 350\text{ m}\ell$$

(T is constant)

$$P_2 = P_1 \times V_{\text{factor}}$$

The volume has increased from 225 to 350 mℓ; hence, the pressure will be decreased, and the volume factor must be written so that the new pressure will be less than the initial pressure. Thus, the volume factor is written as a fraction less than one, and is therefore, 225 mℓ/350 mℓ. Thus, the new pressure is calculated as

$$P_2 = 720\text{ torr} \times \frac{225\text{ m}\ell}{350\text{ m}\ell} = \mathbf{463\text{ torr}}$$

In working gas law problems, keep in mind that units are still important; that is, the units of volume in this problem must be the same so that they divide out.

PROBLEM EXAMPLE 11-2

A sample of gas with a volume of 145 mℓ exerts a pressure of 775 torr. What must the volume be if the pressure is to become 740 torr, at constant temperature?

SOLUTION:

Arranging the data to show what is changing and what we are trying to calculate gives

$$V_1 = 145\text{ m}\ell \qquad\qquad V_2 = ?$$

$$P_1 = 775 \text{ torr} \xrightarrow[\text{volume increases}]{\text{pressure decreases}} P_2 = 740 \text{ torr}$$

(T is constant)

$$V_2 = V_1 \times P_{\text{factor}}$$

Since the pressure is decreasing, the volume must be increasing; thus we want the pressure factor to be a fraction greater than one, or 775 torr/740 torr. Therefore the new volume is calculated as

$$V_2 = 145 \text{ m}\ell \times \frac{775 \text{ torr}}{740 \text{ torr}} = 152 \text{ m}\ell$$

Relationship between Volume and Temperature, at Constant Pressure (Charles' Law)

Another consequence of the kinetic theory is that if the temperature of a fixed amount of gas is increased and the pressure is to remain constant, the volume must also be increased; that is, if the molecules have greater kinetic energy, the volume must be increased in order to keep the pressure constant. Another way of stating this is to note how the temperature must be changed as the volume is changed, if the pressure is to remain constant. For example, if the volume is increased, the only way that the pressure can remain constant is to increase the temperature.

This was first shown in 1787 by Jacques Charles, a French physicist (1746–1823). Thus the relationship between volume and temperature, at constant pressure, is known as **Charles' Law.**

In stating this relationship as a direct mathematical proportion, Charles showed that we must use the kelvin (absolute) temperature scale. On this scale there is a theoretical zero point where all molecular motion would stop. This point is found to correspond to $-273\,°\text{C}$ (or -273.15 to be precise), and is known as *absolute zero.* Using this new scale, Charles showed that if the temperature is doubled (in degrees K, and hence the kinetic energy is twice as great), the volume must be doubled also in order for the pressure to remain constant. Another way of considering the kelvin (absolute) scale is to say that at zero degrees kelvin the volume of a gas would be zero. Since any gas would become a liquid or solid before it reached that temperature, however, this is just a theoretical value. You will recall, from Section 2-4, that to convert from degrees Celsius to kelvin we added 273. This can be stated as

$$\text{K} = °\text{C} + 273$$

Using the kelvin temperature scale, Charles' Law states that *at constant pressure, for a given amount of gas, the volume is directly proportional to the kelvin (absolute) temperature.* For example, if the volume of a given sample of gas is doubled, the temperature must be doubled, in degrees kelvin, if the pressure is to remain constant. Likewise, if the temperature in degrees kelvin is reduced by one-half, the volume must be reduced by one-half in order to keep the pressure constant.

This relationship can be expressed mathematically as

$$V \propto T \quad \text{(at constant pressure)}$$

This states that volume (V) is directly proportional to kelvin temperature (T), at constant pressure. By introducing a proportionality constant this can be written as the equation

$$V = kT \quad \text{or} \quad \frac{V}{T} = k$$

Thus, the volume divided by the kelvin temperature is a constant. The relationship shown earlier in terms of kinetic theory can now be stated in terms of Charles' Law; that is, if the volume is doubled, the temperature (in degrees kelvin) must also be doubled, in order for V/T to remain constant. Figure 11-4 demonstrates Charles' Law.

In doing calculations involving Charles' Law the same procedure is used as with Boyle's Law calculations. To find the new volume, the initial volume is multiplied by some temperature factor (where both temperatures *must* be expressed as degrees kelvin). Or, to find the new temperature the initial temperature (in degrees kelvin) is multiplied by some volume factor. Consider the following problem example.

PROBLEM EXAMPLE 11-3

A sample of gas occupies a volume of 275 mℓ at 20°C and 1 atm pressure. Calculate the volume of the gas at 0°C and 1 atm pressure.

SOLUTION:

$V_1 = 275$ mℓ $\qquad\qquad$ $V_2 = ?$

$$T_1 = 20 + 273 = 293\text{ K} \xrightarrow[\text{volume decrease}]{\text{temperature decreases}} T_2 = 0 + 273 = 273\text{ K}$$

(P is constant)

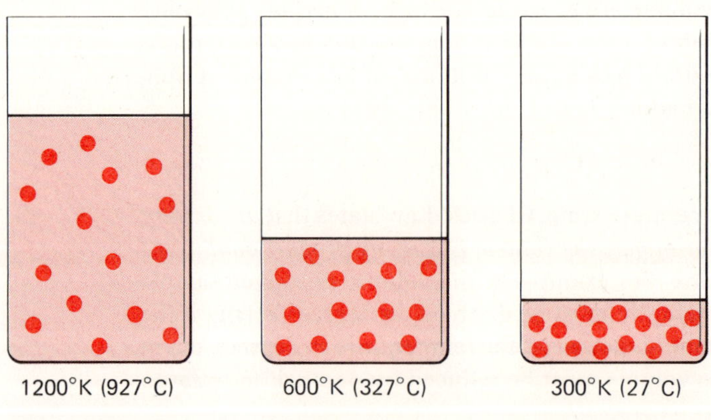

Figure 11-4
Demonstration of Charles' law (pressure is constant)

Since the temperature is decreasing, the volume must also decrease. Therefore, the temperature factor is set up as a fraction less than one, and the new volume is calculated as

$$V_2 = 275\ \text{m}\ell \times \frac{273\ \cancel{\text{K}}}{293\ \cancel{\text{K}}} = \mathbf{256\ m\ell}$$

Relationship between Temperature and Pressure, at Constant Volume (Gay-Lussac's Law)

In 1802 Joseph Gay-Lussac demonstrated that the pressure of a gas increases as the temperature is increased, at constant volume. As with the work of Charles on the relationship between volume and temperature, Gay-Lussac found that if the temperature is measured on the kelvin scale, pressure is directly proportional to temperature. For example, if the temperature, in kelvin, is doubled, the pressure will double. Thus **Gay-Lussac's Law** states that *at constant volume, the pressure of a given sample of gas is directly proportional to the kelvin (absolute) temperature.* This can be expressed mathematically as

$$P \propto T \quad \text{(at constant volume)}$$

or

$$P = kT$$

This can be written as the equation

$$\frac{P}{T} = k$$

This relationship can be explained in terms of kinetic theory by considering what would happen to a gas in a container of constant volume, as the temperature was increased or decreased. If the temperature were increased, the kinetic energy of the molecules would increase and thus the pressure would increase. Likewise, as the kinetic energy decreased, the pressure would decrease. Figure 11-5 demonstrates Gay-Lussac's Law.

To calculate the new temperature as the pressure changes, or the new pressure as the temperature changes, we use the relationship

$$T_2 = T_1 \times P_{\text{factor}}$$

or

$$P_2 = P_1 \times T_{\text{factor}}$$

It must be kept in mind in doing calculations involving this relationship, that the temperature must be expressed in degrees kelvin. Consider the following example.

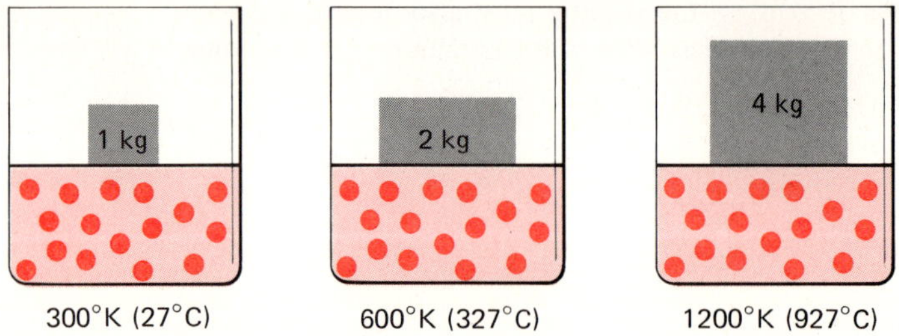

Figure 11-5 Demonstration of Gay-Lussac's law (volume is constant)

PROBLEM EXAMPLE 11-4

A sample of gas at 25°C has a pressure of 1.2 atm. Calculate the final pressure in atmospheres if the temperature is changed to 100°C, at constant volume.

SOLUTION:

$P_1 = 1.20$ atm $\qquad\qquad$ $P_2 = ?$

$T_1 = 25 + 273 = 298$ K $\xrightarrow[\text{pressure increases}]{\text{temperature increases}}$ $T_2 = 100 + 273 = 373$ K

(V is constant)

Since the temperature is increasing, the pressure will also increase. Therefore the temperature factor is set up as a fraction greater than one, and the new pressure is calculated as

$$P_2 = 1.20 \text{ atm} \times \frac{373 \not{K}}{273 \not{K}} = \mathbf{1.64 \text{ atm}}$$

The Combined Gas Laws

We can combine these gas laws to do calculations, using the following relationships:

$$V_2 = V_1 \times T_{\text{factor}} \times P_{\text{factor}}$$
$$P_2 = P_1 \times V_{\text{factor}} \times T_{\text{factor}}$$
$$T_2 = T_1 \times V_{\text{factor}} \times P_{\text{factor}}$$

In doing calculations involving more than one change, we must consider each change individually. For example, to calculate the new volume as the temperature and pressure are both changed, we first consider how the volume would change as the temperature changes, considering the pressure to be constant. Then we consider how the volume would change as the pressure changes, considering the temperature to be constant.

Consider the following problem examples involving changes in two factors.

PROBLEM EXAMPLE 11-5

A sample of gas occupies a volume of 12.5 ℓ at 625 torr pressure and 150°C. What would the pressure be, in torr, if the volume were increased to 15.0 ℓ and the temperature were increased to 300°C?

SOLUTION:

We first set up our list of original and new conditions, as done in previous problems.

$P_1 = 625$ torr $\qquad\qquad\qquad\qquad P_2 = ?$

$V_1 = 12.5\ \ell \quad \xrightarrow[\text{pressure decreases}]{\text{volume increases}} \quad V_2 = 15.0\ \ell$

$T_1 = 150 + 273 = 423\text{ K} \quad \xrightarrow[\text{pressure increases}]{\text{temperature increases}} \quad T_2 = 300 + 273 = 573\text{ K}$

We then consider each factor individually; that is, we consider first what will happen to the pressure if the volume is increased (considering temperature to remain constant) and then we determine what will happen to the pressure if the temperature is increased (considering volume to remain constant). In the first case, if the volume is increased, the pressure will decrease. Thus the volume factor is set up as 12.5 ℓ/15.0 ℓ, in order to give a decrease due to this factor. In the second case, the temperature is increasing, and thus the pressure will increase. Thus the temperature factor is set up as 573 K/423 K, in order to give an increase for this factor. Therefore, the overall change is calculated as

$$P_2 = 625\text{ torr} \times \frac{12.5\ \cancel{\ell}}{15.0\ \cancel{\ell}} \times \frac{573\ \cancel{\text{K}}}{423\ \cancel{\text{K}}} = \mathbf{706\ torr}$$

PROBLEM EXAMPLE 11-6

A sample of gas occupies a volume of 512 mℓ at 20°C and 740 torr. What volume would this gas occupy at STP?

SOLUTION:

The two sets of conditions are

$V_1 = 512\text{ m}\ell \qquad\qquad\qquad\qquad V_2 = ?$

$T_1 = 20 + 273 = 293\text{ K} \quad \xrightarrow[\text{volume decreases}]{\text{temperature decreases}} \quad T_2 = 0 + 273 = 273\text{ K}$

$P_1 = 740\text{ torr} \quad \xrightarrow[\text{volume decreases}]{\text{pressure increases}} \quad P_2 = 760\text{ torr}$

Proceeding as in Problem Example 11-5, considering each change individually, the new volume is calculated as

$$V_2 = 512\text{ m}\ell \times \frac{273\ \cancel{\text{K}}}{293\ \cancel{\text{K}}} \times \frac{740\ \cancel{\text{torr}}}{760\ \cancel{\text{torr}}} = \mathbf{464\ m\ell}$$

In this particular case both changes resulted in a decrease. Nevertheless they must be considered separately: first, how the volume would change as the temperature is decreased and then how the volume would change as the pressure is increased.

11-6 Dalton's Law of Partial Pressures

John Dalton, who formulated the first experimentally based atomic theory (Section 3-1) also worked with gases. In 1801 Dalton found that the total pressure of a mixture of gases is the same as the sum of the individual pressures of each gas in the mixture. This is known as **Dalton's Law of Partial Pressures.**

For example, consider a mixture of gases containing hydrogen and oxygen in a container. If the pressure due to the hydrogen is 450 torr (that is, if the hydrogen were by itself in the container it would exert a pressure of 450 torr), and the pressure due to the oxygen is 320 torr, the total pressure is 770 torr (450 torr + 320 torr).

One application of Dalton's Law of Partial Pressures is the case of a gas that is collected over water. For example, when oxygen gas is prepared in the lab, it is usually collected over water. That is, a reaction is carried out which produces O_2 gas which is collected by the displacement of water from a container, as shown in Fig. 11-6. As the O_2 gas is produced in the test tube, it bubbles into the bottle. Since oxygen is only slightly soluble in water, the oxygen gas displaces the water from the bottle. When all the water has been displaced, we know that the bottle is filled with oxygen; however, the oxygen gas in the bottle is mixed with water vapor, due to the vapor pressure of water. Thus the total pressure inside the flask is equal to the pressure due to the oxygen plus the pressure of the water vapor that is also present in the bottle. To

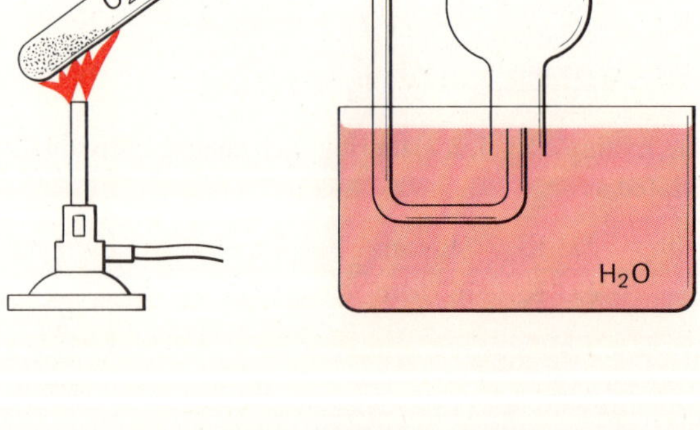

Figure 11-6
Collecting oxygen gas, $O_{2(g)}$, by the displacement of water

find the actual pressure due to the oxygen gas, the vapor pressure of water at the prevailing temperature is subtracted from the total pressure of gas in the bottle. The total pressure in the bottle will be the same as atmospheric pressure if the water level inside the bottle is the same as the level outside.

Consider the following problem example involving the collection of a gas over water.

PROBLEM EXAMPLE 11-7

A sample of oxygen gas collected over water occupies a volume of 210 mℓ at 22°C and 750 torr pressure. Calculate the volume of dry oxygen at STP.

SOLUTION:

Since the total pressure in the bottle is 750 torr and the vapor pressure of water at 22°C is 20 torr (from Appendix V), the pressure due to the oxygen is 750 − 20 = 730 torr. Thus the volume occupied by the dry oxygen at STP is calculated as follows,

$$V_1 = 210 \text{ m}\ell \qquad V_2 = ?$$

$$T_1 = 22 + 273 = 295 \text{ K} \qquad T_2 = 273 \text{ K}$$

$$P_1 = (750 - 20) \text{ torr} = 730 \text{ torr} \qquad P_2 = 760 \text{ torr}$$

$$V_2 = 210 \text{ m}\ell \times \frac{273 \text{ K}}{295 \text{ K}} \times \frac{730 \text{ torr}}{760 \text{ torr}} = \mathbf{187 \text{ m}\ell}$$

11-7 Avogadro's Hypothesis and the Molar Volume of a Gas

Kinetic theory says that the molecules of any gas at the same temperature will have the same average kinetic energy. Therefore, if an equal number of moles of any gas are placed in the same size container at the same temperature the pressure will be the same. This is Avogadro's hypothesis.

From this it can be seen that for a given number of moles of any gas, at the same temperature and pressure, the volume must be the same; that is, the gases would have to be contained in equal volumes in order that the temperature and pressure be the same. It has been found experimentally that 1 mole of any gas at 1 atm (760 torr) pressure and 0°C (273 K) will occupy a volume of 22.4 ℓ; that is, if 1 mole of any gas is placed in a container of 22.4-ℓ volume, at 273 K, the pressure will be 1 atm. Thus 22.4 ℓ is called the molar volume of a gas, at STP.

For gases it is essential that the temperature and pressure be specified when the volume is being considered. Therefore it is not correct to say that 1 mole of any gas will occupy a volume of 22.4 ℓ, but rather we must say that 1 mole of any gas will occupy a volume of 22.4 ℓ at STP.

This dependence of volume of a gas on temperature and pressure must be kept in mind when doing any calculations involving gases. It is especially important however, when doing calculations involving the molar volume of a gas.

We can use this concept to carry out several types of calculations. For example, we can calculate the density of a gas (at STP or any other conditions), the molecular weight (from the density, or the mass of a given volume, at any conditions), the volume of a given mass of gas (or vice versa), or any type of stoichiometry calculations, from a balanced equation involving gases (either at STP or any other conditions). In many of these calculations there are two different things involved: (1) conversions involving the molar volume of a gas, at STP, and (2) gas law calculations, either in going from STP to the given conditions or from the given conditions to STP.

Consider the following problem examples.

PROBLEM EXAMPLE 11-8

Calculate the volume, at STP, occupied by 12.5 g of nitrogen gas.

SOLUTION:

Since nitrogen gas has the formula N_2, the molecular weight is 28.0. Thus, we can convert the given mass to moles and then go from moles to liters, at STP, as follows:

$$12.5\ \cancel{\text{g N}_2} \times \frac{1\ \cancel{\text{mol N}_2}}{28.0\ \cancel{\text{g N}_2}} \times \frac{22.4\ \ell\ \text{(STP)}}{1\ \cancel{\text{mol N}_2}} = \mathbf{10.0\ \ell\ (STP)}$$

PROBLEM EXAMPLE 11-9

Calculate the mass, in grams, of 230 mℓ of Cl_2 gas at 20°C and 740 torr pressure.

SOLUTION:

There are two steps involved in this calculation—first we must find what volume this sample would occupy at STP; then we can use the molar volume of a gas (22.4 ℓ at STP per mole) to finish the calculation.

(a) $V_1 = 230\ \text{m}\ell \qquad V_2 = ?$

$T_1 = 20 + 273 = 293\ \text{K} \qquad T_2 = 273\ \text{K}$

$P_1 = 740\ \text{torr} \qquad P_2 = 760\ \text{torr}$

$$V_2 = 230\ \text{m}\ell \times \frac{273\ \cancel{\text{K}}}{293\ \cancel{\text{K}}} \times \frac{740\ \cancel{\text{torr}}}{760\ \cancel{\text{torr}}} = 209\ \text{m}\ell\ \text{(STP)}$$

(b) $$0.209\ \cancel{\ell\ \text{(STP)}} \times \frac{1\ \cancel{\text{mol Cl}_2}}{22.4\ \cancel{\ell\ \text{(STP)}}} \times \frac{71.0\ \text{g Cl}_2}{1\ \cancel{\text{mol Cl}_2}} = \mathbf{0.662\ g\ Cl_2}$$

PROBLEM EXAMPLE 11-10

Calculate the molecular weight of a gas if 600 mℓ of the gas measured at 30°C and 630 torr has a mass of 0.600 g.

SOLUTION:

(a) Calculate the volume that this sample would occupy at STP.

$V_1 = 600\ \text{m}\ell \qquad V_2 = ?$

$P_1 = 630\ \text{torr} \qquad P_2 = 760\ \text{torr}$

$T_1 = 30 + 273 = 303\ \text{K} \qquad T_2 = 0 + 273 = 273\ \text{K}$

$$V_2 = 600\ \text{m}\ell \times \frac{630\ \cancel{\text{torr}}}{760\ \cancel{\text{torr}}} \times \frac{273\ \cancel{\text{K}}}{303\ \cancel{\text{K}}} = 448\ \text{m}\ell\ (\text{STP}) = 0.448\ \ell\ (\text{STP})$$

(b) Calculate the molecular weight (which can be expressed as grams per mole) as

$$\frac{0.600\ \text{g}}{0.448\ \cancel{\ell\ (\text{STP})}} \times \frac{22.4\ \cancel{\ell\ (\text{STP})}}{1\ \text{mol}} = 30.0\ \text{g/mol}$$

Therefore, the molecular weight is **30.0.**

PROBLEM EXAMPLE 11-11

Calculate the density of O_2 gas at STP.

SOLUTION:

Since 1 mole O_2 = 32.0 g = 22.4 ℓ at STP,

$$\text{density} = \frac{32.0\ \text{g}}{22.4\ \ell} = \mathbf{1.43\ g/\ell\ \text{at STP}}$$

PROBLEM EXAMPLE 11-12

Calculate the density of O_2 gas at 25.0°C and 1.05 atm pressure.

SOLUTION:

To calculate the density at the given conditions we can first calculate the volume of 1 mole of gas at the given conditions and then use the calculated volume with the molecular weight (32.0) to calculate the density.

(a) $V_1 = 22.4\ \ell \qquad V_2 = ?$

$T_1 = 273\ \text{K} \qquad T_2 = 25 + 273 = 298\ \text{K}$

$P_1 = 1.00\ \text{atm} \qquad P_2 = 1.05\ \text{atm}$

$$V_2 = 22.4\ \ell \times \frac{298\ \cancel{\text{K}}}{273\ \cancel{\text{K}}} \times \frac{1.00\ \cancel{\text{atm}}}{1.05\ \cancel{\text{atm}}} = 23.3\ \ell$$

(b) Since we know that the mass of 1 mole of O_2 is 32.0 g and we have just calculated the volume occupied by 1 mole of O_2 at the given conditions, we can calculate the density at the given conditions as

$$\frac{32.0\ \text{g}}{23.3\ \ell} = \mathbf{1.37\ g/\ell}$$

An alternate solution would be to calculate the density at STP and then convert the density to the new conditions, as a gas law problem. In doing this you must keep in mind that density is mass/volume. Therefore, any factor that causes the volume to decrease causes the density to increase, and vice versa. The density at STP is

$$\frac{32.0\ \text{g}}{22.4\ \ell} = 1.43\ \text{g}/\ell\ \text{(STP)}$$

$d_1 = 1.43\ \text{g}/\ell$ $\quad d_2 = ?$

$T_1 = 273\ \text{K}$ $\quad T_2 = 25 + 273 = 298\ \text{K}$

$P_1 = 1.00\ \text{atm}$ $\quad P_2 = 1.05\ \text{atm}$

$$d_2 = 1.43\ \text{g}/\ell \times \frac{273\ \cancel{\text{K}}}{298\ \cancel{\text{K}}} \times \frac{1.05\ \cancel{\text{atm}}}{1.00\ \cancel{\text{atm}}} = \mathbf{1.37\ g/\ell}$$

Notice here that as the temperature increases, the volume increases and hence the density decreases (therefore, the temperature factor is 273 K/298 K). Likewise, as the pressure increases, the volume decreases and thus the density increases (therefore the pressure factor is 1.05 atm/1.00 atm).

We can also use Avogadro's hypothesis and the molar volume of a gas to do stoichiometry calculations involving gases. These are sometimes referred to as mass-volume, and volume-volume stoichiometry. From Avogadro's hypothesis we can see that the balanced equation gives not only the relationships between reactants and products in terms of moles but also in terms of volume. That is, at the same temperature and pressure, if 2 moles of A react with 3 moles of B, we can also say that 2 ℓ of A react with 3 ℓ of B. For example, for the reaction

$$N_{2(g)} + 3\ H_{2(g)} \longrightarrow 2\ NH_{3(g)}$$

the following relationships can be obtained (at the same temperature and pressure):

$$1\ \ell\ N_2 = 3\ \ell\ H_2$$
$$1\ \ell\ N_2 = 2\ \ell\ NH_3$$
$$3\ \ell\ H_2 = 2\ \ell\ NH_3$$

Consider the following problem examples.

PROBLEM EXAMPLE 11-13

Calculate the volume of CO_2, at 27°C and 785 torr, that could be obtained by heating 12.8 g of calcium carbonate.

SOLUTION:

The equation for the reaction is

$$CaCO_3 \xrightarrow{\Delta} CaO + CO_2$$

As with all other stoichiometry calculations, we use moles to find the relationship from the balanced equation. Then we use the molar volume of a gas to calculate the volume of CO_2 at STP. Finally we must do a gas law calculation to determine what the volume of CO_2 would be at the given conditions.

$$12.8 \text{ g } CaCO_3 \times \frac{1 \text{ mol } CaCO_3}{100 \text{ g } CaCO_3} \times \frac{1 \text{ mol } CO_2}{1 \text{ mol } CaCO_3} \times \frac{22.4\ l\ CO_2 \text{ (STP)}}{1 \text{ mol } CO_2} = 2.87\ l\ CO_2 \text{ (STP)}$$

$$V_1 = 2.87\ l \qquad V_2 = ?$$
$$T_1 = 273 \text{ K} \qquad T_2 = 300 \text{ K}$$
$$P_1 = 760 \text{ torr} \qquad P_2 = 785 \text{ torr}$$

$$V_2 = 2.87\ l \times \frac{300 \text{ K}}{273 \text{ K}} \times \frac{760 \text{ torr}}{785 \text{ torr}} \quad \mathbf{3.05\ l}$$

PROBLEM EXAMPLE 11-14

From the equation

$$C_2H_{4(g)} + 3\ O_{2(g)} \longrightarrow 2\ CO_{2(g)} + 2\ H_2O_{(g)}$$

calculate the volume of O_2 necessary and the volume of CO_2 produced from 1.2 l of C_2H_4, all volumes being measured at 350°C and 1.2 atm pressure.

SOLUTION:

Since the gases are all of the same temperature and pressure, the volumes are related to the coefficients from the balanced equation in the same way that moles are related. Thus the volume of O_2 and CO_2 are calculated as

$$1.2\ l\ C_2H_4 \times \frac{3\ l\ O_2}{1\ l\ C_2H_4} = \mathbf{3.6\ l\ O_2}$$

$$1.2\ l\ C_2H_4 \times \frac{2\ l\ CO_2}{1\ l\ C_2H_4} = \mathbf{2.4\ l\ CO_2}$$

11-8 Gas Law Relationships Involving Moles

Up to this point we have only considered changes in volume, pressure, and temperature for a given amount of gas. There are also direct relationships between the number of moles of gas and the volume, pressure, and temperature. The three relationships are as follows:

1. *Volume is directly proportional to the number of moles (n).* That is, if the pressure and temperature are to remain constant as more moles

(and hence more molecules) are added, the volume must increase. This can be expressed as

$$V \propto n \quad \text{(constant } T \text{ and } P\text{)}$$

or

$$V = kn \quad \text{or} \quad \frac{V}{n} = k$$

2. *Pressure is directly proportional to the number of moles of gas.* That is, for the same volume at constant temperature (and hence the same average kinetic energy) if there are more molecules striking the sides of the container, the pressure is increased. This can be expressed as

$$P \propto n \quad \text{(constant } V \text{ and } T\text{)}$$

or

$$P = kn \quad \text{or} \quad \frac{P}{n} = k$$

3. *Temperature is inversely proportional to the number of moles.* That is, if more moles of gas are added and the volume and pressure are to remain constant, the temperature must be decreased. Conversely, if the number of molecules (moles) of gas is decreased and the pressure is to remain constant, the temperature must be increased. This can be expressed as

$$T \propto \frac{1}{n} \quad \text{(constant } V \text{ and } P\text{)}$$

or

$$T = \frac{k}{n} \quad \text{or} \quad Tn = k$$

It must be kept in mind when considering these relationships that the number of moles does not change as the pressure, temperature, or volume change. If two variables, such as temperature and volume, are to remain constant, however, the pressure must change as the number of moles is changed or the number of moles must be changed (that is, some gas removed, or more added) if the pressure is to change.

These relationships can be used in calculations just as the first three relationships that we considered in Section 11-5.

Consider the following problem examples.

PROBLEM EXAMPLE 11-15

A container of constant volume contains 2.4 moles of a gas at 100°C and a pressure of 1.2 atm. If 1.4 moles of gas are added and the temperature remains constant, calculate the new pressure.

SOLUTION:

$$P_1 = 1.2 \text{ atm} \qquad\qquad P_2 = ?$$

$$n_1 = 2.4 \text{ moles} \xrightarrow[\text{pressure increases}]{\text{number of moles increases}} n_2 = 3.8 \text{ moles}$$

$$P_2 = 1.2 \text{ atm} \times \frac{3.8 \text{ mol}}{2.4 \text{ mol}} = \mathbf{1.9 \text{ atm}}$$

PROBLEM EXAMPLE 11-16

If 1.2 moles of a gas are confined in a 10.0-l container at a temperature of 100°C, what is the pressure?

SOLUTION:

It looks as though we only have one set of conditions here; however, we know that 1 mole of gas at STP occupies a volume of 22.4 l. Hence, we can use this as the initial set of conditions as

$$P_1 = 1.0 \text{ atm} \qquad\qquad P_2 = ?$$

$$V_1 = 22.4\ l \xrightarrow[P \text{ increases}]{V \text{ decreases}} V_2 = 10.0\ l$$

$$T_1 = 273 \text{ K} \xrightarrow[P \text{ increases}]{T \text{ increases}} T_2 = 373 \text{ K}$$

$$n_1 = 1 \text{ mole} \xrightarrow[P \text{ increases}]{n \text{ increases}} n_2 = 1.2 \text{ moles}$$

$$P_2 = 1.0 \text{ atm} \times \frac{22.4\ l}{10.0\ l} \times \frac{373 \text{ K}}{273 \text{ K}} \times \frac{1.2 \text{ mol}}{1 \text{ mol}} = \mathbf{3.7 \text{ atm}}$$

11-9 Ideal Gas Law (Equation of State)*

Combining the relationships between volume and temperature ($V \propto T$), volume and pressure ($V \propto 1/P$), and volume and number of moles ($V \propto n$) gives

$$V \propto (T)\left(\frac{1}{P}\right)(n) \quad \text{or} \quad V = R(T)\left(\frac{1}{P}\right)(n)$$

where R is a proportionality constant. This equation is usually rewritten as

$$PV = nRT.$$

This is called the equation of state for an ideal gas.

The value of R will depend on the units used to express the volume and pressure; remember temperature must always be in kelvin and n is always moles. For example, if volume is in liters and pressure in atmospheres,

$$R = \frac{PV}{nT} = \frac{(1.0 \text{ atm})(22.4\ l)}{(1.0 \text{ mol})(273 \text{ K})} = \frac{0.082\ l\text{-atm}}{\text{mol K}}$$

*This section is optional and can be omitted without loss of continuity.

(from the fact that, at STP, 1 mole of any gas = 22.4 ℓ). If volume is in milliliters and pressure in torr,

$$R = \frac{(760 \text{ torr})(22{,}400 \text{ m}\ell)}{(1.00 \text{ mol})(273 \text{ K})} = 6.24 \times 10^4 \frac{\text{m}\ell\text{-torr}}{\text{mol K}}$$

and so forth. Therefore, if this equation is used, make sure that the correct value of R is used.

It is also possible to use this equation to calculate molecular weight by substituting grams divided by molecular weight of the gas for n; that is,

$$PV = \frac{\text{g } RT}{\text{mw}} \quad \text{or} \quad \text{mw} = \frac{\text{g } RT}{PV}$$

This equation of state is not necessary for calculations involving gases. Any problem involving gases can be solved without it, but it is a useful equation and you should learn to use it along with other types of gas law calculations. If this is used, it is essential to understand how it is derived; that is, how it is related to the various gas laws and what the value of R is with the units that are being used.

Consider the following problem example.

PROBLEM EXAMPLE 11-17

Calculate the pressure in atmospheres, in Problem Example 11-16, using $PV = nRT$.

SOLUTION:

If we want to express the pressure in atmospheres, we use 0.082 ℓ-atm/mol-K for R.

$$P = \frac{nRT}{V} = \frac{(1.2 \text{ mol})(373 \text{ K})(0.082) \ell \text{ atm}}{10.0 \ell \text{ mol K}} = \mathbf{3.7 \text{ atm}}$$

This equation is very useful for gas calculations if the correct value of R is used. From this we can also derive another equation to use for any gas law calculation, as follows: Since $PV = nRT$, or $R = PV/nT$, and since R is a constant, we can say that

$$R_1 = R_2 \quad \text{or} \quad \frac{P_1 V_1}{n_1 T_1} = \frac{P_2 V_2}{n_2 T_2}$$

If all but one of these are known, we can solve for the unknown quantity. In some cases, one or more of the quantities may remain constant, in which case that quantity drops out. For example, solving for V_2, for a given amount of gas (that is, n is constant),

$$V_2 = V_1 \times \frac{P_1}{P_2} \times \frac{T_2}{T_1}$$

11-10 Graham's Law of Diffusion*

The average kinetic energy of a particle of matter, such as a gas molecule, in motion is equal to one-half the mass times the velocity squared. This is stated mathematically as

$$\text{ke} = \tfrac{1}{2}mv^2$$

At the same temperature, the average kinetic energy of any molecule is the same and, therefore, in comparing two gases at the same temperature,

$$\text{ke}_1 = \text{ke}_2 \quad \text{or} \quad \tfrac{1}{2}m_1{v_1}^2 = \tfrac{1}{2}m_2{v_2}^2$$

Simplifying this gives

$$\frac{m_1}{m_2} = \frac{{v_2}^2}{{v_1}^2} \quad \text{or} \quad \sqrt{\frac{m_1}{m_2}} = \frac{v_2}{v_1}$$

Since the molecular weight of a gas gives the relative mass of the molecules, we can substitute molecular weight for mass in this equation. Therefore, if the relative velocities of two different gases can be measured and the molecular weight of one of them is known, the molecular weight of the other can be calculated. To measure the relative velocities of two gases, we can compare their rates of diffusion and from this we can calculate the molecular weight. This relationship is called **Graham's Law of Diffusion,** which can be stated as *the relative rates of diffusion of two gases are inversely proportional to the square root of their molecular weights.*

Consider the following problem example.

PROBLEM EXAMPLE 11-18

If a gas diffuses at a rate of $\frac{1}{2}$ as fast as O_2, what is the molecular weight of the gas?

SOLUTION:

If we let the unknown gas be gas 1 and oxygen be gas 2,

$$\frac{\frac{1}{2}}{1} = \sqrt{\frac{32.0}{x}}$$

The easiest way to solve this is to square both sides of the equation. This gives

$$\left(\frac{1}{2}\right)^2 = \frac{32.0}{x}; \qquad \frac{1}{4} = \frac{32.0}{x}; \qquad x = (32.0)(4) = \mathbf{128}$$

In doing calculations of this type, always check your answer to see if it is reasonable. For example, in Problem Example 11-18, we know that the unknown gas should have a molecular weight greater than that of oxygen since it diffuses slower.

*This section is optional and can be omitted without loss of continuity.

SUMMARY

1. Due to the arrangement of the particles in gases, they are easy to compress, have no fixed volume or shape, diffuse very readily, mix completely, and have much lower densities than liquids or solids.
2. *Kinetic theory* states that temperature is an indication of the motion of particles in a substance. For an ideal gas it is assumed that: the volume of the particles is negligible; the molecules are moving so rapidly and are so far apart that the forces of attraction between molecules are negligible; and the average kinetic energy is the same for any size molecule at the same temperature.
3. Real gases act almost like ideal gases at ordinary conditions (that is, not extremely high pressure or extremely low temperature).
4. In making quantitative comparisons of the properties of gases, temperature must always be expressed in degrees kelvin. Pressure is often expressed in centimeters or millimeters of Hg, or torr, where 1 torr equals 1 mm Hg. The amount of gas is always expressed in moles (n). Standard temperature and pressure (STP) is taken as 0°C (273 K) and 1 atm pressure (760 mm Hg or 760 torr).
5. The six gas law relationships can be expressed as follows:
   (a) Boyle's Law

$$P \propto \frac{1}{V} \quad \text{or} \quad P = \frac{k}{V} \quad \text{or} \quad PV = k$$

(at constant T and n)

   (b) Charles' Law

$$V \propto T \quad \text{or} \quad V = kT \quad \text{or} \quad \frac{V}{T} = k$$

(at constant P and n)

   (c) Gay-Lussac's Law

$$P \propto T \quad \text{or} \quad P = kT \quad \text{or} \quad \frac{P}{T} = k$$

(at constant V and n)

   (d)

$$P \propto n \quad \text{or} \quad P = kn \quad \text{or} \quad \frac{P}{n} = k$$

(at constant V and T)

   (e)

$$V \propto n \quad \text{or} \quad V = kn \quad \text{or} \quad \frac{V}{n} = k$$

(at constant P and T)

(f)

$$T \propto \frac{1}{n} \quad \text{or} \quad T = \frac{k}{n} \quad \text{or} \quad Tn = k$$

(at constant P and V)

6. At STP, 1 mole of any gas will occupy a volume of 22.4 ℓ. Thus, we can calculate the molecular weight, the density, or the volume, temperature, or pressure for any known amount of a gas. In general, it is noted that equal numbers of moles of any gas, at the same conditions of temperature and pressure, will occupy the same volume.
7. Since 1 mole of any gas occupies the same volume at the same conditions of temperature and pressure, for gases in balanced chemical equations the relationship in terms of volume is the same as the relationship in moles.
8. Since $V = kT$, $V = k(1/P)$, and $V = kn$, we can combine these into the statement, $V = (R)(T)(1/P)(n)$, where R is a constant. This is usually written as $PV = nRT$ and is known as the equation of state for an ideal gas. This equation can be used to do gas law calculations, either in this form or if gases at different conditions are being compared this can be used to derive the expression

$$\frac{P_1V_1}{n_1T_1} = \frac{P_2V_2}{n_2T_2}$$

9. *Graham's Law of Diffusion* states that the ratio of the diffusion of two gases is inversely proportional to the square root of the ratio of their molecular weights:

$$\frac{R_1}{R_2} = \sqrt{\frac{m_2}{m_1}}$$

Upon completion of this chapter you should be able to do the following:

1. Given any of the properties of gases (expansion, indefinite shape and volume, compressibility, low density, mixability, diffusion, etc.), explain it in terms of the kinetic theory as applied to gases.
2. Given a pressure measurement in any unit, convert it to any other unit (mm Hg, cm Hg, torr, atm, or Pa).
3. Given any one quantity of a gas (volume, temperature, pressure, or number of moles), *explain* the relationship between that quantity and any other quantity, assuming everything else is held constant. For example, explain how pressure varies with temperature, volume with number of moles, etc. (there are six relationships here: *P-V*, *P-T*, *P-n*, *V-T*, *V-n*, *T-n*).

4. Given a fixed amount of gas, at a stated volume, temperature, and pressure, calculate the new volume if the temperature or pressure is changed, the new temperature if the pressure or volume is changed, or the new pressure if the temperature or volume is changed. (1 and 2)
5. Given the partial pressures of a mixture of gases, calculate the total pressure of the gases. Given the total pressure of a mixture of gases, calculate the partial pressure of one gas if the pressure of the other gases in the mixture is known or the fraction of each gas present is known. (3)
6. Given a fixed amount of a given gas, collected over water, at a stated volume, temperature, and pressure, and the vapor pressure of the water at the stated temperature, calculate the volume of the dry gas at any temperature and pressure. (4)
7. Given the name or formula of a gas and a table of atomic weights, calculate the density of the gas at any stated temperature and pressure. (5)
8. Given the mass of a gas and its volume under any stated conditions of temperature and pressure, calculate the molecular weight of the gas. (6)
9. Given the mass or volume (at any conditions) of a sample of a given gas, calculate the volume (at any conditions) or mass of the sample. (7, 8)
10. Given any reaction involving gases, do stoichiometric calculations involving mass-volume or volume-volume relationships. (9)
11. Given any three of the quantities volume, temperature, pressure, and number of moles for any gas, be able to calculate the fourth quantity, using either the molar volume of a gas at STP or the equation of state of an ideal gas ($PV = nRT$). (10)
12. Given the names or formulas of any two gases, calculate the relative rates of diffusion of the gases. (11)
13. Given the relative rates of diffusion of two gases and the molecular weight of one of the gases (or the name or formula and a table of atomic weights), calculate the molecular weight of the other gas. (12)
14. Given any of the terms listed in Exercise 1, give a definition or explanation in your own words.

EXERCISES

1. Define or explain the following terms:
   (a) ideal gas
   (b) STP
   (c) torr
   (d) millimeters of Hg pressure
   (e) molar volume of a gas
   (f) Dalton's Law of Partial Pressures
2. Distinguish between a real gas and an ideal gas.
3. Explain, in terms of kinetic theory, each of the following:
   (a) As the temperature of a sample of gas increases, at constant volume, the pressure increases.
   (b) Temperature and volume are directly proportional for a given sample of gas at constant pressure.
   (c) Pressure and volume are inversely proportional for a given sample of gas at constant temperature.
   (d) As the number of moles of gas is increased, at constant temperature, the volume must be increased in order to keep the pressure constant.
   (e) The number of moles of gas present in a sample is directly proportional to the pressure at constant volume and temperature.
   (f) Number of moles and temperature are inversely proportional at constant pressure and volume.

PROBLEMS

Gas law calculations related to volume, temperature, and pressure

1. A sample of gas occupies a volume of 1.25 *l* at 100°C and 760 torr.
   (a) Calculate the new volume, in liters, if the temperature is changed to 25°C at constant pressure.
   (b) Calculate the new temperature, in kelvin, if the volume is changed to 1.75 *l* and the pressure remains constant.
   (c) Calculate the new pressure, in torr, if the temperature is changed to 20°C at constant volume.
   (d) Calculate the new temperature, in degrees Celsius, if the volume is changed to 0.850 *l* at constant pressure.
   (e) Calculate the new pressure, in atmospheres, if the volume is changed to 1.65 *l* at constant temperature.
   (f) Calculate the new temperature, in kelvin, if the volume is changed to 1.12 *l* at constant pressure.
   (g) Calculate the new volume, in liters, if the temperature is changed to 30°C and the pressure remains constant.
   (h) Calculate the new temperature, in kelvin, if the pressure is changed to 780 torr at constant volume.
   (i) Calculate the new pressure, in torr, if the temperature is changed to 20°C, at constant volume.
   (j) Calculate the new pressure, in atmospheres, if the volume is changed to 0.950 *l* at constant temperature.

Combined gas law problems

2. A sample of gas occupies a volume of 625 ml at 22°C and 1.12 atm.
   (a) Calculate the new pressure, in atmospheres, if the volume is changed to 750 ml and the temperature is changed to 27°C.
   (b) Calculate the volume at STP.
   (c) Calculate the new temperature, in degrees kelvin, if the volume is changed to 535 ml and the pressure is changed to 0.950 atm.
   (d) Calculate the new volume if the temperature is changed to 100°C and the pressure is changed to 775 torr.
   (e) Calculate the new temperature, in degrees Celsius, if the volume is changed to 1.15 l and the pressure is changed to 75.2 cm Hg.
   (f) Calculate the new volume if the pressure is changed to 0.825 atm and the temperature is changed to 35°C.
   (g) Calculate the new temperature, in degrees kelvin, if the volume is changed to 425 ml and the pressure is changed to 1.32 atm.
   (h) Calculate the new pressure, in atmospheres, if the volume is changed to 845 ml and the temperature is changed to 15°C.
   (i) Calculate the new temperature, in degrees Celsius, if the volume is changed to 1.35 l and the pressure is changed to 740 torr.
   (j) Calculate the new pressure, in centimeters of Hg, if the volume is changed to 435 ml and the temperature is changed to 100°C.

Dalton's Law of Partial Pressures

3. Calculate the total pressure of each of the following:
   (a) a mixture of gases, each of which exerts the given partial pressure: H_2, 150 torr; N_2, 210 torr; He, 320 torr.
   (b) a mixture of gases, each of which exerts the given partial pressure: He, 250 torr; Ne, 180 torr; Ar, 310 torr.
4. Each of the following gases is measured over water. Calculate the volume of dry gas, at STP, in each case (use the vapor pressure of water as found in Appendix V).
   (a) 250 ml of oxygen, at 20°C and 745 torr
   (b) 525 ml of nitrogen, at 25°C and 1.15 atm
   (c) 435 ml of hydrogen, at 27°C and 774 torr
   (d) 125 ml of oxygen, at 30°C and 0.920 atm

Molar volume of a gas

5. Calculate the density of each of the following:
   (a) carbon dioxide gas at STP
   (b) nitrogen gas at 20°C and 745 torr
   (c) NO_2 gas at STP
   (d) helium gas at 25°C and 775 torr
6. Calculate the molecular weight of each of the following gases:
   (a) a gas whose density is 1.75 g/l at STP
   (b) a volume of 125 ml of a gas at STP has a mass of 0.420 g
   (c) a gas whose density is 1.25 g/l at 0.75 atm and 20°C
   (d) a volume of 245 ml of gas at 27°C and 740 torr has a mass of 0.825 g
   (e) a volume of 140 ml of a gas at STP has a mass of 0.570 g
   (f) a gas whose density is 1.52 g/l at STP

(g) a volume of 325 ml of a gas at 20°C and 735 torr has a mass of 1.05 g

(h) a gas whose density is 1.45 g/l at 27°C and 1.12 atm pressure.

7. Calculate the mass, in grams, of each of the following:
   (a) 48.6 l of H_2 gas at STP
   (b) 1.75 l of CO_2 gas at 20°C and 740 torr
   (c) 2.82 l of oxygen gas at STP
   (d) 32.5 l of NO_2 gas at 27°C and 1.05 atm

8. Calculate the volume of each of the following:
   (a) 4.25 g of oxygen gas at STP
   (b) 0.530 g of He gas at 100°C and 730 torr
   (c) 0.375 g of nitrogen gas at STP
   (d) 8.25 g of Ne gas at 150°C and 1.45 atm

Mass-volume and volume-volume stoichiometry

9. For the reaction

$$2\,C_2H_{6(g)} + 7\,O_{2(g)} \longrightarrow 4\,CO_{2(g)} + 6\,H_2O_{(g)}$$

   (a) calculate the volume, in liters, of CO_2, at STP, that could be produced from 7.25 g of C_2H_6.
   (b) calculate the mass, in grams, of O_2 necessary to react with 425 ml of C_2H_6 at STP.
   (c) calculate the volume of CO_2, at 20°C and 750 torr, that could be produced from 3.75 moles of O_2.
   (d) calculate the volume of O_2, at 100°C and 1.25 atm, that would react with 7.52 g of C_2H_6.
   (e) calculate the volume of C_2H_6 that will react with 1.35 l of O_2 (at the same temperature and pressure).
   (f) calculate the mass, in grams, of CO_2 that could be produced from 8.75 l of C_2H_6 at STP.
   (g) calculate the volume, in liters, of CO_2 at STP that could be produced from 0.825 g of O_2.
   (h) calculate the volume of O_2, at 35°C and 1.20 atm, that would react with 1.45 g of C_2H_6.
   (i) calculate the volume of CO_2, at 100°C and 820 torr, that could be produced from 12.5 g of C_2H_6.
   (j) calculate the volume of O_2 that will react with 4.82 l of C_2H_6 (at the same temperature and pressure).

Molar volume of a gas (or equation of state)

10. Calculate the number of
   (a) moles of gas occupying 742 ml at STP.
   (b) grams of CO_2 gas occupying 28 l at STP.
   (c) grams of NO_2 gas occupying a volume of 275 ml at 200°C and 748 torr.
   (d) moles of O_2 gas occupying 346 ml at STP.
   (e) grams of NO gas occupying 5.2 l at STP.
   (f) grams of CO_2 gas occupying a volume of 4.20 l at 100°C and 1.28 atm.

Graham's Law of Diffusion

11. Calculate the relative rates of diffusion of each of the following:
   (a) H_2 and N_2
   (b) N_2 and O_2

12. Calculate the molecular weight of the unknown gas in each of the following:
(a) a gas that diffuses one-third as fast as hydrogen
(b) a gas that diffuses one-half as fast as nitrogen

General problems

13. Calculate the final pressure, in torr, for each of the following:
(a) 142 mℓ of a gas measured at 480 torr pressure is allowed to expand to volume of 200 mℓ at constant temperature.
(b) A sample of a gas at -60°C and 1 atm pressure is heated to 200°C at constant volume.
(c) A given sample of gas has a volume of 5.2 ℓ at 27°C and 640 torr pressure. Its volume and temperature are changed to 2.1 ℓ and 100°C, respectively.
(d) 320 mℓ of a gas measured at 600 torr pressure is changed to a volume of 250 mℓ at constant temperature.
(e) A gas at 300°C and 900 torr pressure is cooled to 0°C at constant volume.
(f) A gas measures 250 mℓ at STP. Its volume and temperature are changed to 450 mℓ and 50°C, respectively.

14. Calculate the final temperature in degrees Celsius for each of the following:
(a) 425 mℓ of a gas measured at 20°C is cooled at constant pressure until its volume is 365 mℓ.
(b) A gas measured at 25°C and 600 torr is heated at constant volume until its pressure is 1400 torr.
(c) A gas measured 150 mℓ at STP. Its volume and pressure are changed to 300 mℓ and 950 torr, respectively.
(d) A gas measured at 100°C and 700 torr pressure is heated at constant volume until its pressure is 1200 torr.

15. Calculate the final density, in grams per liter, of each of the following:
(a) A gas has a density of 2.3 g/ℓ at 50°C and 760 torr pressure. Its temperature and pressure are changed to 0°C and 640 mm Hg, respectively.
(b) The density of a certain gas at STP is 1.43 g/ℓ. Its temperature and pressure are changed to 40°C and 600 torr, respectively.

16. Calculate the density of each of the following gases in grams per liter at STP.
(a) NH_3
(b) N_2
(c) CO_2
(d) PH_3

17. Calculate the density of each of the following gases in grams per liter at the temperature and pressures indicated.
(a) CO_2 at 30°C and 3.0 atm
(b) Cl_2O at 50°C and 2.0 atm
(c) C_3H_8 at 100°C and 800 torr
(d) CH_4 at -50°C and 300 torr

18. Calculate the molecular weight of each of the following gases:
(a) The density of a gas at 20°C and 600 torr is 1.8 g/ℓ
(b) A volume of 720 mℓ of a gas at STP weighs 0.20 g
(c) The density of a gas at 30°C and 460 torr is 2.1 g/ℓ.
(d) At STP, 0.40 g of a gas occupies 1.23 ℓ

19. Calculate the volume, in liters, of each of the following:
    (a) 7.25 g of CO_2 gas at 20°C and 0.925 atm
    (b) 12.5 moles of oxygen gas at 120°F and 68.5 cm Hg pressure
    (c) 4.5×10^{24} molecules of H_2O gas at STP
    (d) 0.275 mole of helium gas at 27°C and 825 torr.
    (e) 18.5 g of NO_2 gas at 70°F and 78.4 cm Hg pressure
    (f) 9.2×10^{25} molecules of CO_2 gas at STP
20. Calculate the volume, in liters, of hydrogen gas produced at STP
    (a) from the reaction of 2.35 g of zinc with excess hydrochloric acid.
    (b) from 1.75 g of magnesium and 25.0 mℓ of 15.0% nitric acid, whose density is 1.12 g/mℓ.
    (c) from the reaction of 8.75 g of aluminum with excess nitric acid.
    (d) from 4.37 g of zinc and 42.5 mℓ of 24.0% hydrochloric acid whose density is 1.20 g/mℓ.
21. Calculate the volume, in liters, of H_2 gas produced at the given conditions:
    (a) from 0.150 g of sodium reacting with excess water; the H_2 is measured at 25°C and 1.08 atm.
    (b) from 9.35 g of calcium with excess nitric acid; the H_2 gas is measured at 100°C and 785 torr.
22. For the reaction,

$$N_{2(g)} + 3\ H_{2(g)} \longrightarrow 2\ NH_{3(g)}$$

    (a) how many liters of NH_3, at STP, could be produced from 18.5 g of H_2?
    (b) how many liters of H_2 react with 7.25 ℓ of N_2, all at STP?
    (c) if 4.8 ℓ of N_2 and 9.3 ℓ of H_2 are present to start, how many liters of excess reagent will remain after the reaction is complete (all measured at STP)?
    (d) how many liters of NH_3 could be produced from 7.3 ℓ of N_2, all at STP?
    (e) how many liters of H_2, at STP, would react with 3.25 g of N_2?
    (f) if 0.75 ℓ of N_2 and 0.98 ℓ of H_2 are present to start, how many liters of excess reagent will remain after reaction is complete (all measured at STP)?

REFERENCES

BENSON, SIDNEY W., *Chemical Calculations,* 3rd ed. New York: John Wiley & Sons, Inc., 1971.

PETERS, EDWARD I., *Problem Solving for Chemistry,* 2nd ed. Philadelphia: W. B. Saunders Co., 1977.

Objectives

1. To understand what types of attractive forces hold the particles together in ionic, polar covalent, nonpolar covalent, metallic, and macromolecular liquids and solids and to be able to predict variations in these forces for different substances.
2. To understand the properties of the liquid state of matter and to be able to explain how liquids differ from gases.
3. To understand what is meant by the term miscibility of liquids and to be able to predict which types of liquids will mix and which will not mix.
4. To understand what evaporation and vapor pressure are and to be able to explain these properties in terms of kinetic theory and to explain and to predict variations in various compounds.
5. To understand what boiling point and heat of vaporization are and to be able to explain and predict variations in these properties.
6. To understand what distillation is and to be able to explain what it is used for and how it works.
7. To understand what surface tension is and what it is caused by and to explain how it varies in different compounds.
8. To understand the properties of solids and to be able to explain how solids differ from liquids.
9. To understand what melting (or freezing) point and heat of fusion are and to be able to explain and to predict variations in these properties.
10. To understand what sublimation is.
11. To understand what allotropic forms or allotropes are.
12. To be able to do calculations involving heat of fusion, heat of vaporization, and specific heat.
13. To understand what a phase diagram is and to be able to interpret phase diagrams.
14. To be able to predict variations in properties such as melting point, boiling point, vapor pressure, heat of vaporization, heat of fusion, specific heat, and surface tension from a knowledge of the forces of attraction between the units in a substance.
15. To learn what hydrogen bonding is and how this affects the physical properties of water, such as density, melting point, boiling point, heat of fusion, vaporization, and specific heat.

Twelve

Liquids and Solids and Changes of State

When the temperature of any real gas becomes sufficiently low, or the pressure sufficiently high, the gas will turn to a liquid. When a liquid is cooled, the kinetic energy decreases, the motion of the particles is slowed down, and the liquid becomes solid. These changes take place due to the attractive forces that exist between the particles that make up the substance. You will recall that in the gas phase the molecules are so far apart and moving with such rapid motion that the forces of attraction are negligible. In the liquid and solid state, however, the forces of attraction between particles are very important. We shall see that by considering the variation in these forces for different substances we can explain and predict variations in many of the properties of liquids and solids.

12-1 Attractive Forces in Liquids and Solids

The attractive forces between particles are very important in considering the liquid and solid states and changes of state. There are five types of attractive force that we shall consider: ionic, dipole, van der Waals, metallic, and covalent.

Ionic Attraction

In Section 7-1 we discussed the ionic bond and how it is formed. You will recall that this bond consists of an electrostatic attraction between two oppositely charged ions. This is a rather strong force of attraction, especially when we consider that in an ionic crystal each negative ion is surrounded by positive ions and each positive ion is surrounded by negative ions. An ion with a relatively small size and relatively high charge is said to have a high *charge density.* The greater the charge density is on the ions in a crystal, the stronger are the forces of attraction. We shall see later in this chapter that the stronger the forces of attraction between the particles (ions in this case) in a liquid or solid, the higher are the melting and boiling points and other related properties.

Dipole Attraction

In a polar covalent compound (Section 7-2), one part of the molecule has a partial positive charge and another part has a partial negative charge. This is therefore considered as a *dipole.* Thus, the positive portion of one molecule will be attracted toward the negative portion of another. The more polar the compound is, the greater these forces of attraction will be. Therefore, within this class of polar covalent compounds there is considerable variation between the attractive forces; as we shall see later in the chapter, this causes variation in the properties such as melting and boiling points.

Van der Waals Forces

Even in a completely nonpolar substance, such as oxygen, hydrogen, or helium gas, there are weak forces of attraction between the particles. In the gas state these forces are negligible; but as the gas is cooled down and compressed, it eventually becomes a liquid. In this state there must be some force of attraction in order to hold the particles together. These weak forces of attraction in nonpolar substances are referred to either as **van der Waals forces** or *London dispersion forces.* They are generally considered to be caused by a temporary dipole being formed as the electrons move about in the substance. This causes a very weak force of attraction for a short period of time and then another temporary dipole is formed.

This type of attractive force probably exists in all substances, but it is so much weaker than the ionic attraction in ionic compounds or the dipole-dipole attraction in polar covalent substances that it is negligible in these compounds. In the absence of any other type of attractive forces, however, these very weak ones are important.

Since the number of electrons in a substance increases with the formula weight, the van der Waals forces also increase with increasing formula weight. Thus we shall see that those properties of liquids and solids that depend on the forces of attraction between particles (such as melting and boiling point) vary with formula weight for all substances but that this variation is most noticeable for nonpolar substances.

Metallic Bond

The metallic bond is generally considered to consist of a group of positive ions and a sea of electrons, which are free to move about among the ions. This accounts for the general properties of metals, such as their ability to conduct an electric current. For example, in a piece of sodium metal the "ions" are held in a fixed position in the metal and the valence electrons (one from each sodium atom) are free to move about among the various electron clouds. This is shown in Fig. 12-1. This type of bonding can be considered to be somewhat

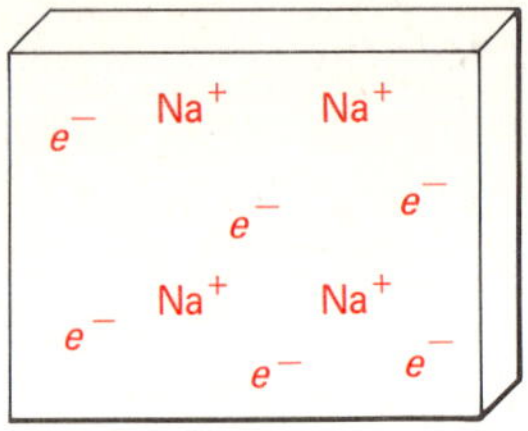

Figure 12-1
Metallic bonding in sodium metal

between ionic and covalent bonding; that is, the valence electrons are not transferred from one atom to another and are not shared between any two atoms in a molecular orbital but are free to move about among the various electron clouds.

Therefore, in metals the forces of attraction that must be overcome to convert from the solid to the liquid state, or from the liquid to the gas state, are quite strong. Of course, these attractive forces vary from one metal to another but in general they are quite strong.

Covalent Bond

The last type of attractive force that we shall consider is the covalent bond between atoms in a *macromolecular* substance, such as a diamond. A macromolecular substance is considered to be one large molecule. The attractive forces that must be overcome to melt or boil one of these substances are the covalent bonds between the atoms. These attractive forces are extremely great and it takes a very high temperature to melt a substance of this type. Along with the diamond, another common macromolecular substance is silicon dioxide (SiO_2), which is the principal constituent of sand.

12-2 The Liquid State

In discussing gases, it was pointed out that in the gas state the molecules are widely separated from each other and are moving at a very rapid rate. Due to this wide separation and rapid motion, the forces of attraction between the molecules are negligible for an ideal gas. At ordinary conditions of temperature and pressure, this is also true for real gases. It was also pointed out, however, that at low temperatures or high pressures real gases begin to deviate from ideal behavior. This is due to the fact that as the motion of molecules is slowed sufficiently or as they are forced closer together by an increase in pressure, the forces of attraction that exist in all real substances become significant.

Therefore, as a real gas is cooled down sufficiently the gas changes to the liquid state. In a liquid the particles are attracted to each other and are held together, even though they are quite free to move about. The liquid state is actually an intermediate condition between a gas and a solid. In a gas each molecule is completely independent of the others and is moving very rapidly, while in a solid the particles are attracted to each other and are held in a definite arrangement with very little, if any, movement.

Since the particles in a liquid are attracted to each other and are held together, even though they are moving quite freely, a sample of liquid has a definite volume, no matter what size container it is placed in. The liquid does not have a definite shape, however, but takes the shape of whatever container it is in. Also, liquids have considerably higher densities than gases and are very difficult to compress. As a comparison of the difference in density between liquids and gases, water in the liquid state at 100°C and 1 atm pressure has a density of 0.96 m/. Water in the gaseous state, under the same conditions, however, has a density of only 0.60 g/l (0.00060 g/ml).

Mixability of Liquids

When two liquids that will mix with each other are placed together, one will diffuse throughout the other but at a much slower rate than is the case with gases. This property is due to the arrangement of the particles in a liquid as opposed to the arrangement of molecules in a gas.

Liquids that are alike will mix, whereas liquids that are unlike will not mix; that is, two polar liquids will mix or two nonpolar liquids will mix, but a polar and a nonpolar will not mix. For example, when a nonpolar substance such as oil is shaken up with a polar substance such as water, as soon as the shaking is stopped, the two liquids will separate out into layers. The oil will be on top and the water on the bottom layer, due to the difference in density between the two substances. The reason that the oil and water will not mix is that the polar water molecules are attracted to each other and force the nonpolar oil molecules out of the way; therefore they form one layer of the polar substance and another layer of the nonpolar substance. When two polar substances are mixed, however, one polar molecule is just as readily attracted to a polar molecule of a different substance as it is to another polar molecule of the same substance; thus the two polar substances will mix. There are some substances, such as some alcohols, that will mix with either polar or nonpolar substances. The reason for this is that part of the alcohol molecule is polar and part of the molecule is nonpolar; hence the alcohol will mix with either a polar or nonpolar substance.

Of course there are intermediate cases of substances that will mix to a slight extent. Two liquids that will mix in any proportion to form a solution are said to be **miscible;** two liquids that do not mix are said to be **immiscible.** When two liquids will mix partially, they are considered to be partially miscible. For example, when two liquids, A and B, that are partially miscible are shaken together and then allowed to separate, they will form two layers. The layer of liquid A will contain a small amount of liquid B, however, and the layer of liquid B will contain a small amount of liquid A.

Evaporation

At a given temperature, the molecules in a particular liquid have a constant average kinetic energy. They do not all have the same kinetic energy, however; some have higher energies and some have lower. The molecules with sufficiently high kinetic energy can escape from a liquid, and therefore the liquid *evaporates.* As the temperature of a liquid is increased, more of the molecules have sufficient energy to escape and thus the rate of **evaporation** increases.

Since the molecules that are escaping have a higher kinetic energy than the average, the remaining molecules will have a lower average kinetic energy and thus a lower temperature. Therefore, evaporation is a cooling process. You have probably noticed this effect when you come out of a swimming pool and feel cool, especially on a windy day when the evaporation process is speeded up. In the case of a beaker of water that is evaporating, however, the water remains at a constant temperature since it will absorb heat from the surroundings. Therefore, the water continues to evaporate at the same rate and eventually the entire beaker of water will evaporate.

The weaker the forces of attraction between the particles in the liquid, the greater the rate of evaporation. For example, a nonpolar liquid such as gasoline will evaporate more rapidly than a polar liquid such as water at the same temperature. This is due to the fact that the molecules of the nonpolar gasoline can escape more readily since they are not attracted to the other molecules as strongly as the polar water molecules are attracted to each other.

Vapor Pressure

If a liquid is placed in a closed container, the molecules that escape from the liquid (evaporate) into the gas phase will cause a pressure on the container. After a very short period of time the rate at which molecules are leaving the liquid become equal to the rate at which other molecules are going back into the liquid phase since the container is closed. This is described as a condition of equilibrium at which the rate of evaporation is equal to the rate of condensation. At this equilibrium point the molecules in the gas phase will exert a certain pressure on the container at a given temperature. This pressure is known as the **vapor pressure** of the liquid at that temperature. This is shown in Fig. 12-2.

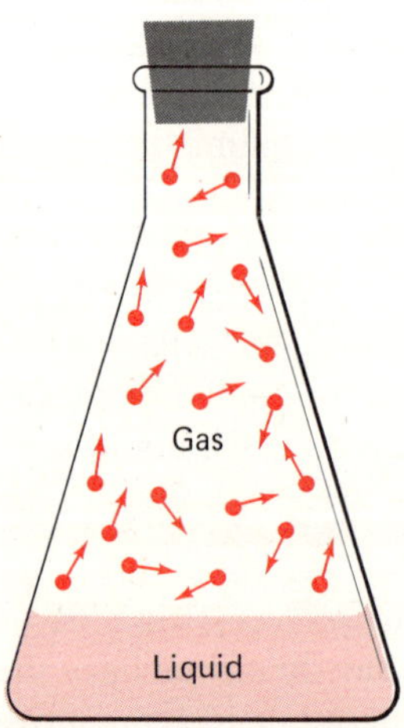

Figure 12-2
Vapor pressure of a liquid

As the temperature of a liquid increases, the vapor pressure increases. Also, as mentioned earlier in discussing evaporation, for liquids at the same temperature the weaker the forces of attraction between particles, the greater the vapor pressure will be. This is shown in Fig. 12-3, which is a graph comparing vapor pressure at various temperatures for the very polar liquid water to the much less polar liquid ethyl alcohol. Notice on this graph that the vapor pressure of each substance increases with temperature. At any given temperature, however, the vapor pressure of the water is lower than the vapor pressure of the alcohol.

Boiling Point

When the vapor pressure of a liquid becomes equal to the pressure of the atmosphere against the liquid, the liquid boils. Therefore, **boiling point** is defined as *the temperature at which the vapor pressure of a liquid is equal to the atmospheric pressure.* As atmospheric pressure decreases, boiling point decreases; as atmospheric pressure increases, boiling point increases. Since standard pressure is taken as 1 atm or 760 torr, the **normal boiling point** is defined as *the temperature at which the vapor pressure of a liquid is equal to 760 torr.* Thus we can see in Fig. 12-3 that the normal boiling point of ethyl alcohol is about 78°C, whereas the normal boiling point of water is 100°C.

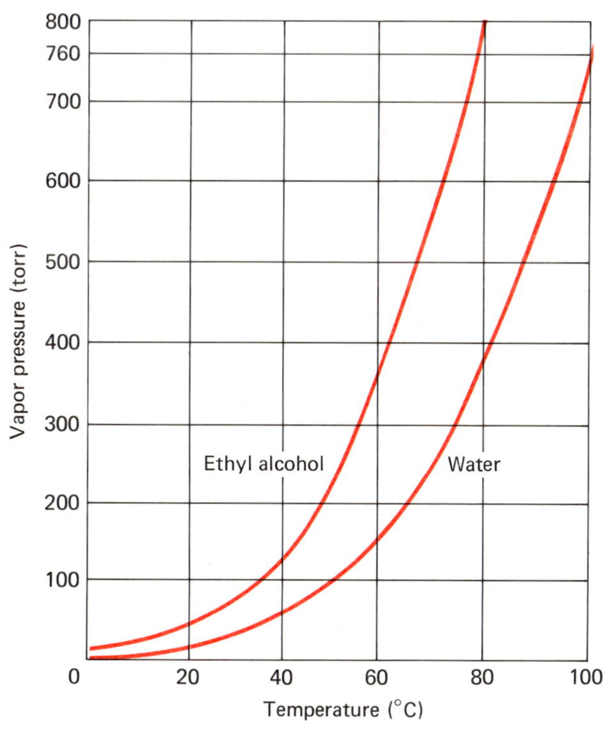

Figure 12-3
Vapor pressure versus temperature for ethyl alcohol and water

Appendix V gives the vapor pressure of water at various temperatures; you will note in this table also that the vapor pressure of water at 100°C is 760 torr.

As with the other properties of liquids that have been discussed, boiling point varies with the forces of attraction between the particles. Hence, in general, polar substances have higher boiling points than nonpolar substances. Likewise, if various polar substances are being compared, the more polar one will have the higher normal boiling point. If nonpolar substances are being compared, the one with the higher molecular (formula) weight will have the higher normal boiling point.

At higher altitudes where the atmospheric pressure is less, the boiling point of a given liquid is lower than the normal boiling point. For example, at an altitude of about 5000 ft, which is approximately the altitude of the city of Denver, Colorado, the boiling point of water is about 95°C. In cooking foods in boiling water, the lower the temperature of the boiling water, the slower the food will cook. Thus, at higher altitudes where the boiling point is decreased, foods will cook more slowly. The same principle is applied in the use of a pressure cooker to cook foods at a more rapid rate. In the pressure cooker when the container is sealed up and heat is applied, pressure builds up inside the container and thus the boiling point of the water present is increased. Therefore, the food is being cooked at a higher temperature and will cook faster.

Heat of Vaporization

In Section 2-5, specific heat was defined as the amount of heat necessary to raise the temperature of a given amount of substance by 1 degree Celsius, usually expressed as cal/g°C; that is, it takes a certain amount of heat to raise the temperature of 1 g of a given substance by 1°C. For water this value is 1.00 cal/g°C since a calorie is defined as the amount of heat necessary to raise the temperature of 1 g of water 1°C. At the boiling point of a liquid it also takes a certain amount of heat to convert a given amount of the liquid to gas. This is defined as the **heat of vaporization** of a liquid and is usually expressed as calories per gram or kilocalories per mole (or kilojoules per mole). Note that there is no change in temperature here. This is simply the amount of heat necessary to change the state from liquid to gas at the normal boiling point. For example, the heat of vaporization of water is 540 cal/g. Therefore, it takes 540 cal of heat to convert 1 g of liquid water to steam at the normal boiling point. The heat of vaporization of liquids varies in the same way as boiling point and other properties.

When a gas is condensed back into the liquid state, the same amount of heat is given off as was necessary to vaporize the liquid originally. Thus, when 1 g of steam is condensed to liquid water, 540 cal of heat are liberated. This explains why very severe burns can occur from steam. When the steam condenses on the skin, this large amount of heat is being put into the skin and thus causes the burn.

Table 12-1 lists the heats of vaporization in various units for several different substances. Note from this table that the ionic substance sodium chloride has the highest heat of vaporization, the very polar liquid water is

TABLE 12-1 *Heat of vaporization of various liquids at their normal boiling point*

LIQUID	NORMAL BOILING POINT (°C)	HEAT OF VAPORIZATION cal/g	kcal/mol	kJ/mol
Sodium chloride	1465	698	40.9	171
Water	100	540	9.72	40.6
Ethyl alcohol (C_2H_5OH)	78.3	204	9.40	39.3
Carbon tetrachloride (CCl_4)	76.7	52.1	8.00	33.4

next highest, the slightly polar liquid ethyl alcohol is next, and the nonpolar liquid carbon tetrachloride is lowest. The reason for the similarity in boiling points between ethyl alcohol and carbon tetrachloride is that the alcohol is slightly polar but has a relatively low molecular weight compared to the carbon tetrachloride.

Distillation

Distillation is a process used to separate liquids due to differences in their boiling points. By heating a mixture, the liquid with the lowest boiling point will boil off first and will thus be separated from the rest of the mixture. The apparatus used to carry out the distillation is shown in Fig. 12-4. The mixture

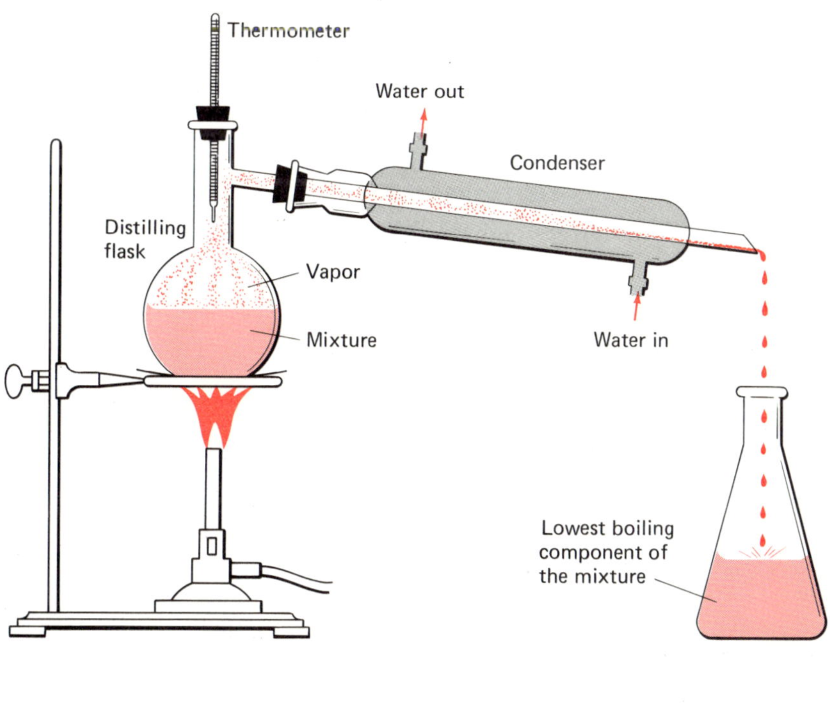

Figure 12-4
Distillation of a mixture

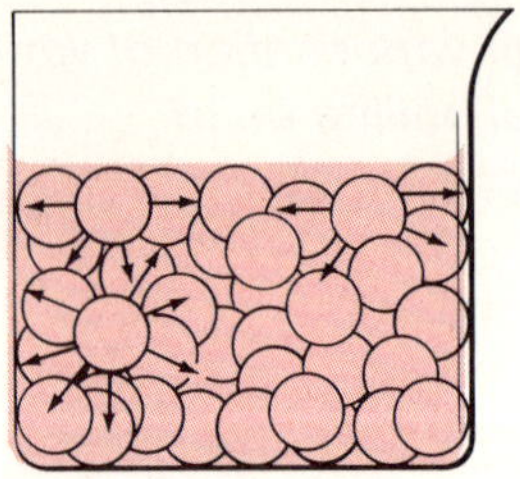

Figure 12-5
Surface tension of a liquid

being distilled is placed in the distilling flask and heated. At the boiling point of the lowest boiling component, vapor of that substance is formed. This passes through the condenser tube where cold water passes around the outside of the tube and causes it to condense back down to the liquid state, which is then collected.

Distillation can also be used to purify water. The impure water is heated in the distilling flask. Pure water vapor forms and is condensed and collected. The impurities remain in the distilling flask.

Surface Tension

The **surface tension** of a liquid is the property that tends to pull the molecules at the surface of a liquid in toward the body of the liquid and thus reduce the surface to a minimum. This is caused by the fact that within the body of the liquid each molecule is attracted to others in all directions, but at the surface they are only attracted in toward the body of the liquid. This is shown in Fig. 12-5.

The stronger the forces of attraction between the particles are, the greater will be the surface tension. For example, mercury has a higher surface tension than water, due to the greater forces of attraction from the metallic bonds in the mercury.

As the temperature of a liquid increases, the surface tension decreases. This is due to the fact that the increased kinetic energy of the molecules tend to overcome the attractive forces between particles and thus the surface tension is lowered. Part of the cleansing action of soaps is due to a lowering of the surface tension of water, although there are other factors involved also.

12-3 The Solid State

As a liquid is cooled down, the molecules move more slowly until a point is reached where most of this motion stops and the particles are held in a definite arrangement. At this point the liquid has turned to a solid. Due to this arrangement of particles in the solid state, solids have relatively high densities, do not mix or diffuse appreciably, and are practically incompressible. Of the three physical states of matter, particles in the gaseous state have the greatest motion, liquids have less, and solids have the least.

Solids are either *crystalline* or **amorphous.** In a crystalline solid the particles are arranged in a definite shape that is distinctive for a given solid. In an amorphous solid, the particles are arranged in an irregular manner and therefore do not have any particular shape or form. Glass is an example of an amorphous solid, although it can also be considered as a very viscous liquid.

Some crystalline solids can exist in more than one form. These various forms are called *allotropic forms* or *allotropes.* One common example of this is carbon, which can exist in two allotropic forms: diamond and graphite.

Crystal Structure

Crystalline solids exist in a definite geometric form. There are seven different forms possible, and several variations within these forms. The seven forms can be described as follows. These are shown in Fig. 12-6.

1. Cubic—all axes the same length and all angles 90°.
2. Rhombohedral—all axes the same length and all angles equal, but not 90°.
3. Tetragonal—two axes the same length and all angles 90°.
4. Orthorhombic—no axes the same length and all angles 90°.
5. Monoclinic—no axes the same length and two angles 90°.
6. Triclinic—no axes the same length, no angles the same, and no angles 90°.

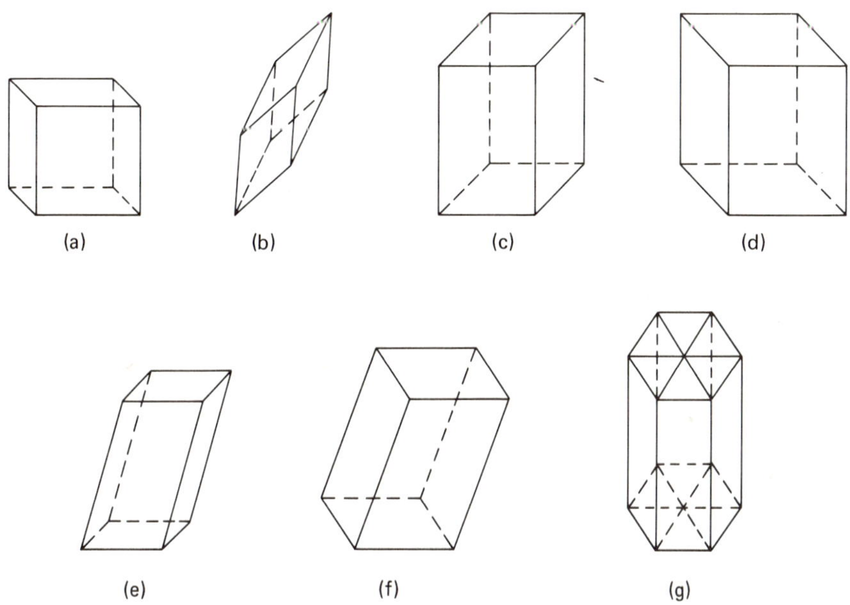

Figure 12-6
Crystal structures: (a) cubic, (b) rhombohedral, (c) tetragonal, (d) orthorhombic, (e) monoclinic, (f) triclinic, and (g) hexagonal

7. Hexagonal—two axes the same length, two angles 90°, and one angle 120°.

In describing these, the lines that show the edges of the crystal are called *axes.*

Melting or Freezing Points

The temperature at which the liquid and solid states of a particular substance can exist in equilibrium is called the *melting point.* At this equilibrium point the rate at which the solid is melting is equal to the rate at which the liquid is freezing. For example, at the melting or freezing point of water (0°C), there will be some constant amount of ice and water if the temperature is held constant. The rate at which ice is melting is equal to the rate at which water is freezing. This is sometimes defined as a dynamic equilibrium, to emphasize the fact that the two processes are taking place at the same rate and not that they have stopped. This is often shown with the double arrows as indicated by the following notation:

$$\text{ice} \rightleftharpoons \text{liquid water}$$

If we are considering the change from the solid to the liquid, this is called the **melting point;** if we are considering the change from liquid to solid, we are considering the **freezing point.** However, it is the same temperature in both cases. As with the properties of liquids that have been discussed previously, the melting points or freezing points also depend on the forces of attraction between the particles; that is, the stronger the forces of attraction holding the particles in a definite arrangement in a solid, the higher the temperature must be in order to melt that solid. Thus, we find that in general ionic substances have higher melting points than covalent substances, and polar covalent substances have higher melting points than nonpolar covalent substances.

A change in pressure affects the melting or freezing point of a substance only slightly, as opposed to the effect on the boiling with changes in pressure. For most substances, an increase in pressure causes an increase in the melting point. Ice is an exception to this, however, in that the melting point is lowered when pressure is increased. This property of water is important in ice skating. The pressure caused by the force of the thin blade on the ice lowers the freezing point of water, and the heat caused by friction causes a thin layer of water to form and enables the blade of the skate to slide smoothly over the ice. We shall consider this change in melting or freezing point and change in pressure in more detail when we consider phase diagrams in Section 12-5.

Heat of Fusion

In discussing changes from the liquid to the gas phase at the boiling point, it was shown that a certain amount of heat is necessary to carry out the conversion due to the attraction between the particles in the liquid. The same thing is true in changing from the solid to the liquid state. It takes a certain amount of heat to convert a given amount of solid to liquid at the freezing point. This amount of heat is called the **heat of fusion** and is usually expressed as calories per gram, or kilocalories per mole (or kilojoules per mole). As with heat of vaporization there is no change in temperature here; we are simply considering the amount of energy necessary for a change of state at the melting point. The heat of fusion of water is 80 cal/g. Thus, it takes 80 cal to convert 1 g of ice to liquid water at the freezing point (0°C). Table 12-2 lists the heat of fusion of several solids.

Sublimation

A substance in the solid state also has a certain vapor pressure, although this is generally much lower than the vapor pressure of a liquid. Therefore, a solid can pass directly from the solid state into the gaseous state, just as in evaporation of liquids the substance goes from the liquid to the gas phase. This process of passing directly from the solid to the gas phase without passing through the liquid state is called **sublimation.** Substances such as dry ice, moth balls, and ice are all examples of solids that undergo sublimation. We shall consider this in more detail in Section 12-5 when considering phase diagrams.

12-4 Heat Transformations in Changing from One State to Another

We have previously considered the following facts: (1) It takes a certain amount of energy to convert from the solid to the liquid state, at the freezing point (heat of fusion). (2) It takes a certain amount of heat to raise the temperature of a given amount of substance by 1 degree Celsius (specific heat). (3) It takes a certain amount of heat to convert a given amount of liquid to gas, at the boiling point (heat of vaporization). These changes can be shown by plotting heat added to a substance versus temperature, as shown in Fig. 12-7. On this graph, as heat is added to a solid, the temperature increases until the melting point is reached. Then the temperature remains constant as more heat is added since the added energy is being used to change state, from solid to liquid. Then once all the solid is melted, the temperature increases again

TABLE 12-2 *Heat of fusion of various solids at their melting point*

SOLID	MELTING POINT (°C)	HEAT OF FUSION		
		cal/g	kcal/mol	kJ/mol
Sodium chloride	804	124	7.25	30.3
Water	0	80	1.44	6.02
Ethyl alcohol	−117	24.9	1.15	4.81
Carbon tetrachloride	−23	4.2	0.64	2.68

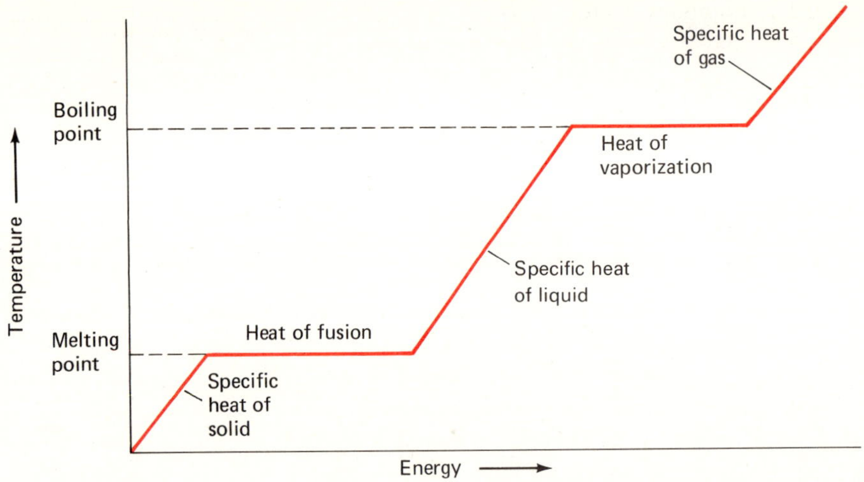

Figure 12-7
Energy, in the form of heat, added to a substance versus temperature change

until the boiling point is reached; at this point the added energy is used to change the state from liquid to gas.

We now want to consider some calculations of the amount of heat necessary to carry out various transformations. Consider the following examples.

PROBLEM EXAMPLE 12-1

How much heat, in kilocalories, is required to convert 50.0 g of ice at 0°C to steam at 100°C?

SOLUTION:

The changes that must be considered are as follows:
(a) Convert solid to liquid, at 0°C (use heat of fusion).
(b) Increase temperature of liquid from 0° to 100°C (use specific heat).
(c) Convert liquid to gas at 100°C (use heat of vaporization).
From Table 12-2, the heat of fusion of water is 80.0 cal/g; hence, for the change in Step (a),

$$50.0\,\cancel{g} \times \frac{80.0\text{ cal}}{1\,\cancel{g}} = 4000\text{ cal} = 4.00\text{ kcal}$$

Since the specific heat of water is 1.00 cal/g °C, we can calculate the change in Step (b), as

$$50.0\,\cancel{g} \times (100 - 0)°\cancel{C} \times \frac{1.00\text{ cal}}{(1\,\cancel{g})(1\,°\cancel{C})} = 5000\text{ cal} = 5.00\text{ kcal}$$

(Review Section 2-5 for the calculation in Step 2, if necessary.) From Table 12-1, the heat of vaporization is 540 cal/g; hence, for the change in Step (c),

$$50.0\,\cancel{g} \times \frac{540\text{ cal}}{1\,\cancel{g}} = 27{,}000\text{ cal} = 27.0\text{ kcal}$$

Therefore, the total amount of heat from all three steps is

Step (a):	4.00 kcal
Step (b):	5.00 kcal
Step (c):	27.0 kcal
Total heat required =	**36.0 kcal**

Note that the amount of heat required in Step (a) (4000 cal) and Step (b) (5000 cal) is nearly equal and relatively small but that over five times this amount of heat is required in Step (c) (27,000 cal) to convert the liquid water (100°C) to vapor (100°C).

The process can be reversed—that is, heat is given off on cooling. Consider the following example:

PROBLEM EXAMPLE 12-2

How much heat, in kilocalories, is liberated when 45.0 g of steam at 100.0°C is changed to water at 20.0°C?

SOLUTION:

The changes that must be considered are as follows:
(a) Convert from steam (gas) to liquid (use heat of vaporization).
(b) Decrease temperature from 100° to 20°C (use specific heat).

Step (a): $$45.0\ \cancel{g} \times \frac{540\ \text{cal}}{1\ \cancel{g}} = 24{,}300\ \text{cal} = 24.3\ \text{kcal}$$

Step (b): $$45.0\ \cancel{g} \times 80.0\ \cancel{°C} \times \frac{1.00\ \text{cal}}{(1\ \cancel{g})(1\ \cancel{°C})} = 3600\ \text{cal} = 3.60\ \text{kcal}$$

The total of the two steps is

Step (a):	24.3 kcal
Step (b):	3.6 kcal
Total heat liberated =	**27.9 kcal**

12-5 Phase Diagrams. Critical Temperature and Pressure

A **phase diagram** is a graph giving the relationship among the solid, liquid, and gaseous states of a given substance as a function of the temperature and pressure. For example, the phase diagram for water is given in Fig. 12-8. This describes the state in which water will exist under any given conditions of temperature and pressure. At point A, water will exist only as a gas. At point B, water exists in two phases in equilibrium; in this case, solid and liquid, and hence point B is the melting point of water of 1 atm pressure, or the normal melting point. At point C, water can exist in three phases in equilibrium. This is called the **triple point** and is 0.01°C at 4.6 torr. Point D corresponds to the normal boiling point of water.

We stated in Section 12-1 that a gas can be converted to a liquid by increasing the pressure, which forces the molecules together so that the forces

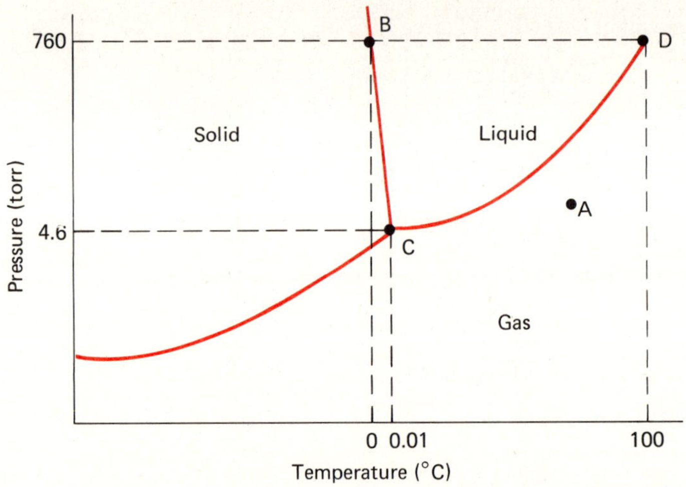

Figure 12-8
Phase diagram of water (line BC is somewhat distorted to show the triple point C, at 0.01°C and 4.6 torr)

of attraction between them can cause the molecules to be attracted to each other and thus form a liquid. For each gas there is a temperature above which the gas cannot be liquified, however, no matter how much pressure is applied. This is due to the fact that above this temperature the molecules have such a high kinetic energy that they cannot be forced together sufficiently to cause the gas to become a liquid. This temperature, which is different for each gas, is called the **critical temperature** and is defined as the temperature above which a gas cannot be liquified, no matter how much pressure is applied.

The pressure that is necessary to liquify the gas when it is at the critical temperature is called the **critical pressure.** Recall that the critical temperature is the temperature *above* which a gas cannot be liquified. It can be liquified *at* the critical temperature, and the amount of pressure that it takes to do this is the critical pressure.

Critical temperature and pressure vary for different gases in the same way that properties such as melting and boiling points vary; that is, the stronger the forces of attraction are for a particular substance, the higher the critical temperature is and the lower the critical pressure.

12-6 Water

Water is probably the most common and most abundant chemical compound on the earth. Water covers approximately 75% of the earth's surface and is necessary for all living organisms.

Physical Properties of Water

Pure water at room temperature is a colorless, odorless, tasteless liquid; however, most water contains various types of impurities, such as substances dissolved in the water from the earth. Also, city water-supplies not only contain minerals from the earth but also substances such as chlorine which are added to purify the water by killing any bacteria in it.

Table 12-3 gives the melting point, normal boiling point, and heat of fusion for water and other hydrogen compounds of the Group VIA elements. You will notice from this table that water has some very unusual properties. For example, the melting point, boiling point, heat of fusion, and heat of vaporization are all much higher for water than for any of these other compounds. You will recall from what was discussed earlier in this chapter that the more polar a substance is, the higher the properties such as melting point and boiling point will be. Due to the high electronegativity of oxygen, as compared to sulfur, selenium, or tellurium, water is much more polar than any of these substances.

TABLE 12-3 *Some physical properties of hydrogen compounds of group VIA*

PROPERTY	H_2O	H_2S	H_2Se	H_2Te
Molecular weight (amu)	18.0	34.1	81.0	129.6
Melting point (°C)	0	−86	−66	−49
Normal boiling point (°C)	100	−61	−41	−2
Molar heat of fusion (kcal/mol	1.44	0.57	0.60	1.0
and kJ/mol)	6.02	2.4	2.5	4.18

Hydrogen Bonding

Due to the high polarity of the water molecules there is a rather strong force of attraction between molecules in any sample of water. The electrostatic attraction between the hydrogen atoms in one water molecule and the oxygen atoms in another is called a **hydrogen bond.** It is a much weaker bond than the covalent bonds within the water molecule but is sufficiently strong to have a considerable effect on the properties of water.

The hydrogen bonding in water not only affects properties such as melting and boiling point but also has a great effect on the density of water. In the solid state (ice) there is a regular arrangement of water molecules, with hydrogen bonds formed between molecules. As the ice is melted, many of these hydrogen bonds are broken, in order that the molecules in the liquid state can move about. Even in the liquid state there is considerable hydrogen bonding, however, with more of these bonds being broken as the water is heated up from 0° to 100°C. This accounts for the high specific heat of water (1 cal/g °C). At the boiling point, in order to change the state from liquid to gas, all remaining hydrogen bonds must be broken, in order for the molecules to be completely independent in the gas phase. This accounts for the high heat of vaporization of water (540 cal/g).

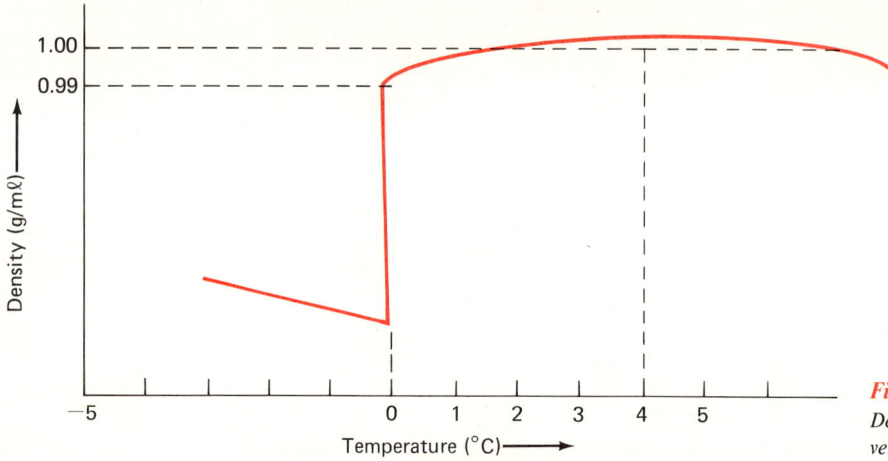

Figure 12-9
Density of water versus temperature

As liquid water is cooled, the volume decreases, as we would expect, and the density increases. As the temperature of the water starts to approach the freezing point, however, more and more hydrogen bonds form. This causes open spaces to form within the various clusters of molecules; hence the volume begins to increase and consequently the density begins to decrease. This happens at 4°C, at which point water has its maximum density. This means that below 4°C, the density decreases.

When ice forms, more hydrogen bonds form, causing even more of an open structure within the clusters of molecules. Thus ice is less dense than water, and the ice floats on the liquid water. This has a great effect on life as we know it. For example, if this were not the case, lakes and ponds would freeze from the bottom up and in cold climates would very likely freeze solid during the winter. Thus fish and other aquatic life could not exist in them.* Figure 12-9 shows the relationship between density and temperature for water.

SUMMARY

1. As real gases are cooled, the motion slows enough to permit the forces of attraction between particles to become significant and the gas first turns to *liquid;* then if it is cooled further, the liquid eventually becomes *solid.* The forces of attraction between particles in the liquid or solid states are therefore very important in determining the physical properties of the substance. These forces of attraction can be
(a) ionic electrostatic attraction (in ionic compounds),

*There is an interesting science fiction type of story called "The Catalyst" concerning this in *Chemistry,* **46,** No. 8 (September 1973), p. 8.

(b) dipole-dipole attractions (in polar covalent compounds),
(c) van der Waals forces (in nonpolar covalent compounds),
(d) metallic electrostatic attraction (in metals),
(e) covalent bond (in macromolecular substances).

2. The stronger the forces of attraction between the units that make up the substance, the higher will be the melting point, heat of fusion, boiling point (due to lower vapor pressure), heat of vaporization, surface tension, and critical temperature.
3. Variations in these properties depend on
   (a) size and charge of ions, for ionic compounds,
   (b) polarity and molecular weight for polar covalent compounds, and
   (c) molecular weight for nonpolar covalent compounds.
4. Table 12-4 summarizes relationships between structure and physical properties for various types of substances by considering the types of units that the substance is composed of, the forces of attraction between these units, and the relative properties of the substances.
5. Liquids that are similar in polarity will mix, and those that are not similar will not mix.
6. Properties such as *evaporation, vapor pressure, boiling point, heat of vaporization,* and *surface tension,* for liquids or *melting (freezing) point,* and *heat of fusion,* for solids vary as follows: ionic substances, (with higher charge density the attractive forces increase); metallic; polar covalent (from most polar to least polar); nonpolar covalent (in order of decreasing molecular weight).
7. In calculating the amount of heat necessary to increase the temperature of a substance, or change the state, the *specific heat* (for temperature changes), *heat of fusion* (to change from solid to liquid), and *heat of vaporization* (to change from liquid to gas) are used.
8. A *phase diagram* is a graph showing the relationship between the solid, liquid, and gas state of a substance as a function of temperature and pressure.
9. The *critical temperature* of a gas is the temperature above which the gas cannot be liquified, no matter how much pressure is applied; the *critical pressure* is the amount of pressure necessary to liquify a gas at the critical temperature.
10. The *hydrogen bonding* in water accounts for many of the physical properties. For example, the high heat of fusion and vaporization and specific heat of water are due to the fact that some hydrogen bonds are being broken to go from the solid to the liquid phase, more are being broken as the liquid is heated, and many hydrogen bonds must be broken in going from the liquid to the gas phase. This also causes open spaces in the structure of liquid and solid water and thus accounts for the fact that water has its maximum density at 4°C and that ice is less dense than water.

TABLE 12-4 *Relationship between structure and properties*

TYPE OF SUBSTANCE	MOLECULAR	IONIC	COVALENT (MACROMOLECULAR)	METALLIC
Units	Molecules	Ions	Atoms	Positive ions in a "sea of electrons"
Forces of attraction between units	Van der Waals for nonpolar; dipole attraction for polar	Electrostatic attraction between oppositely charged particles	Covalent bond (i.e., shared electrons)	Electrostatic attraction between positive "ions" and electrons
Relative properties	Relatively low melting and boiling points; the more polar the molecules, the higher these properties would be for similar molecular weights	Relatively high melting and boiling points; the *higher* the charge and the *smaller* the size, the higher these properties are	Very high melting and boiling points; will not conduct an electric current; not soluble in anything; etc.	
Examples	H_2, Cl_2, CO_2, H_2O	NaCl, CaF_2	C, SiO_2	Na, Fe, Hg

Upon completion of this chapter, you should be able to do the following:

1. Given the type of attractive forces between particles in a liquid or solid, explain the relative strength between these forces (that is, ionic, dipole, van der Waals, metallic, and covalent bonds).
2. Given any of the properties of liquids or solids, such as compressibility, density, shape, and volume, explain in terms of the arrangement and forces of attraction between the particles.
3. Given a group of liquid or solid substances, predict the variations and explain in terms of types of forces of attraction between particles (and molecular weight if necessary) in the following properties:
   (a) evaporation
   (b) vapor pressure
   (c) normal boiling point
   (d) heat of vaporization
   (e) surface tension
   (f) melting (or freezing) point
   (g) heat of fusion
   (1)
4. Given the heat of fusion, specific heat, and heat of vaporization of a substance, calculate the amount of heat (in calories, kilocalories, joules, etc.) necessary to change the temperature by a given amount or change from one state to another for a given amount of the substance. (2–6)
5. Given the phase diagram for any substance, interpret it in terms of finding the state of the substance at any temperature and pressure, including cases where two (or all three) states are at equilibrium.
6. Given any of the physical properties of water, explain them in terms of hydrogen bonding.
7. Given any of the terms in Exercise 1, give a brief definition or explanation in your own words.

EXERCISES

1. Define or explain each of the following:
   (a) evaporation
   (b) condensation
   (c) vapor pressure
   (d) miscible
   (e) immiscible
   (f) sublimation
   (g) distillation
   (h) melting point
   (i) dipole attraction
   (j) hydrogen bond
   (k) van der Waals forces
   (l) boiling point
   (m) normal boiling point
   (n) heat of vaporization
   (o) heat of fusion
   (p) specific heat
   (q) surface tension
   (r) phase diagram
   (s) critical temperature
   (t) critical pressure
   (u) triple point
   (v) macromolecular substance
2. Explain why evaporation is a cooling process.

3. Explain each of the following:
   (a) Liquids are much less compressible than gases.
   (b) Liquids have a surface tension.
   (c) The vapor pressure of a liquid or a solid increases with temperature.
   (d) Heat is absorbed when a liquid vaporizes.
4. Liquid A has a vapor pressure of 30 torr at 20°C and liquid B has a vapor pressure of 200 torr at 20°C. Answer each of the following and explain each answer:
   (a) Which has the higher surface tension?
   (b) Which has the highest normal boiling point?
   (c) Which has the higher critical pressure?
5. Explain why the boiling point of PH_3 is lower than that of NH_3.
6. A sample of water vapor in equilibrium with liquid water is contained in a volume of 2 l at 50°C. If the volume is expanded to 4 l at constant temperature, what change will occur in each of the following (assuming equilibrium is maintained)?
   (a) The number of molecules in the vapor.
   (b) The concentration of molecules in the vapor.
   (c) The pressure exerted by the vapor.
   (d) The average kinetic energy of the molecules in the vapor.
7. Explain how hydrogen bonding in water affects the following physical properties:
   (a) density
   (b) boiling point
   (c) heat of fusion
8. (a) Using the sp^3 orbital approach to the structure of water, what is the predicted bond angle?
   (b) What is the actual bond angle?
   (c) Explain the difference between the two values. (Review Section 7-7 if necessary.)
9. In your own words, explain why ice floats on water.
10. Discuss the consequences to the earth if, on going from liquid water to solid ice, the volume of the solid did not expand.

PROBLEMS

1. Arrange each group in order of increasing melting point, and explain your answer.
   (a) H_2O, NaCl, He, H_2S, Br_2, NH_3
   (b) F_2, $CaCl_2$, SO_2, Br_2, SO_3, H_2
2. Calculate the amount of heat, in kilocalories, necessary for each of the following changes (use Tables 12-1 and 12-2).
   (a) To melt 18.5 g of ice at 0°C.
   (b) To liquefy 0.780 mole of ethyl alcohol at its freezing point.
   (c) To vaporize 8.75 g of carbon tetrachloride at its boiling point.
   (d) To convert 0.750 kg of liquid water to steam at 100°C.

(e) To vaporize 1.25 g of alcohol at its boiling point.
(f) To convert 2.45 g of ice at 0°C to water at 27.5°C.
(g) To convert 125 g of water at 85.0°C to steam at 100°C.
(h) To convert 8.25 g of ice at 0°C to steam at 100°C.
(i) To convert 85.0 g of ethyl alcohol from solid to liquid at the melting point.
(j) To vaporize 1.20 kg of water at 100°C.
(k) To melt 2.45 moles of ice at 0°C.
(l) To raise the temperature of 3.25 g of water from 0°C to 29.3°C.
(m) To convert 1.80 moles of water at 100°C to steam at 100°C.
(n) To convert 1.35 g of water at 53.0°C to steam at 100°C.
(o) To convert 124 g of ice at 0°C to water at 20.0°C.
(p) To convert 2.75 kg of ice at 0°C to steam at 100°C.

3. Calculate the amount of heat given off, in kilocalories, when 3.50 g of steam at 100°C is converted to water at 20°C.
4. Calculate the number of kilocalories of heat given off when 250 mg of steam at 100°C is converted to water at 37°C.
5. Calculate the amount of water, in grams, that could be heated from 0° to 25°C by the heat evolved on cooling 1.00 kg of water from 100° to 25°C.
6. Calculate the kilograms of water that could be heated from 20° to 25°C by the heat evolved on condensing 85 g of water vapor to liquid at the boiling point.

REFERENCES

BERNAL, J. D., "The Structure of Liquids," *Sci. Am.*, Aug. 1960, **203,** p. 124.

BOYD, T. A., "The Wonder of Water," *Chemistry*, June 1974, **47,** p. 6.

CHALMERS, BRUCE, "How Water Freezes," *Sci. Am.*, Feb. 1959, **200,** p. 114.

MOTT, SIR NEVILL, "The Solid State," *Sci. Am.* Sept. 1967, **217,** p. 80.

REVELLE, ROGER, "Water," *Sci. Am.*, Sept. 1963, **209,** p. 92.

RUNNELS, L. K., "Ice," *Sci. Am.*, Dec. 1966, **215,** p. 118.

Objectives

1. To learn the meaning of the terms solution, solute, solvent, solubility, saturated solution, unsaturated solution, supersaturated solution, concentrated versus dilute solution, concentration, percent concentration (by mass and volume), molal concentration, mole fraction, and molar concentration.
2. To be able to list the types of solutions and give an example of each.
3. To understand how the following factors affect solubility: (a) properties of solute and solvent, (b) temperature, and (c) pressure.
4. To understand how the following factors affect rate of solution: (a) particle size, (b) rate of stirring, and (c) temperature.
5. To be able to do calculations involving concentration of solution, using percent (by mass and volume), molal concentration, mole fraction, and molar concentration.
6. To understand what colligative properties of solutions are and to be able to do calculations involving molal freezing point depression and boiling point elevation.
7. To understand what osmosis is.
8. To understand what colloids are and how they differ from solutions.

Thirteen

Solutions and Colloids

In Chapter 3 it was pointed out that a solution is a homogeneous mixture. Also at that time the difference between pure substances and mixtures was considered and a brief introduction to percent and molar concentrations was given. We now want to consider in more detail the properties of solutions and the methods used to express concentrations of solutions. We shall also briefly consider some of the properties of colloids.

In a solution, the particles (molecules or ions) are small and they are bound to the solvent molecules so that they are homogeneously dispersed throughout the solution. In a precipitate, the particles are not bound to solvent molecules and are large enough that eventually they will settle out. Colloids are intermediate between matter in solution and solid matter that will settle out of a mixture. In a colloid, the particles are not bound to solvent molecules but are not large enough to settle out on standing.

13-1 Solutions

A **solution** is a homogeneous mixture composed of a **solute,** dissolved in a **solvent.** In general, the solvent is the component present in a greater amount. For example, in a solution of sugar in water, often referred to as an *aqueous* sugar solution, the sugar is the solute and the water is the solvent. The solute and solvent in a solution can exist either as molecules or ions. For example, when sugar is dissolved in water, the sugar goes into solution in the form of molecules. On the other hand, when sodium chloride dissolves in water, it breaks up into sodium ions and chloride ions. In either case the molecules or ions are surrounded by water molecules in the solution. Thus we say that the molecules or ions in the solution become *hydrated.* This is illustrated in Fig. 13-1.

Solutions may exist in any one of the three physical states of matter; that is, the solute or solvent can be a gas, a liquid, or a solid. For example, a gas,

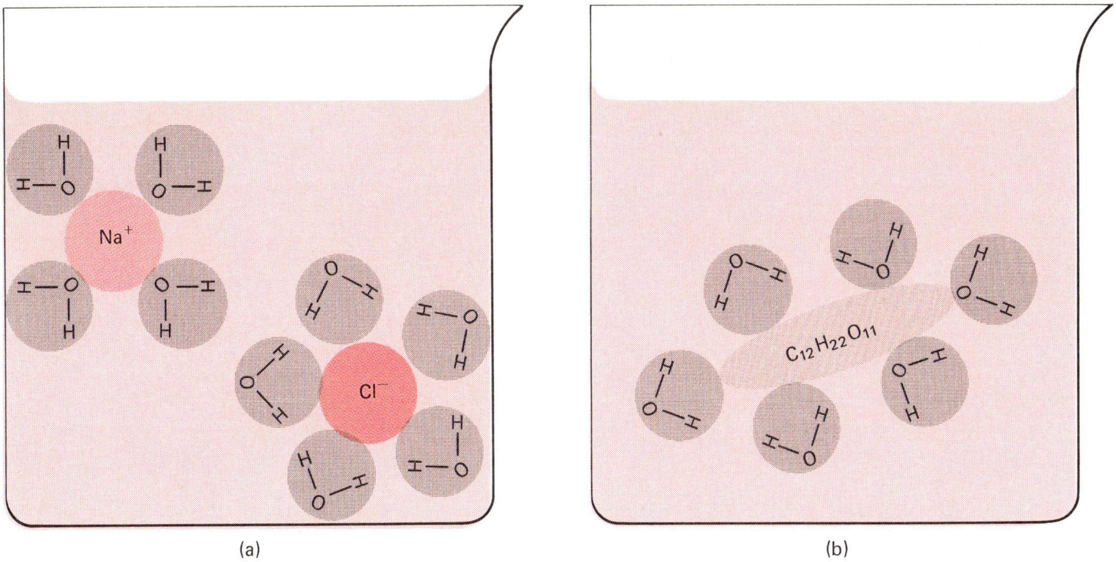

Figure 13-1
Solutions of (a) sodium chloride and (b) sugar. The sodium chloride is in the form of Na^+ and Cl^- ions, whereas the sugar remains as molecules

a liquid, or a solid can be dissolved in a liquid. Also alloys of metals are solutions in the solid state of one metal in another. The three most common types of solutions are gas in liquid, liquid in liquid, and solid in liquid. Examples of these are given in Table 13-1.

A substance that breaks up into ions when it goes into solution is called an **electrolyte** and one that does not ionize is a **nonelectrolyte.** If the electrolyte is completely ionized, it is classified as a **strong electrolyte.** Some electrolytes are only partially ionized, however; these are classified as **weak electrolytes.** Figure 13-2 shows the apparatus that is used to check solutions to determine whether or not the solute is an electrolyte. The solute is dissolved in pure water and the electrodes are dipped into the solution. If the bulb lights, it indicates that the solution contains an electrolyte. If the bulb lights up brightly, the substance is a *strong electrolyte;* if it only glows dimly, the substance is a *weak electrolyte.*

We shall have more to say about electrolytes in Chapter 14, in discussing the difference between weak and strong acids and bases, and again in Chapter 17 with solution equilibrium.

TABLE 13-1 *Some types of solutions*

SOLUTE	SOLVENT	EXAMPLE
Gas	Liquid	HCl gas in water, to form hydrochloric acid
Liquid	Liquid	Alcohol in water
Solid	Liquid	Sodium hydroxide in water

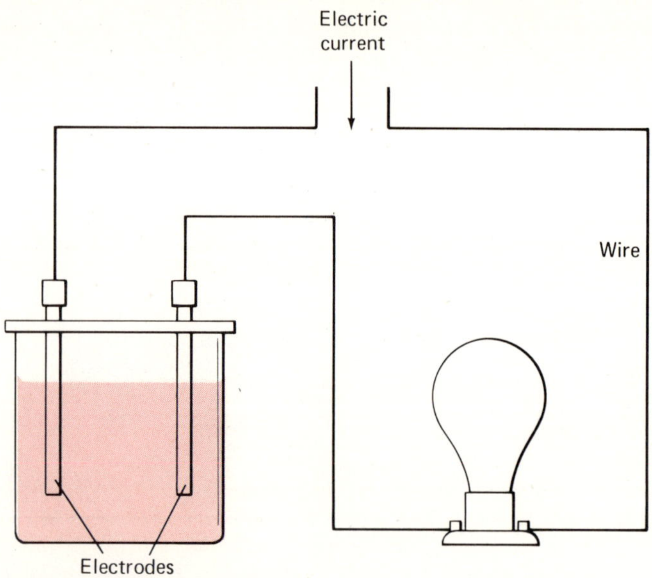

Figure 13-2
Apparatus used to determine if a solution contains an electrolyte

13-2 Factors that Affect Solubility

Properties of Solute and Solvent

You will recall from Chapter 7 that water is a polar substance, due to the shape of the molecule and the difference in electronegativity between hydrogen and oxygen. For this reason, many ionic substances are soluble in water. In forming solutions the polar water molecules are attracted to the ions in the ionic solid and pull them away from each other; however, ionic substances such as sodium chloride generally do not dissolve in nonpolar liquids. For example, carbon tetrachloride (CCl_4) is a nonpolar liquid due to the tetrahedral arrangement of chlorine around the central carbon atom. An ionic solute, such as sodium chloride, will not dissolve in a nonpolar liquid such as carbon tetrachloride since there is no attraction between the liquid molecules and the Na^+ or Cl^- ions. Generally we find that ionic compounds are soluble in polar solvents and covalent compounds are soluble in nonpolar solvents. This is often stated as *like dissolves like.*

Temperature

For solutions of gas in liquid, the solubility decreases with temperature. An example of this is the bubbles that form when a beaker of water is heated. These bubbles form long before the boiling point is reached, due to the dissolved oxygen in the water. As the temperature increases, the solubility of the oxygen gas in the liquid water decreases and therefore the gas is given off.

In the case of solids being dissolved in liquids, for many substances the solubility increases with temperature. There are some substances, however, for which the solubility remains relatively constant with changes in temperature. And there are other substances for which the solubility decreases as the temperature increases. Some examples of this are shown in Fig. 13-3.

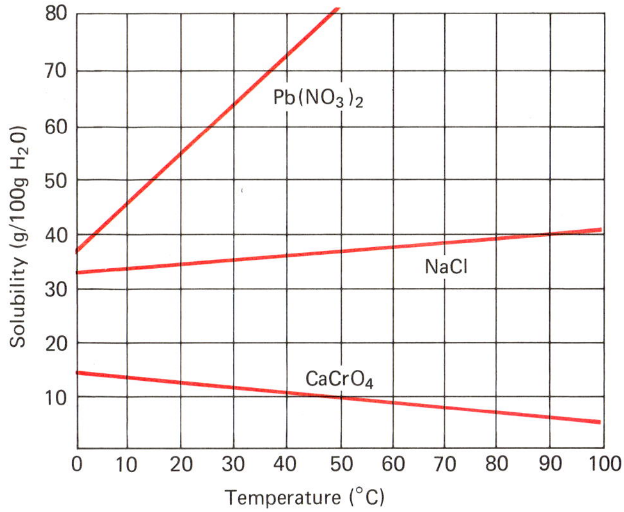

Figure 13-3
Change in solubility of various salts with change in temperature

Pressure

Solutions that contain only liquids and solids are not appreciably affected by changes in pressure. For a solution of a gas in a liquid, however, there is an appreciable affect. This was first shown by William Henry (1774–1836), an English chemist and physician. In 1803, Henry showed that the solubility of a gas in a liquid is directly proportional to the partial pressure of the gas above the liquid. From his results Henry formulated the law, now referred to as Henry's Law, that states this relationship. Since the solubility is directly proportional to the partial pressure above the liquid, if the partial pressure is doubled, the solubility is doubled; if the partial pressure is one-half as great, the solubility is one-half as much. An example of this is the carbon dioxide gas dissolved in carbonated beverages. As long as the bottle is closed, there is considerable partial pressure due to the carbon dioxide above the liquid, and the carbon dioxide gas remains dissolved in the liquid. When the bottle is opened, however, the pressure is released, the solubility of the gas decreases, and the carbon dioxide gas bubbles off of the liquid.

Henry's Law can be written as

$$C = kP$$

where C is the solubility (or concentration), k is a constant, and P is the partial pressure.

Consider the following problem example.

PROBLEM EXAMPLE 13-1

The solubility of nitrogen gas in water is 0.012 mol/ℓ at 0°C when the partial pressure of nitrogen above the water is 1 atm. Calculate the solubility in moles per liter when the partial pressure is 0.35 atm.

SOLUTION:

(a) First calculate the value of k:

$$0.012 \text{ mol}/\ell = (k)(1 \text{ atm})$$
$$k = 0.012 \text{ mol}/\ell\text{-atm}$$

(b) Then use this value to calculate the new solubility:

$$C = 0.012 \text{ mol}/\ell\text{-atm} \times 0.35 \text{ atm} = \mathbf{4.2 \times 10^{-3} \text{ mol}/\ell}$$

That is, since the pressure is about one-third as great, the solubility is about one-third as much.

13-3 Factors that Affect Rate of Solution

Particle Size

Due to the greater surface that is exposed to the solvent, small particles will dissolve more rapidly than larger ones. Therefore, the rate of solution of these is faster than for large particles. For example, a solid lump of sugar would dissolve much more slowly than the same amount in small granular form.

Rate of Stirring

The rate of solution (that is, the rate at which the solute dissolves), can also be increased by stirring. This causes more of the solute particles to come in contact with solvent molecules in a shorter period of time and hence causes the solute to dissolve more rapidly.

Temperature

Increasing the temperature will always cause a solute to dissolve more rapidly. This is true whether the actual solubility increases or decreases at a higher temperature. The reason for this increase is that the kinetic energy of the solute, solvent, and solution all increase. The increased kinetic energy of the solute causes the particles to break away from the crystals more readily. The increased kinetic energy of the solvent causes the molecules to move rapidly and hence come in contact with the solute particles more rapidly, in the same way that stirring causes an increase in rate of solution. In considering this change in rate of solution by heating, it should be kept in mind that temperature is a measurement of the average kinetic energy of the particles in a sample. Hence, at a higher temperature all the particles involved have a higher kinetic energy; therefore the overall rate of solution is increased.

13-4 Saturated, Unsaturated, and Supersaturated Solutions

A **saturated solution** is one that contains as much of the solute as can be dissolved at the given temperature. If there is excess solute present, an equilibrium condition exists wherein the rate at which solute is dissolving is equal to the rate at which excess solute is coming out of solution. The actual concentration of solute in a saturated solution depends on the solubility of the solute in the solvent, as well as the temperature.

To prepare a saturated solution, an excess of solute is added to the solvent and the mixture is stirred or shaken until no more of the solute is dissolved. This process can be speeded up by heating the mixture. This will cause the solute to dissolve faster and will generally cause more of the solute to dissolve than it would at a lower temperature. Then, as the solution cools, the excess solute crystallizes out, leaving a saturated solution.

In an **unsaturated solution** there is less solute than can be dissolved at a given temperature.

A **supersaturated solution** is a rather unstable condition in which there is more solute in solution than can normally exist at a given temperature. If

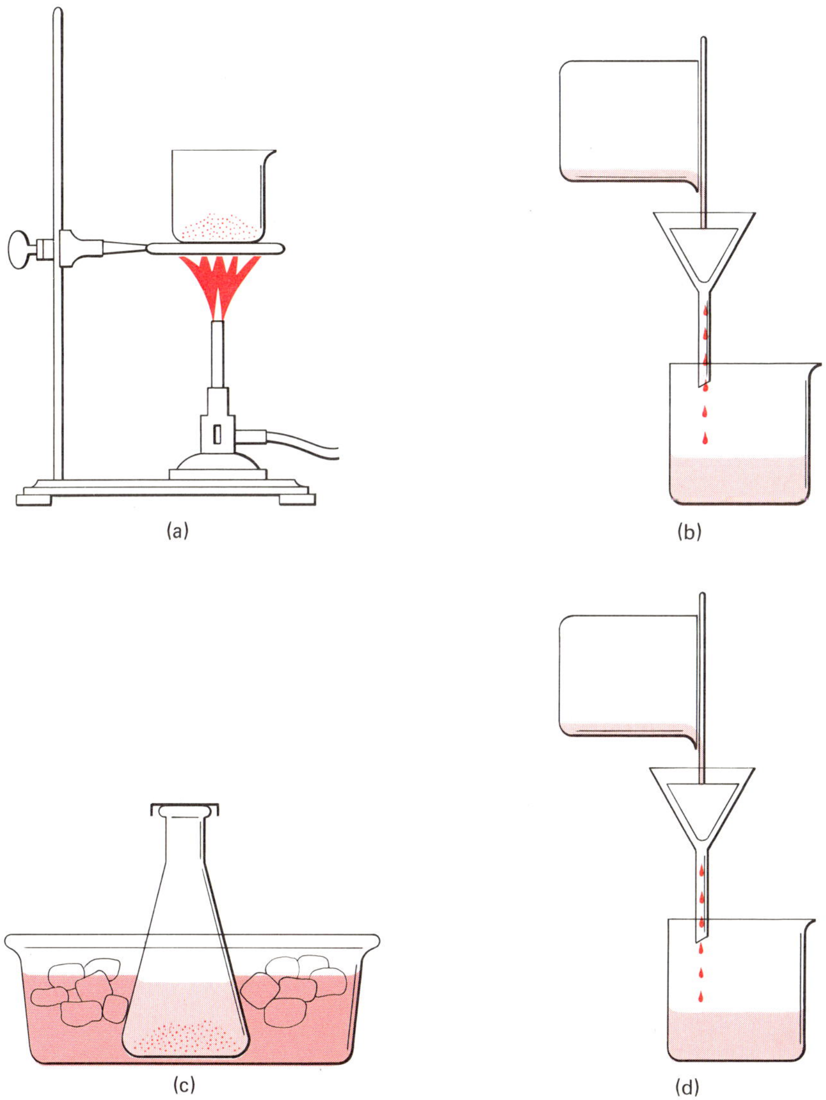

Figure 13-4
Purification of a solid by crystallization: (a) dissolve the solid in water to form a solution; (b) filter the solution to remove water insoluble impurities; (c) cool the filtrate to form crystals of purified solid, leaving the water-soluble impurities in the solution; (d) filter the crystals

anything is done to this solution to begin crystallization, all the excess solute will precipitate out, leaving a saturated solution.

This process of *crystallization* of solute from a supersaturated solution can be used as a method of purifying a solid. For example, a particular solid may contain both water-soluble and water-insoluble impurities. If the solid product that is desired is quite soluble in hot water but insoluble in cold water, which is quite often the case, the following procedure can be used to obtain the product in a pure form. First the impure material is placed in water and heated. This will cause the product and the soluble impurities to dissolve. This can be filtered to remove the insoluble impurities. Then the *filtrate* (the liquid that passes through the filter) is cooled. The product will crystallize out, leaving the water-soluble impurities in solution. Often several of these recrystallizations are necessary to obtain a pure product. This process is shown in Fig. 13-4.

Occasionally a supersaturated solution forms when a precipitate of a particular substance is wanted. When this happens, there are several methods that can be used to begin crystallization. In some cases simply stirring the solution will cause crystallization to begin. If that does not work, sometimes making a small scratch on the container with a stirring rod will cause precipitation. This works by forming a place where crystallization can start; once it starts, all the excess solute will precipitate out. If nothing else works, it is sometimes necessary to add a small crystal, called a *seed crystal.* Preferably the seed crystal should be of the same substance or a similar substance. This method has the disadvantage of adding more material if a quantitative determination is being done or adding an impurity if the substance being crystallized is not known and a different kind of seed crystal is used.

13-5 Concentration of Solutions

The concentration of a solution expresses the amount of solute present in a given amount of solution. The terms *concentrated* and *dilute* are just relative expressions.* A concentrated solution has more solute in it than a dilute solution; however, this does not give any indication of the exact amount of solute present. Therefore, we need more exact, quantitative methods of expressing concentration.

In Chapter 3 we briefly considered percent by mass and molar concentration as two methods of expressing quantitatively the amount of solute

*By standard convention, certain concentrations are used for common *dilute* and *concentrated* acids and bases used in the laboratory. These are as follows:

ACID	DILUTE	CONCENTRATED	BASE	DILUTE	CONCENTRATED
HCl	6 *M*	12 *M*	$NH_{3(aq)}$ (sometimes called ammonium hydroxide)	6 *M*	15 *M*
H_2SO_4	6 *M*	18 *M*			
HNO_3	6 *M*	16 *M*			

present in a given solution. We shall now consider these terms in more detail and also look at some other methods of expressing concentration. We shall consider the following methods of expressing concentration:

1. percent by mass
2. percent by volume
3. mole fraction
4. molal concentration
5. molar concentration

In Chapter 14, after discussing acids and bases, we shall also consider normal concentration, or normality.

Percent by Mass (Weight)

You should recall from Chapter 3 that **percent concentration (by mass)** is simply the mass of solute divided by the mass of solution, all multiplied by 100. This is just like any other percent calculation—a fraction of 100. Thus we are taking the mass of some part of the solution (the solute), divided by the mass of the total solution (solute plus solvent) to give a fraction, and then that fraction is multiplied by 100. For example, consider a solution that contains 20 g of sodium chloride in 80 g of water (to give 100 g of solution). This would be a 20%, by mass, solution of NaCl in water. This could also be expressed as a 20% aqueous NaCl solution. Figure 13-5 shows how this solution would be prepared.

Therefore, percent by mass can be expressed as

$$\text{percent by mass} = \frac{\text{mass of solute}}{\text{mass of solution}} \times 100$$

or

$$\text{percent by mass} = \frac{\text{mass of solute}}{\text{mass of solute} + \text{mass of solvent}} \times 100$$

Consider the following problem examples.

PROBLEM EXAMPLE 13-2

If 28.5 g of calcium hydroxide is dissolved in enough water to make 185 g of solution, calculate the percent concentration of calcium hydroxide in the solution.

SOLUTION:

Since the mass of solute is 28.5 g and the total mass of the solution is 185 g, the percent of calcium hydroxide (the solute) is

$$\frac{28.5 \text{ g Ca(OH)}_2}{185 \text{ g soln}} \times 100 = \mathbf{15.4\%}$$

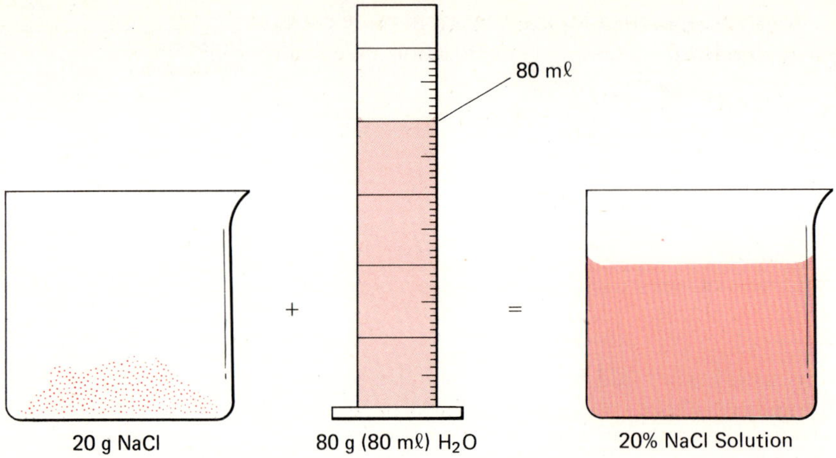

Figure 13-5
Preparation of a twenty percent aqueous solution of sodium chloride by combining twenty grams of sodium chloride with eighty grams (eighty milliliters) of water

PROBLEM EXAMPLE 13-3

If 7.5 g of sodium nitrate is dissolved in 85.0 mℓ of water, calculate the percent concentration of sodium nitrate in the solution.

SOLUTION:

To find the total mass of solution, we must add the mass of solute (7.5 g) and the mass of water. Since the density of water is 1.00 g/mℓ, 85.0 mℓ of water is equivalent to 85.0 g of water. Therefore the total mass of the solution is

$$\begin{array}{r} 7.5 \text{ g } NaNO_3 \\ \underline{85.0 \text{ g } H_2O} \\ 92.5 \text{ g solution} \end{array}$$

and the percent of $NaNO_3$ is

$$\frac{7.5 \text{ g } NaNO_3}{92.5 \text{ g soln}} \times 100 = \mathbf{8.1\%}$$

PROBLEM EXAMPLE 13-4

Calculate the number of grams of magnesium chloride that would be needed to prepare 250 mℓ of a 12% aqueous solution (the density of the solution is 1.1 g/mℓ).

SOLUTION:

This problem says 250 mℓ soln = ? g $MgCl_2$. Since the percent concentration desired is 12%, we want 12 g of $MgCl_2$ in each 100 g of solution. Since the density of the solution is 1.1 g/mℓ, we know that 1.1 g soln = 1 mℓ of soln.* Therefore the calculation is done as follows:

$$250\ \text{m}\ell\ \text{soln} \times \frac{1.1\ \text{g soln}}{1\ \text{m}\ell\ \text{soln}} \times \frac{12\ \text{g MgCl}_2}{100\ \text{g soln}} = \mathbf{33\ g\ MgCl_2}$$

PROBLEM EXAMPLE 13-5

Calculate the number of grams of sodium hydroxide that must be added to 1.00 ℓ of water to make a 15.0% solution.

SOLUTION:

Since 1.00 ℓ of water is equal to 1000 g of water, the problem says

$$1000\ \text{g H}_2\text{O} = ?\ \text{g NaOH (in a 15.0\% solution)}$$

In a 15.0% solution there will be 15.0 g of NaOH for each 85.0 g of H_2O (to give a total of 100 g of solution). This gives the relationship

$$15.0\ \text{g NaOH} = 85.0\ \text{g H}_2\text{O}$$

which can be used as a conversion factor. Thus, the calculation is done as

$$1000\ \text{g H}_2\text{O} \times \frac{15.0\ \text{g NaOH}}{85.0\ \text{g H}_2\text{O}} = \mathbf{176\ g\ NaOH}$$

Percent by Volume†

Occasionally it is convenient to express concentration as percent by volume. This is simply done by dividing the volume of solute by the total volume of solution and multiplying by 100:

$$\text{percent by volume} = \frac{\text{volume of solute}}{\text{volume of solution}} \times 100$$

Consider the following calculation.

PROBLEM EXAMPLE 13-6

A wine contains 12% alcohol by volume. Calculate the number of milliliters of alcohol in 350 mℓ of the wine.

*In using density in calculations of this type, you must keep in mind that it gives the relationship between the mass of solution and the volume of the solution, as in this example. It does *not* give the amount of solute in the solution. This is true for *all* solution calculations and is an important point to remember when doing these calculations. We are simply using density as a conversion factor, to obtain the relationship between mass of solution and volume of solution.

†This portion of Section 13-5 is optional. It can be omitted without loss of continuity.

SOLUTION:

Since the solution is 12% alcohol by volume, we can see that

$$100 \text{ m}\ell \text{ wine} = 12 \text{ m}\ell \text{ of alcohol}$$

(Of course any other volume units could be used here also, but we want milliliters.) The problem says

$$350 \text{ m}\ell \text{ wine} = ?\ \text{m}\ell \text{ alcohol}$$

Therefore the calculation is done as

$$350 \text{ m}\ell \text{ wine} \times \frac{12 \text{ m}\ell \text{ alcohol}}{100 \text{ m}\ell \text{ wine}} = \mathbf{42\ m\ell\ alcohol}$$

Mole Fraction*

The **mole fraction** of a component in a solution is equal to the number of moles of that component divided by the total number of moles of all components present. The sum of the mole fractions of all components in a solution will always be equal to one. For example, in a solution of A and B containing n_A moles of A and n_B moles of B, the concentrations of A and B, expressed as mole fractions, would be

$$\text{Mole fraction of A } (X_A) = \frac{n_A}{n_A + n_B}$$

$$\text{Mole fraction of B } (X_B) = \frac{n_B}{n_A + n_B}$$

The following problem example illustrates the calculations of mole fractions for components in a solution.

PROBLEM EXAMPLE 13-7

Calculate the mole fraction of phosphoric acid in 25% aqueous phosphoric acid solution.

SOLUTION:

The solution contains 25 g H_3PO_4 for every 75 g of water, and the moles of each component present in 100 g of solution are calculated as follows:

$$25 \text{ g } H_3PO_4 \times \frac{1 \text{ mol } H_3PO_4}{98 \text{ g } H_3PO_4} = 0.26 \text{ mol } H_3PO_4$$

$$75 \text{ g } H_2O \times \frac{1 \text{ mol } H_2O}{18 \text{ g } H_2O} = 4.2 \text{ mol } H_2O$$

*This portion of Section 13-5 is optional. It can be omitted without loss of continuity.

The mole fraction of phosphoric acid in the solution is therefore

$$X_{H_3PO_4} = \frac{0.26}{0.26 + 4.2} = \mathbf{0.058}$$

*Molal Concentration**

Molal concentration, which is abbreviated as *m*, is defined as the number of moles of solute per kilogram of solvent. This is expressed as

$$\text{molal concentration } (m) = \frac{\text{moles of solute}}{\text{kilograms of solvent}}$$

This method of expressing concentration is especially useful when considering colligative properties of solutions (Section 13-7).

An example of this would be a solution in which 1 mole of NaOH (40 g) was dissolved in 1 kg (1 *ℓ*) of water to make a one molal solution, as shown in Fig. 13-6.

Consider the following problem examples.

PROBLEM EXAMPLE 13-8

Calculate the molal concentration of a solution that contains 18 g of sodium hydroxide in 100 m*ℓ* of water.

SOLUTION:

Since the density of water is 1 g/m*ℓ*, 100 m*ℓ* of water = 100 g H_2O. The formula weight of NaOH is 40 g. Therefore, the moles of NaOH per kilogram of H_2O is calculated as

$$\frac{18\text{ g NaOH}}{100\text{ g }H_2O} \times \frac{1\text{ mol NaOH}}{40\text{ g NaOH}} \times \frac{1000\text{ g }H_2O}{1\text{ kg }H_2O} = 4.5\,\frac{\text{mol NaOH}}{\text{kg }H_2O} = \mathbf{4.5}\ \boldsymbol{m}$$

PROBLEM EXAMPLE 13-9

Calculate the molal concentration of a 20.0% aqueous solution of calcium chloride.

SOLUTION:

A 20.0% aqueous solution of $CaCl_2$ contains 20.0 g of $CaCl_2$ in each 80.0 g of water; therefore, the molal concentration is calculated as

$$\frac{20.0\text{ g }CaCl_2}{80.0\text{ g }H_2O} \times \frac{1000\text{ g }H_2O}{1\text{ kg }H_2O} \times \frac{1\text{ mol }CaCl_2}{111\text{ g }CaCl_2} = 2.25\,\frac{\text{mol }CaCl_2}{\text{kg }H_2O} = \mathbf{2.25}\ \boldsymbol{m}$$

*This portion of Section 13-5 is optional. It can be omitted without loss of continuity.

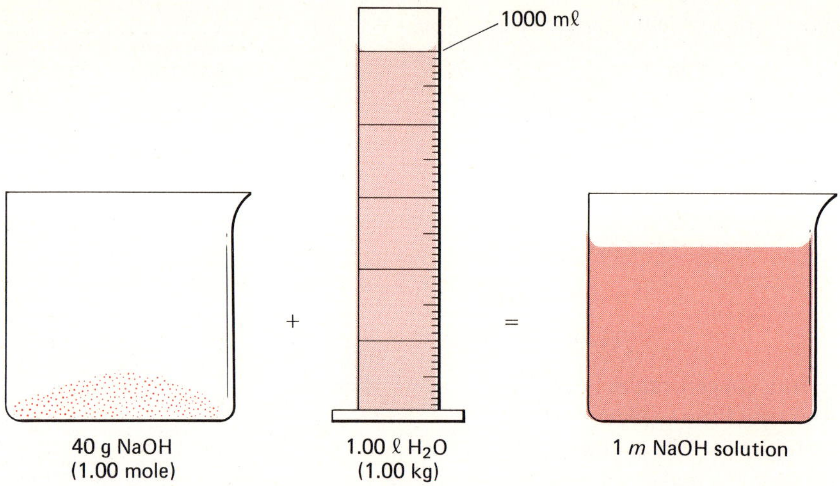

Figure 13-6
Preparation of a one molal (m) solution of sodium hydroxide by combining forty grams of sodium hydroxide with one liter of water

Molar Concentration

In Chapter 3 we defined **molar concentration** (M) as the number of moles of solute per liter of solution:

$$M = \frac{\text{moles of solute}}{\text{liter of solution}}$$

Another term that can be used for this method of expressing concentration is **formal concentration** (F). This gives the same quantity but refers to the formula weight expressed in grams rather than the molecular weight expressed in grams. This term has not gained much popularity, however, and thus *molar concentration* is generally used. Using the correct definition of the mole as the formula weight expressed in grams also eliminates the need to differentiate between substances that exist as molecules from those that exist as formula units. Therefore, in this text we shall use molar concentration exclusively, even though formula weight is used.

Molar concentration is especially useful for doing *stoichiometric* calculations (Chapter 9) involving solutions. In Section 13-6 we shall consider some calculations of this type.

In doing calculations, or preparing solutions, involving molar concentration it is especially important to keep in mind that no information is given as to the amount of solvent that is added. Only the total amount of solution is expressed. For example, to prepare a one molar aqueous solution of calcium bromide ($CaBr_2$), 1 mole of solute (200 g $CaBr_2$) is dissolved in enough water to make a total of 1 *l* of solution, as shown in Fig. 13-7.

Consider the following problem examples.

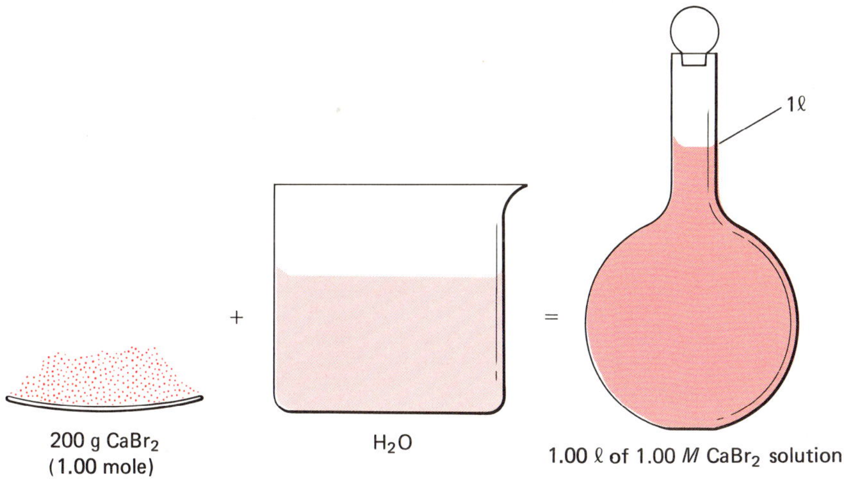

Figure 13-7
Preparation of one liter of a one molar (M) solution of calcium bromide in water by putting two hundred grams of calcium bromide in a one liter volumetric flask and adding enough water to fill to the calibration mark

PROBLEM EXAMPLE 13-10

Calculate the molar concentration of a solution that contains 15 g of potassium hydroxide in 225 mℓ of solution.

SOLUTION:

There are two steps involved here: conversion of grams of KOH to moles using the formula weight of KOH (56 g/mol) and conversion of 225 mℓ to liters. Thus the molar concentration is calculated as

$$\frac{15\ \cancel{\text{g KOH}}}{225\ \cancel{\text{m}\ell\ \text{soln}}} \times \frac{1\ \text{mol KOH}}{56\ \cancel{\text{g KOH}}} \times \frac{1000\ \cancel{\text{m}\ell\ \text{soln}}}{1\ \ell\ \text{soln}} = 1.2\ \frac{\text{mol KOH}}{\ell\ \text{soln}} = \mathbf{1.2}\ \boldsymbol{M}$$

(The last step here can be simplified by recognizing that 225 mℓ = 0.225 ℓ.)

PROBLEM EXAMPLE 13-11

Calculate the number of grams of calcium nitrate necessary to prepare 450 mℓ of 2.25 *M* solution.

SOLUTION:

Since 450 mℓ is equivalent to 0.45 ℓ, the problem says

$$0.450\ \ell\ \text{soln} = ?\ \text{g}\ Ca(NO_3)_2,$$

the fact that 1 ℓ soln = 2.25 mol $Ca(NO_3)_2$ and that the formula weight NO$_3$)$_2$ is 164 [and therefore 164 g $Ca(NO_3)_2$ = 1 mole]. This is then ed as

$$\text{n} \times \frac{2.25 \text{ mol } Ca(NO_3)_2}{1\ \ell \text{ soln}} \times \frac{164 \text{ g } Ca(NO_3)_2}{1 \text{ mol } Ca(NO_3)_2} = 166 \text{ g } Ca(NO_3)_2$$

PROBLEM EXAMPLE 13-12

Calculate the number of milliliters of 3.50 *M* ammonium carbonate solution necessary to provide 12.5 g of ammonium carbonate.

SOLUTION:

The formula weight of ammonium carbonate, $(NH_4)_2CO_3$, is 96.0. In a 3.50 *M* solution of this, 1 ℓ soln = 3.50 moles $(NH_4)_2CO_3$. Therefore the number of liters needed is calculated as

$$12.5 \text{ g } (NH_4)_2CO_3 \times \frac{1 \text{ mol } (NH_4)_2CO_3}{96.0 \text{ g } (NH_4)_2CO_3} \times \frac{1\ \ell \text{ soln}}{3.50 \text{ mol } (NH_4)_2CO_3} \times \frac{1000 \text{ m}\ell \text{ soln}}{1\ \ell \text{ soln}}$$
$$= 37.2 \text{ m}\ell \text{ soln}$$

PROBLEM EXAMPLE 13-13

Calculate the molar concentration of a 45.0% aqueous calcium chloride solution whose density is 1.10 g/mℓ.

SOLUTION:

A 45.0% solution of $CaCl_2$ contains 45.0 g of $CaCl_2$ in 100 g of solution; the formula weight of $CaCl_2$ is 111; the density tells us that 1.10 g of solution is equivalent to 1 mℓ of solution. Thus the molar concentration is calculated as

$$\frac{45.0 \text{ g } CaCl_2}{100 \text{ g soln}} \times \frac{1.10 \text{ g soln}}{1 \text{ m}\ell \text{ soln}} \times \frac{1 \text{ mol } CaCl_2}{111 \text{ g } CaCl_2} \times \frac{1000 \text{ m}\ell \text{ soln}}{1\ \ell \text{ soln}} = \frac{4.46 \text{ mol } CaCl_2}{\ell \text{ soln}}$$
$$= 4.46\ M$$

13-6 Solution Stoichiometry

In Chapter 9 we did several types of calculations using balanced chemical equations. We also did more of these in Chapter 11, using reactions that involved gases. You should recall that in all calculations of this type we used the following steps:

1. Convert from the given units to moles, if not given in moles.
2. Convert from moles of the given quantity to moles of the desired quantity, using the balanced equation.
3. Convert from moles to any other desired units.

We can also do calculations of this type using amount and concentration of a given solution to calculate the number of moles of solute present.

Consider the following problem example.

PROBLEM EXAMPLE 13-14

Calculate the number of grams of calcium nitrate that could be produced by reacting 125 mℓ of 6.00 *M* nitric acid with excess calcium hydroxide.

SOLUTION:

The first step is to write a balanced equation:

$$2\ HNO_3 + Ca(OH)_2 \longrightarrow 2\ H_2O + Ca(NO_3)_2$$

Next we consider what we are going to calculate; that is, this problem says, 125 mℓ HNO_3 soln = ? g $Ca(NO_3)_2$, given the following facts.

(a) The molar concentration of HNO_3 solution.
(b) The balanced equation.
(c) The mass of 1 mole of calcium nitrate in grams (molar mass), which is calculated from the formula to be 164 g $Ca(NO_3)_2$ per mole.

Following the steps outlined for calculations from balanced equations gives the following result:

$$125\ m\ell\ \cancel{HNO_3\ soln} \times \frac{6.00\ \cancel{mol\ HNO_3}}{1000\ \cancel{m\ell\ HNO_3\ soln}} \times \frac{1\ \cancel{mol\ Ca(NO_3)_2}}{2\ \cancel{mol\ HNO_3}} \times \frac{164\ g\ Ca(NO_3)_2}{1\ \cancel{mol\ Ca(NO_3)_2}} = \mathbf{61.5\ g\ Ca(NO_3)_2}$$

13-7 Colligative Properties of Solutions*

Since solutions are mixtures, the properties of solutions depend on the amount of the various substances that make up the solution. Some of the physical properties of solutions depend only on the concentration of particles in the solution, regardless of what these particles are. These properties are known as *colligative* properties and include such things as vapor pressure, freezing point, and boiling point. For solutions of nonionic substances, *Raoult's Law* states that the partial vapor pressure (at a given temperature) of a volatile component in an ideal solution is equal to the product of the mole fraction of this component in the solution and the vapor pressure of this component in the pure state. A sugar solution contains a nonvolatile component (sugar) whose vapor pressure is negligible and a volatile component (water) whose vapor pressure is appreciable. The vapor pressure of the water in this solution (P_A) will be equal to the product of

*This section is optional and can be omitted without loss of continuity.

the mole fraction of the water (X_A) and the vapor pressure of pure water (P_A°) at that temperature.

$$P_A = X_A P_A^\circ$$

The following problem example illustrates the application of Raoult's Law.

PROBLEM EXAMPLE 13-15

Calculate the vapor pressure of a solution containing 600 g of sugar ($C_{12}H_{22}O_{11}$) in 1000 g of water at 25°C. ($P_{H_2O}^\circ$ at 25°C is 23.6 torr.)

SOLUTION:

Since the formula (molecular) weight of sugar is 342 and water is 18.0, the number of moles of sugar and of water in the solution are

$$600\ \text{g sugar} \times \frac{1\ \text{mol sugar}}{342\ \text{g sugar}} = 1.75\ \text{mol sugar}$$

$$1000\ \text{g}\ H_2O \times \frac{1\ \text{mol}\ H_2O}{18.0\ \text{g}\ H_2O} = 55.6\ \text{mol water}$$

The mole fraction of water in the solution is

$$X_{H_2O} = \frac{55.6}{55.6 + 1.75} = 0.969$$

Since the sugar is nonvolatile, the vapor pressure of the solution would be equal to the vapor pressure of the water in the solution, times the mole fraction of water in the solution

$$P_{H_2O} = X_{H_2O} P_{H_2O}^\circ = 0.969 \times 23.6\ \text{torr} = \mathbf{22.9\ torr}$$

One direct consequence of the lowering of the vapor pressure of a solution is an elevation in the boiling point. Depending on what the solvent is, 1 mole of any nonelectrolyte will raise the boiling point of a given amount of a solvent by a constant amount. Also, it is found that one mole of any nonelectrolyte will lower the freezing point of a given amount of a solvent by a constant amount. For example, the value for water is 1.86°C/mol/kg H_2O; that is, 1 mole of any nonelectrolyte will lower the freezing point of 1 kg of H_2O by 1.86°C. This is called the *molal freezing point depression constant* (K_f). Thus, if 1 mole of sugar is dissolved in 1 kg of water, the freezing point of the solution will be −1.86°C (0°C − 1.86°C = −1.86°C).

Likewise, 1 mole of any nonvolatile, nonelectrolyte will elevate the boiling point of a given amount of a solvent by a constant amount, depending on what the solvent is. This value, called the *molal boiling point elevation constant* (K_b), is 0.52°C/mol/kg H_2O, for water. Thus, 1 mole of sugar in 1 g of water will increase the boiling point to 100.52°C (100°C + 0.52°C).

TABLE 13-2 *Molal freezing point depression and boiling point elevation constants for various solvents (°C/mol solute/kg solvent).*

SOLVENT	K_f	K_b
Benzene	4.9	2.5
Acetic acid	3.9	3.1
Chloroform	4.7	3.6
Water	1.86	0.52

Table 13-2 lists freezing point depressions and boiling point elevations for one molal solutions of nonvolatile, nonelectrolytes in various solvents.

From the units used in these constants (°C/mol solute/kg solvent), we can show that the temperature change (Δt) is equal to

$$\Delta t_b = (\text{molality})(K_b) \quad \text{or} \quad \Delta t_f = (\text{molality})(K_f)$$

The following problem examples illustrate calculations involving boiling point elevation and freezing point depression.

PROBLEM EXAMPLE 13-16

Calculate the freezing point and boiling point of a solution containing 4.27 g of sugar ($C_{12}H_{22}O_{11}$) in 50.0 g of water.

SOLUTION:

The molecular weight of sugar ($C_{12}H_{22}O_{11}$) is 342, and the molality of the solution is calculated as follows:

$$\frac{4.27 \text{ g sugar}}{50.0 \text{ g H}_2\text{O}} \times \frac{1000 \text{ g H}_2\text{O}}{1 \text{ kg H}_2\text{O}} \times \frac{1 \text{ mol sugar}}{342 \text{ g sugar}} = \frac{0.250 \text{ mol sugar}}{\text{kg water}}$$

The freezing point depression is calculated as

$$\Delta t_f = 0.250\ m \times \frac{1.86°\text{C}}{m} = 0.465°\text{C}$$

or the freezing point would be **−0.465°C.** The boiling point elevation is calculated in a similar manner:

$$\Delta t_b = 0.250\ m \times \frac{0.52°\text{C}}{m} = 0.13°\text{C}$$

or the boiling point would be

$$100.00 + 0.13 = \mathbf{100.13°C} \quad \text{(at 760 torr)}$$

Another way of solving this is to use the K_f as a type of conversion factor:

$$\frac{0.250\ \cancel{\text{mol sugar}}}{\cancel{\text{kg H}_2\text{O}}} \times \frac{1.86\ (°\text{C})}{\frac{\cancel{(\text{mol sugar})}}{\cancel{\text{kg H}_2\text{O}}}} = 0.465°\text{C change}$$

and therefore the new freezing point is **−0.465°C.**

PROBLEM EXAMPLE 13-17

If 4.35 g of ethylene glycol is dissolved in 100 g of water, the resulting solution has a freezing point of −1.31°C. Calculate the molecular weight of ethylene glycol.

SOLUTION:

Since the freezing point of the solution is −1.31°C, $\Delta t_f = 1.31$, and the molality of the solution may be calculated as

$$\text{molality} = \frac{\Delta t_f}{K_f} = \frac{1.31°\text{C}}{1.86°\text{C}/m} = 0.704\ m$$

Using the units of molality (moles of solute per liter of solvent) and solving for grams per mole (molecular weight), the solution is as follows:

$$\frac{4.35\ \text{g ethylene glycol}}{100\ \cancel{\text{g H}_2\text{O}}} \times \frac{1000\ \cancel{\text{g H}_2\text{O}}}{1\ \cancel{\text{kg H}_2\text{O}}} \times \frac{1\ \cancel{\text{kg H}_2\text{O}}}{0.704\ \text{mol ethylene glycol}} = \frac{\mathbf{61.8\ g}}{\mathbf{mol}}$$

For a volatile solute the boiling point is *lowered* by addition of the solute. The reason for this is that the solute contributes to the vapor pressure, thus increasing the vapor pressure and lowering the boiling point.

For electrolytes the freezing point depression or boiling point elevation is approximately the value given (as in Table 13-2) for that substance times the number of moles of ions formed per mole of electrolyte. For example, a solution of sodium chloride (NaCl) would be approximately twice the value given, calcium chloride ($CaCl_2$) would be about three times the value, and so forth. This is especially true in dilute solutions, which indicates that the more dilute the solution is, the more the individual ions formed can act as separate particles.

13-8 Osmosis and Osmotic Pressure

If a solution is separated from a pure solvent by a semipermeable membrane (which allows the solvent but not the solute to pass through), the solvent tends to flow through the membrane toward the side containing the solution. This is the process of **osmosis.** In doing this the water is tending to equalize the solutions; that is, the water would also pass through the membrane from a more dilute solution to a more concentrated one. Actually the solvent is passing through the membrane in both directions, but the rate of flow is greater

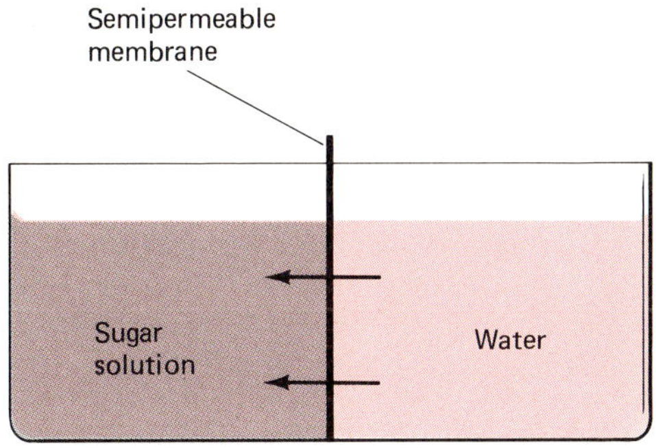

Figure 13-8
Osmosis

from the more dilute to the more concentrated side. The process of osmosis is illustrated in Fig. 13-8.

The force causing this diffusion, or the amount of pressure that is required to stop the diffusion, is called the **osmotic pressure.** Two solutions having the same osmotic pressure are said to be *isotonic.* A solution with a higher osmotic pressure than a second solution is said to be *hypertonic* to the second solution, and the second is *hypotonic* to the first; that is, for two solutions, A and B, if the osmotic pressure of A is greater than B, A is hypertonic to B and B is hypotonic to A.

13-9 Colloids

At the beginning of this chapter it was pointed out that **colloids** are intermediate between solutions and precipitates. The particles in colloids are larger than the molecules or ions that make up solutions. They are not large enough to settle out upon standing, however, as a precipitate will do. The colloidal particles are in the approximate range of 10 to 100 nm (nanometers) in diameter; whereas ions and molecules in solution are generally in the range of about 0.1 to 10 nm. These colloidal particles can consist of individual, very large molecules, such as starch, or aggregates of smaller molecules.

In discussing solutions, we spoke of solutes and solvents. In discussing colloids, we use the terms *dispersed particles* (dispersed phase) and *dispersing medium* (dispersing phase). Dispersed particles are the colloidal particles, comparable to the solute in a solution. The dispersing medium is the substance in which the colloidal particles are distributed, comparable to the solvent in a solution. Milk is an example of a colloid; butterfat constitutes the dispersed particles and water is the dispersing medium.

As with solutions, colloids may exist in any one of the three physical states of matter. These different types of colloids have different names: a *foam* is a gas dispersed in a liquid or solid; a liquid *aerosol* is a liquid dispersed in a gas; a solid aerosol or a *smoke* is a solid dispersed in a gas; an *emulsion* is a liquid dispersed in a liquid; a *gel* is a liquid dispersed in a solid; and a *sol* is a solid dispersed in a liquid or solid. Table 13-3 lists these various types of colloids and gives an example of each.

TABLE 13-3 *Types of colloids*

DISPERSED PARTICLES	DISPERSING MEDIUM	NAME	EXAMPLE
Gas	Liquid	Foam	Whipped cream
Gas	Solid	Solid foam	Styrofoam
Liquid	Gas	Liquid aerosol	Fog
Liquid	Liquid	Emulsion	Mayonnaise
Liquid	Solid	Gel	Jello
Solid	Gas	Solid aerosol	Smoke
Solid	Liquid	Sol	Paint
Solid	Solid	Soild sol	Gems, such as ruby, turquoise, or garnet

Beside particle size, colloids have other identifying properties. Properties peculiar to colloids are (1) optical effect, (2) motion effect, and (3) electrical charge effect.

Optical Effect

When a relatively narrow beam of light is passed through a colloid, such as dust particles in the air, the light is scattered by the dust particles and they appear in the beam as bright, tiny specks of light. In a solution, the light is not scattered, and thus the particles cannot be observed, as shown in Fig. 13-9. The scattering of light in a colloid is due to the reflection of the light by the larger colloidal particles producing a visible beam of light. No observable reflection is caused by the smaller solute particles in a solution. This optical effect is named the *Tyndall effect,* after British physicist John Tyndall (1820–1893), who investigated it in 1860.

Motion Effect

If a colloid is viewed with a special microscope, dispersed colloidal particles appear to move in a zigzag, random motion through the dispersing medium. The special microscope consists of a regular microscope focused on the colloid with a strong light source at right angles to the colloid, viewed against a black background. What are actually seen are reflections from the colloidal particles. The erratic, random motion of the dispersed colloidal particles in

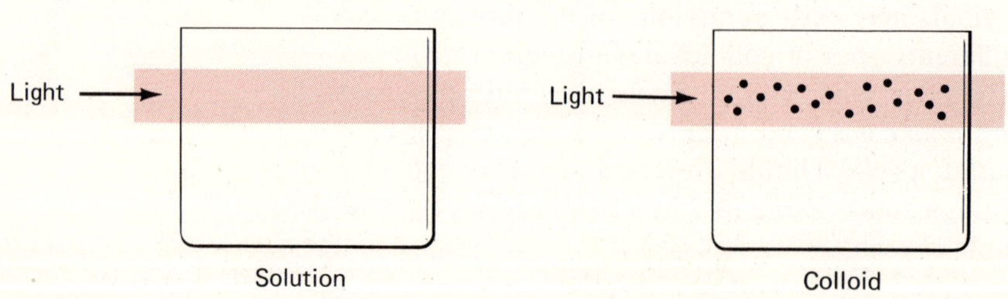

Figure 13-9
Optical (Tyndall) effect of colloids

the dispersing medium is due to bombardment of the dispersed colloidal particles by the medium, creating a zigzag, random motion. Since this reflection of light is due to the colloidal particle size, no such motion is seen in a solution; however, the solute and solvent particles are in continuous random motion. This motion effect of colloids is one reason why colloidal particles do not settle out on standing. The consant bombardment by the dispersing medium particles on the dispersed colloidal particles keeps the dispersed colloidal particles suspended indefinitely. This effect, named after the English botanist Robert Brown (1773–1858), who first studied it in 1827, is called *Brownian movement.*

Electrical Charge Effect

Very often a colloid will have ions adsorbed on its surface. This, along with the Brownian motion of the colloid, prevents colloids from coagulating and precipitating.

In the laboratory this often takes place when a particular substance is being precipitated, such as in the identification of ions in qualitative analysis. To prevent the formation of colloids, due to this charge effect, an electrolyte of some type is added to the solution. This electrolyte must be something that will not interfere with the analysis. Often an excess of the precipitating agent is added, just to be present as an electrolyte. In other cases a solution such as ammonium nitrate, which will not form a precipitate with any common ions, is added. The presence of the electrolyte in the solution tends to neutralize the charge on the colloidal particles and allows them to precipitate.

The coagulation effect of an electrical charge on colloids is used in removing suspended particles from the effluent gases in industrial smokestacks. The device used in this operation is called a *Cottrell precipitator.* In this device the gas being given off, which often contains various pollutants in the form of small particles, is given an electrical charge. Then as the smoke (a colloid containing a solid in a gas) passes up, before going out into the atmosphere, it is passed through charged plates. These attract the charged colloidal particles, which then are held on the plates until the current giving the plates their charge is shut off. When this is done the particles fall to the ground and can be collected, rather than passing out into the atmosphere. This is shown in Fig. 13-10.

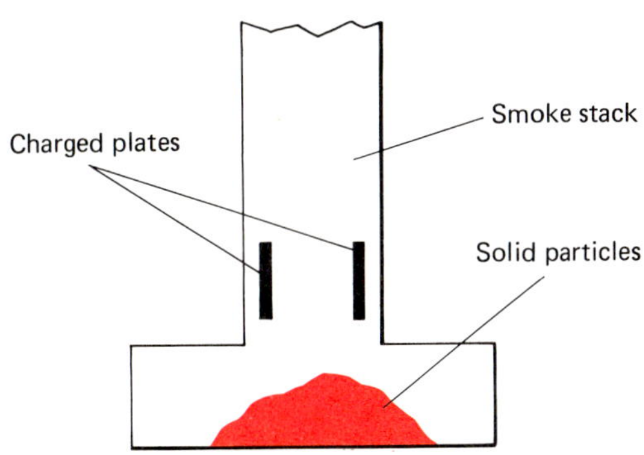

Figure 13-10
Cottrell precipitator

SUMMARY

1. A *solution* is a homogeneous mixture made up of a *solute* and a *solvent,* either of which can be in any of the three states of matter.
2. The factors that affect *solubility* are properties of the solute and solvent (such as polarity and ionic character), temperature, and pressure (only for those involving gases).
3. The factors that affect *rate of solution* are particle size, stirring, and temperature.
4. A solution can be *saturated, unsaturated,* or *supersaturated.*
5. The *concentration* of a solution can be expressed in numerous ways, four of which are as follows:

   (a) $\text{percent by mass} = \dfrac{\text{mass solute}}{\text{mass solution}} \times 100$

   $\left(\text{percent by volume} = \dfrac{\text{volume solute}}{\text{volume solution}} \times 100\right)$

   (b) $\text{molal } (m) = \dfrac{\text{moles solute}}{\text{kilograms solvent}}$

   (c) mole fraction of solute (or solvent)

   $$= \frac{\text{moles solute (or solvent)}}{\text{total moles solute and solvent}}$$

   (d) $\text{molar } (M) \text{ or formal } (F) = \dfrac{\text{moles solute}}{\text{liter solution}}$
6. According to *Raoult's Law,* the vapor pressure of the volatile component of a mixture will be equal to the mole fraction of that component times the initial vapor pressure of the component; that is, $P_A = X_A P_A^\circ$.
7. *Colligative properties* of a solution depend on the number of particles of solute present in a given amount of solvent. For example, the freezing point of a solution is lowered by a given amount (for a particular solvent) for each mole of solute particles that are present in a kilogram of the solvent. Thus, for any solvent there is a constant called the *molal freezing point depression (or boiling point elevation) constant,* which gives the change in freezing point (or boiling point) for a one molal solution of any nonelectrolyte in the solvent.
8. *Colloids* are similar to solutions, except that the particles are larger in a colloid; they are large enough to be detected (unlike solute particles in a solution) but are not large enough to settle out (unlike a precipitate, which will eventually settle out). The colloidal particles are called the *dispersed phase,* and the other part of the mixture is called the *dispersing medium* (equivalent to the solute and solvent in a solution). Colloidal particles can be observed when a beam of light is passed through them, due to the scattering of the light by the particles (this is called the *Tyndall effect*). Also, due to the bombardment of the dis-

persed particles by the molecules of the dispersing medium, a rapid, random motion of the particle can be observed (called *Brownian movement*). Often colloidal particles have a layer of ions surrounding them, which also keeps the particles from settling out.

9. If a solution is separated from a pure solvent by a semipermeable membrane, the solvent tends to flow into the solution, to make it more dilute, due to *osmosis*.

Upon completion of this chapter, you should be able to do the following:

1. Given the partial pressure, calculate the solubility of a gas. (1)
2. Calculate the percent (by mass) and the molal (m), mole fraction, or molar (M) concentration, given any of the following. (Assume in each case that a periodic table with atomic weights is available and that density is given when necessary.) (2, 6, 7, 10)
   (a) mass of solute and mass of solvent
   (b) mass of solute and volume or mass of solution
   (c) moles of solute and volume of solution
   (d) moles of solute and mass of solvent
   (e) percent (by mass) and molal, mole fraction, or molar concentration
3. Calculate the mass in grams, or the number of moles, of solute in a given amount of solution of known concentration or the amount of water that must be added to a given amount of solute to prepare a solution. (3, 4, 5, 8, 9, 11)
4. Calculate the volume of solution, of known concentration, that contains a given amount of solute. (12)
5. Calculate the volume of solution, of known concentration, that would react with a given amount of a substance, or the number of grams of product formed from a given amount of solution, of known concentration. (13, 14)
6. Given the mole fraction (or the quantities necessary to calculate it) of the volatile component of a mixture and the initial vapor pressure of that component, calculate the new vapor pressure. (15)
7. Given any three of the following quantities, calculate the fourth quantity: the moles (or mass) of nonelectrolyte in a solution, the amount of solvent, the new freezing point (or the change in freezing point), and the molal freezing point depression constant. (16, 17, 18)
8. Given any of the terms in Exercise 1, give a definition or explanation in your own words.

EXERCISES

1. Define or explain the following terms:
   (a) filtrate
   (b) heterogeneous mixture
   (c) homogeneous mixture
   (d) solution
   (e) solute
   (f) solvent
   (g) solubility
   (h) electrolyte
   (i) nonelectrolyte
   (j) saturated solution
   (k) unsaturated solution
   (l) supersaturated solution
   (m) percent by mass
   (n) percent by volume
   (o) mole fraction
   (p) molality
   (q) molarity
   (r) colloid
   (s) osmosis
   (t) osmotic pressure
2. Give an example of each of the following types of solutions:
   (a) solid in liquid
   (b) liquid in liquid
   (c) gas in liquid
   (d) gas in gas
   (e) solid in solid
3. Distinguish between
   (a) a solution and a pure substance
   (b) a colloid and a solution
   (c) solubility and rate of solution
   (d) concentrated and dilute solution
   (e) percent by mass and percent by volume
   (f) molality and molarity
   (g) electrolyte and nonelectrolyte
   (h) weak electrolyte and strong electrolyte
4. Describe, in your own words, the process of purification of a solid by crystallization.

PROBLEMS

Henry's Law

1. Calculate the solubility of the gas, in grams per liter, under the new conditions.
   (a) Calculate the solubility of oxygen in water at 3.50 atm and 5°C if the solubility is 0.0132 g oxygen per liter of water at 1.20 atm and 5°C.
   (b) Calculate the solubility of nitrogen in water at 700 torr and 0°C if the solubility is 0.0292 g of nitrogen per liter of water at 0°C and 760 torr.

Percent concentration by mass

2. Calculate the percent by mass of the solute in each of the following solutions:
   (a) 8.60 g of salt in 95.0 g of solution
   (b) 25.0 g of potassium carbonate in 100.0 g of water
   (c) 12.7 g of sodium chloride in 110 g of solution
   (d) 10.4 g of barium chloride in 80.7 g of water

3. Calculate the grams of solute that must be dissolved in
   (a) 350 g of water in the preparation of a 15.0% potassium sulfate solution
   (b) 15.0 g of water in the preparation of a 10.0% sodium chloride solution
4. Calculate the grams of water that must be added to
   (a) 16.0 g of sugar ($C_{12}H_{22}O_{11}$) in the preparation of a 25.0% sugar solution
   (b) 4.00 g of potassium iodide in the preparation of a 2.00% potassium iodide solution
5. Calculate the number of grams of solution necessary to provide the following:
   (a) 68.5 g of sodium chloride from a 12.0% sodium chloride solution
   (b) 1.00 g of sodium bicarbonate from a 6.00% sodium bicarbonate solution

Mole fraction

6. Calculate the mole fraction of the solute in each of the following solutions:
   (a) a solution containing 75 g of ethanol (C_2H_6O) in 450 g of water
   (b) a 20% aqueous hydrochloric acid solution
   (c) a solution containing 2.6 g of sodium sulfate in 20 g of water
   (d) a 40% solution of phenol (C_6H_6O) in benzene (C_6H_6)

Molal concentration

7. Calculate the molality of each of the following solutions:
   (a) 175 g of ethyl alcohol (C_2H_6O) in 650 g of water
   (b) 3.50 g of H_2SO_4 in 13.5 g of aqueous solution
   (c) 37.5% hydrochloric acid solution
   (d) 12.6 g of ethylene glycol ($C_2H_6O_2$) in 480 g of water
   (e) 28.0 g of $CaCl_2$ in 628 g of aqueous solution
   (f) 70.5% nitric acid solution
8. Calculate the number of grams of solute necessary to prepare the following solutions:
   (a) 400 g of a 0.500 *m* solution of ethyl alcohol (C_2H_6O)
   (b) 700 g of a 0.600 *m* solution of sulfuric acid
9. Calculate the number of grams of water that must be added to
   (a) 60.0 g of glucose ($C_6H_{12}O_6$) in the preparation of a 2.00 *m* solution
   (b) 85.0 g of sugar ($C_{12}H_{22}O_{11}$) in the preparation of a 8.00 *m* solution

Molar concentration

10. Calculate the molarity of each of the following solutions:
    (a) 2.50 moles of ethyl alcohol (C_2H_6O) in 450 m*l* of solution
    (b) 2.60 g of sodium chloride in 40.0 m*l* of solution

(c) 20.0 mg of sugar ($C_{12}H_{22}O_{11}$) in 275 μl of solution
(d) 17.5% aqueous sodium hydroxide solution (density = 1.05 g/ml)
(e) 3.25 *m* calcium chloride solution (density = 1.02 g/ml)
(f) 3.75 moles of sodium bromide in 850 ml of solution
(g) 10.0 g of calcium chloride in 640 ml of solution
(h) 16.7 mg of barium bromide in 750 μl of solution
(i) 45.0% aqueous calcium nitrate solution (density = 1.12 g/ml)
(j) 7.50 *m* sulfuric acid solution (density = 1.08 g/ml)

11. Calculate the number of grams of solute necessary to prepare the following solutions:
(a) 500 ml of a 0.10 *M* sodium hydroxide solution
(b) 250 ml of a 0.020 *M* calcium chloride solution
(c) 100 ml of a 0.15 *M* sodium sulfate solution
(d) 135 ml of a 0.80 *M* phosphoric acid solution

12. Calculate the number of milliliters of solution required to provide the following:
(a) 1.00 mole of sulfuric acid from a 6.00 *M* solution
(b) 5.00 g of sodium bromide from a 0.100 *M* solution
(c) 7.65 mg of calcium chloride from a 1.4 *M* solution
(d) 3.50 moles of nitric acid from a 3.0 *M* solution
(e) 8.50 g of magnesium chloride from a 2.5 *M* solution
(f) 10.2 mg of sodium hydroxide from a 6.0 *M* solution

Solution stoichiometry

13. Calculate the number of milliliters of solution required in each of the following:
(a) 2.25 *M* nitric acid to react with 24.2 ml of 3.50 *M* sodium hydroxide solution
(b) 4.35 *M* hydrochloric acid to react with 18.5 g of calcium hydroxide
(c) 12.5% sodium hydroxide solution (density = 1.15 g/ml) to react with 42.5 ml of 6.50 *M* chlorous acid
(d) 3.50 *M* hydrochloric acid to react with 5.45 g of zinc
(e) 1.50 *M* sodium hydroxide solution to react with 35.5 ml of 1.75 *M* nitric acid
(f) 24.5% hydrochloric acid (density = 1.20 g/ml) to react with 3.45 g of calcium oxide

14. Calculate the number of grams of precipitate formed in each of the following.

(a) An excess of silver nitrate is added to 25.0 m*l* of 1.50 *M* hydrochloric acid.
(b) An excess of sodium hydroxide is added to 42.5 m*l* of 15.5% magnesium nitrate solution whose density is 1.12 g/m*l*.
(c) 25.0 m*l* of 1.25 *M* sodium carbonate solution is mixed with 32.5 m*l* of 1.50 *M* aluminum chloride solution.
(d) An excess of sodium carbonate solution is added to 34.5 m*l* of 2.45 *M* calcium nitrate solution.
(e) An excess of sodium phosphate solution is added to 15.5 m*l* of 7.25% barium nitrate solution whose density is 1.08 g/m*l*.
(f) 15.0 m*l* of 2.25 *M* iron (III) nitrate solution is mixed with 18.5 m*l* of 2.30 *M* ammonium carbonate solution.

Colligative Properties of Solutions

15. Assuming Raoult's Law applies, calculate the vapor pressure of
    (a) a solution of 20 g of sucrose ($C_{12}H_{22}O_{11}$) in 80 g of water at 25°C. (the vapor pressure of pure water at 25°C is 24 torr)
    (b) a solution of 45 g of glucose ($C_6H_{12}O_6$) in 95 g of water 25°C. (the vapor pressure of pure water at 25°C is 24 torr)
16. Calculate the molecular weight of the solute in each of the following.
    (a) A solution containing 4.20 g of a nonelectrolyte in 400 g of water freezes at −0.250°C. (K_f for $H_2O = 1.86$)
    (b) A 5.0% solution of a nonvolatile nonelectrolyte in nitrobenzene ($C_6H_5NO_2$) has a boiling point of 212.15°C at 760 torr. The normal boiling point of nitrobenzene at 760 torr is 210.85°C and K_b for nitrobenzene is 5.2. Calculate the molecular weight of the solute.
    (c) A solution containing 2.47 g of a nonelectrolyte in 100 g of acetic acid ($HC_2H_3O_2$) freezes 1.3°C below the normal freezing point for pure acetic acid. (K_f for $HC_2H_3O_2 = 3.9$)
    (d) A solution containing 4.32 g of a nonvolatile nonelectrolyte in 250 g of water has a boiling point of 100.20°C at 760 torr. Calculate the molecular weight of the solute. (K_b for $H_2O = 0.52$.)
17. Calculate the boiling point and freezing point for each of the following solutions of nonelectrolytes:
    (a) a 0.20 molal solution of sugar ($C_{12}H_{22}O_{11}$) in water
    (b) a solution containing 5 g of glycerol ($C_3H_8O_3$) in 20 g of water
    (c) a 20% aqueous sucrose ($C_{12}H_{22}O_{11}$) solution
    (d) a 0.45 molal solution of glycerol ($C_3H_8O_3$) in water

(e) a solution containing 3.5 g of sucrose ($C_{12}H_{22}O_{11}$) in 25 ml of water

(f) a 35% aqueous glucose ($C_6H_{12}O_6$) solution

18. Calculate the freezing point of
    (a) a 9.5 molar solution of ethanol (C_2H_5OH) in water. Density of 9.5 molar ethanol is 0.91 g/ml.
    (b) a 1.2 molar aqueous sugar ($C_{12}H_{22}O_{11}$) solution whose density is 1.1 g/ml.

General problems

19. Calculate the percent, by mass, of solute in each:
    (a) in a solution that contains 3.85 g of calcium chloride in 78.50 g of water
    (b) in a solution that contains 32.3 g of sodium bromide in 49.5 g of water
20. Calculate the molal (m) concentration of the following solutions.
    (a) A solution that contains 2.60 g of glucose ($C_6H_{12}O_6$) in 100 g of water
    (b) A solution that contains 6.25 g of calcium phosphate in 100 g of water
21. Calculate the mass in grams of solute necessary to prepare each of the following solutions:
    (a) 425 g of a 3.0 m solution of ethylene glycol ($C_2H_6O_2$)
    (b) 345 g of a 2.80 m solution of aluminum sulfate
22. Calculate the molar (M) concentration of each:
    (a) a solution that contains 3.80 g of calcium bromate in 125 ml of solution
    (b) a solution that contains 20 g of sugar ($C_{12}H_{22}O_{11}$) in 275 ml of solution
23. Calculate the number of grams of solute necessary to prepare each of the following solutions:
    (a) 100 ml of a 0.15 M sodium sulfate solution
    (b) 225 ml of a 1.15 M glucose ($C_6H_{12}O_6$) solution
24. Calculate the molar (M), molal (m), and percent concentration of each:
    (a) a solution made up of 32.3 g nitric acid and 67.7 g H_2O (specific gravity = 1.20)
    (b) a solution made up of 8.35 g of calcium carbonate and 25.0 g H_2O (specific gravity = 1.12)
25. Calculate the mass in grams of solute necessary to make the following solutions:
    (a) 40 ml of a 0.10 M sulfuric acid solution
    (b) 35 ml of a 0.12 M sodium chloride solution
26. Calculate the number of liters of solution that could be made from each of the following:
    (a) 0.10 M solution of silver nitrate from 20 g of silver nitrate
    (b) 1.2 M solution of magnesium sulfate from 97 g of magnesium sulfate

27. Calculate the molar concentration of each:
    (a) a 20% sulfuric acid solution with a specific gravity of 1.96
    (b) a 35% potassium bromide solution with a specific gravity of 1.15
28. Calculate the mass, in grams, of each:
    (a) calcium oxide to react with 32 mℓ of 1.2 *M* hydrochloric acid
    (b) sodium oxide to react with 8.5 mℓ of 0.18 *M* nitric acid
29. Calculate the number of milliliters of each of the following:
    (a) 6.0 *M* hydrochloric acid to react with 3.5 g of calcium hydroxide
    (b) 3.5 *M* HNO_3 to react with 24 g of aluminum hydroxide
30. Calculate the number of milliliters necessary for each of the following:
    (a) 4.5 *M* sodium hydroxide to react with 25 mℓ of 6.2 *M* sulfuric acid
    (b) 0.850 *M* sodium hydroxide to react with 12.5 mℓ of 1.20 *M* hydrochloric acid
31. Calculate the molal concentration, molar concentration, and mole fraction of each:
    (a) 3.25% solution of aluminum sulfate (density = 1.06 g/mℓ)
    (b) 18.2% solution of sulfuric acid (density = 1.48 g/mℓ)
32. Calculate the percent, molar concentration, and mole fraction of each:
    (a) a 7.25 molal nitric acid solution whose specific gravity is 1.14
    (b) a 3.05 molal sulfuric acid solution whose specific gravity is 1.08
33. Calculate the percent, molal concentration, and mole fraction of each:
    (a) a 1.40 *M* solution of calcium nitrate whose density is 1.03 g/mℓ
    (b) a 6.20 *M* solution of ferrous carbonate whose density is 1.10 g/mℓ

References

REFERENCES

BENSON, SIDNEY, W., *Chemical Calculations,* 3rd ed. New York: John Wiley & Sons, Inc., 1971.

PETERS, EDWARD I., *Problem Solving for Chemistry,* 2nd ed. Philadelphia: W. B. Saunders Co., 1977.

Objectives

1. To recognize the general properties of acids and bases in solution.
2. To learn the Arrhenius definition of acids and bases and to be able to identify substances as Arrhenius acids or bases.
3. To learn to write equations representing the dissociation or ionization of acids and bases in solution.
4. To understand what is meant by the terms *weak* and *strong* acid and base and to learn which common acids and bases are strong.
5. To learn the Brønsted-Lowry definition of acids and bases and to be able to identify substances as Brønsted-Lowry acids or bases or as amphoteric.
6. To learn what conjugate pairs of acids and bases are and to be able to predict strength of any acidic or basic species from this concept.
7. To learn to do stoichiometry and solution calculations using mole relationships for acid-base reactions in aqueous solution.
8. To learn the Lewis definition of acids and bases and be able to identify substances in chemical reactions that are acting as Lewis acids or bases.
9. To learn what equivalent weight is and to be able to calculate the equivalent weight of any acid, base, or salt from a knowledge of the reaction involved.
10. To learn what normality (N) is and to be able to use it in calculations involving acids and bases.
11. To learn to write ionic equations for reactions involving strong and weak electrolytes, insoluble compounds, and covalent or nonionized compounds.

Fourteen

Acids, Bases, and Ionic Equations

The terms *acid* and *base* have been used previously and discussed briefly. For example, in Chapter 6 the terms were used in discussing names and formulas. At that time, it was pointed out that in writing names of substances we recognize an acid by the fact that an H appears first in the formula. A base was recognized as being a substance that contains the OH^- group. In Chapter 8, these substances were again recognized as having particular properties, namely the formation of water along with a salt when they react. In all these cases, we were using the fact that substances that produce H^+ ion are acids and substances that produce OH^- ion are bases.

In Chapter 8, however, it was also pointed out that oxides can act as acids or bases. Metal oxides will form bases in water solution, and thus they react in the same way as bases. Likewise, nonmetal oxides will react with water to form acids, and hence these oxides will react like acids. For this reason, metal oxides were referred to as basic anhydrides and nonmetal oxides were called acid anhydrides.

In this chapter we shall consider further the properties of these substances that we recognize to be acids and bases. Also, we shall see that there are substances that are acids and bases other than those discussed previously.

After we have considered acids and bases and their properties, including the concept of weak and strong acids and bases, we shall consider ionic equations and their meaning and use.

14-1 Definitions and Properties of Acids and Bases

General Properties of Acids and Bases in Aqueous Solution

Aqueous solutions of acids generally have the following properties:

1. *Sour taste.* Citrus fruits such as lemons have a sour taste due to the citric acid found in them; vinegar is a solution of approximately 5%

acetic acid; sour milk contains lactic acid, which gives it its sour taste. Other acids have similar sour tastes. (You should not attempt to taste any of the common strong acids found in the laboratory.)

2. *Ability to turn blue litmus paper red.* Litmus is an *indicator,* which is a substance whose color is changed by acids and bases. Aqueous acids turn blue litmus paper red; whereas aqueous bases turn red litmus blue.

3. *Neutralize bases.* Acids react with bases, as discussed in Section 8-8.

Aqueous solutions of bases, in general, have the following properties:

1. *Bitter taste.*

2. *Soapy or slick feeling.* A dilute solution of sodium hydroxide, for example, has a soapy slick feeling on the skin. This is due to the dissolving of a thin layer of skin. Strong bases will also dissolve hair, and for this reason sodium hydroxide and potassium hydroxide are often found in drain openers.

3. *Turn red litmus paper blue.*

4. *Neutralize acids.*

Arrhenius Concept of Acids and Bases

The **Arrhenius** definition of **acids** and **bases** was proposed in 1894 by Svante August Arrhenius (1859–1927), a Swedish physicist and chemist. Arrhenius defined an acid as a substance that yields hydrogen ions (H^+) when dissolved in water and a base as a substance that yields hydroxide ions (OH^-) when dissolved in water.

The hydrogen ion is a bare proton, but in water solution it is *hydrated* and exists as a **hydronium ion** (H_3O^+),* as shown in the following equation:

$$H^+ + :\ddot{\underset{\cdot\cdot}{O}}:H \longrightarrow H\cdot\ddot{\underset{\cdot\cdot}{O}}\cdot H^+ \quad \text{(with H below each O)} \tag{14-1}$$

For simplicity in this text, however, we shall use the simplified form $H^+_{(aq)}$ or just H^+, with the understanding that it would be hydrated.

Some common examples of acids are hydrochloric acid (HCl), sulfuric acid (H_2SO_4), nitric acid (HNO_3), and acetic acid ($HC_2H_3O_2$). These acids in the pure anhydrous state exist as highly polar covalent molecules. For example, hydrogen chloride is a gas, pure hydrogen sulfate is a viscous liquid, and pure hydrogen nitrate and hydrogen acetate are liquids. When these polar molecules are dissolved in water, they form ions, such as H^+ and Cl^- for hydrochloric acid. This takes place due to the fact that water is very polar and can cause the molecule, such as HCl, to split up into the ions. As in all

*It is now known that there are often several water molecules associated with each H^+ ion, to form ions such as $H_5O_2^+$ and $H_9O_4^+$.

cases, the H^+ ion would be hydrated. This process is called *ionization* and refers to the formation of ions from atoms or molecules. The following equations illustrate this process of ionization.

$$\text{(14-2)} \qquad HCl_{(g)} + H_2O_{(l)} \longrightarrow H_3O^+_{(aq)} + Cl^-_{(aq)}$$

$$\text{(14-3)} \qquad H_2SO_{4(l)} + H_2O_{(l)} \longrightarrow H_3O^+_{(aq)} + HSO_4^-{}_{(aq)}$$

$$\text{(14-4)} \qquad HNO_{3(l)} + H_2O_{(l)} \longrightarrow H_3O^+_{(aq)} + NO_3^-{}_{(aq)}$$

$$\text{(14-5)} \qquad HC_2H_3O_{2(l)} + H_2O_{(l)} \rightleftarrows H_3O^+_{(aq)} + C_2H_3O_2^-{}_{(aq)}$$

Weak versus Strong Acids and Bases

We have previously defined a substance that breaks up into ions in solution as an *electrolyte.* Also, it has been pointed out that if a substance is completely ionized, it is classified as a *strong electrolyte.* A substance that is only partially ionized, however, is classified as a *weak electrolyte.* This same principle also applies to acids and bases; that is, *a* **strong acid** *or* **base,** *is one that is completely ionized,* and *a* **weak acid** *or* **base** *is one that is only partially ionized.*

Hydrochloric, sulfuric, and nitric acids are all completely ionized in aqueous solution and are therefore classified as *strong acids.* Acetic acid undergoes only partial dissociation into ions, however, and thus a dynamic equilibrium is established between the ionic species and the undissociated acetic acid molecules. This is indicated by the double arrow ($\rightleftarrows$).* Therefore, acetic acid is classified as a *weak acid.* Most acids are weak, but the three strong acids listed above (HCl, H_2SO_4, and HNO_3) are very common, and it is therefore important to remember these. Also, along with these three, hydrobromic (HBr), hydroiodic (HI), and perchloric ($HClO_4$) acids are strong. All others are weak.

Acids with more than one replaceable H^+ ion, such as H_2SO_4, are ionized in steps. In all cases of this type the second step of the ionization is much less than the first. Therefore, when we say that H_2SO_4 is a strong acid, we mean for the first step of the ionization, as shown in Eq. (14-3). This will be considered in more detail in Chapter 17, with equilibrium involving weak acids and bases.

Basic or alkaline compounds can also exist as covalent molecules that react with water to form ions in solution. The most common of these is ammonia gas, which reacts with water to form ions, as shown in Eq. (14-6). Note that this reaction gives an equilibrium similar to Eq. (14-5); therefore, this is a *weak base.*

$$\text{(14-6)} \qquad NH_{3(g)} + H_2O_{(l)} \rightleftarrows NH_4^+{}_{(aq)} + OH^-_{(aq)}$$

*In this chapter we shall use a longer arrow in one direction to indicate which way the equilibrium is displaced. After the quantitative aspects of equilibrium are discussed in Chapter 17, we shall use equal-sized arrows and indicate the displacement of the equilibrium by considering the value for the equilibrium constant.

The substance NH_4OH does not exist; ammonia (NH_3) in water gives a few ammonium (NH_4^+) ions and hydroxide (OH^-) ions but no NH_4OH molecules. For convenience, this is sometimes written as NH_4OH, but what is actually meant here is aqueous ammonia, or $NH_3 + H_2O$.

Metal oxides and hydroxides tend to form basic solutions, but most are only very slightly soluble (or often classified as insoluble). The alkakli metal (Group IA) and alkaline earth (Group IIA) hydroxides are in general strong bases, but other hydroxides are weak bases. In fact, some hydroxides are such weak bases that under certain circumstances they can act as acids. Compounds of this type, which can act as either acids or bases, are said to be **amphiprotic** or **amphoteric** (*amphi* means both kinds).*

In Section 8-4 it was shown that metal oxides in water form hydroxides. Therefore, metal oxides are often called *basic anhydrides.* Many of these hydroxides are only slightly ionized, however, and are thus weak bases. The more active metals form stronger bases. This is due to the fact that in the general formula, M—O—H, if M is an active metal (which readily gives up its electrons), the M—OH bond can break quite readily and from OH^- ions. On the other hand, if M is not an active metal, and the MOH is in the presence of a strong base, the O—H bond can break forming H^+ ions. Thus hydroxides of this latter type are classified as being amphoteric. We shall have more to say about this in Chapter 21, when considering periodic variations in properties. In general, the more ionic an oxide is, the more basic the hydroxide formed from it will be.

Brønsted-Lowry Definition of Acids and Bases

The Brønsted-Lowry definition is more general than the Arrhenius concept since it is not limited to water solutions. This concept, proposed in 1923 by Johannes Brønsted (1879–1947) and Thomas Lowry, working independently, defines acids and bases as follows:

> An acid is any substance that can donate a proton (H^+ ion) to another substance; a base is any substance that can accept a proton (H^+) from another substance.

Thus, according to this definition, *an acid is a proton donor, and a base is a proton acceptor.* This can include not only molecules but also ions. Any substance that is an Arrhenius acid or base will also be a Brønsted acid or base. This is due to the fact that the H^+ ion (produced in Arrhenius acids) is the proton donated by a Brønsted acid. And the OH^- ion (produced in Arrhenius bases) is a proton acceptor, according to the equation

$$H^+ + OH^- \longrightarrow H_2O$$

*Amphoteric is the more general term, meaning any substance that acts as an acid or a base, by any definition. Amphiprotic refers strictly to the ability to donate or accept protons (H^+ ions).

All the examples given in Eqs. (14-2) through (14-5) are also examples of Brønsted acids. In each of these cases, water is acting as a Brønsted base since it is accepting a proton to form H_3O^+. In Eq (14-6), however, water is acting as an acid since it is donating a proton to the NH_3 molecule, to form NH_4^+.

Table 14-1 shows several substances reacting as acids or bases according to the Brønsted-Lowry definition. The acid-base pairs (that is, $Acid_1$-$Base_2$) are called **conjugate pairs.** Note here that the reverse reaction, in the ones that go both ways ($\rightleftharpoons$), also represents an acid-base reaction and that both molecules and ions are involved in these reactions. In considering conjugate acid-base pairs, the stronger an acid is, the weaker is its conjugate base, and vice versa. For example, since HCl is a strong acid, the Cl^- ion is a weak base.

Here again substances capable of acting as either acids or bases are said to be amphoteric. Water, for example, behaves as a base (proton acceptor) toward hydrogen chloride [Eq. (14-2)] and as an acid toward ammonia [Eq. (14-6)]. Certain ions, such as hydrogen sulfate (HSO_4^-) and hydrogen carbonate (HCO_3^-), are also amphoteric. In the self-ionization of ammonia [Eq. (14-11)], one molecule of NH_3 acts as a base toward another molecule acting as an acid. The amide ion (NH_2^-) does not exist in water solution since it reacts rapidly and completely with H_2O to form OH^- and NH_3.* Therefore, in water solution, which is the usual form in which NH_3 is used, ammonia would not be considered to be amphoteric.

The definition of **Brønsted-Lowry acids** and **bases** is a more general, and in some ways, a more useful concept of acids and bases. For example, according to the Arrhenius definition, the hydrogen carbonate, or bicarbonate, ion (HCO_3^-) can only act as an acid. But using the Brønsted-Lowry definition, it is obvious that this can also act as a base. For this reason, sodium hydrogen carbonate, or bicarbonate ($NaHCO_3$) is used to neutralize acids or bases if they are spilled on the skin in the laboratory. Also, the Brønsted-Lowry definition of acids and bases includes reactions in solvents other than water and reactions of gases.

TABLE 14-1 *Conjugate pairs of acids and bases*

	$ACID_1$		$BASE_1$		$ACID_2$		$BASE_2$
(14-7)	H_2SO_4	+	H_2O	$\rightarrow$	H_3O^+	+	HSO_4^-
(14-8)	HSO_4^-	+	H_2O	$\rightleftharpoons$	H_3O^+	+	SO_4^-
(14-9)	H_3O^+	+	OH^-	$\rightleftharpoons$	H_2O	+	H_2O
(14-10)	NH_4^+	+	Cl^-	$\rightleftharpoons$	HCl	+	NH_3
(14-11)	NH_3	+	NH_3	$\rightleftharpoons$	NH_4^+	+	NH_2^- (amide ion)

*The oxide ion (O^{2-}) does this also, according to the equation

$$O^{2-} + H_2O \rightarrow 2\,OH^-$$

*Lewis Definition of Acids and Bases**

The concept of acids and bases as proposed by G. N. Lewis in 1923 is the most general definition and includes all substances that are acids or bases according to the Arrhenius and Brønsted-Lowry concepts. According to the Lewis concept, *an acid is any species that can accept a pair of electrons, and a base is any species that can donate a pair of electrons.* Therefore, the hydrogen ion (H^+) is an acid since it can accept a pair of electrons, as shown in Eq. (14-1), and OH^- is a base since it can donate an electron pair, as shown in Eq. (14-9). There are many other substances, however, which can act as acids and bases which are not included in the previous definitions. For example, as shown in Eq. (14-12), calcium oxide can react with carbon dioxide to form calcium carbonate. Or, as another example, in the formation of the complex ion between copper(II) ion (Cu^{2+}) and ammonia, the copper ion is acting as an acid and the ammonia again is acting as a base, as shown in Eq. (14-13).

(14-12) $$CaO_{(s)} + CO_{2(g)} \longrightarrow CaCO_{3(s)}$$

(14-13) $$Cu^{2+}{}_{(aq)} + 4\ NH_{3(aq)} \longrightarrow Cu(NH_3)_4{}^{2+}{}_{(aq)}$$

This is a very useful concept for some purposes. In general, however, when the terms *acid* and *base* are used, they are understood to mean the more common type of acid and base reactions that take place in solution and are thus acids or bases as defined by the Arrhenius or Brønsted-Lowry definitions. If the Lewis concept is being used, the substance is generally referred to as a *Lewis acid* or *Lewis base.*

Neutralization Reactions in Aqueous Solution

In Section 8-8 we considered neutralization reactions as one of the five types of reactions. You will recall that in a neutralization reaction an acid or acid oxide (nonmetal oxide) reacts with a base or basic oxide (metal oxide) to form a salt. Also, in most cases water is formed as a second product. We shall see in discussing ionic equations in Section 14-3 that the formation of water is often the driving force behind a neutralization reaction.

In carrying out a neutralization reaction, an **indicator** is often used to determine when the neutralization point, or **end point,** has been reached. An indicator is a substance that changes color as the solution that it is in changes from acid to base or vice versa. For example, phenolphthalein is colorless in acid solution and red in basic solution. Another commonly used indicator, methyl orange, is red in acid solution and yellow in basic solution. In using these indicators, it is important to note, as precisely as possible, just when the color changes. One way to do this is by a **titration,** in which acid is added to a basic solution, or base is added to acid, until the end point is reached. In doing this, the acid or base is added from a *buret,*

*This portion of Section 14-1 is optional. It can be omitted without loss of continuity.

which is a carefully calibrated piece of equipment (as shown in Fig. 14-1) that is used to measure out an exact amount of solution. Figure 14-1 demonstrates how a titration is carried out by the following steps: (a) A measured amount of acid (or base) is placed in the flask. (b) Next a few drops of indicator are added. (c) Base (or acid) is added slowly from the buret until the end point is reached. From the results of this, the amount of acid (or base) initially present can be calculated, as shown in the following problem examples.

PROBLEM EXAMPLE 14-1

A solution of 25.0 ml of hydrochloric acid was titrated with 3.50 M sodium hydroxide. If it took 38.2 ml of the sodium hydroxide to neutralize the acid, what was the molar concentration of the acid?

SOLUTION:

First we must write a balanced equation for the reaction that takes place:

$$\text{NaOH} + \text{HCl} \longrightarrow \text{H}_2\text{O} + \text{NaCl}$$

From this we can calculate the number of moles of HCl that are equivalent to 38.2 ml of 3.50 M NaOH, as follows:

$$38.2\ \cancel{\text{m}l\ \text{NaOH soln}} \times \frac{3.50\ \cancel{\text{mol NaOH}}}{1000\ \cancel{\text{m}l\ \text{NaOH soln}}} \times \frac{1\ \text{mol HCl}}{1\ \cancel{\text{mol NaOH}}} = 0.134\ \text{mol HCl}$$

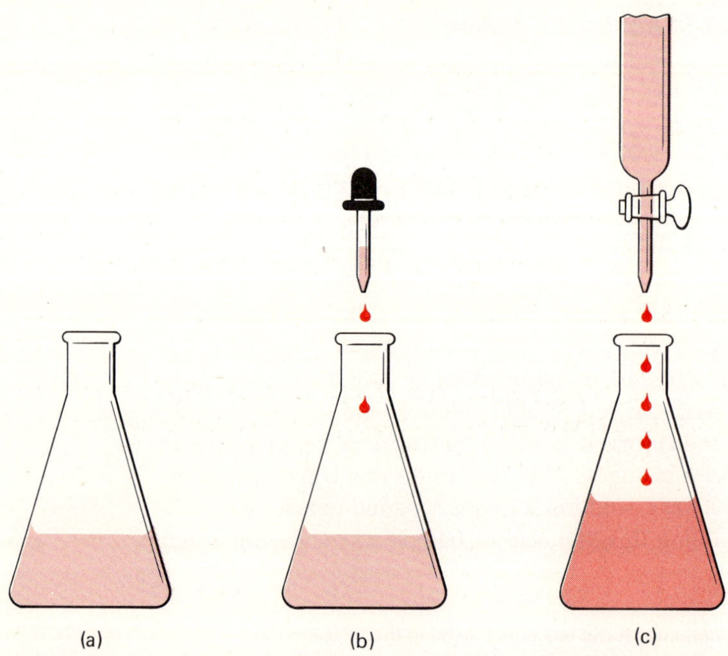

Figure 14-1
Titration. (a) Measured amount of acid (or base) (b) A few drops of indicator are added (c) Acid (or base) added from buret

Then, from the number of moles of HCl and the number of milliliters of HCl solution, we can calculate the molar concentration (moles HCl per liter of solution) as

$$\frac{0.134 \text{ mol HCl}}{25.0 \cancel{\text{m}\ell \text{ soln}}} \times \frac{1000 \cancel{\text{m}\ell \text{ soln}}}{1 \ \ell \text{ soln}} = 5.36 \frac{\text{mol HCl}}{\ell \text{ soln}} = \mathbf{5.36} \ \boldsymbol{M}$$

This could also be solved using

$$M = \frac{\text{millimole (mmol)}}{\text{milliliter (m}\ell)}$$

$$\frac{38.2 \cancel{\text{m}\ell \text{ NaOH soln}}}{25.0 \text{ m}\ell \text{ HCl soln}} \times \frac{3.50 \cancel{\text{mmol NaOH}}}{1 \cancel{\text{m}\ell \text{ NaOH soln}}} \times \frac{1 \text{ mmol HCl}}{1 \cancel{\text{mmol NaOH}}} = \mathbf{5.36} \ \boldsymbol{M} \ \mathbf{HCl}$$

PROBLEM EXAMPLE 14-2

To find the concentration of an acid solution, pure sodium carbonate can be used as a standard. In finding the molar concentration of an HCl solution, 2.15 g of pure sodium carbonate was dissolved in water and then the solution was titrated with the hydrochloric acid solution. If it took 28.5 mℓ of acid to reach the end point, calculate the molar concentration of the acid.

SOLUTION:

The equation is

$$2 \text{ HCl} + Na_2CO_3 \longrightarrow 2 \text{ NaCl} + CO_2 + H_2O$$

The moles of HCl are calculated from the amount of Na_2CO_3 that was used, as follows:

$$2.15 \cancel{\text{g } Na_2CO_3} \times \frac{1 \cancel{\text{mol } Na_2CO_3}}{106 \cancel{\text{g } Na_2CO_3}} \times \frac{2 \text{ mol HCl}}{1 \cancel{\text{mol } Na_2CO_3}} = 0.0406 \text{ mole HCl}$$

and the molar concentration is calculated as

$$\frac{0.0406 \text{ mol HCl}}{28.5 \cancel{\text{m}\ell \text{ soln}}} \times \frac{1000 \cancel{\text{m}\ell \text{ soln}}}{1 \ \ell \text{ soln}} = 1.42 \frac{\text{mol HCl}}{1 \ \ell \text{ soln}} = \mathbf{1.42} \ \boldsymbol{M} \ \mathbf{HCl}$$

14-2 Equivalent Weight and Normality*

The **equivalent weight,** or gram-equivalent weight, of an acid is defined as the mass in grams that will produce 1 mole of H^+ ions in a reaction. Likewise, the equivalent weight of a base is the amount in grams that will furnish 1 mole of OH^- ions. For example, consider a reaction of H_2SO_4 in which both H^+ ions are replaced, such as

$$H_2SO_4 + 2 \text{ NaOH} \longrightarrow Na_2SO_4 + 2 H_2O \qquad (14\text{-}14)$$

*This section is optional and can be omitted without loss of continuity.

In this reaction 1 mole of H_2SO_4 (98 g) furnishes 2 moles of H^+ ion, and thus the amount of H_2SO_4 necessary to furnish 1 mole of H^+ would be 98 g/2 = 49 g; that is, $\frac{1}{2}$ mole of H_2SO_4 (49 g) will furnish 1 mole of H^+ ion, in this reaction. Therefore, we say that the equivalent weight of H_2SO_4 in this reaction is 49 g.

Another way of expressing equivalent weight is to say that 1 equivalent of H_2SO_4 (in a reaction in which both H^+ ions are replaced) is equal to 49 g of H_2SO_4. This is a very useful way of expressing equivalent weight, as we shall see in doing calculations. In using this concept of *an equivalent,* it should be kept in mind that this is a quantity similar to a mole; that is,

1 mole = amount, in grams, of a substance that contains Avogadro's number of units

1 equivalent = amount, in grams, of an acid or base that furnishes Avogadro's number (1 mole) of H^+ or OH^- ions, in a given reaction*

In Eq. (14-14), each mole of NaOH is furnishing 1 mole of OH^- ion, which is always the case for this base since it only contains one OH^- ion per formula unit. Therefore the equivalent weight of NaOH is equal to the formula weight in grams, divided by 1, or 40 g/1 = 40 g.

Therefore, to summarize, we say that the equivalent weight of an acid or base is equal to

$$\frac{\text{formula weight in grams}}{\text{number of } H^+ \text{ or } OH^- \text{ ions transferred}}$$

This can be expressed [using H_2SO_4 in Eq. (14-14) as an example] as

1. equivalent weight = 98 g/2 = 49 g
2. 1 equivalent H_2SO_4 = 49 g H_2SO_4 or
3. 49 g H_2SO_4 per equivalent

It is also sometimes useful to express the relationship between moles and equivalents. In Eq. (14-14), we would say that

$$1 \text{ mole } H_2SO_4 = 2 \text{ equivalents } H_2SO_4$$

and

$$1 \text{ mole NaOH} = 1 \text{ equivalent NaOH}$$

Consider the following problem example.

*We shall see in Chapter 20 that, strictly speaking, an equivalent is related to the transfer of 1 mole of electrons. Since the charge on the hydrogen ion is +1 and on the hydroxide ion is −1, however, it should be obvious that for each mole of H^+ or OH^- ions, 1 mole of electrons has been transferred. This extension of the use of equivalents and equivalent weight is quite useful in doing calculations involving acids and bases as well as other calculations that we shall consider later.

PROBLEM EXAMPLE 14-3

Calculate the equivalent weight of each of the following:
(a) H_2SO_4 in a reaction in which only one H^+ ion is replaced.
(b) $Ca(OH)_2$, if both OH^- ions are replaced.
(c) HCl

SOLUTION:

(a) $$\text{equivalent weight} = \frac{\text{formula weight in grams}}{1} = \frac{98\ \text{g}}{1} = \mathbf{98\ g}$$

(b) $$\text{equivalent weight} = \frac{\text{formula weight in grams}}{2} = \frac{74\ \text{g}}{2} = \mathbf{37\ g}$$

(c) $$\text{equivalent weight} = \frac{\text{formula weight in grams}}{1} = \mathbf{36.5\ g}$$

Note that in this last case that it was not necessary to specify how many H^+ ions are being replaced since there is only one.

This concept of equivalent weight and equivalents can be extended to include salts. In this case the equivalent weight is calculated by dividing the formula weight by the total positive (or negative since they must be the same) charge. It should be obvious that this is directly related to the number of H^+ and OH^- ions transferred when the salt was formed. Consider the following problem example.

PROBLEM EXAMPLE 14-4

Calculate the equivalent weight of each of the following salts:
(a) $AlCl_3$
(b) $CaSO_4$
(c) $Al_2(SO_4)_3$

SOLUTION:

(a) $$\frac{\text{formula weight in grams}}{3} = \frac{133.5\ \text{g}}{3} = \mathbf{44.5\ g}$$

(b) $$\frac{\text{formula weight in grams}}{2} = \frac{136\ \text{g}}{2} = \mathbf{68.0\ g}$$

(c) $$\frac{\text{formula weight in grams}}{6} = \frac{342\ \text{g}}{6} = \mathbf{57.0\ g}$$

The main reason for introducing this concept of equivalent weight and equivalents is for calculations from balanced equations involving acids, bases, and salts; that is, once we know how many H^+ or OH^- ions are being replaced from a particular acid or base, we can simply use the relationship that 1 equivalent of A will react exactly with 1 equivalent of B, where A and B can be acid and base, acid and salt, etc. This is especially useful for solutions in which we use normal concentration or normality, which is considered next.

In using equivalent weights, and normality, it is often assumed that all hydrogen or hydroxide ions are replaced. To be absolutely correct the reaction should be specified in all cases. You will find, however, that in general the equivalent weight of H_2SO_4 is assumed to be $\frac{98}{2}$, H_3PO_4 is assumed to be $\frac{98}{3}$, $Al(OH)_3$ is assumed to be $\frac{78}{3}$, and so forth. If no other information is given, it is proper to assume complete neutralization.

Normal Concentration (N)

Normal concentration or **normality** (N) is defined as the number of equivalents of solute per liter of solution:

$$N = \frac{\text{equivalents of solute}}{\text{liter of solution}}$$

or often stated more simply as

$$N = \frac{\text{equiv}}{\ell}$$

But keep in mind that it is per liter of *solution*, not solvent. For example, to prepare a one normal ($1N$) aqueous solution of sodium sulfate, we would dissolve 1 equivalent (142 g/2 = 71.0 g) of Na_2SO_4 in water. Then add enough water to bring the total volume of the solution to 1 ℓ.

Consider the following problem examples involving normality.

PROBLEM EXAMPLE 14-5

Calculate the normality of a phosphoric acid solution containing 2.50 g of phosphoric acid in 135 mℓ of solution in reactions that replace all three hydrogen ions.

SOLUTION:

Since 1 mole of phosphoric acid is 98.0 g and since 3 moles of hydrogen ions are used per 1 mole of the acid, one equivalent of H_3PO_4 is (98.0 g/3) = 32.7 g.

$$\frac{2.50\ \cancel{\text{g } H_3PO_4}}{0.135\ \ell\ \text{soln}} \times \frac{1\ \text{equiv}\ H_3PO_4}{32.7\ \cancel{\text{g } H_3PO_4}} = \frac{0.566\ \text{equiv}\ H_3PO_4}{\ell\ \text{soln}} = \mathbf{0.566}\ N$$

PROBLEM EXAMPLE 14-6

Calculate the number of grams of sulfuric acid necessary to prepare 225 mℓ of 0.150 N sulfuric acid solution in reactions that replace both hydrogen ions.

SOLUTION:

Since 1 mole of sulfuric acid is 98.1 g and since 2 moles of hydrogen ions are used per 1 mole of the acid, 1 equivalent of H_2SO_4 is (98.1 g/2) = 49.0 g. In a

0.150 N H_2SO_4 solution, there is 0.150 equivalent of H_2SO_4 per liter of solution. Therefore, the number of grams of H_2SO_4 needed to prepare this solution would be

$$225\ \cancel{m\ell\ soln} \times \frac{1\ \cancel{\ell\ soln}}{1000\ \cancel{m\ell\ soln}} \times \frac{0.150\ \cancel{equiv\ H_2SO_4}}{1\ \cancel{\ell\ soln}} \times \frac{49.0\ g\ H_2SO_4}{1\ \cancel{equiv\ H_2SO_4}} = \mathbf{1.65\ g\ H_2SO_4}$$

PROBLEM EXAMPLE 14-7

Calculate the normality of a 1.2 M calcium hydroxide solution.

SOLUTION:

Assuming that both OH^- ions are replaced (which is almost always the case with a base of this type), we can say that

$$1\ \text{mole}\ Ca(OH)_2 = 2\ \text{equivalents}\ Ca(OH)_2$$

Therefore the molar concentration is calculated as

$$\frac{1.2\ \cancel{mol\ Ca(OH)_2}}{\ell} \times \frac{2\ equiv\ Ca(OH)_2}{\cancel{1\ mol\ Ca(OH)_2}} = 2.4\ \frac{equiv}{\ell} = \mathbf{2.4\ N}$$

PROBLEM EXAMPLE 14-8

Calculate the normality of a 15.0% nitric acid solution, with a density of 1.12 g/mℓ.

SOLUTION:

One equivalent of HNO_3 has a mass of 63.0 g since there is only one replaceable H^+. Therefore, the normality is calculated as

$$\frac{15.0\ \cancel{g\ HNO_3}}{100\ \cancel{g\ soln}} \times \frac{1.12\ \cancel{g\ soln}}{1\ \cancel{m\ell\ soln}} \times \frac{1000\ \cancel{m\ell\ soln}}{1\ \ell\ soln} \times \frac{1\ equiv\ HNO_3}{63.0\ \cancel{g\ HNO_3}} = \frac{2.67\ equiv\ HNO_3}{\ell\ soln} = \mathbf{2.67\ N}$$

Use of Normality in Calculations

Once the equations are known so that we can calculate the equivalent weight of an acid or base (or salt), we can use equivalents and normality to do several types of calculations. To do this, we use the relationship that 1 equivalent A = 1 equivalent B, where A and B are acid and base, acid and salt, salt and base, etc. It should be kept in mind that this is just a short way of doing calculations. Any of these could also be done by using the balanced equation and finding the relationship in terms of moles.

Since

$$\text{equivalents A} = \text{equivalents B}$$

or

$$\text{milliequivalents (meq) A} = \text{milliequivalents (meq) B}$$

and from the definition of normality (N) (N = equivalents/liter, or N = meq/mℓ),

$$\text{m}\ell_{\text{A}} \cdot N_{\text{A}} = \text{meq A}$$

Therefore,

$$\text{m}\ell_{\text{A}} \cdot N_{\text{A}} = \text{m}\ell_{\text{B}} \cdot N_{\text{B}}$$

or

$$\text{m}\ell_{\text{A}} \cdot N_{\text{A}} = \text{meq B}, \quad \text{and so forth}$$

Consider the following problem examples involving these equations.

PROBLEM EXAMPLE 14-9

How many milliliters of 0.20 N base would it take to neutralize 0.050 equivalent of acid?

SOLUTION:

$$0.050\,\cancel{\text{equiv A}} \times \frac{1\,\cancel{\text{equiv B}}}{1\,\cancel{\text{equiv A}}} \times \frac{1000\ \text{m}\ell_{\text{B}}}{0.20\,\cancel{\text{equiv B}}} = \mathbf{250\ m\ell_{B}}$$

or, using $\text{m}\ell_{\text{B}} \times N_{\text{B}} = \text{meq}_{\text{A}}$,

$$\text{m}\ell_{\text{B}} = \frac{50\ \text{meq A}}{0.20\ N_{\text{B}}} = 250\ \text{m}\ell_{\text{B}}$$

PROBLEM EXAMPLE 14-10

How many milliliters of 0.50 N HCl will it take to react with 3.5 g of $CaCO_3$?

SOLUTION:

By using equivalents it is not necessary to write out the balanced equation; just use

$$\text{equiv wt CaCO}_3 = \frac{\text{formula wt}}{2} = \frac{100}{2} = 50$$

$$3.5\,\cancel{\text{g CaCO}_3} \times \frac{1\,\cancel{\text{equiv CaCO}_3}}{50\,\cancel{\text{g CaCO}_3}} \times \frac{1\,\cancel{\text{equiv HCl}}}{1\,\cancel{\text{equiv CaCO}_3}} \times \frac{1000\ \text{m}\ell\ \text{HCl}}{0.50\,\cancel{\text{equiv HCl}}} = \mathbf{140\ m\ell\ HCl}$$

or we could first calculate the number of milliequivalents (meq) of $CaCO_3$ by

$$3.5 \text{ g } CaCO_3 \times \frac{1 \text{ meq } CaCO_3}{0.050 \text{ g } CaCO_3} = 70 \text{ meq } CaCO_3$$

Then use

$$m\ell_A = \frac{\text{meq B}}{N_A} = \frac{70}{0.50} = \mathbf{140\ m\ell\ HCl}$$

PROBLEM EXAMPLE 14-11

How many milliliters of 0.150 N HNO_3 will it take to neutralize 25.0 mℓ of 0.230 N $Ca(OH)_2$?

SOLUTION:

$$25.0 \text{ m}\ell\ Ca(OH)_2 \text{ soln} \times \frac{0.230 \text{ meq } Ca(OH)_2}{\text{m}\ell\ Ca(OH)_2 \text{ soln}} \times \frac{1 \text{ meq } HNO_3}{1 \text{ meq } Ca(OH)_2} \times \frac{1 \text{ m}\ell\ HNO_3}{0.150 \text{ meq } HNO_3} = \mathbf{38.3\ m\ell\ HNO_3}$$

or, using $m\ell_A \times N_A = m\ell_B \times N_B$,

$$\frac{(25.0\ m\ell_B)(0.230\ N_B)}{0.150\ N_A} = \mathbf{38.3\ m\ell_A}$$

PROBLEM EXAMPLE 14-12

If it takes 24.5 mℓ of 0.102 N base to neutralize 0.490 g of acid, what is the equivalent weight of the acid?

SOLUTION:

To calculate this, we need to calculate the number of equivalents of acid present and then to divide this into the number of grams of acid (to find grams per equivalent, or equivalent weight):

$$(24.5\ m\ell_A)(0.102\ N_A) = 2.50 \text{ meq B} = 0.00250 \text{ equiv B}$$

$$\text{equiv wt} = \frac{0.490 \text{ g}}{0.0025 \text{ equiv}} = \mathbf{196\ g/equiv}$$

14-3 Ionic Equations

In discussing chemical equations in Chapter 8, it was pointed out that equations may be written as molecular equations or as ionic equations. We now want to consider the steps involved in writing **ionic equations.** In general, this type of equation gives a better representation of a chemical change in aqueous solution than a molecular equation. One of the reasons for this is that an ionic equation shows which ionic substances (electrolytes) are in solution and what weak electrolytes or nonelectrolytes are present. In many cases this shows why the reaction takes place. For example, when a solution of sodium chloride ($Na^+ + Cl^-$) is mixed with a solution of silver nitrate ($Ag^+ + NO_3^-$), a precipitate of solid silver chloride (AgCl) forms, as shown in Eq. (14-15). Or, in a common neutralization reaction, such as that between sodium hydroxide solution and hydrochloric acid, the ionic equation shows that the nonionized substance H_2O is formed, as in Eq. (14-16).

$$Na^+ + Cl^- + Ag^+ + NO_3^- \longrightarrow AgCl_{(s)} + Na^+ + NO_3^- \quad (14\text{-}15)$$

$$Na^+ + OH^- + H^+ + Cl^- \longrightarrow H_2O_{(l)} + Na^+ + Cl^- \quad (14\text{-}16)$$

The following are some general rules for writing ionic equations:

1. Follow the general rules for writing equations, as outlined in Chapter 8—write correct formulas for reactants if necessary, decide what the product(s) will be, write the correct formula for the products, and balance.

2. In determining what the products of a reaction are going to be, consider the formation of (a) insoluble substances (see general solubility rules on the inside back cover of this text), (b) weak acids and bases, and (c) nonelectrolytes (such as H_2O, NH_3, CO_2, organic compounds, and gases such as H_2, O_2, and N_2). Also, if necessary, consider displacement reactions (see *activity series* of metals on the inside back cover of this text).

3. Write the formulas of all strong electrolytes, such as salts in solution and strong acids and bases, in ionic form. Write the formulas of precipitates and weak electrolytes (most acids and bases and some salts such as $HgCl_2$) and nonelectrolytes in molecular form.

4. When writing compounds in the ionic form, use subscripts only in polyatomic ions. For example, sodium sulfate in solution is written as $2\ Na^+ + SO_4^{2-}$.

5. In balancing the equation, make sure that both the atoms and the total net charge are exactly the same on each side of the equation.

The **net ionic equation** shows only those ions that have actually undergone a chemical change. The ions appearing on both sides of the equation that have not undergone a change are crossed out and are not included in the net ionic equation. These ions, which do not undergo any change, are often referred to as *spectator ions.* Also, in the net ionic equation, be sure to use the smallest coefficients possible. Consider the following list of ionic equations. In each of these we shall write the reactants first and give a brief explanation of why the reaction takes place. Then the ionic equation will be written, applying the criteria listed in rule 3 above. Finally, the net ionic equation is written, by eliminating all the spectator ions and using the smallest coefficients possible.

1. Aqueous calcium chloride solution plus aqueous silver nitrate solution

Since these are both salts in solution, they are completely dissociated into ions and are therefore written as individual ions.

$$Ca^{2+} + 2\ Cl^- + Ag^+ + NO_3^- \longrightarrow\ ?$$

Note that the Cl^- is written as 2 Cl^- (since the formula for calcium chloride is $CaCl_2$).
Next, by considering the solubility rules, we must decide what the products are going to be. In this case, since silver chloride is insoluble, this will form and precipitate out. The other ions (Ca^{2+} and NO_3^-) will remain in solution as ions since calcium nitrate is soluble.

$$Ca^{2+} + 2\ Cl^- + Ag^+ + NO_3^- \longrightarrow AgCl_{(s)} + Ca^{2+} + NO_3^-$$
(not balanced)

To balance the equation we need 2 Ag^+, 2 AgCl, and 2 NO_3^- (on both sides).

$$Ca^{2+} + 2\ Cl^- + 2\ Ag^+ + 2\ NO_3^- \longrightarrow 2\ AgCl_{(s)} + Ca^{2+} + 2\ NO_3^-$$

This is the complete ionic equation. Note that we did not indicate (aq) solution for each ion. It is understood that in order for $CaCl_2$ to exist as individual ions, it would have to be in solution. We do, however, indicate the state for any solid, weak electrolyte, or gas that is produced in the reaction. Also, it should be mentioned at this point that these are actually equilibrium reactions, as we shall see in Chapter 17.
For the net ionic equation we leave out the Ca^{2+} and the NO_3^- ions since they are just spectator ions. Also, we use the smallest coefficients possible. This gives

$$Cl^- + Ag^+ \longrightarrow AgCl_{(s)}$$

2. Sodium acetate solution + hydrochloric acid

Since Na $C_2H_3O_2$ is a soluble salt and HCl is a strong acid, they are both written in ionic form. In this case a weak acid, acetic acid, is formed. This is therefore written in molecular form. The other product, sodium chloride, is a soluble salt and is therefore written in ionic form. This gives

$$Na^+ + C_2H_3O_2^- + H^+ + Cl^- \longrightarrow HC_2H_3O_{2(aq)} + Na^+ + Cl^-$$

The net ionic equation is

$$C_2H_3O_2^- + H^+ \longrightarrow HC_2H_3O_{2(aq)}$$

3. Sodium chloride solution + potassium nitrate solution

Here the sodium chloride and potassium nitrate are both soluble salts and would therefore be written as ions. In this case, however, there is no insoluble substance, weak acid or base, or nonionized substance formed, therefore, there is no reaction;

$$\cancel{Na^+} + \cancel{Cl^-} + \cancel{K^+} + \cancel{NO_3^-} \longrightarrow \cancel{Na^+} + \cancel{Cl^-} + \cancel{K^+} + \cancel{NO_3^-}$$

4. Solid calcium oxide + hydrochloric acid

Here the CaO is written in molecular form (since it is a solid) and the HCl in ionic form (since it is a strong acid). This is an acid-base reaction since calcium oxide is a basic oxide (or basic anhydride). Therefore, water (written in molecular form since it is a nonelectrolyte) and the salt calcium chloride (written in ionic form since it is a soluble strong electrolyte) are formed. Thus the overall ionic equation is

$$CaO_{(s)} + 2\,H^+ + 2\,Cl^- \longrightarrow H_2O_{(l)} + Ca^{2+} + 2\,Cl^-$$

and the net ionic equation is

$$CaO_{(s)} + 2\,H^+ \longrightarrow H_2O_{(l)} + Ca^{2+}$$

5. Aluminum metal + nitric acid

This is a displacement reaction; that is, aluminum will displace hydrogen gas from an acid. When this takes place, the aluminum becomes an Al^{3+} ion since it forms the soluble salt $Al(NO_3)_3$. Therefore, the overall ionic equation is

$$Al_{(s)} + H^+ + NO_3^- \longrightarrow H_{2(g)} + Al^{3+} + NO_3^- \quad \text{(not balanced)}$$

To balance this, we need 3 NO_3^- [for the $Al(NO_3)_3$]; but this means that there would also be 3 H^+ and hence $1\frac{1}{2}$ H_2. To eliminate this fraction, we go through and multiply everything by 2 to give

$$2\,Al_{(s)} + 6\,H^+ + 6\,NO_3^- \longrightarrow 3\,H_{2(g)} + 2\,Al^{3+} + 6\,NO_3^-$$

For the net ionic equation, we can only eliminate the NO_3^- ions. Note that on the left side of the equation the Al is in the free state (often indicated as Al^0) and on the right it is an Al^{3+} ion. These are not the same and therefore must be left in the net ionic equation, which is

$$2\,Al_{(s)} + 6\,H^+ \longrightarrow 3\,H_{2(g)} + 2\,Al^{3+}$$

Note that in this case we must leave the coefficients 2 Al and 2 Al^{3+} and 6 H^+ and 3 H_2, in order to balance the charge.

SUMMARY

1. There are several ways to define *acids* (produces H^+ in solution, proton donor, or electron pair acceptor) and *bases* (produces OH^- in solution, proton acceptor, or electron pair donor). But in considering acids and bases in solution it is only necessary to (a) recognize what kind of reaction is taking place and (b) know what H^+ or OH^- producers are completely ionized (strong acids and bases) or only partially ionized (weak acids and bases).
2. In writing *ionic equations* these weak acids and bases (as well as insoluble and nonionized compounds) are written in molecular form, and strong acids and bases (as well as practically all soluble salts) are written in the form of ions.

3. The *equivalent weight* of an acid or base is the amount in grams that will furnish 1 mole of H^+ or OH^- ions; thus, 1 *equivalent* is this mass expressed in grams. For a salt the equivalent weight is the amount formed from 1 equivalent of acid or base (or equiv wt = mol wt/total + charge).
4. *Normal concentration (N)* is the number of equivalents of solute per liter of solution. Since 1 equivalent of any substance will react exactly with 1 equivalent of another substance, we can say

$$\text{equiv A} = \text{equiv B}$$

or

$$N_A \times \ell_A = N_B \times \ell_B = \text{equiv}_A = \text{equiv}_B$$

or

$$N_A \times \text{m}\ell_A = N_B \times \text{m}\ell_B = \text{meq}_A = \text{meq}_B$$

Summary

Upon completion of this chapter you should be able to do the following:

1. Given the name or formula of any substance, classify it as an Arrhenius acid or base, Brønsted-Lowry acid or base, or Lewis acid or base. (1)
2. Given the name or formula of any acid or base, classify it as a weak acid, a weak base, both of these, a strong acid, or a strong base. (2)
3. Given the volume of any acid or base solution and the amount of base or acid it will react with, calculate the concentration of the acid or base. (3)
4. Given any acid, base, or salt, calculate the equivalent weight. (4)
5. Given the mass or number of moles of solute and the volume of solution (or mass and density of solution), calculate the normal (N) concentration of the solution. (5a, b, c)
6. Given the percent, molal, or molar concentration of any solution (and the density if necessary), calculate the normality (N) of the solution or vice versa. (5d, e, f)
7. For any titration, given the normality and volume of A, or grams of A and the volume of B used, calculate the normality of B, or given the normality of B, calculate the volume of B necessary for neutralization. (6, 7)
8. Given the volume and normality of A that will react with a given mass of B, calculate the equivalent weight of B. (8)
9. Given the reactants, write a balanced molecular equation, ionic equation, and net ionic equation (using knowledge of weak acids and bases, general solubility rules, and nonionized substances such as CO_2 and H_2O). (9)
10. Given any of the terms in Exercise 1, give an explanation or definition in your own words.

EXERCISES

1. Define or explain each of the following:
   (a) Arrhenius acid
   (b) Arrhenius base
   (c) Brønsted-Lowry acid
   (d) Brønsted-Lowry base
   (e) Lewis acid
   (f) Lewis base
   (g) conjugate pairs of acids and bases
   (h) ionic equation
   (i) net ionic equation
   (j) weak acid or base
   (k) strong acid or base
   (l) indicator
   (m) end point
   (n) amphiprotic substance
   (o) amphoteric substance
   (p) equivalent weight of acid, base, or salt
   (q) titration
   (r) hydronium ion
   (s) normality
2. Write the formulas for six strong acids.
3. Write the formulas for eight strong bases.

PROBLEMS

Acid-base concepts

1. Classify each of the following as (i) Arrhenius acid, (ii) Arrhenius base, (iii) Brønsted-Lowry acid, (iv) Brønsted-Lowry base, (v) Lewis acid, (vi) Lewis base. (The answer could be none of these, or one, or more.)
   (a) SO_2
   (b) HNO_3
   (c) HCO_3^-
   (d) NH_4^+
   (e) KOH
   (f) $C_2H_3O_2^-$
   (g) SO_3
   (h) $HClO_4$
   (i) HSO_4^-
   (j) H_3O^+
   (k) NaOH
   (l) PO_4^{3-}
2. Classify each of the following as a weak acid, or base, or strong acid, or base. (The answer could be none of these, or one, or more.)
   (a) HCl
   (b) $HC_2H_3O_2$
   (c) KOH
   (d) NH_3
   (e) HSO_4^-
   (f) $Al(OH)_3$
   (g) HNO_3
   (h) HClO
   (i) NaOH
   (j) $Be(OH)_2$
   (k) HCO_3^-
   (l) $Zn(OH)_2$

Neutralization reactions

3. For each of the following, calculate the molar concentration of the acid.
   (a) 35.0 m*l* of HCl are neutralized by 42.5 m*l* of 0.250 *M* NaOH.
   (b) 27.8 m*l* of HNO_3 are needed to just neutralize 1.85 g of CaO.
   (c) 24.3 m*l* of HNO_3 are neutralized by 37.2 m*l* of 1.25 *M* KOH.
   (d) 18.5 m*l* of HCl are needed to just neutralize 1.23 g of MgO.

Equivalent weight and normality

4. Calculate the equivalent weight of each of the following:
   (a) phosphoric acid [for reactions that replace all three (3) hydrogen ions]
   (b) calcium hydroxide
   (c) calcium phosphate
   (d) phosphoric acid, in the reaction $H_3PO_4 + 2\ NaOH \longrightarrow Na_2HPO_4 + 2\ H_2O$
   (e) sulfuric acid (for reactions that replace both hydrogen ions)
   (f) magnesium hydroxide
   (g) aluminum sulfate
   (h) sulfuric acid, in the reaction $H_2SO_4 + NaOH \longrightarrow NaHSO_4 + H_2O$
5. Calculate the normality of each of the following solutions:
   (a) 3.75 g of sodium hydroxide in 135 ml of solution
   (b) 4.32 g of phosphoric acid in 85.5 ml of solution in reactions that replace all three hydrogen ions
   (c) 14.2 g of sodium sulfate in 625 ml of solution
   (d) 8.0 *M* hydrochloric acid solution
   (e) 25% calcium nitrate solution (density = 1.2 g/ml)
   (f) 4.0 *m* sulfuric acid solution (density = 1.2 g/ml)
   (g) 0.45 g of barium hydroxide in 125 ml of solution in reactions that replace both hydroxide ions
   (h) 1.32 g sulfuric acid in 85.0 ml of solution in reactions that replace both hydrogen ions
   (i) 11.1 g of calcium chloride in 700 ml of solution
   (j) 2.5 *M* nitric acid solution
   (k) 20% sodium hydroxide solution (density = 1.2 g/ml)
   (l) 2.1 *m* sugar ($C_{12}H_{22}O_{11}$) solution (density = 1.1 g/ml)
6. Calculate the normality of the base in each of the following solutions:
   (a) 45.0 ml of NaOH solution reacts with 32.8 ml of 1.50 *N* HCl
   (b) 37.2 ml of KOH solution neutralizes 41.5 ml of 0.350 *N* H_2SO_4
   (c) 18.5 ml of base solution reacts with 24.8 ml of 2.25 *N* acid
   (d) 38.5 ml of NaOH solution reacts with 9.25 g of benzoic acid ($C_7H_6O_2$—only one replaceable hydrogen)
   (e) 24.2 ml of potassium hydroxide solution reacts with 4.32 g of sodium hydrogen carbonate
   (f) 18.4 ml of KOH solution reacts with 23.7 ml of 0.750 *N* HNO_3
   (g) 42.3 ml of NaOH solution neutralizes 38.7 ml of 1.25 *N* H_2SO_4
   (h) 24.3 ml of base solution reacts with 36.2 ml of 0.350 *N* acid
   (i) 15.2 ml of KOH solution reacts with 3.72 g of benzoic acid ($C_7H_6O_2$—only one replaceable hydrogen)
   (j) 47.5 ml of sodium hydroxide solution reacts with 6.42 g of potassium hydrogen sulfate
7. Calculate the volume of base necessary to react in each of the following:
   (a) 0.370 *N* KOH reacts with 24.4 ml of 1.25 *N* HCl
   (b) 2.42 *N* sodium hydroxide reacts with 37.0 ml of 1.85 *N* sulfuric acid

(c) 1.45 N sodium hydroxide reacts with 8.75 g of sodium hydrogen sulfate
(d) 2.12 N NaOH reacts with 36.5 mℓ of 2.75 N HNO_3
(e) 0.750 N potassium hydroxide reacts with 23.2 mℓ of 0.385 N sulfuric acid
(f) 2.15 N sodium hydroxide reacts with 3.23 g of sodium hydrogen carbonate

8. Calculate the equivalent weight of each of the following bases:
(a) 27.2 mℓ of 2.15 N acid reacts with 5.42 g of base
(b) 13.5 mℓ of 0.350 N H_2SO_4 reacts with 0.285 g of base
(c) 42.8 mℓ of 1.23 N acid reacts with 4.85 g of base
(d) 26.3 mℓ of 0.825 N H_2SO_4 reacts with 1.04 g of base

Ionic equations

9. Write an ionic and a net ionic equation for each of the following:
(a) barium chloride solution + sodium carbonate solution
(b) silver nitrate solution + hydrochoric acid
(c) ammonium chloride solution + barium acetate solution
(d) nitric acid + calcium carbonate
(e) nitric acid + ammonia
(f) sodium hydroxide solution + ammonium chloride solution
(g) silver nitrate solution + sodium chloride solution
(h) hydrochloric acid + lead(II) carbonate
(i) ammonium nitrate solution + sodium hydroxide solution
(j) barium carbonate + nitric acid
(k) ammonia + acetic acid
(l) calcium nitrate solution + sodium chloride solution

General problems

10. Carry out the following calculations from titrations.
(a) If 18.8 mℓ of 1.25 M sodium hydroxide solution is necessary to neutralize 25.0 mℓ of a hydrochloric acid solution to a phenolphthalein end point, calculate (i) the molarity, and (ii) the percent by mass (density of solution = 1.02 g/mℓ) of the hydrochloric acid solution.
(b) If 1.25 g of pure sodium carbonate was dissolved in water and the solution titrated with 22.7 mℓ of hydrochloric acid to a methyl orange end point, calculate (i) the molarity and (ii) the percent by mass (density of solution = 1.02 g/mℓ) of the hydrochloric acid solution.
(c) Vinegar is a dilute solution of acetic acid. In the titration of 10.0 mℓ of vinegar, 19.5 mℓ of 4.20 M sodium hydroxide solution was required to neutralize the vinegar to a phenolphthalein end point. Calculate (i) the molarity and (ii) the percent by mass (density of solution = 1.01 g/mℓ) of acetic acid in the vinegar.
(d) If 0.825 g of pure sodium hydroxide is titrated with 24.7 mℓ of hydrochloric acid solution to a methyl orange end point, calculate (i) the molarity and (ii) the percent by mass (density of solution = 1.02 g/mℓ) of the hydrochloric acid solution.

11. Calculate the number of grams of solute necessary to prepare the following solutions:
    (a) 25.0 mℓ of a 0.200 N sulfuric acid solution in reactions that replace both hydrogen ions
    (b) 50.0 mℓ of a 1.25 N phosphoric acid solution in reactions that replace all three hydrogen ions
12. Calculate the number of milliliters of solution required to provide the following:
    (a) 17.5 g of sulfuric acid from a 2.50 N solution in reactions that replace both hydrogen ions
    (b) 24.2 g of calcium chloride from a 1.25 N solution in reactions that replace both chloride ions
13. Calculate the percent concentration of each of the following solutions:
    (a) 2.50 N sodium hydroxide solution (density = 1.25 g/mℓ)
    (b) 4.80 N potassium hydroxide solution (density = 1.30 g/mℓ)
14. Calculate the molar concentration of each of the following:
    (a) 6.0 N sulfuric acid (in reactions that replace both hydrogens)
    (b) 3.5 N potassium hydroxide
    (c) 4.0 N phosphoric acid (in reactions that replace two hydrogen ions)
    (d) 6.2 N sodium hydroxide
15. How many milliliters of 0.250 N hydrochloric acid are necessary to react with 20.2 mℓ of 0.480 N calcium hydroxide?
16. How many milliliters of 0.35 N sulfuric acid are necessary to react with 30.2 mℓ of 0.52 N sodium hydroxide?
17. How many grams of magnesium hydroxide are there in 125 mℓ of 3.50 N solution?
18. How many grams of calcium hydroxide are there in 230 mℓ of 2.8 N solution?
19. How many grams of aluminum hydroxide will react with 25 mℓ of 4.2 N acid?
20. How many grams of barium hydroxide will react with 45 mℓ of 2.8 N acid?
21. If it takes 24.8 mℓ of 1.20 N acid to neutralize 3.25 g of a base, what is the equivalent weight of the base?
22. If it takes 32.5 mℓ of 1.82 N acid to neutralize 4.80 g of a base, what is the equivalent weight of the base?

REFERENCES

JENSEN, WILLIAM B., "Lewis Acid-Base Theory," *Chemistry,* March–May 1974, **47.**

KING, EDWARD, *Ionic Reactions and Separations; Experiments in Qualitative Analysis,* New York: Harcourt, Brace and World, Inc., 1973.

VANDERWERF, CALVIN A., *Acids, Bases, and the Chemistry of the Covalent Bond.* New York: Reinhold Publishing Corp., 1961.

Objectives

1. To understand what is meant by the terms *reaction rate* and *kinetics.*
2. To understand what a rate law tells about a reaction, what the rate determining step in a reaction is, and what the Law of Mass Action is.
3. To understand what is meant by order of reaction and to be able to tell, from the rate law for a reaction, what the order of reaction is in terms of a specific reactant or for the overall reaction.
4. To be able to calculate the increase in the rate of a reaction with a specific increase in the concentration(s) of the reactant(s), given the specific rate law.
5. To understand what the collision theory of reaction rates is and to be able to explain how the nature of the reactants, temperature, a catalyst, and concentration of reactants affect the reaction rate, in terms of the collision theory.
6. To understand how the mechanism of a reaction is determined and to be able to tell from a series of possible mechanisms and the experimentally determined rate law which is the correct mechanism.
7. To understand how reaction rates are determined experimentally.

Fifteen

Kinetics

In previous chapters we have considered the products formed in a chemical reaction as well as the stoichiometric (chemically equivalent) relationships from the balanced equation. We now want to consider the speed with which a reaction takes place, or the *reaction rate.* This shows the rate or speed at which the products are produced or the reactants consumed. We shall also consider the path or *mechanism* by which a reaction takes place. These two topics constitute the study of *chemical kinetics.*

The study of reaction rates has many practical applications. For example, in the chemical industry it is important to find conditions that give a desired product at a reasonable speed in order to be economically feasible. Also, reaction rates are important in areas such as the growth of plants and the refrigeration of food to prevent it from spoiling.

The study of reaction mechanisms is useful as a means of formulating theories about why reactions take place. It also provides information that is useful in carrying out similar reactions and thus developing new products.

We shall also find the study of reaction rates useful in understanding the next topic that we shall consider—chemical equilibrium.

15-1 Reaction Rates and Rate Laws

The **rate of a reaction** is a quantity that tells how the concentration of a given reactant or product changes with time. This can be expressed as

$$\text{rate} = \frac{\Delta_{\text{concentration}}}{\Delta_{\text{time}}}$$

which is read as change (Δ) in concentration per unit change of time. If the concentration of a particular reactant A is expressed as moles per liter (M), this can be expressed as

$$\text{rate} = \frac{d[\text{A}]}{dt}$$

which is read as rate of change of A with respect to time. The brackets, [ ], are always used to indicate molar concentration.

The *rate law* for a particular reaction is a mathematical expression that relates the rate of the reaction to the concentration of the reactants. For example, for the general reaction,

$$a\text{A} + b\text{B} \longrightarrow \text{products}$$

the rate of the overall reaction is proportional to the concentration of A and B raised to certain powers. This is expressed as

$$\text{rate of reaction} \propto [\text{A}]^x[\text{B}]^y$$

The concentration of A and B is expressed as mole per liter, except for gases where pressure units are often used. The values of x and y are whole or fractional numbers or zero. The value of these must be determined experimentally.

Many reactions go in two or more steps. When this is the case, there is usually one slow step, which determines the rate of the reaction. For example, the reaction $2\text{A} + \text{B} \longrightarrow \text{C}$ may go in the following steps:

$$\text{A} + \text{B} \longrightarrow \text{AB}$$
$$\text{AB} + \text{A} \longrightarrow \text{C}$$

If the first step goes slowly and the second goes fast, the first step will determine the rate at which the reaction takes place. This is due to the fact that the overall reaction cannot proceed any faster than the slowest step.

This could be compared to an assembly line involving several steps in the production of a product. The rate at which the final product is turned out cannot be any greater than the slowest step in the process, no matter how fast the other steps proceed. The slow step in a process of this type is called the **rate determining step.**

In 1867 two Norwegian chemists, Cato Guldberg (1836–1902) and Peter Waage (1833–1900), proposed the **Law of Mass Action.** This law states that for the rate determining step in a reaction the rate of the reaction is proportional to the product of the concentrations of the reactants, each raised to the power of its coefficient in the balanced equation. For example, for a reaction in which the rate determining step is

$$n\text{A} + m\text{B} \longrightarrow \text{C}$$

the rate is

$$\text{rate} \propto [\text{A}]^n[\text{B}]^m$$

By putting in a proportionality constant k, this can be written as

$$\text{rate} = k[\text{A}]^n[\text{B}]^m$$

This proportionality constant is called the *specific rate constant.*

For some reactions the overall reaction corresponds to the rate determining step, and for these the rate law can be written directly from the balanced equation. It must be kept in mind, however, that the overall rate law must be determined experimentally. The Law of Mass Action only applies to the rate determining step.

Consider the following example.

PROBLEM EXAMPLE 15-1

For a particular reaction the rate determining step is

$$2\text{A} + \text{B} \longrightarrow \text{C}$$

Write the expression for the rate law for this reaction.

SOLUTION:

According to the Law of Mass Action the rate law is

$$\textbf{rate} = k[\text{A}]^2[\text{B}]$$

We shall see later that there are two other factors besides concentration of the reactants that affect the rate of a reaction. These are temperature and a catalyst. These will be discussed in Section 15-4 after we consider the collision theory.

15-2 Order of Reaction

From the rate law for a reation we can determine the **order of reaction.** For a particular species in a reaction, the order is equal to the exponent for that species in the rate law. For example, in a reaction for which the rate law is

$$\text{rate} = k[\text{A}]^2[\text{B}]$$

the reaction is said to be second order with respect to A and first order with respect to B. This is often stated as second order in A and first order in B.

The overall order of reaction is equal to the sum of all the individual orders. Therefore, the reaction given above is third order overall.

From the order of reaction we can determine how the rate of a reaction will change as the concentration of each of the reactants is changed. For example, for the general reaction, $2\text{M} + \text{N} \longrightarrow \text{M}_2\text{N}$, the rate law is determined experimentally to be

$$\text{rate} = k[\text{M}]^2[\text{N}]$$

Thus, we say that this reaction is second order in M and first order in N. Therefore, if the concentration of M is doubled, the reaction will go four times as fast ($2^2 = 4$). If the concentration of N is doubled, the reaction will go twice as fast ($2^1 = 2$). If both M and N are doubled, the reaction will go eight times as fast. For gases, the partial pressure of a particular gas can be used in place of molar concentration. This is due to the fact that the partial pressure of a gas in a mixture is directly proportional to the concentration of that gas (Section 11-6).

Consider the following problem example.

PROBLEM EXAMPLE 15-2

For the reaction

$$2\ NO_{2(g)} \longrightarrow 2\ NO_{(g)} + O_{2(g)}$$

the rate law has been determined experimentally to be

$$\text{rate} = k[NO_2]^2$$

(a) What is the order of reaction with respect to NO_2 and what is the overall order?
(b) How much faster will the reaction proceed when the partial pressure of NO_2 is 3 atm and when the partial pressure of NO_2 is 1 atm?

SOLUTION:

(a) From the rate law, the reaction is second order in NO_2 and second order overall (since NO_2 is the only reactant).
(b) Since the reaction is second order in NO_2, the reaction will proceed at a rate $3^2 = 9$ times as fast when the partial pressure (or concentration) of NO_2 is three times as great.

15-3 Collision Theory of Reaction Rates

One way of considering what happens in a chemical reaction is to think of a reaction as taking place due to the collision of particles. In some reactions almost all collisions result in reaction, and in others only a very few are effective. The only collisions that are effective are those in which the colliding particles have sufficient energy. This energy that is necessary for effective collisions is called the **activation energy.** The graph in Fig. 15-1 gives a representation of this activation energy for the reaction $A + B \longrightarrow C + D$. Note that on this graph the total potential energy of the products (C + D) is lower than that of the reactants (A + B). Thus, this is an exothermic reaction. There is an energy barrier that must be overcome, however, before reaction can take place; that is, the colliding particles must have sufficient energy to overcome this barrier before they can react. This is true for both exothermic and endothermic reactions. Next we shall consider the factors that affect reaction rates, in relation to the collision theory and the potential energy curve.

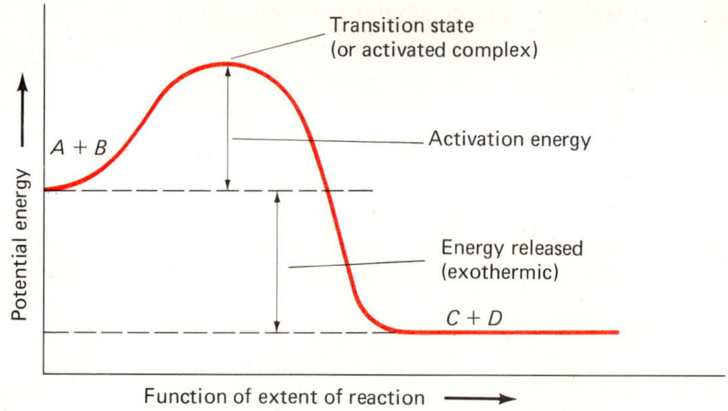

Figure 15-1
Potential energy versus function of extent of reaction

15-4 Factors that Control Reaction Rates

We have seen that the rate of a reaction increases as the concentration of the reactants increases, according to the Law of Mass Action. Also, we can show that for almost all reactions the rate of reaction increases with an increase in temperature. Finally, it is found that a catalyst also alters the rate of a reaction. In this section we shall consider these three factors in relation to the collision theory and the potential energy curve.

Before considering these three factors that can alter the rate of a reaction, it should be mentioned that the nature of the reactants is probably the most fundamental property in determining reaction rates. This is a factor that cannot be controlled, however, and thus is generally not considered as one of the factors that control reaction rates. The nature of the reactants simply determines the nature of the activation energy, or the height of the energy barrier that must be overcome in order for a reaction to take place.

Some reactions are very rapid due to a low activation energy and others are very slow due to a high activation energy. For example, reactions of ions are usually very fast and practically all collisions are effective. This is due to the fact that the ions have an attraction for each other and therefore do not need additional energy to cause them to react.

Concentration

The frequency of collision between molecules that are reacting is first of all dependent on the number of molecules present in a given volume. As the concentration increases, more reactant molecules collide with other reactant molecules in a given period of time, and more product molecules are formed. Thus the concentration of reactants is the first factor to be considered.

The Law of Mass Action was actually stated in terms of *active masses* of the reactants; however, the active mass is directly related to the concentration of the reactants, in moles per liter. For gases, the active mass is also related to the partial pressure of a particular gas in a reaction mixture. Therefore, for reactions involving gases we can use either molar concentration or partial pressure.

Temperature

Since temperature is a measure of the average kinetic energy of the molecules in a gas, as the temperature increases, the average kinetic energy increases. This is shown in Fig. 15-2, which represents the distribution of energy among the molecules in a sample of gas at two different temperatures.

In this figure, temperature T_2 is greater than T_1. It can be seen that for any particular kinetic energy, such as that indicated by x, at temperature T_2 a greater percentage of the molecules have this energy. For example, if x represents the activation energy of a particular reaction, there will be more molecules (a versus b) having this energy at temperature T_2 than there are at T_1. Some of the additional energy goes into increased rotational and vibrational energy of the molecules, but here we are concerned only with the increased velocity. Since the molecules have a higher average kinetic energy, more of them have sufficient energy to make their collisions effective and therefore the rate of reaction is more rapid. As a general approximation, the rate roughly doubles for each 10°C rise in temperature.

When the temperature of a reaction is changed, the rate law does not change but the value of the rate constant does change.

Catalyst

A catalyst has been defined previously (Section 8-1) as a substance that alters the speed of a reaction without being consumed. Generally a catalyst is used to speed up a reaction, but there are also catalysts that slow down a reaction. A substance of this type is sometimes called a *negative catalyst* or an **inhibitor.** A catalyst alters the speed of the reaction by changing the activation energy, or the energy barrier. Figure 15-3 shows how a catalyst that speeds up a reaction works by decreasing the activation energy, while a negative catalyst increases the activation energy.

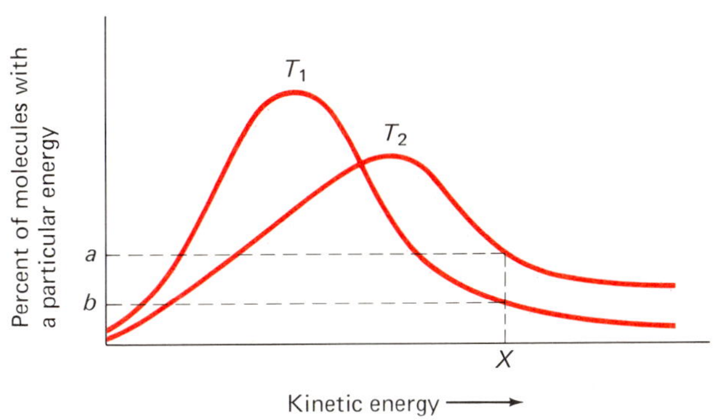

Figure 15-2
Distribution of energy among the molecules at different temperatures

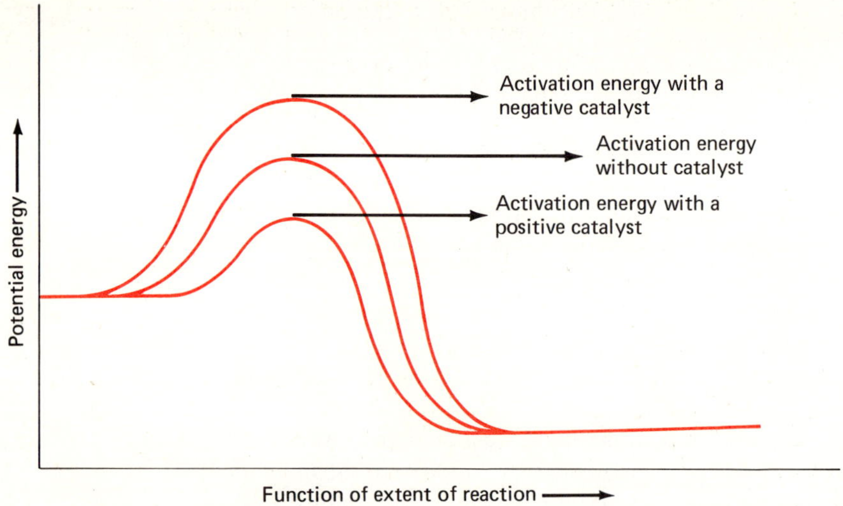

Figure 15-3
Effect of a catalyst on the activation energy of a reaction

15-5 Reaction Mechanisms

The **mechanism of a reaction** shows the path or steps by which the reaction takes place. The mechanism of a reaction is determined by the experimentally determined rate law or by the kinetics of the reaction. The way that this is done is to propose several possible mechanisms that could possibly take place. In doing this, each proposed mechanism must be consistent with the known properties of the substances involved. Also the total of all the steps included in any proposed mechanism must agree with the known, overall reaction. For example, for the general reaction 2 A + B $\longrightarrow$ C, the following two possible mechanisms could be proposed.

1. A + B $\longrightarrow$ AB (slow step)

   AB + A $\longrightarrow$ C (fast step)

   From this mechanism the rate law (using the Law of Mass Action, based on the slow or rate determining step) would be

   $$\text{rate} = k[\text{A}][\text{B}]$$

   that is, the reaction would be first order in A and first order in B and second order overall.

2. A + B $\longrightarrow$ AB (fast step)

   AB + A $\longrightarrow$ C (slow step)

   In this case, the rate law would be

   $$\text{rate} = [\text{AB}][\text{A}]$$

   The concentration of AB, [AB], is proportional to [A][B], however, since AB is formed from A and B much faster in the first step than it is used up in the second step. Thus the rate law is

   $$\text{rate} = \text{k}[\text{A}][\text{B}][\text{A}] = [\text{A}]^2[\text{B}]$$

   This is second order in A and first order in B and third order overall.

If the experimentally determined rate law agrees with that given in item 1 above, this must be the correct mechanism; if it agrees with that given in item 2, that would be the correct mechanism. Note that in both of these if the two steps are added together, they give the overall equation; that is,

$$\begin{array}{rcl} A + B & \longrightarrow & AB \\ AB + A & \longrightarrow & C \\ \hline A + B + AB + A & \longrightarrow & AB + C \end{array}$$

which can be simplified to

$$2\,A + B \longrightarrow C$$

Experimental Determination of Reaction Rate Laws

To determine the rate of a reaction experimentally, it is necessary to have some method of following the progress of the reaction. There are several ways of doing this, depending on the reaction. For example, in a reaction in which H^+ ion is one of the products, the reaction can be followed by measuring the change in pH as the reaction proceeds. For substances that contain a species with a distinctive color, the progress can be followed by measuring the absorbency of light passing through the solution, in a *colorimeter* or other type of spectrophotometer. If one of the products is a gas, the progress can be followed by measuring the amount of gas given off.

SUMMARY

1. *Kinetics* is concerned with the rate at which reactions take place; that is, how the concentration of a given reactant or product changes with time.
2. One of the main purposes of studying reaction rates is to be able to determine the path or steps by which a reaction takes place; this is called the *mechanism* of the reaction.
3. The *rate law* of a particular reaction is a mathematical expression that relates the rate of a reaction to the concentration of the reactants. The rate of a reaction is proportional to the concentration of the reactants raised to certain powers. The rate law must be determined experimentally.
4. For the reactions that go in more than one step, there is always a slow *rate determining* step.
5. The *Law of Mass Action* states that for the rate determining step in a reaction, the rate is proportional to the product of the concentrations of the reactants, each raised to the power of the coefficient in the balanced equation. For example, for a reaction in which the rate determining step is

$$2\,A + B \longrightarrow C$$

$$\text{rate} = k[A]^2[B]$$

where k is a proportionality constant, called the *specific rate constant.*

6. The *order of reaction* for a particular species is equal to the exponent for that species, in the rate law. The overall order is the sum of the individual orders. For example, for the equation given in item 5 above, the reaction is second order in A, first order in B, and third order overall.
7. From the order of reaction for a particular species we can calculate how much faster the reaction would take place if the concentration (or partial pressure for a gas) is changed by a given amount. For example, in the reaction given in item 5, if the concentration of A is doubled, the reaction will go four times as fast ($2^2 = 4$).
8. The *collision theory* of reaction rates is based on the fact that in order for reactions to take place the reacting molecules must collide and must have sufficient energy to overcome the energy barrier, called the *activation energy.*
9. Besides the concentration of reactants there are two other factors that affect reaction rates; these are temperature and a catalyst. Each of these three factors can be explained in terms of collision theory, as follows:
   (a) As the concentration of reactants increases, there are more collisions and hence more reaction.
   (b) As the temperature is increased, a larger percentage of the colliding particles have sufficient energy to overcome the activation energy and hence react.
   (c) A *catalyst* lowers the activation energy and hence increases the rate of reaction (a negative catalyst, or inhibitor, increases the activation energy and slows down the reaction).
10. To determine the mechanism of a reaction, all possible mechanisms are considered; the one that agrees with the experimentally determined rate law is the correct one.

Upon completion of this chapter, you should be able to do the following:

1. Given the rate determining step in a reaction, write the rate law for that reaction. (1)
2. Given the rate law for any reaction, tell what the order of reaction is for any given reactant, or what the overall order of reaction is, and calculate how much the rate would be increased or decreased by any given change in concentration of any of the reactants. (1, 3)
3. Given the order of reaction in terms of each reactant, write the rate law, and explain what this tells about the mechanism for the reaction. (2)
4. Given any of the factors—temperature, catalyst, or concentration—explain in terms of collision theory and activation energy what effect that factor has on rate of reaction. (Exercise 2)
5. Given any of the terms in Exercise 1, give a definition or explanation, in your own words.

EXERCISES

1. Define or explain each of the following:
   (a) reaction rate
   (b) rate law
   (c) rate determining step
   (d) Law of Mass Action
   (e) order of reaction
   (f) collision theory
   (g) activation energy
   (h) catalyst
   (i) negative catalyst
   (j) inhibitor
   (k) mechanism of a reaction
2. Explain, in terms of collision theory and activation energy, how each of the following factors affects the rate of a reaction:
   (a) increase in concentration
   (b) increase in temperature
   (c) catalyst

PROBLEMS

1. Assuming that each of the following reactions go in one step, write the rate law and give the order of reaction for each reactant and the overall order.
   (a) $N_2O_{4(g)} \longrightarrow 2\,NO_{2(g)}$
   (b) $C_2H_{4(g)} + H_{2(g)} \longrightarrow C_2H_{6(g)}$
   (c) $N_{2(g)} + O_{2(g)} \longrightarrow 2\,NO_{(g)}$
   (d) $C_3H_{6(g)} + H_{2(g)} \longrightarrow C_3H_{8(g)}$
2. Each of the following reactions is first order. Write the rate law for each and explain what this tells about the mechanism of the reaction.
   (a) $2\,N_2O_{5(g)} \longrightarrow 4\,NO_{2(g)} + O_{2(g)}$
   (b) $2\,A_2B_3 \longrightarrow 4\,AB + B_2$
3. Assuming that the reaction: $2\,NO_{(g)} + O_{2(g)} \longrightarrow 2\,NO_{2(g)}$ is second order in NO and first order in O_2, tell how much faster the reaction will go under the following conditions as compared to the rate when all pressures are 1 atm.
   (a) The partial pressure of NO is 2 atm and the O_2 remains the same.
   (b) The partial pressure of both NO and O_2 are doubled.
   (c) The partial pressure of NO is 3 atm and the partial pressure of O_2 is $\frac{1}{2}$ atm.
   (d) The partial pressure of NO is 3 atm and the O_2 remains the same.
   (e) The partial pressure of both gases are reduced by one-half.
   (f) The partial pressure of NO is 2 atm and the partial pressure of O_2 is 3 atm.

REFERENCES

BENSON, SIDNEY W., *Chemical Calculations,* 3rd ed. New York: John Wiley & Sons, Inc., 1971.

EYRING, HENRY, and EDWARD M. EYRING, *Modern Chemical Kinetics,* New York: Reinhold Publishing Corp., 1963.

Objectives

1. To understand what is meant by the terms *irreversible* and *reversible* reactions.
2. To be able to predict what reactions tend to go toward completion, by the formation of a precipitate, a gas, or some other nonionized or slightly ionized substance such as water or a weak acid or base.
3. To be able to write the expression for the equilibrium constant for any equilibrium reaction.
4. To understand what the equilibrium constant tells about a given reaction.
5. To understand what is meant by heterogeneous equilibrium and to be able to write the expression for K_{eq} for any equilibrium reactions of this type.
6. To understand how the factors of temperature change and a change in concentration (or pressure for gases) causes the equilibrium to shift, and to be able to predict the shift for a given change in concentration of products or reactants or for a given temperature change.
7. To be able to predict the change in concentration of products and reactants when a temperature or concentration change causes the equilibrium to shift.
8. To be able to calculate equilibrium constants from concentrations or pressures (for gases) or, using equilibrium constants, to be able to calculate the equilibrium concentrations of products and reactants.

Sixteen

Introduction to Chemical Equilibrium

We have previously considered some examples of physical equilibrium, such as the equilibrium between solid ice and liquid water at the freezing point. We now want to consider *chemical equilibrium,* which involves reversible reactions representing chemical changes. Some chemical reactions appear to go only in one direction or are said *to go to completion.* For many reactions, however, there is also a reversible reaction that takes place, and an equilibrium condition is established. Equilibrium is defined as the point at which the rate of the forward reaction is equal to the rate of the reverse reaction. This is indicated by the double arrow ($\rightleftarrows$).

In this chapter we shall apply the principles of reaction rates and rate laws to calculate equilibrium constants for reversible reactions. We shall also do calculations involving equilibrium constants and learn to predict how the position of equilibrium can be shifted in these reactions. Then in the following two chapters we shall consider a very important type of equilibrium involving equilibrium reactions that take place in solution.

16-1 Reversible Reactions and Chemical Equilibrium

All chemical reactions are theoretically reversible. There are cases however, in which the reverse reaction is so slight, or it takes place at such a slow rate, that for all practical purposes the reaction is considered to be irreversible. For example, the reaction of hydrogen gas with oxygen gas to form water is generally considered an irreversible reaction. Also, for a reaction in which one of the products is given off as a gas, the reaction is generally considered to be irreversible.

The reactions that we want to consider now are cases where there is a definite equilibrium established between the forward and the reverse reactions. A very common example of this involves the formation of ammonia from nitrogen gas and hydrogen gas by the Haber process. This process was developed by the German chemist Fritz Haber (1869–1934) in 1914. This involves an equilibrium reaction in the gas phase, as shown by the following equation:

(16-1) $$N_{2(g)} + 3\ H_{2(g)} \rightleftarrows 2\ NH_{3(g)} + 22\ \text{kcal}$$

Notice here that heat is given off in the reaction, and thus this is an exothermic reaction. The reaction is carried out at a temperature of about 500°C and a pressure of 200 to 600 atm, in the presence of a catalyst.

16-2 Equilibrium Constants

We now want to consider the quantative aspects of the equilibrium established between the forward and reverse reactions in a reversible equation. We shall start with a general case involving two substances A and B that react to form two products C and D according to the following equation:

$$A + B \longrightarrow C + D$$

As A and B begin to react, some of the products, C and D, are formed. We find that as more of the two reactants are used up, the rate of the reaction decreases due to the change in the concentration of the reactants. As the two products C and D begin to form, it is found that they react with each other to form A and B, according to the following equation:

$$C + D \longrightarrow A + B$$

As more C and D is formed, the rate of this reaction increases due to the increase in the concentration of C and D. Eventually these two reactions reach a point where the rate at which A and B are reacting to form C and D is equal to the rate at which C and D are reacting to form A and B. This is the condition of *equilibrium.* It must be kept in mind that this is a dynamic equilibrium since the two reactions have not stopped but are continually taking place at an equal rate. This is the same type of process as we considered earlier with the dynamic equilibrium between water and ice at the freezing point of 0°C. At the point of equilibrium, the rate at which water freezes is equal to the rate at which ice melts.

These two opposing reactions involving A and B in one case and C and D in the other case are shown to be in equilibrium in the following equation:

(16-2) $$A + B \rightleftarrows C + D$$

Fig. 16-1 represents the attainment of equilibrium between these two reactions by showing a graph in which reaction rate is plotted against time. Notice at the top part of the graph on the left side, which represents A and B going to form C and D, the reaction rate decreases with time. Likewise for the bottom part of the graph on the left side, the rate at which C and D react to form A and B increases with time. Once equilibrium is established, there is no change in reaction rate for either the forward or the reverse reaction. It is important to keep in mind when considering these reversible reactions that

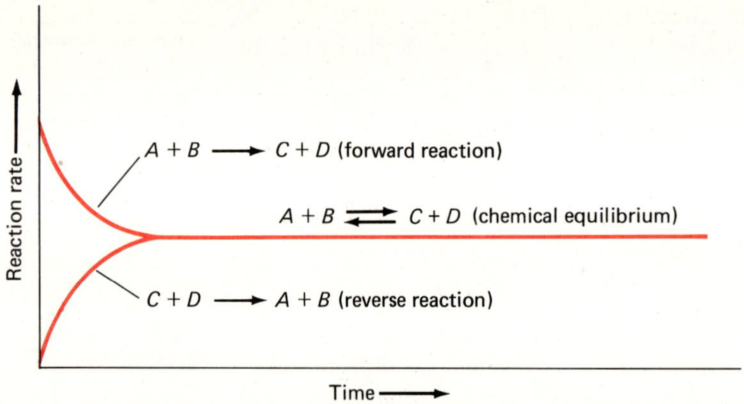

Figure 16-1
Chemical equilibrium is reached when the rate of the forward reaction is equal to the rate of the reverse reaction

at equilibrium we are not saying that the amount of products and reactants are the same but only that the rate at which the two opposing reactions are taking place is equal.

For any equilibrium reaction we can derive an **equilibrium constant,** using the principles of reaction rates and rate laws as considered in Chapter 15. It can be shown that for any reaction at equilibrium the rate of the forward reaction will be directly proportional to the concentration of the reactants raised to the power of the coefficient in the balanced equation. (We shall show why this is true later in this section, after considering equilibrium constants.) For example, using the general reaction show in Eq. (16-2), the rate of the forward reaction is proportional to the molar concentration of A times the molar concentration of B, which can be written as

$$\text{rate}_f = k_f[\text{A}][\text{B}]$$

where k_f is the proportionality constant, or specific rate law constant, for the forward reaction. Likewise, for the reverse reaction, the rate is expressed as

$$\text{rate}_r = k_r[\text{C}][\text{D}]$$

where k_r is the specific rate constant for the reverse reaction.

At equilibrium the rate of the forward reaction is equal to the rate of the reverse reaction and, therefore,

$$\text{rate}_f = \text{rate}_r \quad \text{or} \quad k_f[\text{A}][\text{B}] = k_r[\text{C}][\text{D}]$$

Rearranging this gives

$$\frac{k_f}{k_r} = \frac{[\text{C}][\text{D}]}{[\text{A}][\text{B}]}$$

Since k_f and k_r are both constant, k_f/k_r is also a constant, called the equilibrium constant, K_{eq}. Therefore,

$$K_{eq} = \frac{[\text{C}][\text{D}]}{[\text{A}][\text{B}]}$$

These equilibrium constants are always written as products/reactants.

For the general reaction,

(16-3) $$2\,A \rightleftharpoons B + C$$

the equilibrium constant can be derived in the same way. If we first write the equation as

$$A + A \rightleftharpoons B + C$$

we can see that the equilibrium constant is

$$K_{eq} = \frac{[B][C]}{[A][A]} = \frac{[B][C]}{[A]^2}$$

For the very general reaction,

$$aA + bB \rightleftharpoons cC + dD$$

where A, B, C, and D represent different molecular species and *a*, *b*, *c*, and *d* are coefficients in the balanced equation, the equilibrium constant is written as

$$K_{eq} = \frac{[C]^c[D]^d}{[A]^a[B]^b}$$

Thus, the coefficients in the chemical equation become exponents for the concentrations in the expression for the equilibrium constant, K_{eq}.

At a given temperature, K_{eq} remains constant. As the temperature changes, the value of K_{eq} also changes.

By standard convention these constants are always written as products/reactants; that is, in Eq. (16-3) we could have written

$$\frac{k_r}{k_f} = \frac{[A][B]}{[C][D]}$$

and still have a constant. But it would *not* be the equilibrium constant for Eq. (16-2) as written there. This would be the reciprocal of the equilibrium constant for Eq. (16-2).

These equilibrium constants may or may not have units. For example, for Eqs. (16-2) and (16-3), the constant would be a pure number since the units would be

$$\frac{(\text{mol}/\ell)(\text{mol}/\ell)}{(\text{mol}/\ell)(\text{mol}/\ell)}$$

However, for the equation

$$2\,A \rightleftharpoons B + 2\,C$$

the units would be

$$\frac{(\text{mol}/\ell)(\text{mol}/\ell)^2}{(\text{mol}/\ell)^2} = \text{mol}/\ell$$

In this text we shall not put in units for equilibrium constants, but it is essential to keep in mind that all concentrations must be expressed as moles per liter (*M* concentration).

For equilibrium reactions involving gases, the equilibrium constant can also be written in terms of the partial pressures of the gases rather than as molar concentration. For example, for the equilibrium reaction between nitrogen gas and hydrogen gas forming ammonia,

$$N_{2(g)} + 3\,H_{2(g)} \rightleftarrows 2\,NH_{3(g)}$$

The equilibrium constant can be written as either

$$K_{eq} = \frac{[NH_3]^2}{[N_2][H_2]^3} \quad \text{or} \quad K_p = \frac{P_{NH_3}{}^2}{P_{N_2} \cdot P_{H_2}{}^3}$$

(where P represents partial pressure).

The value for a particular equilibrium constant can be determined by measuring the concentrations of all reactants and products at equilibrium, at a given temperature. The larger the value of the equilibrium constant, the greater the concentration of products at equilibrium. And likewise, the smaller the value of the equilibrium constant, the lower the concentration of the products. For example, a very small equilibrium constant would indicate that there is not much forward reaction taking place and therefore only a small amount of the products is being formed at equilibrium. Likewise, a large constant would mean that the forward reaction takes place to a great extent and thus there is a larger concentration of products than reactants.

For any reaction that is at equilibrium it can be shown that the equilibrium constant is always equal to the concentrations of the products over the concentrations of the reactants, all raised to the power of the coefficients from the balanced equation. For example, for the general reaction

$$2\,A + B \rightleftarrows C + D$$

if the forward reaction went in two steps,

1. $A + B \longrightarrow AB$
2. $AB + A \longrightarrow C + D$

then the reverse reactions would be

3. $C + D \longrightarrow AB + A$
4. $AB \longrightarrow A + B$

Therefore, the four rate laws for these four reactions would be

$$R_1 = k_1[A][B]$$
$$R_2 = k_2[AB][A]$$
$$R_3 = k_3[C][D]$$
$$R_4 = k_4[AB]$$

At equilibrium, since the overall rates of all forward reactions must equal the overall rates of all reverse reactions,

$$R_1 \cdot R_2 = R_3 \cdot R_4$$

or

$$k_1[A][B] \cdot k_2[AB][A] = k_3[C][D] \cdot k_4[AB]$$

and therefore

$$\frac{k_1 \cdot k_2}{k_3 \cdot k_4} = \frac{[C][D][AB]}{[A][B][AB][A]} = \frac{[C][D]}{[A]^2[B]}$$

or

$$K_{eq} = \frac{[C][D]}{[A]^2[B]}$$

Homogeneous and Heterogeneous Equilibria

If all the reactants and products are in the same state, the equilibrium is an example of a *homogeneous equilibrium.* For example, the reaction of nitrogen gas with hydrogen gas in equilibrium with ammonia gas is an example of a homogeneous equilibrium. On the other hand, if one or more of the reactants or products is not homogeneously dispersed (in the same phase as the other species), the equilibrium is classified as a **heterogeneous equilibrium.** An example of this would be a solid in equilibrium with a gas, such as shown by the following equation:

$$2\,KNO_{3(s)} \rightleftharpoons 2\,KNO_{2(s)} + O_{2(g)}$$

In a heterogeneous equilibrium of this type only the species in the homogeneous phase, that is, the gas phase in this example, are included in the equilibrium constant. Therefore, the equilibrium constant for this reaction is simply

$$K_{eq} = [O_2] \quad \text{or} \quad K_p = P_{O_2}$$

Consider the following problem example, involving both homogeneous and heterogeneous equilibrium.

PROBLEM EXAMPLE 16-1

Write the expression for the equilibrium constant for each of the following reactions:

(a) $H_{2(g)} + Cl_{2(g)} \rightleftharpoons 2\,HCl_{(g)}$
(b) $CaCO_{3(s)} \rightleftharpoons CaO_{(s)} + CO_{2(g)}$

SOLUTION:

(a) $K_{eq} = \frac{[HCl]^2}{[H_2][Cl_2]}$

(b) This is an example of a heterogeneous equilibrium. Therefore, only the $CO_{2(g)}$ is involved in the constant.

$$K_{eq} = [CO_2]$$

16-3 Factors that Cause a Shift in Equilibrium

The equilibrium constant for a given reaction at equilibrium must remain constant at a given temperature. Therefore, if any change in conditions causes a change in the concentration of any of the products or reactants, the others must also change in such a way as to compensate for this; that is, the position of equilibrium will shift in order to compensate for the change. This means that if more of one of the reactants is added to the equilibrium system, the forward reaction will speed up momentarily and for a short period of time the system is not in equilibrium. As more of the products are formed, however, the rate of the reverse reaction also increases. It soon esablishes a new position of equilibrium in which there is a somewhat higher concentration of products then there was originally. Thus, we say that the equilibrium has *shifted* to the right.

Another way of stating this is by the **Principle of Le Châtelier,** which was formulated by the French chemist Henry Louis Le Châtelier in 1888. This principle states that if a stress is applied to a system at equilibrium, the system will shift, if possible, in such a way as to reduce the stress. In the example given above, the stress can be reduced by using up some of the added reactant and forming more product.

There are two factors that can cause the equilibrium to shift. The first of these is concentration and the second is temperature. The concentration can be changed either by changing the amount of a particular substance or the volume that it is contained in. This latter change will, of course, cause a change in pressure, and often this is discussed in terms of changes in pressure for a gas. A change in temperature not only causes the equilibrium to shift but also changes the value of the equilibrium constant.

A catalyst will not cause a shift in equilibrium since it causes both the forward and reverse reactions to proceed faster to the same degree.

Concentration

When the concentration of one of the substances in a system at equilibrium is changed, the equilibrium will shift in such a way as to compensate for this change. For example, if the concentration of one of the reactants is increased, the equilibrium will shift to the right, or toward the products side. There are several ways of explaining why this happens.

1. One way is simply to think in terms of reaction rates. As the concentration of one or more of the reactants is increased, the forward reac-

tion will speed up momentarily until a new position at equilibrium is reached. At this new position of equilibrium, however, there will be different amounts of products and reactants than there were before the equilibrium was shifted.

2. This can also be seen by considering the equilibrium constant. As long as the temperature did not change, the value of the equilibrium constant will not change. Therefore, if the concentration of one or more of the reactants is increased, the other concentrations would have to change in such a way as to keep the value of the equilibrium constant the same.

3. Considering this from the viewpoint of Le Châtelier's Principle, we would simply say that increasing the concentration of one of the reactants would cause a stress on the system, and the system would shift in such a way as to relieve that stress. To do this, the equilibrium would have to shift to the right, to use up more of the reactants and to form more of the products.

For example, in the reaction considered for the Haber process of the preparation of ammonia,

$$N_{2(g)} + 3\ H_{2(g)} \rightleftharpoons 2\ NH_{3(g)} + \text{heat}$$

if the concentration of either nitrogen or hydrogen is increased, the equilibrium will shift to the right, causing the concentration of ammonia to increase and the concentration of the other reactant to decrease. Similarly, decreasing the concentration of either hydrogen or nitrogen will cause the equilibrium to shift to the left (the reactant's side), thus decreasing the concentration of ammonia.

If there is a change in the volume in which the reaction mixture is contained the concentration of each reactant and product will also change since concentration is amount per unit volume. For example, in an equilibrium mixture between gases, if the volume is decreased, the pressure will increase. According to Le Châtelier's Principle, the equilibrium will then shift in such a way as to relieve this stress that is placed on the system. In order to do this, the equilibrium will shift toward the side that has the least number of moles. You will recall that in discussing gases it was shown that 1 mole of any gas at the same temperature will exert a certain pressure. Therefore, the less moles that are present, the less the pressure. For example, using the ammonia reaction again, if the pressure were increased, the equilibrium would shift to the right to form more ammonia and use up more of the nitrogen and hydrogen. This is due to the fact that on the left side of the equation (reactants) there are 4 moles and on the right side (products) there are only 2 moles. Therefore, by shifting to the right the stress on the system will be relieved. For a reaction like hydrogen gas plus chlorine gas forming hydrogen chloride gas, according to the following equilibrium reaction,

$$H_{2(g)} + Cl_{2(g)} \rightleftharpoons 2\ HCl_{(g)}$$

it can be seen that an increase or decrease in pressure would not cause the equilibrium to shift. Since there are the same number of moles on each side of the equation, an increase or decrease in pressure would not favor one side or the other.

Temperature

The shift in equilibrium caused by a change in temperature will depend on whether the reaction as written is exothermic or endothermic. For an exothermic reaction, heat can be considered as one of the products. Likewise for an endothermic reaction, heat could be considered as a reactant. Therefore, if the temperature is increased for an exothermic reaction, it will have the same effect as increasing one or more of the products; the equilibrium will shift to the left in order to relieve the stress on the system. Conversely, a decrease in temperature for an exothermic reaction would cause the equilibrium to shift to the right to form more heat and thus relieve the stress on the system. For an endothermic reaction, the changes would be just the opposite of this; that is, an increase in temperature would cause a shift to the right and a decrease in temperature would cause a shift to the left. With temperature changes, however, it is important to keep in mind that along with the shift in equilibrium the equilibrium constant will also be changed. An equilibrium constant has a particular value only at one given temperature and at other temperatures it would have a different value.

For example, consider the ammonia equilibrium reaction again, which is an exothermic reaction.

$$N_{2(g)} + 3\,H_{2(g)} \rightleftharpoons 2\,NH_{3(g)} + \Delta$$

The following changes would take place with temperature changes. If the temperature is increased, the equilibrium will shift to the left, the equilibrium constant will change, the concentration of ammonia will decrease, and the concentration of hydrogen and nitrogen will increase. If the temperature is decreased, the equilibrium will shift to the right, the concentration of ammonia will increase, the concentrations of nitrogen and hydrogen will decrease, and the equilibrium constant will change.

Consider the following problem example.

PROBLEM EXAMPLE 16-2

For the reaction

$$N_{2(g)} + 3\,H_{2(g)} \rightleftharpoons 2\,NH_{3(g)} + \Delta$$

(a) Which way will the equilibrium shift if more N_2 is added to the system?
(b) Which way will the equilibrium shift if the pressure is increased?
(c) Which way will the equilibrium shift if the temperature is decreased?
(d) What will happen to the concentration of N_2 if H_2 is removed from the system?

(e) What will happen to the concentration of NH_3 if the pressure is decreased (by increasing the volume)?
(f) What will happen to the concentration of H_2 if the temperature is increased?

SOLUTION:

(a) An increase in the concentration of N_2 will cause the equilibrium to shift to the right (to use up more of the N_2).
(b) An increase in pressure in this case will cause the equilibrium to shift toward the side with the least number of moles; therefore it will shift to the right.
(c) Since this is an exothermic reaction (where heat can be thought of as a product), if heat is removed (that is, temperature is decreased), the equilibrium will shift to the right.
(d) If H_2 is removed, the equilibrium will shift to the left; therefore the concentration of N_2 will increase.
(e) If the pressure is decreased, the equilibrium will shift to the left; therefore the concentration of NH_3 will decrease.
(f) If the temperature is increased (heat put into the system), the equilibrium will shift to the left; therefore the concentration of H_2 will increase. Likewise, the N_2 concentration will increase and the NH_3 will decrease—thus along with the shift in equilibrium in this case there is also a change in the value of the equilibrium constant.

16-4 Calculations Involving Equilibrium Constants

Consider the following calculations involving equilibrium constants.

PROBLEM EXAMPLE 16-3

For the reaction

$$SO_{2(g)} + \tfrac{1}{2}O_{2(g)} \rightleftharpoons SO_{3(g)}$$

$$K_{eq} = \frac{[SO_3]}{[SO_2][O_2]^{1/2}} = 25.0 \text{ at } 600°C$$

Calculate the value of K_{eq} for each of the following reactions, at the same temperature:

(a) $2\,SO_{2(g)} + O_{2(g)} \rightleftharpoons 2\,SO_{3(g)}$
(b) $SO_{3(g)} \rightleftharpoons SO_{2(g)} + \tfrac{1}{2}O_{2(g)}$

SOLUTION:

(a) $K_{eq} = \dfrac{[SO_3]^2}{[SO_2]^2[O_2]} = (25.0)^2 = \mathbf{625}$

Since the reaction is the same as the original, but each factor is multiplied by two, the equilibrium is just the original raised to the second power (or squared).

(b) Since this is just the reverse of the original equation, the equilibrium constant is the reciprocal of the original. Recall that the K_{eq} is always written as products over reactants; that is,

$$K_{eq} = \frac{[SO_2][O_2]^{1/2}}{[SO_3]} = \frac{1}{25.0} = 0.0400$$

PROBLEM EXAMPLE 16-4

For the reaction

$$N_{2(g)} + 3\,H_{2(g)} \rightleftarrows 2\,NH_{3(g)}$$

at a given temperature it is found that an equilibrium mixture in a 10-ℓ container contains 2.0 moles of NH_3, 0.20 mole of N_2, and 6.0 moles of H_2. What is the value of the equilibrium constant at this temperature?

SOLUTION:

$$K_{eq} = \frac{[NH_3]^2}{[N_2][H_2]^3}$$

Remember that these are all molar concentrations and therefore the number of moles of each substance must be divided by the total volume of the container.

$$K_{eq} = \frac{(2.0/10)^2}{(0.20/10)(6.0/10)^3} = \frac{(0.20)^2}{(0.020)(0.60)^3} = 9.3$$

PROBLEM EXAMPLE 16-5

For the reaction

$$H_{2(g)} + I_{2(g)} \rightleftarrows 2\,HI_{(g)}$$

at 500°C, $K_{eq} = 62.5$. If 5.0 moles of H_2 and 5.0 mol of I_2 are placed in a 10-ℓ container at 500°C and allowed to come to equilibrium, calculate the final concentration of H_2, I_2, and HI.

SOLUTION:

$$K_{eq} = \frac{[HI]^2}{[H_2][I_2]} = 62.5$$

Let x equal the number of moles of H_2 and I_2 that react (since they will be equal). Therefore, the number of moles of HI that are formed is $2x$, and the number of moles of H_2 and I_2 that remain at equilibrium are each $(5.0 - x)$. Therefore,

$$\frac{(2x/10)^2}{[(5.0 - x)/10]^2} = 62.5$$

Taking the square root of both sides gives

$$\frac{(2x\ /10)}{[(5.0 - x)/10]} = 7.9 \qquad \frac{2x}{5.0 - x} = 7.9$$

and solving for x gives

$$2x = 39.5 - 7.9x \qquad 9.9x = 39.5 \qquad x = 4.0$$

Therefore, at equilibrium, keeping in mind that we want moles/liters,

$$[HI] = \frac{2x}{10} = \frac{(2)(4.0)}{10} = \mathbf{0.80}$$

$$[H_2] = [I_2] = \frac{5.0 - x}{10} = \frac{5.0 - 4.0}{10} = \mathbf{0.10}$$

SUMMARY

1. For a *reversible reaction,* there is an *equilibrium* established in which the rate of forward reaction is equal to the rate of the reverse reaction.
2. The *equilibrium constant,* called K_{eq} (or K_p for gases), is always written as concentration of products raised to the power of the coefficient in the balanced equilibrium reaction, divided by concentration of reactants, raised to the proper power from the balanced equation.
3. For heterogeneous systems, only those substances in the homogeneous phase are included in the equilibrium constant.
4. The equilibrium can be shifted (and therefore the various concentrations changed) by changing concentration of any of the reactants or products, changing the pressure (but only for those reactions involving gases where there are an unequal number of moles on either side of the equation), or changing the temperature (but this also gives a different value for the constant—the way that the equilibrium shifts depends on whether the reaction is exothermic or endothermic).
5. The shift in equilibrium can be explained in terms of change in the rate of the forward or reverse reaction, until a new position of equilibrium is reached; changes in the various concentrations in order to keep the K_{eq} constant; or *Le Châtelier's Principle,* which states that if a stress is placed on the system, it will shift in such a way as to relieve the stress.

Upon completion of this chapter, you should be able to do the following:

1. Given any equilibrium reaction, derive the expression for the equilibrium constant.
2. Given any equilibrium reaction (including those involving heterogeneous equilibrium), write the expression for the equilibrium constant in terms of molar concentration, or pressure for gases. (1)
3. Given any equilibrium reaction, and the equilibrium concentrations of all products and reactants, calculate the value of the equilibrium constant. (2)
4. Given any equilibrium reaction, tell which way the equilibrium would shift and how any of the concentrations (or partial pressures for gases) would change for any change in temperature or concentration (or pressure for those involving gases). (3, 4)
5. Given the initial concentration(s) of reactant(s), and concentration(s) of product(s) after equilibrium is established, calculate the value of the equilibrium constant. (5)
6. Given the value of the equilibrium constant for a particular reaction at a given temperature and the initial concentrations of reactants, calculate the equilibrium concentration of all reactants and products. (6)
7. Given the value of any equilibrium constant, calculate the value for the constant for any related reaction (that is, for the reverse of the original, for the original divided or multiplied through by any number, and so on). (7, 8)
8. Given any of the terms in Exercise 1, give a definition or explanation in your own words.

EXERCISES

1. Define or explain each of the following:
   (a) reversible reaction
   (b) chemical equilibrium
   (c) equilibrium constant
   (d) Le Châtelier's principle
   (e) heterogeneous equilibrium
2. Explain (using an example in each case) how the equilibrium shifts in a reaction when
   (a) the temperature is changed.
   (b) the concentration (or pressure for gases) of any of the reactants or products is changed.

PROBLEMS

Equilibrium constants

1. Write the expression for the equilibrium constant (K_{eq}) for each of the following reactions.
   (a) $H_{2(g)} + Cl_{2(g)} \rightleftharpoons 2\,HCl_{(g)}$
   (b) $S_{(s)} + O_{2(g)} \rightleftharpoons SO_{2(g)}$
   (c) $CH_{4(g)} + 2\,O_{2(g)} \rightleftharpoons CO_{2(g)} + 2\,H_2O_{(g)}$
   (d) $2\,SO_{2(g)} + O_{2(g)} \rightleftharpoons 2\,SO_{3(g)}$
   (e) $SnO_{2(s)} + 2\,CO_{(g)} \rightleftharpoons Sn_{(s)} + 2\,CO_{2(g)}$
   (f) $2\,N_{2(g)} + O_{2(g)} \rightleftharpoons 2\,NO_{2(g)}$
   (g) $H_{2(g)} + F_{2(g)} \rightleftharpoons 2\,HF_{(g)}$
   (h) $2\,CO_{(g)} + O_{2(g)} \rightleftharpoons 2\,CO_{2(g)}$
   (i) $CuO_{(s)} + H_{2(g)} \rightleftharpoons Cu_{(s)} + H_2O_{(g)}$
   (j) $C_2H_{4(g)} + 3\,O_{2(g)} \rightleftharpoons 2\,CO_{2(g)} + 2\,H_2O_{(g)}$
2. Given the concentrations at equilibrium, calculate the value of the equilibrium constant.
   (a) For the reaction,
   $2\,A + B \rightleftharpoons 3\,C + D$
   $[A] = 0.30\,M$
   $[B] = 0.60\,M$
   $[C] = 0.20\,M$
   $[D] = 0.50\,M$
   (b) For the reaction,
   $A + 3\,B \rightleftharpoons 2\,C + D$
   $[A] = 1.2\,M$
   $[B] = 0.50\,M$
   $[C] = 3.4\,M$
   $[D] = 0.75\,M$
3. For the equilibrium reaction
   $$2\,NO_{2(g)} \rightleftharpoons 2\,NO_{(g)} + O_{2(g)} + \text{heat}$$
   (a) Write the expression for the equilibrium constant.
   (b) Which way will the equilibrium shift if the temperature is decreased?
   (c) Which way will the equilibrium shift if NO is removed from the system?
   (d) Which way will the equilibrium shift if more NO_2 is added?
   (e) What will happen to the concentration of NO_2 if the temperature is increased?
   (f) What will happen to the concentration of O_2 if more NO is added?
   (g) Which way will the equilibrium shift if the pressure is decreased (by increasing the volume)?
   (h) What will happen to the concentration of NO_2 if the pressure is increased?
4. For the equilibrium reaction
   $$2\,SO_{2(g)} + O_{2(g)} \rightleftharpoons 2\,SO_{3(g)} + \text{heat}$$
   (a) Write the expression for the equilibrium constant.
   (b) Which way will the equilibrium shift if more O_2 is added to the system?
   (c) Which way will the equilibrium shift if SO_3 is removed from the system?

(d) What will happen to the concentration of O_2 if more SO_2 is added to the system?
(e) Which way will the equilibrium shift if the temperature of the system is increased?
(f) What will happen to the concentration of O_2 if the temperature of the system is decreased?
(g) Which way will the equilibrium shift if the pressure of the system is increased (by decreasing the volume)?
(h) What will happen to the concentration of O_2 if the pressure of the system is decreased (by increasing the volume)?

Calculations involving equilibrium constants

5. For the reaction

$$2\,NO_{2(g)} \rightleftharpoons N_2O_{4(g)}$$

(a) At a given temperature, 1 mole of NO_2 is placed in a 10-l container After equilibrium is established, there are 0.20 mole of N_2O_4 present. Calculate K_{eq} at this temperature.
(b) At a given temperature, 3.2 moles of N_2O_4 is placed in a 4.0-l container. After equilibrium is established, there is 0.15 mole of NO_2 present. Calculate K_{eq} at this temperature.

6. For the reaction

$$2\,HI_{(g)} \rightleftharpoons H_{2(g)} + I_{2(g)} \qquad K_{eq} = 0.016 \text{ at } 500°C$$

(a) If 1.0 mole of HI is placed in a 2.0-l container and allowed to come to equilibrium at 500°C, calculate the equilibrium concentrations of H_2, I_2, and HI.
(b) If 0.35 mole of HI is placed in a 0.75-l container and allowed to come to equilibrium at 500°C, calculate the equilibrium concentrations of H_2, I_2, and HI.

7. Using the value given for K_{eq} for the equation in Problem 6, at 500°C, calculate
(a) K_{eq} for the reaction $H_{2(g)} + I_{2(g)} \rightleftharpoons 2\,HI_{(g)}$
(b) K_{eq} for the reaction $HI_{(g)} \rightleftharpoons \frac{1}{2}\,H_{2(g)} + \frac{1}{2}\,I_{2(g)}$
(c) K_{eq} for the reaction $\frac{1}{2}\,H_{2(g)} + \frac{1}{2}\,I_{2(g)} \rightleftharpoons HI_{(g)}$
(d) K_{eq} for the reaction $4\,HI_{(g)} \rightleftharpoons 2\,H_{2(g)} + 2\,I_{2(g)}$

8. For the reaction

$$2\,A_{(g)} + B_{(g)} \rightleftharpoons 2\,C_{(g)} \qquad K_{eq} = 0.18 \text{ at } 200°C$$

calculate
(a) K_{eq} for the reaction $2\,C_{(g)} \rightleftharpoons A_{(g)} + B_{(g)}$
(b) K_{eq} for the reaction $A_{(g)} + \frac{1}{2}\,B_{(g)} \rightleftharpoons C_{(g)}$
(c) K_{eq} for the reaction $C_{(g)} \rightleftharpoons A_{(g)} + \frac{1}{2}\,B_{(g)}$
(d) K_{eq} for the reaction $4\,A + 2\,B \rightleftharpoons 4\,C$

REFERENCES

BERLINE, STEVEN, and CLARK BRICKER, "The Law of Mass Action," *J. Chem Educ.*, 1969, **46,** p. 499.

MORRIS, KELSO B., *Principles of Chemical Equilibrium,* New York; Reinhold Publishing Corp., 1965.

Objectives

1. To understand equilibria involving slightly soluble salts, in terms of
   (a) understanding the meaning of the terms soluble and insoluble, the general solubility rules, and the three methods that can be used to dissolve precipitates;
   (b) understanding what a solubility product constant, K_{sp}, represents and being able to tell how it can be used to predict when a precipitate will form; and
   (c) being able to do calculations involving solubility product constants, such as calculating K_{sp} from solubility, solubility from K_{sp}, and solubility when a common ion is present.
2. To understand equilibria involving the ionization of weak acids and bases in terms of ionization constants and what they represent and tell about the strength of an acid or base and to be able to do calculations involving K_i (or K_a or K_b).
3. To understand the ionization of water and the meaning and use of the term pH in terms of
   (a) being able to write the expression for and knowing the value of the ionization constant of water (K_W), and from this to calculate and use the relationship between $[H^+]$ and $[OH^-]$ in any water solution;
   (b) understanding what pH and pOH represent and being able to do calculations involving the relationship between pH, pOH, $[H^+]$, and $[OH^-]$; and
   (c) being able to calculate pH (or pOH) of any acid or base solution, including both weak and strong acids and bases.
4. To understand what a buffer solution is and how it works and to be able to do calculations involving buffer solutions, such as calculating the pH of a buffer, the pH after acid or base has been added to the buffer, and the preparation of a buffer of a given pH.

Seventeen

Equilibria in Solution I–Solubility, Ionization of Weak Acids and Bases, and pH

In Chapters 15 and 16, we have considered reaction rates and chemical equilibrium. We now want to look at chemical equilibrium involving reactions that take place in solution. We shall begin with the dissociation of slightly soluble compounds, then consider weak electrolytes such as weak acids and bases, and then look at the ionization of water and the pH of solutions. Several of these things have been considered qualitatively in previous chapters, but now we shall consider the quantitative aspects; that is, we shall calculate the actual amount of a slightly soluble compound that dissolves, or the H^+ ion concentration or pH of a solution of a weak acid.

17-1 Solubility Product Equilibria

There are three things that cause a reaction in solution to go toward completion. These are formation of an *insoluble* compound, formation of a weak electrolyte, and formation of a nonelectrolyte. In this section we shall consider the first of these. We now want to consider not only the formation of the slightly soluble compound, however, but also the equilibrium between the slightly soluble electrolyte and its ions in solution. This involves a dynamic equilibrium between the constant dissolving of the electrolyte and reprecipitation of it at the surface of the solid crystals. At equilibrium the solution is saturated—the rate of solution equals the rate of **precipitation** and, therefore, at equilibrium the concentration of the ions remains constant. As with other types of equilibria this can be shifted by changing the concentration of one of the ions present in the solution.

For example, consider a slightly soluble electrolyte of the type AB in water. The electrolyte will dissolve until the solution is saturated. At this point there is an equilibrium established between the solid and the ions in solution. This is represented by the equation

(17-1) $$AB_{(s)} \rightleftharpoons A^+ + B^-$$

In this equation it is understood that the A^+ and B^- ions are in solution. Therefore, it is not necessary to indicate this by the notation (aq).

The equilibrium constant for this equation would be written as

(17-2) $$K_{eq} = \frac{[A^+][B^-]}{[AB]}$$

Since the AB is in the solid state and is thus not part of the solution, however, its concentration remains constant. Therefore, the [AB] term is included with the K_{eq}, and a special type of equilibrium constant, called the **solubility product constant** (K_{sp}), is written as

(17-3) $$K_{sp} = [A^+][B^-]$$

This is defined as the molar concentration of the ions present, raised to the power of their coefficients in the equilibrium expression.

For a substance of the type A_2B, which ionizes according to the equation

(17-4) $$A_2B \rightleftharpoons 2\,A^+ + B^{2-}$$

the solubility product constant is written as

(17-5) $$K_{sp} = [A^+]^2[B^{2-}]$$

Consider the following problem example, of writing solubility product constant expressions.

PROBLEM EXAMPLE 17-1

Write the expression for the K_{sp} for each of the following slightly soluble electrolytes.
(a) AgCl
(b) $Mg(OH)_2$
(c) $Ca_3(PO_4)_2$

SOLUTION:

(a) The equilibrium equation for the dissociation is

$$AgCl_{(s)} \rightleftharpoons Ag^+ + Cl^-$$

Therefore, the solubility product constant is written as

$$\mathbf{K_{sp} = [Ag^+][Cl^-]}$$

Recall that this is heterogeneous equilibrium; therefore, the [AgCl] is constant and is included as part of the K_{eq}.

(b) The equilibrium equation is

$$Mg(OH)_2 \rightleftharpoons Mg^{2+} + 2\,OH^-$$

and the K_{sp} is written as

$$K_{sp} = [Mg^{2+}][OH^-]^2$$

(c) The equilibrium equation is

$$Ca_3(PO_4)_2 \rightleftharpoons 3\,Ca^{2+} + 2\,PO_4^{3-}$$

and the K_{sp} is written as

$$K_{sp} = [Ca^{2+}]^3[PO_4^{3-}]^2$$

When the **ion product** (the concentration of the ions raised to the proper power) is greater than the value of the K_{sp}, a supersaturated solution results. Normally, the excess will precipitate out and equilibrium will be established between the solid precipitate and the ions in solution. In some cases, precipitation does not begin immediately but, nevertheless, *once precipitation begins and equilibrium is established, the ion product is always equal to the* K_{sp}. Therefore, we can calculate from the K_{sp} what concentration of ions is necessary to begin precipitation for a particular compound.

In comparing values of K_{sp}, the electrolytes must all be of the same type, such as all AB or all A_2B_3. The larger the value of K_{sp} for an electrolyte of a given type, the more soluble the electrolyte in water. For example, AgCl with a K_{sp} of 1.8×10^{-10} is more soluble than AgBr whose K_{sp} is 7.0×10^{-13}.

To compare the solubility of electrolytes of different types, we must use the K_{sp} values and calculate the actual molar solubility. For example, the K_{sp} for silver oxalate, $Ag_2C_2O_4$, is 1.0×10^{-11}; the K_{sp} for zinc oxalate, ZnC_2O_4 is 1.5×10^{-9}. Since the first case is A_2B type and the second is AB type, however, the calculated solubilities turn out to be 1.4×10^{-4} *M* for $Ag_2C_2O_4$ and 3.9×10^{-5} *M* for ZnC_2O_4. We shall carry out some calculations of this type in the following problem examples.

The table of K_{sp} values does give an idea of what compounds are considered to be *insoluble,* or more accurately, slightly soluble, however, and the value of the K_{sp} gives at least a qualitative indication of how *insoluble* a subtance is. The values for the solubility product constants of several compounds are given in Table 1, Appendix VI. We shall use these values in the following calculations.

Consider the following problem examples involving solubility product constants.

PROBLEM EXAMPLE 17-2

Calculate the solubility of AgCl in (a) moles per liter and (b) grams per 100 m*l*.

SOLUTION:

(a) The equilibrium reaction is

$$AgCl_{(s)} \rightleftharpoons Ag^+ + Cl^-$$

The solubility product constant (K_{sp}), as found in Appendix VI, is

$$K_{sp} = [Ag^+][Cl^-] = 1.8 \times 10^{-10}$$

If we let x = the solubility in moles/l (that is, the molar solubility), then x = moles of AgCl that dissolve per liter and, therefore,

$$[Ag^+] = [Cl^-] = x$$

that is, for each mole of AgCl that dissolves, 1 mole of Ag^+ and 1 mole of Cl^- is formed.

Putting this into the expression for K_{sp} gives

$$(x)(x) = 1.8 \times 10^{-10}$$
$$x^2 = 1.8 \times 10^{-10}$$
$$x = \sqrt{1.8 \times 10^{-10}} = 1.3 \times 10^{-5}$$

Since all concentrations are molar, the solubility is **1.3×10^{-5} *M*.**

(b) To calculate the solubility in grams per 100 ml, we use the formula weight of AgCl (143 g/mole) as follows:

$$0.10\ l \times \frac{1.3 \times 10^{-5}\ \text{mol AgCl}}{l} \times \frac{143\ \text{g AgCl}}{1\ \text{mol AgCl}} = 1.9 \times 10^{-4}\ \text{g}$$

Therefore, the solubility is **1.9×10^{-4} g/100 ml.**

PROBLEM EXAMPLE 17-3

Calculate the solubility, in moles per liter, of $Hg(OH)_2$.

SOLUTION:

The equilibrium reaction is

$$Hg(OH)_{2(s)} \rightleftharpoons Hg^{2+} + 2\,OH^-$$

and the solubility product constant (using the value given in Appendix VI) is

$$K_{sp} = [Hg^{2+}][OH^-]^2 = 3.2 \times 10^{-26}$$

Let x = molar solubility of $Hg(OH)_2$, and therefore

$[Hg^{2+}] = x$

$[OH^-] = 2x$ (the coefficients in the equilibrium equation show that 1 mole of $Hg(OH)_2$ yields 2 moles of OH^- ion)

Putting these values into the expression for K_{sp} gives

$$(x)(2x)^2 = 3.2 \times 10^{-26}$$

There are two things to keep in mind when doing calculations of this type:

(a) The $[OH^-]$ is $2x$; you are *not* taking two times the concentration.

(b) Raising the $[OH^-]$ to the second power (that is, squaring it) is done when substituting into the K_{sp} expression; that is, $[OH^-]$ is *not* $(2x)^2$; this comes in when using the K_{sp}.

Solving for x gives

$$(x)(4x^2) = 3.2 \times 10^{-26}$$

$$4x^3 = 3.2 \times 10^{-26}$$

$$x^3 = 0.80 \times 10^{-26} = 8.0 \times 10^{-27}$$

$$x = \sqrt[3]{8 \times 10^{-27}} = 2.0 \times 10^{-9}$$

Therefore, the solubility is **2.0×10^{-9} moles/ℓ.**

PROBLEM EXAMPLE 17-4

The solubility of the electrolyte AB is 2.0×10^{-6} M. Calculate the value of K_{sp} for AB.

SOLUTION:

The equilibrium reaction is

$$AB_{(s)} \rightleftharpoons A^+ + B^-$$

and the solubility product constant is

$$K_{sp} = [A^+][B^-]$$

Since the solubility is 2.0×10^{-6} mol/ℓ

$$[A^+] = [B^-] = 2.0 \times 10^{-6}$$

Putting these values into the expression for K_{sp} gives

$$K_{sp} = (2.0 \times 10^{-6})^2 = \mathbf{4.0 \times 10^{-12}}$$

PROBLEM EXAMPLE 17-5

The solubility of $Pb(OH)_2$ is 1.0×10^{-5} M at 20°C. Calculate the value of K_{sp} for $Pb(OH)_2$ at this temperature.

SOLUTION:

The equilibrium reaction is

$$Pb(OH)_2 \rightleftharpoons Pb^{2+} + 2\ OH^-$$

and the solubility product constant is

$$K_{sp} = [Pb^{2+}][OH^-]^2$$

Since the solubility is 1.0×10^{-5},

$$[Pb^{2+}] = 1.0 \times 10^{-5}$$

$$[OH^-] = (2)(1.0 \times 10^{-5}) = 2.0 \times 10^{-5}$$

and therefore

$$K_{sp} = (1.0 \times 10^{-5})(2.0 \times 10^{-5})^2 = (1.0 \times 10^{-5})(4.0 \times 10^{-10})$$
$$= \mathbf{4.0 \times 10^{-15}}$$

PROBLEM EXAMPLE 17-6

The solubility of PbF_2 is 0.049 g/100 mℓ at 20°C. Calculate the value of K_{sp} for PbF_2 at this temperature.

SOLUTION:

First we must find the solubility in moles per liter.

$$\frac{0.049 \text{ g } PbF_2}{0.10\ \ell} \times \frac{1 \text{ mol } PbF_2}{245 \text{ g } PbF_2} = 2.0 \times 10^{-3}\ M$$

The reaction for the ionization of PbF_2 is

$$PbF_2 \rightleftharpoons Pb^{2+} + 2\,F^-$$

Since the solubility is $2.0 \times 10^{-3}\ M$,

$$[Pb^{2+}] = 2.0 \times 10^{-3}$$
$$[F^-] = (2)(2.0 \times 10^{-3}) = 4.0 \times 10^{-3}$$

Putting these values into the expression for the K_{sp} gives,

$$K_{sp} = [Pb^{2+}][F^-]^2 = (2.0 \times 10^{-3})(4.0 \times 10^{-3})^2 = (2.0 \times 10^{-3})(16 \times 10^{-6})$$
$$= 32 \times 10^{-9} = \mathbf{3.2 \times 10^{-8}}$$

Common Ion Effect

If an ion common to one of the ions present at equilibrium is added to the solution, the solubility will decrease. For example, consider the equilibria between solid silver chloride and its ions:

$$AgCl_{(s)} \rightleftharpoons Ag^+ + Cl^-$$

If an excess of Cl^- ion is added, the equilibrium will shift to the left and therefor the solubility of AgCl will decrease.

Consider the following calculation.

PROBLEM EXAMPLE 17-7

Calculate the molar solubility of AgCl in 0.10 M $AlCl_3$ solution.

SOLUTION:

In a 0.10 M solution of $AlCl_3$, there are 0.30 mol/ℓ of Cl^- ion; that is, each mole of $AlCl_3$ gives 1 mole of Al^{3+} and 3 moles of Cl^-. The equilibrium reaction that we are concerned with is

$$AgCl_{(s)} \rightleftharpoons Ag^+ + Cl^-$$

and the solubility product constant is

$$K_{sp} = [Ag^+][Cl^-] = 1.8 \times 10^{-10}$$

Let $x = M$ solubility, and therefore

$$[Ag^+] = x$$
$$[Cl^-] = x + 0.30$$

(that is, x mol/ℓ from the AgCl and 0.30 mol/ℓ from the $AlCl_3$ solution). Therefore,

$$(x)(x + 0.30) = 1.8 \times 10^{-10}$$

To make the calculations simpler, however, we can assume that x is probably going to be small compared to 0.30* and just use

$$(x)(0.30) = 1.8 \times 10^{-10}$$
$$x = 6.0 \times 10^{-10} \quad \text{(which is small compared to 0.30)}$$

Therefore, the solubility of AgCl in this solution is **6.0×10^{-10} *M*.** Notice that this is less than the solubility of AgCl in pure water, which would be

$$AgCl_{(s)} \rightleftharpoons Ag^+ + Cl^-$$
$$K_{sp} = [Ag^+][Cl^-] = 1.8 \times 10^{-10}$$

Let x = solubility. Therefore,

$$[Ag^+] = [Cl^-] = x$$
$$x^2 = 1.8 \times 10^{-10}$$
$$x = 1.3 \times 10^{-5} = \text{solubility in pure water}$$

This is what we would predict qualitatively—that is, adding Cl^- ions to this equilibrium causes a shift to the left, which means that the solubility is decreased. This principle is used in the quantitative determination of ions such as Ag^+; an excess of Cl^- is added to decrease the solubility of AgCl to a negligible amount.

Dissolving Precipitates

There are four methods that may be used to dissolve a precipitate.

1. Formation of a weak electrolyte—if a slightly soluble electrolyte contains an ion that will form a weak acid, addition of acid (H^+ ion) to the solution will cause the precipitate to dissolve. For example, if any

*As a general rule, if the result using this approximation is less than 5% of the number that you are adding to or subtracting from, the approximation is acceptable. For example, in this calculation, since the answer using this approximation is much less than 0.30 (6.0×10^{-10}), the approximation is close enough. Even if the $(x + 0.30)$ factor had been used and the problem solved by the quadratic equation, the answer to two significant digits would still be 6.0×10^{-10}.

strong acid is added to a solution containing silver acetate in equilibrium with its ions, the precipitate will dissolve because the acetate ion ($C_2H_3O_2^-$) will combine with the hydrogen ion to form the weak acid HC_2H_3O according to the equation

$$AgC_2H_3O_{2(s)} + H^+_{(aq)} \longrightarrow Ag^+_{(aq)} + HC_2H_3O_{2(aq)}$$

The acetic acid ($HC_2H_3O_2$) is still in solution but it exists mainly in the form of molecules. If H^+ (from nitric acid) is added to solid silver chloride, however, no reaction takes place since HCl is a strong acid.

$$AgCl_{(s)} + H^+ \longrightarrow \text{no reaction}$$

2. Formation of a complex ion–for example, a precipitate of AgCl can be dissolved by adding aqueous ammonia. The NH_3 forms a stable *complex* ion with the Ag^+ according to the equation

$$Ag^+ + 2\,NH_3 \rightleftharpoons Ag(NH_3)_2^+$$

This therefore removes Ag^+ ion from the equilibrium and causes the AgCl to dissolve by shifting the equilibrium to the right in the equation

$$AgCl \rightleftharpoons Ag^+ + Cl^-$$

We will consider complex ions in more detail in Chapter 24.

3. Formation of a less soluble compound–for example, adding potassium chromate, K_2CrO_4, to a precipitate of lead(II) chloride, $PbCl_2$, causes the $PbCl_2$ to dissolve and $PbCrO_4$ precipitates. This reaction can be observed by noting the color change since $PbCl_2$ is white whereas $PbCrO_4$ is yellow.

$$\underset{\text{(white)}}{PbCl_{2(s)}} + K_2CrO_{4(aq)} \longrightarrow \underset{\text{(yellow)}}{PbCrO_{4(s)}} + 2\,KCl_{(aq)}$$

4. Oxidation-reduction reaction–for example, a sulfide can be dissolved by *oxidizing* the sulfide ion S^{2-} to free sulfur, as in the following equation:

$$3\,NiS_{(s)} + 8\,HNO_{3(aq)} \longrightarrow 3\,Ni(NO_3)_{2(aq)} + 2\,NO_{(g)} + 3\,S_{(s)} + 4\,H_2O_{(l)}$$

We shall consider reactions of this type in Chapters 19 and 20.

Use of Solubility Product Constants to Predict When a Precipitate Will Form

We can use K_{sp} values to predict when a precipitate will form by comparing the ion product in a particular case with the value of the K_{sp} for the substance being considered. If the ion product is greater than the K_{sp}, a precipitate should form; if it is less than the K_{sp}, a precipitate will not form. Consider the following problem examples:

PROBLEM EXAMPLE 17-8

If silver nitrate solution is added to a solution containing equal concentrations of Cl^- and Br^-, which will precipitate first, AgCl or AgBr?

SOLUTION:

By comparing the K_{sp} values for AgCl (1.8×10^{-10}) and AgBr (5.0×10^{-13}), we can see that AgBr will precipitate first since the K_{sp} is smaller; that is, it will not take as much Ag^+ ion to begin to precipitate AgBr as it will AgCl, if there are equal amounts of Br^- and Cl^- ion present.

PROBLEM EXAMPLE 17-9

If two solutions are mixed so that the final solution contains 1.0×10^{-3} *M* Pb^{2+} and 2.0×10^{-2} *M* Cl^-, will $PbCl_2$ precipitate?

SOLUTION:

The equilibrium equation is

$$PbCl_2 \rightleftharpoons Pb^{2+} + 2\,Cl^-$$

and therefore the ion product is

$$\begin{aligned}(Pb^{2+})(Cl^-)^2 &= (1.0 \times 10^{-3})(2.0 \times 10^{-2})^2 \\ &= (1.0 \times 10^{-3})(4.0 \times 10^{-4}) = 4.0 \times 10^{-7}\end{aligned}$$

The value of the K_{sp} for $PbCl_2$ is 1.6×10^{-5}. Therefore, a precipitate will not form since the ion product (4.0×10^{-7}) is less than the K_{sp} (1.6×10^{-5}).

17-2 Equilibria Involving Weak Acids and Bases

You will recall that a weak acid or base is one that is only slightly ionized. Therefore, a solution of one of these weak electrolytes will contain unionized molecules and ions, both in solution. Notice that this differs from solubility product equilibrium in that the molecules of weak electrolyte are part of the solution, as are the ions that are formed. Therefore, the equilibrium constant, called K_i (or K_a for a weak acid and K_b for a weak base) for the general weak acid HA would be

(17-6) $$HA_{(aq)} \rightleftharpoons H^+ + A^-$$

(17-7) $$K_a = \frac{[H^+][A^-]}{[HA]}$$

Recall that this is a simplified form of the equation for an equilibrium of this type. The actual equation is

(17-8) $$HA + H_2O \rightleftharpoons H_3O^+ + A^-$$

For this reaction the equilibrium constant would be

(17-9) $$K_{eq} = \frac{[H_3O^+][A^-]}{[HA][H_2O]}$$

The concentration of water, in a water solution, can be calculated from the density of water as

$$\frac{1.00 \text{ g } H_2O}{m\ell} \times \frac{1000 \text{ } m\ell}{\ell} \times \frac{1 \text{ mole } H_2O}{18.0 \text{ g } H_2O} = 55.6 \text{ mole } H_2O/\ell$$

In the dilute solutions that we shall be considering, where the concentration of acid or base is in the range of about 0.01 *M* to 1 *M*, the concentration of water will remain practically constant at about 55.6 *M*. Thus, this is combined with the K_{eq} to give a new constant called K_i (or K_a or K_b).

The strength of an acid or base can be found experimentally either by conductivity measurements, which give a measure of the percent of ionization, or by measuring the pH of the solution, which gives a measure of the $[H^+]$ or $[OH^-]$ of the solution (pH will be discussed in Section 17-3).

As with other equilibrium constants the value of the K_i gives an indication of the strength of the weak electrolyte; that is, the smaller the value of K_a or K_b, the weaker the acid or base. Since an ionization constant is a special type of equilibrium constant, there must be an equilibrium reaction. But a strong acid, such as HCl, is considered to be completely ionized and therefore there is no equilibrium, and no ionization constant. That is, for hydrochloric acid the reaction

(17-10) $$HCl + H_2O \longrightarrow H_3O^+ + Cl^-$$

or

(17-11) $$HCl_{(aq)} \longrightarrow H^+ + Cl^-$$

goes to completion. Thus for a solution of HCl, the $[H^+]$ is equal to the concentration of the acid. For example, in a 0.10 *M* solution of HCl, the $[H^+] = 0.10$ *M*. For a weak acid, however, we must calculate the $[H^+]$ from the K_i.

Consider the following calculations involving weak acids and bases in solution.

PROBLEM EXAMPLE 17-10

For a 0.10 *M* solution of acetic acid ($HC_2H_3O_2$), calculate
(a) the $[H^+]$
(b) the percent ionization

SOLUTION:

(a) The equilibrium equation representing the ionization is

$$HC_2H_3O_{2(aq)} \rightleftharpoons H^+ + C_2H_3O_2^-$$

and the ionization constant (from Appendix VI) is

$$K_i = \frac{[H^+][C_2H_3O_2^-]}{[HC_2H_3O_2]} = 1.8 \times 10^{-5}$$

Note here that the $HC_2H_3O_2$ is in solution but in the form of molecules. Therefore, this is not like a solubility product constant where the solid is not included in the K_{sp} expression.

Let x = the amount of ionization. Therefore, at equilibrium,

$$[HC_2H_3O_2] = 0.10 - x$$
$$[H^+] = [C_2H_3O_2^-] = x$$

Assuming that x is going to be small compared to 0.10 (as done in Problem 17-7),

$$\frac{(x)(x)}{0.10} = 1.8 \times 10^{-5}$$
$$x^2 = 1.8 \times 10^{-6}$$
$$x = \sqrt{1.8 \times 10^{-6}} = 1.3 \times 10^{-3}$$

(which is small compared to 0.10; that is, 1.3×10^{-3} is less than 5% of 0.10.) Therefore,

$$[H^+] = \mathbf{1.3 \times 10^{-3}}\ \boldsymbol{M}$$

(b) The percent of ionization is

$$\frac{\text{amount of acid ionized } (M)}{\text{amount of acid initially } (M)} \times 100$$

In (a) the amount of acid ionized was calculated to be 1.3×10^{-3} M, and the initial amount of acid present was 0.10 M. Therefore, the percent of ionization is

$$\text{percent ionization} = \frac{1.3 \times 10^{-3}}{0.10} \times 100 = \mathbf{1.3\%}$$

PROBLEM EXAMPLE 17-11

Calculate the $[H^+]$ of a 0.10 M solution of acetic acid in 0.25 M sodium acetate.

SOLUTION:

The equilibrium reaction is simply the ionization of the acetic acid:

$$HC_2H_3O_2 \rightleftharpoons H^+ + C_2H_3O_2^-$$

and the ionization constant is

$$K_i = \frac{[H^+][C_2H_3O_2^-]}{[HC_2H_3O_2]} = 1.8 \times 10^{-5}$$

This involves a common ion, however; that is, there is more acetate ion ($C_2H_3O_2^-$) present than just the amount from the ionization of the acetic acid. This common ion will cause the equilibrium to shift to the left and therefore cause the $[H^+]$ to be less than it would be without the common ion (in Problem Example 17-10 this was calculated to be 1.3×10^{-3}). Let x = amount ionized. Therefore, at equilibrium,

$$[HC_2H_3O_2] = 0.10 - x$$
$$[C_2H_3O_2^-] = x + 0.25$$

(that is, x from the ionization of the acid and 0.25 from the sodium acetate.)

$$[H^+] = x$$

Assuming that x is going to be small compared to 0.10 (and also compared to 0.25),

$$\frac{(x)(0.25)}{(0.10)} = 1.8 \times 10^{-5}$$

$$x = \frac{(1.8 \times 10^{-5})(0.10)}{(0.25)} = 0.72 \times 10^{-5} = 7.2 \times 10^{-6}$$

This is small compared to 0.10, and therefore

$$[H^+] = \mathbf{7.2 \times 10^{-6}}\ \boldsymbol{M}$$

Notice that this is less than 1.3×10^{-3}, as predicted.

PROBLEM EXAMPLE 17-12

A 0.020 M solution of the weak acid HA is 0.60% ionized. Calculate the value of the ionization constant, K_i (or K_a) for this acid.

SOLUTION:

The equilibrium equation is

$$HA_{(aq)} \rightleftharpoons H^+ + A^-$$

and the expression for the ionization constant is

$$K_a = \frac{[H^+][A^-]}{[HA]}$$

If this 0.020 M solution is 0.60% ionized, the concentrations of H^+ and A^- are

$$(0.020)(0.0060) = (2.0 \times 10^{-2})(6.0 \times 10^{-3}) = 12 \times 10^{-5} = 1.2 \times 10^{-4}$$

and the

$$[HA] = 0.020 - 1.2 \times 10^{-4}$$

which, to two significant digits, is equal to 0.020.

Putting these values into the expression for K_a gives

$$K_a = \frac{(1.2 \times 10^{-4})^2}{2.0 \times 10^{-2}} = \frac{1.44 \times 10^{-8}}{2.0 \times 10^{-2}} = 0.72 \times 10^{-6} = \mathbf{7.2 \times 10^{-7}}$$

PROBLEM EXAMPLE 17-13

If 100 mℓ of 0.20 M aqueous ammonia solution is mixed with 200 mℓ of 0.50 M $MgCl_2$ solution, will a precipitate of $Mg(OH)_2$ form?

SOLUTION:

This problem involves three steps: (a) finding the new concentration of NH_3 and $MgCl_2$ after mixing (that is, as each one is diluted), (b) calculating the $[OH^-]$ from the aqueous NH_3 solution, and (c) using the K_{sp} of $Mg(OH)_2$ to tell if a precipitate should form.

(a) To obtain the new molar concentration of each solution we must find the number of moles present and divide by the total number of liters of solution, which is 300 mℓ or 0.30 ℓ.

$$100\ \text{m}\ell\ \text{NH}_3\ \text{soln} \times \frac{0.20\ \text{mol NH}_3}{1000\ \text{m}\ell\ \text{NH}_3\ \text{soln}} = 0.020\ \text{mol NH}_3$$

$$200\ \text{m}\ell\ \text{MgCl}_2\ \text{soln} \times \frac{0.50\ \text{mol MgCl}_2}{1000\ \text{m}\ell\ \text{MgCl}_2\ \text{soln}} = 0.10\ \text{mol MgCl}_2$$

Therefore, the molar concentrations are

$$\frac{0.020\ \text{mol NH}_3}{0.30\ \ell\ \text{soln}} = \mathbf{0.067\ \mathit{M}\ NH_3}$$

$$\frac{0.10\ \text{mol MgCl}_2}{0.30\ \ell\ \text{soln}} = 0.33\ M\ \text{MgCl}_2 = \mathbf{0.33\ \mathit{M}\ Mg^{2+}}$$

(b) The equilibrium reaction for the ionization of NH_3 in H_2O is

$$NH_3 + H_2O \rightleftharpoons NH_4^+ + OH^-$$

and the K_i (or K_b) for this base is

$$K_b = \frac{[NH_4^+][OH^-]}{[NH_3]} = 1.8 \times 10^{-5}$$

Let $x =$ the amount of ionization and, therefore,

$$[NH_3] = 0.067 - x$$
$$[NH_4^+] = [OH^-] = x$$

Putting these values into the expression for K_b (again assuming that x is going to be small compared to 0.067) gives

$$\frac{x^2}{0.067} = 1.8 \times 10^{-5}$$

$$x^2 = 0.12 \times 10^{-5} = 1.2 \times 10^{-6}$$

$$x = \sqrt{1.2 \times 10^{-6}} = 1.1 \times 10^{-3}$$

(which is less than 5% of 0.067). Therefore,

$$[OH^-] = \mathbf{1.1 \times 10^{-3}}$$

(c) The solubility product constant of $Mg(OH)_2$ is

$$K_{sp} = [Mg^{2+}][OH^-]^2 = 8.9 \times 10^{-12}$$

Since the $[Mg^{2+}]$ is 0.33 and the $[OH^-]$ is 1.1×10^{-3}, the ion product is

$$(0.33)(1.1 \times 10^{-3})^2 = 0.40 \times 10^{-6} = \mathbf{4.0 \times 10^{-7}}$$

This is greater than the K_{sp} for $Mg(OH)_2$, and therefore a precipitate would form.

17-3 Ionization of Water

We have seen earlier that water is only very slightly ionized. In fact, it is sometimes referred to as a nonionized substance; you may have found in the lab that pure water does not conduct an electric current. It can be shown with more sensitive instruments than the simple light bulb apparatus which was discussed in Chapter 3 however, that there is a very slight amount of conductance. The following equations illustrate this slight ionization:

(17-12) $$H_2O + H_2O \rightleftharpoons H_3O^+ + OH^-$$

or simply

(17-13) $$H_2O \rightleftharpoons H^+ + OH^-$$

The equilibrium constant for this would be

(17-14) $$K_{eq} = \frac{[H^+][OH^-]}{[H_2O]}$$

But since $[H_2O]$ remains relatively constant, as with weak acids and bases, we can combine K_{eq} and $[H_2O]$ to write the **ionization constant for water,** called K_W

(17-15) $$K_W = [H^+][OH^-]$$

At room temperature (about 25°C), the value of K_W is 1.0×10^{-14}. Therefore, in pure water,

$$[H^+][OH^-] = 1.0 \times 10^{-14}$$

Since,

$$[H^+] = [OH^-]$$

$$[H^+]^2 = 1.0 \times 10^{-14}$$

$$[H^+] = \sqrt{1.0 \times 10^{-14}} = 1.0 \times 10^{-7}$$

In any aqueous solution, at room temperature,* the ion product $[H^+][OH^-]$ will always equal 1.0×10^{-14}. Therefore, if the concentration of one of these is changed, the other must change also in order to have the K_W remain constant. Therefore, if one value is known, the other can be calculated using the equation

$$[H^+][OH^-] = 1.0 \times 10^{-14} \tag{17-16}$$

If a solution has a hydrogen ion concentration larger than 1.0×10^{-7} (that is, a smaller negative exponent), the solution is acidic since the $[H^+]$ is greater than the $[OH^-]$. For example, if the $[H^+]$ is 1.0×10^{-5}, the $[OH^-]$ is calculated to be

$$(1 \times 10^{-5})[OH^-] = 1.0 \times 10^{-14}$$

$$[OH^-] = \frac{1.0 \times 10^{-14}}{1.0 \times 10^{-5}} = 1.0 \times 10^{-9}$$

Therefore, $[H^+]$ is greater than $[OH^-]$ and hence the solution is acidic.

Likewise, if the $[OH^-]$ is larger than 1×10^{-7}, the $[H^+]$ must be less than 1×10^{-7} and hence the solution is basic. For example, if the $[OH^-]$ of a solution is 2.0×10^{-6}, the $[H^+]$ is calculated to be,

$$[H^+](2.0 \times 10^{-6}) = 1.0 \times 10^{-14}$$

$$[H^+] = \frac{1.0 \times 10^{-14}}{2.0 \times 10^{-6}} = 0.50 \times 10^{-8} = 5.0 \times 10^{-9}$$

Therefore, the $[H^+]$ is less than the $[OH^-]$ and hence the solution is basic. Also, of course, in the first of these two examples, $[H^+]$ is greater than 1×10^{-7} (and hence the solution is acidic) and in the second example $[H^+]$ is less than 1×10^{-7} (and hence the solution is basic).

17-4 pH and pOH

The **pH** of a solution is a measure of the acidity of a solution or, more accurately, the $[H^+]$ of the solution. The notation p means $1/\log$ or $-\log$ and, therefore,

$$pH = -\log [H^+]$$

Likewise,

$$\text{pOH} = -\log [\text{OH}^-]$$

A logarithm (log) is defined as the exponent, or the power to which 10 must be raised to give the number. For example, $\log 10^2 = 2$, $\log 10^{-3} = -3$,

*As with any equilibrium constant the value of this varies with temperature. At 100°C, K_W is about 1×10^{-12}. For present purposes, however, where we shall be working with temperature around 20° to 25°C, we shall use the value of 1.0×10^{-14}, without specifying the temperature.

and log 1 = 0 (since any number raised to the zero power is equal to 1). For logs of numbers that are not integers, a table of logarithms (see Appendix VII) must be used. In using a table of this type, it is easiest to write the number in correct scientific notation (one number before the decimal, with an exponent). Then the log is always 0.______. For example, to find the log of 2.25, you look in the table (go down the N column to 2.2, then to the 5 column) and find that it is 0.3522. To find the log of 2.25×10^{-5}, you look up the log of 2.25 and add it to -5 (since logs are simply exponents). To multiply numbers, the logs are added; conversely, to divide one number by another, the log of the denominator is subtracted from the log of the numerator. Thus, the log of 2.25×10^{-5} is $0.352 - 5 = -4.648$. To find the log of 0.0225, the number would first be written as 2.25×10^{-3} and then the log would be found to be $0.352 - 3 = -2.648$. In considering significant digits of logarithms, only the numbers to the right of the decimal are counted since the number before the decimal only indicates the power of 10 in the scientific notation; thus it just tells where the decimal point in the number is.

To find the antilog of a number, the reverse process is used. The given log is found in the log table and then the number is found. For example, to find the antilog of 0.352, you would go to the log table and find that this is the log of 2.25. Here also, the numbers are always written as 0.______ and then the antilog will have one number before the decimal. To find the antilog of the number 3.352, we want to find what number has this log. The number is $10^{3.352}$ but this is not a convenient way of expressing a number. Therefore, it is written as $10^3 \times 10^{0.352}$, or $10^{0.352} \times 10^3$. The antilog of 0.352 is found from the log table to be 2.25 and therefore the number is 2.25×10^3. To find the antilog of the number $10^{-4.62}$, we write it as $10^{0.48} \times 10^{-5}$ (remember that we add the exponents to multiply numbers, and $-5 + 0.48 = -4.62$). Then we find the antilog of 0.48, which is 3.02. Thus, the number is 3.02×10^{-5}.

For any water solution, the sum of the pH and the pOH will always equal 14. This can be derived as follows.

Since, for any water solution,

$$[H^+][OH^-] = 1 \times 10^{-14}$$

$$\log [H^+] + \log [OH^-] = \log (1 \times 10^{-14}) = \log 1 + \log 10^{-14}$$
$$= 0 + (-14) = -14.$$

Multiplying both sides by -1 gives

$$(-\log [H^+]) + (-\log [OH^-]) = -(-14)$$

and hence, from the definition of pH and pOH,

(17-17) $$pH + pOH = 14$$

Also, from the discussion of the relationship between $[H^+]$ and acidity of solutions in Section 17-3 it can be seen that a pH of 7 is neutral, below 7 is acidic, and above 7 is basic. This can be summarized as

pH below 7	pH = 7	pH above 7
← More Acidic	Neutral	More Basic →

Consider the following calculations involving pH, pOH, $[H^+]$, and $[OH^-]$.

PROBLEM EXAMPLE 17-14

Calculate the pH of each of the following solutions.

(a) $[H^+] = 2.5 \times 10^{-5}$
(b) $[OH^-] = 2.0 \times 10^{-8}$
(c) pOH = 4.2
(d) 0.020 *M* HCl
(e) 0.020 *M* NaOH
(f) 0.10 *M* HCN
(g) 0.10 *M* NH_3 solution

SOLUTION:

(a) $pH = -\log [H^+] = -(-5 + 0.40) = \mathbf{4.60}$

(b) This can be solved in two ways:

(i) Convert from $[OH^-]$ to $[H^+]$ and then calculate pH.

$$[H^+] = \frac{1.0 \times 10^{-14}}{[OH^-]} = \frac{1.0 \times 10^{-14}}{2.0 \times 10^{-8}} = 0.50 \times 10^{-6} = 5.0 \times 10^{-7}$$

$$pH = -(-7 + 0.70) = \mathbf{6.30}$$

(ii) Calculate the pOH and subtract from 14.

$$[OH^-] = 2.0 \times 10^{-8}$$
$$pOH = -(-8 + 0.30) = 7.70$$
$$pH = 14.00 - 7.70 = \mathbf{6.30}$$

(c) $pH = 14.0 - pOH = 14.0 - 4.2 = \mathbf{9.8}$

(d) Since HCl is a strong acid, in a 0.020 *M* HCl solution, the $[H^+] = 0.020 = 2.0 \times 10^{-2}$.

$$pH = -(-2 + 0.30) = \mathbf{1.70}$$

(e) Since NaOH is a strong base, in a 0.020 *M* NaOH solution

$$[OH^-] = 0.020\ M = 2.0 \times 10^{-2}$$

Therefore,

$$[H^+] = \frac{1 \times 10^{-14}}{2.0 \times 10^{-2}} = 0.50 \times 10^{-12} = 5.0 \times 10^{-13}$$

$$pH = -(-13 + 0.70) = \mathbf{12.30}$$

or

$$pOH = -(-2 + 0.30) = 1.70$$
$$pH = 14.00 - 1.70 = \mathbf{12.30}$$

(f) Since HCN is a weak acid, we must first calculate the $[H^+]$ using the K_a, as found in Appendix VI-2. The equilibrium reaction is

$$HCN_{(aq)} \rightleftharpoons H^+ + CN^-$$

and the ionization constant is

$$K_a = \frac{[H^+][CN^-]}{[HCN]} = 4.0 \times 10^{-10}$$

Let x = the number of moles per liter of HCN that ionize. Thus, at equilibrium,

$$[HCN] = 0.10 - x$$
$$[H^+] = [CN^-] = x$$

Putting these values into the expression for K_a gives (again assuming that x will be small compared to 0.10)

$$\frac{x^2}{0.10} = 4.0 \times 10^{-10}$$
$$x^2 = 4.0 \times 10^{-11} = 40 \times 10^{-12}$$

(To take the square root, the exponent must be an even number.)

$$x = \sqrt{40 \times 10^{-12}} = 6.3 \times 10^{-6} = [H^+]$$
$$pH = -(-6 + 0.80) = \mathbf{5.20}$$

(g) Since NH_3 is a weak base, we must first calculate the $[OH^-]$ using the K_b for NH_3, as found in Appendix VI-2. The equilibrium reaction is

$$NH_3 + H_2O \rightleftharpoons NH_4^+ + OH^-$$

and the ionization constant is

$$K_b = \frac{[NH_4^+][OH^-]}{[NH_3]} = 1.8 \times 10^{-5}$$

Let x = the number of moles per liter of NH_3 that ionize. Thus, at equilibrium,

$$[NH_3] = 0.10 - x$$
$$[NH_4^+] = [OH^-] = x$$

Putting these values into the expression for K_b gives (assuming x is small compared to 0.10)

$$\frac{x^2}{0.10} = 1.8 \times 10^{-5}$$

$$x^2 = 1.8 \times 10^{-6}$$

$$x = \sqrt{1.8 \times 10^{-6}} = 1.3 \times 10^{-3} = [OH^-]$$

$$pOH = -(-3 + 0.11) = 2.89$$

$$pH = 14.00 - 2.89 = \mathbf{11.11}$$

PROBLEM EXAMPLE 17-15

Calculate the $[H^+]$ of a solution whose pH = 5.72.

SOLUTION:

If the pH = 5.72,

$$[H^+] = 10^{-5.72} = 10^{0.28} \times 10^{-6} = \mathbf{1.9 \times 10^{-6}}$$

PROBLEM EXAMPLE 17-16

The pH of a saturated solution of $Fe(OH)_2$ is 9.30. Calculate the K_{sp} of $Fe(OH)_2$.

SOLUTION:

From the pH we can find the $[H^+]$, then we can obtain $[OH^-]$ from this, and finally we can calculate K_{sp} from $[OH^-]$.

If pH = 9.3,

$$[H^+] = 10^{-9.3} = 10^{0.7} \times 10^{-10} = 5.0 \times 10^{-10}$$

$$[OH^-] = \frac{1.0 \times 10^{-14}}{5.0 \times 10^{-10}} = 0.20 \times 10^{-4} = 2.0 \times 10^{-5}$$

From the equilibrium reaction for the dissociation of $Fe(OH)_2$, which is

$$Fe(OH)_{2(s)} \rightleftharpoons Fe^{2+} + 2\,OH^-$$

we see that

$$[Fe^{2+}] = \tfrac{1}{2}[OH^-]$$

Since

$$[OH^-] = 2.0 \times 10^{-5} \qquad [Fe^{2+}] = 1.0 \times 10^{-5}$$

Using the expression for the solubility product constant for $Fe(OH)_2$, which is

$$K_{sp} = [Fe^{2+}][OH^-]^2$$

and putting in these values for $[Fe^{2+}]$ and $[OH^-]$ gives

$$K_{sp} = (1.0 \times 10^{-5})(2.0 \times 10^{-5})^2 = \mathbf{4.0 \times 10^{-15}}$$

17-5 Buffer Solutions

A **buffer** solution is a solution of a weak acid and a salt of the acid, or a weak base and a salt of the base. A buffer acts to prevent rapid changes in the pH of the solution as more acid or base is added. An example of a buffer solution would be a solution containing 0.1 mole per liter of acetic acid and 0.1 mole per liter of sodium acetate (although the concentrations of acid and salt do not have to be the same). To find the pH of this solution, we use the K_a for acetic acid, which gives

(17-18) $$HC_2H_3O_2 \rightleftharpoons H^+ + C_2H_3O_2^-$$

(17-19) $$K_a = [H^+]\frac{[C_2H_3O_2^-]}{[HC_3H_3O_2]}$$

Therefore,

(17-20) $$[H^+] = K_a\frac{[HC_2H_3O_2]}{[C_2H_3O_2^-]}$$

Since

$$[HC_2H_3O_2] = [C_2H_3O_2^-]$$

(omitting the small amount of ionization)

$$[H^+] = K_a = 1.8 \times 10^{-5}$$

and

$$pH = -(-5 + 0.26) = 4.74$$

For any buffer solution, the hydrogen ion or hydroxide ion concentration is equal to the ionization constant, K_a or K_b, times the ratio of

$$\frac{[\text{acid or base}]}{[\text{ion}]}$$

As acid is added to the acetic acid–sodium acetate buffer just discussed, the H^+ ion combines with the acetate ion present to shift the equilibrium to the left [in Eq. (17-18)]. This increases the concentration of acid and decreases the concentration of acetate ion and thus changes the ratio of [acid]/[ion]. The $[H^+]$ is equal to the ratio times the K_a, however, which is a small value (1.8×10^{-5}). Thus even though the ratio is changed, the $[H^+]$ changes only slightly; hence the pH remains relatively constant.

As base is added to this buffer, the OH^- reacts with the acetic acid present to form more salt and hence more acetate ion. This causes the equilibrium in Eq. (17-18) to shift to the right and thus decreases the concentration of acid and increases the concentration of acetate ion. As in the previous case, this simply changes the ratio of [acid]/[ion] somewhat, but doesn't change the $[H^+]$ much. Some specific problem examples should help to show how a buffer works.

PROBLEM EXAMPLE 17-17

To the acetic acid–sodium acetate buffer discussed above is added an amount of NaOH that would make the solution 0.020 *M* in NaOH, if the buffer was not present. Calculate the new pH of the solution.

SOLUTION:

The NaOH added will react with the acetic acid present to form more sodium acetate, according to the equation

$$NaOH + HC_2H_3O_2 \longrightarrow H_2O + NaC_2H_3O_2$$

Since 1 mole of NaOH will react with 1 mole of $HC_2H_3O_2$, the 0.020 mol/*l* of NaOH added will react with 0.020 mol/*l* of $HC_2H_3O_2$, to form 0.020 mol/*l* of $NaC_2H_3O_2$ (or $C_2H_3O_2^-$ ion since sodium acetate is a soluble electrolyte). Let x = the number of moles per liter of $HC_2H_3O_2$ that ionize. Thus, at equilibrium, $[HC_2H_3O_2]$ = the initial amount (0.10 *M*) minus the amount that is neutralized by the base (0.020), minus the amount that ionizes (x). Thus,

$$[HC_2H_3O_2] = (0.10 - 0.020) - x = 0.080 - x$$

$[C_2H_3O_2^-]$ = the initial amount (0.10 *M*), plus the amount formed by the reaction of NaOH with $HC_2H_3O_2$ (0.020 *M*), plus the amount of $HC_2H_3O_2$ that ionizes (x)

$$= (0.10 + 0.020) + x = 0.12 + x$$

Assuming that x is going to be small compared to 0.080 and 0.12 and putting these values into Eq. (17-20) gives

$$[H^+] = (1.8 \times 10^{-5})\left(\frac{0.080}{0.12}\right) = 1.2 \times 10^{-5}$$

Since this is small compared to 0.080 or 0.12, the assumption made was valid.

$$pH = -(-5 + 0.079) = \mathbf{4.92}$$

Notice here that this is greater than the pH of the buffer (4.74) since base has been added; however, it is only a small change in pH. If 0.020 mol/*l* of NaOH had been added to pure water, the pH would have changed from 7.0 to 12.3 [See Problem Example 17-14(e)].

PROBLEM EXAMPLE 17-18

To this same buffer solution is added an amount of HCl that would make the solution 0.020 *M* in HCl, if the buffer was not present. Calculate the new pH.

SOLUTION:

The equilibrium is still the ionization of acetic acid:

$$HC_2H_3O_2 \rightleftharpoons H^+ + C_2H_3O_2^-$$

The 0.020 mol/l of H^+ (from the strong acid HCl) that is added will combine with 0.020 mol/l of $C_2H_3O_2^-$ ion that is present, to form 0.020 mol/l of $HC_2H_3O_2$ (which will then ionize slightly).
Let x = number of moles of acetic acid that ionize. Therefore, at equilibrium,

$$[HC_2H_3O_2] = (0.10 + 0.020) - x = 0.12 - x$$
$$[C_2H_3O_2^-] = (0.10 - 0.020) + x = 0.080 - x$$

Assuming that x is going to be small compared to 0.12 or 0.080 and putting these values into Eq. (17-20) gives

$$[H^+] = K_a \frac{[HC_2H_3O_2]}{[C_2H_3O_2^-]} = (1.8 \times 10^{-5})\left(\frac{0.12}{0.080}\right) = 2.7 \times 10^{-5}$$

(which is small compared to 0.12 or 0.080).

$$pH = -(-5 + 0.43) = \mathbf{4.57}$$

Notice that the pH does decrease, as you would expect when acid is added, but only slightly (from 4.74 to 4.57). If the 0.020 mol/l of HCl had been added to pure water, the pH would have changed from 7.0 to 1.7 [See Problem Example 17-14(d)].

PROBLEM EXAMPLE 17-19

To determine the K_a of a weak acid, a student took 20.00 ml of a solution of the acid and titrated it with a base. He found that it took 32.00 ml of base to neutralize the acid. He then took a new 20.00 ml of acid and added exactly 10.00 ml of the same base to it. He measured the pH of this solution and found it to be 4.2. What is the value of K_a for the weak acid?

SOLUTION:

The equilibrium reaction for the ionization of HA is

$$HA_{(aq)} \rightleftharpoons H^+ + A^-$$

and the ionization constant is

$$K_a = \frac{[H^+][A^-]}{[HA]}$$

Therefore, if we know the $[H^+]$ and the ratio of $[A^-]/[HA]$, we can calculate the value of K_a, without knowing the actual value of either $[A^-]$ or $[HA]$. Since we know that it takes 32.00 ml of base to neutralize all the acid, the fraction neutralized (which is equal to the fraction of salt formed) is 10.00/32.00, and the fraction of acid remaining is 22.00/32.00. Therefore, the ratio of

$$\frac{[A^-]}{[HA]} = \frac{10.00}{22.00}$$

If pH = 4.2,

$$[H^+] = 10^{-4.2} = 10^{0.8} \times 10^{-5} = 6.3 \times 10^{-5}$$

Thus,

$$K_a = (6.3 \times 10^{-5})\left(\frac{10.00}{22.00}\right) = \mathbf{2.9 \times 10^{-5}}$$

SUMMARY

1. The *solubility product constant,* K_{sp}, is a special type of equilibrium constant that gives a quantitative measurement of the solubility of slightly soluble electrolytes. Since this is an example of heterogeneous equilibria, the concentration of solid is not included in the expression of K_{sp}.
2. When the product of the concentrations of the ions, each raised to their respective powers, is equal to the value of the solubility product constant, then a saturated solution exists. When the product of these concentrations of the ions, each raised to their respective powers, is less than the value of the solubility product constant, then an unsaturated solution exists; when this product is greater than the value of the solubility product constant, a supersaturated solution exists and precipitation normally occurs. Therefore, the solubility product constant gives the limit to the solubility of the electrolyte in water at a given temperature.
3. Once equilibrium is established between the solid and its ions in solution, K_{sp} is equal to the *ion product.* Therefore, using K_{sp} values, the solubility of a substance can be calculated, or if the solubility can be found, the K_{sp} can be calculated. If the concentration of one of the ions present at equilibrium is changed, the equilibrium will shift. For example, if a common ion is added, the solubility is decreased. If something is added that causes the concentration of one of the ions to be decreased, such as by the formation of a complex ion or a weak acid or base, the solubility will increase and the precipitate can be dissolved.
4. For weak acids or bases, an equilibrium is established between the molecules in solution and the ions formed. As with other types of equilibrium constants, the smaller the constant, called K_i (or K_a or K_b), the less reaction there is to the right and hence the weaker the acid or base. In dilute solution, the $[H_2O]$ remains relatively constant and is therefore included with the K_{eq} (that is, $K_{eq}[H_2O] = K_i$) and is not part of the expression for the ionization constant. From the value of the K_i, and the initial concentration of acid or base, the $[H^+]$ or $[OH^-]$ can be calculated, as well as the percent of ionization. Conversely, if the initial concentration of acid or base is known, along with the percent of ionization, the value of K_i can be calculated. Also, these calculations can be combined with ion concentrations in solution and K_{sp} values to predict if precipitates would form for a given hydroxide in a solution of a weak base.

5. In any water solution, $[H^+][OH^-] = 1 \times 10^{-14}$; therefore, if one of these is known, the other can be calculated. In a neutral solution, $[H^+] = [OH^-] = 1 \times 10^{-7}$. If the $[H^+]$ is greater than 1×10^{-7} (that is, a smaller negative exponent), the solution is acidic since $[H^+] > [OH^-]$; if $[H^+]$ is less than 1×10^{-7} (that is, a larger negative exponent), the solution is basic since $[OH^-] > [H^+]$. The pH of a solution, which gives a measure of the acidity, is equal to $-\log [H^+]$. In any water solution, pH + pOH = 14. Therefore, in any solution for which the $[H^+]$ or $[OH^-]$ is known, or can be calculated, the pH or pOH can be found. In a neutral solution, pH = 7. If pH is less than 7, the solution is acidic; if the pH is greater than 7, it is basic.

Summary

6. A *buffer solution*, which consists of a weak acid or base and a salt of the acid or base, holds the pH of a solution relatively constant as more acid or base is added to the solution. Calculations involving buffer solutions therefore involve the ionization of the weak acid or base with a common ion. In some cases, however, it is only necessary to consider the ratio of [ion] to [acid or base] since

$$K_i = \frac{[H^+][A^-]}{[HA]}$$

for a weak acid. The $[A^-]$ is simply equal to the concentration of salt in the buffer mixture. If the salt is formed during neutralization, the $[A^-]$ is the concentration of acid that has been neutralized, and the [HA] is the concentration of acid that remains.

Upon completion of this chapter, you should be able to do the following:

1. Given a list of slightly soluble substances and the K_{sp} values for these substances, tell which is the most soluble or which is the least soluble. (Exercise 3)
2. Given a slightly soluble substance, predict how addition of another substance (such as a common ion, or a substance that will form a weak acid, a less soluble substance, or a complex ion) would affect the solubility. (Exercise 4)
3. Given the name or formula of any slightly soluble substance, write the equilibrium reaction representing dissociation, and the expression for the solubility product constant. (1)
4. Given the K_{sp}, calculate the solubility (in any units), including cases where a common ion is added. (2, 3)
5. Given the solubility of a slightly soluble substance (in any units), calculate the K_{sp}. (4)
6. Given the concentration of ions in a mixture, and K_{sp} values, tell from the ion product if you would expect a precipitate to form. (5)

7. Given a list of weak acids or bases and the values of the ionization constant, choose the strongest or weakest acid or base. (Exercise 5)
8. Given the concentration of a weak acid or base and the percent of ionization, calculate the value of the ionization constant. (6)
9. Given the ionization constant for any weak acid or base and the concentration of the acid or base (in any units), calculate the $[H^+]$ or $[OH^-]$, including cases where a common ion is added. (7, 8, 9)
10. Given the concentration of any weak acid or base (and the ionization constant), calculate the percent of ionization. (10)
11. Given the $[H^+]$ of a weak acid solution (or the $[OH^-]$ of a weak base solution) or the percent ionization, calculate the amount of acid (or base) necessary to give that $[H^+]$ (or $[OH^-]$) or percent ionization. (11 to 15)
12. Given any of the following, tell if the solution is acidic, basic, or neutral: $[H^+]$, $[OH^-]$, pH or pOH. (Exercise 6)
13. Given the $[H^+]$, $[OH^-]$, or concentration of a solution of any acid or base, calculate the pH of the solution. (17)
14. Given any one of the following, calculate any of the others: $[H^+]$, $[OH^-]$, pH, and pOH. (16, 18)
15. Given any acid or base, tell how the pH of the solution will change upon addition of a given salt, acid, or base. (Exercise 7)
16. Given any buffer mixture, calculate
    (a) the pH of the mixture, (19a)
    (b) the pH after a given amount of water has been added, (19b)
    (c) the pH after a given amount of strong acid or strong base has been added. (19c, d)
17. Given any of the terms listed in Exercise 1, give a definition or explanation in your own words.

EXERCISES

1. Define or explain each of the following:
   (a) K_{sp}
   (b) K_i
   (c) K_a
   (d) K_b
   (e) K_W
   (f) pH
   (g) pOH
   (h) pK_a
   (i) precipitate
   (j) solubility
   (k) ion product
   (l) buffer
2. Distinguish between:
   (a) K_{sp} and K_i
   (b) K_a and K_b
   (c) K_i and K_W
   (d) pH and pOH
3. Choose the most soluble and the least soluble one:
   (a) PbF_2
   (b) $PbCl_2$
   (c) $PbBr_2$
   (d) PbI_2
   (e) $Pb(OH)_2$
4. Which of the following are more soluble in acid than in water?
   (a) $CaCO_3$
   (b) $PbCl_2$
   (c) CaF_2
   (d) $Fe(OH)_2$
   (e) AgCl

5. Choose the strongest acid and the weakest acid:
(a) HF
(b) $HC_2H_3O_2$
(c) HNO_2
(d) HCN
(e) $HClO_2$

6. Classify each of the following as acid, basic, or neutral.
(a) $[H^+] = 1 \times 10^{-5}$
(b) $[OH^-] = 1 \times 10^{-8}$
(c) pH = 9.5
(d) pOH = 8.2

7. Tell how addition of each of the following will affect the pH of a solution of acetic acid:
(a) sodium acetate
(b) hydrochloric acid
(c) sodium hydroxide
(d) sodium chloride

PROBLEMS (Use tables of constants in Appendix VI)

Solubility product constant (K_{sp})

1. Write the expression for the solubility product constant (K_{sp}) for each of the following salts:
(a) AgCl
(b) $PbCl_2$
(c) Bi_2S_3
(d) $Ba_3(PO_4)_2$
(e) HgS
(f) Ag_2CrO_4
(c) CaF_2
(h) $Pb_3(AsO_4)_2$

2. From the solubility product constant (K_{sp}) for each of the following compounds, calculate the solubility of the compound in moles per liter.
(a) FeS
(b) $BaSO_4$
(c) BaF_2
(d) $Mn(OH)_2$
(e) $CaCO_3$
(f) HgS
(g) $Fe(OH)_2$
(h) Ag_2CrO_4

3. From the K_{sp} for each, calculate the molar solubility (solubility in moles per liter) in the solution indicated.
(a) AgI in 0.010 *M* KI
(b) AgBr in 0.20 *M* Ag_2SO_4
(c) $Pb(OH)_2$ in 0.10 *M* $Pb(NO_3)_2$
(d) AgCl in 0.010 *M* HCl
(e) AgI in 0.010 *M* CaI_2
(f) $Mg(OH)_2$ in 0.20 *M* $MgCl_2$

4. From the following solubilities, calculate the value of the K_{sp} for each:
(a) 4.0×10^{-5} mol $BaCO_3/l$
(b) 1.3×10^{-4} mol $Mg(OH)_2/l$
(c) 1.36×10^{-17} mmol Ag_2S/ml
(d) 1.30×10^{-6} mg $Cr(OH)_3/ml$
(e) 1.35×10^{-5} g AgBr/100 ml
(f) 2.0×10^{-13} mol PbS/l
(g) 3.7×10^{-5} mol $Mn(OH)_2/l$
(h) 2.0×10^{-4} mmol CaF_2/ml
(i) 4.4 mg $PbCl_2/ml$
(j) 9.0×10^{-4} g $BaSO_4$/100 ml

5. For each of the following, tell if you would expect a precipitate to form. Explain each answer.
(a) Two solutions are mixed so that the final concentration of Ag^+ ion is 1.0×10^{-5} and Cl^- ion is 1.0×10^{-3}.
(b) To a solution containing 0.010 *M* Al^{3+} is added enough base to make the $[OH^-]$ 1.0×10^{-12} *M*.
(c) 100 ml of 2.0×10^{-5} *M* Ag^+ is mixed with 100 ml of 2.0×10^{-8} *M* Br^-.
(d) Two solutions are mixed so that the final concentration of Ba^{2+} is 2.0×10^{-5} *M* and CO_3^{2-} is 1.0×10^{-9}.

(e) To a solution containing 0.010 *M* Mg^{2+} is added enough base to make the $[OH^-]$ 2.0×10^{-8}.

(f) 100 m*l* of 2.0×10^{-6} *M* Ag^+ is mixed with 100 m*l* of 1.0×10^{-5} *M* Br^-.

Ionization constants (K_i)

6. Calculate the value of the ionization constant (K_i) for each of the following:
   (a) 0.10 *M* acetic acid is 1.35% ionized.
   (b) 0.010 *M* aqueous ammonia is 4.3% ionized.
   (c) 0.10 *M* hydrofluoric acid is 7.9% ionized.
   (d) 0.010 *M* acetic acid is 4.3% ionized.
   (e) 0.10 *M* aqueous ammonia is 1.35% ionized.
   (f) 0.10 *M* nitrous acid is 6.5% ionized.
7. Calculate the $[H^+]$ of each of the following:
   (a) 0.10 *M* $HC_2H_3O_2$
   (b) 0.050 *M* HCN
   (c) 0.010 *M* $HC_2H_3O_2$
   (d) 0.020 *M* HCN
8. Calculate the $[OH^-]$ of each of the following:
   (a) 0.010 *M* NH_3
   (b) 0.10 *M* NH_3
9. Calculate the $[H^+]$ of each of the following solutions:
   (a) A solution containing 0.020 *M* acetic acid and 0.010 *M* sodium acetate.
   (b) A solution containing 0.010 *M* hydrofluoric acid and 0.010 *M* calcium fluoride.
   (c) A solution made by mixing 150 m*l* of 0.020 *M* HCN and 250 m*l* of 0.030 *M* NaCN.
   (d) A solution containing 0.0250 *M* HF and 0.010 *M* NaF.
   (e) A solution containing 0.10 *M* acetic acid and 0.010 *M* calcium acetate.
   (f) A solution made by mixing 350 m*l* of 0.250 *M* HCN and 250 m*l* of 0.40 *M* NH_4CN.
10. Calculate the percent of ionization of each of the following in the given solution:
    (a) 0.10 *M* NH_3
    (b) 0.010 *M* $HC_2H_3O_2$
    (c) 0.010 *M* HCN
    (d) 0.0120 *M* NH_3
11. Calculate the number of moles of HCN that must be present in 1 *l* of solution to give a $[H^+]$ of
    (a) 1.0×10^{-5}
    (b) 2.50×10^{-6}
12. Calculate the number of grams of $HC_2H_3O_2$ that must be added to 100 m*l* of water to give a $[H^+]$ (assuming constant volume) of
    (a) 2.50×10^{-3}
    (b) 1.0×10^{-2}
13. Calculate the number of liters of NH_3, at STP, that must be added to 1 *l* of water (assuming constant volume) to give a $[OH^-]$ of
    (a) 2.40×10^{-3}
    (b) 1.0×10^{-2}
14. Calculate the number of grams of ammonium chloride that must be added to 200 m*l* of 0.20 *M* aqueous ammonia to give a $[OH^-]$ of
    (a) 8.0×10^{-6}
    (b) 1.20×10^{-5}

15. Calculate the molar concentration of each of the following solutions:
  (a) acetic acid that is 1.3% ionized
  (b) HCN that is 0.0250% ionized
  (c) aqueous ammonia that is 1.8% ionized
  (d) nitrous acid that is 4.5% ionized

Ionization of water and pH

16. Calculate the $[H^+]$ of each of the following:
  (a) $[OH^-] = 1.0 \times 10^{-4}$
  (b) $[OH^-] = 2.7 \times 10^{-9}$
  (c) pH = 5.0
  (d) pH = 8.3
  (e) pOH = 8.0
  (f) pOH = 4.5
  (g) $[OH^-] = 1.0 \times 10^{-8}$
  (h) $[OH^-] = 7.3 \times 10^{-4}$
  (i) pH = 8.0
  (j) pH = 4.7
  (k) pOH = 5.0
  (l) pOH = 8.3

17. Calculate the pH of each of the following solutions:
  (a) $[H^+] = 1.0 \times 10^{-5}$
  (b) $[H^+] = 2.30 \times 10^{-9}$
  (c) 0.020 *M* HCl
  (d) 0.020 *M* $HC_2H_3O_2$
  (e) $[OH^-] = 1.0 \times 10^{-4}$
  (f) $[OH^-] = 3.5 \times 10^{-8}$
  (g) 0.15 *M* NaOH
  (h) 0.15 *M* aqueous ammonia
  (i) $[H^+] = 1.0 \times 10^{-8}$
  (j) $[H^+] = 4.25 \times 10^{-4}$
  (k) 0.035 *M* HCl
  (l) 0.450 *M* NaOH
  (m) 0.125 *M* acetic acid
  (n) 0.72 *M* NH_3
  (o) $[OH^-] = 1.0 \times 10^{-9}$
  (p) $[OH^-] = 4.2 \times 10^{-5}$

18. Calculate the pH of each of the following solutions:
  (a) saturated $Mg(OH)_2$
  (b) 0.20 *M* HCN in 0.10 *M* NaCN
  (c) 0.12 *M* NH_3 in 0.25 *M* NH_4Cl
  (d) 4.2 g NaOH in 450 mℓ of solution
  (e) 250 mℓ 3.0 *M* HCl + 250 mℓ 3.0 *M* NaOH
  (f) 45 mℓ 0.10 *M* sodium hydroxide + 50 mℓ 0.150 *M* acetic acid
  (g) saturated $Ca(OH)_2$
  (h) 2.5 *M* aqueous ammonia in 4.2 *M* ammonium chloride
  (i) 1.25 *M* HCN in 1.00 *M* NaCN
  (j) 0.35 g of potassium hydroxide in 55 mℓ of solution
  (k) 45 mℓ of 0.25 *M* NaOH + 85 mℓ of 0.30 *M* $HC_2H_3O_2$
  (l) 450 mℓ 2.50 *M* hydrochloric acid + 250 mℓ 1.50 *M* sodium hydroxide

Buffer solutions

19. A buffer is made by mixing 200 mℓ of 1.50 *M* acetic acid and 150 mℓ of 2.0 *M* sodium acetate. Calculate the pH of the solution in each of the following:
  (a) The buffer solution.
  (b) The buffer solution after 600 mℓ of water is added to the mixture.
  (c) The solution after 50 mℓ of 0.40 *M* HCl is added to the solution.
  (d) The solution after 50 mℓ of 0.40 *M* NaOH is added to the solution.

(e) The mixture before anything else is added.

(f) The solution after 150 ml of 0.50 M HCl is added.

(g) The solution after 150 ml of 0.50 M NaOH is added.

(h) The solution after 400 ml of water is added.

General problems

20. Calculate the number of moles of HCN that must be present in 1 l of solution to give a $[H^+]$ of
(a) 1.0×10^{-2}
(b) 1.2×10^{-3}

21. Calculate the number of grams of NH_4NO_3 that must be added to 350 ml of 0.42 M NH_3 solution to give a $[OH^-]$ of
(a) 2.0×10^{-6}
(b) 9.5×10^{-7}

22. Calculate the molar solubility of each of the following:
(a) calcium carbonate
(b) barium fluoride
(c) plumbous chloride in 0.25 M hydrochloric acid
(d) chromium(II) hydroxide in a solution with a pH of 9.0
(e) iron(II) hydroxide in 0.15 M aqueous ammonia
(f) barium carbonate
(g) calcium fluoride
(h) lead(II) chloride in 1.5 M hydrochloric acid
(i) stannous hydroxide in 0.32 M aqueous ammonia
(j) mercury(II) hydroxide in a solution with a pH of 8.0

23. Calculate the pH of each of the following:
(a) 0.35 M nitric acid
(b) 4.82 g of potassium hydroxide in 34.5 ml of solution
(c) 0.72 M hydrofluoric acid
(d) 2.45 M aqueous ammonia that contains 1.84 g of ammonium chloride in 750 ml of solution
(e) 150 ml of 0.12 M NaOH solution is added to 200 ml of 0.18 M acetic acid
(f) 0.15 M hydrochloric acid
(g) 0.320 g of sodium hydroxide in 18.5 ml of solution
(h) 1.30 M formic acid
(i) 200 ml 0.20 M HCl is added to 350 ml of 0.32 M aqueous ammonia
(j) 0.35 M acetic acid that contains 0.82 g of sodium acetate in 45 ml of solution

24. Calculate the $[OH^-]$ of each of the following solutions:
(a) 1.8 M HCl
(b) 0.342 g of KOH in 85.0 ml of solution
(c) 1.25 M formic acid
(d) a solution with a pOH of 8.20
(e) a solution with a pH of 9.60
(f) 2.5 M HNO_3
(g) 3.20 g of NaOH in 650 ml of solution
(h) 0.72 M hydrocyanic acid
(i) a solution with a pOH of 4.50
(j) a solution with a pH of 3.20

25. Calculate the minimum concentration of Pb^{2+} ion necessary to just begin precipitation of
(a) $PbBr_2$ from a solution that contains 0.012 M Br^-.
(b) PbF_2 from a solution that contains 0.025 M F^-.

26. A 0.010 M solution of the weak acid HA has a pH of 4.20; calculate the value of K_i (K_a) for HA.

27. A 0.025 M solution of the weak acid HX has a pH of 5.70; calculate the value of K_i (K_a) for HX.

References

REFERENCES

GUYON, JOHN C., and BERWYN E. JONES, *Introduction to Solution Equilibrium*, Boston: Allyn and Bacon, 1969.

HOGNESS, T. R., WARREN C. JOHNSON, and ALFRED R. ARMSTRONG, *Qualitative Analysis and Chemical Equilibrium*, 5th ed., New York: Holt, Rinehart and Winston, Inc., 1966.

Objectives

1. To understand the ionization of H_2S (as an example of a polyprotic acid) in terms of the relationships between the ionization constant, $[H^+]$, and $[S^{2-}]$, and to be able to do calculations involving $[H^+]$, $[S^{2-}]$, and K_{sp} of sulfides.
2. To understand what hydrolysis is and to be able to
   (a) write ionic equations to represent hydrolysis of salts,
   (b) predict whether solutions of salts will be acidic, basic, or neutral,
   (c) derive hydrolysis constants for the salt of a weak acid and strong base, a weak base and strong acid, or a weak acid and weak base, and
   (d) do calculations involving hydrolysis reactions.

Eighteen

Equilibria in Solution II—Polyprotic Acids and Hydrolysis of Salts

In this chapter* we shall consider two other types of equilibrium in solution. The first of these deals with acids that contain more than one replaceable hydrogen ion and, therefore, ionize in steps. The second type that we shall consider here involves the reaction of a salt with water to form either a weak acid, or a weak base, or both. This is known as *hydrolysis.* These are somewhat more complex examples of equilibrium in solution. We shall see, however, that the basic principles and the calculations are very similar to the material considered in Chapter 17.

18-1 Polyprotic Acids—Ionization of Hydrosulfuric Acid and Precipitation of Sulfides

It was pointed out in Chapter 14 that any **polyprotic acid** (acid with more than one replaceable hydrogen ion) ionizes in steps. The second step of the ionization is always much weaker than the first, usually by a power of about 10^{-5} or 10^{-6}.

One common example of this is the compound hydrogen sulfide, H_2S, which forms the weak acid hydrosulfuric acid when in aqueous solution and ionizes according to the equations

(18-1) $$H_2S \rightleftharpoons H^+ + HS^-$$

(18-2) $$HS^- \rightleftharpoons H^+ + S^{2-}$$

Since many sulfides are very insoluble, this is a useful compound for the qualitative analysis of ions. Even though only a small amount of S^{2-} ion is produced here, many sulfides will precipitate out upon addition of H_2S due to the small K_{sp} values of the sulfides.

*The material in this chapter is optional. It can be omitted without loss of continuity.

The ionization constants for these two reactions are

(18-3) $$K_1 = \frac{[H^+][HS^-]}{[H_2S]} = 1.0 \times 10^{-7}$$

and

(18-4) $$K_2 = \frac{[H^+][S^{2-}]}{[HS^-]} = 1.3 \times 10^{-13}$$

The equation for the overall ionization is the sum of these two steps:

$$\begin{array}{rcl} H_2S & \rightleftharpoons & H^+ + HS^- \\ HS^- & \rightleftharpoons & H^+ + S^{2-} \\ \hline H_2S + HS & \rightleftharpoons & H^+ + HS^- + H^+ + S^{2-} \end{array}$$

This can be simplified to

(18-5) $$H_2S \rightleftharpoons 2\,H^+ + S^{2-}$$

The overall ionization constant, which is obtained by multiplying $K_1 \times K_2$, is

$$K_i = K_1 \times K_2 = \frac{[H^+][HS^-][H^+][S^{2-}]}{[H_2S][HS^-]} = (1.0 \times 10^{-7})(1.3 \times 10^{-13})$$
$$= 1.3 \times 10^{-20}$$

or, more simply,

(18-6) $$K_i = \frac{[H^+]^2[S^{2-}]}{[H_2S]} = 1.3 \times 10^{-20}$$

In using this overall ionization (Eq 18-5) and the ionization constant (Eq. 18-6), it is essential to keep in mind that the $[S^{2-}]$ is not equal to one-half the $[H^+]$; that is, even though we can write an overall equilibrium expression and calculate a value of the ionization constant for it, the ionization goes in two steps and the second step is much smaller than the first. In fact, looking at the two individual ionization constants it can be shown that the $[S^{2-}]$ is approximately equal to the second ionization constant; that is, in the expression

$$K_2 = \frac{[H^+][S^{2-}]}{[HS^-]}$$

since the second step of ionization is very slight compared to the first, $[H^+]$ and $[HS^-]$ are about equal and therefore $[S^{2-}]$ is approximately equal to K_2.

Using the overall equilibrium constant, we can calculate the sulfide ion concentration if the $[H^+]$, or pH, is known. A saturated solution of H_2S, at room temperature and 1 atm pressure, contains 0.10 mol/l of H_2S. Using this fact, along with the ionization constant for H_2S and the known $[H^+]$, we can calculate the $[S^{2-}]$. Then we can use this to predict what sulfides would precipitate out of a particular solution with a given concentration of various metal ions. Consider the following calculation.

PROBLEM EXAMPLE 18-1

A solution containing 0.10 mol/l of Cu^{2+} ion and 0.10 mol/l of Mn^{2+} is made 0.020 M in H^+ and saturated with H_2S.
(a) Calculate the $[S^{2-}]$.
(b) Will CuS precipitate out?
(c) Will MnS precipitate out?

SOLUTION:

(a) Using the equilibrium reaction

$$H_2S \rightleftharpoons 2\,H^+ + S^{2-}$$

and the ionization constant

$$K_i = \frac{[H^+]^2[S^{2-}]}{[H_2S]} = 1.3 \times 10^{-20}$$

If we know that

$$[H^+] = 0.020\,M = 2.0 \times 10^{-2}\,M$$
$$[H_2S] = 0.10\,M = 1.0 \times 10^{-1}\,M$$

we can calculate $[S^{2-}]$ by putting these values into the expression for the ionization constant and solving for $[S^{2-}]$.

$$\frac{(2.0 \times 10^{-2})^2\,[S^{2-}]}{(1.0 \times 10^{-1})} = 1.3 \times 10^{-20}$$

$$[S^{2-}] = \frac{1.3 \times 10^{-21}}{4.0 \times 10^{-4}} = 0.32 \times 10^{-17} = \mathbf{3.2 \times 10^{-18}}$$

(b) The K_{sp} for CuS (from Table 1, Appendix VI) is

$$K_{sp} = [Cu^{2+}][S^{2-}] = 4.0 \times 10^{-36}$$

The ion product in this solution (found by taking $[Cu^{2+}][S^{2-}]$) is

$$(1.0 \times 10^{-1})(3.2 \times 10^{-18}) = 3.2 \times 10^{-19}$$

Since this is greater than the K_{sp}, **CuS will precipitate.**

(c) The K_{sp} for MnS is

$$K_{sp} = [Mn^{2+}][S^{2-}] = 4.0 \times 10^{-14}$$

The ion product for this solution (found by taking $[Mn^{2+}][S^{2-}]$) is

$$(1.0 \times 10^{-1})(3.2 \times 10^{-18}) = 3.2 \times 10^{-19}$$

Since this is less than the K_{sp}, **MnS will not precipitate.**

Using this ionization constant for H_2S, we can also calculate how to control the pH in such a way as to prevent the precipitation of certain sulfides from a solution while precipitating other sulfides. Consider the following example.

PROBLEM EXAMPLE 18-2

What is the maximum pH that a solution can have if it contains 0.010 *M* Mn^{2+} ion and is saturated with H_2S and yet not precipitate out MnS?

SOLUTION:

From the K_{sp} for MnS

$$K_{sp} = [Mn^{2+}][S^{2-}] = 4.0 \times 10^{-14}$$

the maximum $[S^{2-}]$ that this can have without having the ion product exceed the K_{sp} is

$$(1 \times 10^{-2})[S^{2-}] = 4.0 \times 10^{-14}$$

$$[S^{2-}] = 4.0 \times 10^{-12}$$

To find a sulfide concentration of 4.0×10^{-12} mol/*l* in a saturated solution of H_2S, the $[H^+]$ would be (using the K_i for H_2S)

$$\frac{[H^+]^2(4.0 \times 10^{-12})}{1.0 \times 10^{-1}} = 1.3 \times 10^{-20}$$

$$[H^+]^2 = \frac{1.3 \times 10^{-21}}{4.0 \times 10^{-12}} = 0.32 \times 10^{-9} = 3.2 \times 10^{-10}$$

$$[H^+] = \sqrt{3.2 \times 10^{-10}} = 1.8 \times 10^{-5}$$

The pH corresponding to this $[H^+]$ is

$$pH = -(-5 + 0.26) = 4.74$$

Therefore, if the pH were any greater than 4.74, the $[H^+]$ would be less than 1.8×10^{-5}, the $[S^{2-}]$ would be greater than 4.0×10^{-2}, and MnS would precipitate out.

18-2 Hydrolysis of Salts

The term *hydrolysis* simply means reaction with water. What we are considering in this case is the hydrolysis of salts which form weak acids, weak bases, or both. Since most acids and bases are weak, we can therefore say that most ions in solution are hydrolyzed. For example, the acetate ion in water will hydrolyze to form acetic acid, according to the equation

(18-7) $$C_2H_3O_2^- + H_2O \rightleftharpoons HC_2H_3O_2 + OH^-$$

Another example would be the ammonium (NH_4^+) ion, which hydrolyzes to form ammonia (NH_3) and hydronium ion, according to the equation

$$NH_4^+ + H_2O \rightleftarrows NH_3 + H_3O^+$$

In this case, we must include the water and use the H_3O^+ ion or the reaction would simply be

(18-8) $$NH_4^+ \rightleftarrows NH_3 + H^+$$

Therefore, an ion that forms a weak acid or a weak base will undergo hydrolysis. There are three types that we must consider here: salts that are composed of the ions of a weak acid and a strong base, salts of a strong acid and a weak base, and finally salts of a weak acid and a weak base. The anions that form strong acids (that is, NO_3^-, ClO_4^-, Cl^-, Br^-, and I^-*) and the cations that form strong bases (ions of Group IA and Ca^{2+}, Sr^{2+}, and Ba^{2+}) do not undergo hydrolysis. Therefore, the salt of a strong acid and a strong base does not react with water, other than simply to dissociate into its ions.

Hydrolysis of the Salt of a Weak Acid and a Strong Base

The salt of a weak acid and strong base will hydrolyze to form a basic (alkaline) solution. For example, sodium acetate hydrolyzes according to the equation

(18-9) $$\cancel{Na^+} + C_2H_3O_2^- + H_2O \rightleftarrows HC_2H_3O_2 + \cancel{Na^+} + OH^-$$

that is, the $[OH^-]$ of the solution is increased and thus the $[H^+]$ must be decreased, and therefore the solution is basic (alkaline). This can be thought of as causing a shift in the $H_2O \rightleftharpoons H^+ + OH^-$ equilibrium by removing H^+.

Recall that salts in solution and strong acids and bases are written in ionic form, whereas weak acids or bases and water are written in molecular form in an ionic equation. It is only when we are specifically concerned with the ionization of a weak acid, for example, that we show its ionization—which is actually what the reverse reaction in Eq. (18-9) shows. What is actually happening here is that the acetate ion is reacting with water to form acetic acid but as soon as the acetic acid and hydroxide ion is formed (forward reaction), they start to neutralize each other (reverse reaction) and very quickly an equilibrium is established.

*Even sulfate, SO_4^{2-}, undergoes some hydrolysis due to the fact that sulfuric acid ionizes in two steps, and the second step is rather weak ($K_i = 1.2 \times 10^{-2}$). Therefore, there is some hydrolysis according to the equation

$$SO_4^{2-} + H_2O \rightleftarrows HSO_4^- + OH^-$$

The equilibrium constant for this reaction would be

$$K_{eq} = \frac{[HC_2H_3O_2][OH^-]}{[C_2H_3O_2^-][H_2O]}$$

But again, the $[H_2O]$ remains practically constant and is therefore included with the K_{eq} to give a new constant, called the *hydrolysis constant, K_h:*

(18-10) $$K_h = \frac{[HC_2H_3O_2][OH^-]}{[C_2H_3O_2^-]}$$

Notice that the Na^+ is not included here since it is not part of the net ionic equation. Even if it were put into the expression for K_h, it would cancel out.

The **hydrolysis constant** is related to the ionization constant for acetic acid by the equation

$$K_h = \frac{K_W}{K_a}$$

that is, the hydrolysis constant for any salt of the cation of a strong base and acetate ion is the same and is related to the K_a for acetic acid. This relationship can be shown quite simply by substituting into Eq. (18-10) the expression

$$[OH^-] = \frac{K_W}{[H^+]}$$

which gives

$$K_h = \frac{[HC_2H_3O_2]K_W}{[C_2H_3O_2^-][H^+]} = \frac{[HC_2H_3O_2]}{[H^+][C_2H_3O_2^-]} \times K_W = \frac{1}{K_a} \cdot K_W = \frac{K_W}{K_a}$$
$$= \frac{1.0 \times 10^{-14}}{1.8 \times 10^{-5}} = 5.6 \times 10^{-10}$$

Thus, the general equation for the hydrolysis constant for a salt of a strong base and a weak acid is

(18-11) $$K_h = \frac{K_W}{K_a}$$

Hydrolysis of the Salt of a Weak Base and a Strong Acid

A salt composed of the ions of a weak base and a strong acid will hydrolyze to form an acidic solution. For example, ammonium chloride hydrolyzes according to the equation

(18-12) $$NH_4^+ + \cancel{Cl^-} + H_2O \rightleftarrows NH_3 + H_3O^+ + \cancel{Cl^-}$$

In this case the ammonium ion reacts with water to form some ammonia (NH_3) and hydronium ion (forward reaction), but as soon as they are formed, they begin to neutralize each other (recall that NH_3 is a base) to form the ammonium ion and water (reverse reaction) and again an equilibrium is established. The hydrolysis constant for this is

(18-13) $$K_h = \frac{[NH_3][H_3O^+]}{[NH_4^+]}$$

Substituting $K_W/[OH^-]$ for $[H_3O^+]$ (from the ionization constant for water) gives

(18-14) $$K_h = \frac{[NH_3]K_W}{[NH_4^+][OH^-]} = \frac{K_W}{K_b} = \frac{1.0 \times 10^{-14}}{1.8 \times 10^{-5}} = 5.6 \times 10^{-10}$$

That is, the hydrolysis constant for any salt of a weak base and a strong acid is K_W divided by the ionization constant of the weak base (K_b);

(18-15) $$K_h = \frac{K_W}{K_b}$$

Hydrolysis of the Salt of a Weak Base and a Weak Acid

A salt composed of the ions of a weak base and a weak acid will hydrolyze to form a solution that may be acidic, basic, or neutral, depending on the strength (that is, the ionization constant) of the acid and base formed. For example, the salt ammonium acetate will hydrolyze according to the equation

(18-16) $$NH_4^+ + C_2H_3O_2^- + H_2O \rightleftharpoons NH_3 + H_2O + HC_2H_3O_2$$

or, more simply

(18-17) $$NH_4^+ + C_2H_3O_2^- \rightleftharpoons NH_{3(aq)} + HC_2H_3O_{2(aq)}$$

Note that the water is not included in the net ionic equation. In this particular case the solution would be neutral since the ionization constant for acetic acid and aqueous ammonia are the same; however, this does not mean that no hydrolysis takes place. In fact, we shall see when we calculate the value of K_h for this salt that there is more hydrolysis here than in either of the two previous cases; that is, K_h is a larger value for this salt than for either sodium acetate or ammonium chloride. This indicates that more forward reaction, or more hydrolysis, takes place.

The hydrolysis constant for this, from Eq. (18-17), would be

(18-18) $$K_h = \frac{[NH_3][HC_2H_3O_2]}{[NH_4^+][C_2H_3O_2^-]}$$

To find this in terms of K_a and K_b, we must multiply the equation by $[H^+][OH^-]/[H^+][OH^-]$ to give

$$K_h = \frac{[NH_3][HC_2H_3O_2][H^+][OH^-]}{[NH_4^+][C_2H_3O_2^-][H^+][OH^-]} = \frac{[NH_3]}{[NH_4^+][OH^-]} \times \frac{[HC_2H_3O_2]}{[H^+][C_2H_3O_2^-]}$$

$$\times [H^+][OH^-] = \frac{1}{K_b} \times \frac{1}{K_a} \times K_W = \frac{K_W}{K_a \cdot K_b}$$

$$= \frac{1.0 \times 10^{-14}}{(1.8 \times 10^{-5})(1.8 \times 10^{-5})} = 3.1 \times 10^{-5}$$

That is, the hydrolysis constant for the salt of a weak acid and a weak base is K_W divided by the ionization constant of the weak acid (K_a) times the ionization constant of the weak base (K_b)

(18-19) $$K_h = \frac{K_W}{K_a K_b}$$

To determine whether a solution of a salt of a weak acid and a weak base would be acidic, basic, or neutral, we must consider the K_a of the acid and the K_b of the base. In the case of ammonium acetate, the solution turns out to be neutral since the K_a and K_b just happen to be the same (1.8×10^{-5} in each case). This means that the acid and the base are of equal strength. For a salt like ammonium cyanide, however, the solution would be basic. The K_b for aqueous ammonia (1.8×10^{-5}) is greater than the K_a for hydrocyanic acid (4.0×10^{-10}). This means that the base is stronger than the acid, and thus the solution would be basic.

For a salt of a strong acid and a strong base there is no hydrolysis (no reaction with water to form a weak acid or weak base). Therefore, solutions of this type would be neutral. For example, when sodium chloride is dissolved in water, the solution is neutral and the pH is 7.0, just as it is in pure water.

Calculations Involving Hydrolysis

These calculations are essentially the same as we have been doing with weak acids and bases, except that the value of the hydrolysis constant must be calculated from K_W and K_a or K_b. As with all the other equilibrium problems that we have done, the first step is to write out the equilibrium reaction and the expression for the equilibrium constant, K_h in this case. Consider the following problem examples.

PROBLEM EXAMPLE 18-3

Calculate the $[H^+]$ of a 0.10 *M* solution of sodium acetate.

SOLUTION:

The equilibrium reaction is

$$Na^+ + C_2H_3O_2^- + H_2O \rightleftharpoons HC_2H_3O_2 + OH^- + Na^+$$

or, more simply,

$$C_2H_3O_2^- + H_2O \rightleftharpoons HC_2H_3O_2 + OH^-$$

The expression for the hydrolysis constant and the calculated value is

$$K_h = \frac{[HC_2H_3O_2][OH^-]}{[C_2H_3O_2^-]} = \frac{K_W}{K_a} = \frac{1.0 \times 10^{-14}}{1.8 \times 10^{-5}} = 5.6 \times 10^{-10}$$

Letting X be the amount of hydrolysis gives

$$[C_2H_3O_2^-] = 0.10 - X$$
$$[HC_2H_3O_2] = [OH^-] = X$$

Putting these values into the equation (assuming that X is small compared to 0.10) gives

$$\frac{X^2}{0.10} = 5.6 \times 10^{-10}$$
$$X^2 = 5.6 \times 10^{-11} = 56 \times 10^{-12}$$
$$X = \sqrt{56 \times 10^{-12}} = 7.5 \times 10^{-6}$$

In this case X was equal to the $[OH^-]$ and therefore the $[H^+]$ is

$$[H^+] = \frac{1.0 \times 10^{-14}}{[OH^-]} = \frac{1.0 \times 10^{-14}}{7.5 \times 10^{-6}} = 0.13 \times 10^{-8} = \mathbf{1.3 \times 10^{-9}}$$

PROBLEM EXAMPLE 18-4

Calculate the pH of a solution made by mixing 100 m*l* of 0.10 *M* aqueous ammonia and 100 m*l* of 0.10 *M* hydrochloric acid.

SOLUTION:

Here we have equivalent amounts of acid and base, which react according to the equation

$$NH_{3(aq)} + HCl_{(aq)} \rightleftharpoons NH_4Cl_{(aq)}$$

The salt formed does not give a neutral solution in this case (that is, the pH is not 7) due to the hydrolysis of the NH_4^+ ion, according to the equation

$$NH_4^+ + H_2O \rightleftharpoons NH_3 + H_3O^+$$

To calculate the $[H_3O^+]$ and then the pH, we use the hydrolysis constant, which is

$$K_h = \frac{[NH_3][H_3O^+]}{[NH_4^+]} = \frac{K_W}{K_b} = \frac{1.0 \times 10^{-14}}{1.8 \times 10^{-5}} = 5.6 \times 10^{-10}$$

Since equal volumes of 0.10 *M* acid and base were mixed, the concentration of salt, NH_4Cl, formed is 0.10 $M/2$ = 0.050 *M*. Letting X equal the amount of salt that is hydrolyzed gives

$$[NH_4^+] = 0.050 - X$$
$$[NH_3] = [H_3O^+] = X$$

Putting these values into the hydrolysis constant (again assuming that X is small compared to 0.050) gives

$$\frac{X^2}{0.050} = 5.6 \times 10^{-10}$$
$$X^2 = 0.28 \times 10^{-10} = 28 \times 10^{-12}$$
$$X = \sqrt{28 \times 10^{-12}} = 5.3 \times 10^{-6} = [H_3O^+]$$

Therefore, the pH is

$$pH = -(-6 + 0.72) = \mathbf{5.28}$$

PROBLEM EXAMPLE 18-5

How many grams of sodium acetate would need to be added to 1 l of water to give a pH of 9? (Assume constant volume.)

SOLUTION:

There are three steps involved in this problem:

1. From the pH, calculate the $[OH^-]$.
2. From the equation for the hydrolysis of $NaC_2H_3O_2$ and the calculated value of K_h, calculate $[C_2H_3O_2^-]$ that would be necessary to give this $[OH^-]$.
3. From the calculated, molar concentration of $C_2H_3O_2^-$, which is equal to the concentration of sodium acetate, use conversion factors to calculate the grams of $NaC_2H_3O_2$ that this is equivalent to.
   (a) If pH = 9, $[H^+] = 1.0 \times 10^{-9}$ and $[OH^-] = 1.0 \times 10^{-5}$ (or, if pH = 9, pOH = 5, and $[OH^-] = 1.0 \times 10^{-5}$).
   (b) The net ionic equilibrium reaction for the hydrolysis of sodium acetate is

$$C_2H_3O_2^- + H_2O \rightleftharpoons HC_2H_3O_2 + OH^-$$

and the hydrolysis constant is

$$K_h = \frac{[HC_2H_3O_2][OH^-]}{[C_2H_3O_2^-]} = \frac{K_W}{K_a} = \frac{1.0 \times 10^{-14}}{1.8 \times 10^{-5}} = 5.6 \times 10^{-10}$$

Letting X be the molar concentration of the amount of $C_2H_3O_2^-$ added gives, $[C_2H_3O_2^-] = X - 1 \times 10^{-5}$ (due to the amount that hydrolyzes to form 1×10^{-5} M OH^-). In this case we can probably assume that 1×10^{-5} is going to be small compared to X and therefore use

$$[C_2H_3O_2^-] = X$$

From the equilibrium reaction and the calculated $[OH^-]$,

$$[HC_2H_3O_2] = [OH^-] = 1.0 \times 10^{-5}$$

Putting these values into the expression for the hydrolysis constant gives

$$\frac{(1.0 \times 10^{-5})^2}{X} = 5.6 \times 10^{-10}$$

Solving for X gives

$$X = \frac{(1.0 \times 10^{-5})^2}{5.6 \times 10^{-10}} = \frac{1.0 \times 10^{-10}}{5.6 \times 10^{-10}} = 0.18$$

Therefore, we need 0.18 mole of sodium acetate per liter.

(c) To find the number of grams of sodium acetate we simply convert from moles per liter to grams per liter using the molecular weight of sodium acetate:

$$\frac{0.18 \text{ mol } NaC_2H_3O_2}{\ell \text{ soln}} \times \frac{82 \text{ g } NaC_2H_3O_2}{1 \text{ mol } NaC_2H_3O_2} = \mathbf{\frac{15 \text{ g } NaC_2H_3O_2}{\ell \text{ soln}}}$$

PROBLEM EXAMPLE 18-6

A 0.20 M solution of the salt NaX has a pH of 10.96. Calculate the value of K_a for the weak acid HX.

SOLUTION:

The hydrolysis equation is

$$X^- + H_2O \rightleftarrows HX + OH^-$$

If the pH of the solution is 10.96, the pOH is 3.04 (14.00 − 10.96 = 3.04). Therefore, $[OH^-] = 10^{-3.04} = 10^{0.96} \times 10^{-4} = 9.12 \times 10^{-4}$. For each mole of OH^- produced, 1 mole of HX is produced. Therefore, [HX] is also 9.12×10^{-4}.

$$[X^-] = 0.20 - 9.12 \times 10^{-4} \approx 0.20$$

The expression for the hydrolysis constant is

$$K_h = \frac{[HX][OH^-]}{[X^-]} = \frac{K_W}{K_a}$$

Solving this for K_a gives

$$K_a = \frac{K_W[X^-]}{[HX][OH^-]}$$

And putting in the known values gives

$$K_a = \frac{(1.0 \times 10^{-14})(0.20)}{(9.12 \times 10^{-4})^2} = \frac{2.0 \times 10^{-15}}{83 \times 10^{-8}} = 0.024 \times 10^{-7} = \mathbf{2.4 \times 10^{-9}}$$

PROBLEM EXAMPLE 18-7

In a titration 20.0 m*l* of 0.12 *M* acetic acid is being titrated with 0.15 *M* sodium hydroxide. Calculate the pH after 10.0 m*l* of NaOH has been added.

SOLUTION:

We must first calculate the number of moles (or millimoles) of acid and base that are present.

$$20.0 \text{ m}l \text{ acid} \times \frac{0.12 \text{ mmol acid}}{1 \text{ m}l \text{ acid}} = 2.4 \text{ mmol acid}$$

$$10.0 \text{ m}l \text{ base} \times \frac{0.15 \text{ mmol base}}{1 \text{ m}l \text{ base}} = 1.5 \text{ mmol base}$$

One mole of acid reacts with 1 mole of base according to the equation

$$HC_2H_3O_2 + NaOH \longrightarrow H_2O + NaC_2H_3O_2$$

Therefore, 1.5 mmol of NaOH will react with 1.5 mmol of $HC_2H_3O_2$ to form 1.5 mmol of $NaC_2H_3O_2$, leaving 0.90 mmol $HC_2H_3O_2$ (2.4 − 1.5 = 0.90). The solution thus contains 1.5 mmol of $NaC_2H_3O_2$ and 0.90 mmol of $HC_2H_3O_2$ in 30.0 m*l* of solution.
Therefore, the $[C_2H_3O_2^-]$ and $[HC_2H_3O_2]$ are calculated to be

$$[C_2H_3O_2^-] = \frac{1.5 \text{ mmol}}{30.0 \text{ m}l} = 0.050\ M$$

$$[HC_2H_3O_2] = \frac{0.90 \text{ mmol}}{30.0 \text{ m}l} = 0.030\ M$$

From the hydrolysis reaction,

$$C_2H_3O_2^- + H_2O \rightleftharpoons HC_2H_3O_2 + OH^-$$

and the hydrolysis constant

$$K_h = \frac{[HC_2H_3O_2][OH^-]}{[C_2H_3O_2^-]} = \frac{K_W}{K_a} = \frac{1.0 \times 10^{-14}}{1.8 \times 10^{-5}} = 5.6 \times 10^{-10}$$

we can calculate the $[OH^-]$.
Let X = amount of hydrolysis. Therefore,

$$[OH^-] = X$$

$$[HC_2H_3O_2] = 0.030 + X \qquad \text{(0.030 from the excess acid)}$$

$$[C_2H_3O_2^-] = 0.050 - X \qquad \text{(0.050 from the salt formed)}$$

Assuming that X is going to be small compared to 0.030 and 0.050, we can calculate $[OH^-]$ by putting these values into the expression for the hydrolysis constant

$$\frac{(0.030)(X)}{(0.050)} = 5.6 \times 10^{-10}$$

$$X = \frac{(0.050)(5.6 \times 10^{-10})}{(0.030)} = 9.3 \times 10^{-10}\ \text{mol}/\ell = [OH^-]$$

(This is small compared to 0.030 or 0.050.)

$$\text{pOH} = -\log 9.3 \times 10^{-10} = -(-10 + 0.97) = 9.03$$

$$\text{pH} = 14.00 - \text{pOH} = 14.00 - 9.03 = \mathbf{4.97}$$

Since we have excess acid present, we would expect the pH to be less than 7, which it is.

Hydrolysis of Metal Ions

Since most metals form weak bases (except the IA and most of the IIA metals), these ions in solution undergo hydrolysis to form acidic solutions. This can best be shown by first considering the ionization of these weak acids in steps, just like weak polybasic (polyprotic) acids. For example, $Fe(OH)_3$ ionizes in three steps as

(18-20) $$Fe(OH)_3 \rightleftharpoons Fe(OH)_2{}^+ + OH^-$$

(18-21) $$Fe(OH)_2{}^+ \rightleftharpoons Fe(OH)^{2+} + OH^-$$

(18-22) $$Fe(OH)^{2+} \rightleftharpoons Fe^{3+} + OH^-$$

As with polybasic acids, each step is weaker than the preceding one and, therefore, the last step, Eq. (18-22), is the weakest. Thus, when Fe^{3+} ion undergoes hydrolysis, the main product is $Fe(OH)^{2+}$; that is, the Fe^{3+} ion is removing OH^- ion from water, according to the equation

(18-23) $$Fe^{3+} + H_2O \rightleftharpoons Fe(OH)^{2+} + H^+$$

Actually, this is a somewhat simplified view of what is happening. This is more accurately described by looking at the hydrated Fe^{3+} ion (most ions in solution have several H_2O molecules attached to them, due to the polar properties of H_2O). Thus, Fe^{3+} in aqueous solution is actually $Fe(H_2O)_6{}^{3+}$. The water molecules are arranged around the Fe^{3+} in an octrahedral arrangement, as shown in Fig. 18-1.

Since the Fe^{3+} ion has a rather high charge density, the electron cloud around the oxygen atom is pulled toward the Fe^{3+} ion. This makes the O—H bond weak and allows an H_2O molecule to remove an H^+ from the $Fe(H_2O)_6{}^{3+}$ to form H_3O^+.

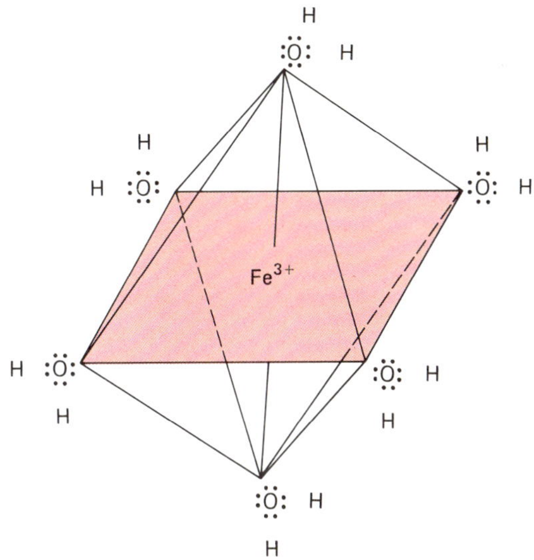

Figure 18-1
Hydrated iron ion, Fe^{3+}

SUMMARY

1. The weak, polybasic (polyprotic) acid H_2S ionizes in two steps to produce a small amount of S^{2-} ion. As with any polybasic acid, the second step of ionization is much smaller than the first. Therefore, using the overall equation ($H_2S \rightleftarrows 2\ H^+ + S^{2-}$), the $[H^+]$ and $[S^{2-}]$ cannot both be calculated. In fact, it can be shown that the $[S^{2-}]$ is approximately equal to the second ionization constant. Using the overall equation and the K_i for this equation, however, if the $[H^+]$ (or the pH) is known, the $[S^{2-}]$ can be calculated, using the fact that in a saturated solution of H_2S at room temperature and about 1 atm pressure, the $[H_2S]$ is 0.10 M. These calculations can then be used along with solubility product constants to calculate what ions will precipitate as sulfides at a given pH when the solution is saturated with H_2S.
2. Many salts, when in water solution, *hydrolyze* (react with the water) to form a weak acid, or weak base, or both. These include any salt that contains the cation of a weak base or the anion of a weak acid, or both of these. The three types of salts that undergo hydrolysis, and the hydrolysis constant for each are
   (a) salt of a weak acid and strong base

$$K_h = \frac{K_W}{K_a}$$

   (b) salt of a strong acid and a weak base

$$K_h = \frac{K_W}{K_b}$$

(c) salt of a weak acid and a weak base

$$K_h = \frac{K_W}{K_a \cdot K_b}$$

It can be seen that in each case the hydrolysis constant is equal to the K_W divided by the ionization constant of the weak acid or weak base or both. (The salt of a strong acid and a strong base is not hydrolyzed.)

3. Using these hydrolysis constants (which must first be calculated), the $[H^+]$, $[OH^-]$, pH, and so on, can be calculated for any solution containing one of these salts or for any reaction in which a salt of this type is formed; conversely, the amount of one of these salts that is necessary to give a particular pH can be calculated.
4. Any metal ion that forms a weak base is also hydrolyzed in water solution. This is probably best described as H^+ ion being removed from the hydrated metal ion.

Upon completion of this chapter, you should be able to do the following (assume tables of ionization constants and solubility product constants are available):

1. Given the $[H^+]$ (or pH, concentration of acid, and so on) for any solution that is saturated with H_2S, calculate the $[S^{2-}]$. (1)
2. Given a particular $[S^{2-}]$, calculate the $[H^+]$ (or pH, concentration of acid, and so on) necessary to obtain this $[S^{2-}]$ in a solution that is saturated with H_2S. (2, 3)
3. Given the concentration of any metal ion in a solution saturated with H_2S, predict whether or not the metal sulfide will precipitate out. (4)
4. Given any weak dibasic acid (H_2A), calculate the approximate $[H^+]$ and $[A^{2-}]$. (5)
5. Given any salt of a weak acid and strong base, strong acid and weak base, or weak acid and weak base, calculate the values of the hydrolysis constant (K_h) for that salt. (6)
6. Given the concentration of any salt in solution (or the amount of acid and base mixed to form a salt), calculate the $[H^+]$ or pH. (7)
7. Given a particular $[H^+]$ or pH, calculate the concentration of a given salt that will give that $[H^+]$ or pH. (8)
8. Given the $[H^+]$ or pH of any salt solution of given concentration, calculate K_a or K_b for the weak acid or base formed by that salt in water solution. (9)
9. Given any of the terms listed in Exercise 1, give a definition or explanation in your own words.

EXERCISES

1. Define or explain the following terms:
   (a) polyprotic acid
   (b) dibasic acid
   (c) hydrolysis of salts
   (d) K_h
2. Derive the expression for the hydrolysis constant for each of the following types of salts:
   (a) for the salt of a weak acid and a strong base,
   (b) for the salt of a strong acid and a weak base,
   (c) for the salt of a weak acid and a weak base.
3. Show that for any dibasic acid, such as H_2A, the $[A^{2-}]$ is approximately equal to the second ionization constant for the acid.
4. Show that for a dibasic acid, such as H_2S, the overall ionization constant is equal to the K_i for the first step of ionization times the K_i for the second step of ionization.

PROBLEMS

Polyprotic acids—H_2S

1. The following solutions are saturated with H_2S. Calculate the $[S^{2-}]$ of each one:
   (a) $[H^+] = 1.0 \times 10^{-4}$
   (b) pH = 6.27
   (c) 0.010 *M* HCl
   (d) 1.0×10^{-4} *M* NaOH
   (e) 0.10 *M* $HC_2H_3O_2$
   (f) 0.010 *M* NH_3
   (g) a solution containing 0.020 *M* acetic acid and 0.010 *M* sodium acetate
   (h) a solution that contains 5.0 g of ammonium chloride in 200 m*l* of 0.20 *M* NH_3
   (i) $[H^+] = 1.0 \times 10^{-5}$
   (j) pH = 5.85
   (k) 0.0230 *M* HNO_3
   (l) 3.5×10^{-6} *M* NaOH
   (m) 0.150 *M* acetic acid
   (n) 0.015 *M* aqueous ammonia
   (o) a solution that contains 0.30 *M* ammonia and 0.20 *M* ammonium chloride
   (p) a solution that contains 2.4 g of sodium acetate in 300 m*l* of 0.020 *M* acetic acid
2. For a solution saturated with H_2S, calculate the $[H^+]$ necessary to give a $[S^{2-}]$ of
   (a) 1.0×10^{-12}
   (b) 2.5×10^{-9}
3. Calculate the pH necessary to give a $[S^{2-}]$ (in a saturated solution of H_2S) of
   (a) 3.2×10^{-15}
   (b) 1.5×10^{-17}
4. The following solutions, all with a pH of 1.20, are saturated with H_2S. Tell if a precipitate will form.
   (a) 0.10 *M* Fe^{2+}
   (b) 4.25 g lead(II) nitrate in 225 m*l* of solution
   (c) 0.10 *M* Cu^{2+}
   (d) 6.75 g $MnCl_2$ in 350 m*l* of solution

5. Calculate the approximate $[H^+]$ and $[CO_3^{2-}]$ for each of the following solutions:
   (a) 0.010 M H_2CO_3
   (b) 0.015 M H_2CO_3

Hydrolysis

6. For each of the following salts in solution, do the following: write the equilibrium reaction (if any), tell if the solution would be acidic, basic, or neutral; write the expression for the hydrolysis constant (K_h); calculate K_h; and calculate the $[H^+]$ of a 0.10 M solution.
   (a) sodium acetate
   (b) ammonium nitrate
   (c) sodium chloride
   (d) ammonium acetate
   (e) ammonium chloride
   (f) sodium cyanide
   (g) potassium acetate
   (h) sodium nitrate
7. Calculate the pH of each of the following solutions:
   (a) 0.250 M $NaC_2H_3O_2$
   (b) 0.020 M NH_4NO_3
   (c) 0.3750 M KCl
   (d) 150 mℓ of 0.10 M HCl + 152 mℓ of 0.10 M NaOH
   (e) 200 mℓ of 0.15 M acetic acid + 200 mℓ of 0.15 M sodium hydroxide
   (f) 150 mℓ of 0.30 M NH_3 + 100 mℓ of 0.45 M HCl
   (g) 250 mℓ of 0.20 M HCl + 400 mℓ of 0.25 M NH_3
   (h) 100 mℓ of 0.20 M NaOH + 200 mℓ of 0.30 M $HC_2H_3O_2$
   (i) 0.320 M NH_4Cl
   (j) 0.120 M $KC_2H_3O_2$
   (k) 1.850 M NaCl
   (l) 200 mℓ of 0.42 M NaOH + 200 mℓ of 0.42 M HCl
   (m) 100 mℓ of 0.21 M sodium hydroxide + 100 mℓ of 0.21 M acetic acid
   (n) 150 mℓ of 0.18 M NH_3 + 250 mℓ of 0.25 M HCl
   (o) 300 mℓ of 0.45 M NH_3 + 250 mℓ of 0.32 M HCl
   (p) 200 mℓ of 0.35 M NH_3 + 150 mℓ of 0.25 M HCl
8. Calculate the number of moles of solute necessary in each of the following:
   (a) potassium acetate in a liter of solution to give a pH of 8.50
   (b) ammonium nitrate in a liter of solution to give a pH of 5.30
   (c) ammonium chloride in a liter of solution to give a pH of 4.80
   (d) sodium acetate in a liter of solution to give a pH of 9.20.
9. Calculate the value of K_i (K_a or K_b) in each of the following:
   (a) A 0.15 M solution of the salt NaX has a pH of 9.80. Calculate K_i for HX.
   (b) A 0.23 M solution of the salt KX has a pH of 10.20. Calculate K_i for HX.

General problems

10. Calculate the molar concentration of potassium acetate necessary to precipitate
    (a) $Mg(OH)_2$ from 0.10 M $MgCl_2$
    (b) $Mn(OH)_2$ from 0.10 M $MnCl_2$
11. Calculate the number of grams of NaCN that must be added to 1 ℓ of solution to give a pH of
    (a) 10.80
    (b) 11.60

12. Calculate the $[S^{2-}]$ of each of the following solutions, when saturated with H_2S.
    (a) 0.15 *M* HCl
    (b) 0.10 *M* $HC_2H_3O_2$
    (c) 0.24 *M* HNO_3
    (d) 0.10 *M* HCN
13. What concentration of hydrogen ion will reduce the sulfide ion concentration in a saturated solution of H_2S to
    (a) 2.0×10^{-18} *M*?
    (b) 9.2×10^{-19} *M*?
14. Calculate the pH of each of the following solutions:
    (a) A solution saturated with H_2S in which the $[S^{2-}]$ is 3.25×10^{-12}.
    (b) A 0.020 *M* solution of H_2CO_3.
    (c) A solution made by mixing 100 m*l* of 0.50 *M* acetic acid and 200 m*l* of 0.10 *M* sodium hydroxide.
    (d) 5.0 g of NH_4Cl is added to 100 m*l* of 0.10 *M* aqueous ammonia.
    (e) A 0.10 *M* solution of KCN.
    (f) A 0.20 *M* solution of NH_3 that contains 0.20 *M* $(NH_4)_2SO_4$.
    (g) A solution made by mixing 100 m*l* of 0.20 *M* HCl with 200 m*l* of 0.10 *M* NaOH.
    (h) A solution made by mixing 200 m*l* of 0.20 *M* NH_3 with 200 m*l* of 0.20 *M* HCl.
    (i) A solution saturated with H_2S in which the $[S^{2-}]$ is 1.12×10^{-11}.
    (j) A 0.025 *M* solution of $NaC_2H_3O_2$.
    (k) A solution made by mixing 200 m*l* of 0.52 *M* aqueous ammonia with 120 m*l* of 0.35 *M* HCl.
    (l) A 0.32 *M* solution of acetic acid that contains 0.12 *M* calcium acetate.
    (m) 6.25 g of ammonium nitrate is added to 325 m*l* of 0.15 *M* aqueous ammonia.
    (n) A solution made by mixing 200 m*l* of 0.30 *M* HCl with 300 m*l* of 0.20 *M* NaOH.
    (o) A solution made by mixing 225 m*l* of 0.35 *M* HCN with 180 m*l* of 0.28 *M* NaOH.
    (p) A 0.015 *M* solution of H_2CO_3.
15. If a 0.015 *M* solution of the salt NaA has a pH of 11.00, calculate the value of K_i for the weak acid HA.
16. If a 0.035 *M* solution of the salt KA has a pH of 12.50, calculate the value of K_i for the weak acid HA.
17. A 0.18 *M* solution of the weak acid HA has a pH of 3.80. Calculate the pH of a 0.25 *M* solution of the salt NaA.
18. A 0.25 *M* solution of the weak base MOH has a pH of 9.30. Calculate the pH of a 0.45 *M* solution of the salt MCl.

REFERENCES

GUYON, JOHN C., and BERWYN E. JONES, *Introduction to Solution Equilibrium*, Boston: Allyn and Bacon, 1969.

HOGNESS, T. R., WARREN C. JOHNSON, and ALFRED R. ARMSTRONG, *Qualitative Analysis and Chemical Equilibrium*, 5th ed., New York: Holt, Rinehart and Winston, Inc., 1966.

Objectives

1. To learn what is meant by the terms oxidation, reduction, oxidizing agent, and reducing agent and to be able to identify all four of these for any oxidation-reduction (redox) equation.
2. To learn the oxidation number method of balancing redox equations and to be able to balance equations by this method.
3. To learn to balance half-reactions and to be able to balance redox equations by the ion-electron method, or the method of half-reactions.

Nineteen

Balancing Oxidation-Reduction Equations

The chemical equations in which there are changes in oxidation number (or oxidation state) of the reactants are called oxidation-reduction or, sometimes more simply, redox equations. Some of the simple types of reactions considered in Chapter 8 (combination, decomposition, and replacement reactions) actually involved oxidation and reduction. In general, however, these redox equations are not so simple as the cases considered in Chapter 8 and thus cannot be balanced by inspection. Therefore, special techniques are needed to balance these more complex oxidation-reduction equations.

We shall first consider the oxidation number method, for which you may need to review the general rules for determining oxidation numbers, as given in Section 5-8. Then we shall learn to balance half-reactions and use this as a second method of balancing redox equations. This method is sometimes referred to as the ion-electron method.

19-1 Definitions of Oxidation and Reduction. Oxidizing and Reducing Agents.

Before we consider the definitions of oxidation and reduction, you may need to review oxidation numbers (oxidation states) as discussed in Section 5-8. Recall that we defined oxidation number as an apparent valence, or apparent charge, on an atom in a compound. This is a way of accounting for where the various electrons are in the compound. Therefore, if the oxidation state is to change, the electrons must be transferred. If electrons are lost by an atom, the oxidation state increases since an electron has a negative charge. This is due to the fact that the atom then has more protons than electrons and thus has a higher charge or oxidation number. Conversely, if electrons are gained, the oxidation state decreases. At one time the term *oxidation* referred only to reactions with oxygen, but this has been expanded to include any chemical reaction in which a substance increases in oxidation state by losing electrons. Therefore, we define oxidation and reduction as follows:

1. *Oxidation* is any chemical change in which a substance loses electrons and thus increases in oxidation state. For example, when a Zn atom loses two electrons to become a Zn^{2+} ion, the zinc has increased in oxidation state (from 0 to +2):

$$Zn \longrightarrow Zn^{2+} + 2\,e^-$$

2. *Reduction* is a chemical change in which a substance gains electrons and thus decreases in oxidation state. For example, when a sulfur atom gains two electrons to become a sulfide (S^{2-}) ion, it decreases in oxidation state (from 0 to −2),

$$S + e^- \longrightarrow S^{2-}$$

In a given reaction, whenever one substance is oxidized, it loses electrons to another substance, which is thereby reduced; that is, oxidation is always accompanied by reduction. Hence these equations are called oxidation-reduction equations or often, more simply, redox.

In an oxidation-reduction equation, the substance that is oxidized is called the *reducing agent* since it causes something else to be reduced. Likewise, the substance that is reduced causes something else to be oxidized and is thus called the *oxidizing agent.*

In summary, we can say that oxidation can be defined either as an increase in oxidation number or as a loss of electrons and that reduction is a decrease in oxidation number or a gain of electrons. For example, in the combination of zinc with sulfur, we can show the changes as follows:

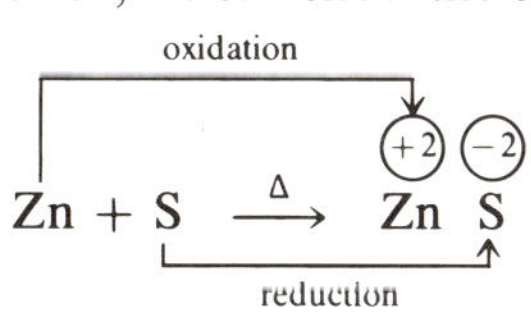

The Zn is the reducing agent and the S is the oxidizing agent. This is therefore an example of an oxidation-reduction equation. It is important to keep in mind when considering equations of this type that there must be both oxidation and reduction taking place. Electrons cannot be lost (oxidation) by one substance unless they are being gained (reduction) by something else.

19-2 Balancing Oxidation-Reduction Equations by the Oxidation Number Method

This method of balancing redox equations can be used for both molecular and ionic equations. We shall first suggest some guidelines to follow and then do some examples, pointing out things to watch for in using this method.

1. Determine which elements undergo a change in oxidation number. At first this may involve calculating the oxidation number for each

element in the equation, but after balancing a few of these you should readily be able to recognize which elements undergo a change in oxidation number.

2. Write the oxidation number above each element that is oxidized or reduced.

3. Determine the change in oxidation number (and hence the number of electrons transferred) for each element undergoing a change in oxidation number. This is best shown by drawing an arrow from the element in the reactant to the element in the product and indicating the increase or decrease of oxidation number above the arrow. (You can also use loss and gain of electrons here. Use one or the other consistently; do not mix up the two.)

4. Balance the increase and decrease of oxidation number by placing the necessary coefficients before the reactants that are changing, using the smallest numbers possible; only put these coefficients on the left side of the equation at this point. The total increase in oxidation number must equal the total decrease; that is, the total number of electrons lost must equal the total number gained.

5. Complete the balancing by inspection. First balance those substances that changed in oxidation state and then any atoms other than hydrogen and oxygen, finally balance hydrogen and oxygen by putting H_2O wherever needed.

6. For ionic equations, just before H_2O is added (that is, the last step above), put in H^+ if in acid or OH^- if in base, on either side of the equation, to balance the charge.

7. Check the equation to see that the coefficients are the lowest possible ratio. For ionic equations, the net charges on both sides of the equation must be exactly the same (that is, $+2$ on both sides or -2 on both sides, *not* $+2$ on one side and -2 on the other).

Consider the following examples. Since the products of these reactions are often difficult to predict, you will be given the reactants and products. Thus, you will be asked only to balance the equations. Also, to simplify writing the equation, the physical states will normally not be included in the equation.

1. $K_2Cr_2O_7 + H_2O + S \longrightarrow SO_2 + KOH + Cr_2O_3$

$-3/\text{atom} \times 2\ \text{atom} = -6$ (Cr: $+6 \rightarrow +3$)

$$\overset{+6}{K_2Cr_2O_7} + H_2O + \overset{0}{S} \longrightarrow \overset{+4}{SO_2} + KOH + \overset{+3}{Cr_2O_3}$$

$+4/\text{atom} \times 1\ \text{atom} = +4$ (S: $0 \rightarrow +4$)

Put a 2 before the $K_2Cr_2O_7$ and a 3 before the S. Since

$$(2)(-6) = -12$$
$$(3)(+4) = +12$$

(Be sure to use the smallest numbers possible.)

$$2\ K_2Cr_2O_7 + H_2O + 3\ S \longrightarrow SO_2 + KOH + Cr_2O_3$$

Balance the S and Cr on the right side of the equation by putting a 3 before the SO_2 and a 2 before the Cr_2O_3.

$$2\ K_2Cr_2O_7 + H_2O + 3\ S \longrightarrow 3\ SO_2 + KOH + 2\ Cr_2O_3$$

Placing a 4 before the KOH to balance the K gives 16 O on the right side and 14 (plus the H_2O) on the left. Therefore, 2 must be placed before the H_2O. Check the H to be sure that it is balanced (we could have used H for the last step instead of O).

$$2\ K_2Cr_2O_7 + 2\ H_2O + 3\ S \longrightarrow 3\ SO_2 + 4\ KOH + 2\ Cr_2O_3$$

The S is oxidized (0 to +4) and is thus the reducing agent; the $K_2Cr_2O_7$ is reduced (Cr goes from +6 to +3) and is thus the oxidizing agent.

2. $HCl + KMnO_4 \longrightarrow MnCl_2 + KCl + Cl_2 + H_2O$

+1/atom × 1 atom (on the left side) = +1

$$\overset{-1}{5\ HCl} + 1\ K\overset{+7}{Mn}O_4 \longrightarrow 1\ \overset{+2}{Mn}Cl_2 + KCl + 2\tfrac{1}{2}\ \overset{0}{Cl_2} + H_2O$$

−5/atom × 1 atom = −5

Multiply everything done so far by 2 (to eliminate the fraction):

$$10\ HCl + 2\ KMnO_4 \longrightarrow 2\ MnCl_2 + KCl + 5\ Cl_2 + H_2O$$

Balance K by putting a 2 before the KCl and then add 6 to HCl on the left side (*add* 6 HCl) to account for the Cl^- that did not change; finally, put an 8 before the H_2O to balance the H (the O will also balance at this point).

$$16\ HCl + 2\ KMnO_4 \longrightarrow 2\ MnCl_2 + 2\ KCl + 5\ Cl_2 + 8\ H_2O$$

The HCl is oxidized (Cl goes from −1 to 0) and is the reducing agent; $KMnO_4$ is reduced (Mn goes from +7 to +2) and is the oxidizing agent.

3. $HNO_2 \longrightarrow NO_2 + NH_3$

In this case, HNO_2 is being both oxidized and reduced; therefore the best way to approach this is probably to write it twice before starting (since it will take at least 2 HNO_2 to balance it in any case). Also, watch NH_3 (and NH_4^+) in these equations—remember that the H is still +1 and that N is −3.

$$\overset{+3}{1\ HNO_2} + \overset{+3}{6\ HNO_2} \longrightarrow \overset{+4}{6\ NO_2} + \overset{-3}{1\ NH_3}$$

(+1: from 6 HNO_2 to 6 NO_2; −6: from 1 HNO_2 to 1 NH_3)

Then we must add 2 H_2O to the right side to balance the H. This gives

$$HNO_2 + 6\ HNO_2 \longrightarrow 6\ NO_2 + NH_3 + 2\ H_2O$$

or

$$7\ HNO_2 \longrightarrow 6\ NO_2 + NH_3 + 2\ H_2O$$

In this case, HNO_2 is both oxidized and reduced; this is sometimes called *auto oxidation-reduction or disproportionation.*

4. $Ag + NO_3^- \longrightarrow Ag^+ + NO$ (acid solution)

This is an ionic equation, which can also be balanced by the oxidation number method. Place the oxidation numbers of elements that change in oxidation number above the elements.

$$\overset{0}{Ag} + \overset{+5}{NO_3^-} \longrightarrow \overset{+1}{Ag^+} + \overset{+2}{NO}$$

Determine the change in oxidation number for each element undergoing a change in oxidation number.

$$\overset{0}{Ag} + \overset{+5}{NO_3^-} \longrightarrow \overset{+1}{Ag^+} + \overset{+2}{NO}$$

(+1/atom × 1 atom: from Ag to Ag^+; −3/atom × 1 atom: from NO_3^- to NO)

Place the coefficients in front of the formulas on the left to make the increase equal the decrease and balance the atoms that are changing on the right:

$$3\ Ag + NO_3^- \longrightarrow 3\ Ag^+ + NO$$

Add 4 H^+ on the left to balance the charge (since it is in acid solution, H^+ can be added to *either* side of the equation). Then add

2 H_2O to the right to balance the H and O. Check the charge to make sure that it balances also.

$$3\ Ag + 4\ H^+ + NO_3^- \longrightarrow 3\ Ag^+ + NO + 2\ H_2O$$

$$\text{Charges:}\quad (+4) + (-1) = +3 \qquad\qquad 3(+1) = +3$$

Since the Ag increases its oxidation number and hence loses electrons, it is oxidized and thus is the reducing agent. In NO_3^-, N decreases in oxidation number and hence gains electrons; therefore, NO_3^- is reduced and is the oxidizing agent.

5. $CrO_2^- + ClO^- \longrightarrow CrO_4^{2-} + Cl^-$ (in alkaline, or basic, solution)

Proceed as in the earlier examples, to give

$$\overset{+3}{2\ CrO_2^-} + \overset{+1}{3\ ClO^-} \longrightarrow \overset{+6}{2\ CrO_4^{2-}} + \overset{-1}{3\ Cl^-}$$

(Cr: $+3 \rightarrow +6$, change $+3$; Cl: $+1 \rightarrow -1$, change -2)

Now add 2 OH^- to the left side to balance the charge (since it is in basic solution, OH^- can be added to either side of the equation) and H_2O to the right side to balance the H and O.

$$2\ OH^- + 2\ CrO_2^- + 3\ ClO^- \longrightarrow 2\ CrO_4^{2-} + 3\ Cl^- + H_2O$$

Check all atoms to see that they are balanced, and also check to see that the charge is exactly the same on both sides of the equation $(-2 - 2 - 3 = -7; 2(-2) - 3 = -7)$.
The CrO_2^- is oxidized (and thus is the reducing agent) and the ClO^- is reduced (and thus is the oxidizing agent).

19-3 Half-Reactions. Balancing Equations by the Ion-Electron Method (Method of Half-Reactions).

A second method that can be used to balance redox equations is the ion-electron method, or the method of *half-reactions.* In this section we shall first consider the balancing of half-reactions and then apply these to the ion-electron method of balancing equations.

In writing these half-reactions it must be kept in mind that a single one of these cannot take place by itself; oxidation is always accompanied by reduction and vice versa. We shall find that these partial equations (half-reactions) are very useful in electrochemistry, which we shall consider in Chapter 20.

In this method of balancing redox equations, two partial equations are written, representing half-reactions. One equation describes the oxidation half-reaction and the other describes the reduction. To balance each half-reaction, it is simply necessary to find the same number of each atom on each side of the equation, using H_2O and then H^+ where needed (and OH^- if in basic solution), and then balance the charge by putting electrons in where needed on either side of the equation. Consider the following examples of half-reactions:

1. $Cr_2O_7^{2-} \longrightarrow Cr^{3+}$ (in acid solution)

   Place a 2 before the Cr^{3+} to balance the 2 Cr on the left (in $Cr_2O_7^{2-}$); then 7 H_2O on the right to balance the 7 O in $Cr_2O_7^{2-}$.

   $$Cr_2O_7^{2-} \longrightarrow 2\,Cr^{3+} + 7\,H_2O$$

   Then, put 14 H^+ on the left side to balance the 14 H in 7 H_2O.

   $$14\,H^+ + Cr_2O_7^{2-} \longrightarrow 2\,Cr^{3+} + 7\,H_2O$$

   This balances all the atoms, but the charge is not balanced [$+14 - 2 = +12$, and $2(+3) = +6$]. This is done by adding electrons on either side of the equations. In this case, 6 e^- are needed on the left side to make the charge exactly the same on both sides of the equation ($+6$). This gives

   $$6\,e^- + 14\,H^+ + Cr_2O_7^{2-} \longrightarrow 2\,Cr^{3+} + 7\,H_2O$$

2. $NO_3^- \longrightarrow NH_3$ (in basic solution)

   For basic solution, OH^- can be added on either side of the equation, as needed. One method of doing this is to proceed just as with acid solution; that is, including H^+ to balance H atoms and then, before balancing the charge, put as many OH^- ions on both sides of the equation as there are H^+ ions. Then simplify by canceling common H_2O. Therefore, to balance this half-reaction the following steps are used:

   (a) 3 H_2O on right (to balance O).

   $$NO_3^- \longrightarrow NH_3 + 3\,H_2O$$

   (b) 9 H^+ on left (to balance H).

   $$9\,H^+ + NO_3^- \longrightarrow NH_3 + 3\,H_2O$$

   (c) 9 OH^- on both sides (since it is in basic solution the H^+ would combine with OH^- to form H_2O).

   $$9\,H_2O + NO_3^- \longrightarrow NH_3 + 3\,H_2O + 9\,OH^-$$

   (c) Cancel 3 H_2O on both sides.

   $$6\,H_2O + NO_3^- \longrightarrow NH_3 + 9\,OH^-$$

   (e) Add 8 e^- on left to balance the charge (-1 on left and -9 on right).

   $$8\,e^- + 6\,H_2O + NO_3^- \longrightarrow NH_3 + 9\,OH^-$$

Remember that these just represent half-reactions, and therefore they do have electrons as part of the balanced equation. A complete redox equation cannot have any excess electrons on either side.

To balance equations by the method of half-reactions, the following steps are recommended.

1. Balance each half-reaction as done in the two preceding examples.
2. Multiply each half-reaction by the smallest number possible to make the total number of electrons lost in the oxidation half-reaction equal to the number gained in the reduction half-reaction. Keep in mind that by multiplying through by any number each half-reaction is still balanced as long as *each* part is multiplied by the same number.
3. Add the two half-reactions.
4. Cancel all common species.

Consider the following example:

$$H_2S + NO_3^- \longrightarrow NO + S \quad \text{(acid solution)}$$

Balancing the two half-reactions gives

$$H_2S \longrightarrow S + 2\,H^+ + 2\,e^- \text{ (add 2 } H^+\text{, then 2 } e^-\text{)}$$

$$3\,e^- + 4\,H^+ + NO_3^- \longrightarrow NO + 2\,H_2O$$

(The steps followed here are to add 2 H_2O, to balance oxygen; add 4 H^+, to balance H; add 3 e^- to balance the charge.)

Multiply the first equation by 3 and the second equation by 2.

$$3\,H_2S \longrightarrow 3\,S + 6\,H^+ + 6\,e^-$$

$$6\,e^- + 8\,H^+ + 2\,NO_3^- \longrightarrow 2\,NO + 4\,H_2O$$

Add the two half-reactions:

$$3\,H_2S + 6\,e^- + 8\,H^+ + 2\,NO_3^- \longrightarrow 3\,S + 6\,H^+ + 6\,e^- + 2\,NO + 4\,H_2O$$

Cancel common species, to give

$$3\,H_2S + 2\,H^+ + 2\,NO_3^- \longrightarrow 3\,S + 2\,NO + 4\,H_2O$$

SUMMARY

1. In a *redox equation,* electrons are lost by one of the reactants (*oxidation*) and gained by another (*reduction*). The total number lost must equal the total number gained—this is the basis for balancing redox equations, by any method.

2. In the electron transfer (or oxidation state) method, the total increase in oxidation state (or total number of electrons lost) is made to equal the total decrease in oxidation state (or total number of electrons gained) by putting in the proper coefficients.
3. In an ionic equation, H^+ (if in acid solution) or OH^- (if in basic solution) is added to either side of the equation to balance the charge. just before H_2O is added as the last step.
4. In a *half-reaction,* all atoms are balanced by inspection and then electrons are added (as the very last step) to either side of the equation to balance the charge.
5. To balance equations by the method of half-reactions, or the ion-electron method, each balanced half-reaction is multiplied by the smallest number possible to give the same number of electrons lost in the oxidation half-reaction and gained in the reduction half-reaction. Then the two half-reactions are added together and simplified if possible.
6. In an ionic equation, both the atoms and the charge must be balanced; that is, there must be the same number of each kind of atom on each side of the equation and exactly the same charge on each side.
7. The substance oxidized is the *reducing agent* and the substance reduced is the *oxidizing agent.*

Upon completion of this chapter you should be able to do the following:

1. Given the products and reactants for any redox equation (molecular or ionic), balance the equation and tell what is oxidized, what is reduced, what is the oxidizing agent, and what is the reducing agent. (1, 2, 3, 4)
2. Given any half-reaction, in either acid or basic solution, write a complete, balanced equation for the half-reaction. (5)
3. Given the terms *oxidation, reduction, oxidizing agent,* and *reducing agent,* define, both in terms of change in oxidation state (number) and transfer of electrons.

EXERCISES

1. Define oxidation and reduction, both in terms of change in oxidation state and gain or loss of electrons.
2. Define the terms *oxidizing agent* and *reducing agent.*
3. Explain what a half-reaction (or partial equation) is.

PROBLEMS

Oxidation Number and Transfer of Electrons

1. For each of the following, tell what is oxidized, what is reduced, the change in oxidation number for each element, and the total number of electrons transferred in the balanced equation (for example, in the equation 2 Na + S $\longrightarrow$ Na_2S, Na is oxidized, S is reduced, the oxidation number of Na goes from 0 to +1, the oxidation number of S goes from 0 to −2, and there are two electrons transferred.)
   (a) $H_2 + Cl_2 \longrightarrow HCl$
   (b) $K + O_2 \longrightarrow K_2O$
   (c) $Zn + HCl \longrightarrow H_2 + ZnCl_2$
   (d) $Al + O_2 \longrightarrow Al_2O_3$
   (e) iron + sulfur $\longrightarrow$ ferric sulfide
   (f) $H_2 + O_2 \longrightarrow H_2O$
   (g) $Na + O_2 \longrightarrow Na_2O$
   (h) $Mg + HNO_3 \longrightarrow H_2 + Mg(NO_3)_2$
   (i) $Al + Cl_2 \longrightarrow AlCl_3$
   (j) sulfur + tin $\longrightarrow$ stannous sulfide
2. For each part of Problem 1, tell what the oxidizing agent is and what the reducing agent is.

Balancing equations

3. Balance the following molecular equations:
   (a) $FeCl_3 + H_2S \longrightarrow FeCl_2 + S + HCl$
   (b) $HNO_3 + I_2 \longrightarrow NO_2 + HIO_3 + H_2O$
   (c) $Bi(OH)_3 + K_2SnO_2 \longrightarrow Bi + K_2SnO_3 + H_2O$
   (d) $I_2O_5 + CO \longrightarrow I_2 + CO_2$
   (e) $HNO_3 + Fe \longrightarrow Fe(NO_3)_3 + NO + H_2O$
   (f) $H_2SO_4 + Ca \longrightarrow CaSO_4 + SO_2$
   (g) $Mn(NO_3)_2 + NaBiO_3 + HNO_3 \longrightarrow$ $HMnO_4 + Bi(NO_3)_3 + NaNO_3$
   (h) $Na_2Cr_2O_7 + FeCl_2 + HCl \longrightarrow CrCl_3 + FeCl_3 + NaCl$
   (i) $NiS + HCl + HNO_3 \longrightarrow NiCl_2 + NO + S + H_2O$
   (j) $Bi(OH)_3 + Na_2SnO_2 \longrightarrow Na_2SnO_3 + Bi + H_2O$
   (k) $MnS + HCl + HNO_3 \longrightarrow MnCl_2 + NO + S + H_2O$
   (l) $HNO_3 + HI \longrightarrow NO + I_2 + H_2O$
   (m) $KI + H_2SO_4 \longrightarrow H_2S + I_2 + K_2SO_4 + H_2O$
   (n) $Cu + HNO_3 \longrightarrow Cu(NO_3)_2 + NO$
   (o) $Cl_2 + KNO_3 \longrightarrow KClO_3 + KCl + HNO_3$
   (p) $FeSO_4 + KI + KIO_3 \longrightarrow Fe(OH)_2 + K_2SO_4 + I_2$
4. Balance the following ionic equations:
   (a) $Sn^{2+} + IO_3^- \longrightarrow Sn^{4+} + I^-$ (acid)
   (b) $AsO_2^- + MnO_4^- \longrightarrow AsO_3^- + Mn^{2+} + H_2O$ (acid)
   (c) $C_2O_4^{2-} + MnO_4^- \longrightarrow CO_2 + Mn^{2+}$ (acid)
   (d) $Mn^{2+} + BiO_3^- \longrightarrow MnO_4^- + Bi^{3+}$ (acid)
   (e) $Fe + NO_3^- \longrightarrow Fe^{3+} + NO$ (acid)
   (f) $Cl_2 \longrightarrow ClO_3^- + Cl^-$ (base)

(g) $MnO_2 + O_2 \longrightarrow MnO_4^{2-}$ (base)
(h) $MnO_4^- + ClO_2^- \longrightarrow MnO_2 + ClO_4^-$ (base)
(i) $S^{2-} + NO_3^- \longrightarrow S + NO$ (acid)
(j) $SO_3^{2-} + MnO_4^- \longrightarrow SO_4^{2-} + Mn^{2+}$ (acid)
(k) $AsO_3^{3-} + Br_2 \longrightarrow AsO_4^{3-} + Br^-$ (acid)
(l) $AsO_4^{3-} + I^- \longrightarrow AsO_3^{3-} + I_2$ (acid)
(m) $Bi(OH)_3 + SnO_2^{2-} \longrightarrow SnO_3^{2-} + Bi$ (base)
(n) $Cr_2O_7^{2-} + C_2O_4^{2-} \longrightarrow Cr^{3+} + CO_2$ (acid)
(o) $Cl_2 \longrightarrow ClO^- + Cl^-$ (base)
(p) $Mn^{2+} + H_2O_2 \longrightarrow MnO_2$ (base)

Half-reactions

5. Write a complete, balanced half-reaction for each of the following:
(a) $MnO_4^- \longrightarrow Mn^{2+}$ (acid)
(b) $SnO_2^{2-} \longrightarrow SnO_3^{2-}$ (base)
(c) $Cr_2O_7^{2-} \longrightarrow Cr^{3+}$ (acid)
(d) $ClO^- \longrightarrow Cl^-$ (acid)
(e) $NO_3^- \longrightarrow NH_4^+$ (base)
(f) $CrO_4^- \longrightarrow CrO_4^{2-}$ (acid)
(g) $NO_3^- \longrightarrow NO$ (base)
(h) $ClO^- \longrightarrow Cl_2$ (acid)
(i) $NH_3 \longrightarrow NO_3^-$ (base)
(j) $S_2O_3^{2-} \longrightarrow SO_2$ (acid)

General problems

6. Tell how many electrons are transferred in each of the following when the equation is balanced:
(a) $Cu + HNO_3 \longrightarrow Cu(NO_3)_2 + NO$
(b) aluminum displaces hydrogen from nitrous acid
(c) $Fe^{2+} + Cr_2O_7^{2-} \longrightarrow Cr^{3+} + Fe^{3+}$ (in acid)
(d) $NH_3 + O_2 \longrightarrow N_2O_5$
(e) $Na_2TeO_3 + NaI + HCl \longrightarrow NaCl + Te + I_2$
(f) $MnO_4^- + SO_2 \longrightarrow Mn^{2+} + SO_4^{2-}$ (in acid)
(g) $Sb + HNO_3 \longrightarrow Sb_2O_5 + NO$
(h) barium displaces hydrogen from hydrochloric acid
(i) $KI + KMnO_4 + HCl \longrightarrow KCl + MnCl_2 + I_2$
(j) $NH_3 + O_2 \longrightarrow NO$

7. Balance the following equations (all in aqueous solution):
(a) $FeSO_4 + KMnO_4 + H_2SO_4 \longrightarrow Fe(SO_4)_3 + MnSO_4 + K_2SO_4$
(b) $Cu + H_2SO_4 \longrightarrow CuSO_4 + SO_2$
(c) $KI + H_2SO_4 \longrightarrow H_2S + I_2 + K_2SO_4$
(d) $K_2Cr_2O_7 + FeSO_4 + H_2SO_4 \longrightarrow$
$Cr_2(SO_4)_3 + Fe_2(SO_4)_3 + KHSO_4$
(e) $KMnO_4 + HNO_2 + H_2SO_4 \longrightarrow MnSO_4 + HNO_3 + K_2SO_4$
(f) $Mn(NO_3)_2 + NaBiO_3 + HNO_3 \longrightarrow$
$HMnO_4 + Bi(NO_3)_3 + NaNO_3$
(g) $Zn + K_2Cr_2O_7 + H_2SO_4 \longrightarrow ZnSO_4 + Cr_2(SO_4)_3 + K_2SO_4$
(h) $NaOH + S \longrightarrow Na_2S_2O_3 + Na_2S$

(i) $KMnO_4 + SO_2 \longrightarrow K_2SO_4 + MnSO_4 + H_2SO_4$
(j) $Na_2SO_3 \longrightarrow Na_2S + Na_2SO_4$
(k) $KIO_3 + H_2SO_3 \longrightarrow KHSO_3 + H_2SO_4 + I_2$
(l) $K_2Cr_2O_7 + HCl \longrightarrow KCl + CrCl_3 + Cl_2$
(m) $KMnO_4 + HCl \longrightarrow MnCl_2 + Cl_2 + KCl$
(n) $K_2CrO_4 + KI + HCl \longrightarrow CrCl_3 + KCl + I_2$
(o) $FeCl_3 + KI \longrightarrow KCl + FeCl_2 + I_2$
(p) $KIO_4 + KI + HCl \longrightarrow KCl + I_2$
(q) $Ag + H_2SO_4 \longrightarrow Ag_1SO_4 + SO_2$
(r) $K_2Cr_2O_7 + KI + H_2SO_4 \longrightarrow Cr_2(SO_4)_3 + K_2SO_4 + I_2$
(s) $PbCrO_4 + KI + HCl \longrightarrow PbCl_2 + CrCl_3 + KCl + I_2$
(t) $As_2S_5 + HNO_3 \longrightarrow H_3AsO_4 + H_2SO_4 + NO_2$

8. Balance the following ionic equations:
(a) $Cu + SO_4^{2-} \longrightarrow Cu^{2+} + SO_2$ (acid)
(b) $I_2 + CdS \longrightarrow HI + S + Cd^{2+}$ (acid)
(c) $Zn + Cr_2O_7^{2-} \longrightarrow Zn^{2+} + Cr^{3+}$ (acid)
(d) $Cr_2O_7^{2-} + Fe^{2+} \longrightarrow Cr^{3+} + Fe^{3+}$ (acid)
(e) $Al + NO_3^- \longrightarrow NH_3 + AlO_2^-$ (base)
(f) $CrO_2 + ClO^- \longrightarrow CrO_4^{2-} + Cl^-$ (base)
(g) $I^- + MnO_4^- \longrightarrow Mn^{2+} + I_2$ (acid)
(h) $MnO_4^- + SO_2 \longrightarrow Mn^{2+} + SO_4^{2-}$ (acid)
(i) $H_2S + Cr_2O_7^{2-} \longrightarrow S + Cr^{3+}$ (acid)
(j) $Cr_2O_7^{2-} + Cl^- \longrightarrow Cr^{3+} + Cl_2$ (acid)
(k) $CrO_4^{2-} + HSnO_2^- \longrightarrow HSnO_3^- + CrO_2^-$ (base)
(l) $P_4 + H_2O \longrightarrow H_2PO_2^- + P_2H_4$ (base)

9. For each part of Problem 8, tell what is oxidized, what is reduced, what is the oxidizing agent, and what is the reducing agent.

REFERENCES

BENSON, SIDNEY W., *Chemical Calculations,* 3rd ed. New York: John Wiley & Sons, Inc, 1971.

SORUM, C. H., and ROBERT S. BOIKESS, *How to Solve Chemistry Problems,* 5th ed., Englewood Cliffs, N.J.: Prentice-Hall, Inc., 1976.

SORUM, C. H., *Introduction to Semimicro Qualitative Analysis,* 4th ed, Englewood Cliffs, N.J.: Prentice-Hall, Inc., 1967 (Chapter 2).

Objectives

1. To understand the basic principles of electrochemistry, including transport of electrical energy through matter, conversion of electrical energy to chemical energy (in electrolysis cells), and conversion of chemical energy into electrical energy (in galvanic or voltaic cells).
2. To understand the reactions that take place in an electrolysis cell, including anode reactions, cathode reactions, and overall reactions, and to be able to write equations for any of these reactions.
3. To understand the quantitative relationships in electrolysis cells, in terms of coulombs, amperes, and faradays and to be able to do calculations involving these quantities as well as equivalent weights.
4. To understand what a galvanic or voltaic cell is and to be able to show for such a cell what the half-reactions are, which is the anode and which is the cathode reaction, and which direction electrons are flowing in the cell.
5. To understand what oxidation or reduction potentials represent, and using these to be able to predict whether a reaction will go spontaneously as written.
6. To understand what effect concentration has on oxidation or reduction potentials, in terms of the Nernst equation, and to be able to do calculations involving the Nernst equation.
7. To understand what a redox equilibrium constant represents and to be able to calculate the constant or use the constant to calculate any of the equilibrium concentrations.
8. To understand the basic principles of the dry cell, the lead storage battery, and the fuel cell.

Twenty

Electrochemistry

In order to have an electric current, which is transported through matter by the conduction of electric charge from one point to another, there must be a force and a charge carrier. The force is usually a battery or some other source of electrical energy. The charge carrier can be either electrons (as in the case of metals) or ions (as in the case of ionic solutions or molten salts). The first case is called *metallic conduction* and the second is *electrolytic conduction.*

In metallic conduction, electrons are "put in" at one end of the wire and "drain out," just as fast, at the other end; the whole wire is always electrically neutral. Since the outer electrons in the metal have very high kinetic energy and are not strongly attached to any particular atom, metallic conduction should be extremely high. But there is some kind of resistance present, probably due to the vibration of the "positive ions" in the metallic bond. At higher temperatures this vibration increases and the conductance decreases.

In electrolytic conduction there are no free electrons, but ions that carry the electric charge. As in metallic conductance there must be electric neutrality in each part of the solution at all times. This can be accomplished either by adding positive or negative ions as fast as they are removed or by removing positive and negative ions at the same rate. Unlike metallic conductance, an increase in temperature increases electrolytic conductance because as the temperature is increased, the ions move faster.

There are two types of electrolytic conductance that we shall be concerned with in this chapter. First we shall consider electrolytic cells, in which electrical energy causes a chemical reaction to take place. Then we shall consider voltaic or galvanic cells, in which a chemical reaction produces electrical energy.

20-1 Electrolytic Cells

In an electrolytic cell, electrical energy is used to cause a chemical reaction to take place. An example of an **electrolytic cell** is shown in Fig. 20-1. The substance that conducts the current is molten sodium chloride. The two electrodes (called the *anode* and *cathode*) are placed into the molten sodium chloride and an electric current is passed through it. This electric current consists of a flow of electrons. Thus electrons are being forced into the cell. The cation (Na^+) moves toward the electrode called the **cathode.** At this electrode, Na^+ ions accept electrons to form sodium metal according to the following reaction:

(20-1) $$Na^+ + e^- \longrightarrow Na$$

Therefore, *reduction (gain of electrons or decrease in oxidation number) takes place at the cathode.*

At the other electrode, called the **anode,** the anion (Cl^-) loses electrons, to form chlorine gas (Cl_2) according to the equation

(20-2) $$2\,Cl^- \longrightarrow Cl_2 + 2\,e^-$$

Therefore, *oxidation (loss of electrons or increase in oxidation number) takes place at the anode.*

In the cell, electrons move from the anode through the external circuit to the cathode. The anions move toward the anode, where oxidation takes

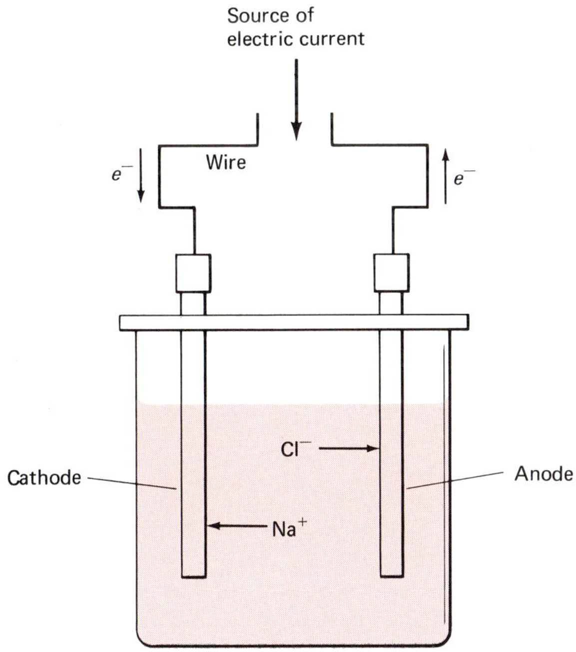

Figure 20-1
A simple electrolytic cell

place. The cations move toward the cathode where reduction takes place. It must be kept in mind that these reactions only take place because of the electrical energy being put into the system. Also, notice that both oxidation and reduction must take place in the cell. The electrons lost by one substance must be gained by another.

To write the overall reaction, the cathode reaction must be doubled (or the anode reaction could be written as $Cl^- \longrightarrow \frac{1}{2} Cl_{2(g)} + e^-$), and then the two half-reactions are added together.

(20-3) Cathode: $2\ Na^+ + 2\ e^- \longrightarrow Na_{(s)}$ (reduction)

(20-4) Anode: $2\ Cl^- \longrightarrow Cl_{2(g)} + 2\ e^-$ (oxidation)

(20-5) Overall: $2\ Na^+ + 2\ Cl^- \longrightarrow 2\ Na_{(s)} + Cl_{2(g)}$

Besides simple cells like the electrolysis of a cation to a free metal at the cathode and an anion to a nonmetal at the anode, there are other types of reactions that can take place at each electrode. Of course, in each case any reaction at the cathode is reduction and at the anode is oxidation. Consider the following examples.

Anode Reactions (Oxidation)

1. Oxidation of an anion to an element, such as

   (20-6) $2\ Cl^- \longrightarrow Cl_{2(g)} + 2\ e^-$

2. Oxidation of an anion or cation in solution to an ion of higher oxidation state, such as

   (20-7) $Fe^{2+} \longrightarrow Fe^{3+} + e^-$

3. Oxidation of a metal anode, such as

   (20-8) $Cu_{(s)} \longrightarrow Cu^{2+} + 2\ e^-$

4. Oxidation of H_2O to produce oxygen:

   (20-9) $2\ H_2O_{(l)} \longrightarrow O_{2(g)} + 4\ H^+ + 4\ e^-$

Cathode Reactions (Reduction)

1. Reduction of a cation to a metal, such as

   (20-10) $Zn^{2+} + 2\ e^- \longrightarrow Zn_{(s)}$

2. Reduction of an anion or cation in solution to an ion of lower oxidation state, such as

   (20-11) $Fe^{3+} + e^- \longrightarrow Fe^{2+}$

3. Reduction of a nonmetal to an anion, such as

$$\text{(20-12)} \qquad Cl_{2(g)} + 2\,e^- \longrightarrow 2\,Cl^-$$

4. Reduction of H_2O to produce hydrogen gas:

$$\text{(20-13)} \qquad 2\,H_2O_{(l)} + 2\,e^- \longrightarrow H_{2(g)} + 2\,OH^-$$

In a particular electrolysis cell there are often several possible reactions that could take place. For example, in any aqueous solution there is always the possibility of the water being oxidized or reduced more readily than the cation or anion present. The electrolysis of aqueous sodium chloride is one case where this takes place. It is found that hydrogen gas is produced at the cathode [see Eq. (20-13)] and chlorine gas at the anode. Thus, the overall reaction for this is

$$\text{(20-14)} \qquad 2\,Na^+ + 2\,Cl^- + 2\,H_2O \longrightarrow H_{2(g)} + Cl_{2(g)} + 2\,OH^- + 2\,Na^+$$

This is one method that can be used for the commercial preparations of hydrogen and chlorine, with an aqueous solution of sodium hydroxide as a third product.

As another example, it is found that, upon electrolysis of aqueous sodium sulfate, hydrogen gas is produced at the cathode [Eq. (20-13)] and oxygen gas at the anode [Eq. (20-9)]. Therefore, the overall reaction here is simply

$$2\,H_2O_{(l)} \longrightarrow 2\,H_{2(g)} + O_{2(g)}$$

Electrolysis of water by this method can be accomplished with any solution as long as the ions in the solution (which must be present in order for electrolysis to take place) are not oxidized or reduced more readily than water.

Another application of electrolysis cells is in the final purification to obtain pure copper.* In this purification the impure copper acts as the anode and a thin plate of pure copper acts as the cathode, with the electrolyte being copper(II) sulfate. The reactions that take place in this electrolysis cell are

Anode: $\underset{\text{(from the impure sample)}}{Cu} \longrightarrow Cu^{2+} + 2\,e^-$

Cathode: $\underset{\text{(formed at the anode)}}{Cu^{2+}} + 2\,e^- \longrightarrow \underset{\text{(pure)}}{Cu_{(s)}}$

The copper that forms on the cathode is better than 99.9% pure. Other metals found in copper ores, such as silver, gold, and platinum, are deposited

*Other metals besides copper that are prepared in nearly pure form using an electrolytic cell are aluminum, sodium, potassium, magnesium, and calcium.

as a sludge at the bottom of the cell and are further refined, and more active metals such as nickel remain in solution and are later recovered, making the entire operation quite economical. This process does require considerable amounts of electricity, however, and thus cheap electrical power must be available near the electrolytic cell plant site.

20-2 Quantitative Relationships in Electrolytic Cells

In electrolysis cells, one very important consideration is the amount of electrical energy necessary for a given amount of material to react. One of the most commonly used units in expressing amount of electrical energy is the **coulomb** (coul). Another very important unit is the **faraday** (F), which is the amount of energy required for the flow of 1 mole of electrons. To three significant digits, 1 faraday equals 96,500 coulombs. The unit commonly used to measure current flow is the **ampere** (A), which is a rate such that 1 coulomb passes a given point in the circuit in 1 second. Therefore, A = coul/s, or coulombs = (amperes)(seconds). Keep in mind that a coulomb is an *amount* of electrical energy and an ampere is a *rate* at which it flows.

The **equivalent weight,** or gram-equivalent weight, of a substance is the mass in grams that is equivalent to 1 mole of electrons; that is, it is the amount that reacts with or is produced by 1 mole of electrons. The faraday is therefore the amount of electrical energy necessary to produce 1 equivalent of pure substance. From the definitions above it can be seen that to determine the equivalent weight of a substance it is necessary to know how many moles of electrons are transferred per mole of substance. Therefore, to calculate the equivalent weight of a substance the formula weight is divided by the number of electrons transferred. Consider the following problem examples.

PROBLEM EXAMPLE 20-1

Calculate the equivalent weight of each of the following:

(a) aluminum
(b) Fe, in the reaction $Fe \longrightarrow Fe^{2+} + 2\,e^-$
(c) Fe, in the reaction $Fe^{3+} + 3\,e^- \longrightarrow Fe$
(d) Fe, in the reaction $Fe^{3+} + e^- \longrightarrow Fe^{2+}$

SOLUTION:

(a) Since Al will always lose three electrons to form Al^{3+}, the equivalent weight is 27.0/3 = **9.00 g.**
(b) In this case, the equivalent weight of Fe is, 55.8/2 = **27.9 g.**
(c) 55.8/3 = **18.6 g**
(d) 55.8/1 = **55.8 g**

Since 1 faraday is the amount of electrical energy necessary for the transfer of 1 mole of electrons, it is the amount of electrical energy necessary to produce 1 equivalent of any substance. For example, referring to Problem Example 20-1, it would take 1 F or 96,500 coul to produce 9.00 g of Al, 27.9 g of Fe from $FeCl_2$, and 18.6 g of Fe from $FeCl_3$. We can also use relationships directly between equivalents and moles, as follows:

$$1 \text{ mole Al} = 3 \text{ equivalents Al}$$

$$1 \text{ mole Fe} = 2 \text{ equivalents Fe (from } FeCl_2)$$

$$1 \text{ mole Fe} = 3 \text{ equivalents Fe (from } FeCl_3)$$

Consider the following problem examples of calculations involving electrolysis cells:

PROBLEM EXAMPLE 20-2

How many coulombs would it take to deposit 1.00 g of silver from solution?

SOLUTION:

Since Ag is always +1 in compounds, the equivalent weight of Ag is always 108/1 = 108.

$$1.00 \text{ g Ag} \times \frac{1 \text{ equiv Ag}}{108 \text{ g Ag}} \times \frac{96{,}500 \text{ coul}}{1 \text{ equiv Ag}} = \textbf{894 coul}$$

PROBLEM EXAMPLE 20-3

How many grams of Zn could be deposited by 96.5 coul?

SOLUTION:

Equivalent weight of Zn = atomic weight/2 = 65.4/2 = 32.7.

$$96.5 \text{ coul} \times \frac{1 \text{ equiv Zn}}{96{,}500 \text{ coul}} \times \frac{32.7 \text{ g Zn}}{1 \text{ equiv Zn}} = \mathbf{3.27 \times 10^{-2} \text{ g Zn}}$$

PROBLEM EXAMPLE 20-4

How long would it take a current of 100 A to deposit 10.0 g of Fe from a solution of $FeCl_2$?

SOLUTION:

$$100 \text{ A} = 100 \text{ coul/s}$$

Equivalent weight of Fe in $FeCl_2$ = 55.8/2 = 27.9.

$$10.0 \text{ g Fe} \times \frac{1 \text{ equiv Fe}}{27.9 \text{ g Fe}} \times \frac{96{,}500 \text{ coul}}{1 \text{ equiv Fe}} \times \frac{1 \text{ s}}{100 \text{ coul}} = \textbf{346 s}$$

PROBLEM EXAMPLE 20-5

How many grams of Al could be obtained by a current of 20 A flowing for 1 hr?

SOLUTION:

$$1\ \text{hr} \times \frac{60\ \text{min}}{1\ \text{hr}} \times \frac{60\ \text{s}}{1\ \text{min}} = 3600\ \text{s}$$

Since coul = A × s, the number of coulombs is

$$(20\ \text{A})(3600\ \text{s}) = 72{,}000\ \text{A-s} = 72{,}000\ \text{coul}$$

$$72{,}000\ \text{coul} \times \frac{1\ \text{equiv Al}}{96{,}500\ \text{coul}} \times \frac{9.00\ \text{g Al}}{1\ \text{equiv Al}} = \mathbf{6.7\ g\ Al}$$

20-3 Voltaic or Galvanic Cells

In a **voltaic** or **galvanic cell,** a chemical reaction produces electrical energy. In a cell of this type, the oxidation-reduction reaction is carried out in such a way that the electrons being transferred from the reducing agent to the oxidizing agent travel through a wire and thus produce an electric current.

One example of a voltaic cell involves zinc metal in a solution containing Cu^{2+} ions. The overall reaction that takes place is

$$Zn_{(s)} + Cu^{2+}{}_{(aq)} \longrightarrow Zn^{2+}{}_{(aq)} + Cu_{(s)}$$

If the reaction is carried out in such a way that the zinc metal does not come in contact with the Cu^{2+} ions, the energy given off may be obtained as electricity. This is done in a cell like that shown in Fig. 20-2. In this cell, the solution containing the Cu^{2+} ions is separated from the solution in which

Figure 20-2
A simple voltaic cell using (a) a porous partition, and (b) a salt bridge

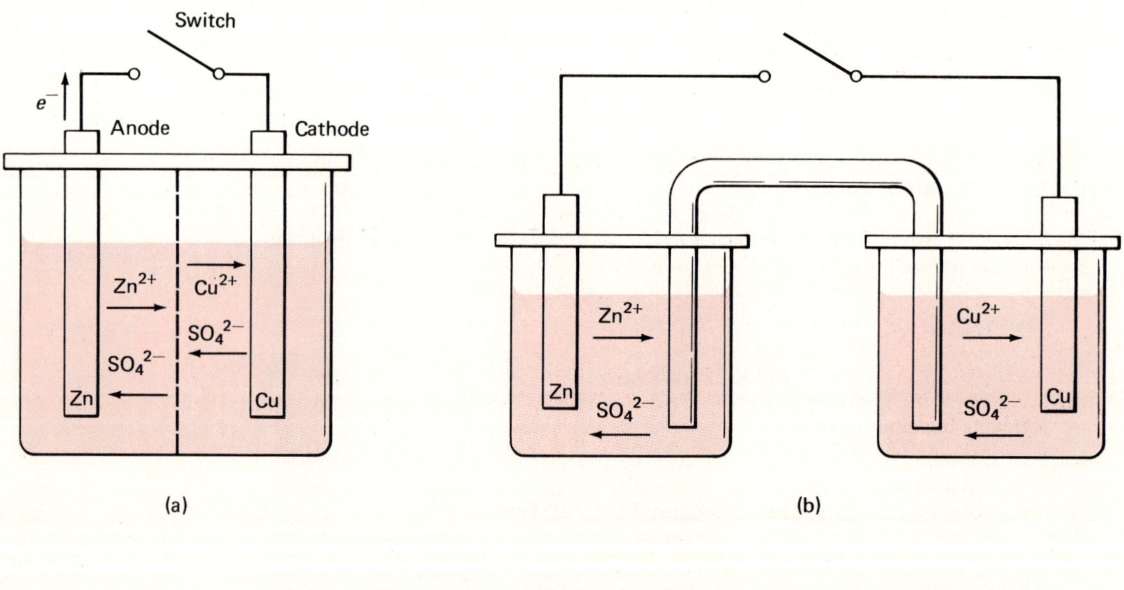

the zinc metal is placed. The two solutions are separated by a porous plate, as shown in Fig. 20-2(a). They can also be separated by a salt bridge, which contains an ionic substance and allows ions to pass from a solution in one container to a solution in another container, as shown in Fig. 20-2(b). When the switch in the external circuit is closed, the reaction will begin to take place since electrons can then pass through from the anode to the cathode. The two half-reactions that take place are

$$\text{Anode:} \qquad Zn \longrightarrow Zn^{2+} + 2\,e^- \quad \text{(oxidation)}$$

$$\text{Cathode:} \qquad Cu^{2+} + 2\,e^- \longrightarrow Cu \quad \text{(reduction)}$$

20-4 Standard Electrode Potentials

In Chapter 8 we discussed the electromotive series of metals. In this series, sometimes called the *activity* series, any metal will displace a lower metal from a solution of its ions. The basis of the series is the *standard electrode potentials (E^0)*, as given in Table 20-1. These give the potential, in volts, for a particular half-reaction to take place. The potentials given in Table 20-1 are written as **reduction potentials,** where each half-reaction is written as a reduction reaction. If these were written as **oxidation potentials,** the reaction

TABLE 20-1 *Standard reduction potentials for some common metals—for aqueous solutions of ions at 1.0 M concentration at 25°C and H_2 gas at 1.0 atm pressure and 25°C.*

REDUCTION HALF-CELL EQUATION	E^0 (V)
$Li^+ + 1\,e^- \rightleftarrows Li$	−3.04
$K^+ + 1\,e^- \rightleftarrows K$	−2.92
$Ba^{2+} + 2\,e^- \rightleftarrows Ba$	−2.91
$Ca^{2+} + 2\,e^- \rightleftarrows Ca$	−2.87
$Na^+ + 1\,e^- \rightleftarrows Na$	−2.71
$Mg^{2+} + 2\,e^- \rightleftarrows Mg$	−2.36
$Al^{3+} + 3\,e^- \rightleftarrows Al$	−1.66
$Zn^{2+} + 2\,e^- \rightleftarrows Zn$	−0.76
$Cr^{3+} + 3\,e^- \rightleftarrows Cr$	−0.74
$Fe^{2+} + 2\,e^- \rightleftarrows Fe$	−0.44
$Cd^{2+} + 2\,e^- \rightleftarrows Cd$	−0.40
$Ni^{2+} + 2\,e^- \rightleftarrows Ni$	−0.25
$Sn^{2+} + 2\,e^- \rightleftarrows Sn$	−0.14
$Pb^{2+} + 2\,e^- \rightleftarrows Pb$	−0.13
$2\,H^+ + 2e^- \rightleftarrows H_2$	0.00*
$Cu^{2+} + 2\,e^- \rightleftarrows Cu$	+0.34
$Hg_2^{2+} + 2\,e^- \rightleftarrows 2\,Hg$	+0.79
$Ag^+ + 1\,e^- \rightleftarrows Ag$	+0.80
$Hg^{2+} + 2\,e^- \rightleftarrows Hg$	+0.85
$Au^{3+} + 3\,e^- \rightleftarrows Au$	+1.42

*Hydrogen is included since it is used as the standard.

would be the reverse of that given, and the sign would be opposite. For example, for the oxidation half-reaction, $Li \longrightarrow Li^+ + e^-$, the potential ($E^0$) is + 3.04 volts ($V$).

It is not possible to measure the potential for any half-reaction by itself, it can only be measured as compared to something else. Thus, each of these is compared to the $2\ H^+ + 2\ e^- \longrightarrow H_2$ half-reaction; that is, this half-reaction is considered to have a standard potential of 0.00 (at 25°C and 1 atm pressure). Then all the other half-reactions are compared to this. The cell used to carry this out is shown in Fig. 20-3.

The metal in a solution containing its ions is used as one electrode and the hydrogen gas being passed over platinum metal is the other electrode. Either the reaction of metal ion going to metal ($M^{n+} + n\ e^- \longrightarrow M$) or metal going to metal ion ($M \longrightarrow M^{n+} + n\ e^-$) will take place, and the voltage is measured. For example, if the Zn–Zn^{2+} half-cell is used for the electrode, the Zn goes to Zn^{2+} and the potential is 0.76 V; therefore, the potential for the reverse reaction ($Zn^{2+} + 2\ e^- \longrightarrow Zn$) is −0.76 V, (as compared to the hydrogen electrode) as given in Table 20-1. If Cu and Cu^{2+} ion are used as the other electrode, along with the hydrogen electrode, the reduction half-reaction ($Cu^{2+} + 2\ e^- \longrightarrow Cu$) takes place and produces a potential of 0.34 V (as compared to the hydrogen electrode). In all these cells the concentration of the ions is 1.0 M, which is taken as the standard. Thus, the potentials given here are called *standard electrode potentials* or, for the particular values given in Table 20-1, *standard reduction potentials.* If the concentrations are not 1.0 M, the potential can be calculated using the Nernst equation, discussed in Section 20-5.

To determine if an oxidation-reduction reaction will take place spontaneously, the two half-reactions are written and the potentials added. If

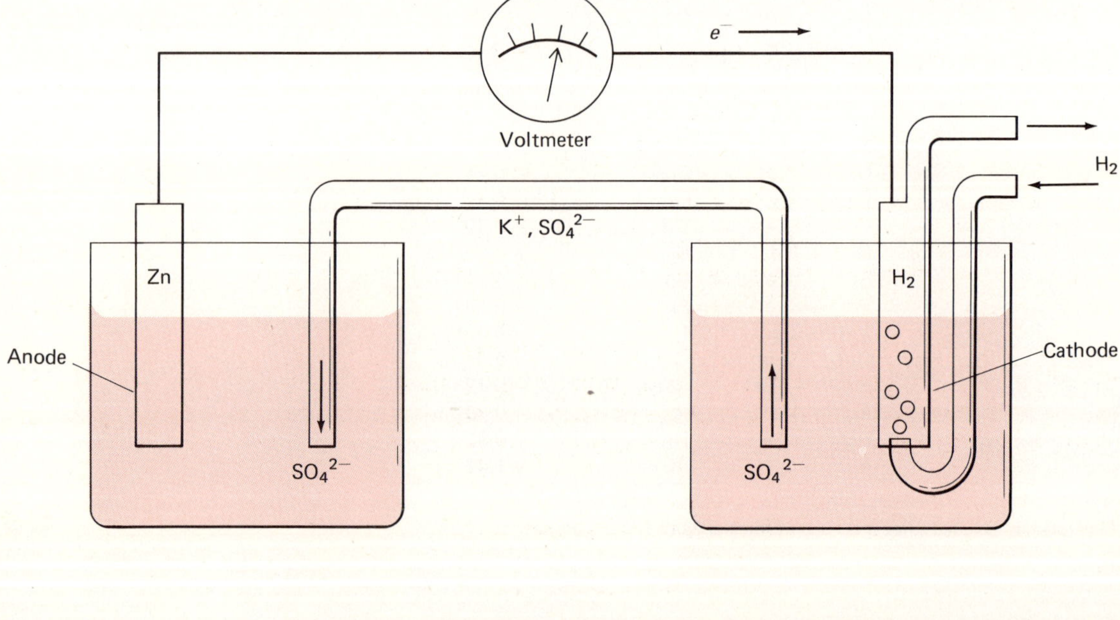

Figure 20-3
Cell used to measure standard potentials

the potential is positive the reaction will go spontaneously; if it is negative, it will not. Consider the following examples (all concentrations 1.00 *M*):

1. $Mg^{2+} + Zn \longrightarrow Zn^{2+} + Mg$

$$\begin{array}{rcll} Mg^{2+} + 2\,e^- & \longrightarrow & Mg & E^0 = -2.36\ \text{V} \\ Zn & \longrightarrow & Zn^{2+} + 2\,e^- & E^0 = +0.76\ \text{V} \\ \hline Mg^{2+} + Zn & \longrightarrow & Mg + Zn^{2+} & E^0 = -1.60\ \text{V} \end{array}$$

Therefore, this reaction will not go spontaneously as written (the reverse reaction would since the potential would be +1.60 V).

2. $3\,Mg + 2\,Al^{3+} \longrightarrow 3\,Mg^{2+} + 2\,Al$

$$\begin{array}{rcll} 3(Mg & \longrightarrow & Mg^{2+} + 2\,e^-) & E^0 = +2.36\ \text{V} \\ 2(Al^{3+} + 3\,e^- & \longrightarrow & Al) & E^0 = -1.66\ \text{V} \\ \hline 3\,Mg + 2\,Al^{3+} & \longrightarrow & 3\,Mg^{2+} + 2\,Al & E^0 = +0.70\ \text{V} \end{array}$$

This reaction will go spontaneously as written. (The potential is *not* multiplied by 3 or 2 in the half-reactions. The reason for this will be shown in Section 20-5 with the Nernst equation.)

Calculations of this type not only tell what reactions will go spontaneously but also tell us something about the reverse reaction; that is, in the first example the given reaction will not go spontaneously but the reverse reaction would. This gives the same information as used to predict replacement reactions in Chapter 8; Zn metal will not displace Mg^{2+} ion from solution but Mg metal will displace Zn^{2+} ion.

20-5 The Nernst Equation*

All the potentials used in the previous section involved solutions in which the concentration of the ions was 1 *M*. This is known as the *standard state*. As the concentration changes, the potential for the half-reaction also changes according to the Nernst equation, which states that

$$E = E^0 - \frac{0.059}{n}\log\frac{[\text{products}]^x}{[\text{reactants}]^y} \quad (20\text{-}15)$$

where E^0 = the standard oxidation or reduction potential,
n = the number of electrons transferred,
x and y = the coefficients in the balanced equation, and
0.059 = a constant at 25°C.

In using this equation it is very important always to write the log factor as the concentration of products over reactants, just as in equilibrium constants, for the reaction as written; that is, if E^0 is an oxidation potential (from the reduced to the oxidized state, often written as $E^0_{\text{red, ox}}$) it is [oxidized state]/

*This section is optional and can be omitted without loss of continuity.

[reduced state]. And if E^0 is a reduction potential ($E^0_{\text{ox, red}}$) it is [reduced state]/[oxidized state].

For anything in the free state, such as a free metal (M^0), the concentration is taken as one. It is actually the activity that is one but for most purposes molar concentration of one is practically the same.

Consider the following problem examples.

PROBLEM EXAMPLE 20-6

Calculate the half-cell potential for the reaction

$$Cu^{2+} + 2\,e^- \longrightarrow Cu^0$$

when the $[Cu^{2+}]$ is 1×10^{-4}.

SOLUTION:

Since the $[Cu^{2+}]$ is less than one, we would expect the potential to decrease (whether it is a positive or a negative value, it should become smaller). The value of E^0 for this half-reaction is +0.34 V (from Table 20-1). The Nernst equation is

$$E = E^0 - \frac{0.059}{2}\log\frac{[Cu^0]}{[Cu^{2+}]}$$

Putting the known values into this (recall that $[Cu^0] = 1$) gives

$$E = 0.34 - \frac{0.059}{2}\log\frac{1}{1 \times 10^{-4}}$$

$$= 0.34 - \left[\left(\frac{0.059}{2}\right)(\log 1 \times 10^4)\right] = \left[0.34 - \left(\frac{0.059}{2}\right)(4)\right]$$

$$= 0.34 - (0.059)(2) = 0.34 - 0.12 = \mathbf{0.22\ V}$$

This value is less than 0.34, as predicted.

PROBLEM EXAMPLE 20-7

For the reaction $2\,Al + 3\,Cu^{2+} \longrightarrow 2\,Al^{3+} + 3\,Cu$, calculate the potential if the $[Cu^{2+}]$ is 1×10^{-4} and the $[Al^{3+}]$ is 1×10^{-3}.

SOLUTION:

This can be solved either as two separate steps or as the overall reaction.

(a) $$E_{Al, Al^{3+}} = 1.66 - \frac{0.059}{3}\log(1 \times 10^{-3}) = 1.66 + 0.059 = 1.72\ V$$

$$E_{Cu^{2+}, Cu} = 0.34 - \frac{0.059}{2}\log\frac{1}{1 \times 10^{-4}} = 0.34 - (0.059)(2) = 0.22\ V$$

$$E_{\text{overall}} = (1.72 + 0.22)\ V = \mathbf{1.94\ V}$$

In this case, it is not necessary to multiply each of the half-reactions by 2 and 3 since we are using the individual cases and simply adding the two new potentials. It should be obvious, from your knowledge of logarithms, that even if the Al, Al^{3+} half-reactions were multiplied through by 2, for example, the answer would have come out the same since the 2 would then show up both in the n term and as a second power in the log term, as

$$1.66 - \frac{(0.059)}{(3)(2)} \log (1 \times 10^{-3})^2 = 1.66 - \frac{0.059}{(3)\cancel{(2)}}(-3)\cancel{(2)}$$

(b) Using the overall reaction, from the sum of the two half-reactions, it can be seen that there are six electrons involved:

$$\begin{array}{rcl} 2\,\text{Al} & \longrightarrow & 2\,\text{Al}^{3+} + 6\,e^- \\ 3\,\text{Cu}^{2+} + 6\,e^- & \longrightarrow & 3\,\text{Cu} \\ \hline 2\,\text{Al} + 3\,\text{Cu}^{2+} & \longrightarrow & 2\,\text{Al}^{3+} + 3\,\text{Cu} \end{array}$$

$$E^0_{\text{overall}} = 1.66 + 0.34 = 2.00 \text{ V (from Table 20-1)}$$

Since the overall reaction involves six electrons,

$$E_{\text{overall}} = 2.00 - \frac{0.059}{6} \log \frac{[\text{Al}^{3+}]^2}{[\text{Cu}^{2+}]^3}$$

$$= 2.00 - \frac{0.059}{6} \log \left(\frac{10^{-6}}{10^{-12}}\right) = 2.00 - \frac{0.059}{6} \log (1 \times 10^6)$$

$$= 2.00 - \frac{0.059}{\cancel{6}} \cancel{(6)} = 2.00 - 0.059 = \mathbf{1.94\ V}$$

20-6 Equilibrium Constants from Redox Equations*

In Section 20-4 it was pointed out that if the overall potential (E) is positive, the reaction takes place spontaneously; if E is negative, the reverse reaction takes place spontaneously. Also, in Section 20-5 it was shown that the potential varies with concentration of the ions involved according to the Nernst equation, Eq. (20-15).

We now want to consider the case where the potential is zero. This means that neither the forward nor reverse reactions will take place spontaneously; that is, the reaction has reached a point of equilibrium. Therefore, at equilibrium the $[\text{products}]^x/[\text{reactants}]^y$ will be constant, called k_{eq} or often k_{redox}. Putting this into the Nernst equation gives

(20-16) $$E = E^0 - \frac{0.059}{n} \log k_{\text{redox}}$$

Since E must equal zero,

(20-17) $$E^0 = \frac{0.059}{n} \log k_{\text{redox}}$$

*This section is optional and can be omitted without loss of continuity.

or

$$\log k_{redox} = \frac{nE^0}{0.059} \qquad (20\text{-}18)$$

Consider the calculations in the following problem example.

PROBLEM EXAMPLE 20-8

Calculate the k_{redox} for the reaction

$$Zn + Cu^{2+} \rightleftharpoons Zn^{2+} + Cu$$

SOLUTION:

$$E^0 = +0.76 + 0.34 = 1.10 \quad \text{(from Table 20-1)}$$

Since there are two electrons transferred in this reaction, using Eq. (20-18) gives

$$\log k_{redox} = \frac{(2)(1.10)}{0.059} = 37.3$$

$$k_{redox} = 10^{37.3} = 10^{0.3} \times 10^{37} = 2.0 \times 10^{37}$$

This large value for the equilibrium constant means that the reaction goes practically to completion.

These constants can then be used to calculate equilibrium concentrations. Consider the following problem example.

PROBLEM EXAMPLE 20-9

Calculate the concentration of Zn^{2+} in equilibrium with 0.30 M Cr^{3+}, from the reaction

$$3\,Zn^{2+} + 2\,Cr \rightleftharpoons 2\,Cr^{3+} + 3\,Zn$$

SOLUTION:

$$E^0 = -0.76 + 0.74 = -0.02$$

In this reaction there are six electrons transferred and therefore

$$\log k_{redox} = \frac{(6)(-0.02)}{0.059} = -2$$

and

$$k_{redox} = 1 \times 10^{-2}$$

Therefore, at equilibrium

$$\frac{[Cr^{3+}]^2}{[Zn^{2+}]^3} = 1 \times 10^{-2}$$

or

$$\frac{(0.30)^2}{[Zn^{2+}]^3} = 1 \times 10^{-2}$$

$$[Zn^{2+}]^3 = \frac{(0.30)^2}{1 \times 10^{-2}} = \frac{9 \times 10^{-2}}{1 \times 10^{-2}} = 9$$

$$[Zn^{2+}] = \sqrt[3]{9} = \mathbf{2.1}\ M$$

This can also be used to determine in which direction a reaction will proceed in order to reach equilibrium. Consider the following problem example.

PROBLEM EXAMPLE 20-10

If a solution is made up that contains 0.20 mol/l of Fe^{2+}, 0.30 mol/l of Cd^{2+} and solid Fe and Cd, which way will the reaction proceed (that is, as $Fe + Cd^{2+} \rightleftharpoons Cd + Fe^{2+}$ or $Cd + Fe^{2+} \rightleftharpoons Fe + Cd^{2+}$)?

SOLUTION:

To solve this, we shall take one of the possible reactions, calculate the k_{redox}, and then calculate the [products]/[reactants]. If this value is less than k_{redox}, the reaction will go forward so as to reach the point where this is equal to k_{redox}. If it is greater than k_{redox}, the reaction will proceed in the reverse direction. Taking

$$Fe + Cd^{2+} \rightleftharpoons Cd + Fe^{2+}$$

$$E^0 = +0.44 - 0.40 = +0.04$$

$$\log k_{redox} = \frac{(2)(0.04)}{0.059} = 1.4 \quad \text{[from Eq. (10-18)]}$$

$$k_{redox} = 10^{1.4} = 10^{0.4} \times 10^1 = 2.5 \times 10^1 = 25$$

and

$$\frac{[Fe^{2+}]}{[Cd^{2+}]} = \frac{0.20}{0.30} = 0.67$$

Therefore, since 0.67 is less than 25, the reaction will proceed **to the right** in order to reach equilibrium.

In Problem Example 20-10, we could also calculate what the concentrations would be at equilibrium, as follows:

As this reaction proceeds, $[Cd^{2+}]$ will decrease by some amount (x) and $[Fe^{2+}]$ will increase by an equivalent amount. Therefore, at equilibrium,

$$\frac{0.20 + x}{0.30 - x} = 25$$

$$(0.30)(25) - 25x = 0.20 + x$$

$$7.5 - 25x = 0.20 + x$$

$$7.3 = 26x$$

$$x = \frac{7.3}{26} = 0.28$$

Therefore,

$$[Cd^{2+}] = 0.30 - 0.28 = 0.02$$

and

$$[Fe^{2+}] = 0.20 + 0.28 = 0.48$$

20-7 Practical Cells

Any combination of half-reactions that gives a positive potential could be used to produce electrical energy; however, there are often difficulties that must be overcome in setting up such a cell. Three common types of practical voltaic or galvanic cells are the dry cell, the lead storage battery, and the fuel cell.

The common *dry cell* (potential, 1.5 V) consists of a zinc container that acts as the anode and a carbon cathode, as shown in Fig. 20-4. The electrolyte is a moist paste of ammonium chloride, manganese(IV) oxide (manganese dioxide), and carbon. The half-cell reaction at the anode (oxidation) is

$$Zn \longrightarrow Zn^{2+} + 2\,e$$

The half-cell reaction at the cathode (reduction) is much more complex; however, it appears to be

$$2\,MnO_2 + 2\,NH_4^+ + 2\,H_2O + 2\,e^- \longrightarrow 2\,Mn(OH)_3 + 2\,NH_3$$

If the manganese(IV) oxide were absent, the cathode reaction would be

$$2\,NH_4^+ + 2\,e^- \longrightarrow 2\,NH_3 + H_2$$

The hydrogen liberated would insulate the carbon cathode from the mixture of ammonium chloride and carbon, and the reaction would soon stop. The presence of manganese(IV) oxide prevents the formation of hydrogen, making the dry cell a practical, continuous source of electricity.

The overall reaction for the dry cell is obtained by adding the two half-cell reactions.

$$Zn + 2\,MnO_2 + 2\,NH_4^+ + 2\,H_2O \longrightarrow Zn^{2+} + 2\,Mn(OH)_3 + 2\,NH_3$$

In the *lead storage battery*, the half-cells are lead and lead(IV) oxide. The lead is the anode and the lead(IV) oxide is the cathode, as shown in Fig. 20-5.

$$Pb \longrightarrow Pb^{2+} + 2\,e^-$$

$$4\,H^+ + PbO_2 + 2\,e^- \longrightarrow Pb^{2+} + 2\,H_2O$$

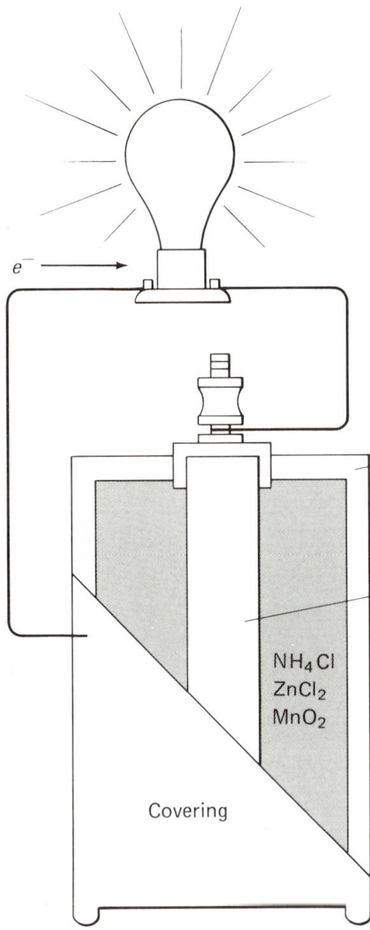

Figure 20-4
The dry cell

The sulfate ion present forms lead(II) sulfate at both electrodes. This insoluble compound adheres to each electrode. The overall reaction for the storage battery is

$$Pb + PbO_2 + 2\,SO_4^{2-} + 4\,H^+ \underset{\text{charge}}{\overset{\text{discharge}}{\rightleftharpoons}} 2\,PbSO_4 + 2\,H_2O$$

To charge a lead storage battery, it must be converted into an electrolytic cell with an outside source of energy. Each cell of a lead storage battery has a voltage of about 2 V. Usually three units are placed in series, giving 6 V.

The *fuel cell* is another type of voltaic cell, which is becoming increasingly important. In one type of fuel cell, hydrogen reacts with oxygen to form water and electrical energy. The two half-reactions are

Anode: $H_{2(g)} + 2\,OH^- \longrightarrow 2\,H_2O_{(l)} + 2\,e^-$

Cathode: $O_{2(g)} + 2\,H_2O_{(l)} + 4\,e^- \longrightarrow 4\,OH^-$

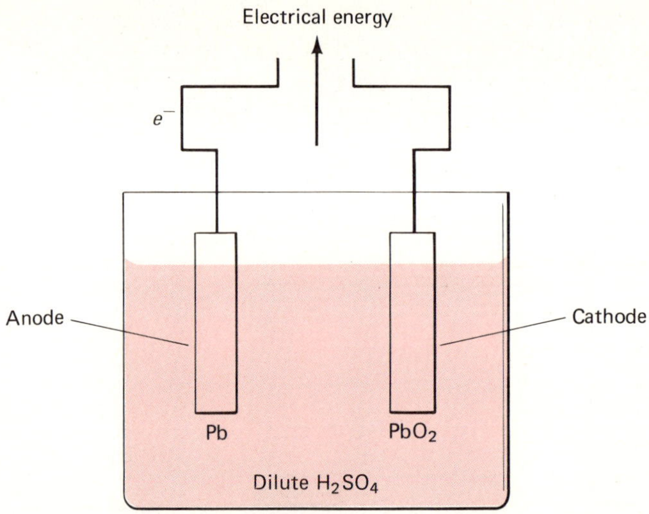

Figure 20-5
One cell of a lead storage battery

These reactions are carried out in basic solution since it is less corrosive than acid. A catalyst must be present in order for the H_2 and O_2 to react to form water (the overall reaction is $2\ H_{2(g)} + O_{2(g)} \longrightarrow 2\ H_2O_{(l)}$).

Fuel cells have been used quite extensively in spacecraft and show promise of being used more for other sources of electrical energy also.

SUMMARY

1. *Electrolysis, or electrolytic, cells* involve chemical reactions that take place when electrical energy is put into the system. In these cells *oxidation takes place at the anode* and *reduction at the cathode.* The anode reactions may involve oxidation of an anion to a free element, oxidation of an anion or cation to an ion of higher oxidation state, oxidation of a metal to a cation, or the oxidation of water to give O_2 gas and H^+ ion. The cathode reactions may involve reduction of a cation to the free metal, reduction of an ion to a lower oxidation state, reduction of a nonmetal to an anion, or reduction of water to give H_2 gas and OH^- ion.
2. In doing quantitative calculations involving electrolysis cells, the units *coulomb* (an amount of electrical energy) and *ampere* (a rate at which it flows) are frequently used. The relationships between these are

$$\text{coul} = (\text{A})(\text{s}) \quad \text{or} \quad \text{A} = \text{coul/s}$$

3. The *equivalent weight* of a substance is the formula weight divided by the number of electrons transferred in the particular reaction under consideration. Since a *faraday* is the amount of electrical energy

necessary to cause the transfer of 1 mole of electrons, it takes 1 faraday (96,500 coul) to cause the reaction of 1 equivalent of a substance; that is, it takes 96,500 coul to plate out 1 equivalent of a metal (such as $\frac{27}{3} = 9$ g of Al) or form 1 equivalent of a gas (such as $\frac{35.5}{1}$ or $\frac{71}{2} = 35.5$ g Cl_2).

4. A *voltaic or galvanic cell* is one in which a spontaneous chemical reaction generates an electric current. For example, the reaction

$$Zn + Cu^{2+} \longrightarrow Zn^{2+} + Cu$$

takes place spontaneously, and if this is set up as a cell and the potential is measured, it is found to be 1.10 V (at 25°C with all concentrations 1.00 *M*). Using a table of *standard potentials*, the overall potential for two half-reactions can be calculated; if this is positive, the reaction will take place spontaneously; if negative, it will not. These standard potentials are just relative values (since there is no way of measuring the absolute value of a half-reaction) taking the reaction $2\ H^+ + 2\ e^- \longrightarrow H_2$ as being zero and comparing all others to this.

5. If the concentrations are not 1 *M* in the solutions involved in a voltaic cell, the *Nernst equation* can be used to calculate the potential with the given concentration. The Nernst equation is

$$E = E^0_{M^0,\ M^{n+}} - \frac{0.059}{n} \log \frac{[M^{n+}]}{[M^0]}$$

Where E is the potential at the new concentrations, E^0 is the standard potential, 0.059 is a constant, n is the number of electrons transferred in the reaction, and the concentration of anything in the free state (M^0) is taken as zero.

6. The *equilibrium constant* for any redox reaction can be calculated from the relationship

$$\log k_{redox} = \frac{nE^0}{0.059}$$

7. Three common practical voltaic cells are the *dry cell*, the *lead storage battery*, and the *fuel cell*.

Upon completion of this chapter you should be able to do the following:

1. Given any half-reaction, tell if it is an oxidation or a reduction, and identify at which electrode (anode or cathode) the reaction would take place. (Exercise 3)
2. Given any electrolysis reaction, write equations for the two half-reactions that take place, including cases where water is more readily oxidized or reduced than the anion or cation present. (Exercise 4)

3. Given any redox equation, determine the equivalent weight of any of the reactants. (1)
4. Given any electrolysis cell, calculate the coulombs, amperes, or time (given the other two quantities) to cause a given amount of substance to react. (2, 3, 4, 5)
5. Given a table of standard oxidation or reduction potentials, calculate the overall standard potential for a given reaction in which the concentrations of all ions are one molar and tell if the reaction will occur spontaneously as written. (6, 7)
6. Given a table of standard potentials, calculate the potential for a given reaction (overall or half-reaction), at any given concentration, using the Nernst equation. (8, 9)
7. Given any redox equation and a table of E^0 values, calculate the value of the redox equilibrium constant (k_{redox}) for the reaction. (10)
8. Given any two half-reactions, either for an electrolysis cell or a galvanic (voltaic) cell, diagram the cell, labeling the anode and cathode, showing which way the ions and electrons are moving and writing the oxidation and reduction half-reactions.
9. Given any of the terms in Exercise 1, give a definition or explanation in your own words.

EXERCISES

1. Define or explain the following terms:
   (a) oxidation
   (b) reduction
   (c) anode
   (d) cathode
   (e) ampere
   (f) coulomb
   (g) faraday
   (h) equivalent weight
   (i) electrolytic cell
   (j) voltaic or galvanic cell
   (k) salt bridge
   (l) E^0
   (m) positive E^0
   (n) negative E^0
   (o) oxidation potential
   (p) reduction potential
2. Distinguish between
   (a) oxidation half-reaction and reduction half-reaction
   (b) cathode and anode, in regard to oxidation and reduction
   (c) electrolytic cells and voltaic or galvanic cells
   (d) negative E^0 values and positive E^0 values for standard reduction potentials
   (e) standard reduction potentials and standard oxidation potentials
   (f) fuel cell and dry cell
3. Classify each as oxidation or reduction and tell which electrode (anode or cathode) it would take to place at
   (a) $Ni^{2+} + 2e^- \rightleftharpoons Ni$
   (b) $Fe \rightleftharpoons Fe^{3+} + 3e^-$
   (c) $Zn \rightleftharpoons Zn^{2-} + 2e^-$
   (d) $Pb^{2+} + 2e^- \rightleftharpoons Pb$
4. In the electrolysis of aqueous sodium chloride, Cl_2 gas is produced at the anode and H_2 gas at the cathode. Write the half-reactions that take place.

PROBLEMS

Electrolytic cells

1. Calculate the equivalent weight of each of the following:
   (a) Zn
   (b) H_2
   (c) Al
   (d) Sn, for the reaction
   $Sn \longrightarrow Sn^{2+} + 2\,e^-$
   (e) I_2, for the reaction
   $2\,I^- \longrightarrow I_2 + 2\,e^-$
   (f) Mg
   (g) O_2
   (h) Ag
   (i) Fe, for the reaction
   $Fe \longrightarrow Fe^{3+} + 3\,e^-$
   (j) Br_2, for the reaction
   $2\,Br^- \longrightarrow Br_2 + 2\,e^-$

2. Calculate the number of grams of aluminum that could be deposited by each of the following:
   (a) 5000 coul
   (b) a current of 100 A flowing for 10 minutes
   (c) 6500 coul
   (d) a current of 220 A flowing for 25 minutes

3. Calculate the time, in seconds, that it would take to carry out each of the following:
   (a) for a current of 50.0 A to deposit 185 g of zinc
   (b) for a current of 18.5 A to form 42.0 g of iron from a solution of iron(II) chloride
   (c) for a current of 100 A to release 15.0 *l* of hydrogen gas (at STP) in the electrolysis of water
   (d) to deposit all the copper from 135 m*l* of 0.450 *M* $CuSO_4$, with a current of 220 A
   (e) for a current of 75.0 A to deposit 88.5 g of aluminum
   (f) for a current of 150 A to deposit 15 g of copper from a solution of copper(II) chloride
   (g) for a current of 50.0 A to release 25.0 *l* of oxygen gas (at STP) in the electrolysis of water
   (h) to deposit all the nickel from 85.0 m*l* of 2.25 *M* $NiSO_4$, with a current of 125 A

4. Calculate the number of amperes necessary to carry out each of the following:
   (a) to deposit 125 g of silver in 3.50 hours
   (b) to deposit 84.5 g of nickel from a solution of $NiCl_3$ in 45.0 minutes
   (c) to form 350 m*l* of chlorine gas (at STP) in 4.50 minutes, in the electrolysis of molten sodium chloride
   (d) to deposit 18.2 g of silver in 25.0 minutes
   (e) to deposit 42.5 g of iron from a solution of $FeCl_3$ in 1.25 hours
   (f) to form 12.5 *l* of hydrogen gas (at STP) in 45.0 minutes, in the electrolysis of water

5. Calculate the number of coulombs necessary for each of the following:
   (a) to plate out 18.5 g of chromium from a solution of $Cr_2(SO_4)_3$
   (b) to plate out 245 mg of tin from a solution of $Sn(NO_3)_2$

(c) to plate out 125 g of copper from a solution of $CuSO_4$

(d) to plate out 385 mg of nickel from a solution of $NiCl_2$

Voltaic cells

6. Calculate the voltage that would be developed by cells using the following reactions (all concentrations 1 *M*). Use Table 20-1.
   (a) $Zn + Ni^{2+} \rightleftharpoons Zn^{2+} + Ni$
   (b) $Pb + Zn^{2+} \rightleftharpoons Pb^{2+} + Zn$
   (c) $2\ Al + 3\ Cd^{2+} \rightleftharpoons 2\ Al^{3+} + 3\ Cd$
   (d) $Hg + Pb^{2+} \rightleftharpoons Hg^{2+} + Pb$
   (e) $Ni + 2\ Ag^{+} \rightleftharpoons Ni^{2+} + 2\ Ag$
   (f) $2\ Cr + 3\ Cu^{2+} \rightleftharpoons 2\ Cr^{3+} + 3\ Cu$
7. From the standard reduction potentials for metals (Table 20-1), determine whether the following reactions will occur:
   (a) $Zn + Ni^{2+} \rightleftharpoons Zn^{2+} + Ni$
   (b) $Pb + Zn^{2+} \rightleftharpoons Pb^{2+} + Zn$
   (c) $2\ Al + 3\ Cd^{2+} \rightleftharpoons 2\ Al^{3+} + 3\ Cd$
   (d) $Hg + Pb^{2+} \rightleftharpoons Hg^{2+} + Pb$
   (e) $Ni + 2\ Ag^{+} \rightleftharpoons Ni^{2+} + 2\ Ag$
   (f) $2\ Cr + 3\ Cu^{2+} \rightleftharpoons 2\ Cr^{3+} + 3\ Cu$

Nernst equation

8. Calculate the potential for each of the following half-reactions with the given concentration (use Table 20-1):
   (a) $Cu^{2+} + 2\ e^{-} \rightleftharpoons Cu$
   $[Cu^{2+}] = 1.0 \times 10^{-3}$
   (b) $Cu \rightleftharpoons Cu^{2+} + 2\ e^{-}$
   $[Cu^{2+}] = 1.0 \times 10^{-4}$
   (c) $Mg^{2+} + 2\ e^{-} \rightleftharpoons Mg$
   $[Mg^{2+}] = 1.0 \times 10^{-4}$
   (d) $Mg \rightleftharpoons Mg^{2+} + 2\ e^{-}$
   $[Mg^{2+}] = 1.0 \times 10^{-5}$
9. Calculate the potential for each of the following reactions, at the given concentrations (use Table 20-1):
   (a) $Cu^{2+} + Zn \rightleftharpoons Zn^{2+} + Cu$
   $[Cu^{2+}] = 1.0 \times 10^{-3}$
   $[Zn^{2+}] = 1.0 \times 10^{-4}$
   (b) $2\ Na + Fe^{2+} \rightleftharpoons 2\ Na^{+} + Fe$
   $[Na^{+}] = 1.0 \times 10^{-2}$
   $[Fe^{2+}] = 1.0 \times 10^{-3}$
   (c) $2\ Al^{3+} + 3\ Ni \rightleftharpoons 2\ Al + 3\ Ni^{2+}$
   $[Al^{3+}] = 1.0 \times 10^{-4}$
   $[Ni^{2+}] = 1.0 \times 10^{-2}$
   (d) $Mg + Fe^{2+} \rightleftharpoons Mg^{2+} + Fe$
   $[Mg^{2+}] = 1.0 \times 10^{-2}$
   $[Fe^{2+}] = 1.0 \times 10^{-3}$
   (e) $Zn^{2+} + 2\ K \rightleftharpoons 2\ K^{+} + Zn$
   $[Zn^{2+}] = 1.0 \times 10^{-3}$
   $[K^{+}] = 1.0 \times 10^{-4}$
   (f) $2\ Al + 3\ Mg^{2+} \rightleftharpoons 2\ Al^{3+} + 3\ Mg$
   $[Mg^{2+}] = 1.0 \times 10^{-5}$
   $[Al^{3+}] = 1.0 \times 10^{-3}$
10. Calculate the value of K_{redox} for each of the following reactions (use Table 20-1):
    (a) $Fe + Ni^{2+} \rightleftharpoons Ni + Fe^{2+}$
    (b) $Sn^{2+} + Pb \rightleftharpoons Pb^{2+} + Sn$
    (c) $Fe^{2+} + Cd \rightleftharpoons Cd^{2+} + Fe$
    (d) $Ni + Sn^{2+} \rightleftharpoons Ni^{2+} + Sn$

General problems

11. For each of the following, tell what is oxidized and what is reduced, and write the anode half-reaction and the cathode half-reaction.
    (a) electrolysis of molten sodium chloride
    (b) electrolysis of molten potassium bromide
12. Calculate the voltage of each of the following cells, and tell if the reaction would go spontaneously or not (all concentrations 1 *M*).
    (a) $Cu^{2+} + 2\,Hg \rightleftharpoons Cu + Hg_2^{2+}$
    (b) $Zn^{2+} + Sn \rightleftharpoons Zn + Sn^{2+}$
13. How many liters of Cl_2 gas, at STP, will be given off at the anode in the electrolysis of molten calcium chloride if 10 g of calcium are produced at the cathode?
14. How many grams of sodium are produced at the cathode in the electrolysis of molten sodium if 42.5 *l* of Cl_2 are given off at the anode?
15. How many amperes flowing for 12.5 minutes would it take to deposit 100 g of Mg from a solution of magnesium chloride?
16. How many grams of Al could be deposited by 8.5×10^3 coul from a solution of aluminum chloride?
17. How many coulombs would it take to deposit 49.5 g of aluminum from a solution of aluminum chloride?
18. How many minutes would it take to deposit all the Al from 225 m*l* of 0.10 *M* aluminum nitrate with a current of 220 A?
19. How many grams of hydrogen could be liberated by a current of 100 A flowing for 20 minutes?
20. How many seconds would it take for a current of 50 A to deposit 20 g of zinc?
21. Calculate the potential for the reaction $Fe + Mg^{2+} \rightleftharpoons Mg + Fe^{2+}$ when the concentration of the Fe^{2+} ion is 0.010 *M* and Mg^{2+} concentration is 0.0010 *M*.
22. Calculate the potential for the reaction $Zn + Ni^{2+} \rightleftharpoons Zn^{2+} + Ni$ when the $[Zn^{2+}]$ is 1.0×10^{-4} and the $[Ni^{2+}]$ is 1.0×10^{-3}.

REFERENCES

HOFFMAN, GEORGE A., "The Electric Automobile," *Sci. Am.,* Oct. 1966, **215,** p. 34.

LAUREN, PAUL M., "Sir Humphrey Davy's Battle with the Sea," *Chemistry,* Sept. 1977. **50,** p. 14.

SHEN, SAMUEL, "What Is the Real Value of a Faraday?," *Chemistry,* Feb. 1966, **39,** p. 8.

WEISSBART, JOSEPH, "Fuel Cells–Electrochemical Converters of Chemical to Electrical Energy," *J. Chem. Educ.,* 1961, **38,** p. 267.

WEISSMAN, EUGENE Y., "Batteries–The Workhorses of Chemical Energy Conversion," *Chemistry,* Nov. 1972, **45,** p. 6.

Objectives

1. To review the electronic structure of the atom and the arrangement of the periodic table, in terms of types of elements and orbitals occupied, and to be able to predict and explain variations in atomic size, ionization energy, electronegativity, differentiating electrons, valence electrons, oxidation states, metallic properties, and properties of ions.
2. To understand the periodic variations in acid and base strength and to be able to predict relative strength from position in the periodic table for (a) hydrogen compounds (binary acids and oxyacids) and (b) oxides.
3. To become familiar with the physical characteristics of metals; to learn some of the more common chemical properties of metals; and to be able to write equations for the reactions of metals, using the activity series (electromotive series), as well as the reactions that take place in the metallurgy of metals.

Twenty One

The Periodic Table-Periodic Variations

In this chapter we shall consider some aspects of the variations in properties within the periodic table. Some of this will be review of topics covered in earlier chapters. We shall be considering several new topics, however, and some of those covered earlier will now be considered in more detail. This should help to tie together many of the facts and principles that we have considered. It will also lead into the discussion of individual elements and groups of elements, which we shall consider in the next two chapters.

Section 21-1 is a very brief review of the material covered in detail in Chapter 5. If necessary, you may want to refer back to that chapter, especially to various tables of values given there.

21-1 Review of Electronic Structure and Related Topics

Atomic size depends on the electronic configuration of the atom and the nuclear charge. Going across a period in the periodic table, there is a general decrease in the size due to the increased nuclear charge and valence electrons in the same energy level. Going down a group, the electron configuration is the predominant thing, and there is a general increase in size. (In each family the first member is somewhat different due to its small size compared to the rest of the group.) There are some deviations in these general trends, due to things such as screening and *effective nuclear charge.*

Ionization energy (energy required to remove an electron from an atom) and *electronegativity* (which just refers to a general tendency for an atom in a compound to attract electrons toward itself) both increase across a period and decrease going down a group.

You should recall that, in general, for representative elements the *differentiating electron* is in an *s* or *p* orbitals, for transition elements it is in a *d* orbital, and for inner transition elements it is in an *f* orbital. There are exceptions to this but it does give the general relationship between types of orbital for the differentiating electron and position in the periodic table. The *valence electrons* (those available for bonding) are the ones in the outer energy level for representative elements, and hence for these elements the number of valence electrons corresponds to the group number. For the transition elements the outer electrons plus some of the electrons in the inner *d* orbitals are available for bonding.

In general, the group number corresponds to the maximum *oxidation state,* except for Group IB and possibly Group VIII. Also, for the nonmetals the minimum oxidation state is equal to the group number minus 8; for example, for the halogens (VIIA) it is minus 1 (7 − 8), for Group VIA it is minus 2 (6 − 8), and for nitrogen and phosphorus (VA) it is minus 3 (5 − 8). The representative metals only exhibit one or two oxidation states, but the transition metals and the nonmetals generally have several different ones.

The majority of the elements are *metals,* with about 17 of those in the upper right side of the table being *nonmetals;* there are a few intermediate ones called *metalloids* or semimetals. Metals are, in general, much better conductors of heat and electricity and more *electropositive* than the nonmetals. Metals can lose electrons to form positive ions; nonmetals, being more *electronegative,* can gain electrons to form negative ions. In polar covalent molecules, or the extreme case of ionic compounds, the electron density is shifted away from the more metallic element and toward the more nonmetallic. In general then, metallic properties decrease going across a period and increase down a group.

In discussing periodic variations in properties of ions, there are two terms that are important; these are **charge density** and *polarizability.* Charge density (or ionic potential) is the charge on an ion divided by the radius of the ion and is therefore small when the ion has a low charge and a large size. For cations the charge density increases across a periodic (charge increases and radius decreases) and decreases down a group (charge is constant and radius increases), while for anions it decreases across a period (charge decreases and radius decreases slightly) and down a group (charge is constant and radius increases). Polarizability refers to the ability of electrons in ions or atoms to be shifted, or the electron cloud distorted. Ions or atoms that are small and compact would have little tendency to shift electron density away from themselves and are therefore nonpolarizable. On the other hand, ions or atoms that are large have more tendency to shift electron density away from themselves and are therefore classified as polarizable.

21-2 Periodic Variations in Acid–Base Strength of Compounds.

Hydrogen Compounds

Binary Acids You will recall from Chapter 14 that the strength of an acid is determined by how readily it ionizes. A strong acid is completely ionized, whereas a weak acid is only partially ionized. For binary acids, of the type HX, acid strength depends on the size of X and the electronegativity of X.

Within a period, size doesn't change much, but electronegativity does. Therefore, for a series of binary acids, acid strength increases across a period, from left to right. For example, in terms of acid strength, $CH_4 < NH_3 < H_2O < HF$. Due to the high electronegativity of fluorine, the electron cloud is shifted more toward the F in the HF molecule, and hence the H^+ ion is more readily produced.

Within a group, such as in the series HF-HCl-HBr-HI, size is the main factor. Acid strength increases going down the group, due to the larger size of the atoms. Hence, in terms of acid strength, $HF < HCl < HBr < HI$. There is less overlap between the electron clouds of the hydrogen atom and the halogen atom with a larger atom. Thus, the bond is weaker and the acid stronger. This is shown in Fig. 21-1, where HF and HI are compared.

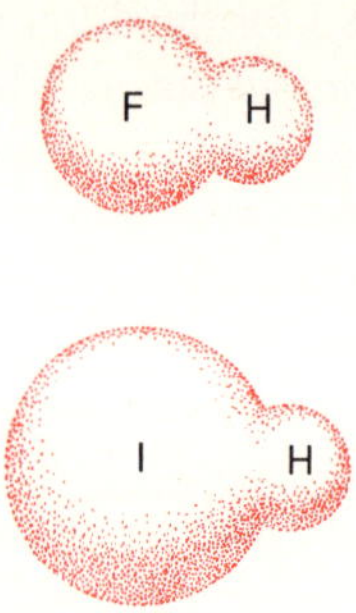

Figure 21-1
Comparison of overlap of electron clouds in HF, and HI

Oxyacids In an oxyacid the hydrogen is always connected to an oxygen atom. For example, HClO is more correctly written as HOCl and $HClO_4$ as $HOClO_3$. Within a series like HClO-$HClO_2$-$HClO_3$-$HClO_4$, there is an increase in acid strength with increasing amounts of oxygen. Thus, in terms of acid strength, $HClO < HClO_2 < HClO_3 < HClO_4$. Likewise, nitrous acid (HNO_2) is a weaker acid than nitric acid (HNO_3), and sulfurous (H_2SO_3) is weaker than sulfuric (H_2SO_4). Since the oxygen has a high electronegativity, the more oxygen that is present, the more it tends to pull electron density away from the H—O bond. This makes the bond weaker and hence the acid is stronger. This is shown in Fig. 21-2 for HNO_3 and HNO_2.

For series of oxyacids across a period (left to right) or up a group, acid strength increases, due to the greater electronegativity of the central atom. For example, in terms of acid strength, $H_3PO_4 < H_2SO_4 < HClO_4$, and $HIO_3 < HBrO_3 < HClO_3$. This is shown in Fig. 21-3 where HIO_3 is compared to $HClO_3$.

Oxides

The acidity of oxides is related to the metallic character and the oxidation state of the element other than oxygen. Oxides of strongly electropositive

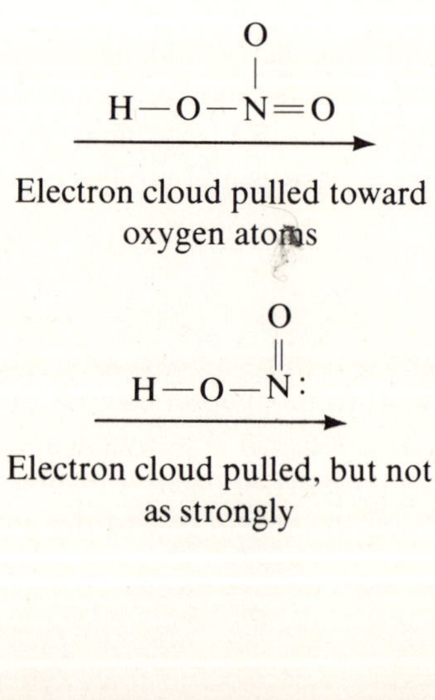

Figure 21-2
Comparison of acid strength of HNO_3, and HNO_2

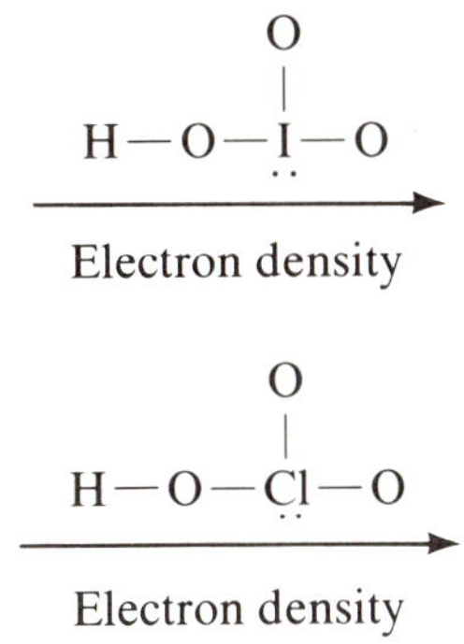

Figure 21-3
Comparison of acid strength of HIO_3 and $HClO_3$. In the $HClO_3$, the electron cloud is pulled away from the H—O bond more strongly, since chlorine has a greater electronegativity than iodine

elements react with water to form bases and hence are often referred to as *basic anhydrides.* Oxides of highly electronegative elements react with water to form acids and hence are often referred to as *acid anhydrides.* Oxides of elements of intermediate metallic properties (intermediate electronegativity) show both acidic and basic properties in water and are thus said to be *amphoteric.* Therefore, due to periodic variations in electronegativity, oxides generally increase in acidity going across a period and decrease in acidity going down a group. For example, Na_2O is basic, Al_2O_3 is amphoteric, and SO_2 is acidic. Likewise, B_2O_3 is acidic, Ga_2O_3 is amphoteric, and Tl_2O_3 is basic.

When a metal forms two or more oxides, the one with the lower oxidation state is the most basic; that is, there is an increase in acidic character (or decrease in basic character) with increase in oxidation state.

When an acid or base is formed from the anhydride (oxide), the central atom will always have the same oxidation number in the acid or base as it had in the oxide. For example, SO_3 reacts with water to form H_2SO_4 (sulfur is +6 in both cases), SO_2 forms H_2SO_3 (sulfur is +4 in both cases), CaO forms $Ca(OH)_2$ (calcium is +2 in both cases), and Fe_2O_3 forms $Fe(OH)_3$ (iron is +3 in both cases).

Another way of comparing the acidic or basic properties of oxides is to consider how ionic the oxide is. The more ionic the oxide, the more basic it is. Conversely, the more covalent the oxide, the more acidic it is. The reason for this is that the oxide ion, O^{2-}, is not stable in water. It reacts rapidly and completely to form OH^-, according to the equation

$$O^{2-} + H_2O \longrightarrow 2\,OH^-$$

For example, sodium oxide is ionic and is therefore strongly basic. On the other hand, due to the small size and large charge density of the Be^{2+} ion, there is considerable covalent character in BeO. Thus, $Be(OH)_2$ is a weak base and is even somewhat amphoteric. Even $Mg(OH)_2$ is a weaker base than $Ca(OH)_2$. Keep in mind that we are considering strength of the base, not solubility. Most oxides and hydroxides are not very soluble (except for those of the IA metals), and therefore any solution of these compounds will be quite dilute. But the strength of the base is a measure of the amount of ionization of the substance once it is in solution.

If a metal forms two or more oxides, the one with the lower oxidation state would tend to be more ionic and hence more basic. For example, if we compare the two oxides PbO and PbO_2, we can see that the PbO could be ionic since it is possible to form a Pb^{2+} ion. PbO_2 would have to be covalent, however, since there could not be a Pb^{4+} ion. Hence, we would expect PbO to be more basic (a stronger base) than PbO_2. For metals that form more than two oxides (that is, some of the transition metals) there is a regular decrease in ionic character and hence in basic character, with increase in oxidation state; that is, for different oxides of the same element there is an increase in acidity with increasing oxidation state.

In terms of the hydrated form of the oxide (that is, the hydroxide) the acidic or basic properties can be compared by looking at the strength of the metal-oxygen bond (M—O) versus the oxygen-hydrogen bond (O—H) in the general formula M—O—H. The more electropositive M is, the greater the tendency for the M—O bond to break, to give off OH^- ions. The more electronegative M is, the greater the tendency for the O—H bond to break to give off H^+ ions.

Oxides of elements that are intermediate in electronegativity are amphoteric (can act as either acids or bases). For example, ZnO [or $Zn(OH)_2$] can form either Zn^{2+} or ZnO_4^{2-} (zincate ion) depending on the conditions.

Consider the following examples:

$$Zn(OH)_2 + 2\ HCl \rightleftharpoons ZnCl_2 + 2\ H_2O$$

$$H_2ZnO_2 + 2\ NaOH \rightleftharpoons Na_2ZnO_2 + 2\ H_2O$$

This second reaction can be considered as the formation of a complex ion:*

$$Zn(OH)_2 + 2\ NaOH \rightleftharpoons 2\ Na^+ + Zn(OH)_4^{2-}$$

In summary, it can be said that, in general, oxides increase in acidity (or decrease in basicity) going across a period and decrease in acidity (or increase in basicity) going down a group in the periodic table. This can be explained in terms of metallic character, electronegativity, or ionic character.

21-3 Metals

Physical Properties of Metals

Some general physical characteristics of metals are listed below:

1. Metallic luster
2. High electrical conductivity
3. High thermal conductivity
4. Malleability (ability to be shaped)

*Complex ions will be considered in Chapter 24. The $Zn(OH)_4^{2-}$ shown here is just the hydrated form of ZnO_2^{2-}; that is, $ZnO_2 \cdot 2\ H_2O^{2-}$ is the same as $Zn(OH)_4^{2-}$.

5. Ductility (ability to be drawn out to a thin wire)
6. Hardness
7. High density
8. Low volatility

These properties vary from metal to metal. For example, mercury is a liquid at room temperature while tungsten (wolfram) melts at 3370°C. The alkali metals can be cut with a knife, but most metals are too hard to be cut in this way. One of the most important properties of metals is that of electrical conductivity. Apparently the valence electrons are mobile enough to be displaced by a potential applied to the metal and can readily pass from atom to atom. The electrical conductivity decreases as the temperature is raised because of increased vibration of the atoms, which interferes with the electron flow.

Occurrence of Metals

Depending on their reactivity, metals are found in nature in the combined state or as the free element. Those in the combined state occur with a variety of anions and require chemical reduction to produce the metals. Those metals occurring in the free state must be concentrated and refined before the metal can be obtained in a pure state. Only the most inactive metals are found in nature in the elemental state. This group includes gold, silver, copper, and the platinum metals (Ru, Rh, Pd, Os, Ir, and Pt).

Many of the metals occur as sulfides, oxides, halides, silicates, sulfates, carbonates, and phosphates. The important ores for each of the metals will be considered under the appropriate group discussion in Chapter 22.

Metallurgy

The extraction of a metal from its ore depends on the chemical properties of the metal, the nature of the ore, and the impurities present in the ores. In most cases if the ore is in the form of the sulfide or carbonate, it is first converted to the oxide. This process is known as *roasting* and is carried out in the presence of air (especially for sulfide). Examples of roasting typical sulfides and carbonates are

(21-1) $$2\,ZnS + 3\,O_2 \xrightarrow{\Delta} 2\,ZnO + 2\,SO_2$$

(21-2) $$ZnCO_3 \xrightarrow{\Delta} ZnO + CO_2$$

A *reduction step* is necessary to obtain the pure metal from the oxide. Very often carbon and carbon monoxide are used as reducing agents in metallurgical processes. The metal to be reduced by the carbon is usually in the form of an oxide.

(21-3) $$SnO_2 + 2\,C \xrightarrow{\Delta} Sn + 2\,CO$$

(21-4) $$Fe_2O_3 + 3\,CO \xrightarrow{\Delta} 2\,Fe + 3\,CO_2$$

Hydrogen is an expensive reducing agent and is not commonly used. Some reductions cannot be carried out by using carbon, however, because of carbide formation. In these reductions hydrogen may be a feasible substitute. Tungsten (wolfram) is an example of a metal obtained by hydrogen reduction of the oxide, WO_3.

(21-5) $$WO_3 + 3\,H_2 \xrightarrow{\Delta} W + 3\,H_2O$$

One metal can be used to obtain another metal from one of its compounds. Aluminum and sodium metals have been used for this purpose.

(21-6) $$2\,Al + Cr_2O_3 \xrightarrow{\Delta} Al_2O_3 + 2\,Cr$$

(21-7) $$TiCl_4 + 4\,Na \xrightarrow{\Delta} Ti + 4\,NaCl$$

Some metals may be obtained by the *thermal decomposition* of certain of their compounds. Mercury is an example of this.

(21-8) $$HgS + O_2 \xrightarrow{\Delta} Hg + SO_2$$

Electrolysis is one of the most important methods of obtaining a pure metal from its compounds. Examples are known in which a crude metal obtained by a carbon reduction is refined by electrolysis. Electrolytic procedures are divided into two classes: aqueous and nonaqueous.

Aqueous If an aqueous solution is used in electrolysis, the metal reduced (cathodic reduction) must be relatively inactive in water. Otherwise hydrogen will be discharged at the cathode rather than the metal. Copper, nickel, and zinc are often refined by such electrolytic methods.

Nonaqueous The nonaqueous electrolytic procedures are important for obtaining the more active metals such as magnesium, aluminum, sodium, and calcium. The electrolyte can either be the compound reduced, such as $MgCl_2$ in the preparation of magnesium, or it can be an inert compound such as Na_3AlF_6 in the electrolytic reduction of aluminum oxide. Usually a high temperature is required to melt the electrolyte, and the terms *fused* or *molten* are used.

(21-9) $$MgCl_2 \xrightarrow{\text{fused}} Mg + Cl_2$$

(21-10) $$2\,Al_2O_3 \xrightarrow[\text{electrolyte}]{\text{fused } Na_3AlF_6} 4\,Al + 3\,O_2$$

Chemical Properties of Metals

Many of the more important reactions of metals can be summarized in the activity series (or electromotive series), as shown in Table 21-1. As well as showing simple displacement reactions, this also includes reactions with H_2, O_2, and H_2O, and some properties of metal oxides.

TABLE 21-1 *Activity series of metals*

Metals			
Li, K, Ba, Sr, Ca, Na	Combine with H_2 to form hydrides	React with H_2O to form H_2 and hydroxides	Combine directly with O_2
Mg, Al, Zn, Cr		React with steam to form H_2 and hydroxides	Combine directly with O_2
Fe, Ni, Sn, Pb, H, Cu, Bi, Sb	Oxides reduced by H_2 (at high temperature) to free metals and H_2O		Combine directly with O_2
Hg	Oxides reduced by H_2 (at high temperature) to free metals and H_2O	Oxides decompose when heated	Combine directly with O_2
Ag, Pt, Au	Oxides reduced by H_2 (at high temperature) to free metals and H_2O	Oxides decompose when heated	

↑ Increasing tendency to go to oxidized state (i.e., to lose electrons)

SUMMARY

1. In general there are regular variations within the periodic table, across a *period* or down a *group*, in *atomic size, ionization energy, electronegativity, differentiating electrons, valence electrons, oxidation states, metallic properties,* and *acid-base strength* of hydrogen compounds (binary acids and oxyacids) and oxides.
2. With binary acids the size of the atom other than hydrogen is the major factor in determining acid strength—the smaller the size, the weaker the acid—and with oxyacids the electronegativity or number of oxygen atoms present are the determining factors—the greater the electronegativity or the more oxygen atoms present, the stronger the acid.
3. The more ionic an oxide is, the more basic it is; thus, in general, metal oxides are basic and nonmetal oxides are acidic. Oxides become more acidic (less basic) going across a period and up a group. If an element forms more than one oxide, the lower the oxidation state is, the more ionic (and hence the more basic) the oxide is. Oxides with intermediate electronegativity are *amphoteric.*

4. The properties of metals vary considerably, but in general they are good conductors, hard, ductile, and malleable and have high densities. Some metals are found in the free state but most occur as oxides, sulfides, halides, silicates, sulfates, carbonates, and phosphates. In general, they are prepared by roasting (heating in air) to form oxides. These oxides are then reduced to the free metal, by using carbon, hydrogen, or a more active metal; by heating; and by electrolysis (either in aqueous solution or in the molten or fused state).
5. The *activity (electromotive) series* of metals can be used not only to predict displacement reactions but also to predict which metals will react with water or steam, which will combine with hydrogen to form hydrides, and which will combine directly with oxygen; which oxides will be reduced by hydrogen; and which oxides decompose when heated.

Upon completion of this chapter you should be able to do the following:

1. Given any of the following, predict how the property would vary within a period and a group in the periodic table: atomic size, ionization energy, electronegativity, differentiating electron, valence electrons, possible oxidation states, metallic properties, and charge density. (1–5)
2. Given any group of acids, either binary or oxyacids, predict and explain variations in acid strength. (6)
3. Given the name or formula of any oxide, predict whether it would be acidic or basic, and tell what acid or base (hydroxide) it would form in water solution; or given any group of oxides, predict which would be the most basic or least basic. (7, 8)
4. Given the name or formula of any of the following, write a complete balanced equation: acid + base; metal oxide + acid; nonmetal oxide + base; metal oxide + nonmetal oxide. (9)
5. Given any metal compound commonly found in the ore of that metal (such as a sulfide), write an equation to represent the roasting of the compound in air and the reduction of the oxide formed, using carbon, hydrogen, or electrolysis. (10)
6. Given an activity series of metals, write equations for displacement reactions, reactions of metals with water or steam, reactions of metals with hydrogen to form hydrides, reactions of metals with oxygen, and reactions of metal oxides with hydrogen. (11)
7. Given any of the terms in Exercise 1, give a definition or explanation, in your own words.

EXERCISE

1. Define or explain each of the following:
   (a) ionization energy
   (b) electronegativity
   (c) differentiating electron
   (d) valence electron
   (e) oxidation state (oxidation number)
   (f) charge density
   (g) polarizability
   (h) binary acid
   (i) oxyacid
   (j) acidic oxide
   (k) basic oxide
   (l) amphoteric substance
   (m) roasting
   (n) electrolysis
   (o) hydride

PROBLEMS

Electronic structure and related properties

1. Choose the correct answer to each part from the following list:

   Mg, P, Cl, Al, S, K, F, Ca

   (a) the one with the smallest atomic size
   (b) the one with the highest ionization energy
   (c) the one with the lowest electronegativity
   (d) the one with the highest electronegativity
   (e) the one with the largest atomic size
   (f) the one with the lowest ionization energy
2. Tell how many valence electrons there are in each of the following:
   (a) P
   (b) Ca
   (c) Br
   (d) K
   (e) S
   (f) Cl
3. Tell what type of orbital (1*s*, 3*p*, and so on) the differentiating electron would be in for each of the following:
   (a) S
   (b) Fe
   (c) Sn
   (d) P
   (e) Pb
   (f) Ni
4. List the possible oxidation states for each of the following:
   (a) Ba
   (b) Al
   (c) Pb
   (d) Mg
   (e) Sn
   (f) Ca
5. In each group, choose the one with the highest charge density and the one with the lowest charge density:
   (a) Ca^{2+}, K^{+}, Al^{3+}, Mg^{2+}
   (b) Sr^{2+}, Ag^{+}, Cs^{+}, Sn^{2+}
   (c) Pb^{2+}, Hg^{2+}, Cs^{+}, Ba^{2+}
   (d) Na^{+}, K^{+}, Ca^{2+}, Mg^{2+}

Acid strength

6. In each group, choose the strongest acid and the weakest acid.
   (a) H_2O, NH_3, HF
   (b) HCl, HF, HBr
   (c) H_2S, HBr, HCl
   (d) HBrO, $HBrO_4$, $HBrO_2$
   (e) $HClO_2$, HIO_2, $HBrO_2$

(f) sulfurous acid, sulfuric acid
(g) HCl, HF, H_2S
(h) ammonia and water
(i) HI, HCl, HBr, HF
(j) phosphoric acid, phosphorous acid
(k) HClO, $HClO_2$, $HClO_3$, $HClO_4$
(l) chloric acid, bromic acid, iodic acid

Oxides

7. Tell if each of the following is acidic or basic:
   (a) CaO
   (b) SnO
   (c) SO_2
   (d) Cs_2O
   (e) P_2O_5
   (f) NO_2
   (g) K_2O
   (h) P_2O_3
   (i) PbO
   (j) BaO
8. In each group, choose the most basic and the least basic one:
   (a) CuO, K_2O, SO_2
   (b) SnO, CaO, SnO_2
   (c) MgO, PbO, PbO_2
   (d) NO_2, ZnO, CaO
9. Write a balanced equation for each of the following:
   (a) $HNO_3 + Ca(OH)_2$
   (b) $Na_2O + HCl$
   (c) $MgO + SO_2$
   (d) hydrochloric acid + ferrous oxide
   (e) aluminum hydroxide + dinitrogen trioxide
   (f) KOH + HCl
   (g) $HNO_3 + CaO$
   (h) $CO_2 + Na_2O$
   (i) stannic oxide + nitric acid
   (j) diphosphorus pentoxide + chromium(II) hydroxide

Metallurgy

10. Write a balanced equation for each of the following reactions:
    (a) roasting of copper(II) sulfide
    (b) reduction of copper(II) oxide with carbon
    (c) electrolysis of aqueous copper(II) chloride
    (d) roasting of zinc sulfide
    (e) reduction of zinc oxide with hydrogen
    (f) electrolysis of fused sodium chloride

Reactions of metals

11. Write a balanced equation for each of the following reactions:
    (a) aluminum + aqueous ferrous chloride
    (b) stannic oxide + H_2 + heat
    (c) calcium + hydrogen
    (d) magnesium + hydrochloric acid
    (e) aluminum + steam
    (f) zinc + aqueous stannic chloride
    (g) barium + hydrogen
    (h) magnesium + steam
    (i) zinc + hydrochloric acid
    (j) plumbous oxide + H_2 + heat

General problems

12. Explain each of the following:
    (a) A sodium atom is larger than an aluminum atom.
    (b) Chlorine has a higher electronegativity than iodine.
    (c) Potassium has a lower ionization energy than sodium.
    (d) Na_2O is more basic (a stronger base) than BeO.
    (e) H_3PO_4 is a stronger acid than H_3PO_3.
    (f) Bromous acid is a weaker acid than bromic acid.
    (g) Calcium nitrite undergoes hydrolysis in water but calcium nitrate does not.
    (h) Hypochlorous acid has a lower ionization constant than chloric acid.
    (i) A phosphorus atom is larger than a nitrogen atom.
    (j) Calcium has a higher ionization energy than potassium.
    (k) Carbon has a lower electronegativity than oxygen.
    (l) Sulfuric acid is a stronger acid than sulfurous acid.
    (m) MgO is a weaker base than BaO.
    (n) K_2O is strongly basic but PbO_2 is very weakly basic.
    (o) $HBrO_3$ has a higher ionization constant than HBrO.
    (p) Sodium chlorite undergoes hydrolysis in water but sodium perchlorate does not.

13. Write a balanced equation for each of the following reactions:
    (a) aluminum + zinc chloride solution
    (b) calcium + water
    (c) barium + hydrogen + heat
    (d) aluminum oxide + nitric acid
    (e) aluminum oxide + sodium hydroxide
    (f) dinitrogen trioxide + calcium hydroxide solution
    (g) chloric acid + potassium hydroxide
    (h) chlorous acid + calcium metal
    (i) zinc hydroxide + sodium hydroxide
    (j) potassium + water
    (k) magnesium + zinc chloride solution
    (l) hydrochloric acid + iron(III) oxide
    (m) calcium + hydrogen + heat
    (n) diphosphorus pentoxide + sodium oxide
    (o) zinc oxide + sodium hydroxide
    (p) bromic acid + calcium oxide
    (q) perchloric acid + zinc
    (r) carbon dioxide + barium oxide solution

REFERENCES

HABASHI, FATHI, "Chemistry & Metallurgy," *Chemistry*, Oct. 1972, **45**, p. 6.

RICH, RONALD, *Periodic Correlations*, New York: W. A. Benjamin, Inc., 1965.

SANDERSON, R. T., *Chemical Periodicity*, New York: Reinhold Publishing Corp., 1960.

Objectives

1. To be able to explain and to predict the more common properties of the groups of representative elements.
2. To learn and to be able to correlate with position in the periodic table what oxidation states the elements in each group can form.
3. To learn some of the more common properties of some elements and their compounds and to be able to write equations for the reactions of these elements, including the following types:
   (a) Reaction with hydrogen, oxygen, and water
   (b) Reactions of metals with nonmetals
   (c) Reactions of oxides and hydroxides of the elements
   (d) Reactions that take place in the preparation of some of the more important compounds
   (e) Reactions of the more common ions and compounds.

Twenty Two

Properties of the Representative Elements

In this chapter we shall consider some of the groups of representative elements. In doing this we shall be concerned first of all with the relationship between position in the table (and hence electron structure) and properties. With each group of elements we shall first consider some general trends and variations within the group. Then we shall look at some specific elements and compounds and their reactions.

The material included here is not intended to be a thorough study of inorganic chemistry. It is a brief introduction to the chemistry of some elements and compounds. Rather than being presented in detail, the material is condensed as much as possible, in order to point out trends and major aspects. There are many excellent references available where further information on all of these topics can be found. Those listed at the end of this chapter should be useful for obtaining any of this information.

22-1 Hydrogen

Very little hydrogen is found in the free state; however, it does occur in many compounds such as water and all living material. Hydrogen is a colorless, odorless, tasteless gas, which will burn very readily in oxygen. It is the least dense of all known substances and is partically insoluble in water.

At room temperature hydrogen is rather inert; however, at elevated temperatures it combines with most nonmetals and with active metals to form hydrides. It can be used to reduce some metal oxides to the free metal.

Table 22-1 summarizes the properties of hydrogen.

Consider the following problem example involving some reactions of hydrogen.

TABLE 22-1 *Properties of hydrogen*

GENERAL PROPERTIES	Colorless, odorless, tasteless gas; will burn in oxygen; least dense of all known substances; insoluble in water
OCCURRENCE	In water and most organic compounds
PREPARATION	Lab—metal plus water, steam, or acid (depending on activity of metal—see activity series in Table 21-1); hydride + water ($\rightarrow H_2 + OH^-$); amphoteric metal + base (e.g., Zn) Commercial—electrolysis of water
CHEMICAL PROPERTIES	Rather inert at room temperature; at elevated temperatures reacts with nonmetals ($\rightarrow H_2O$, H_2S, NH_3, and so on), active metals ($\rightarrow$ hydride), and metal oxides ($\rightarrow$ metal + H_2O)

PROBLEM EXAMPLE 22-1

Write a balanced equation for each of the following reactions:
(a) hydrogen + nitrogen (with heat)
(b) hydrogen + sodium (with heat)
(c) hydrogen + copper(II) oxide (with heat) to form copper metal

SOLUTION:

(a) Hydrogen will combine directly with nitrogen to form ammonia:

$$3\ H_{2(g)} + N_{2(g)} \longrightarrow 2\ NH_{3(g)}$$

(Recall that hydrogen and nitrogen in the free state are always H_2 and N_2.)

(b) Since sodium is an active metal, hydrogen will combine with it to form a hydride:

$$H_{2(g)} + 2\ Na_{(s)} \longrightarrow 2\ Na\ H_{(s)}$$

(c) Hydrogen will reduce the copper in copper(II) oxide to the free metal. The hydrogen must be oxidized (to the +1 oxidation state) and therefore forms H_2O:

$$CuO_{(s)} + H_{2(g)} \longrightarrow Cu_{(s)} + H_2O_{(g)}$$

22-2 Group IA—The Alkali Metal Group: Lithium (Li), Sodium (Na), Potassium (K), Rubidium (Rb), Cesium (Cs), and Francium (Fr)

Since these elements have only one valence electron, their chemistry is relatively simple and the chemical properties of the elements are quite similar. They form hydroxides that are strongly basic, and they are very good reducing agents since they readily lose their one valence electron. These elements are thus classified as very active metals. The reactivity increases with increasing atomic weight (that is, going down the group). The densities of these elements are quite low for metals. For example, lithium, sodium, and potassium are less dense than water. Also, they are unusually soft and have low melting points. The melting points decrease going down the group.

Since these metals are so active, they are never found in nature in the free state. They always occur in compounds, generally with a +1 oxidation state, and usually as salts. Sodium is more abundant than potassium but the two are usually found together in the form of salts. Lithium, rubidium, and cesium are somewhat rare, occurring as impurities to some extent with various sodium salts. Francium exists only as an unstable radioactive species (see Chapter 25).

The free metals are usually prepared by electrolysis of the salt, to reduce the +1 ion to the free metal. Since these metals are so active, the electrolysis cannot be done in aqueous solutions but must be done with the fused (molten) salt. Sodium is produced in much larger quantities than any of the other alkali metals since sodium chloride is very plentiful and there are very few uses of alkali metals in which sodium does not work so well as the others.

Almost all the compounds of these metals are white solids that are soluble in water. The one exception to this is lithium, which is more similar to magnesium in some ways than it is to the other members of this family, mainly due to the small size of the atom. Therefore, the fluoride, carbonate, and phosphate of lithium are rather insoluble.

The alkali metals are too expensive and too reactive to be used for their typical metallic properties; however, they do have some special uses. Sodium is a very good heat conductor and is therefore used as a heat transfer medium, especially in nuclear reactors. Sodium is also used as a reducing agent, such as in the preparation of other metals by the reduction of their chlorides. For example, chromium is formed by the following reaction:

$$3\ Na + CrCl_3 \longrightarrow Cr + 3\ NaCl$$

All the alkali metals react with water to form hydrogen and the metal hydroxide. They also combine with hydrogen to form hydrides. Upon reaction with oxygen, lithium forms the normal oxide (Li_2O); sodium forms the peroxide (Na_2O_2); and potassium, rubidium, and cesium form *superoxides* (MO_2) in which the oxygen has an oxidation state of $-\frac{1}{2}$. These metals also combine with the halogens (VIIA elements) to form halides.

The hydrides of these elements react with water to form hydrogen gas and the hydroxide, such as in the reaction

$$NaH + H_2O \longrightarrow H_2 + NaOH$$

The oxides all react with water to form hydroxide; peroxides and superoxides also form oxygen, along with the hydroxide. The hydroxides are all strong bases, except LiOH, which is quite weak, due to the small size of the Li atom and hence the somewhat covalent character of the compound.

Since almost all compounds of sodium and potassium are soluble, the Na^+ and K^+ ions are identified by flame tests. The Na^+ ion gives a long-lasting, yellow-orange flame. The K^+ ion produces a rather brief, purple color in the flame.

Table 22-2 summarizes the properties of the **alkali metals.**

TABLE 22-2 *Properties of the IA Metals—Li, Na, K, Rb, Cs, Fr*

GENERAL PROPERTIES	Active metals; good conductors, and so on
OCCURRENCE	Never found in free state (too active); Na and K as salts; Li, Rb, and Cs rather rare; Fr only exists as radioactive species
PREPARATION	Reduced by electrolysis of molten salts
PHYSICAL PROPERTIES	Soft, low melting point, low density; most compounds soluble in water (except Li compounds, due to high charge density)
CHEMICAL PROPERTIES	Very reactive (good reducing agents); form only +1 compounds; oxides are strong bases (except Li_2O); react with H_2O ($\rightarrow H_2 + MOH$), H_2 ($\rightarrow$ hydride), O_2 ($\rightarrow Li_2O$, Na_2O_2, KO_2), and halogens ($\rightarrow$ halides)
SOME IMPORTANT COMPOUNDS	Halides—NaCl, $CaCl_2$, KCl, and so on (NaCl is used to produce Na, Cl_2, and NaOH) Hydrides—react with H_2O for form H_2 + hydroxide Oxides—react with H_2O to form hydroxide (peroxides and superoxides also form O_2 along with hydroxide) Hydroxides—strong bases (except LiOH, which is a weak base); soluble in water Carbonates—Na_2CO_3 is used in the preparation of many things, such as paper, and glass Hydrogen carbonates (bicarbonates)—will neutralize acid or base; used in fire extinguisher, with H_2SO_4 (forms CO_2)
IDENTIFICATION OF IONS	Na^+ and K^+ are identified by flame tests since almost all compounds are soluble (Na^+ = orange, K^+ = purple)

Consider the following problem example involving reactions of the alkali metals:

PROBLEM EXAMPLE 22-2

Write a balanced equation for each of the following reactions:

(a) sodium + water
(b) potassium + hydrogen (with heat)
(c) lithium + oxygen
(d) sodium + oxygen
(e) potassium + oxygen
(f) sodium + chlorine
(g) lithium oxide + water
(h) sodium peroxide + water
(i) sodium hydrogen carbonate + sulfuric acid

SOLUTION:

(a) Since sodium is an active metal, it displaces hydrogen gas from water:

$$2\,Na_{(s)} + 2\,H_2O_{(l)} \longrightarrow H_{2(g)} + 2\,NaOH_{(aq)}$$

(b) Potassium combines with hydrogen to form the hydride:

$$2\,K_{(s)} + H_{2(g)} \longrightarrow 2\,KH_{(s)}$$

(c) Lithium combines with oxygen to form the normal oxide:

$$4\,Li_{(s)} + O_{2(g)} \longrightarrow 2\,Li_2O_{(s)}$$

(d) Sodium combines with oxygen to form the peroxide:

$$2\,Na_{(s)} + O_{2(g)} \longrightarrow Na_2O_{2(s)}$$

(e) Potassium combines with oxygen to form the superoxide:

$$K_{(s)} + O_{2(g)} \longrightarrow KO_{2(s)}$$

(f) Sodium combines with chlorine to form the chloride:

$$2\,Na_{(s)} + Cl_{2(g)} \longrightarrow 2\,NaCl_{(s)}$$

(g) Lithium oxide reacts with water to form the hydroxide:

$$Li_2O_{(s)} + H_2O_{(l)} \longrightarrow 2\,LiOH_{(aq)}$$

(h) Sodium peroxide reacts with water to form the hydroxide and oxygen:

$$2\,Na_2O_{2(s)} + 2\,H_2O_{(l)} \longrightarrow 4\,NaOH_{(aq)} + O_{2(g)}$$

(i) Sodium hydrogen carbonate (bicarbonate) reacts with the acid to form carbonic acid (H_2CO_3) which immediately breaks up into carbon dioxide and water:

$$2\,NaHCO_3 + H_2SO_{4(ag)} \longrightarrow 2\,H_2O_{(l)} + 2\,CO_{2(g)} + Na_2SO_{4(aq)}$$

22-3 Group IIA—The Alkaline Earth Metals: Beryllium (Be), Magnesium (Mg), Calcium (Ca), Strontium (Sr), Barium (Ba), and Radium (Ra)

The elements in this group show definite metallic properties, with beryllium being the least metallic. These are all good conductors of heat and electricity, although magnesium is the only one used much as a metal; beryllium is quite rare and therefore costly; calcium, strontium, and barium are too reactive. These metals are all strong reducing agents (with the exception of Be) but not so strong as the IA metals. In their compounds they all exhibit the +2 oxidation state, and generally these metals lose their two valence electrons to form +2 ions (again with the exception of Be). Calcium, strontium, and barium resemble each other more closely in their chemical properties than do beryllium and magnesium.

Magnesium is found in a number of minerals and also in seawater. Calcium is also quite abundant, occurring as the carbonate, sulfate, and phosphate. Strontium and barium are less abundant and also occur as the sulfate and carbonate. Radium only occurs as a radioactive species.

As with the alkali metals, due to the activity of these metals they are usually prepared by the electrolysis of the fused salts.

Beryllium is a very light metal and is used mainly in alloys with copper and nickel and in trace amounts for hardening other metals. Magnesium, calcium, strontium, and barium are relatively soft metals, and their softness increases as the atomic weight increases. Magnesium is the only one of these four having satisfactory structural properties. It is often used in alloys, which are light and strong. Calcium, strontium, and barium are all used as reducing agents in the preparation of the various metals.

The activity of these metals increases going down the group. Beryllium is the least active, and calcium, strontium, and barium are similar in activity to the alkali metals.

All these metals combine directly with oxygen, halogens, nitrogen, and sulfur; Ca, Sr, and Ba react with cold water, and Mg reacts with steam; Ca, Sr, and Ba react with H_2. Examples are shown in the following general equations:

$$2\,M + O_2 \longrightarrow 2\,MO \text{ (oxide)}$$
$$M + X_2 \text{ (halogen)} \longrightarrow MX_2 \text{ (halide)}$$
$$3\,M + N_2 \longrightarrow M_3N_2 \text{ (nitride)}$$
$$M + S \longrightarrow MS \text{ (sulfide)}$$
$$M + H_2 \longrightarrow MH_2 \text{ (hydride; } M = \text{Ca, Sr, or Ba)}$$
$$M + 2\,H_2O \longrightarrow H_2 + M(OH)_2 \text{ (not Be)}$$

Beryllium oxide is amphoteric but the other oxides are all basic. The hydroxides increase in strength as bases going down the group; $Mg(OH)_2$ is weak, $Ca(OH)_2$ is somewhat stronger, and $Sr(OH)_2$ and $Ba(OH)_2$ are strong. The oxides and hydroxides are not very soluble in water but the solubility increases going down the group. Calcium oxide (lime) is easily prepared by heating limestone (calcium carbonate), according to the following equation:

$$CaCO_{3(s)} \xrightarrow{\Delta} CaO_{(s)} + CO_{2(g)}$$

This reacts with water to form the hydroxide:

$$CaO_{(s)} + H_2O_{(l)} \longrightarrow Ca(OH)_{2(aq)}$$

The carbonates of this group are very insoluble in water although the bicarbonates are soluble. Large natural deposits of calcium carbonate are leached by natural waters containing carbon dioxide, thus forming calcium hydrogen carbonate (bicarbonate), which is a constituent of hard water:

$$CaCO_3 + H_2O + CO_2 \longrightarrow Ca(HCO_3)_2$$

The chief components of hard water are calcium and magnesium cations associated with sulfate and hydrogen carbonate (bicarbonate) anions. Hardness may be temporary (bicarbonate salts) or permanent (sulfate salts). Temporary hardness may be eliminated by heating.

$$Ca(HCO_3)_{2(aq)} \xrightarrow{\Delta} CaCO_{3(s)} + H_2O_{(l)} + CO_{2(g)}$$

Hardness due to sulfate salts is not removed by heating. These can be removed by ion-exchange resins, such as in water softeners.

Many of the salts of these ions are soluble, and thus in the qualitative identification they are generally about the last ones tested for. For example, most sulfides are insoluble, except IA and IIA metal sulfides. The ions of this group do form insoluble carbonates, sulfates, and oxalates, however, and can thus be identified in this way.

The Ca^{2+}, Sr^{2+}, and Ba^{2+} ions can also be identified by the color of the flame produced in a flame test. Ca^{2+} gives a red-orange flame; Sr^{2+}, a red flame; and Ba^{2+} gives a green color. However, these ions are generally separated by precipitation first, and then a flame test is used to confirm the presence.

Table 22-3 summarizes the properties of the **alkaline earth metals.**

Consider the following problem example involving reactions of the alkaline earth metals:

PROBLEM EXAMPLE 22-3

Write a balanced equation for each of the following reactions:
(a) calcium + nitrogen (with heat)
(b) barium + hydrogen (with heat)
(c) magnesium nitrate solution + sodium hydroxide solution

SOLUTION:

(a) Calcium will combine with nitrogen to form a nitride:

$$3\ Ca_{(s)} + N_{2(g)} \xrightarrow{\Delta} Ca_3N_{2(s)}$$

The oxidation state of nitrogen in a binary compound of this type is always -3; hence the formula, Ca_3N_2.

(b) Barium is an active metal and therefore forms a hydride:

$$Ba_{(s)} + H_{2(g)} \xrightarrow{\Delta} BaH_{2(s)}$$

(c) Since magnesium hydroxide is not very soluble, it will precipitate out:

$$Mg(NO_3)_{2(aq)} + 2\ NaOH_{(aq)} \longrightarrow Mg(OH)_{2(s)} + 2\ NaNO_{3(aq)}$$

22-4 Group IIIA—Boron (B), Aluminum (Al), Gallium (Ga), Indium (In), and Thallium (Tl)

The Group IIIA elements vary from nonmetallic for boron to metallic for the others. Aluminum is a rather active metal, which only exhibits the +3 oxidation state. Gallium, indium, and thallium are quite rare; they are soft, low melting metals, and form +1 and +3 compounds. These last three elements are not so active as aluminum. They are generally found as impurities in metals near them in the periodic table, such as zinc, lead, and copper. Their chemistry is similar to that of zinc, cadmium, and mercury.

Aluminum is a very abundant element, occurring in many types of compounds, such as the oxide, as well as in complex compounds like clays.

TABLE 22-3 *Properties of the IIA Metals—Be, Mg, Ca, Sr, Ba, Ra*

GENERAL PROPERTIES	Good conductors, and so on (Be least metallic); less active than IA metals (increasingly active going down the group); only Mg is used as a metal (Ca, Sr, and Ba are too reactive); generally lose their two valence electrons to form +2 ions (except Be); oxides are all basic (BeO is amphoteric), with base strength increasing down the group
OCCURRENCE	Generally as carbonates and sulfates; Mg found in several minerals and in seawater; Ca also abundant; Sr and Ba are rather rare; Ra only occurs as radioactive species
PREPARATION	Electrolysis of molten salts
USES	Be used in alloy; Mg used as light metal; Ca, Sr, and Ba used as reducing agents
CHEMICAL PROPERTIES	React with O_2 (→ oxide), halogens (→ halide), N_2 (→ nitride), S (→ sulfide), cold H_2O for Ca, Sr, Ba, and steam for Mg (→ H_2 + hydroxide), and H_2 (→ hydride, but only Ca, Sr, Ba)
SOME COMPOUNDS	Oxides—basic; not very soluble (increase from Mg to Ba) Hydroxides—increase in basic character from Mg to Ba Carbonates—insoluble; when heated form oxide + CO_2 Sulfates—not very soluble (unlike most other sulfates)
IDENTIFICATION OF IONS	Do not form insoluble sulfides and therefore the last group in the qualitative analysis scheme; precipitate as carbonates, sulfates, and oxalates—then flame tests are run (Ba^{2+}—green; Ca^{2+}—orange red; Sr^{2+}—dark red)

Since it is an active metal, it is generally prepared by the electrolysis of the oxide. Since aluminum is a light, strong metal, it is used for many structural purposes.

Aluminum will react with either acid or base to form hydrogen gas.

$$2\,Al_{(s)} + 6\,HCl_{(aq)} \longrightarrow 3\,H_{2(g)} + 2\,AlCl_{3(aq)}$$

$$2\,H_2O + 2\,Al_{(s)} + 2\,NaOH_{(aq)} \longrightarrow 2\,NaAlO_{2(s)} + 3\,H_{2(g)}$$

Boron is a nonmetal and resembles silicon and carbon more than it does the other Group 3 elements. It is brittle and is not a good conductor of heat or electricity. Boron is used in small quantities in steel, but otherwise the free element is not important commercially.

Boron and aluminum react rather vigorously with most electronegative elements, such as oxygen and halogens, especially if warmed. They only show oxidation states of +3. The compounds of these two elements are all covalent, with the possible exception of AlF_3, which shows considerable ionic character. The salts of aluminum ionize in water however, to form the Al^{3+} ion.

Aluminum hydroxide is amphoteric, and thus it is soluble in either acid or base. Also, aluminum salts are hydrolyzed in water, since $Al(OH)_3$ is a weak base.

The Al^{3+} ion is precipitated out of solution as the hydroxide in a weakly basic solution. Since this hydroxide is amphoteric, it will dissolve in a strong base. This is used in qualitative analysis to separate it from other ions, such as Fe^{3+}, that do not dissolve. The $Al(OH)_3$ will reprecipitate upon addition of NH_4^+, which decreases the OH^- ion concentration by forming ammonia and water.

Table 22-4 summarizes the properties of the elements of Group IIIA. Consider the following problem example involving the IIIA elements.

PROBLEM EXAMPLE 22-4

Write a balanced equation for each of the following reactions:
(a) electrolysis of fused aluminum oxide
(b) boron + oxygen
(c) aluminum hydroxide + hydrochloric acid
(d) aluminum hydroxide + sodium hydroxide

SOLUTION:

(a) The aluminum is reduced to the free metal, and oxygen is oxidized to the free element:

$$2\ Al_2O_{3(l)} \xrightarrow{\text{electrolysis}} 4\ Al_{(s)} + 3\ O_{2(g)}$$

TABLE 22-4 *Properties of the IIIA elements—B, Al, Ga, In, Tl*

GENERAL PROPERTIES	B is least metallic (a semiconductor); others are typical metals with aluminum being the most active; Al is a light, strong metal; Ga, In, and Tl are soft, low melting metals
PREPARATION	B_2O_3 and Al_2O_3 are reduced to the free metal—B_2O_3 with an active metal, and Al_2O_3 by electrolysis
CHEMICAL PROPERTIES	B and Al are active elements, reacting readily with O_2 and the halogens (always form +3 compounds); Al will react with either acid or base to form H_2 (and Al^{3+} or AlO_2^- salts); Ga, In, and Tl form +1 and +3 compounds
COMPOUNDS	B_2O_3 forms a weak acid in solution; Al_2O_3, Ga_2O_3 and In_2O_3 are amphoteric; Tl_2O_3 is strongly basic. Aluminum hydroxide is amphoteric and therefore will dissolve in acid or base (to form Al^{3+} or AlO_2^-); since it is a weak base, salts of Al^{3+} readily hydrolyze
IDENTIFICATION OF IONS	Al^{3+} precipitates as $Al(OH)_3$ in weakly basic solution; will dissolve [along with $Zn(OH)_2$ and $Cr(OH)_3$] in strong base; will reprecipitate if NH_4^+ is added

(b) Both boron and aluminum will combine directly with oxygen; the oxidation state is always +3:

$$4\,B_{(s)} + 3\,O_{2(g)} \longrightarrow 2\,B_2O_{3(s)}$$

(c) This is a typical acid-base reaction, forming water plus a salt:

$$Al(OH)_{3(s)} + 3\,HCl_{(aq)} \longrightarrow 3\,H_2O_{(l)} + AlCl_{3(aq)}$$

(d) Since aluminum hydroxide is amphoteric, this is also an acid-base reaction. Aluminum hydroxide could be written as H_3AlO_3, to emphasize its amphoteric nature. It is known, however, that only one hydrogen can be replaced and therefore it is usually thought of as $HAlO_2 \cdot H_2O$ to emphasize this:

$$\underset{[\text{or } HAlO_2 \cdot H_2O]}{Al(OH)_{3(s)}} + NaOH_{(aq)} \longrightarrow 2\,H_2O_{(l)} + NaAlO_{2(aq)}$$

22-5 Group IVA—Carbon (C), Silicon (Si), Germanium (Ge), Tin (Sn), and Lead (Pb)

As with the Group IIIA elements, this group of elements goes from nonmetals (C and Si) to metals (Sn and Pb), with Ge being classified as a metalloid or semimetal. The metallic elements, tin and lead, show variable oxidation states (+2 and +4), just as the metals beyond Al in Group IIIA do. This is due to the transition series that precede these metals.

The occurrence of these elements is widespread and in many different types of compounds. Carbon occurs in inorganic compounds such as carbonates and carbon dioxide and in a very large number of organic compounds (see Chapter 26). Silica (SiO_2) and metal silicates are found everywhere on the earth's crust, but silicon never occurs in the free state. Germanium is a rare element and occurs in minute quantities in many metal sulfides. Tin occurs chiefly as the oxide, SnO_2, and lead occurs mainly as the sulfide, PbS, called galena.

Carbon is used as a reducing agent in the preparation of many metals; silicon is used in steel; germanium is used as a semiconductor in transistors; tin and lead have a variety of uses as metals (they are both soft, low melting metals and lie above hydrogen in the activity series).

Carbon is relatively inert at room temperature; at elevated temperatures, however, it combines directly with oxygen and the halogens as well as with most other nonmetals. Germanium, tin, and lead are rather active metals. They will displace hydrogen gas from acids and combine with most nonmetals. The product formed (such as SnO or SnO_2) depends on the conditions.

The oxidation states commonly shown by these elements are +4 for carbon and silicon (although carbon shows a variety of oxidation states, especially in organic compounds) and either +2 or +4 for Ge, Sn, and Pb. The stability of the +2 state increases as the atomic weight increases, and the stability of the +4 state decreases in the same order.

The common oxide for carbon and silicon is the dioxide, CO_2 and SiO_2, respectively. Both of these occur in nature. There is a tremendous difference in their physical properties. Carbon dioxide is a gas that sublimes at approximately $-80°C$ at atmospheric pressure, while silicon dioxide is a solid, melting at approximately 1600°C. Both of these oxides are acidic.

Tin and lead form the common oxides MO and MO_2, SnO_2 being the stable tin oxide and PbO the stable lead oxide. These oxides are formed by direct combination of the free elements. The oxides and hydroxides of tin and lead are amphoteric. Tin(II) oxide and hydroxide form stannite ion (SnO_2^{2-}) upon reaction with a base, and lead(II) oxide and hydroxide form plumbite ion (PbO_2^{2-}). The +4 oxides (SnO_2 and PbO_2) form stannates (SnO_3^{2-}) and plumbates (PbO_3^{2-}) upon reaction with base. The plumbate ion is a powerful oxidizing agent since the +2 oxidation state is much more stable than the +4 state for lead. The following equations illustrate the formation of these ions:

$$Sn(OH)_2 + 2\,NaOH \longrightarrow Na_2SnO_2 + 2\,H_2O$$
$$Pb(OH)_4 + 2\,NaOH \longrightarrow Na_2PbO_3 + 3\,H_2O$$

Identification of Ions

The carbonate ion (CO_3^{2-}) forms CO_2 gas upon addition of acid to either a solid carbonate or an aqueous solution. To confirm that the gas is CO_2, it can be bubbled into limewater (a saturated solution of calcium hydroxide) to form the insoluble calcium carbonate:

$$CO_{2(g)} + Ca(OH)_{2(aq)} \longrightarrow CaCO_{3(s)} + H_2O_{(l)}$$

The Pb^{2+} ion can be precipitated out of solution as the chloride; however, $PbCl_2$ is quite soluble, especially in hot water. It is also precipitated out as a sulfide, along with Sn^{2+}, in acid solution. These sulfides have a very low solubility and are precipitated from solution when H_2S is added, even when the S^{2-} ion concentration is quite low, which it is in an acid solution.

Table 22-5 summarizes the properties of the Group IVA elements.

Consider the following problem example involving the IVA elements:

PROBLEM EXAMPLE 22-5

Write a balanced equation for each of the following reactions:
(a) silicon + oxygen
(b) lead + oxygen
(c) tin(II) hydroxide + sodium hydroxide solution
(d) lead(IV) hydroxide + potassium hydroxide solution
(e) tin(II) oxide + sodium hydroxide solution

TABLE 22-5 *Properties of the IVA Elements—C, Si, Ge, Sn, Pb*

GENERAL PROPERTIES	Go from nonmetal (C, Si) to metalloid (Ge), to metal (Sn, Pb)
OCCURRENCE	Carbon occurs in carbonates and CO_2, as well as organic compounds; Si is very abundant as the oxide SiO_2; Ge is rare; Sn occurs as SnO_2 and Pb as PbS
PREPARATION	Generally by reduction of oxides, with carbon or active metal, such as in the reaction $SnO_2 + 2\,C \rightarrow Sn + 2\,CO$
CHEMICAL PROPERTIES	React with O_2, halogens, and other nonmetals; form compounds with oxidation states of +2 and +4 (in organic compounds, C exhibits many oxidation states), with the +2 becoming more stable going down the group; carbon is often used as a reducing agent, where it forms CO
COMPOUNDS	Oxides—CO_2 and SiO_2 are weakly acidic; SiO_2 is a macromolecular substance; SnO, SnO_2, PbO, and PbO_2 are amphoteric (PbO is most basic); in OH^-, the Sn and Pb oxides form MO_2^{2-} (stannite and plumbite) and MO_3^{2-} (stannate and plumbate)
IDENTIFICATION OF IONS	CO_3^{2-} forms CO_2 gas upon addition of acid Sn^{2+} and Pb^{2+} precipitate as sulfides with H_2S in acid solution ($PbCl_2$ also precipitates with Cl^- but is rather soluble especially in hot water)

SOLUTION:

(a) Silicon will combine with oxygen to form $SiO_{2(s)}$ since +4 is the only oxidation state for silicon:

$$Si_{(s)} + O_{2(g)} \longrightarrow SiO_{2(s)}$$

(b) Since the most stable oxidation state of lead is +2, the most probable compound formed is PbO:

$$2\,Pb_{(s)} + O_{2(g)} \longrightarrow 2\,PbO_{(s)}$$

(c) Since tin(II) hydroxide is amphoteric, it reacts with sodium hydroxide by a typical acid-base reaction:

$$\underset{(\text{or } H_2SnO_2)}{Sn(OH)_{2(s)}} + 2\,NaOH_{(aq)} \longrightarrow 2\,H_2O_{(l)} + Na_2SnO_{2(aq)}$$

(d) Lead(IV) hydroxide is amphoteric but only two hydrogens are replaced to form the plumbate ion (PbO_3^{2-}):

$$\underset{(\text{or } H_2PbO_3 \cdot H_2O)}{Pb(OH)_{4(s)}} + 2\,KOH_{(aq)} \longrightarrow 3\,H_2O_{(l)} + K_2PbO_{3(aq)}$$

(e) Tin(II) oxide is amphoteric, forming the stannite ion (SnO_2^{2-}):

$$SnO_{(s)} + 2\,NaOH_{(aq)} \longrightarrow H_2O_{(l)} + Na_2SnO_{2(aq)}$$

22-6 Group VA—Nitrogen (N), Phosphorus (P), Arsenic (As), Antimony (Sb), and Bismuth (Bi)

This group of elements is often referred to as the nitrogen family. Only the last two elements show definite metallic properties, the others being distinctly nonmetallic. All these elements exhibit a variety of oxidation states.

Air is approximately 80% nitrogen. Although nitrogen does occur in many compounds, the only commercial source of the element is from air. The nitrogen is obtained by the distillation of liquid air. This separates the nitrogen from the oxygen and inert gases in the air. Nitrogen is a colorless, odorless, tasteless gas, with a boiling point of −196°C. Its solubility in water is very low. Since nitrogen is quite inert, liquid nitrogen is often used in the laboratory to obtain low temperatures. The element is used in large quantities to form compounds such as ammonia and nitric acid. Nitrogen is essential to life but must be in the form of a compound in order to be used by plants. Compounds of nitrogen have many uses, an important one being as fertilizers.

Phosphorus occurs chiefly as calcium phosphate, and the other elements in the group generally occur as sulfides.

Phosphorus exists in two allotropic forms. White phosphorus is volatile, flammable, and toxic. It is very reactive and when it is heated in the absence of air, it is converted to the red form, which is relatively nonvolatile, nontoxic, and unreactive. Antimony and bismuth are soft, low melting solids.

Arsenic, antimony, and bismuth are used in small amounts in alloys. Arsenic and antimony are used to harden lead, and bismuth is used to make low melting alloys.

The elements in this group vary from nonmetallic to metallic going down the group, mainly due to the increasing size of the atoms. This change from nonmetallic to metallic behavior can be seen by looking at the trioxides. N_2O_3 and P_2O_3 are strongly acidic; As_2O_3 and Sb_2O_3 are amphoteric; Bi_2O_3 is definitely basic.

Each of the elements in Group VA has five valence electrons. Therefore, it is unlikely that they would completely lose five electrons or gain three electrons. In most of their reactions they form covalent bonds, forming compounds with a maximum oxidation state of +5 (with more electronegative atoms) or a minimum oxidation sate of −3 (with less electronegative atoms). Both of these oxidation states are found for all the Group VA elements, although the stability of the −3 oxidation state decreases going down the group. In addition to these two oxidation states, a +3 state is also found for all these elements. Nitrogen and phosphorus form compounds with oxidation states from −3 to +5. In the free state none of the Group VA elements is very reactive. In the +5 oxidation state all the elements except phosphorus form compounds that are very strong oxidizing agents.

Nitrogen reacts with hydrogen to form ammonia. With active metals it forms nitride and with oxygen it forms a variety of oxides—N_2O, NO, N_2O_3, $NO_2(N_2O_4)$, and N_2O_5.

Phosphorus, arsenic, antimony, and bismuth react with oxygen to form oxides of the type P_2O_5 and P_4O_6 (dimer of P_2O_3). They react with chlorine to form halides of the type PCl_3 and PCl_5. They can also exist in a -3 oxidation state, in the compounds phosphine (PH_3), arsine (AsH_3), stibine (SbH_3), and bismuthine(BiH_3). These are analogous to ammonia but are all unstable.

Ammonia is one of the most important compounds of nitrogen. It is produced from N_2 and H_2 by the Haber process. Ammonia is a gas with a high heat of vaporization and is used in mechanical refrigeration systems. It is highly soluble in water, with which it reacts to form the weak base ammonium hydroxide (or aqueous ammonia):

$$NH_3 + H_2O \longrightarrow NH_4^+ + OH^-$$

Ammonia is a base and thus reacts with acids to form ammonium salts. It also readily forms complexes (Chapter 24) with certain metal ions, such as Ag^+.

$$NH_3 + HCl \longrightarrow NH_4Cl$$
$$2\,NH_3 + Ag^+ \longrightarrow Ag(NH_3)_2^+$$

Another important compound of nitrogen is nitric acid. Nitric acid is manufactured chiefly by the oxidation of ammonia to nitrogen dioxide and treating the oxide with water. If nitrogen dioxide reacts with cold water, a mixture of nitric acid and nitrous acid results.

$$H_2O + 3\,NO_2 \xrightarrow{\text{warm}} 2\,HNO_3 + NO$$
$$H_2O + 2\,NO_2 \xrightarrow{\text{cold}} HNO_2 + HNO_3$$

Note that both of these are oxidation-reduction reactions. They are not simple combinations involving an acid anhydride with water, in which there is no change in oxidation number. There is no simple anhydride of nitric acid.

Nitric acid is a strong acid; in addition, it is a powerful oxidizing agent both in dilute and concentrated solutions. A number of reduction products are possible, making the chemistry of nitric acid as an oxidizing agent quite complicated. More than one reduction product may be formed but in general one predominates. If concentrated nitric acid is used, the reduction product is always NO_2. If dilute acid is used, several products are possible, such as NO_2, NO, N_2O, N_2, or NH_3, depending on the strength of the reducing agent. The reduction product never has a $+5$ oxidation state.

Identification of Ions

The ammonium ion (NH_4^+) is identified by adding a strong base, to form NH_3 gas according to the following equation:

$$NH_4^+ + OH^- \longrightarrow NH_{3(g)} + H_2O_{(l)}$$

The NH_3 gas is given off, especially if heated, and can be identified by its odor or by holding a piece of moist red litmus paper in the gas; the paper turns blue, due to the basic nature of ammonia in water.

The phosphate ion (PO_4^{3-}) is identified by the formation of a yellow precipitate of phosphomolybdate when ammonium molybdate solution is added.

The nitrate ion (NO_3^-) is identified by adding $FeSO_4$ and H_2SO_4 to the solution. The formation of a brown ring, due to the $Fe(NO)^{2+}$ ion, at the interface of the two layers is a positive test for NO_3^-.

Table 22-6 summarizes the properties of the Group VA elements.

Consider the following problem example involving the Group VA elements:

PROBLEM EXAMPLE 22-6

Write a balanced equation for each of the following reactions:
(a) magnesium + nitrogen (with heat)
(b) nitrogen + chlorine (with heat)
(c) diphosphorus pentoxide + water
(d) dinitrogen trioxide + potassium hydroxide solution
(e) bismuth(III) oxide + hydrochloric acid

SOLUTION:

(a) Since magnesium is an active metal, it will combine with nitrogen to form a nitride, in which nitrogen has a −3 oxidation state:

$$3\ Mg_{(s)} + N_{2(g)} \longrightarrow Mg_3N_{2(s)}$$

(b) Since chlorine is more electronegative than nitrogen, it will combine to form a compound in which nitrogen has a positive oxidation state; however, it can only be NCl_3 (and not NCl_5, like PCl_5) due to the Lewis structure (NCl_5 would have more than eight valence electrons around the N):

$$N_{2(g)} + 3\ Cl_{2(g)} \longrightarrow 2\ NCl_{3(g)}$$

(c) P_2O_5 is the anhydride of phosphoric acid:

$$P_2O_5 + 3\ H_2O_{(l)} \longrightarrow 2\ H_3PO_{4(aq)}$$

(d) N_2O_3 is the anhydride of HNO_2, and thus this is a typical acid base reaction:

$$N_2O_{3(g)} + 2\ KOH_{(aq)} \longrightarrow H_2O_{(l)} + 2\ KNO_{2(aq)}$$

(e) Bi_2O_3 is the anhydride of $Bi(OH)_3$ and therefore this is also an acid-base reaction:

$$Bi_2O_{3(s)} + 6\ HCl_{(aq)} \longrightarrow 3\ H_2O_{(l)} + 2\ BiCl_{3(aq)}$$

TABLE 22-6 *Properties of the VA Elements—N, P, As, Sb, Bi*

GENERAL PROPERTIES	Vary from nonmetals to metals (Sb and Bi) going down the group
OCCURRENCE	Air is about 80% N_2, and many compounds contain this element; P occurs mainly as phosphates; As, Sb, and Bi occur mainly as sulfides
PHYSICAL PROPERTIES AND USES	Nitrogen is a colorless, odorless, tasteless gas, which is not very soluble in water and is used in the form of its compounds (NH_3, ammonium, and nitrate salts) in fertilizers and in the preparation of other compounds Phosphorus exists in two allotropic forms—white (very reactive) and red (unreactive) Sb and Bi are soft, low melting solids Bi is used in low melting alloys; As and Sb are used in small amounts to harden lead
CHEMICAL PROPERTIES	Nitrogen forms compounds with oxidation states from -3 to $+5$; it reacts with active metals to form nitrides; with O_2 it forms a series of oxides; with H_2 it forms NH_3; with chlorine it forms NCl_3 Phosphorus (as well as As, Sb, and Bi) exhibits oxidation states of -3 (PH_3), $+3$ (P_4O_6) and $+5$ (P_2O_5); they form halides of the type PCl_3 and PCl_5; with hydrogen they form compounds analogous to ammonia, of the type PH_3; the stability of the $+5$ state decreases going down the group; compounds in the $+5$ oxidation state are strong oxidizing agents (except P)
COMPOUNDS	Oxides—N_2O_3, P_4O_6, and P_2O_5 are acidic; As_2O_3 and Sb_2O_3 are amphoteric; Bi_2O_3 is basic NH_3—produced by the Haber process; forms a weak base in water; readily forms complex ions with many metal ions HNO_3—a strong acid and powerful oxidizing agent; formed from NO_2 and H_2O Phosphate salts—most are insoluble; all hydrolyze readily (HPO_4^{2-} is a very weak acid)
IDENTIFICATION OF IONS	NH_4^+ is identified by adding strong base (forms NH_3); PO_4^{3-} is precipitated as the yellow phosphomolybdate; NO_3^- is identified by the formation of brown $Fe(NO)^{2+}$ when $FeSO_4$ and H_2SO_4 is added to the solution

22-7 Group VIA—Oxygen (O), Sulfur (S), Selenium (Se), Tellurium (Te), and Polonium (Po)

The elements in this group are primarily nonmetallic, although Se, Te, and Po do show some metallic properties.

Oxygen is by far the most abundant element in the earth's crust, on a basis of both mass and number of atoms; by mass, 49.5% is oxygen. There are more atoms of oxygen than all other elements combined. Air is 21% oxygen by mass and 20% by volume. In the combined state, oxygen is found in plants, animals, minerals, and water. Oxygen readily reacts with most other elements and generally exists in the −2 oxidation state, except in peroxides (O_2^{2-}) and superoxides (O_2^-). It is usually prepared by the electrolysis of water, or the fractional distillation of liquid air, along with nitrogen. Small amounts of oxygen are prepared in the laboratory by heating compounds such as mercury(II) oxide or potassium chlorate, according to the following equations:

$$2\,HgO_{(s)} \longrightarrow 2\,Hg_{(l)} + O_{2(g)}; \qquad 2\,KClO_{3(s)} \longrightarrow 2\,KCl_{(s)} + 3\,O_{2(g)}$$

Oxygen is a colorless, odorless, tasteless gas. It is slightly soluble in water but more soluble than nitrogen; therefore, the oxygen dissolved in water is purer than in air.

Oxygen will not burn, but it will support combustion. It is not very active at ordinary temperatures but is active at elevated temperatures. All the elements form oxides. Oxygen combines directly with *most* metals (see Table 21-1) to form metal oxides. With the heaviest elements of Group IA (K, Rb, and Cs), compounds of the type MO_2, called *superoxides,* are formed. These are ionic solids that contain the cation M^+ and the anion O_2^- (hence, the oxygen has an oxidation state of $-\frac{1}{2}$). Many metals, such as aluminum, combine with oxygen in the air. Oxygen also combines directly with most nonmetals.

Sulfur is also a very common element and occurs in the earth's crust in many forms, including that of the free element. Sulfur is also commonly found combined in the form of the metal sulfides and sulfates.

In the free state, sulfur is obtained by the Frasch process. In this procedure, steam under pressure is forced into the ground containing the sulfur. This melts the sulfur, and compressed air then forces it to the surface.

Two common crystalline forms of sulfur are known. The rhombic form can be crystallized from carbon disulfide and is stable at temperatures below 95.5°C; at higher temperatures the monoclinic form is stable. Sulfur just above its melting point forms a clear yellow liquid but as the temperature is increased, the liquid darkens and becomes increasingly viscous. Finely divided sulfur obtained by condensing the vapor is called flowers of sulfur.

Selenium and tellurium occur much less abundantly than sulfur. They are found along with sulfur in metal sulfides and are recovered in some quantity from lead and copper refining processes. They are relatively unimportant compared with sulfur. There are general variations in their properties going down the group. Polonium occurs only in the form of radioisotopes as a result of the decay of the actinide elements. Its properties can be predicted by considering the general properties of the other elements of this group.

With the exception of oxygen, the elements in this group will assume positive oxidation states (as well as the −2 state) in the presence of more negative elements. The most important positive oxidation states are 4 and 6 but +2 compounds exist, such as in thiosulfate ion, $S_2O_3^{2-}$. The formation of polyatomic anions with oxygen is a common occurrence, as represented by sulfite, sulfate, selenite, selenate, tellurite, and tellurate radicals. It is thus possible for these elements to act either as oxidizing or reducing agents, depending on conditions. In the presence of most nonmetals (oxidizing agents) they tend to go to a +4 or +6 state, depending on the oxidizing power of the nonmetal. Sulfur does not react with the most inactive of elements (noble gases, noble metals) but combines with most others. Some common reactions of sulfur are illustrated by the following equations:

$$Mg + S \longrightarrow MgS$$
$$S + O_2 \longrightarrow SO_2$$
$$S + 3\,F_2 \longrightarrow SF_6$$

Some Important Compounds

Sulfur dioxide, which is the anhydride of sulfurous acid, is formed when sulfur is burned in oxygen. This can be converted to sulfur trioxide by heating in the presence of a catalyst. In the commercial preparation of sulfuric acid, sulfur is burned in air to sulfur dioxide, which is oxidized to sulfur trioxide. The sulfur trioxide is converted slowly to sulfuric acid by the addition of water. It is converted more rapidly if first dissolved in sulfuric acid to form pyrosulfuric acid ($H_2S_2O_7$), which is then diluted with water to form H_2SO_4. Sulfuric acid is a strong acid and, when concentrated, is a strong oxidizing agent. Sulfuric acid is widely used in industry because of these properties and because it is relatively economical to produce.

The sulfide ion (S^{2-}) can be identified by adding acid to the solution. This forms the weak, insoluble acid H_2S, which is given off and can be identified by its odor. The sulfate ion ($SO_4{}^{2-}$) is identified by adding Ba^{2+} ion to precipitate $BaSO_4$. The solution is first made acidic to decrease the concentration of other ions that would precipitate with the Ba^{2+} ion, such as sulfide, carbonate, or phosphate (all these ions form weak acids with the H^+ added and are thus essentially no longer present in the solution as ions).

Table 22-7 summarizes the properties of the elements of Group VIA.

Consider the following problem example involving the Group VIA elements.

PROBLEM EXAMPLE 22-7

Tell which of these acids, selenous or selenic, is the stronger, and explain why. Also, give the name and formula of the anion formed from each, upon replacement of both hydrogen ions.

SOLUTION:

The formulas for these acids are H_2SeO_3 and H_2SeO_4. Selenic acid (H_2SeO_4) is the stronger, due to the greater number of oxygen atoms around the central Se atom (see Section 21-2).
Selenous acid would form the selenite ion, $SeO_3{}^{2-}$, analogous to sulfite, $SO_3{}^{2-}$, and selenic acid would form the selenate ion, $SeO_4{}^{2-}$, analogous to sulfate, $SO_4{}^{2-}$.

TABLE 22-7 *Properties of the VIA elements—O, S, Se, Te, Po*

GENERAL PROPERTIES	All nonmetals, but Se, Te, and Po do show some metallic properties; oxygen is a colorless, odorless, tasteless gas at room temperature but all the others are solids
OCCURRENCE AND PREPARATION	Oxygen is the most abundant element on the earth's crust; found in the air, water, plants, animals, and minerals; it is produced by electrolysis of H_2O, or distillation of liquid air Sulfur also occurs in the free state, as well as in sulfides and sulfates; free sulfur is mined by the Frasch process; Se and Te are not abundant; Po exists only as a radioactive species
CHEMICAL PROPERTIES	Oxygen usually exists in the -2 oxidation state (but also -1 in peroxide and $-\frac{1}{2}$ in superoxide); it will not burn but is necessary for combustion; combines directly with most metals and with most nonmetals, usually to form several oxides Sulfur (and Se, Te, and Po) exhibits oxidation states of -2, $+4$, and $+6$ (for example, in sulfide, sulfite, sulfate, tellurite, tellurate, selenite, and selenate)
COMPOUNDS	Oxides—all oxides are acidic, with acid strength decreasing down the group Sulfides—most metal sulfides are insoluble (except IA and IIA metals), and thus H_2S (weak acid, forms a small amount of S^{2-}) is used in the analysis of metal ions (by controlling the pH of the solution the S^{2-} can be controlled); some sulfides can be dissolved by adding acid (forms H_2S), but for the very insoluble ones an oxidizing agent is necessary (forms free S) SO_2 and H_2SO_3—when sulfur is burned in air SO_2 is formed; this reacts with H_2O to form H_2SO_3 SO_3 and H_2SO_4—SO_2 will react with more O_2, in the presence of a catalyst, to form SO_3, which reacts with H_2O to form H_2SO_4 (a strong acid, which in concentrated solution is a strong oxidizing agent; it is also a good dehydrating agent)
IDENTIFICATION OF IONS	SO_4^{2-} is identified by precipitation of $BaSO_4$ upon addition of Ba^{2+} in acid (to decrease the concentration of anions of weak acids such as CO_3^{2-} or PO_4^{3-}, which would also precipitate with the Ba^{2+}) S^{2-} can be identified by adding a strong acid, which produces H_2S gas

22-8 Group VIIA—The Halogens: Fluorine (F), Chlorine (Cl), Bromine (Br), Iodine (I), and Astatine (At)

The **halogens** constitute the most active family or group of nonmetals in the periodic table. Their high reactivity is due to their position in the periodic table next to the noble gases. They can achieve a noble gas electronic configuration by a reaction in which they add one electron. In the free state these elements exist as diatomic molecules.

Due to their activity the halogens are never found in the free state. Fluorine occurs chiefly as fluorite (CaF_2) and cryolite (Na_3AlF_6). Chlorine is found in large quantities as sodium chloride in seawater and in inland salt deposits. Bromine is often found with sodium chloride as the sodium, potassium, and magnesium bromides, while iodine's most important commercial source is sodium iodate, $NaIO_3$.

Since these elements occur mainly as halide ions (X^-) in metal halides, their preparation will require the use of an oxidizing agent more powerful than the halogen in question.

$$2X^- \xrightarrow[\text{agent}]{\text{oxidizing}} X_2 + 2\,e^-$$

In the case of fluorine there is no more powerful chemical oxidizing agent and electrolytic oxidation is required. The reactivity of fluorine is so great that it is usually prepared as needed and used immediately.

There are a number of chemical oxidizing agents more powerful than chlorine so it can be prepared rather easily. The most important commercial preparation is through electrolysis of an aqueous sodium chloride solution.

$$2\,NaCl_{(aq)} + 2\,H_2O_{(l)} \xrightarrow{\text{electrolysis}} H_{2(g)} + Cl_{2(g)} + 2\,NaOH_{(aq)}$$

Bromine can be prepared by using the same oxidizing agents used to prepare chlorine. Since it is a poorer oxidizing agent than chlorine, milder oxidizing agents may also be used. Industrially bromine is displaced from bromides by chlorine.

$$Cl_2 + 2\,Br^- \longrightarrow Br_2 + 2\,Cl^-$$

Iodine is the weakest oxidizing agent of the halogen family. It is readily prepared from iodides by employing the same oxidizing agents used for the chlorine and bromine preparations. In addition, chlorine and bromine will displace iodine from iodides.

$$Cl_2 + 2\,I^- \longrightarrow I_2 + 2\,Cl^-$$

$$Br_2 + 2\,I^- \longrightarrow I_2 + 2\,Br^-$$

In going down the family there is a change in the physical properties as can be seen in Table 22-8.

The halogens all readily form compounds in which they exhibit the -1 oxidation state, as would be expected from the electron configuration. With

TABLE 22-8 *Physical properties of the halogens*

ELEMENT	STATE	COLOR	MELTING POINT (°C)	NORMAL BOILING POINT (°C)
F	Gas	Yellow	−223	−187
Cl	Heavy gas	Greenish yellow	−101.6	−34.6
Br	Liquid	Brownish red	−7.2	+58.7
I	Solid	Purplish black	+113.5	+184.3

the exception of fluorine, which is the most electronegative element in the periodic table, they also form compounds in which they exhibit positive oxidation states–namely, +1, +3, +5, and +7. For example, chlorine forms the series of oxyacids (and corresponding salts): $HClO$, $HClO_2$, $HClO_3$, and $HClO_4$. Many of the corresponding compounds of Br and I are also known. The salts of these acids are all strong oxidizing agents. Sodium hydrochlorite is a common household bleach.

The halogens are all very active chemically; however, the tendency of the elements to form stable negative ions decreases as their atomic weight increases. They are all oxidizing agents (being reduced to X^- in the process), but the oxidizing power decreases going down the group. Fluorine is the most powerful oxidizing agent known. Fluorine and chlorine will react directly with all metals to form metal halides. Due to the oxidizing power of fluorine and chlorine, the metal in the compound generally has its highest oxidation state. Bromine and iodine are less reactive than chlorine and fluorine and require much more strenuous conditions before reacting with inactive metals; iodine does not react with certain inactive metals under any conditions.

The halogens will react directly with most nonmetals to form covalent compounds. Again fluorine reacts with more nonmetals and in a more violent manner than the other halogens. Fluorine will react directly with all nonmetals except nitrogen, oxygen, and the noble gases. Chlorine, in addition to the elements mentioned above, does not react directly with carbon. Bromine and iodine are much less reactive toward nonmetals. With the exception of the noble gases, compounds between halogens and all other nonmetals can be prepared indirectly if not by direct union.

The *hydrogen halides* are among the most important compounds formed by the halogens. In the anhydrous state these compounds are covalent but they are highly ionized in aqueous solution, with the exception of HF, to form hydrohalic acids, which are thus strong acids (with the exception of hydrofluoric acid). These acids show the properties characteristic of acids as described previously. They are not oxidizing agents (since the halide cannot be reduced further). Therefore when they react with active metals, the salt formed generally has the metal in its lowest oxidation state.

When Ag^+ ion is added to a solution containing Cl^-, Br^-, or I^-, all of them will precipitate as the silver halide. The AgCl will dissolve in ammonia by forming the silver ammonia complex ion $Ag(NH_3)_2^+$. If an oxidizing agent such as NaOCl (bleach) is then added, the iodide and bromide ions are oxidized to free iodine and bromine. The I_2 can be extracted by adding CCl_4 and shaking. This causes a violet color to form in the carbon tetrachloride layer, and the Br_2 gives the water layer a yellow-brown color.

Table 22-9 summarizes the properties of the **halogens.**

Consider the following problem example involving the halogens:

PROBLEM EXAMPLE 22-8

Write a balanced equation for each of the following reactions:
(a) chlorine + iron
(b) chlorine + sodium bromide
(c) iodine + sodium bromide

TABLE 22-9 *Properties of the VIIA elements—F, Cl, Br, I, At*

GENERAL PROPERTIES	The most active group of nonmetals; they never occur free in nature; vary from yellow gas (F) to green gas (Cl), to low boiling red liquid (Br), to purple solid (I)
OCCURRENCE AND PREPARATION	Occur as salts, in which the halide ion must be oxidized to the free state Fluorine must be prepared electrolytically, and Cl_2 is usually prepared this way also; Cl_2 can oxidize Br^- and I^-, and Br_2 can oxidize I^-
CHEMICAL PROPERTIES	All readily form compounds with the -1 oxidation state; Cl, Br, and I can also be $+1$, $+3$, $+5$, and $+7$; they react less readily with metals going down the group; can also react with most nonmetals (again with fluorine and chlorine reacting most readily) All the oxides are acidic, forming acids of the type HXO, HXO_2, HXO_3, and HXO_4 (fluorine, of course, does not form these)
IDENTIFICATION OF IONS	Chloride, bromide, and iodide ions are usually identified by reaction with Ag^+; Cl^-, Br^-, and I^- all precipitate out, but AgCl will dissolve in dilute NH_3 [forms the complex ion $Ag(NH_3)_2^+$]; I^- and Br^- can be oxidized to I_2 and Br_2 and the I_2 extracted with CCl_4; the I_2 gives a purple color to the CCl_4 layer, and Br_2 gives a yellow-brown color to the water layer

(d) $Cl_2O_3 + H_2O$
(e) Br_2O_5 + NaOH solution

SOLUTION:

(a) Chlorine will readily combine with iron. In the compound formed, iron will have its highest oxidation state since chlorine is a strong oxidizing agent:

$$3\ Cl_2 + 2\ Fe \longrightarrow 2\ FeCl_3$$

(b) Since chlorine is a more active nonmetal than bromine (that is, it has a greater tendency to attract electrons toward itself and form a Cl^- ion), the chlorine will displace the Br^- ion:

$$Cl_2 + 2\ NaBr \longrightarrow Br_2 + 2\ NaCl$$

(Recall that all the halogens exist as diatomic molecules.)

(c) Since iodine is less active than bromine, there is no reaction.

(d) This will form an acid of the type HClO, $HClO_2$, $HClO_3$, or $HClO_4$ with the same oxidation state for Cl as in Cl_2O_3. Hence, the acid is $HClO_2$, in which Cl has an oxidation number of +3:

$$Cl_2O_3 + H_2O \longrightarrow 2\ HClO_2$$

(e) Since Br_2O_5 is the anhydride of $HBrO_3$ (using the same reasoning as in part d), the reaction is a typical acid base reaction in which $NaBrO_3$ is the salt formed.

$$Br_2O_5 + 2\ NaOH \longrightarrow H_2O + 2\ NaBrO_3$$

SUMMARY

1. Hydrogen occurs in many compounds, such as water and all living material. It is colorless, odorless, tasteless gas that will burn, and combines with most nonmetals at elevated temperatures. It also combines with active metals to form hydrides. It is the least dense of all known substances. It is produced commercially by the electrolysis of water; in the lab it is produced by an active metal with acid (or water for very active metals).
2. The Group IA elements, called the *alkali metals,* are all active metals and are quite similar in their properties. They are light, soft, low melting metals and form only +1 compounds. They are found in the form of their salts and usually are prepared by electrolysis. The oxides are all strong bases, except Li_2O.
3. The Group IIA elements, called the *alkaline earth metals,* are less active than the IA metals, with the activity increasing down the group. BeO is amphoteric, but the other oxides are all basic. They also occur as salts and are prepared by electrolysis. Oxides and hydroxides increase in basicity and solubility going down the group.

4. The Group IIIA elements vary from nonmetallic for B to metallic for the others. Al is a rather active metal that only exhibits the +3 oxidation state. Ga, In, and Tl are quite rare; they are soft, low melting metals and form +1 and +3 compounds. B_2O_3 is weakly acidic; Al_2O_3 and $Al(OH)_3$ are amphoteric.
5. The Group IVA elements go from nonmetals (C and Si) to metalloid (Ge) to metals (Sn and Pb). They form compounds with +2 and +4 oxidation states (as well as others for C in organic compounds) with the +2 becoming more stable going down the group. All oxides of Sn and Pb are amphoteric (PbO is most basic).
6. The Group VA elements also vary from nonmetals to metals (Sb and Bi) going down the group. Nitrogen is a colorless, odorless, tasteless, inert gas and not very soluble in water. It is used in the form of ammonia and ammonium and nitrate salts. The elements of this group have variable oxidation states, (−3 to +5 for N; −3, +3, and +5 for the others), with the +5 becoming less stable from P to Bi. Oxides vary from acidic (N and P) to amphoteric (As and Sb) to basic (Bi).
7. The Group VIA elements are all nonmetals. Oxygen is a colorless, odorless, tasteless gas at room temperature that reacts readily with most other elements; the other members of this family are solids. Oxygen usually exists in the −2 oxidation state; the others also form +4 and +6 compounds.
8. The Group VIIA elements, called the *halogens,* are all rather active nonmetals. They are quite similar in their chemical properties and show regular variations in their physical properties going down the group. They all readily form −1 compounds, and Cl, Br, and I can also be +3, +5, and +7. All the oxides are acidic. They decrease in strength as oxidizing agents going down the group.

Upon completion of this chapter, you should be able to do the following:

1. Given any group of representative elements, list the elements in the group, in order of increasing atomic weight, and tell
   (a) the principal properties, both physical and chemical.
   (b) the variations in properties going down the group, such as metallic properties, reactivity, oxidation states, and properties of the oxides.
   (c) the principal source of the more common elements and how the free elements are prepared.
   (d) what some of the more common compounds are, why they are important, and how they are prepared.
   (e) how the more common ions in the group are identified.
2. Given the reactants, write a balanced equation for any of the more common reactions of the elements and compounds in any group of representative elements. (1–8)
3. Given any of the terms in Exercise 1, give a definition or explanation in your own words.

EXERCISES

1. Define or explain each of the following:
   (a) representative elements
   (b) alkali metals
   (c) alkaline earth elements
   (d) active metal
   (e) acid anhydride
   (f) basic anhydride
   (g) halogens
   (h) hydride
   (i) nitride
   (j) halide
   (k) electrolysis
   (l) amphoteric
   (m) thiosulfate
   (n) hydrohalic acid
   (o) Haber process
   (p) Frasch process
2. List the main properties and principal source of each of the following elements and tell briefly how it is prepared:
   (a) hydrogen
   (b) sodium
   (c) calcium
   (d) aluminum
   (e) carbon
   (f) lead
   (g) nitrogen
   (h) oxygen
   (i) sulfur
   (j) chlorine
3. Tell if each of the following oxides is acidic, basic, or amphoteric:
   (a) Na_2O
   (b) BeO
   (c) CaO
   (d) B_2O_3
   (e) Al_2O_3
   (f) CO_2
   (g) GeO
   (h) SnO_2
   (i) N_2O_3
   (j) P_2O_5
   (k) SO_3
   (l) Cl_2O_3

PROBLEMS (In each case if there is no reaction, indicate no reaction.)

Hydrogen

1. Write a balanced equation for each:
   (a) hydrogen + oxygen
   (b) hydrogen + nitrogen
   (c) hydrogen + sodium
   (d) sodium hydride + water
   (e) zinc + hydrochloric acid
   (f) hydrogen + phosphorus
   (g) hydrogen + sulfur
   (h) water + sodium
   (i) hydrogen + potassium
   (j) potassium hydride + water

Group IA

2. Write a balanced equation for each:
   (a) potassium + water
   (b) cesium + hydrogen
   (c) sodium + oxygen
   (d) lithium + oxygen
   (e) sodium oxide + nitric acid
   (f) rubidium + hydrogen
   (g) sodium + sulfur
   (h) cesium + water
   (i) potassium oxide + hydrochloric acid
   (j) potassium + oxygen

Group IIA

3. Write a balanced equation for each:
   (a) calcium + oxygen
   (b) magnesium + nitrogen
   (c) calcium + hydrogen
   (d) barium + water
   (e) beryllium + hydrogen
   (f) barium + hydrogen
   (g) magnesium + oxygen
   (h) calcium + nitrogen
   (i) magnesium + hydrogen
   (j) strontium + water

Group IIIA

4. Write a balanced equation for each:
   (a) aluminum + oxygen
   (b) aluminum + nitric acid
   (c) boron + chlorine
   (d) aluminum hydroxide + sodium hydroxide
   (e) gallium + hydrogen
   (f) aluminum + chlorine
   (g) boron + oxygen
   (h) indium + hydrogen
   (i) aluminum + sodium hydroxide solution
   (j) aluminum oxide + hydrochloric acid

Group IVA

5. Write a balanced equation for each:
   (a) carbon + excess oxygen
   (b) carbon dioxide + sodium hydroxide solution
   (c) tin(II) oxide + potassium hydroxide
   (d) lead(II) hydroxide + sodium hydroxide
   (e) sodium carbonate + nitric acid
   (f) sodium oxide + carbon dioxide
   (g) silicon + oxygen
   (h) lead(II) oxide + potassium hydroxide
   (i) tin(II) hydroxide + sodium hydroxide
   (j) hydrochloric acid + potassium carbonate

Group VA

6. Write a balanced equation for each:
   (a) nitrogen + magnesium
   (b) dinitrogen trioxide + water
   (c) nitrogen + hydrogen
   (d) diphosphorus trioxide + sodium hydroxide
   (e) Bi_2O_3 + nitric acid
   (f) nitrogen + calcium
   (g) $P_2O_3 + H_2O$
   (h) diphosphorus pentoxide + potassium hydroxide
   (i) nitrogen + chlorine
   (j) bismuth(III) oxide + perchloric acid

Group VIA

7. Write a balanced equation for each:
   (a) sulfur + oxygen
   (b) sulfur dioxide + potassium hydroxide
   (c) $SeO_3 + H_2O$
   (d) sodium sulfide solution + zinc nitrate solution
   (e) potassium sulfide solution + hydrochloric acid
   (f) oxygen + aluminum
   (g) sulfur trioxide + sodium hydroxide
   (h) $SeO_2 + H_2O$
   (i) sodium sulfide solution + nitric acid
   (j) potassium sulfide solution + tin(II) chloride solution

Group VIIA

8. Write a balanced equation for each:
   (a) chlorine + iron
   (b) sodium bromide + chlorine
   (c) potassium bromide + iodine
   (d) Cl_2O_3 + KOH

(e) silver nitrate solution + ammonium chloride solution
(f) copper + chlorine
(g) sodium iodide + bromine
(h) potassium chloride + bromine
(i) sodium bromide solution + silver nitrate solution
(j) Cl_2O_5 + NaOH

General problems

9. Explain each of the following, using equations where applicable:
   (a) Group IA metals are never found in the free state.
   (b) Lithium oxide is less basic than the other Group IA oxides.
   (c) The sodium ion is identified by a flame test.
   (d) Magnesium is somewhat similar to lithium.
   (e) Aluminum salts readily hydrolyze in water.
   (f) CO_2 is acidic and PbO_2 is amphoteric.
   (g) SO_2 reacts with water to form sulfurous acid.
   (h) Fluorine never exists in a positive oxidation state.
   (i) An aqueous solution of NaCl is neutral, but an aqueous solution of NaF is basic.
   (j) Lime is prepared from limestone.
   (k) A hydrogen carbonate will neutralize either acid or base.
   (l) Cl_2O_3 is acidic.
   (m) The sulfide ion concentration is lower in an acid solution that is saturated with H_2S than it is in a basic solution.
   (n) The potassium ion is identified by a flame test.
   (o) Calcium hydroxide is prepared from lime.
   (p) Sodium bicarbonate will neutralize either acid or base.
   (q) BeO is amphoteric.
   (r) The halogens are never found in the free state.
   (s) Barium is more like the IA metals than magnesium is.
   (t) Fluorides hydrolyze but chlorides do not.
   (u) Cl_2O_5 is acidic.
   (v) Beryllium salts hydrolyze in water.
   (w) B_2O_3 is acidic and Al_2O_3 is amphoteric.
   (x) The sulfide ion concentration is higher in basic solution that is saturated with H_2S than it is in acidic solution.
   (y) There are no compounds of fluorine like ClO^-, ClO_2^-, and so on.
   (z) SO_3 reacts with water to form H_2SO_4.

10. Complete and balance the following equations (if no reaction takes place, indicate *no reaction*).
   (a) $Al_{(s)} + HCl_{(aq)}$
   (b) $Zn_{(s)} + NaOH_{(aq)}$
   (c) $Mg_{(s)} + H_{2(g)}$
   (d) $Na_{(s)} + H_{2(g)}$
   (e) $NaHCO_{3(s)} + HCl_{(aq)}$
   (f) $K_{(s)} + H_2O_{(l)}$
   (g) $Na_2CO_{3(aq)} + CaCl_{2(aq)}$
   (h) $FeS_{(s)} + HCl_{(aq)}$
   (i) $Mg_{(s)} + H_2O_{(l)}$
   (j) $Zn_{(s)} + HCl_{(aq)}$
   (k) $Al_{(s)} + NaOH_{(aq)}$
   (l) $Al_{(s)} + Br_{2(g)}$
   (m) $Sn_{(s)} + Cl_{2(g)}$
   (n) $Pb(OH)_{2(aq)} + HNO_{3(aq)}$
   (o) $CaCO_{3(s)} + HCl_{(aq)}$
   (p) $NaOH_{(aq)} + MgCl_{2(aq)}$

11. Write a net ionic equation for each of the following (if no reaction takes place, indicate *no reaction*). Assume each is in aqueous solution.

(a) Bi^{3+} ion + hydrogen sulfide
(b) ammonia + acetic acid
(c) zinc sulfide + hydrochloric acid
(d) ammonium chloride solution + sodium hydroxide solution
(e) solid calcium oxide + acetic acid
(f) zinc chloride solution + aluminum
(g) perchoric acid + sodium oxide
(h) bicarbonate ion + nitric acid
(i) calcium chloride solution + ammonium carbonate solution
(j) lead(II) oxide + sodium hydroxide solution
(k) gallium(I) ion is oxidized by chlorine gas
(l) sulfur dioxide + sodium hydroxide solution
(m) iron(II) sulfide + nitric acid
(n) ammonium nitrate solution + potassium hydroxide solution
(o) H_2S + Cu^{2+}
(p) acetic acid + sodium hydroxide solution
(q) ammonia + hydrochloric acid
(r) calcium + zinc nitrate solution
(s) tin(II) oxide + sodium hydroxide solution
(t) sodium carbonate solution + barium chloride solution
(u) magnesium oxide + acetic acid
(v) hydrochloric acid + hydrogen carbonate ion
(w) calcium oxide + perchloric acid
(x) nitric acid + sodium acetate solution
(y) hydrochloric acid + sodium hydroxide solution
(z) ferrous ion is oxidized by chlorine gas

REFERENCES

BIGELOW, JEROME M., *The Representative Elements*, Croton-on-Hudson: Bogden and Quigley, Inc., 1970.

GOULD, EDWIN S., *Inorganic Reactions and Structure*, New York: Holt, Rinehart, and Winston, 1962.

HAMPEL, CLIFFORD A. (editor), *The Encyclopedia of the Chemical Elements*, New York: Reinhold, 1968.

LATIMER, WENDELL M., and JOEL H. HILDEBRAND, *Reference Book of Inorganic Chemistry*, 3rd ed., New York: The Macmillan Co., 1951.

NECHAMKIN, HOWARD, *The Chemistry of the Elements*, St. Louis: McGraw-Hill Book Company, 1968.

ROCHOW, EUGENE G., *The Metalloids*, Chicago: D. C. Heath and Company, 1966.

ROCHOW, EUGENE G., *Modern Descriptive Chemistry*, Philadelphia: W. B. Saunders and Co., 1977.

WEEKS, MARY ALVIRA, *Discovery of the Elements*, 7th ed., Springfield, Pa.: Journal of Chemical Education Publications, 1977.

Objectives

1. To be able to explain and predict the more common properties of the groups of transition elements.
2. To learn and be able to correlate with position in the periodic table what oxidation state the elements in each group can form and which compounds are good oxidizing or reducing agents.
3. To be able to predict variations in basic character of the oxides of the more common transition elements.
4. To learn some of the more important properties of some common transition elements, such as how they occur, how they react, what they are used for, and how their ions are identified.

Twenty Three

Properties of the Transition Elements

In this chapter we shall consider briefly some of the more important properties of the transition elements. We shall look at each group of elements but will be mainly concerned with the first member of each series since these are by far the most common elements in this section of the periodic table.

23-1 Introduction

These elements are all metals and in general they have higher densities, melting points, and boiling points than the representative metals. They form many complex ions, due in part, at least, to vacant orbitals in the metal atom.

In most of the compounds of the transition metals, electrons are readily elevated from $(n - 1)d$ to ns orbitals. This is especially true if there are unpaired electrons, in which case the absorption of visible light provides sufficient energy to cause this. Also, this absorption of visible light causes a particular color to appear. Thus, most of these compounds that have unpaired electrons are colored, such as in Fe^{2+}, Fe^{3+}, Ni^{2+}, and Cu^{2+}.

The transition metals form compounds with a variety of oxidation states, due to the fact that the $(n - 1)d$ and ns energy levels differ very little in energy and, therefore, electrons from both of these energy levels are available for bonding. In general, it is not possible to predict what oxidation state is most stable, but as a general tendency the maximum oxidation state becomes more stable and the lower oxidation states less stable with increasing atomic number. Through Group VIIB the maximum oxidation state corresponds to the group number. There are compounds known with oxidation states of 8, in Group VIII, but these are very rare.

The transition metals can lose up to three electrons to form monatomic cations, but the higher the oxidation state, the more covalent the compound is, and all compounds with oxidation states greater than three are covalent.

The last two members of each family are quite similar in size, due to the *lanthanide contraction;* that is, the elements following the first inner transition series are smaller than would be normally expected due to the series of elements that precede them. This series causes an increase in the nuclear charge, and hence, a contraction of the electron clouds. This similarity in size causes the last two elements in each family to be more alike in their properties than the first two.

For Groups IIIB–VIIB there is a direct correlation between the oxidation states formed by the elements and the electron configuration (the one underlined is the most stable one; those in parentheses are known but are quite rare).

IIIB (s^2d^1)—usually $\underline{+3}$ (also +2)

IVB (s^2d^2)—+2, +3, $\underline{+4}$

VB (s^2d^3)—+2, +3, +4, $\underline{+5}$

VIB (s^1d^5)—+2, $\underline{+3}$, (+4, +5) +6

VIIB (s^2d^5)—$\underline{+2}$, +3, +4, (+5), +6, +7

After this the electrons in *d* orbitals are not so readily available for bonding, especially in the first series of transition elements. For example, iron exhibits oxidation states of +2 or +3; cobalt is also +2 or +3 (but the +3 state is quite rare); nickel is practically always +2 (+3 is very rare); copper is only +1 or +2; and zinc is only +2.

23-2 Group IIIB— The Scandium Subgroup

The elements of this group include not only scandium (Sc), yttrium (Y), lanthanum (La), and actinium (Ac), but also the two inner transition series, called the *lanthanum series* and the *actinium series.* The elements in the La series are all very similar and are often called the **rare earth elements.** Since the 4*f* and 5*d* energy levels are very close in energy, there is considerable variation in the electron configurations of these elements. There are even cases where the configuration differs for a particular element when the environment in which it is found changes. All these elements form compounds with the +3 oxidation state, and also the +2 state is known.

All the lanthanide elements (La through Lu) occur in mixtures, which are hard to separate due to the similarity in chemical properties. They are all quite reactive, being similar to magnesium in reactivity.

All the actinide elements (Ac through Lr) beyond uranium are prepared synthetically and only exist as radioactive isotopes, which are usually not very stable.

The elements in this subgroup have physical properties that are typical of metals (high luster, good conductivity, and so on). They are all quite reactive chemically, and the oxides are basic.

Scandium is found in nature in small amounts in some minerals but it is never found in the free state. All its compounds are colorless, and Sc always has a +3 oxidation state. It reacts quite readily with nonmetals, acids, and oxidizing agents. The Sc^{3+} ion precipitates as an oxalate.

23-3 Group IVB—The Titanium Subgroup: Ti, Zr, Hf

The elements in this subgroup—titanium, zirconium, and hafnium—are more common that those of the scandium subgroup, but titanium is the only one that has much commercial use. These elements can form compounds with oxidation states of +2, +3, and +4. The +4 oxidation state is the most common one; the +2 compounds are good reducing agents.

Titanium is the tenth most abundant element in the earth's crust but is not found in very large deposits. It is difficult to prepare the pure metal, due to its great affinity for carbon, nitrogen, oxygen, and hydrogen.

The most important compound is titanium dioxide (TiO_2). It is used as a pigment since it has good covering power and is quite inert. The main uses of titanium are for hardening and toughening steel and as a structural material. It is very resistant to corrosion due to the surface coating of oxide and nitride that are formed when it is exposed to the air.

Zirconium and hafnium have very similar chemical properties, mainly because the atomic radii are almost equal, and their outer electron configurations are the same. Like titanium, these metals are very difficult to prepare in the pure state. Zirconium, like titanium, finds some use as a strong, corrosion-resistant metal.

Titanium can be identified by the red color produced by hydrogen peroxide in acid solution.

23-4 Group VB—The Vanadium Subgroup: V, Nb, Ta

The elements of this subgroup—vanadium, niobium, and tantalum—are less abundant than the elements of the titanium subgroup. They form compounds with oxidation states of +2 through +5, with +5 being the most stable. The oxides are amphoteric.

Vanadium is found in very small amounts in some minerals. It is obtained as a by-product from the uranium mineral carnotite ($KUVO_6$). The compounds of vanadium with the lower oxidation states (+2 and +3) are good reducing agents. The oxide V_2O_5 is probably the most important compound of vanadium; it is used as a catalyst in many oxidation reactions where O_2 is the oxidizing agent. The main use of vanadium is in steel, where it increases the tensile strength, toughness, and elasticity of the steel. It is quite unreactive at ordinary temperatures but will combine with nonmetals at high temperature.

Niobium and tantalum are also very similar in their chemical properties (like Zr and Hf). Both of these elements are quite rare. Tantalum is used in small amounts as a metal that is very resistant to corrosion, for example, in apparatus used for handling acids. Also, it is used for pins, plates, and wire that can be left in the human body, such as those used to strengthen a broken bone.

23-5 Group VIB—The Chromium Subgroup: Cr, Mo, W

The elements of this group—chromium, molybdenum, and tungsten—are all metals with high melting points, great hardness, and excellent resistance to corrosion.

Chromium is one of the less abundant metals on the earth's crust; however, it is very useful due to its good resistance to corrosion. Small amounts of chromium (about 1%) in steel make it very strong and hard. Stainless steel contains about 30% chromium. It is also used as a plating material, both for its decorative and protective effect. It retains its high luster due to the formation of an invisible oxide coating.

Chromium forms compounds with oxidation states of +2, +3, and +6. Molybdenum and tungsten form compounds in which they exhibit oxidation states of +2 through +6. For chromium the most stable oxidation state is +3. Compounds in which chromium exhibits a +2 oxidation state are good reducing agents (since they are readily oxidized to +3); compounds such as dichromate ($Cr_2O_7^{2-}$), in which chromium has a +6 oxidation number, are readily reduced to the +3 state and are thus good oxidizing agents.

Chromium(III) hydroxide is amphoteric, as illustrated by the following equations:

$$Cr(OH)_3 + 3\ HCl \longrightarrow CrCl_3 + 3\ H_2O$$

$$Cr(OH)_3 + NaOH \longrightarrow NaCrO_2 + 2\ H_2O$$

Note in this second equation that only one hydrogen is given off as an H^+ ion and therefore this is often written as $HCrO_2 \cdot H_2O$.

The Cr^{3+} ion is identified by oxidizing it to the yellow CrO_4^{2-}, which can then be precipitated out with Ba^{2+}:

$$Ba^{2+} + CrO_4^{2-} \longrightarrow BaCrO_{4(s)}$$

23-6 Group VIIB—The Manganese Subgroup: Mn, Tc, Re

Manganese is the only important element in this group. Technetium is radioactive and is not found in nature. Rhenium is so rare that it is actually a chemical curiosity.

Manganese occurs mostly as the oxide MnO_2. It is difficult to prepare in the pure state. For example, when carbon is used as a reducing agent, the manganese reacts with it to form manganese carbide. Since manganese is used mainly in alloy steels, however, it can be used without extensive purification.

Manganese exhibits oxidation states of +2 to +7. The +2 compounds are often called manganous and the +3 are called manganic. The +2 compounds are good reducing agents. The permanganate ion (MnO_4^-) is a strong oxidizing agent since manganese is in the +7 oxidation state and is readily reduced. In acid solution the manganese is reduced to the +2 state and in basic solution the manganese is reduced to manganese(IV) oxide, such as in the following equations:

$$2\,KMnO_4 + 16\,HCl \longrightarrow 2\,KCl + 2\,MnCl_2 + 5\,Cl_2 + 8\,H_2O$$

$$2\,KMnO_4 + H_2O + KI \longrightarrow 2\,MnO_2 + 2\,KOH + KIO_3$$

Manganese is softer than iron and has a lower melting point (1250°C). It is more active than zinc and will react slowly with cold water. It reacts readily with electronegative elements to form Mn^{2+} compounds, except with fluorine, which forms MnF_3.

Mn^{2+} is identified by precipitating out the pink MnS when H_2S is added to alkaline solution. Also, any compound in which manganese exists as an oxidation state lower than 7 can be oxidized to the purple permanganate ion (MnO_4^-) by sodium bismuthate ($NaBiO_3$).

23-7 Group VIII—The Iron Triad: Fe, Co, and Ni

In the next three subgroups of transition elements the chemical properties are more similar in the horizontal sequence than in the vertical. Therefore, these will be considered as the *iron triad* (Fe, Co, and Ni) and the *platinum metals* (Ru, Rh, and Pd; and Os, Ir, and Pt). The elements of the iron triad are rather reactive; whereas the platinum metals are rather inert. The platinum metals, along with silver, gold, and mercury, are sometimes called the *noble metals.*

The metals in the iron triad do not occur free in nature. Iron is very abundant and its occurrence is widespread; however workable deposits are not common. The chief source of iron in industry is hematite (Fe_2O_3). The sulfide, iron pyrite (FeS_2), occurs with a number of other sulfide ores but the extraction of iron from this mineral is not economical. This compound looks similar to gold and is often called "fools gold." Ores of cobalt and nickel are much less abundant than those of iron; they are generally obtained from sulfide ores.

These three metals are all silvery in appearance. Iron is soft and malleable, while cobalt and nickel are somewhat harder.

Commercially, iron is the most important metal. It can be obtained inexpensively and a great variation in properties can be obtained when alloyed with small amounts of other elements. Pure iron has very little use.

Iron, cobalt, and nickel are above hydrogen in the activity series. Cobalt and nickel are less reactive than iron; they all dissolve slowly in most acids and they all react with the active nonmetals (halogens, sulfur, and oxygen). In their compounds they all exhibit oxidation states of +2 and +3. The stability of the +2 state increases from iron to nickel, and the stability of the +3 state decreases in the same order. Iron(II) hydroxide (ferrous hydroxide) is a relatively strong base; whereas iron(III) hydroxide (ferric hydroxide) is a very weak base. Therefore, iron(III) salts are readily hydrolyzed in water solution. All the ions of this group precipitate out as sulfides in basic solution of H_2S. Fe^{3+} will form an insoluble hydroxide with aqueous ammonia; Co^{2+} and Ni^{2+} will not since they form complex ions. Iron is identified by the red-colored complex ion formed when Fe^{3+} reacts with thiocyanate (SCN^-). Co^{2+} precipitates as the potassium hexanitritocobaltate(III) salt, and Ni^{2+} precipitates as the red nickel dimethylglyoxime.

23-8 Group VIII—The Platinum Metals: Ru, Rh, Pd, Os, Ir, Pt

These metals are very unreactive and occur chiefly in the free state. They occur together and are difficult to separate. They are corrosion-resistant metals and are good catalysts, especially when hydrogen is involved. Platinum and paladium resemble each other and are soft and malleable. The others are harder than platinum, with some degree of brittleness. They all have high melting points.

The oxidation states of these elements are listed below:

Ruthenium	+2	+3	+4	+6	+7	+8
Rhodium		+3	+4			
Palladium	+2		+4			
Osmium		+3	+4	+6		+8
Iridium		+3	+4	+6		
Platinum	+2		+4			

23-9 Group IB—The Copper Subgroup: Cu, Ag, Au

All the elements in this group occur free in nature because of their low reactivity. Copper is the most reactive, and therefore it is also often found in the combined state, as sulfides, carbonates, silicates, and oxides. Silver occurs in the combined state as the telluride (Ag_2Te) and also in small amounts with copper, lead, and zinc ores. Gold generally occurs in the free state.

The elements of this group show definite metallic properties; they are excellent conductors of heat and electricity and are quite malleable and ductile. These metals are relatively soft and have intermediate melting points (950°–1100°C). Copper and gold are the only colored metals.

As mentioned earlier, these elements are inactive chemically, with the activity decreasing going down the group. Copper and silver react with the halogens and other electronegative elements reasonably well, but gold is very slow in its reaction with these elements and in many cases does not react at all.

Copper forms compounds in which it exhibits oxidation states of +1 or +2 with the +2 being the most stable. Silver is generally considered to exhibit only the +1 oxidation state; there are compounds known in which it is +3 but these are very rare. Gold can be either +1 or +3, with the

+3 being the most stable oxidation state. In general, the oxides of these metals are basic.

The ions of these metals form a number of complex ions. For example, the silver-ammonia complex is formed when aqueous ammonia is added to solid silver chloride, and the blue copper-ammonia complex is used to identify the Cu^{2+} ion. These form according to the following equations:

$$AgCl_{(s)} + 2\,NH_{3(aq)} \longrightarrow Ag(NH_3)_2{}^+{}_{(aq)} + Cl^-{}_{(aq)}$$

$$Cu^{2+} + 4\,NH_{3(aq)} \longrightarrow \underset{\text{(blue)}}{Cu(NH_3)_4{}^{2+}{}_{(aq)}}$$

Cu^{2+} precipitates with H_2S even in acid solution (that is, with very low S^{2-} concentration); the Cu^{2+} can be identified by the bright blue $Cu(NH_3)_4{}^{2+}$ complex ion formed with aqueous ammonia. Ag^+ is one of the few ions (Ag^+, $Hg_2{}^{2+}$, Pb^{2+}) that form insoluble chlorides and is thus precipitated with HCl, usually in the first group in qualitative analysis. AgCl will dissolve in aqueous ammonia, by forming the $Ag(NH_3)_2{}^+$ complex ion.

23-10 Group IIB—The Zinc Subgroup: Zn, Cd, Hg

Since the elements of this group have a filled set of *d* orbitals, they differ somewhat from the other transition elements. Therefore, they are sometimes classified as *terminal* elements rather than transition elements.

Zinc and cadmium occur in nature only in the combined state. Mercury mainly occurs as the sulfide HgS, called cinnabar, but is also found as the free element. Zinc occurs chiefly as the sulfide (ZnS) but is also found as the carbonate and oxide. Cadmium is found as an impurity in zinc ores; there is no important ore of cadmium.

Zinc and cadmium are soft metals and form an adherent oxide or basic carbonate coating when exposed to air, which prevents further oxidation of the metal. Therefore, zinc and cadmium are often used as protective coatings for iron objects. Mercury has the lowest melting point of any metal and exists as a liquid at room temperature. All these metals are used in alloys; alloys of mercury are called *amalgams.* They all form compounds with +2 oxidation states. Mercury also forms +1 compounds, which actually exist as $Hg_2{}^{2+}$ (since Hg^+ would have an s^1 electron configuration). They are much more reactive than the metals of the copper group. Zinc and cadmium are both above hydrogen in the activity series. Mercury, while classed as a noble metal, is more active than gold.

These metals will react with the halogens, nitric acid, and other common oxidizing agents, according to the following general equations:

$$M + X_2 \longrightarrow MX_2$$

$$2M + O_2 \longrightarrow 2\,MO$$

$$M + 4\,HNO_3\,(conc) \longrightarrow M(NO_3)_2 + 2\,NO_2 + 2\,H_2O$$

The oxides and hydroxides of these metals are weak bases, with base strength increasing going down the group. Zinc hydroxide is amphoteric, reacting with acid or base, such as in the following equations:

$$Zn(OH)_2 + 2\,HCl \longrightarrow ZnCl_2 + 2\,H_2O$$

$$Zn(OH)_2 + 2\,NaOH \longrightarrow Na_2ZnO_2 + 2\,H_2O$$

Zinc metal reacts with warm sodium hydroxide solution, forming sodium zincate and hydrogen.

$$Zn + 2\,NaOH \longrightarrow Na_2ZnO_2 + H_2$$

The Zn^{2+} ion is precipitated as ZnS from basic solutions of H_2S. It is separated from the other ions in the analytical group (generally Al, Cr, Mn, Fe, Co, and Ni) by the fact that Zn^{2+} forms stable complexes with both OH^- and NH_3. $Hg_2{}^{2+}$ precipitates as Hg_2Cl_2 in the first group of the qualitative analysis scheme; it turns black upon addition of NH_3 (whereas AgCl dissolves). The Hg^{2+} ion reacts with $SnCl_2$ to form Hg_2Cl_2 and $SnCl_4$ (this is used both as a test for Hg^{2+} and for Sn^{2+}).

SUMMARY

1. In general, the *transition elements* (a) are all metals; (b) have higher densities, melting points, and boiling points than the representative elements; (c) form compounds in which the metal exhibits a variety of oxidation states, with the maximum corresponding to the group number (except for Groups VIII and IB); (d) form complex ions; and (e) often form ions that have distinctive color, especially if there are unpaired electrons in the ion.
2. The IIIB subgroup includes not only Sc, Y, La, and Ac but the two inner transition series as well. The elements of the La series are all quite similar and are usually referred to as the *rare earth elements;* the elements of the Ac series beyond U (92) are all synthetically produced and generally are unstable radioactive species; thus, in many cases not much is known about their chemical properties. The other elements of the IIIB subgroup are all typical metals, whose oxides are basic. They are quite reactive, being similar to Mg in reactivity.
3. The IVB subgroup (Ti, Zr, Hf) is more common than the IIIB elements, but only Ti has practical use (as structural material, used to harden steel—it is also very resistant to corrosion). The metals are difficult to prepare in the pure state. They form compounds with oxidation states of +2, +3, and +4, with the +4 being the most common. The +2 compounds are good reducing agents.
4. The elements of the VB subgroup (V, Nb, Ta) are not very abundant, although V is obtained as a by-product from uranium ores and is used in steel. Small amounts of Ta are used in alloys that find use in the

human body, due to its good resistance to corrosion. These elements form compounds in which they have oxidation states of +2 through +5, with +5 being the most common. Compounds with +2 or +3 oxidation states are good reducing agents. The oxides of these elements are amphoteric.

5. The metals of the VIB subgroup (Cr, Mo, W) are typical metals with good resistance to corrosion and are thus used in alloys. They exhibit oxidation states of +2 through +6 with CrO being most basic and CrO_3 most acidic (with the +3 oxide or hydroxide of Cr being amphoteric). Cr in the +2 state is a good reducing agent and the +6 state is a good oxidizing agent (such as in $Cr_2O_7^{2-}$).
6. The only important element of Group VIIB is Mn; it forms compounds with oxidation states of +2 to +7 with +2 compounds being good reducing agents. The permanganate ion (MnO_4^-) is a powerful oxidizing agent, being reduced to Mn^{2+} in acid solution and MnO_2 in basic solution. Manganese is a rather active metal, being more active than Zn.
7. The Group VIII metals are usually divided into the iron triad (Fe, Co, Ni) and the Pt metals. The first of these are rather active metals, all being above hydrogen in the activity series, whereas the Pt metals are very inactive.
8. Iron is quite abundant, whereas Co and Ni are much less so. The +3 oxidation state is more stable for Fe than the +2, whereas for Co and Ni the reverse is true. The ferrous (Fe^{2+}) state of iron is a good reducing agent. Nickel exists almost entirely in the +2 state. Pure iron has little use, but with varying amounts of other metals and with small amounts of carbon it has many uses. $Fe(OH)_2$ is a rather strong base, but $Fe(OH)_3$ is very weak; thus, Fe^{3+} salts are readily hydrolyzed.
9. Since the Pt metals are very unreactive, they are found in the free state. These metals form a variety of oxidation states, from +2 to +8. They are good catalysts, especially for reaction of hydrogen.
10. The Group IB metals (Cu, Ag, Au) are also relatively unreactive and occur free in nature; Cu is most reactive and thus often occurs as a sulfide, oxide, or carbonate. Copper occurs as +2 and +1 compounds (+2 most stable); Ag occurs only as +1; and Au occurs as +3 and +1 (+3 most stable). They all show typical metallic properties, and the ions very readily form complex ions.
11. The elements of Group IIB (Zn, Cd, Hg) have a filled set of *s* and *d* orbitals and thus do not show much variation in their oxidation states. All form +2 compounds, and Hg also exhibits a +1 oxidation state. Zinc and mercury occur mainly as sulfides but Hg is also found in the free state. Zinc is amphoteric but the others are not. They all have low melting points, with Hg being a liquid at room temperature. They are all used in alloys; alloys of mercury are called amalgams.

Upon completion of this chapter, you should be able to do the following:

1. Given any group of transition elements, tell (a) the name and symbol for the first member of the group; (b) what oxidation states you would predict the group to exist in and which oxidation states you would predict to be good oxidizing or reducing agents; (c) which of the oxides of the group would be most basic; (d) the general properties of the group, including reactivity, uses of the more common metals, and how the more common ions are identified.
2. Given any transition metal ion, tell how many unpaired electrons it has; predict whether it would be colored or colorless. (Exercise 3)
3. Given any of the terms in Exercise 1, give a definition or explanation in your own words.

EXERCISES

1. Define or explain the following:
   (a) transition elements
   (b) lanthanum series
   (c) actinide elements
   (d) rare earth elements
   (e) iron triad
   (f) platinum metals
   (g) noble metals
2. Explain each of the following:
   (a) The rare earth elements are all very similar in their properties.
   (b) The last two members of each B subgroup are often more similar to each other than they are to the first one, due to the lanthanide contraction.
   (c) The +2 oxide of Cr is basic, the +6 oxide is acidic, and the +3 oxide is amphoteric.
   (d) Dichromate ion and permanganate ion are both strong oxidizing agents.
   (e) The platinum metals, along with Ag, Au, and Hg, are often found in the free state.
   (f) The mercury(I) ion exists as Hg_2^{2+}.
3. Tell how many unpaired electrons there are in each of the following ions, and tell if the ions are colored or colorless:
   (a) Ag^+ (b) Fe^{2+} (c) Cr^{3+} (d) Zn^{2+} (e) Ni^{2+} (f) Cu^{2+} (g) Fe^{3+} (h) Mn^{2+} (i) Co^{2+} (j) Hg^{2+}

REFERENCES

GOULD, EDWIN S., *Inorganic Reactions and Structure*, New York: Holt, Rinehart and Winston, 1962.

MOELLER, THERALD, *The Chemistry of the Lanthanides*, New York: Reinhold Publishing Corporation, 1963.

NECHAMKIN, HOWARD, *The Chemistry of the Elements*, St. Louis: McGraw-Hill Book Company, 1968.

Objectives

1. To understand what a coordination compound or complex is and to be able to tell what the central atom is and what the ligands are.
2. To have a general understanding of what atoms form complexes and what type of ions and molecules act as ligands.
3. To be able to tell what the coordination number is for a given compound.
4. To be able to define the terms monodentate, polydentate, and chealate and to identify these types of species.
5. To be able to explain or to predict the possible geometry of complexes with coordination numbers of 2, 4, and 6, using the valence bond theory and the concept of hybridization of orbitals.
6. To understand what is meant by the terms isomer and geometric isomer (stereoisomer) and to be able to tell what types of complexes can form these isomers.
7. To be able to identify *cis* and *trans* isomers.
8. To understand some of the simpler methods used to determine the formula of a complex experimentally and to be able to choose the correct formula from experimental evidence.
9. To understand the basic ideas of the valence bond and ligand field (crystal field) theories of bonding for coordination compounds and to be able to explain and to predict bonding in complexes using these.
10. To learn to write correct formulas and names for these compounds.

Twenty Four

Coordination Compounds

In Chapter 17 we referred to complex ions and in Chapter 22 it was mentioned that transition metals often form coordination compounds or complexes. In this chapter we shall briefly consider these compounds. In discussing these compounds we shall be using several fundamental principles considered earlier, such as oxidation states, bonding, and electron structure.

24-1 Introduction

Coordination compounds, which can be molecules, positive ions, or negative ions, are made up of a *central atom* surrounded by ions or molecules called **ligands.** The central atom is generally a transition metal, and the ligands always contain a lone pair of electrons that can form a coordinate covalent bond.

Compounds of this type were first discovered around 1800. At that time, however, no one could explain how they were held together or why they formed. One of the first ones to be prepared was a yellow substance composed of cobalt(III) chloride and ammonia. It was found that there were 6 moles of ammonia associated with each mole of cobalt(III) chloride. Since both cobalt(III) chloride and ammonia are stable substances, however, it could not be explained why they would combine to form this new type of substance. It took nearly 100 years of study and research to find out why these compounds formed and to propose theories for their bonding. In the meantime, they became known as *complex substances,* or since many of them were found to be ionic, the name *complex ion* developed. One of the earliest theories of bonding, known as the valence bond theory (which is one of the currently accepted models for the bonding in these compounds), uses the concept of a coordinate covalent bond being formed between the ligand and the central atom. Hence the name **coordination**

compound has come into use. There is some question about this theory of bonding, however, and a somewhat more recent theory considers the bond to be an electrostatic attraction. Therefore, the name *coordination compound* is perhaps also somewhat incorrect. The best term for these, and one widely used, is to simply refer to them as *complexes.* In this text, we shall use the terms coordination compound, complex, and complex ion interchangeably.

Metal ions with a high charge density tend to form complexes quite readily. An example of this is the +2 and +3 ions of the VIB through IIB transition metals. The higher oxidation states of these metals form more complexes than the lower ones. This includes species such as Cu^{2+}, Fe^{3+}, Co^{3+}, and so on. The IA and IIA metal ions do not tend to form complexes since they have lower charge densities. For example, the IA metals only form +1 ions and the IIA metals that are definitely ionic, such as calcium, strontium, and barium, also have rather low charge densities due to their relatively large size. There are exceptions to this general rule, such as with the Ag^+ ion, which has a low charge density, but nevertheless forms complexes.

Nonmetal atoms with high electronegativities tend to form good ligands if they exist as ions, in molecules, or in polyatomic anions, with a lone pair of electrons. Due to the presence of this lone pair of electrons, the species that act as ligands are all polar and therefore a ligand will always be either a polar molecule or an ion. Some common examples of ligands are H_2O, NH_3, OH^-, F^-, Cl^-, Br^-, CN^-, NO_2^-, $C_2O_4^{2-}$, and organic molecules containing nitrogen such as ethylenediamine ($H_2NCH_2CH_2NH_2$).

Some examples of coordination compounds or complex ions that you have probably worked with in the lab include the blue complex formed between the copper(II) ion and ammonia (which is used in the qualitative identification of the copper(II) ion), the colorless silver ammonia complex ion (which is used to dissolve silver chloride in the qualitative test for silver ion), and the compounds formed between amphoteric hydroxides and hydroxide ion. An example of the last case is the formation of a complex ion when excess hydroxide ion is added to a precipitate of aluminum hydroxide. Also, many metal ions form complexes with water molecules, such as the light blue aquo complex of the copper(II) ion.

The **coordination number** of a complex is equal to the number of bonds formed around the central atom. If each of the ligands only forms one bond with the central atom, the coordination number is equal to the number of ligands. Some ligands, however, such as the oxalate ion ($C_2O_4^{2-}$), have more than one position on the group that can form a bond with the central atom (see Table 24-1). In this case, each of the end oxygens in the group can form a bond; therefore, in this ligand there are two points of attachment. Ligands of this type are called **polydentate** ligands or, more specifically, bidentate, tridentate, and so on. The **bidentate** ones are the most common. Those ligands that only have one point of attachment are referred to as **monodentate.** Table 24-1 shows the structures of some common monodentate and polydentate ligands.

TABLE 24-1 *Some common ligands*

NAME	FORMULA	LEWIS STRUCTURE
ammonia	NH_3	H:N̈:H (monodentate) H
chloride ion	Cl^-	:C̈l:⁻ (monodentate)
oxalate ion	$C_2O_4^{2-}$	:Ö:C : C:Ö:²⁻ (bidentate) :O::O:
ethylenediamine (en)	$NH_2(CH_2)_2NH_2$	H H H H H:N:C:C:N:H (bidentate) H H

Complexes with polydentate ligands are called **chelates,** from the Greek word *chele,* meaning *claw.* Thus a chelating agent is any substance that can form a chelate.

By far the most common coordination numbers found in complexes are 2, 4, and 6. There are complexes known with other coordination numbers but these three are by far the most common.

24-2 Geometry of Coordination Compounds

Since complexes with coordination numbers 2, 4, and 6 are by far the most common ones, we shall just consider the possible geometries with complexes of these coordination numbers.

1. Coordination number 2—complexes of this type are always linear.
2. Coordination number 4—these can be either tetrahedral, in which case apparently an sp^3 hybrid orbital is used around the central atom, or square planar, in which case it appears as though a dsp^2 hybrid orbital is formed. (Review Chapter 7 for hybrid orbitals if necessary.) See Fig. 24-1.
3. Coordination number 6—these are octahedral, with d^2sp^3 or sp^3d^2 hybrid orbitals being formed around the central atom (according to the valence bond theory of bonding). See Fig. 24-1.

Some complexes with coordination number 4 (with a square planer structure) or coordination number 6 can form different *isomers.* These are compounds with the same formula but different structural arrangements of the atoms. The type of isomers that we are considering here are actually

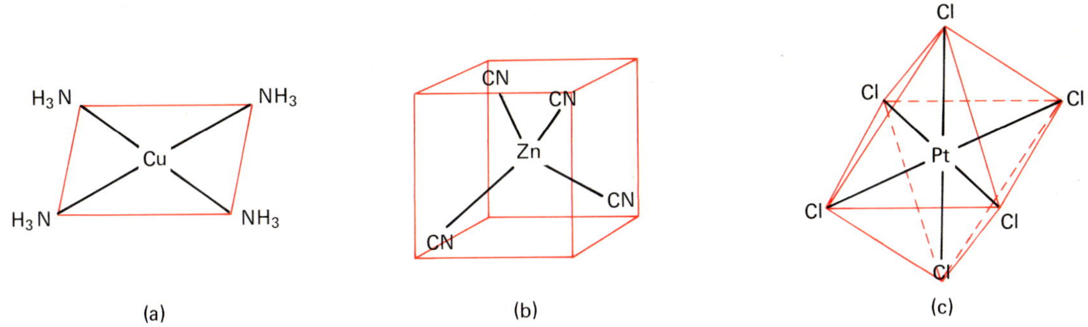

Figure 24-1
Examples of coordination compounds. (a) $Cu(NH_3)_4^{2+}$ with a square planar arrangement (b) $Zn(CN)_4^{2-}$ with a tetrahedral arrangement (c) $PtCl_6^{2-}$ with an octahedral arrangement

geometric isomers or stereoisomers. Figure 24-2 shows some possible isomers of this type.

A *cis-* isomer has groups that are alike next to each other; whereas a *trans-* isomer has the like groups opposite. If there are more than two different ligands in a complex, there are often several different isomers possible, especially with octahedral complexes. You may find it useful to build some models of various types to help to see the differences in isomers.

Keep in mind that for the octahedral arrangement, any two positions that are opposite can be used for the *trans-* isomer. For coordination number 4 with a tetrahedral arrangement there can be no stereoisomers since each position is adjacent to all three of the others. If there are *cis-* and *trans-* isomers possible for a particular complex, the one being considered must be indicated in the name since these are often quite different compounds.

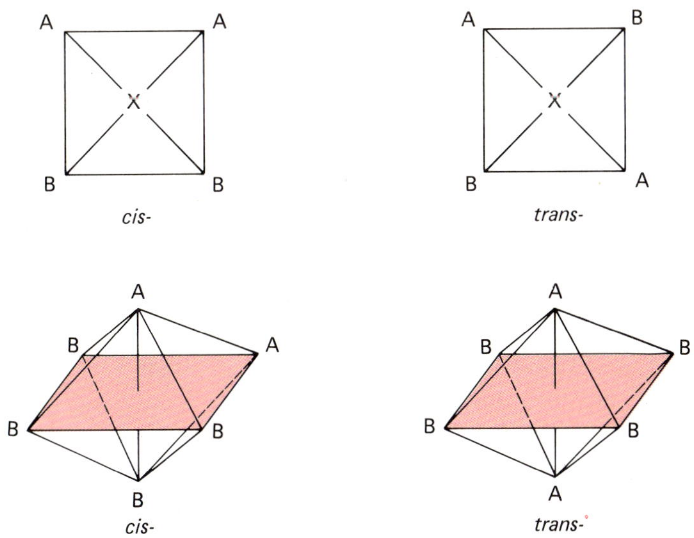

Figure 24-2
cis- *and* trans- *Isomers*

24-3 Experimental Determination of the Formula of a Complex

Very often there are several different coordination compounds that have the same molecular formula. For example, Table 24-2 shows four possible different compounds with the same formula. The first one is simply a hydrate of the neutral complex, tetrachlorplatinum(IV). The second one is a cation complex, dichlorodiaquoplatinum(IV), with 2 Cl^- ions associated with it (not as ligands). The third is the neutral complex, tetrachlorodiaquoplatinum(IV). And the fourth is the cation complex, diaquoplatinum(IV), with 4 Cl^- ions associated with it. To determine experimentally which of these four a particular substance might be, there are two methods that could be used. First, the substance could be dissolved in water and the conductivity measured; if it conducts a current, it must be either number 2 or number 4, which are the only ionic species. Simple chemical tests could be run to determine what the actual formula is. For example, if heating the compound drives off 2 moles of water for each mole of compound, it must be the simple hydrate as shown in formula 1. If reacting with silver nitrate gives 2 moles of silver chloride for each mole of the substance, it must be as shown in formula 2, which contains two chloride ions per mole. If 4 moles of silver chloride are formed for each mole of complex, it must be formula 4. If no water is given off upon heating and no chloride precipitate is formed upon addition of silver nitrate, it must be as shown in formula 3. This is just one simple example of some of the types of tests that can be run to determine what the actual formula is for a particular complex.

TABLE 24-2 *Different complex compounds having the same formula*

COMPOUND	LIGANDS
(1) $PtCl_4 \cdot 2\,H_2O$	4 Cl^- ions (no H_2O)
(2) $[Pt(H_2O)_2Cl_2]Cl_2$	2 H_2O, 2 Cl^-
(3) $[Pt(H_2O)_2Cl_4]$	2 H_2O, 4 Cl^-
(4) $[Pt(H_2O)_2]Cl_4$	2 H_2O (no Cl^-)

24-4 Theories of Bonding*

There are currently three theories that are used to describe the bonding in coordination compounds. These are the valence bond theory, the ligand field (or crystal field) theory, and the molecular orbital theory.

The *valence bond theory* is what was used in discussing covalent bonds in Chapter 7. As applied to complexes, this theory simply says that a coordinate covalent bond is formed, with the ligand furnishing the electron pair and hybrid orbitals being formed around the central atom. The type of

*This section is optional and can be omitted without loss of continuity.

hybrid formed determines the shape of the complex; that is, *sp* or *dp* hybrids give a linear shape for coordination number 2; sp^3 or dsp^2 giving a tetrahedral or square planar, respectively, shape for coordination number 4; and d^2sp^3 or sp^3d^2 giving an octahedral shape for coordination number 6. In this last case, the *d* orbital used in forming the hybrid can be either of a lower energy level than the *s* and *p* orbitals, such as the 3*d* in an iron atom or ion, or of the same energy level as the *s* and *p* orbitals such as the 4*d* in the case of iron. If it is the 3*d*, they are written as d^2sp^3 and called *inner complexes;* if 4*d* is used, it is written as sp^3d^2 and referred to as an *outer complex.* This can be determined by measuring the magnetic properties, which gives a measurement of the number of unpaired electrons. A detailed explanation of this is beyond the scope of this text but can be found in the references listed at the end of the chapter.

The *ligand field theory* or the somewhat more mathematical and slightly more accurate modification of this, called the *crystal field theory,* is mainly an electrostatic model; that is, the ligand, which is always an ion or a polar molecule, is attracted to the central atom or ion. In the presence of the ligand, which acts as a magnetic field, the *d* orbitals around the central atom are distorted from their usual arrangement and some of them become lower in energy than they were originally while others become higher. The somewhat simpler ligand field theory just explains what happens to the *d* orbitals; the crystal field theory allows for calculations of differences in stability by calculating the energy required to break bonds and so forth, due to different types of metal ions and ligands. Again a detailed study of these theories of bonding is beyond the scope of this book but can be found in the references at the end of the chapter.

The *molecular orbital theory* is just an expansion of the basic ideas discussed regarding this theory in Chapter 7. Due to the more complex nature of these substances than the simple diatomic molecules which we consider briefly in Chapter 7, however, this becomes quite complicated for coordination compounds. It is a more mathematical description and although more accurate than the others is not as useful in visualizing the geometry of these substances.

24-5 Nomenclature

The names of these coordination compounds or complexes often sound quite complicated, but when they are broken down to the individual parts they are really quite simple. We shall only consider some rather simple examples in this text and therefore shall not be concerned with a complete detailed study of the nomenclature for all compounds of this type. By learning a few general rules, however, you will recognize what these names mean when you see them, and you will also be able to write the correct name or formula for the majority of the compounds of this type that you will be using or have reference to in the laboratory. Keep in mind that the complex can be a positive ion, a negative ion, or a neutral molecule. If it is a positive ion, it may have an anion associated with it such as a chloride or sulfate ion. Likewise, if it is a complex anion, it may have a positive ion associated with it in its formula, such as a sodium ion, a calcium ion, or an ammonium ion.

The *formula* for the complex part of any substance is always *written in the order of central atom followed by neutral ligands and finally anion ligands.* Usually brackets are put around it to separate the complex part from the rest of the formula.

In the name *the complex is named in the reverse order from the way the formula is written;* that is, it is written as follows:

1. First the anion ligands are named, and they take an -o ending, such as chloro (Cl^-), hydroxo (OH^-), nitro (NO_2^-), cyano (CN^-), and oxalato ($C_2O_4^{2-}$). If there is more than one of the same ligand present, the prefixes di-, tri-, tetra-, penta-, and hexa- are used.
2. Next the neutral ligands are named. Several of the more common ones have common names, such as aquo for H_2O and ammine for NH_3. As with the anion ligands, if there is more than one of the same kind, the prefixes are used.
3. Finally the central atom is named, with a Roman numeral in parentheses to indicate the oxidation state. If the *complex* is an anion, the central atom takes an -ate ending, and the Latin names are used for Fe, Cu, Sn, Pb, and Ag.

When writing the name, the complex part is written all as one word, with the oxidation state as part of the word; that is, they are not separated by any spaces.

Consider the following examples.

1. $Na_2[ZnCl_4]$ = sodium tetrachlorozincate(II)
   The sodium is not part of the complex but is a cation associated with the complex anion. It is not necessary to indicate that there are two sodium ions, just as with simple compounds such as Na_2S. Notice that the complex part, $[ZnCl_4^{2-}]$, is an anion, and thus the central atom takes the name zincate, followed by the Roman numeral to indicate the oxidation state of zinc.
2. $[Co(NH_3)_6]Cl_3$ = hexamminecobalt(III) chloride
   This is a positive complex and thus the central atom is just named as *cobalt.* Notice that a space is left between the name of the complex and the anion (Cl^-), but the complex is written all as one word, including the Roman numeral. The oxidation state of cobalt here is determined from the fact that NH_3 is neutral and each chloride ion has a -1 charge.
3. $Pt(NH_3)_2Cl_2$ = dichlorodiammineplatinum(II)
   This is a neutral complex. If there are no brackets, it is assumed that the whole thing is part of the complex. Anion ligands are named before neutral ligands.

4. $Cu(NH_3)_4^{2+}$ = tetraaminecopper(II) ion
5. $Ca[Pt(C_2O_4)_2]$ = calcium dioxalatoplatinate(II)
   Since the oxalate ion has a −2 charge and calcium is always +2, the platinum here must be +2. Note that the coordination number of Pt in this case is 4 since the oxalate ion is a bidentate ligand.

SUMMARY

1. Coordination compounds consist of a *central atom* and one or more *ligands,* where the ligand generally funishes the pair of electrons used in forming the bond. The transition metals are the most common central atoms but not the only ones.
2. The *coordination number* is the number of bonds formed around the central atom. For *monodentate ligands* the coordination number is equal to the number of ligands. A *polydentate ligand* has more than one point of attachment, and complexes of this type form ring structures that are called *chelates.*
3. By far the most common coordination numbers are 2 (which form linear structures), 4 (which can be either square planar or tetrahedral), and 6 (which are octahedral). For square planar and octahedral structures, *stereoisomers* are possible (*cis*- and *trans*- forms).
4. There are three theories, or models, used to describe the bonding in these compounds—*valence bond, ligand field (or crystal field), and molecular orbital.*
5. The formulas of compounds containing coordination compounds, or complex ions, are written as

(+ion, if any)[central atom-neutral ligand(s)-anion ligand(s)](−ion, if any)

6. The names are written as:

(+ion)[anion ligands with o- ending-neutral ligands-central atom (Roman numeral for oxidation state)](−ion)

Use the prefixes di-, tri-, and so on for more than one ligand of same type; for anion complex, central atom takes -ate ending, and Latin names are used for Fe, Cu, Sn, Pb, and Ag.

Upon completion of this chapter, you should be able to do the following:

1. Given any complex, or any substance containing a complex, identify the central atom and the ligands, tell what the coordination number is (including cases with bidentate ligands), predict the possible shape, and tell if there could be stereoisomers. Also, given the formula of

any complex, write the name, or given the name, write the formula. (1, 2)

2. Given any of the terms in Exercise 1, give a definition or explanation in your own words.

EXERCISES

1. Define the following terms:
   (a) coordination compound
   (b) ligand
   (c) monodentate
   (d) polydentate
   (e) bidentate
   (f) chelate
   (g) coordination number
   (h) stereoisomers
   (i) *cis*-, *trans*-, isomers
2. Distinguish between
   (a) monodentate and bidentate ligands
   (b) square planar and tetrahedral complexes
   (c) *cis*- and *trans*- isomers

PROBLEMS

1. For each of the following, tell the coordination number, what the ligands are, the possible shape, if stereoisomers are possible, and the correct name:
   (a) $K_2[ZnCl_4]$
   (b) $[Pt(NH_3)_4F_2]$
   (c) $[Cr(NH_3)_6]Cl_3$
   (d) $[Co(NH_3)_5H_2O]I_3$
   (e) $[Co(H_2O)_4(NH_3)_2]Cl_3$
   (f) $[Pt(H_2O)_4]Cl_2$
   (g) $[Pt(H_2O)_4Cl_2]$
   (h) $[Co(en)F_2]$
   (i) $Ca[Ni(NH_3)_2Cl_4]$
   (j) $[Cr(NH_3)_2(H_2O)_2](ClO_4)_3]$
   (k) $Na_2[PtCl_4]$
   (l) $[Co(NH_3)_2(H_2O)_4]Cl_2$
   (m) $[Pt(NH_3)_4Cl_2]$
   (n) $[Cr(H_2O)_4](ClO_3)_3$
   (o) $[Ni(en)Cl_2]NO_3$
   (p) $Na_2[NiCl_4]$
   (q) $[Cr(H_2O)_6]Cl_3$
   (r) $[Pt(NH_3)_2]Cl_2$
   (s) $[Pt(NH_3)_2Cl_2]$
   (t) $Ca[Pt(NH_3)_2Cl_4]$
2. For each of the following, tell the coordination number, what the ligands are, the possible shape, if stereoisomers are possible, and the correct formula:
   (a) sodium hexabromoplatinate(IV)
   (b) potassium tetrachlorozincate(II)
   (c) tetraaquochromium(III) chlorate
   (d) hexaamminechromium(III) chloride
   (e) pentaammineaquo cobalt(III) iodide
   (f) potassium trichlorotriamminenickelate(II)
   (g) dichloroethylenediaminecobalt(II)
   (h) sodium tetrachloroplatinate(II)
   (i) dichlorotetraaquoplatinum(II)
   (j) potassium tetrafluorodiamminenickelate(II)
   (k) calcium tetrachlorodiaquoplatinate(II)
   (l) tetraamminechromium(III) nitrate

(m) amminepentaaquonickel(III) bromide
(n) potassium hexachloroplatinate(IV)
(o) tetrachlorodiaquoplatinum(IV)
(p) sodium trichlorotriamminenickelate(II)
(q) diaquoethylenediaminecobalt(II)
(r) hexaamminechromium(III) chloride
(s) trichlorohexaamminechromium(III)
(t) difluoroethylenediaminecobalt(II)

REFERENCES

BASOLO, FRED, and RONALD C. JOHNSON, *Coordination Chemistry,* New York: W. A. Benjamin, Inc., 1964.

MURMANN, KENT R., *Inorganic Complex Compounds,* New York: Reinhold Publishing Corporation, 1964.

SCHLESSINGER, GERT C., "A Summer Short Course in Coordination Chemistry," *Chemistry,* June, July, & August 1966, **39.**

Objectives

1. To understand what nuclear reactions and radioactivity are.
2. To learn the nature and properties (charge and mass) of alpha (α) and beta (β) particles, gamma (γ) rays, positrons (β^+), and neutrons and to be able to write nuclear equations involving these particles.
3. To understand the difference between natural and artificial radioactivity and to be able to write nuclear equations representing transmutation of elements and artificially produced radioactive species.
4. To understand what half-life is and to be able to do calculations involving half-life.
5. To understand what nuclear fission and fusion are and to be able to distinguish between these.
6. To understand the basic principles of nuclear reactors.
7. To understand the basic principles of the ionization counter, such as the Geiger counter, and to be able to explain briefly how this works.
8. To understand the relationship between matter and energy and to be able to do calculations involving the conversion of matter to energy in nuclear reactions and calculation of the binding energy of atoms.
9. To learn some of the applications of nuclear chemistry in industry, medicine, preservation of food, dating of objects, biology, and the various branches of chemistry such as reaction mechanisms and structure studies.

Twenty Five

Introduction to Nuclear Chemistry

In most of the topics that we have considered up to this point we have been concerned with the valence electrons in atoms and how they behave in chemical reactions. In Chapters 3 and 4 we briefly considered the nucleus of the atom and the particles (protons and neutrons) that are found there. We now want to consider some of the more important aspects of the topic of nuclear chemistry.

In this chapter we shall consider the changes that take place in the nucleus during a *nuclear reaction.* Along with these nuclear reactions we shall consider what radioactivity is and the various types of nuclear radiation that are produced. Also accompanying nuclear reactions we shall see that there are rather large energy changes that take place.

25-1 Nuclear Radiations

Certain isotopes of some elements are unstable, giving off various types of radiation. These are known as *radioactive isotopes.* Some of these occur naturally and others are formed in nuclear reactions. We shall consider these two types of **radioactivity** (natural and synthetic) later (Sections 25-3 and 25-5). But first we want to look at the various types of radiation that are given off and learn to write nuclear reactions.

There are several types of nuclear radiations; however, the most common are (1) alpha particles, (2) beta particles, (3) gamma rays, (4) positrons, and (5) neutrons. Only the first three of these occur in naturally radioactive species. It is found that if a small sample of a naturally radioactive species is placed in a lead block and the radiation emitted is passed through an electric field and then recorded on a photographic plate, the gamma rays will not be deflected but the alpha and beta particles will be. This is shown in Fig. 25-1. It is found that the alpha particles are attracted toward the negative side of

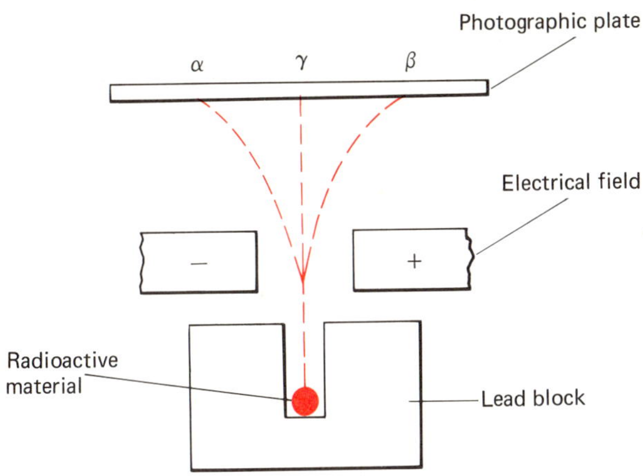

Figure 25-1 Radiation from a radioactive substance, under the influence of an electric field

the field and the beta particles are attracted toward the positive side. Thus the alpha particles must have a positive charge and the beta particles must have a negative charge.

Alpha (α) Particles

It is found that an **alpha particle** (α) is identical to a helium nucleus. Thus the alpha particle has a mass number of 4 and an atomic number of 2. It consists of two protons and two neutrons and has a charge of +2. The symbol usually used for this type of particle is ${}^{4}_{2}He$. When an alpha particle is discharged from the nucleus of an atom, it rapidly picks up two electrons from some other atom or from electrons in the atmosphere. Thus the alpha particle quickly becomes a helium atom.

Alpha particles are emitted from a nucleus with a rather low velocity, only about one-tenth the speed of light. Due to the mass, charge, and speed of alpha particles, they do not have very high penetrating power and can be stopped by a thin sheet of paper. Thus they do not penetrate the body. If a radioactive substance that emits alpha particles does get into the body, however, it is very harmful. The alpha particles injure normal cells in the body. They are more harmful to the body than beta particles or gamma rays. Alpha particles also cause gases to ionize very readily. This is the basis of detection of radiation in the Geiger counter, which we shall consider in Section 25-8.

Beta (β) Particles

The **beta particle** is found to be identical to the electron; that is, it has a charge of −1 and a negligible mass. The symbol used for the beta particle is ${}^{0}_{-1}e$, to represent a mass of almost zero and a charge of −1.

Even though beta particles are emitted from the nucleus of some radioactive substances, it is known that there are no electrons in the nucleus. The beta particles are believed to be produced when a neutron is transformed into a proton in the reaction.

$$\text{neutron } ({}^{1}_{0}n) \longrightarrow \text{proton } ({}^{1}_{1}p) + \text{beta particle } ({}^{0}_{-1}e)$$

It may seem contradictory to say that an electron (beta particle) can be produced from the nucleus when there are no electrons in the nucleus. As an analogy, however, if we consider that a piece of iron can be made into nails, we would not conclude that the piece of iron was made up of nails.

Beta particles are emitted from the nucleus at a much greater velocity than alpha particles. A beta particle is emitted with a speed about nine-tenths the speed of light. Thus beta particles have greater penetrating power than alpha particles. Even they can be stopped, however, by a sheet of aluminum about 1 cm thick. Beta particles are also harmful if they get into the body.

Gamma (γ) Rays

Gamma radiation consists of rays of very short wavelength. Thus they have no charge or mass. The symbol used to represent these, since they do not have any mass or charge, is simply the lowercase Greek letter gamma (γ). These rays result from a difference in energy between reactants and products in a nuclear reaction.

Gamma rays are emitted from the nucleus with a velocity equal to the velocity of light. They have very high penetrating power and thus readily penetrate the body. They are harmful to the body since they affect the genes, causing mutations.

Positrons (β^+)

A **positron** is found to be a positive electron; that is, it is a particle with negligible mass and a charge of +1. The symbol used to represent the positron is ${}^{0}_{+1}e$. Positrons are believed to be produced in the nucleus of certain radioactive substances when a protron is converted to a neutron, according to the equation

$$\text{proton } ({}^{1}_{1}p) \longrightarrow \text{neutron } ({}^{1}_{0}n) + \text{positron } ({}^{0}_{+1}e)$$

Positrons are similar to beta particles in velocity but have very low penetrating power.

Neutrons (n)

Neutrons are emitted from certain radioactive species. They have velocities that vary from about 10^{-5} times the speed of light to about 10^{-1} times the speed of light. They have relatively high penetrating power that varies with the velocity. The symbol used for the neutron is ${}^{1}_{0}n$.

A proton is also sometimes involved in a nuclear equation—not as a type of radiation produced but as a reacting particle. This has a mass of about 1 amu and a charge of +1. The symbol used for this is ${}^{1}_{1}\text{H}$, or ${}^{1}_{1}p$.

TABLE 25-1 *Summary of properties of nuclear radiations*

NUCLEAR RADIATION	SYMBOL	MASS (amu)	CHARGE	PENETRATING POWER (Rank — highest = 1)
Alpha (α)	${}^{4}_{2}He$	4	+2	4
Beta (β)	${}^{0}_{-1}e$	Negligible	−1	3
Gamma (γ)	γ	0	0	1
Positron (β^{+})	${}^{0}_{+1}e$	Negligible	+1	5
Neutron	${}^{1}_{0}n$	1	0	2

25-2 Nuclear Equations

Since a **nuclear reaction** involves a change in the nucleus of the atom, there is a change into either an atom of a different element or a different isotope of the original element. In writing nuclear equations, the sum of all the mass numbers of the products must equal those of the reactants and the sum of the atomic numbers in the products must equal those in the reactants.

Consider the following problem examples of nuclear reactions.

PROBLEM EXAMPLE 25-1

Write the nuclear equation for the reaction in which ${}^{221}Fr$ decays by alpha emission.

SOLUTION:

Using the periodic table, we find that francium (Fr) is element number 87. Since an alpha particle is written as ${}^{4}_{2}He$, we can begin by writing

$$ {}^{221}_{87}Fr \longrightarrow {}^{4}_{2}He + X $$

From the atomic numbers and mass numbers, we can determine that the mass number of X must be 217 (221 − 4) and the atomic number must be 85 (87 − 2). Therefore, from the periodic table we see that X must be At and therefore the complete nuclear equation is

$$ {}^{221}_{87}Fr \longrightarrow {}^{4}_{2}He + {}^{217}_{85}At $$

PROBLEM EXAMPLE 25-2

Write the nuclear equation for the reaction in which ${}^{40}K$ decays by beta emission.

SOLUTION:

From the periodic table, and using the fact that the beta particle is ${}^{0}_{-1}e$, we write

$$ {}^{40}_{19}K \longrightarrow {}^{0}_{-1}e + X $$

The mass number of X is 40 (40 − 0) and the atomic number is 20 (19 + 1 = 20, or 20 − 1 = 19). Therefore, X is calcium and the complete equation is

$$^{40}_{19}K \longrightarrow {}^{0}_{-1}e + {}^{40}_{20}Ca$$

PROBLEM EXAMPLE 25-3

Write the nuclear equation for the reaction in which ^{37}Ar reacts by capturing an electron. (This type of nuclear reaction is generally referred to as *electron capture* or sometimes as K capture since the electron would be from the first energy level or the *K* shell.)

SOLUTION:

$$^{37}_{18}Ar + {}^{0}_{-1}e \longrightarrow {}^{37}_{17}Cl$$

25-3 Nuclear Stability and Natural Radioactivity

It was mentioned earlier that some nuclei emit various types of radiation. A substance in nature that is radioactive is called a *naturally radioactive isotope.* It is still not understood exactly why some nuclei are stable and others are not. There are several observations that have been made, however, and some theories as to why certain isotopes are radioactive. First of all, it seems the neutron to proton ratio has an effect on stability. Since protons are positively charged, they apparently repel each other in the nucleus. Therefore, for elements with higher atomic numbers and hence more protons, there must also be more neutrons present in order to have a stable nucleus; that is, if the neutron to proton ratio is too low (not enough neutrons for the number of protons), the nucleus is unstable and therefore radioactive. There are several types of reactions that can take place if there are too many protons present. For example, the nucleus might emit alpha particles, which decreases the number of protons present, or it could emit a positron, which causes a proton to change into a neutron ($^{1}_{1}p \longrightarrow {}^{1}_{0}n + {}^{0}_{+1}e$); or the nucleus could capture an electron, to cause a proton to form a neutron ($^{1}_{1}p + {}^{0}_{-1}e \longrightarrow {}^{1}_{0}n$).

It is also found that all nuclei with more than 92 protons are unstable, and there appears to be certain numbers of protons and neutrons that give stable nuclei, just as there are certain numbers of electrons, in the noble gas elements, that are particularly stable. These numbers, often referred to as the *magic numbers* of protons or neutrons, are 2, 8, 20, 50, 82, and 126. It is also known that nuclei with even numbers of protons and neutrons are more stable than those with odd numbers; that is, the 274 stable isotopes known fit into the following approximate classification:

1. Even number of both protons and neutrons—162 stable isotopes.

2. Even number of protons, odd number of neutrons or odd number of protons, even number of neutrons—108 stable isotopes.
3. Odd number of both protons and neutrons—4 stable isotopes.

One theory of nuclear structure is that the nucleons (protons and neutrons) are arranged in energy levels, just as the electrons are. This would account for the greater stability of nuclei with certain numbers of nucleons, as well as explaining the release of gamma rays in nuclear reactions; that is, this energy of short wavelength would come from nucleons going from one energy level (or shell) to another.

Using the generalizations given here, we can begin to see why some nuclei are stable and others are radioactive. We can also begin to predict which are the more stable nuclei. Consider the following problem examples.

PROBLEM EXAMPLE 25-4

Arrange the following in order of decreasing stability and explain:

$$^{35}Cl,\ ^{4}He,\ ^{22}Na$$

SOLUTION:

^{4}He would be most stable since it has an even number of protons (2) and an even number of neutrons (2), plus the fact that 2 is one of the magic numbers.
^{35}Cl would be next—it has an odd number of protons (17) but an even number of neutrons (18).
^{22}Na would be least stable—it has an odd number of both protons and neutrons (11 of each).

PROBLEM EXAMPLE 25-5

Explain each of the following in terms of nuclear stability:
(a) ^{212}Po decays by alpha emission.
(b) ^{214}Bi decays by beta emission.

SOLUTION:

(a) When ^{212}Po decays according to the equation

$$^{212}_{84}Po \longrightarrow {}^{208}_{82}Pb + {}^{4}_{2}He$$

the new product formed has a magic number of protons (82).
(b) When ^{214}Bi decays, according to the equation

$$^{214}_{83}Bi \longrightarrow {}^{214}_{84}Po + {}^{0}_{-1}e$$

the odd-odd nucleus (83 protons, 231 neutrons) goes to an even-even nucleus (84 protons, 130 neutrons).

25-4 Half-Life

The rate of disintegration of radioactive isotopes is expressed in terms of the **half-life,** which is the time that it takes for half of a given amount to disintegrate. This rate is independent of external conditions, such as temperature and pressure, and depends only on the nature of the substance. It only gives the probability of decay, and this probability is always the same for a given type of nucleus, no matter how it was formed, what it is combined with, and so forth. The number of nuclei decaying per second is the probability per second times the number of nuclei present. Since the number present decreases as a result of the decays, the number decaying per second decreases steadily. By the time half the nuclei have decayed (one half-life), the number decaying per second has dropped in half. The remaining nuclei have the same probability per second of decaying and therefore it takes another time interval of one half-life for half of these to decay, and so forth.

For isotopes with relatively short half-lives it is possible to find the half-life by simply taking counts per minute and measuring the time it takes for this to be reduced by one-half, for a given sample of the radioactive substance. This will be considered in Section 25-8.

Consider the following problem examples:

PROBLEM EXAMPLE 25-6

The radioactive isotope ^{137}Cs has a half-life of 30.0 years. Starting with 1.00 mg of ^{137}Cs, how much would remain after 90.0 years?

SOLUTION:

At this time (0 years), we have 1.00 mg of the isotope; after 30.0 years, we shall have one-half of the original, or 0.500 mg; after 60.0 years, we shall have one-half of 0.500 mg, or 0.250 mg. Finally, after 90.0 years, we shall have one-half of 0.250 mg, or 0.125 mg:

0 years—1.00 mg

30.0 years—0.500 mg

60.0 years—0.250 mg

90.0 years—**0.125 mg**

An alternate solution to this problem is to count the number of half-lives (90.0/30.0 = 3) that have passed and multiply $(\frac{1}{2})^3$ by the original amount, to give

$$(1.00\ \text{mg})(\tfrac{1}{2})^3 = (1.00\ \text{mg})(\tfrac{1}{8}) = 0.125\ \text{mg}$$

PROBLEM EXAMPLE 25-7

At a given time there are 1.2 mg of a radioactive substance present. After 6 hours there is only 0.3 mg remaining. What is the half-life?

SOLUTION:

Since there is one-fourth of the initial amount present, two half-lives must have passed ($\frac{1}{2} \times \frac{1}{2} = \frac{1}{4}$). Therefore, the half-life must be **3 hours.**

25-5 Artificial Radioactivity. Transmutation of Elements and Preparation of New Elements.

Along with naturally occurring radioactive species, there are also some that are synthetically radioactive. Radioactivity of this type is known as *artificial* radioactivity. It has been found that by bombarding a nucleus with a particle (such as an alpha particle, a proton, a neutron, and so on) a new nuclear species can be formed. The first example of this was carried out by Ernest Rutherford (1871–1937) in 1919. He reacted an alpha particle with a nitrogen nucleus to cause the following reaction to take place:

$$^{4}_{2}He + ^{14}_{7}N \longrightarrow ^{1}_{1}H + ^{17}_{8}O$$

This was the first example of the change of one element into another, called *transmutation* of an element.

Since that time many other reactions of this type have been carried out. In doing this the particle that is to bombard the nucleus must be accelerated to very high speeds, in order to penetrate the nucleus. The instrument used to do this is called an *accelerator.* In the accelerator small alternating voltages are used to increase the energy of the particle. One of the first types of accelerator was the *cyclotron,* shown in Fig. 25-2. The particle passes

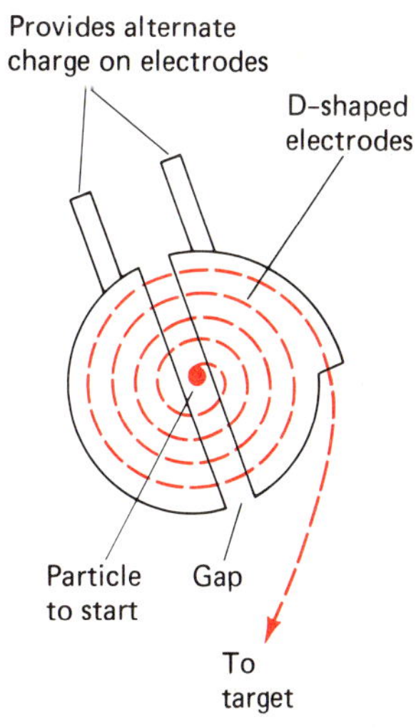

Figure 25-2
Sketch of a cyclotron

between the accelerating electrodes, in the presence of the magnetic field. The magnet causes the particle to move in a spiral path. By alternating the current in the electrodes, the particle is given many small increases in energy as it spirals through the accelerator, until it finally is expelled, with very high energy.

In a *linear accelerator* the particle travels along a straight path. Many accelerating electrodes are used, and the particle gains energy as it moves through the accelerator. These accelerators can be compared to a group of people striking a ball, either as it spins around or as it travels along a straight line. Each time the ball is struck, it gains a little more energy until it is finally released toward the target with very high energy.

In many transmutation reactions the new isotope formed is radioactive. For example, when ^{238}U is bombarded with a neutron, it forms ^{239}U, which is radioactive and gives off a beta particle. The two reactions are

$$^{238}_{92}U + ^{1}_{0}n \longrightarrow ^{239}_{92}U$$

$$^{239}_{92}U \longrightarrow ^{0}_{-1}e + ^{239}_{93}Np$$

The synthesis of new elements is another example of transmutation of the elements followed by artificial radioactivity. Element 105, with the proposed name of hahnium (Ha)* in honor of Otto Hahn, is formed by the bombardment of californium-248 with nitrogen-15 nuclei to form element 105 (mass number 260) and the emission of four neutrons. The newly formed element has a half-life of 1.6 s and disintegrates to lawrencium-256 and an alpha particle. The nuclear equations for these nuclear reactions are as follows:

$$^{249}_{98}Cf + ^{15}_{7}N \longrightarrow ^{260}_{105}(Ha) + 4\,^{1}_{0}n$$

$$^{260}_{105}(Ha) \longrightarrow ^{256}_{103}Lr + ^{4}_{2}He \text{ (half-life} = 1.6 \text{ s)}$$

Consider the following examples of nuclear reactions involving transmutation of elements and artificial radioactivity.

PROBLEM EXAMPLE 25-8

Write the nuclear equation for the reaction in which ^{16}O is bombarded with a neutron to form a new element plus an alpha particle.

SOLUTION:

$$^{16}_{8}O + ^{1}_{0}n \longrightarrow X + ^{4}_{2}He$$

From the atomic number and mass number it can be seen that the atomic

*The official name for this element has not yet been determined. A group of Russian scientists claim to have prepared this element by a different process and have proposed the name nielsborium (Ns).

number of the new element must be 6 (8 + 0 = 6 + 2) and the mass number must be 13 (16 + 1 = 13 + 4). Therefore, the nuclear equation is

$$^{16}_{8}O + ^{1}_{0}n \longrightarrow ^{13}_{6}C + ^{4}_{2}He$$

Reactions of this type are sometimes written as $^{16}O(n, \alpha)^{13}C$.

PROBLEM EXAMPLE 25-9

Write the complete nuclear equation for the reaction

$$^{44}Ca\ (p, n)\ X$$

SOLUTION:

$$^{44}_{20}Ca + ^{1}_{1}H \longrightarrow ^{1}_{0}n + X$$

X must have a mass number of 44 (44 + 1 = 1 + 44) and an atomic number of 21 (20 + 1 = 0 + 21). Therefore, X is $^{44}_{21}Sc$, and the equation is

$$^{44}_{20}Ca + ^{1}_{1}H \longrightarrow ^{1}_{0}n + ^{44}_{21}Sc$$

PROBLEM EXAMPLE 25-10

Write a complete nuclear equation for each of the reactions when ^{238}U reacts with a neutron to form a new atom plus a beta particle; the new atom formed decays by beta emission.

SOLUTION:

$$^{238}_{92}U + ^{1}_{0}n \longrightarrow ^{239}_{93}Np + ^{0}_{-1}e$$

$$^{239}_{93}Np \longrightarrow ^{239}_{94}Pu + ^{0}_{-1}e$$

25-6 Nuclear Fission

Some very heavy nuclei react by splitting into two smaller nuclei. This is called **nuclear fission.** In this process, a small amount of mass is converted to a very large amount of energy. One example of this is when ^{235}U is bombarded with neutrons to form ^{94}Sr and ^{139}Xe, plus three neutrons according to the equation

$$^{235}_{92}U + ^{1}_{0}n \longrightarrow ^{94}_{38}Sr + ^{139}_{54}Xe + 3\,^{1}_{0}n$$

Since three neutrons are formed in the reaction, they can each cause another reaction to take place, and this can set off a *chain reaction.* An example of a chain reaction is shown in Fig. 25-3 with an array of dominos standing on end. In order to do this there must be enough material present for the neutrons to react with it, rather than just escaping. This minimum amount of material is known as the *critical mass.* In an atomic bomb two

Figure 25-3
Dominos stacked on end. When the first one falls, it knocks over the next two, which in turn knock over the next three, and so forth.

samples of this type of less than the critical mass (called a *noncritical* or *subcritical* mass) are brought together in such a way as to produce an uncontrolled chain reaction.

25-7 Nuclear Fusion

Nuclear fusion involves the combination of two or more nuclei to form a heavier nucleus. One example of nuclear fusion is the reaction that takes place on the sun. The nuclear reaction on the sun possibly involves the fusion of two isotopes of hydrogen—deuterium, $^{2}_{1}H$, and tritium, $^{3}_{1}H$—to form helium, $^{4}_{2}He$, according to the following overall nuclear equation:

$$^{2}_{1}H + ^{3}_{1}H \longrightarrow ^{4}_{2}He + ^{1}_{0}n$$

During the fusion process there is considerable loss in mass, which is converted to energy, primarily in the form of heat. But a considerable amount of heat energy must be supplied for the occurrence of one of these nuclear fusion reactions. This type of heat energy is available on the sun since the sun's interior temperature is estimated to be 2×10^{7} °C.

25-8 Detection of Radiation

One common type of detector of radiation is the Geiger-Mueller counter (Geiger counter). This consists of a gas-filled tube, with a positive and negative electrode. The basic parts of this are shown in Fig. 25-4. The radiation enters the tube and causes the gas present to ionize. When it does, the positive ions move to the negative electrode and the electrons move to the

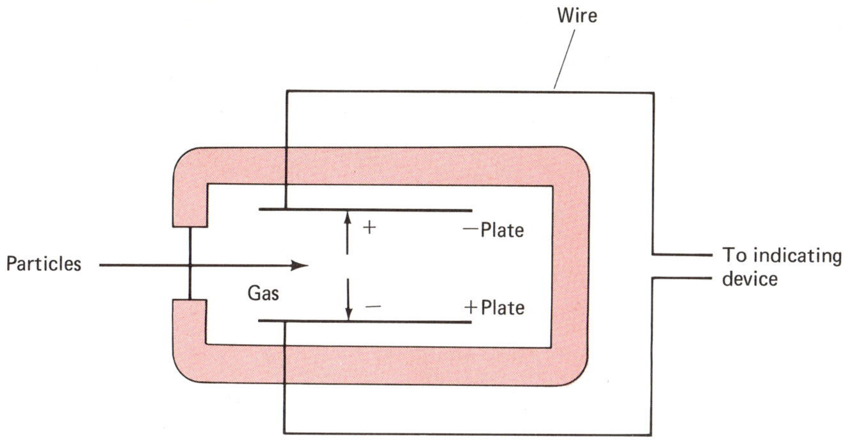

Figure 25-4
Sketch of the main parts of a Geiger counter

positive electrode. As each hits the electrode, a pulse is noted on the indicating device, and a "count" is noted.

One of the units used to measure the quantity of radiation is the **curie** (Ci). This is defined as the amount of radiation produced from a source that decays at a rate of 3.7×10^{10} disintegrations per second. A substance with a radiation of more than 1 Ci is considered to have high activity.

Consider the following problem example involving the use of a Geiger counter to determine the half-life of a radioactive substance:

PROBLEM EXAMPLE 25-11

From the following data, calculate the half-life:

TIME (MIN)	COUNTS/MIN
0	1000
5	800
10	650
15	500
20	350
30	250

SOLUTION:

After 15 minutes, the counts were reduced from 1000 to 500. Since this is one-half of the original counts, it means that only one-half of the original sample remains. Thus the half-life is **15 minutes.**

As mentioned earlier, the energy resulting from nuclear reactions is due to the small change in mass. These quantities are related by the Einstein equation, which says that $E = mc^2$, or $\Delta E = \Delta mc^2$. If m is expressed in grams and c (speed of light) as 3.00×10^{10} cm/s, E is in ergs.

Using the values n = 1.00898 amu, e^- = 0.000549 amu, p = 1.00759 amu, ${}^4_2He$ = 4.00387 amu, ${}^7Li$ = 7.08122, the energy released in the reaction

$$ {}^7_3Li + {}^1_1H \longrightarrow 2\,{}^4_2He $$

is calculated as follows: 7.08122 amu + 1.00759 amu − 2(4.00387) amu = 0.08107 amu; 1.00 g = 6.02×10^{23} amu; therefore, the energy is

$$ \frac{0.08107\ \text{amu}}{\text{atom}} \times \frac{1.00\ \text{g}}{6.02 \times 10^{23}\ \text{amu}} \times \frac{(3.00 \times 10^{10})^2\ \text{cm}^2}{\text{s}^2} = 1.21 \times 10^{-4}\ \text{ergs/atom} $$

For one mole of reactants, and converting to kilocalories (where 1 erg = 2.39×10^{-11} kcal), this would be

$$ \frac{1.21 \times 10^{-4}\ \text{erg}}{\text{atom}} \times \frac{6.02 \times 10^{23}\ \text{atoms}}{\text{mol}} \times \frac{2.39 \times 10^{-11}\ \text{kcal}}{\text{erg}} = 1.74 \times 10^{9}\ \text{kcal/mol} $$

(Compared to a chemical reaction where the energy changes are in general less than about 300 kcal/mole of reactants, it can be seen that this is very large.)

This also accounts for the difference in mass between the total number of protons, neutrons, and electrons and the mass of a given isotope; that is, the mass of six electrons, six protons, and six neutrons is not exactly 12 amu, which is the mass of a ${}^{12}C$ atom. This mass difference again can be converted to energy, and this is referred to as the *binding energy*. For example, using the values given above, the binding energy of a helium nucleus would be

$$ E = \frac{[(2)(1.00759) + (2)(1.00898) - 4.00387]\ \text{amu}}{\text{atom}} \times \frac{1\ \text{g}}{6.02 \times 10^{23}\ \text{amu}} $$

$$ = \frac{4.86 \times 10^{-26}\ \text{g}}{\text{atom}} $$

$$ \frac{4.86 \times 10^{-26}\ \text{g}}{\text{atom}} \times \frac{(3.00 \times 10^{10})^2\ \text{cm}^2}{\text{s}^2} = \frac{4.37 \times 10^{-5}\ \text{erg}}{\text{atom}} $$

or

$$ \frac{4.37 \times 10^{-5}\ \text{erg}}{\text{atom}} \times \frac{6.02 \times 10^{23}\ \text{atom}}{\text{mol}} \times \frac{2.39 \times 10^{-11}\ \text{kcal}}{\text{erg}} = 6.29 \times 10^{8}\ \text{kcal/mol} $$

25-10 Applications of Nuclear Reactions and Radioactive Isotopes

As well as the destructive uses of nuclear power there are many peaceful uses of nuclear reactions and radioactive isotopes. One of the most important ones currently, and also one of the most controversial, is the use of nuclear reactors to produce power. This is potentially a source of vast amounts of energy, in place of the use of petroleum. There are questions about the safety, however. There are two aspects to this. This first simply has to do with the possibility of a nuclear explosion, which could be very disastrous. The other aspect concerns the disposal of the radioactive waste materials that would be generated. Many of these are quite active and have half-lives as long as thousands of years. These must be safely stored until the radioactivity has decreased to a negligible amount, which could be a very long time.

Other applications of nuclear reactions include the preparation of new elements (which was discussed in Section 25-5); the use of radioactive tracers, which can trace an element through a chemical reaction, including reactions that take place in the body; in cancer therapy; for the preservation of food by radioactive sterilization; and the determination of the age of objects by measuring the amount of the radioactive isotope ^{14}C that is present in them.

Radioactive species are used to trace the course of a chemical reaction by using a particular radioactive isotope and checking to see where it is in the final product, by checking the radioactivity of the various products. This same procedure can be used in the body. For example, radioactive iodine can be used to map the thyroid, by taking small amounts of the iodine into the body. The radioactive iodine acts exactly the same as other iodine in the body but it can be traced by following its radioactivity.

Certain radioactive species can be used to prevent the growth of cancerous tissue in the body. The cancerous tissue is more readily affected by the radioactive species; therefore by carefully controlled use the radiation can be used to destroy the cancerous tissue without greatly affecting normal tissue.

Bacteria in food can be killed by subjecting it to heavy amounts of radiation. Food treated in this way can be sealed and kept for long periods of time without deteriorating. There are still some difficulties with this, however, since in some cases the texture of the food is altered and it is also possible to form radioactive isotopes in the food.

By using the half-life of the radioactive isotope ^{14}C, it is possible to determine the age of various objects. In living material, the ratio of ^{14}C to ^{12}C (which is not radioactive) remains relatively constant. When the tissue in an animal or plant dies, the ^{14}C decreases because the intake and utilization of ^{14}C does not occur. Therefore, in dead tissue the ratio of ^{14}C to ^{12}C would decrease, depending on the age of the tissue. The age of the dead tissue is determined in the following way. A sample of the dead tissue is burned to carbon dioxide, and the carbon dioxide is analyzed for the ratio of ^{14}C to ^{12}C. The ratio in the dead tissue is compared to that in normal living tissue by considering a decrease of the ratio in dead tissue by one-half for every 5770 years (the half-life for ^{14}C); from this, the age of the dead tissue can then be determined.

SUMMARY

1. Some nuclei are unstable and emit various types of *radiations* in order to become stable. This is an atomic process and is thus independent of the chemical state of the particular radioactive isotope. These radiations include alpha (^4_2He) and beta ($_{-1}^{0}e$) particles, gamma rays, positrons ($_{+1}^{0}e$), and neutrons (^1_0n).
2. *Radioactive species* may occur naturally or may be artificially produced. Synthetically produced radioactivity includes not only the radiations listed above but also electron capture, protons as bombarding or emitted particles, and nuclear fission.
3. In a nuclear reaction, whether it represents a natural radioactive species or a synthetically produced one, the total of all charges and mass numbers of reactants must equal the sum of all charges and mass numbers of products. Therefore, to write correct *nuclear equations* it is only necessary to remember the charge and mass of the various radiations mentioned above.
4. The rate of decay of radioactive species is expressed in terms of the *half-life,* which is the amount of time it takes for half of a given amount of any sample to decay; that is, after n half-lives, the amount remaining equals (original amount)$(\frac{1}{2})^n$.
5. Radiation can be detected in an *ionization counter* such as the Geiger counter in which the radiation causes ions and free electrons to form. These are attracted to charged plates in the counter. As these particles strike the plates, a count is recorded by the counting device.
6. Even though the total number of nuclear species must be retained in the products of a nuclear reaction, the total mass of all products is always less than the total mass of reactants. Therefore, some mass is always converted to energy, according to the equation $E = \Delta mc^2$. The difference between the total mass of protons, neutrons, and electrons in a given isotope and the actual mass of that isotope is the *mass defect.* The amount of energy equivalent to this mass can be calculated by the equation above and is referred to as the *binding energy.*
7. Radioactive isotopes are used in industry, in the preservation of food, to find the age of rocks and other objects, to trace chemical reactions and biological processes, and in medicine, to mention only a few of the many present and possible future uses.

Upon completion of this chapter, you should be able to do the following:

1. Given any nuclear reaction, representing either a naturally radioactive species or a synthetically produced one, involving alpha particles, beta particles, gamma rays, positrons, neutrons, protons, or those that react by electron capture, write a complete nuclear equation for the reaction. (1)

2. For a nuclear reaction in which an isotope is bombarded with a particle and a new isotope plus a particle or other type of radiation is emitted, given either all reactants and products, or all but one reactant or product, write a complete nuclear equation to represent the reaction. (2)
3. Given the half-life of any radioactive species, calculate how much of a given sample will remain after any whole number of half-lives. (3)
4. Given the amount of a radioactive species to start, and the amount remaining after a given amount of time, calculate the half-life of the substance. (4)
5. Briefly explain how the ionization counter works and how it can be used to determine the half-life of a short-lived radioactive species.
6. Given the exact mass of the proton, neutron, and electron, plus the exact mass of any isotope, calculate the mass defect and convert this to energy, expressed in the units ergs per atom, or kilocalories (or kilojoules) per mole. (5)
7. List and briefly explain several uses of radioactive isotopes.
8. Given any of the terms listed in Exercise 1, give a brief definition or explanation in your own words.

EXERCISES

1. Define or explain the following terms:
   (a) nuclear reactions
   (b) radioactivity
   (c) alpha particles
   (d) beta particles
   (e) gamma rays
   (f) positron particles
   (g) natural radioactivity
   (h) half-life
   (i) transmutation of an element
   (j) artificial radioactivity
   (k) nuclear fission
   (l) nuclear fusion
   (m) electron capture
   (n) isotopes
2. Distinguish between
   (a) an alpha particle and a beta particle
   (b) an alpha particle and a gamma ray
   (c) an alpha particle and a positron
   (d) a beta particle and a positron
   (e) a proton and a deuteron
   (f) natural and artificial radioactivity
3. In your own words, explain the operation of a cyclotron.
4. In your own words, briefly explain the operation of a Geiger counter.

PROBLEMS

Nuclear equations

1. Complete the following nuclear equations:
   (a) Krypton-87 (Kr) decays by beta emission.
   (b) Uranium-232 (U) decays by alpha emission.
   (c) Praseodymium-140 (Pr) decays by positron emission.
   (d) Oxygen-16 plus a neutron results in the formation of another element and the release of an alpha particle.
   (e) Beryllium-9 plus a proton results in the formation of another element and the release of an alpha particle.
   (f) Lutetium-176 (Lu) plus a bombarding particle results in the formation of lutetium-177 and gamma rays.
   (g) Silicon-32 decays by beta emission.
   (h) Americium-243 (Am) decays by alpha emission.
   (i) Neon-18 decays by positron emission.
   (j) Boron-10 plus a neutron results in the formation of another element and the release of an alpha particle.

(k) Lithium-7 plus a proton results in the formation of another element and the release of a neutron.

(l) Gold-197 plus a bombarding particle results in the formation of mercury-197 and a neutron.

2. Give the atomic number, mass number, and symbol for the new element formed in each of the following reactions:
   (a) Nickel-58 is bombarded with protons, and an alpha particle is emitted in the transmutation process.
   (b) Palladium-108 (Pd) is bombarded with high-speed alpha particles, and a proton is emitted in the transmutation process.

Half-life

3. Carry out the following calculations using the given half-lives:
   (a) Thallium-201 has a half-life of 73.0 hr. If 4.00 mg of thallium-201 disintegrates over a period of 6.00 days and 2.00 hr, how many milligrams of thallium-201 will remain?
   (b) The half-life of a radioactive isotope, X, is 2.00 yr. How many years would it take a 4.00 g sample of X to decay and have only 0.500 g of X remain?
   (c) Sodium-25 has a half-life of 60.0 s. If 5.00 mg of sodium-25 disintegrates over a period of 3.0 min, how many milligrams will remain?
   (d) Selenium-83 has a half-life of 25.0 min. How many hours would it take a 10.0-mg sample of selenium-83 to decay and have only 1.25 mg of it remain?
4. Calculate the half-life of each:
   (a) 1.20 mg of a radioactive species decays to 0.30 mg in 40 min.
   (b) 0.80 mg of a radioactive species decays to 0.20 mg in 80 s.

Binding energy

5. Calculate the binding energy (in kilocalories per mole) for each of the following isotopes:
   (a) ^{9}Be whose atomic weight is 9.0103
   (b) ^{19}F whose atomic weight is 19.0045

General problems

6. Following a nuclear fission explosion, the sodium in glass containers formed sodium-24. Sodium-24 decays with emission of a beta particle and has a half-life of 15.0 hr.
   (a) Write the nuclear equation for this reaction.
   (b) If 5.00×10^{-6} g of sodium-24 is formed in a glass container, how many grams will remain in 2.50 days?

7. Zinc-65 accounts for 50% or more of the total radioactivity in fish exposed to a nuclear fission explosion. Zinc-65 decays by positron emission and has a half-life of 245 days.
   (a) Write a nuclear equation for this reaction.
   (b) If 1.00 μg of zinc-65 is formed in a fish, how many days will it take to have just 0.250 μg of zinc-65? Note that the 0.250 μg of zinc-65 remains if the fish lives or dies, but if it is caught and eaten by people, the zinc-65 resides in the people who ate the fish.
8. A 4.00×10^{-6}% solution of sodium-24 chloride is prepared at 9:00 a.m. on Tuesday. What is the percent of sodium-24 chloride at 3:00 p.m. on Wednesday if sodium-24 has a half-life of 15.0 hr?
9. A 1.50×10^{-6} M solution of cesium-137 chloride was prepared by a research chemist in 1948. What is the molar concentration of cesium-137 chloride in 1978 if the half-life of cesium-137 is 30.0 yr?
10. For each of the following radioactive isotopes, determine the mass number (atomic weight), atomic number, and symbol for the element formed when the isotope decays as indicated.
    (a) ^{32}Si decays by beta emission.
    (b) ^{240}Cm decays by alpha emission.
    (c) ^{18}Ne decays by positron emission.
    (d) ^{87}Kr decays by beta emission.
    (e) ^{232}U decays by alpha emission.
    (f) ^{240}Pu decays by positron emission.
11. Write a complete nuclear reaction, including all atomic numbers and atomic weights for each of the following:
    (a) e^- capture by ^{7}Be
    (b) ^{208}Po emits an alpha particle
    (c) ^{66}Cu emits a beta particle
    (d) $^{16}O(n, \alpha)\underline{\ ?\ }$
    (e) $^{9}Be(p, \alpha)\underline{\ ?\ }$
    (f) $^{10}B(n, 2\alpha)\underline{\ ?\ }$
    (g) $^{6}Li(\underline{\ ?\ }, \alpha)^{3}H$
    (h) $^{60}Co \longrightarrow ? + \beta$
    (i) ^{22}Na emits a positron
    (j) $^{140}Pr \longrightarrow ? +$ positron
    (k) $^{197}Au + \underline{\ ?\ } \longrightarrow {}^{197}Hg + n$
    (l) successive decay of ^{211}Pb by beta, alpha, and beta emission
    (m) $^{108}Pd(\alpha, p)Ag$
    (n) $^{9}Be(\alpha, n)C$

REFERENCES

CHOPPIN, GREGORY R., *Nuclei and Radioactivity,* New York: W. A. Benjamin, Inc., 1964.

FERMI, LAURA, *The Story of Atomic Energy,* New York: Random House, 1961.

HARVEY, BERNARD G., *Nuclear Chemistry,* Englewood Cliffs, N.J.: Prentice-Hall, Inc., 1965.

Overman, Ralph T., *Basic Concepts of Nuclear Chemistry,* New York: Reinhold Publishing Corporation, 1963.

Seaborg, Glenn T., "Some Recollections of Early Nuclear Age Chemistry," *Journal of Chemical Education,* 1968, **45,** p. 278.

Szabadvary, Ferenc, "Discovery of Radioactivity," Chemistry, May 1974, **47,** p. 19.

Objectives

1. To review the principles of bonding and to relate this to the properties of organic compounds.
2. To learn what a structural formula is and what isomers are and to be able to recognize what structural formulas represent isomers of the same compound.
3. To be able to distinguish between a saturated and an unsaturated hydrocarbon as well as between an aliphatic and an aromatic compound.
4. To learn the names of the first ten (10) alkanes.
5. To be able to recognize the functional group and to write correct names and formulas for the following types of compounds:
   (a) Alkanes, alkenes, alkynes (including cyclic ones), and substituted hydrocarbons—including use of the common systems using the iso-, and primary, secondary, and tertiary system.
   (b) Aromatic compounds, including the use of the common prefixes *o*-, *m*-, and *p*- for disubstituted benzene compounds.
   (c) Alcohols and phenols—including both the common and IUPAC systems.
   (d) Ethers—using common names.
   (e) Carboxylic acids (including common names for the first three) and the derivatives of these: esters, anhydrides, and amides.
   (f) Alkehydes and ketones—including both the common and IUPAC systems.
   (g) Amines—including the notation primary (1°), secondary (2°), and tertiary (3°).
6. To be able to write a complete, balanced equation for the reaction of a hydrocarbon with oxygen (that is, burning hydrocarbon in oxygen) or an alkene or alkyne with H_2, Cl_2, Br_2, HCl, HBr, or HI.

Twenty Six

Introduction to Organic Chemistry

As pointed out in Chapter 1, organic chemistry is the study of compounds of carbon, with the exception of carbon monoxide, carbon dioxide, carbonates, hydrogen carbonates (bicarbonates), and cyanides (CN^-). There are many more organic compounds than there are inorganic. There are over 2 million organic compounds and approximately 100,000 inorganic compounds known. The primary source of organic compounds is oil, although all living matter is composed of organic compounds.

26-1 General Principles

Review of Atomic Structure and Bonding in Carbon Compounds

In all organic compounds there are four bonds around each carbon atom. These could be (1) four single bonds, in which the four electron pairs would occupy sp^3 hybrid orbitals having a tetrahedral shape, and all the bonds would be sigma (σ) bonds; (2) three sigma bonds and one pi (π) bond (one double bond) in which case the electron pairs would occupy sp^2 hybrid orbitals for the sigma bonds and the unhybridized p orbital would overlap with another p orbital to form the pi bond; and (3) two sigma bonds, in which sp hybrid orbitals would be occupied, and two pi bonds (to form a triple bond) using the two unhybridized p orbitals.

In the tetrahedral (sp^3) structure all bond angles are approximately 109°. In the sp^2 structure, the shape is a plane triangle and the bond angles are 120°. In the sp structure, the angles are 180°, to give a linear structure. The material in Chapter 7 on hybridization and geometry of molecules should be reviewed at this point, if necessary

Properties of Organic Compounds

In general, organic compounds are covalent and not very polar, although some do have some ionic bonds within the compound and many are somewhat polar. Therefore, they are generally (1) nonelectrolytes, (2) insoluble in water but soluble in nonpolar solvents, (3) liquids or low melting solids, and (4) quite volatile.

As with other compounds, the higher the molecular weight, the higher the melting and boiling points will be for similar compounds. For example, low molecular weight hydrocarbons (Section 26-2) are gases. Those with higher molecular weights are liquids or solids. Similarly, low molecular weight organic acids, such as acetic acid (vinegar), are liquids; whereas high molecular weight acids, such as steric acid (found in candles), are solids.

Structural Formulas and Isomers

Structural formulas show the arrangement of atoms within a molecule. For organic compounds it is not only important to know the molecular formula (as opposed to the empirical formula) but also essential to know the structural formula. For example, the compound with the molecular formula C_5H_{12} could be any of the following three compounds:

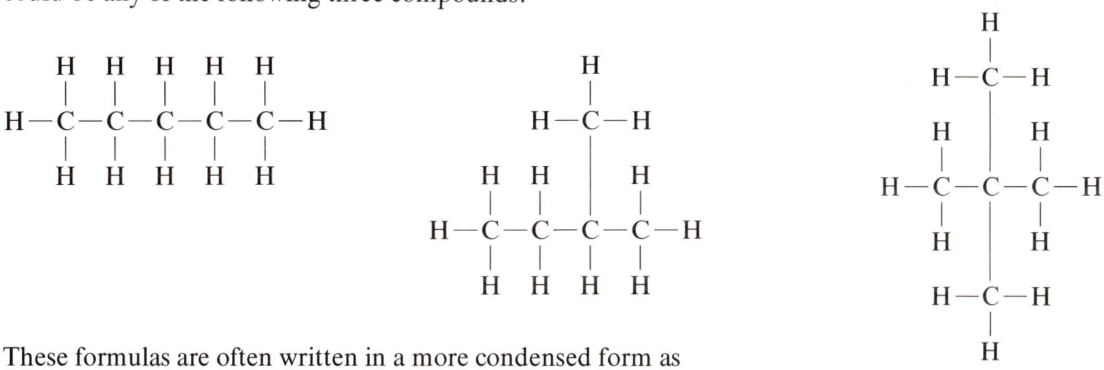

These formulas are often written in a more condensed form as

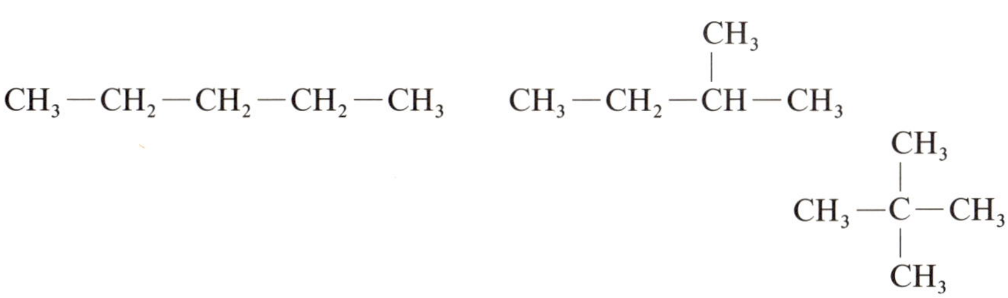

These different structural formulas are referred to as *isomers.* Recall that isomers are compounds with the same molecular formula but different structural formulas. To be more specific, these should be called *structural isomers* since there are also other types of isomerism possible. For our purposes we need only consider this one type.

There are two points that should be noted about these isomers.

1. The carbons are not actually in a straight chain since the angles in the compound (C_5H_{12}) shown above, for example, are actually 109°. Therefore, we often refer to the longest continuous chain in a compound (which would be five carbons in the first formula, four carbons in the second, and three in the third) but you shouldn't think of these as a straight chain.

2. In considering these isomers it is important that you can distinguish which are actually different isomers. The best way to determine this is to note whether the structural formula has been changed simply by turning the molecule around or bending it (in which case the two forms are not isomers) or whether a bond would need to be broken to change from one form to the other. Consider the following example:

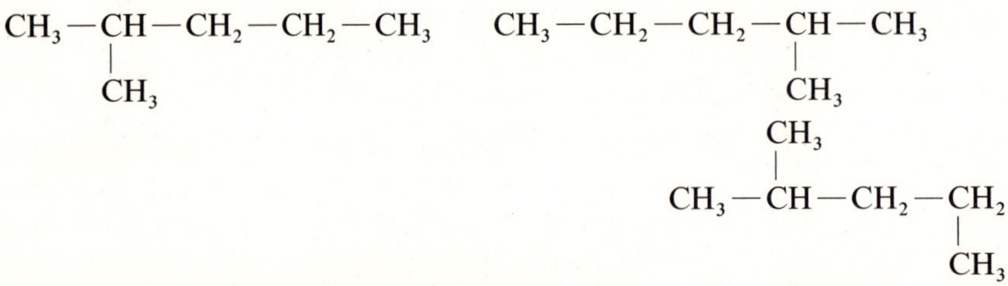

are not different isomers; however,

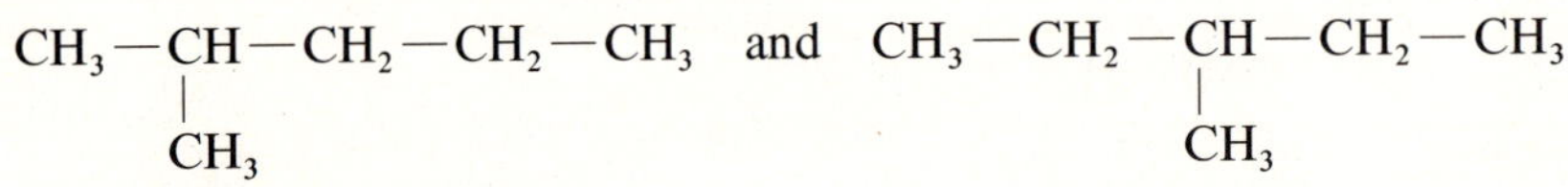

are different isomers.

It is often helpful to make some models of these in order to become familiar with the concept. Consider the following problem example involving structural formulas.

PROBLEM EXAMPLE 26-1

Which of the following represent the same compound?

(a) 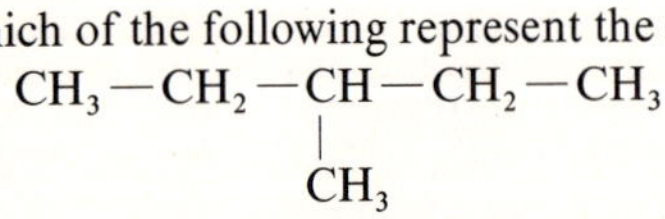

(b) 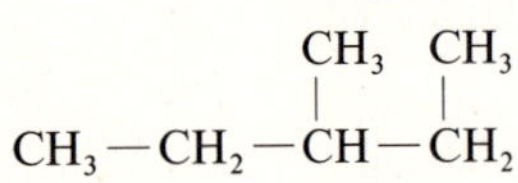

(c) 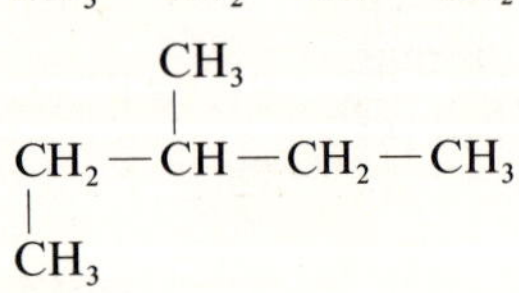

(d) $CH_3-\underset{\displaystyle CH_3}{\underset{|}{CH}}-CH_2-CH_2-CH_3$

SOLUTION:

(a), (b), and (c) are all the same since it is possible to change from one to the other without breaking any bonds. Also, all these have five carbons in the longest continuous chain with another one attached on the third carbon. However, (d) has the sixth carbon attached to the second one in the continuous chain. Notice that in all these structural formulas there are always four bonds around each carbon atom.

Functional Groups

It would be impossible even to begin to remember all the reactions for even a small fraction of all the organic compounds presently known. Fortunately, there are only a small number of types of organic compounds and therefore, by classifying them by the type of *functional group* that they contain, we can make a systematic study of organic chemistry. The functional group in general determines the chemical properties of an organic compound. For example, organic molecules that contain the —OH group are classified as alcohols, and they all undergo reactions typical of this particular group.

We shall consider compounds that contain the following functional groups (the notation in parentheses represents the functional group):

1. *Hydrocarbons*—These contain only carbon and hydrogen. There are four different types here:
   (a) *alkanes,* which contain all single bonds (—).
   (b) *alkenes,* which contain double bonds (=).
   (c) *alkynes,* which contain triple bonds (≡).
   (d) *aromatic* compounds, which contain the benzene ring (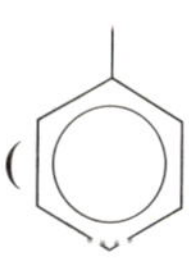).
2. *Alcohols*—These contain the —OH group (—OH). Aromatic alcohols are called *phenols.*
3. *Ethers*—These contain an oxygen, joining two other organic groups (—O—).
4. *Acids*—These contain the carboxyl group $\left(-\overset{\displaystyle O}{\overset{\|}{C}}-OH\right)$.
5. *Aldehydes and Ketones*—These contain the carbonyl group $\left(-\overset{\displaystyle O}{\overset{\|}{C}}-\right)$.
6. *Amines*—These contain nitrogen $\left(-\underset{|}{N}-\right)$ and are often considered as organic derivatives of ammonia.

26-2 Hydrocarbons

Hydrocarbons are compounds that contain only carbon and hydrogen. Hydrocarbons that contain only single bonds are called *saturated hydrocarbons* and those that contain double or triple bonds are referred to as being *unsaturated.* Saturated hydrocarbons are called *alkanes.* Unsaturated hydrocarbons that have double bonds are called *alkenes* and those with triple bonds are called *alkynes.* Compounds that contain the benzene ring (Section 26-7) are referred to as *aromatic* compounds and all others are called *aliphatic.* Table 26-1 summarizes this information.

The reactions of hydrocarbons vary considerably, according to which of the types above they are; that is, unsaturated hydrocarbons are much more reactive than alkanes and aromatic hydrocarbons. We shall consider a few of the more common reactions of hydrocarbons in Sections 26-3 through 26-7. Any hydrocarbon when burned in an excess of oxygen gives CO_2 and H_2O as the products of the reaction. In a limited amount of oxygen, however, some CO and even C is formed, and therefore when hydrocarbons are burned in a car engine (remember that air is only about 20% oxygen), all four of these products are formed—H_2O, CO_2, CO, and C.

26-3 Alkanes

Alkanes are hydrocarbons that have all single bonds and they are thus often referred to as *saturated hydrocarbons.* They have the general formula C_nH_{2n+2}.

In general, the naming of organic compounds is more systematic than for inorganic compounds, even though common names are sometimes used in place of the preferred systematic nomenclature. The convention followed in the systematic nomenclature of organic compounds was set up by the International Union of Pure and Applied Chemistry and is called the IUPAC, or sometimes just the IUC, system. Table 26-2 gives the names of the first 10 alkanes. Since this is the basis for all organic compounds, according to the IUPAC system, these must be learned. Notice that all of these end in *-ane;* this is the ending used for any alkane.

TABLE 26-1 *Classification of hydrocarbons*

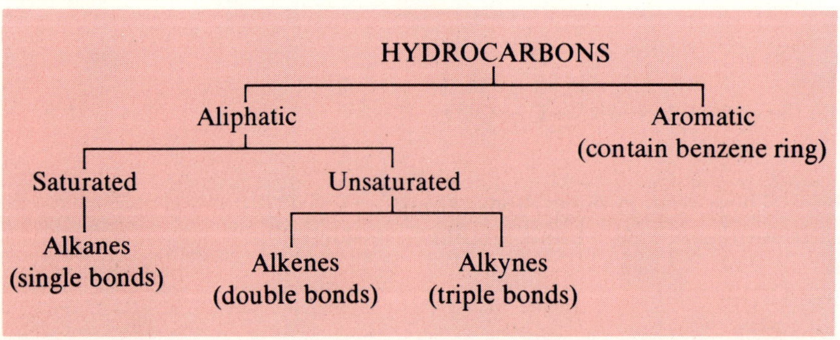

TABLE 26-2 *Names of first ten alkanes*

FORMULA	NAME
CH_4	Methane
C_2H_6	Ethane
C_3H_8	Propane
C_4H_{10}	Butane
C_5H_{12}	Pentane
C_6H_{14}	Hexane
C_7H_{16}	Heptane
C_8H_{18}	Octane
C_9H_{20}	Nonane
$C_{10}H_{22}$	Decane

The alkanes are quite unreactive but they do burn in oxygen to form carbon dioxide and water. For example, when butane is burned in an excess of oxygen, the reaction is

$$2\,C_4H_{10} + 13\,O_2 \longrightarrow 8\,CO_2 + 10\,H_2O$$

In the presence of sunlight, they will undergo a substitution reaction with Cl_2 or Br_2, in which one or more chlorine or bromine atoms are substituted for hydrogen atoms. The reactions take place in several steps and generally a mixture of products is obtained. For example, when Cl_2 is reacted with methane, one product formed is CCl_4, commonly called *carbon tetrachloride.* This reaction goes in steps, however, and several products are formed, according to the following equations:

$$CH_4 + Cl_2 \xrightarrow{\text{sunlight}} \underset{\substack{\text{methyl chloride}\\\text{(or chloromethane)}}}{CH_3Cl} + HCl$$

$$CH_3Cl + Cl_2 \xrightarrow{\text{sunlight}} \underset{\substack{\text{methylene chloride}\\\text{(or dichloromethane)}}}{CH_2Cl_2} + HCl$$

$$CH_2Cl_2 + Cl_2 \xrightarrow{\text{sunlight}} \underset{\substack{\text{chloroform}\\\text{(or trichloromethane)}}}{CHCl_3} + HCl$$

$$CHCl_3 + Cl_2 \xrightarrow{\text{sunlight}} \underset{\substack{\text{carbon tetrachloride}\\\text{(or tetrachloromethane)}}}{CCl_4} + HCl$$

Methyl chloride is used in paint remover; chloroform is a common biological preservative; and carbon tetrachloride is used as a cleaning agent (although it is quite toxic).

If the alkane is a ring structure, the same base is used for the name, and *cyclo-* is added. For example,

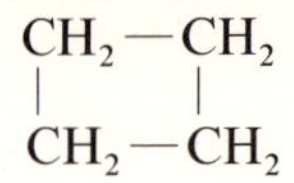

is cyclobutane.

26-4 Alkenes

Alkenes are hydrocarbons with double bonds. They have the general formula C_nH_{2n}.

According to the IUPAC system of nomenclature for alkenes, the root of the name is the same as for the saturated hydrocarbons, with an *-ene* ending. The carbon atoms in the chain are numbered in such a way that the carbon atoms of the double bond have the lowest numbers possible. The position of the double bond is indicated by the number of the lower numbered carbon atom (that is, the first carbon of the double bond). The number is placed before the name of the compound, with a hyphen between the number and the name. Consider the following examples:

FORMULA	NAME
$CH_2{=}CH{-}CH_2{-}CH_3$	1-Butene
$CH_3{-}CH{=}CH{-}CH_3$	2-Butene
$CH_3{-}CH_2{-}CH_2{-}CH{=}CH{-}CH_3$	2-Hexene (start at the right to give the smallest number)

If there is more than one double bond, a number is used for each one, with the prefix di-, tri-, and so on placed just before the -ene ending; for example, $CH_3{-}CH{=}CH{-}CH_2{-}CH{=}CH{-}CH_3$ is 2,5-heptadiene.

Common names are sometimes used for the simplest member of the series, such as $CH_2{=}CH_2$ (ethylene) and $CH_3{-}CH{=}CH_2$ (propylene).

Due to the pi bonds in alkenes, these are much more reactive than alkanes. For example, H_2 (in the presence of a catalyst), X_2 (where X is Cl or Br), and HX (where X is Cl, Br, or I) readily react with alkenes. Consider the following examples:

1. 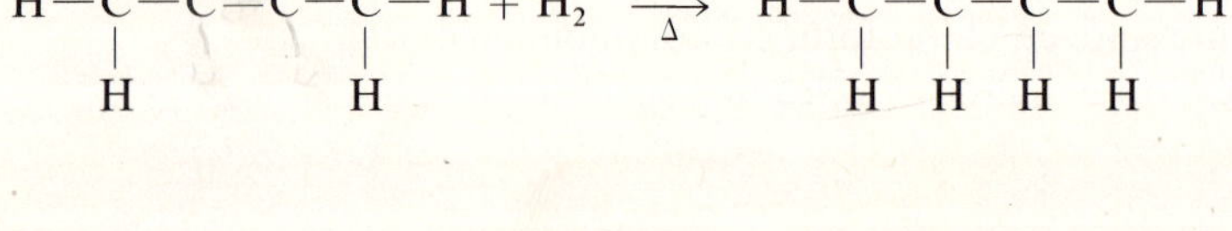

2. 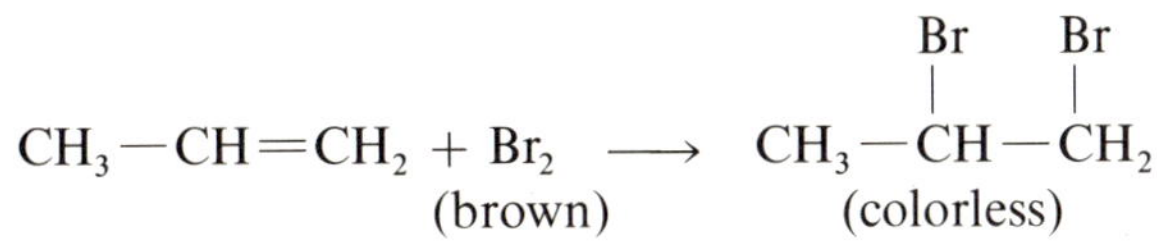

This reaction is often used to test for the presence of double (or triple—see Section 26-5) bonds, due to the color change from brown to colorless.

3. 

26-5 Alkynes

Alkynes are hydrocarbons with triple bonds. They have the general formula C_nH_{2n-2} and are often referred to as acetylenes. This name comes from the common name of the first member of this series, acetylene, for $H{-}C{\equiv}C{-}H$.

In writing names and formulas for alkynes, the same general rules are followed as with alkenes, except that the ending in all cases is *-yne*. The first member of this series, $CH{\equiv}CH$, is known almost exclusively by its common name acetylene.

Consider the following examples of alkynes:

FORMULA	NAME
$CH_3{-}C{\equiv}CH$	Propyne (no number necessary)
$CH_3{-}C{\equiv}C{-}CH_2{-}CH_3$	2-Pentyne

The alkynes are similar in their properties to the alkenes, except that there are two π bonds rather than one. Therefore, in their reactions, 1 mole of alkyne will react with 2 moles of H_2, X_2, and HX (where X is the same as in the reactions of alkenes). Consider the following examples:

(1) $$CH_3{-}C{\equiv}CH + 2\,H_2 \xrightarrow[\Delta]{Pt} CH_3{-}CH_2{-}CH_3$$

(2) 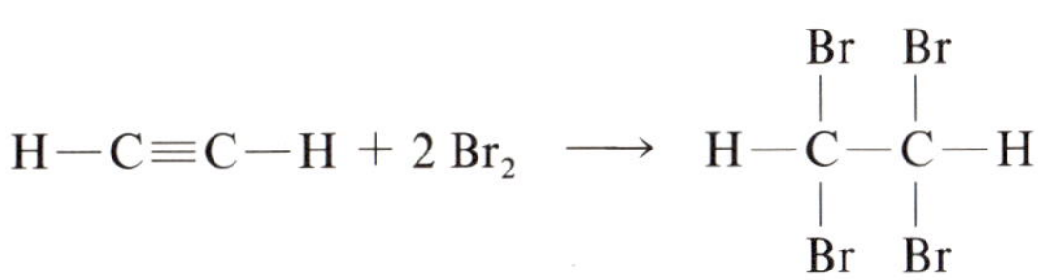

(3) $$\mathrm{H-C\equiv C-H + 2\,HCl \longrightarrow H-\overset{\displaystyle Cl}{\underset{\displaystyle Cl}{C}}-\overset{\displaystyle H}{\underset{\displaystyle H}{C}}-H^*}$$

26-6 Substituted Hydrocarbons

In any type of hydrocarbon, one or more of the hydrogens may be replaced by a different atom or group of atoms. Table 26-3 lists the *alkyl groups* that correspond to the first 10 alkanes. The names of these are directly related to the name of the alkane, with a *-yl* ending. Keep in mind that these are not compounds but rather groups that are part of a compound. For example, the methyl group (CH_3^-) has the structural formula $\mathrm{H-\overset{\displaystyle H}{\underset{\displaystyle H}{C}}-}$ and must be attached to something else in order to form a compound. These groups are often referred to as *radicals.*

When halogen atoms are substituted into the various types of hydrocarbons, the halogen takes an *-o* ending, such as *chloro*. The $-NO_2$ group is called *nitro.*

In writing correct names and formulas for hydrocarbons with various atoms or groups attached to them, the same general rules are followed as given in Sections 26-3 through 26-5, with the following additions:

1. The longest continuous chain is used for the stem of the name of substituted alkanes; for alkenes and alkynes, the longest chain containing the multiple bond is used.
2. A number is used to tell where *each* substituent atom or group is on the chain; number in such a way that the smallest numbers possible are used. For alkenes and alkynes, the smallest number *for the multiple bond* is used.
3. If there is more than one substituent group, they are written in alphabetical order.
4. If more than one of the same atom or group is on the chain, the prefixes di-, tri-, and so on are used to tell how many there are, with a number for each one. In considering alphabetical order, the prefix is not considered. That is, dibromo would come before chloro.

*Notice that both Cl atoms go on the same carbon. This goes in two steps, and in the second step the C with the most H tends to get the H, although actually a mixture of products is obtained. This is known as *Markownikoff's Rule,* and can be remembered as "them that has, gets."

TABLE 26-3 *Names of first ten alkyl groups or radicals*

RADICAL	NAME	RADICAL	NAME
CH_3-	methyl	$C_6H_{13}-$	hexyl
C_2H_5-	ethyl	$C_7H_{15}-$	heptyl
C_3H_7-	propyl	$C_8H_{17}-$	octyl
C_4H_9-	butyl	$C_9H_{19}-$	nonyl
$C_5H_{11}-$	pentyl (or amyl)	$C_{10}H_{21}-$	decyl

Consider the following examples.

FORMULA	NAME
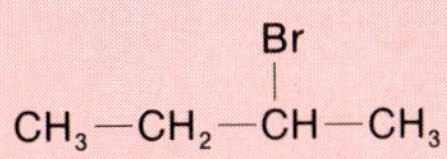	2-Bromobutane (start numbering from the end that gives the smallest numbers)
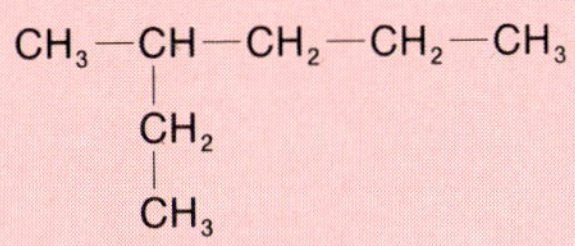	3-Methylhexane (use the longest *continuous* chain)
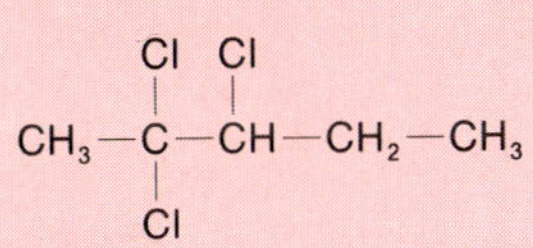	2,2,3-Trichloropentane (a number must be included for *each* Cl)
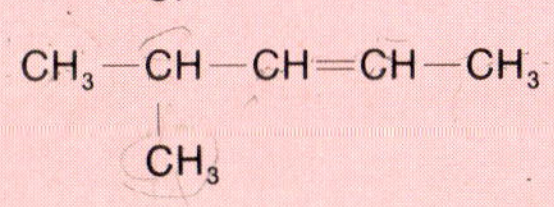	4-Methyl-2-pentene (use smallest number *for double bond*)

Often common names are used for organic compounds. Two of the most frequently used are the following:

1. *Iso* for a branching on the end of a chain and *normal* (*n*) if there is no branching. This is only used for rather simple compounds but is quite commonly used for many of these. In this system, the radical is named and then the substituent. For example,

$$CH_3-CH_2-CH_2-CH_2-Br$$

would be called *n*-butyl bromide and

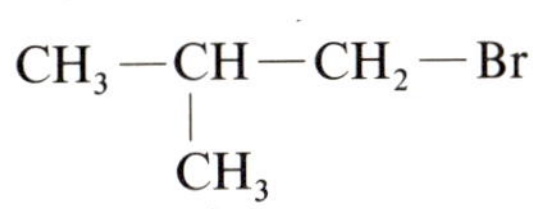

would be called isobutyl bromide.

2. Classification of alkyl groups into three types, as follows (in these, R stands for any alkyl group):

$R-CH_2-$ Primary–The carbon to which the substituent is attached is only joined to one other C atom. This is often written as 1°.

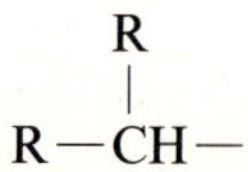

Secondary (*sec*)–The carbon to which the substituent is attached is joined to two other carbons. This is written as 2°.

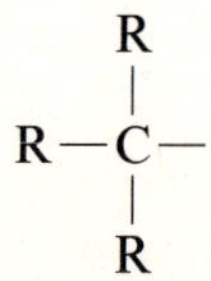

Tertiary (*tert*)–The carbon to which the substituent is attached is joined to three other carbons. This is written as 3°.

In this system also the radical is named first and then the substituent.

For example, 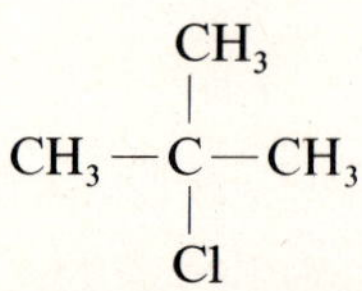 would be called *tert*-butyl chloride.

Consider the following examples, in which both the common name and the IUPAC name are given:

FORMULA	COMMON NAME	IUPAC NAME
$CH_3-CH_2-CH_2-CH_3$	*n*-Butane	Butane
$CH_3-CH-CH_3$ with CH_3 on the CH	Isobutane (all carbons are counted)	2-Methylpropane
$CH_3-CH-CH_2-CH_2-Br$ with CH_3 on the CH	Isopentyl bromide	1-Bromo-3-methylbutane
$CH_3-CH_2-CH_2-Cl$	*n*-Propyl chloride	1-Chloropropane

Alkynes are sometimes named as derivatives of acetylene. For example,

$$CH_3-C\equiv C-CH_3$$

could be called dimethylacetylene and

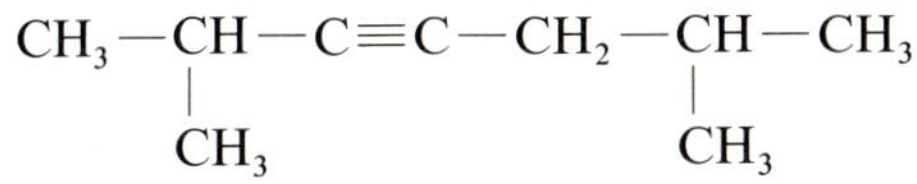

could be called isopropylisobutylacetylene. These are common names, however, and not the correct IUPAC system nomenclature.

26-7 Aromatic Compounds

The compound *benzene* has the molecular formula C_6H_6 and the structural formula:

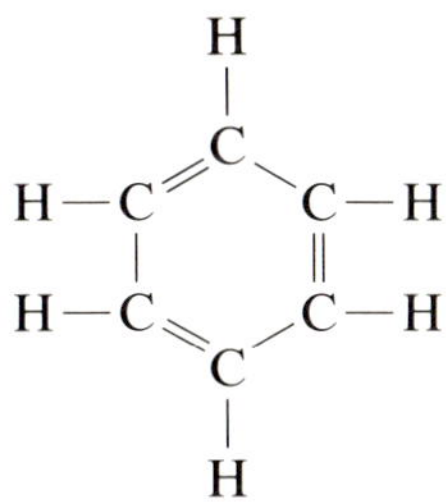

This is often abbreviated as

The benzene ring can be considered as consisting of two contributing *resonance* forms

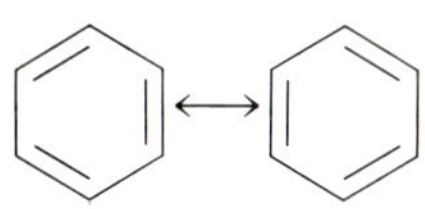

or as three *delocalized π bonds.* Therefore, the formula is often written as

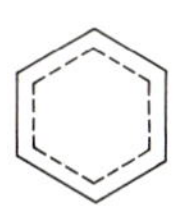

or

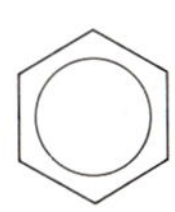

In this text we shall use this last notation for the benzene ring.

Any compound that contains this benzene ring is referred to as an **aromatic compound.** This name was originally used because many of the first compounds known that were derivatives of benzene had rather fragrant odors.

The multiple bonds in the benzene ring are not reactive like double bonds in aliphatic compounds. For example, Cl_2 will not break one of the 'double' bonds in the benzene ring like it will in an alkene. Chlorine can be made to react with benzene but the reaction is a substitution type, where a —Cl replaces a —H on the benzene, not an addition where a double bond is broken.

These compounds are generally named by one of two systems. In the first of these, the benzene ring is considered as a substituent of another compound. In this case the group

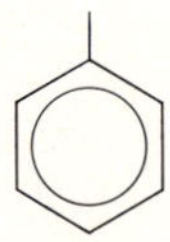

(i.e., the benzene ring with one H removed) is called a *phenyl* group. For example,

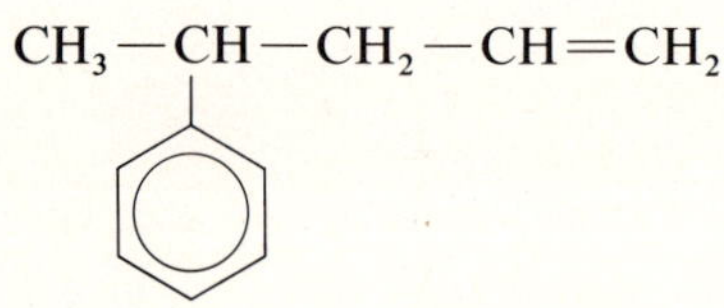

could be named as 4-phenyl-1-pentene.

In the second system the compound is named as a derivative of benzene. In this system if there is more than one substituent on the benzene ring, numbers are used to indicate where the substituents are (the numbering is done in such a way that the smallest numbers possible are used).

For disubstituted benzene derivatives, the prefixes ortho- (*o*), meta (*m*), and para- (*p*) are also used, as follows:

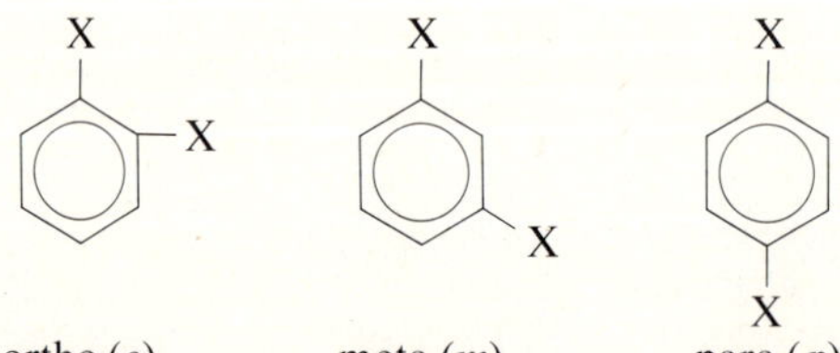

ortho (*o*) (substituents on adjacent carbons)	meta (*m*) (separated by one carbon)	para (*p*) (separated by two carbons)

Compounds that contain a methyl group on the benzene ring, along with other groups, are sometimes named as derivatives of toluene.

Consider the following examples:

FORMULA	NAME
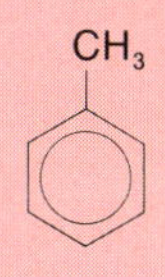	Methylbenzene (no number necessary) This is often called toluene.
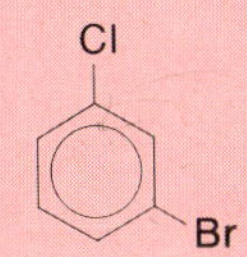	1-Chloro-3-bromobenzene or 1-bromo-3-chlorobenzene (these are equally correct—in either case the smallest numbers possible are used)
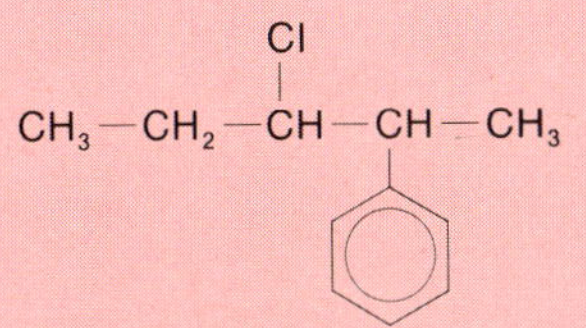	3-Chloro-2-phenylpentane (substituents are written in alphabetical order)
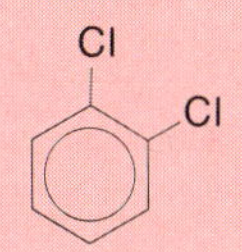	*o*-Dichlorobenzene (or, 1,2-dichlorobenzene)
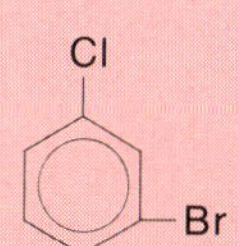	*m*-Bromochlorobenzene (or, 1-bromo-3-chlorobenzene)
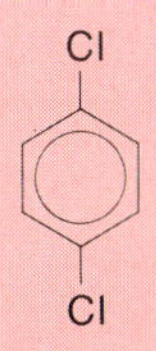	*p*-Dichlorobenzene (or, 1,4-dichlorobenzene)
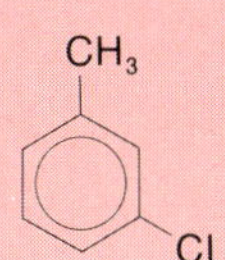	1-Methyl-3-chlorobenzene, (or, *m*-chlorotoluene)

The functional group of **alcohols** and **phenols** is —OH. A phenol is an aromatic alcohol, and the name comes from the simplest one, which is

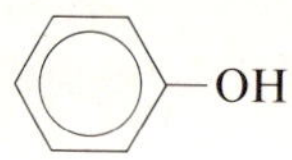

26-8 Alcohols and Phenols

The general formula for alcohols is R—OH, where R— can be any kind of aliphatic group—it could even contain other functional groups, such as multiple bonds. The general formula for phenols in Ar—OH, where the Ar— group is any aromatic group; that is, any group containing a benzene ring. The alcohols are *not* bases, as might be expected by the presence of the —OH group, but are actually very weak acids.

In naming alcohols, common names are often used for the simpler ones. In this system the alkyl group is just named, followed by alcohol. According to the IUPAC system of naming alcohols, the longest chain of carbon atoms that contains the hydroxyl (—OH) group is used as the stem of the name, with an *-ol* ending. A number is used to indicate the position of the hydroxyl group (the number is omitted if it is not necessary, as in methanol or ethanol). Phenols are usually just named as derivatives of phenol. Consider the following examples.

FORMULA	COMMON NAME	IUPAC NAME
CH_3-OH	Methyl alcohol	Methanol
$CH_3-CH_2-CH_2-CH_2-OH$	*n*-Butyl alcohol	1-Butanol
$CH_3-CH-CH_2-OH$ | CH_3	Isobutyl alcohol	2-Methyl-1-propanol
$CH_3-CH-CH_2-CH_3$ | OH	*sec*-Butyl alcohol	2-Butanol
CH_3 | CH_3-C-CH_3 | OH	*tert*-Butyl alcohol	2-Methyl-2-propanol
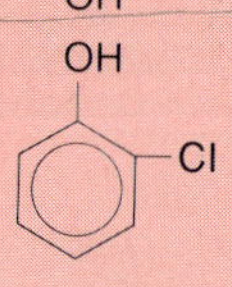	*o*-Chlorophenol	2-Chlorophenol

26-9 Ethers

Ethers have the general formula R—O—R. They can be either aliphatic or aromatic. A simple ether is one in which both groups attached to the oxygen are the same, and a mixed ether is one in which they are different. Ethers are relatively inert and are therefore often used as solvents in organic reactions.

Common names are frequently used for ethers. A system is followed, however, and thus there is usually no difficulty. According to this common system, the two groups that are attached to the oxygen are named, followed by *ether.* In a simple ether it is not necessary to use the prefix di-, but this is sometimes done to avoid any confusion. For example, CH_3-O-CH_3 could be called methyl ether or dimethyl ether.*

Consider the following examples.

FORMULA	NAME
$CH_3-O-CH_2-CH_2-CH_3$	Methyl propyl ether
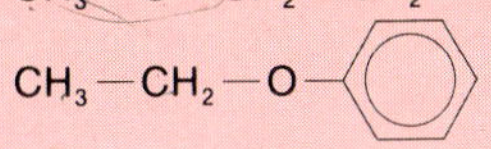	Ethyl phenyl ether
$\underset{\displaystyle CH_3}{\underset{\vert}{CH_3-CH}}-O-\underset{\displaystyle CH_3}{\underset{\vert}{CH}}-CH_3$	Isopropyl ether

26-10 Carboxylic Acids and Their Derivatives

Organic compounds that contain the carboxyl group

$$-\overset{\displaystyle O}{\overset{\|}{C}}-OH$$

are called **carboxylic acids.** Therefore, the general formula for carboxylic acids is

$$R-\overset{\displaystyle O}{\overset{\|}{C}}-OH\left(\text{or, } Ar-\overset{\displaystyle O}{\overset{\|}{C}}-OH\right).$$

These are all weak acids and only have one acidic hydrogen; that is, only the hydrogen on the carboxyl group can be given off as an H^+ ion.

Since these compounds are quite common in many natural products, they were among the first organic compounds studied. Many of them have common names, derived from a Latin or Greek name for the substance in which they were originally found. The IUPAC system of nomenclature uses all the rules discussed earlier, with an *-oic acid* ending. For example,

$$CH_3-CH_2-\overset{\displaystyle O}{\overset{\|}{C}}-OH$$

is propanoic acid; note in this name that all carbons are counted, including the one in the carboxyl group. The following examples list three common

*According to the IUPAC system, ethers are named as alkoxy derivatives of hydrocarbons. For example, dimethyl ether would be called methoxymethane.

carboxylic acids, with their IUPAC names, common names and what they are derived from.

FORMULA	IUPAC NAME	COMMON NAME	WHERE FOUND	DERIVATION OF NAME
$H-\overset{\overset{\large O}{\|}}{C}-OH$	Methanoic acid	Formic acid	Ants	Latin—formica (ant)
$CH_3-\overset{\overset{\large O}{\|}}{C}-OH$	Ethanoic acid	Acetic acid	Vinegar	Latin—acetum (vinegar)
$CH_3-CH_2-CH_2-\overset{\overset{\large O}{\|}}{C}-OH$	Butanoic acid	Butyric acid	Rancid butter	Latin—butyrum (butter)

The compound

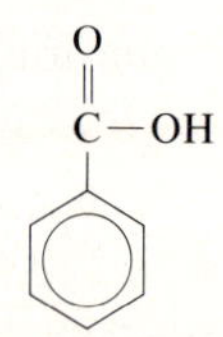

is benzoic acid.

Sometimes the carbon next to the carboxyl group is referred to as the alpha (α) carbon and the next one, the beta (β) carbon. This is the basis for another common system of nomenclature. For example,

$$CH_3-CH_2-\underset{\underset{\large Cl}{|}}{CH}-\overset{\overset{\large O}{\|}}{C}-OH$$

could be called α-chlorobutyric acid; however, the IUPAC name would be 2-chlorobutanoic acid. In the IUPAC system, the carbon of the carboxyl group is always the number 1 carbon.

Esters

Esters are derivatives of carboxylic acids that are formed when an acid is combined with an alcohol. They are named by taking the name of the alcohol and the name of the acid with an *-ate* ending (like a salt of an oxyacid), such as

$$CH_3-CH_2-OH + CH_3-\overset{\overset{\displaystyle O}{\|}}{C}-OH \longrightarrow CH_3-\overset{\overset{\displaystyle O}{\|}}{C}-O-CH_2-CH_3 + H_2O$$

ethyl alcohol + acetic acid ⟶ ethyl acetate

or

ethanol + ethanoic acid ⟶ ethyl ethanoate

Many esters have very pleasant, sweet smelling odors. For example, the ester, amyl acetate (or pentyl ethanoate) has the odor of bananas. Even though there are several compounds that contribute to the aroma and flavor of fruits, such as the banana, this ester is mainly responsible for the odor.

Acid Anhydrides

An **acid anhydride** is formed when one molecule of water is removed from two molecules of acid, such as

$$R-\overset{\overset{\displaystyle O}{\|}}{C}-OH + HO-\overset{\overset{\displaystyle O}{\|}}{C}-R \longrightarrow R-\overset{\overset{\displaystyle O}{\|}}{C}-O-\overset{\overset{\displaystyle O}{\|}}{C}-R + H_2O$$

The R groups here can be the same or different and can also be aromatic.

These are named by taking the name of the acid (or acids) and adding *anhydride.* Consider the following examples.

FORMULA	COMMON NAME	IUPAC NAME
$CH_3-\overset{\overset{\displaystyle O}{\|}}{C}-O-\overset{\overset{\displaystyle O}{\|}}{C}-CH_3$	Acetic anhydride	Ethanoic anhydride
$CH_3-\overset{\overset{\displaystyle O}{\|}}{C}-O-\overset{\overset{\displaystyle O}{\|}}{C}-H$	Acetic formic anhydride	Ethanoic methanoic anhydride

When water is added to an anhydride, the original acid is formed. In the case where the R groups are different, a mixture of two acids is formed.

Amides

An **amide** is a derivative of a carboxylic acid with the general formula

$$R-\overset{\overset{\displaystyle O}{\|}}{C}-NH_2.$$

These are usually named by a common system in which the name of the acid is used, the -ic acid ending is dropped, and *-amide* is added. Either common or IUPAC names can be used for the acid. Consider the following examples.

FORMULA	NAME
$CH_3-\overset{\overset{\displaystyle O}{\|}}{C}-NH_2$	Acetamide* or ethanamide
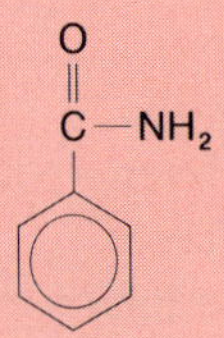	Benzamide

26-11 Aldehydes and Ketones

Aldehydes and **ketones** are organic compounds that contain the carbonyl group, $-\overset{\overset{\displaystyle O}{\|}}{C}-$. If two organic groups are attached to the carbon (either alkyl or aryl), the compound is a ketone; that is,

$$R-\overset{\overset{\displaystyle O}{\|}}{C}-R \quad \text{or} \quad R-\overset{\overset{\displaystyle O}{\|}}{C}-Ar \quad \text{or} \quad Ar-\overset{\overset{\displaystyle O}{\|}}{C}-Ar$$

If one organic group and one H are attached to the carbon, the compound is an aldehyde; that is,

$$R-\overset{\overset{\displaystyle O}{\|}}{C}-H \quad \text{or} \quad Ar-\overset{\overset{\displaystyle O}{\|}}{C}-H$$

Aldehydes and ketones can be formed from the oxidation of alcohols. A primary alcohol gives an aldehyde upon oxidation and a secondary alcohol

*Thioacetamide, below, is used in qualitative analysis to produce H_2S.

$$CH_3-\overset{\overset{\displaystyle S}{\|}}{C}-NH_2$$

It reacts with water according to the following equation:

$$CH_3-\overset{\overset{\displaystyle S}{\|}}{C}-NH_2 + H_2O \rightarrow H_2S + CH_3-\overset{\overset{\displaystyle O}{\|}}{C}-NH_2$$

gives a ketone. Consider the following examples:

$$CH_3-CH_2-OH \xrightarrow{(O)} CH_3-\overset{\overset{\displaystyle O}{\|}}{C}-H$$

$$CH_3-\overset{\overset{\displaystyle OH}{|}}{C}H-CH_3 \xrightarrow{(O)} CH_3-\overset{\overset{\displaystyle O}{\|}}{C}-CH_3$$

These are not complete equations but only represent the main product. The other product formed would depend on what the oxidizing agent was; the symbol (O) simply represents an oxidizing agent. Notice also that this is not necessarily a reaction with oxygen. It is a general type of oxidation in which the oxidation number of the carbon is increased, from zero in the alcohol to $+2$ in the aldehyde and $+4$ in the ketone. In general, when considering oxidation numbers for organic compounds, it is easiest to consider any alkyl group to be -1 since it has lost one H; oxygen of course is still -2 and hydrogen $+1$.

Aldehydes can be oxidized further to acids. You should be able to show that the oxidation number of the carbon is increased further when the aldehyde forms an acid.

The simplest and most common aldehyde is different in that it has two hydrogens attached to the carbonyl group, to give formaldehyde, $H-\overset{\overset{\displaystyle O}{\|}}{C}-H$.

The common names for aldehydes use the common name of the acid that the aldehyde is derived from; the -ic acid is dropped and *aldehyde* is added; that is, formic acid gives formaldehyde. As with acids, the carbon adjacent to the carbonyl group is sometimes called the alpha (α) carbon. Therefore, the compound $CH_3-CH_2-\overset{\overset{\displaystyle Cl}{|}}{C}H-\overset{\overset{\displaystyle O}{\|}}{C}-H$ could be called α-chlorobutyraldehyde.

According to the IUPAC system, the aldehyde takes an -al ending and the carbonyl carbon is always number 1.

The common names of ketones use the common name for the two R (or Ar) groups with *ketone*. The IUPAC system takes an -one ending, with a number to indicate the position of the carbonyl group.

Consider the following examples.

FORMULA	COMMON NAME	IUPAC NAME
$CH_3-\overset{\overset{\displaystyle O}{\|}}{C}-H$	Acetaldehyde	Ethanal

FORMULA	COMMON NAME	IUPAC NAME
$CH_3-CH_2-\overset{\overset{\large O}{\Vert}}{C}-CH_2-CH_2-CH_3$	Ethyl *n*-propyl ketone	3-Hexanone
$CH_3-\overset{\overset{\large O}{\Vert}}{C}-CH_3$	Dimethyl ketone (also, more commonly known as acetone)	Propanone (the 2 is not needed here)

26-12 Amines

Amines can be considered as organic derivatives of ammonia in which one or more of the hydrogens has been replaced by alkyl or aryl groups. If only one of the hydrogens has been replaced, the amine is a primary (1°) amine. If two hydrogens have been replaced, it is a secondary (2°) amine; if all three hydrogens have been replaced, it is a tertiary (3°) amine. These compounds thus have the following general formulas:

$$\underset{1^\circ\ \text{amine}}{R-NH_2} \qquad \underset{2^\circ\ \text{amine}}{R-\underset{\underset{\large R}{|}}{N}H} \qquad \underset{3^\circ\ \text{amine}}{R-\underset{\underset{\large R}{|}}{N}-R}$$

The R groups can also be aromatic, or aryl (Ar) groups.

These compounds are generally named according to the groups that have replaced the hydrogen(s), followed by *amine.* Consider the following examples.

FORMULA	NAME
CH_3-NH_2	Methylamine
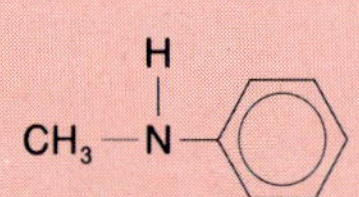	Methylphenylamine
$CH_3-\underset{\underset{\large CH_3}{|}}{N}-CH_3$	Trimethylamine

The simplest aromatic amine is aniline

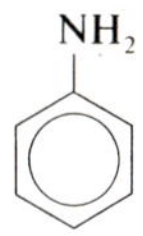

Other aromatic amines are often named as derivatives of this, such as 3-methylaniline (or *m*-methylaniline) for the compound

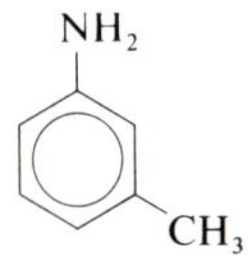

SUMMARY

Table 26-4, on the following pages, summarizes the types of organic compounds that have been considered.

TABLE 26-4 *Summary of organic compounds*

TYPE OF COMPOUND	GENERAL FORMULA	CHARACTERISTICS	IUPAC NAMES	COMMON NAMES
Alkane	C_nH_{2n+2}	Saturated hydrocarbon; not very reactive (will react by substitution)	-ane ending (basic names of all organic compounds)	iso- for branch on end of chain; all carbons counted (e.g., isobutane)
Alkene	C_nH_{2n}	Unsaturated hydrocarbon—double bonds; quite reactive—by addition reaction	Stem from basic alkanes; -ene ending; number (smallest possible) for first carbon of double bond (e.g., 2-butene)	Radical with -ene ending (e.g., ethylene)
Alkyne	C_nH_{2n-2}	Unsaturated hydrocarbon—triple bonds; reactive—like alkene but two pi bonds to add on to	Like alkene but -yne ending (e.g., 3-heptyne)	Radical(s) plus acetylene (e.g., methylethylacetylene)
Substituted hydrocarbon		Alkyl group, halo group, nitro (—NO_2) group, or phenyl 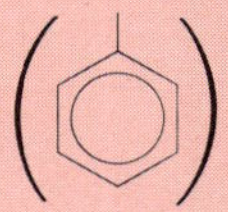 group on chain	Stem of longest continuous chain (or chain that contains multiple bond); number and name for each group on chain (e.g., 2,3-dimethyl-2-butene)	(a) iso- for branch on end of chain, or normal (*n*); (b) primary, secondary, and tertiary carbons (e.g., *n*-butyl, isobutyl, *sec*-butyl, and *tert*-butyl)
Aromatic	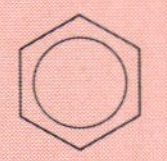	Contain the benzene ring; *not* like alkene (react by addition)	Derivatives of benzene; number for each substituent group, (e.g., 1-chloro-3-methylbenzene)	Toluene (and derivatives) for 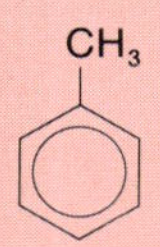 *o*-, *m*-, and *p*- for disubstituted (e.g., *p*-dichlorobenzene)
Alcohol	R—OH	Aromatic alcohols are called phenols; the —OH group can be on a 1°, 2°, or 3°C	Stem from alkane, with -ol ending and number to show where —OH group is (e.g., 2-butanol)	Radical + alcohol; including iso-, *sec*-, and *tert*- (e.g., *sec*-butyl alcohol)

Ether	R—O—R	Can be simple or mixed	Stem of alkane with -oxy ending, + radical (e.g., methoxy methane)	Radical(s) + ether (e.g., isopropyl *n*-butyl ether)
Acid	R—C(=O)—OH	H on carboxyl group (—COOH) is acidic; weak acids	Stem of alkane (including C in carboxyl group), with -oic acid ending (e.g., pentanoic acid)	From name of substance acid is derived from (e.g., formic, and acetic)
Ester	R—C(=O)—O—R	Formed from acid + alcohol (removal of a molecule of water)	Radical from alcohol + acid with -ate ending (e.g., ethyl pentanoate)	Common name of acohol + common name of acid anion (e.g., isopropyl acetate)
Anhydride	R—C(=O)—O—C(=O)—R	Formed by removal of a molecule of water from two molecules of acid (same acid or different)	Name of acids(s) + anhydride (e.g., methanoic propanoic anhydride)	Common name of acid(s) + anhydride (e.g., acetic anhydride)
Amide	R—C(=O)—NH_2	Derived from carboxylic acid	Acid (drop -oic acid) + amide (e.g., ethanamide)	Like IUPAC, but common name of acid (e.g., acetamide)
Aldehyde	R—C(=O)—H	Formed from oxidation of primary alcohol (R—OH $\xrightarrow{(O)}$ R—C(=O)—H); can be oxidized further to acid	Stem of alkane, with -al ending (e.g., propanal)	Common name of corresponding acid; drop -ic acid and add aldehyde (e.g., acetaldehyde)
Ketone	R—C(=O)—R	Formed from oxidation of secondary alcohol R—CH(OH)—F $\xrightarrow{(O)}$ R—C(=O)—R	Stem of alkane, with -one ending, and number to indicate which C is the carbonyl group (e.g., 3-hexanone)	Common name(s) of radicals, + ketone (e.g., methyl isopropyl ketone); use di- if two of the same radical.
Amine	R—NH_2, R—N(R)—H, or R—N(R)—R (R can also be an aromatic or aryl[Ar] group)	Organic derivative of NH_3; can be 1°, 2°, or 3°	Name(s) of radical(s) + amine (e.g., dimethyl amine)	Common name(s) of radical(s) + amine (e.g., *sec*-butyl amine)

Upon completion of this chapter, you should be able to do the following:

1. Given any hydrocarbon describe the bonding in the compound. (1)
2. Given any molecular formula, write all possible structural isomers. (2)
3. Given the formula of any organic compound, identify the functional group and classify a substance as one of the following types: alkane, alkene, alkyne, aromatic, alcohol, ether, acid, ester, anhydride, amide, aldehyde, ketone, or amine. (3)
4. Given the IUPAC or the common name, be able to write the correct formula; or given the formula, be able to write the correct name for any hydrocarbon, substituted hydrocarbon, alcohol, ether, acid, ester, anhydride, amide, aldehyde, ketone, or amine. (4, 5, 7 through 26)
5. Given any of the following reactants, write a balanced equation to represent the reaction that takes place: any hydrocarbon upon burning in an excess of oxygen; alkene or alkyne + Br_2 or Cl_2; alkene or alkyne + HBr or HCl; alkene or alkyne + H_2 (in the presence of a catalyst and heat). (6)
6. Given the name or formula of any alcohol or amine, classify it as primary, secondary, or tertiary. (10, 25)
7. Given any of the terms in Exercise 1, give a brief definition or explanation in your own words.

EXERCISES

1. Define or explain each of the following:
   (a) hydrocarbon
   (b) alkane
   (c) alkene
   (d) alkyne
   (e) aromatic compound
   (f) alcohol
   (g) phenol
   (h) ether
   (i) acid
   (j) ester
   (k) anhydride
   (l) amide
   (m) aldehyde
   (n) ketone
   (o) amine
2. Distinguish between
   (a) alkane and alkene
   (b) alkene and alkyne
   (c) aliphatic and aromatic compound
   (d) primary (1°), secondary (2°), and tertiary (3°) alcohol
   (e) amide and amine
   (f) common and IUPAC names for alcohols

PROBLEMS

General principles

1. Describe the bondng in each of the following:
   (a) $CH_3-CH_2-CH_3$
   (b) $CH_3-CH=CH_2$
   (c) $CH_3-C\equiv CH$
   (d) benzene
2. Write the structural isomers for each of the following (the number in parentheses tells how many there are).
   (a) C_4H_{10} (2)
   (b) C_6H_{14} (5)
   (c) C_5H_{12} (3)
   (d) C_7H_{16} (7)
3. Classify each of the following as alkane, alkene, alkyne, aromatic hydrocarbon, alcohol, ether, acid, aldehyde, ketone, or amine:
   (a) $CH_3-CH=CH_2$
   (b) 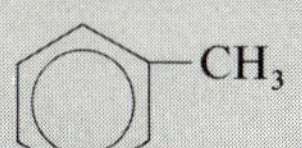
   (c) CH_3-O-CH_3
   (d) $CH_3-C\equiv CH$
   (e) $CH_3-C(=O)-OH$
   (f) $CH_3-CH_2-NH_2$
   (g) $CH_3-C(=O)-H$
   (h) CH_3-CH_2-OH
   (i) $CH_3-CH_2-CH_3$
   (j) $CH_3-C(=O)-CH_3$
   (k) $CH_3-O-CH_2-CH_3$
   (l) $CH_3-CH_2-C(=O)-H$
   (m) $CH_3-CH=CH-CH_3$
   (n) CH_3-NH_2
   (o) $CH_3-CH_2-C(=O)-OH$
   (p) $CH_3-C\equiv C-CH_3$
   (q) CH_3-OH
   (r) $CH_3-CH_2-CH_2-CH_3$
   (s) $CH_3-CH_2-C(=O)-CH_3$
   (t) 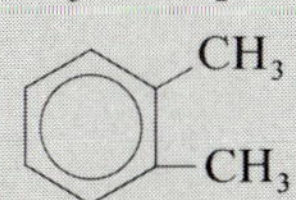

Hydrocarbons

4. Write the correct name for each of the following:
   (a) $CH_3-CH_2-CH=CH_2$
   (b) $CH_3-C\equiv C-CH_2-CH_3$
   (c) $CH_3-CH(CH_3)-CH_2-CH_2-CH_3$
   (d) $CH_3-CH(Br)-CH_2-CH_3$
   (e) $CH_3-CH(CH_3)-CH(CH_3)-CH_3$
   (f) $CH_2-CH-CH_3$ (ring: CH_2 and CH each bonded below to CH_2-CH_2)

(g) 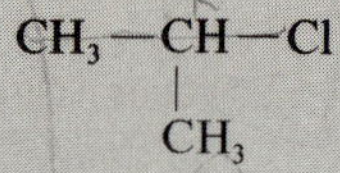

(h) 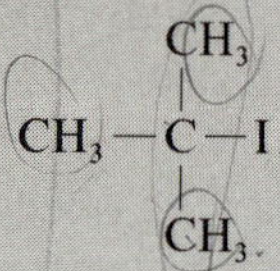

(i) 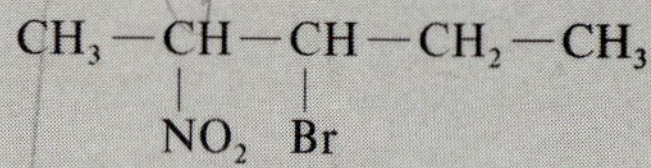

(j) 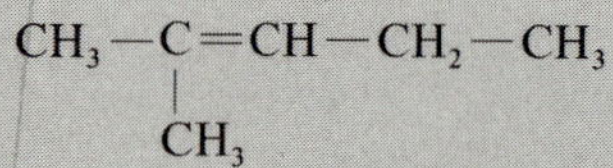

(k) 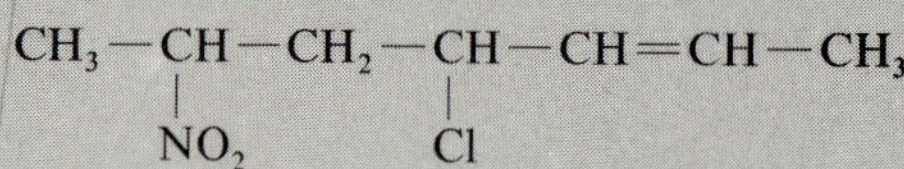

(l) 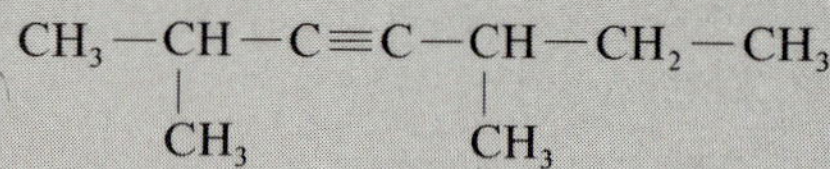

(m) $CH_3-CH=CH_2$

(n) $CH_2=CH-CH_2-CH_2-CH_2-CH_3$

(o) 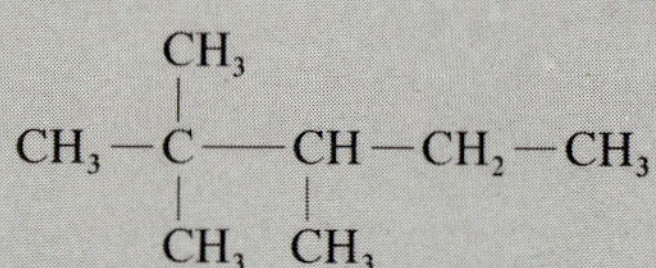

(p) 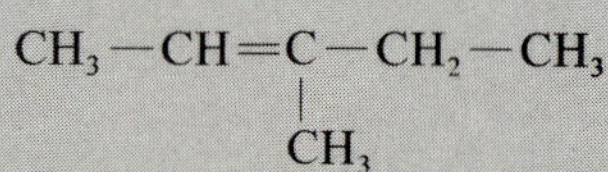

(q) 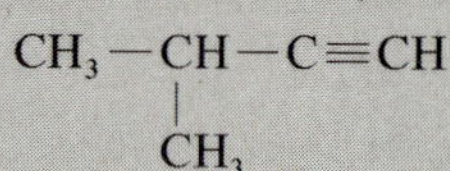

(r) 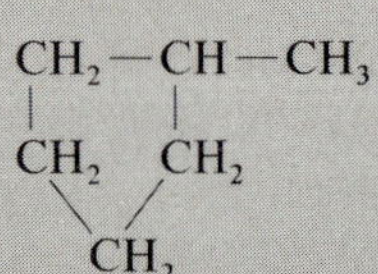

(s) 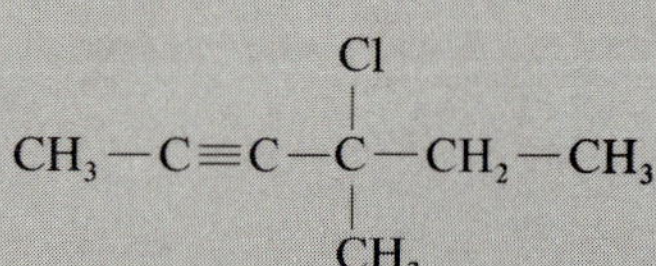

(t) 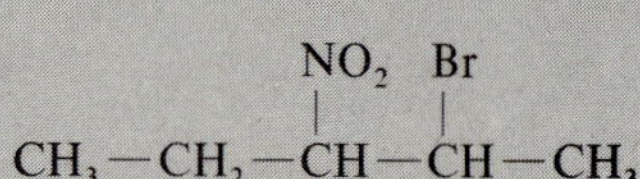

(u) 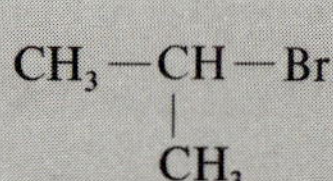

(v) 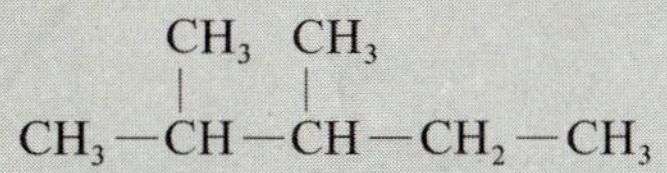

(w) 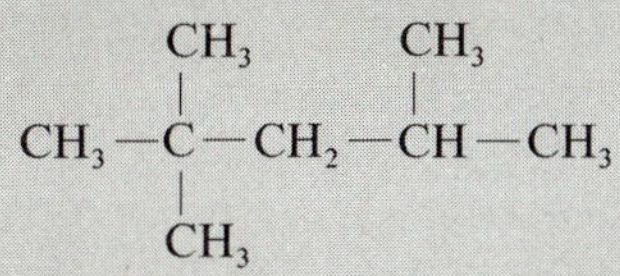

(x) 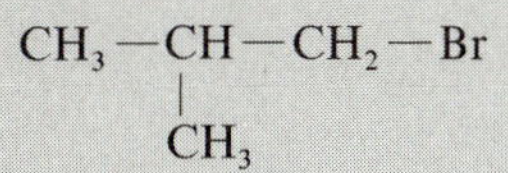

5. Write the correct formula for each of the following:
   (a) 2-nitropropane
   (b) 2-bromo-3,3-dimethylhexane
   (c) 3-ethylpentane
   (d) ethylcyclopentane
   (e) 4-methyl-2-hexene
   (f) 3,3-dimethyl-1-heptyne
   (g) 2-methyl-3-ethylheptane
   (h) 3-nitrohexane
   (i) methylcyclobutane
   (j) 2-methyl-4,4-dichlorohexane
   (k) 3-bromo-4-methyl-2-pentene
   (l) 2-nitro-3-hexyne

6. Write a balanced equation for each of the following:
   (a) propene + hydrogen
   (b) 1-butyne + 2 moles bromine
   (c) ethene + hydrogen chloride
   (d) pentane + oxygen (and heat)
   (e) 2-methyl-3-octene + chlorine
   (f) 2,2-dimethylheptane is burned in an excess of oxygen
   (g) propene + chlorine
   (h) ethene + hydrogen bromide
   (i) 2-butene + hydrogen (with catalyst)
   (j) 1-pentyne + 2 moles chlorine

7. Write the correct IUPAC name for each of the following:
   (a) dimethyacetylene
   (b) isobutane
   (c) ethylchloride
   (d) isopentyl bromide
   (e) carbon tetrachloride
   (f) *tert*-butyl bromide
   (g) 3-chloro-5-octene
   (h) *sec*-propyl chloride
   (i) propyl bromide
   (j) carbon tetrabromide
   (k) ethylpropylacetylene
   (l) isopentane
   (m) isobutyl bromide
   (n) *sec*-butyl bromide
   (o) *tert*-pentyl chloride
   (p) 2-methyl-4-heptyne

Aromatic compounds

8. Write the correct name for each of the following:

(a) 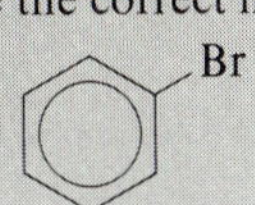

(b) 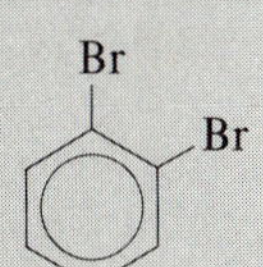

(c) 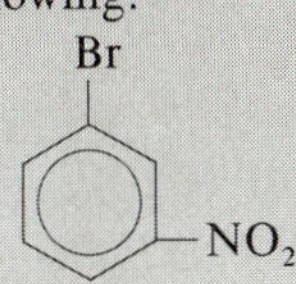

(d) 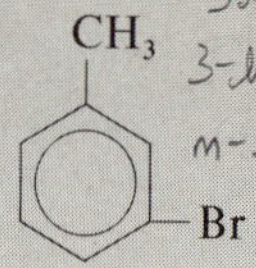

(f) 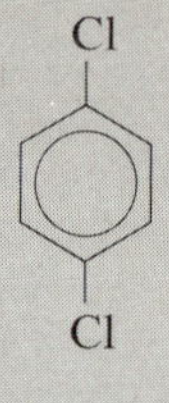

(e) 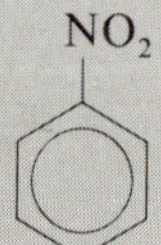

9. Write the correct formula for each of the following:
   (a) iodobenzene
   (b) 1,4-dichlorobenzene
   (c) *m*-iodonitrobenzene
   (d) *o*-dichlorobenzene
   (e) *p*-nitrotoluene
   (f) *p*-dichlorobenzene
   (g) bromobenzene
   (h) *o*-bromonitrobenzene
   (i) *m*-chlorotoluene
   (j) 2,4-dichlorotoluene

Alcohols and phenols

10. Write the correct IUPAC name for each of the following and classify as primary, secondary, or tertiary for the aliphatic alcohols:
   (a) $CH_3{-}CH_2{-}CH_2{-}CH_2{-}OH$
   (b) 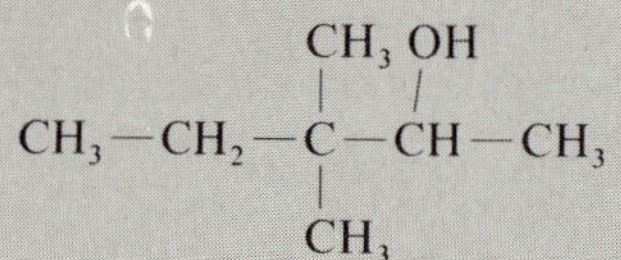
   (c) 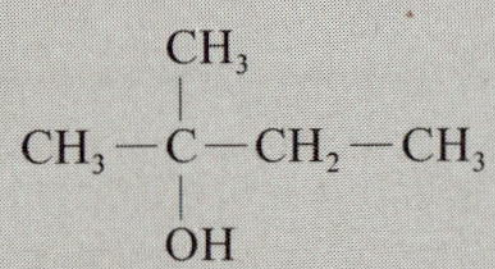
   (d) 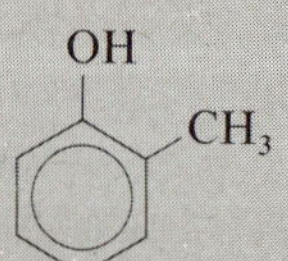
   (e) 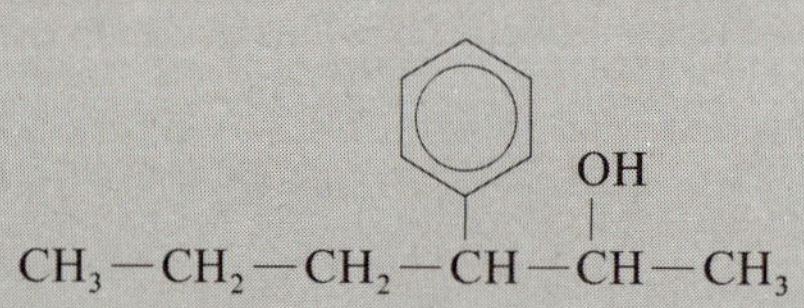
   (f) $CH_3{-}CH_2{-}CH_2{-}OH$
   (g) 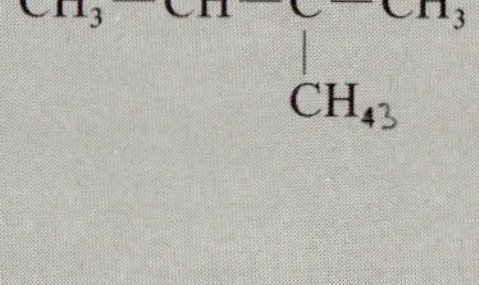

(h) 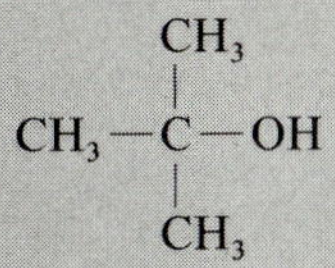

(i) 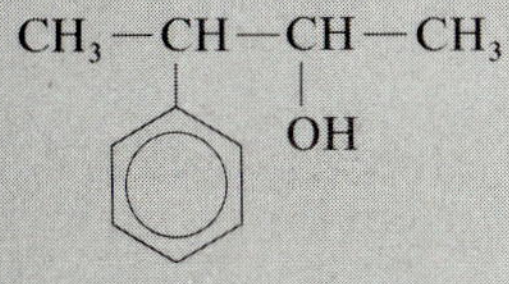

(j) 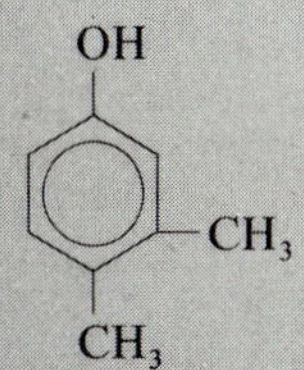

11. Write the formula for each of the following:
    (a) 2-hexanol
    (b) 3-methyl-2-pentanol
    (c) *n*-propyl alcohol
    (d) *tert*-butyl alcohol
    (e) *sec*-propyl alcohol
    (f) 2-chlorophenol
    (g) *p*-nitrophenol
    (h) 2-butanol
    (i) *n*-butyl alcohol
    (j) 2,3-dimethyl-2-hexanol
    (k) *sec*-butyl alcohol
    (l) 3-nitrophenol
    (m) isobutyl alcohol
    (n) *p*-chlorophenol
12. Write the correct IUPAC name for each of the following:
    (a) *tert*-butyl alcohol
    (b) isobutyl alcohol
    (c) *sec*-propyl alcohol
    (d) isopropyl alcohol
    (e) *sec*-butyl alcohol
    (f) *n*-hexyl alcohol

Ethers

13. Name the following:
    (a) $CH_3-O-C_2H_5$
    (b) 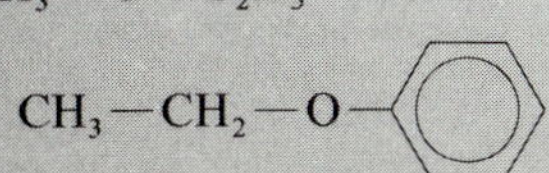
    (c) 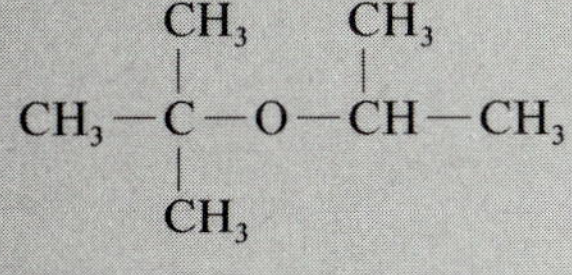
    (d) $CH_3-CH_2-O-CH_2-CH_3$
    (e) 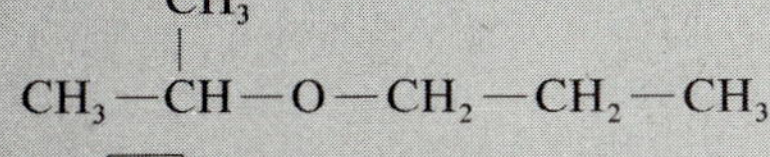
    (f) 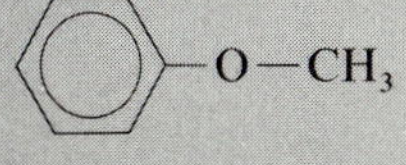

14. Write the correct formula for each of the following
    (a) dimethyl ether
    (b) *sec*-butyl isobutyl ether
    (c) phenyl propyl ether
    (d) diethyl ether
    (e) ethyl phenyl ether
    (f) *tert*-butyl isopropyl ether

Acids

15. Write the correct IUPAC name for each of the following:

(a)
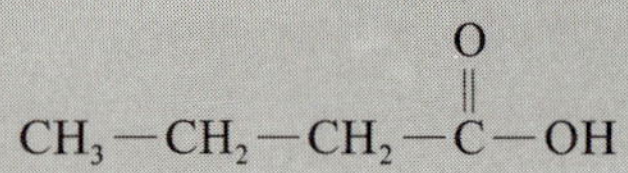

(b)
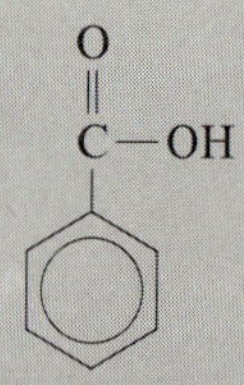

(c)
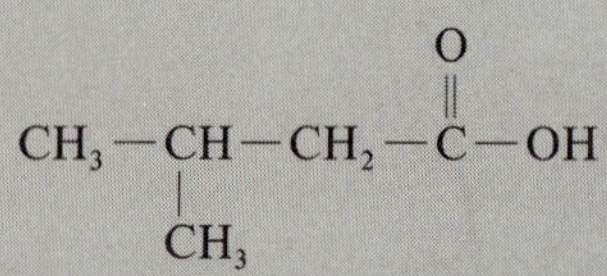

(d)
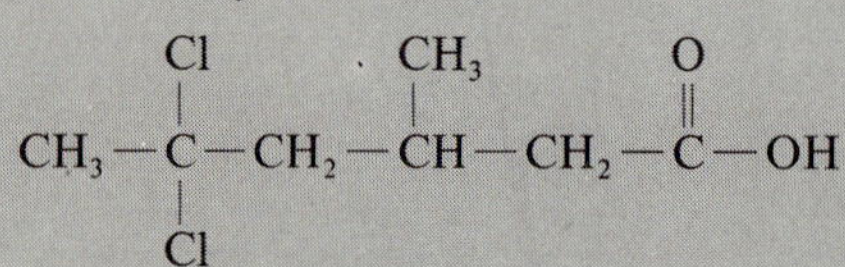

(e)
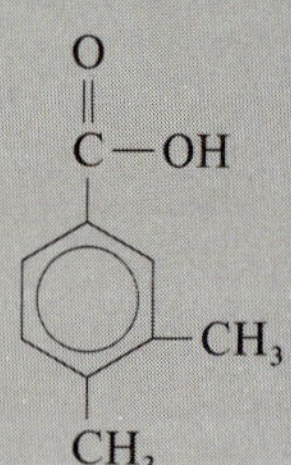

(f)
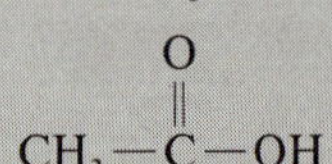

(g)
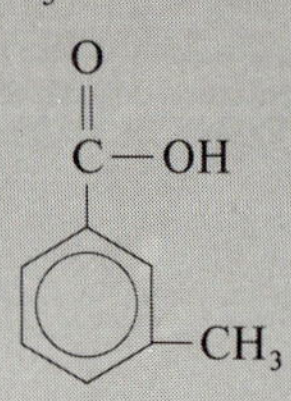

(h)
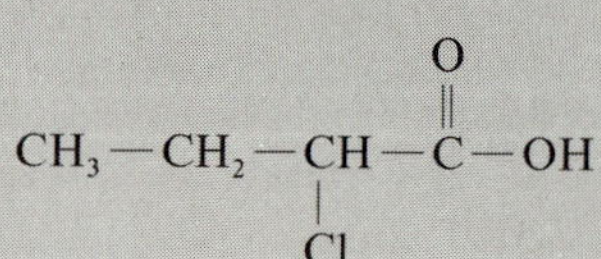

(i)

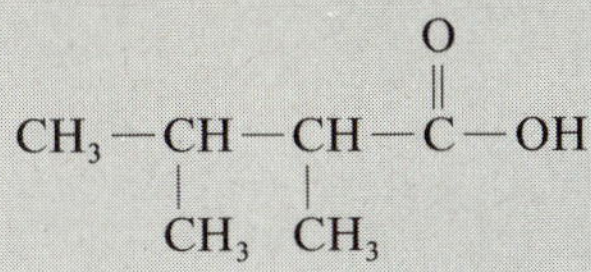

(j)

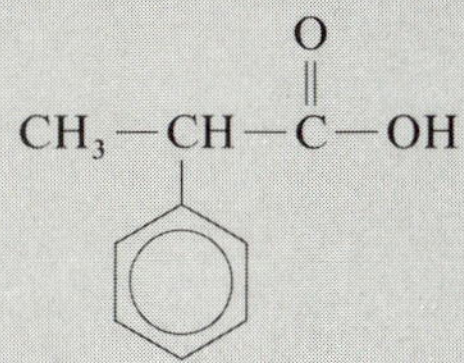

16. Write the correct formula for each of the following:
    (a) acetic acid
    (b) pentanoic acid
    (c) 2-methyl-3,4-dichloroheptanoic acid
    (d) 3-bromobenzoic acid
    (e) α-bromobutyric acid
    (f) formic acid
    (g) hexanoic acid
    (h) 2,3-dimethyloctanoic acid
    (i) α-methylpentanoic acid
    (j) *o*-chlorobenzoic acid

Esters

17. Write the correct IUPAC name for each of the following:

(a)

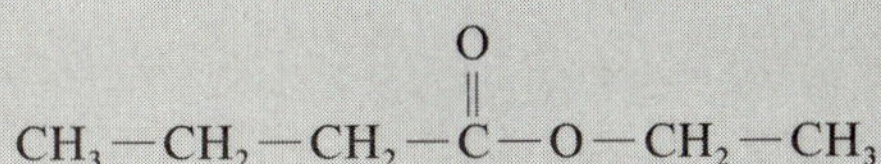

(b)

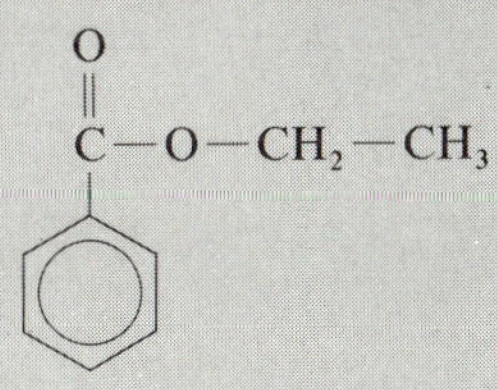

(c)

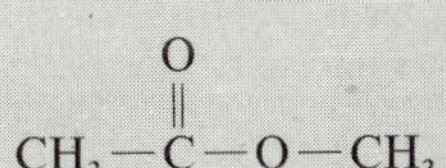

(d)

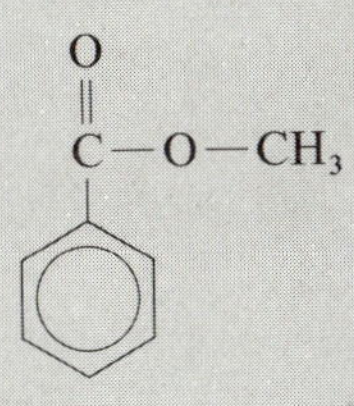

18. Write the structural formula for each of the following:
    (a) isobutyl acetate
    (b) *sec*-butyl formate
    (c) hexyl benzoate
    (d) *sec*-propyl formate
    (e) ethyl benzoate
    (f) methyl-3-chloropentanoate

Acid anhydrides

19. Write the correct IUPAC name for each of the following:

(a) 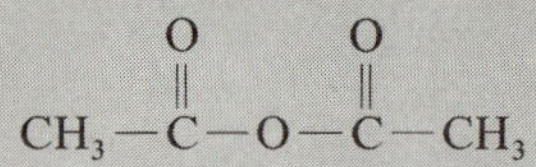

(b) 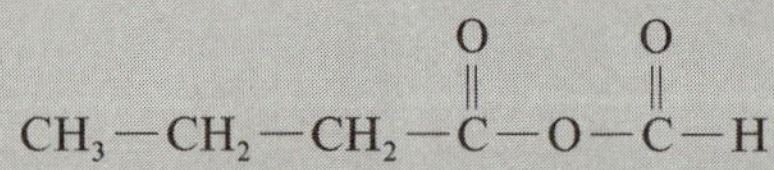

(c) 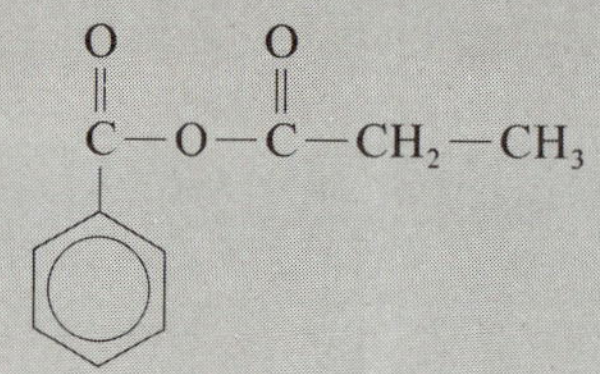

(d) 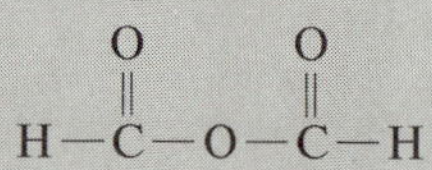

(e) 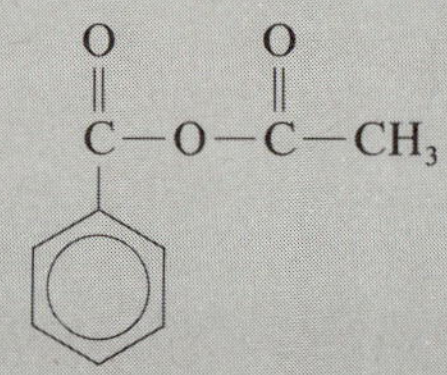

(f) 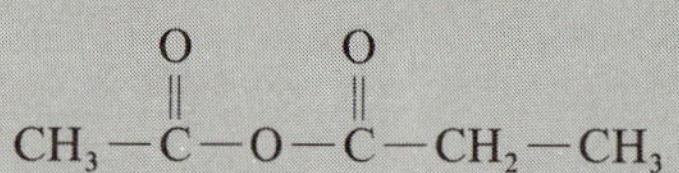

20. Write the correct formula for each of the following:
    (a) benzoic anhydride
    (b) butanoic pentanoic anhydride
    (c) benzoic formic anhydride
    (d) acetic anhydride
    (e) benzoic hexanoic anhydride
    (f) pentanoic anhydride

Amides

21. Write the correct IUPAC name for each of the following:

(a) 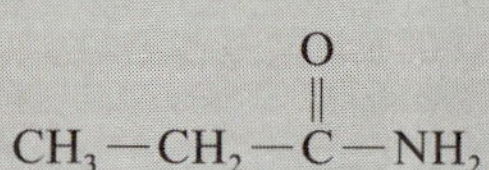

(b) 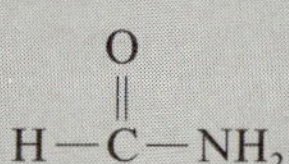

(c) 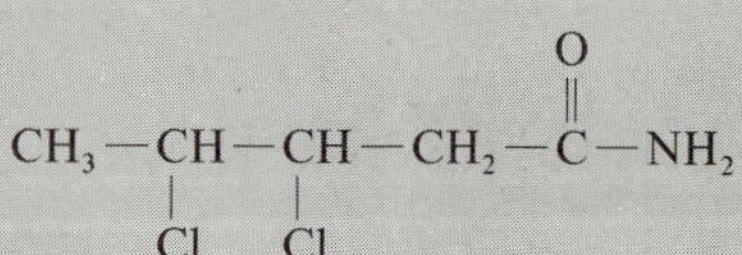

(d)

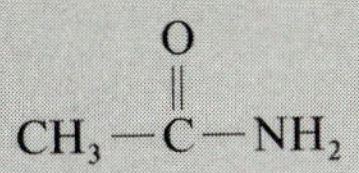

(e)

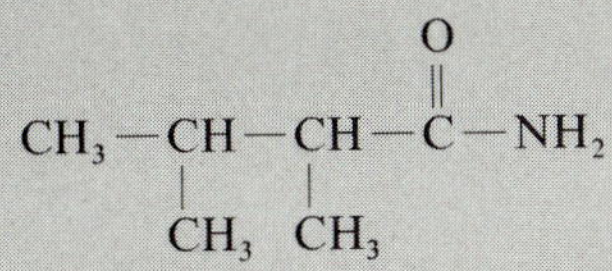

(f)

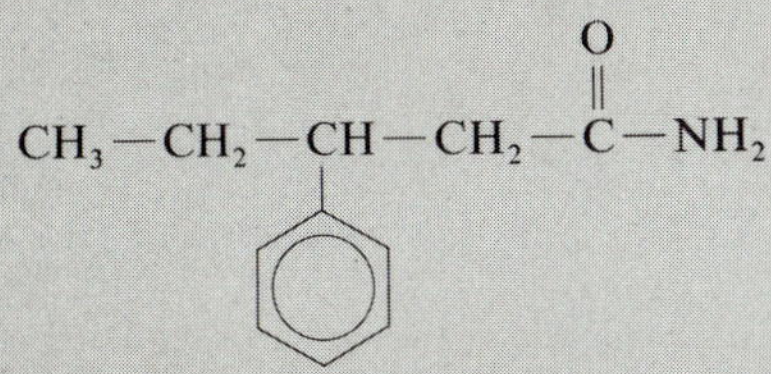

22. Write the structural formula for each of the following:
   (a) acetamide
   (b) butanamide
   (c) 2,3-dimethylhexanamide
   (d) formamide
   (e) hexanamide
   (f) 2-chloro-3-methylpentanamide

Aldehydes and ketones

23. Write the correct IUPAC name for each of the following:

(a)

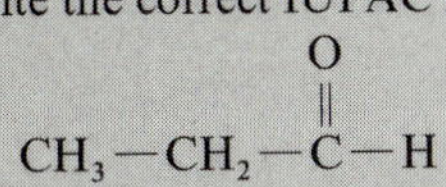

(b)

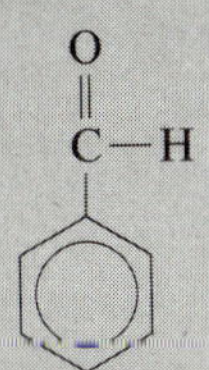

(c)

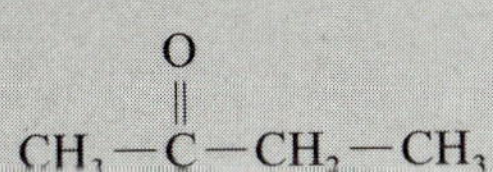

(d)

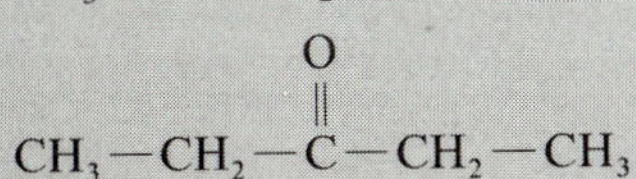

(e)

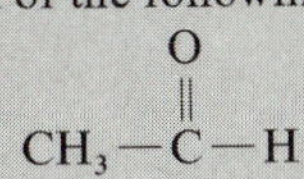

(f)

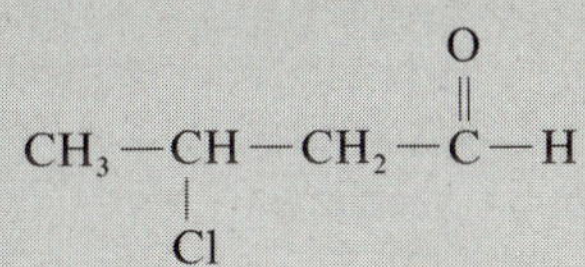

(g)

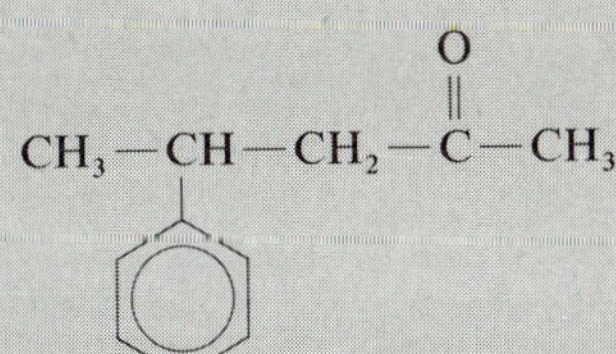

(h)

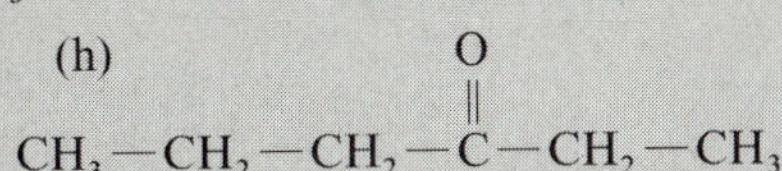

24. Write the structural formula for each of the following:
   (a) formaldehyde
   (b) hexanal
   (c) α-chlorobutyraldehyde
   (d) diphenyl ketone
   (e) methyl ethyl ketone
   (f) 3-heptanone
   (g) pentanal
   (h) acetaldehyde
   (i) 2-methylheptanal
   (j) ethyl phenyl ketone
   (k) 2-butanone
   (l) 3-chloro-2-pentanone

Amines

25. Write the common name for each of the following, and classify as 1°, 2°, or 3°:

(a) CH_3-NH_2

(b) 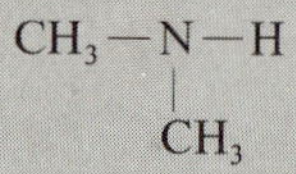

(c) 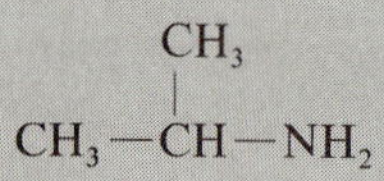

(d) 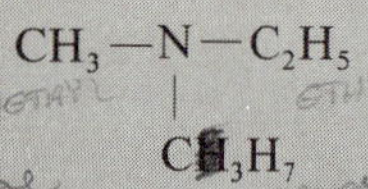

(e) 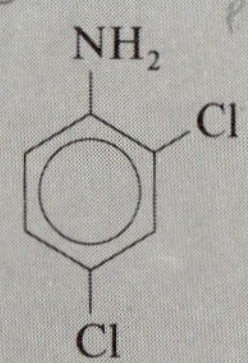

(f) $CH_3-CH_2-NH_2$

(g) 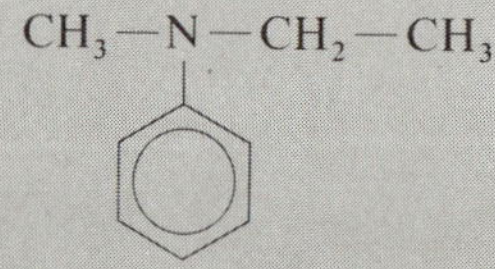

(h) 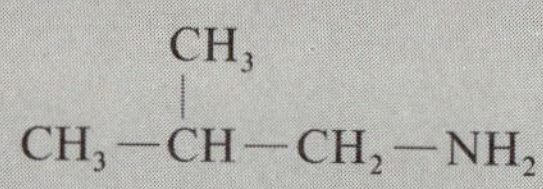

(i) 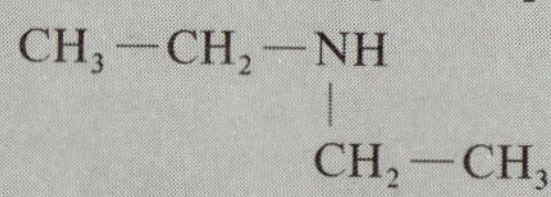

(j) 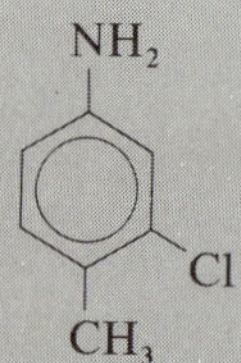

26. Write the structural formula for each of the following:
   (a) diethylamine
   (b) methylethylpropylamine
   (c) isobutylamine
   (d) *p*-chloroaniline
   (e) 2,4,6-trimethylaniline
   (f) methylamine
   (g) ethylphenylamine
   (h) *sec*-butylethylpropylamine
   (i) isopropylamine
   (j) *o*-chloroaniline

General problems

27. Tell if each of the following is an alkane, alkene, or alkyne:
   (a) 1 mole of a substance reacts immediately with 1 mole of Br_2 to form a colorless solution.
   (b) 1 mole of a substance reacts with 2 moles of Br_2 to form a colorless substance.
   (c) A substance only reacts with Br_2 in the presence of a catalyst, such as sunlight.
   (d) 1 mole of a substance reacts with 2 moles of HBr.
   (e) A substance does not react with HBr.
   (f) 1 mole of a substance reacts with 1 mole of HCl.

28. For each of the following, write the correct formula and classify as alkane, alkene, alkyne, aromatic compound, alcohol, ether, acid, ester, anhydride, amide, aldehyde, ketone, primary amine, secondary amine, or tertiary amine.
   (a) 3-methyl-2-hexene
   (b) dimethylacetylene
   (c) methanal
   (d) phenol
   (e) 2-methyl-3-pentanol
   (f) formic acid
   (g) 1,3-octadiene
   (h) dimethylamine

(i) 1,3,5-tribromobenzene
(j) aniline
(k) *p*-dichlorobenzene
(l) 2-hexanone
(m) *tert*-butylamine
(n) ethyl acetate
(o) pentanoic acid
(p) isopropyl alcohol
(q) toluene
(r) 1,2-butadiene
(s) 3-pentanone
(t) ethyl propyl ether
(u) *p*-chloroaniline
(v) methyl ethyl ketone
(w) 3-methylheptanal
(x) isopropylbenzoate

29. Write the correct IUPAC name for each of the following:
(a) isopropyl alcohol
(b) 4,5-dimethyltoluene
(c) methyl ethyl ketone
(d) 2,3-dimethyl-4-hexene
(e) acetaldehyde
(f) *sec*-butyl alcohol
(g) *tert*-butyl alcohol
(h) ethylformate
(i) formaldehyde
(j) isobutyl ethyl ketone
(k) *n*-butyl alcohol
(l) methyl acetate
(m) 5,6-dichlorobenzoic acid
(n) 2-chloro-3-methyl-5-heptyne
(o) isobutyl alcohol
(p) methyl propyl acetylene

REFERENCES

HART, HAROLD, and ROBERT D. SCHUETZ, *Organic Chemistry,* 5th ed., Boston: Houghton Mifflin, 1978.

Objectives

1. To learn what a carbohydrate is and what the difference is between an aldose and a ketose and between a monosaccharide and a polysaccharide.
2. To learn what hydrolysis reactions are, in regard to biochemical processes, and to understand why this is such an important type of reaction.
3. To learn what an amino acid is, to be able to write the general formula for alpha (α) amino acids, and to understand what an essential amino acid is.
4. To learn what proteins are and what functions they serve.
5. To learn how amino acids are joined together to form proteins by a peptide linkage.
6. To learn the difference between the primary and secondary structure of proteins and to understand what the tertiary and quaternary structures refer to.
7. To learn some of the more important properties of proteins and to understand the importance of hydrogen bonds, why proteins are amphoteric and act as buffers, what a zwitterion is, what the isoelectric point is, how proteins undergo hydrolysis and can be denatured, and what the products of hydrolysis are.
8. To learn the difference between a simple and conjugated protein and the classification of proteins according to their function.
9. To learn what lipids, fatty acids, and fats and oils are and what the difference is between a fat and an oil.
10. To learn what enzymes are, why there are so many different ones in the body, why they are essential to life, and how they are classified according to the type of reaction that they catalyze.
11. To learn where nucleic acids occur in the body and what the functions of DNA and RNA are.
12. To learn the difference between digestion and metabolism and between catabolism and anabolism.
13. To understand the basic processes involved in the production of energy by the metabolism of carbohydrates in the body.
14. To learn the two functions of the amino acids that are produced when proteins undergo metabolism in the body and what happens to the ammonia formed in one of these processes.
15. To learn what is meant by nitrogen balance and why this takes place in adults but not in children.
16. To understand how fats and oils produce energy in the body and what purpose stored fats serve.
17. To learn what the study of nutrition involves and the two methods by which nutrients may be classified.
18. To learn the three functions of nutrients and the six chemical types of nutrients.
19. To learn the three main food types that provide energy to the body, what the three body-building materials in the body are, and some of the more important elements necessary to regulate body processes.
20. To consider briefly some important drugs and to begin to understand the general methods by which drugs are used to prevent and to cure diseases and alleviate pain.

Twenty Seven

Introduction to Biochemistry

In this chapter we shall look at some biologically active organic substances. Most of these are very complex compounds but if you watch for the various functional groups that you are familiar with, it will make it easier to understand the structure and some of the simpler reactions of these substances. We shall not attempt to undertake any kind of thorough study of the branch of chemistry referred to as *biochemistry.* Rather, we shall simply consider some common types of biologically important organic compounds and look briefly at how they affect living organisms.

27-1 Carbohydrates

The name *carbohydrate* comes from the fact that, in general, these compounds have formulas that suggest that they are hydrates of carbon; that is, the majority of them have the general formula $C_n(H_2O)_n$. There are some carbohydrates which do not fit this general formula, however, and some compounds which do fit it are not carbohydrates. Therefore, carbohydrates are best defined as naturally occurring polyhydroxy aldehydes (called *aldoses*) and ketones (called *ketoses*) or compounds that yield these substances when reacted with water (hydrolysis).* These compounds include sugars, starches, and cellulose.

The two simplest carbohydrates are 2,3-dihydroxypropanal, more commonly known as glyceraldehyde, and 1,3-dihydroxypropanone, more

*These reactions with water are very common and important for the biologically active compounds that we shall be studying in this chapter. Thus, we shall consider many hydrolysis reactions and the products of hydrolysis.

commonly known as dihydroxyacetone. The structural formulas for these are as follows:

glyceraldehyde (2,3-dihydroxypropanal) dihydroxyacetone (1,3-dihydroxypropanone)

Carbohydrates are generally classified as *monosaccharides* (containing only one sugar unit), *disaccharides* (containing two units), and *polysaccharides* (containing several units, although this term can be used for anything other than monosaccharides). The polysaccharides can be broken down, usually by hydrolysis, to two or more monosaccharide units.

Monosaccharides

Since the monosaccharides contain only one unit, they do not undergo hydrolysis to break down into the component parts as the disaccharides and polysaccharides do. They can be classified either according to the number of carbon atoms that they contain, such as trioses (three carbon atoms), tetroses (four carbon atoms), pentoses (five carbon atoms), hexoses (six carbon atoms), and so forth, or on the basis of the carbonyl group, such as aldoses or ketoses. Often both things are indicated in the classification, such as in an aldotriose. Glyceraldehyde is the simplest aldose (a triose, or aldotriose) and dihydroxyacetone is the simplest ketose (a ketotriose).

Two of the most important monosaccharides are glucose and fructose, which have the following structural formulas:

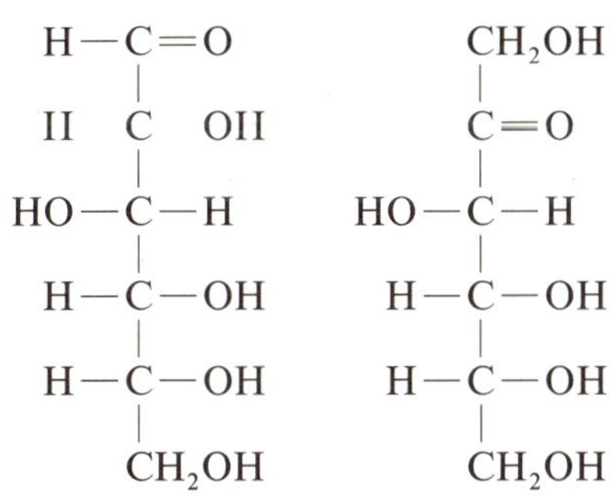

glucose (an aldohexose) fructose (a ketohexose)

Both of these sugars are found in fruits and honey. These two monosaccharides are the units that make up the disaccharide sucrose (table sugar). Glucose is the sugar that is in the blood, and in the urine of people with diabetes. Fructose is the sweetest of all sugars and is found in honey.

Disaccharides

The disaccharides yield two simple units upon hydrolysis. Consider the following examples.

Sucrose This is by far the most common one and is made up of a glucose unit and a fructose unit.

Maltose This contains two glucose units. It is obtained from starch.

Polysaccharides

The most common polysaccharides are made of hexoses and are called *hexosans.* These include starch and cellulose, which are both made up of repeating glucose units, in different arrangements. Starch is an important nutrient for man and is hydrolyzed in the body by enzymes (Section 27-4). Cellulose is the main structural constituent of plants. It is not hydrolyzed by enzymes in man but certain animals called grazing animals do contain the enzymes that will hydrolyze it. Cellulose is important to man in other ways since it is the main constituent of wood, cotton, paper, and so forth.

Formation of Carbohydrates

Carbohydrates are formed from carbon dioxide and water in the cells of plants in a process called *photosynthesis.* In this reaction the energy from sunlight causes the reaction to take place, in the presence of chlorophyll. Chlorophyll is a complex compound, with magnesium as the central atom. There is a series of reactions in photosynthesis, but the overall reaction can be shown as follows, using the molecular formula $C_6H_{12}O_6$ for the monosaccharide, glucose, as an example.

$$6\,CO_2 + 6\,H_2O \xrightarrow[\text{chlorophyll}]{\text{sunlight}} C_6H_{12}O_6 + 6\,O_2$$

Reactions of Carbohydrates

Hydrolysis of polysaccharides forms the simple units that the polysaccharide is made up of. For example, hydrolysis of sucrose gives glucose and fructose, and hydrolysis of starch produces glucose.

Fermentation of carbohydrates, in the presence of a catalyst such as yeast, forms alcohol and carbon dioxide. For example, sucrose undergoes fermentation to produce ethyl alcohol.

Oxidation of sugars causes the carbonyl group $\left(\overset{\displaystyle O}{\overset{\|}{-C-}}\right)$ of the sugar molecule to be oxidized in the Benedict's and Fehling's test for reducing sugars (those that can be oxidized). Since the Cu^{2+} ion is blue in solution, the basis of the test is the disappearance of the blue color upon reaction with a reducing sugar and the formation of a red precipitate of Cu_2O. Since glucose is a reducing sugar, this is the test that is used for sugar in the urine.

Another test for reducing sugars is the Tollens test in which Ag^+ is reduced to metallic silver. All the common monosaccharides and disaccharides are reducing sugars except sucrose. This is due to the fact that the carbonyl $\left(\begin{array}{c} \mathrm{O} \\ \| \\ -\mathrm{C}- \end{array}\right)$ groups from the two units (glucose and fructose) are eliminated when they join together to form the sucrose unit.

27-2 Amino Acids and Proteins

Proteins are the main constituents of living cells and serve many functions—as catalysts to control the rate of biochemical reactions, as regulators for various body functions, to transport oxygen within the body, and as defense mechanisms in the body. They are made up of carbon, hydrogen, oxygen, nitrogen, and usually sulfur and have molecular weights between 10,000 and 50 million or more. In general, they are polymers (many molecules) of amino acids, linked through the nitrogen. The term *peptide* linkage is sometimes used for this and thus proteins are sometimes called *polypeptides*. As with the carbohydrates they can be hydrolyzed (that is, broken down) to the individual amino acids that make them up.

Amino acids are organic compounds that contain both an amino ($-NH_2$) and a carboxyl ($-COOH$) group. All the amino acids that have been isolated from proteins are *alpha amino acids;* that is, the amino group is on the carbon adjacent to the carboxyl group. These have the general formula

$$\begin{array}{ccccccc} & & \mathrm{NH_2} & & \mathrm{O} & & \\ & & | & & \| & & \\ \mathrm{R} & - & \mathrm{CH} & - & \mathrm{C} & - & \mathrm{OH} \end{array}$$

They can have either one or two amino groups and one or two carboxyl groups. The R group can be either aliphatic or aromatic. Several amino acids also have hydroxyl ($-OH$) groups and sulfur atoms in them. There are over 20 different amino acids known.

The simplest amino acid is glycine, in which the R is simply a hydrogen atom. Thus, the formula for glycine is

$$\begin{array}{ccccccc} & & \mathrm{NH_2} & & \mathrm{O} & & \\ & & | & & \| & & \\ \mathrm{H} & - & \mathrm{CH} & - & \mathrm{C} & - & \mathrm{OH} \end{array}$$

In alanine the R group is the methyl group (see Table 27-1), in phenylalanine it is

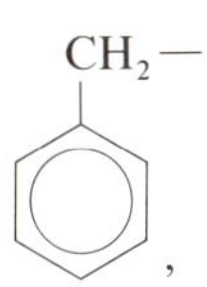

,

in leucine it is the isobutyl group

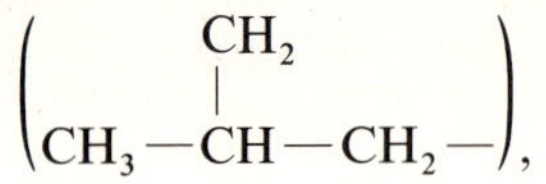

and in isoleucine it is the *sec*-butyl group

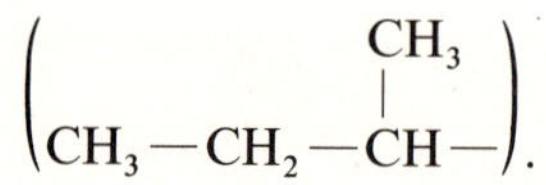

PROBLEM EXAMPLE 27-1

Write the structural formula for isoleucine.

SOLUTION:

Since R is the *sec*-butyl group, the formula is

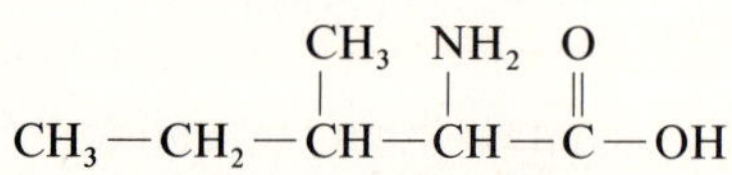

Amino acids can be classified according to the number of amino groups, the number of carboxyl groups, or the other groups they contain, such as hydroxy or sulfur. Table 27-1 gives the formula and name of several common amino acids, showing examples of each possible type.

The **essential amino acids** are those that are required by man but cannot be synthesized by the body (those that can be synthesized by the body are often called *nonessential amino acids*). Therefore it is necessary for the essential amino acids to be included in the diet.

Structure of Proteins

A protein consists of many amino acid molecules joined together by the amino group in one acid and the carboxyl group in another, by splitting off one molecule of water, as shown in the following equation:

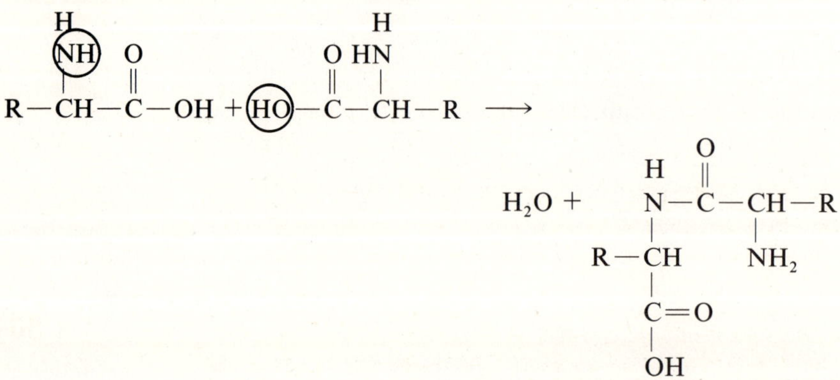

TABLE 27-1 *Classification, name, and structural formula of some amino acids*

CLASSIFICATION	FORMULA	NAME
Monoamino monocarboxylic	$CH_3-\overset{NH_2}{\overset{\vert}{C}}H-\overset{O}{\overset{\Vert}{C}}-OH$	Alanine
Monoamino dicarboxylic	$HO-\overset{O}{\overset{\Vert}{C}}-CH_2-\overset{NH_2}{\overset{\vert}{C}}H-\overset{O}{\overset{\Vert}{C}}-OH$	Aspartic acid
Diamino monocarboxylic	$\overset{NH_2}{\overset{\vert}{C}}H_2-CH_2-CH_2-CH_2-\overset{NH_2}{\overset{\vert}{C}}H-\overset{O}{\overset{\Vert}{C}}-OH$	Lysine
Hydroxy	$\overset{OH}{\overset{\vert}{C}}H_2-\overset{NH_2}{\overset{\vert}{C}}H-\overset{O}{\overset{\Vert}{C}}-OH$	Serine
Sulfur	$\overset{SH}{\overset{\vert}{C}}H_2-\overset{NH_2}{\overset{\vert}{C}}H-\overset{O}{\overset{\Vert}{C}}-OH$	Cysteine

As mentioned at the beginning of this section, this is called a *peptide linkage,* which consists of the $\left(-\overset{H}{\overset{\vert}{N}}-\overset{O}{\overset{\Vert}{C}}-\right)$ part of the molecule. Different proteins consist of various amino acids joined together through many of these linkages; hence, the name *polypeptides* is sometimes used.

There are generally considered to be four parts to the structure of proteins. These are as follows:

1. The kind of amino acids and the order in which they are arranged determines what is often referred to as the *primary structure.*

2. The chains of acids in the protein are often coiled, and the hydrogen in the $-NH_2$ group of one part of the chain forms a hydrogen bond with the oxygen in the carboxyl group ($C{=}O$) of another part of the chain, as $-C{=}O-H\cdots N-$. This is often referred to as the *secondary structure.*

3. The *tertiary structure* has to do with the overall shape of the molecule, determined by the folding of the chain to accommodate other intramolecular interactions.

4. The final determination of the structure of a protein has to do with interactions between protein units to form large molecular aggregates. This is called the *quaternary structure.*

Properties of Proteins

Due to the fact that amino acids contain both an amino group, which is basic, and a carboxyl group, which is acidic, proteins are amphoteric and act as buffers. At low pH the acid group is more active, and at high pH the amino group is more active. In solution the amino acid molecules appear to react with themselves to form a type of double ion, called a *zwitterion;* that is, one part of the molecule becomes positively charged and another part becomes negatively charged. Due to this, amino acids and, hence, proteins generally have high melting points and are soluble in polar solvents, like ionic substances. In acid solution the zwitterion becomes positively charged, and in basic solution it becomes negatively charged according to the following equations:

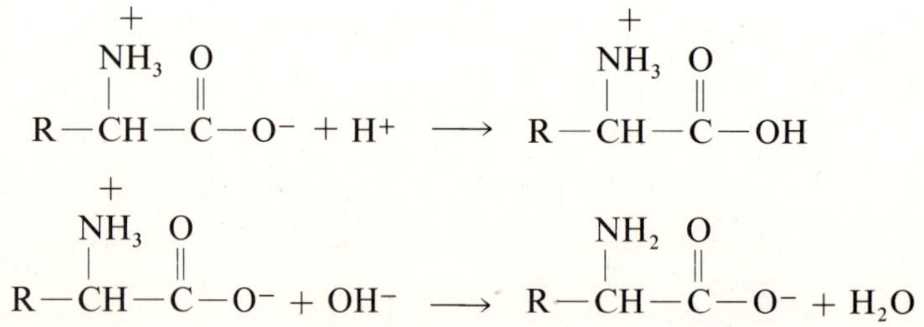

The pH at which the molecule is electrically neutral is called the *isoelectric point* of the protein. At this isoelectric point the protein is least soluble.

Proteins undergo *hydrolysis* in the presence of acids, bases, and certain enzymes to yield the amino acids that the protein is composed of.

Denaturation of proteins is the process whereby the hydrogen bonds in the protein are broken, and thus the structure and reactivity is changed; however, the peptide linkages are not broken. This can be done by heat or by the addition of chemicals such as acids, bases, or alcohols. The cooking of an egg is an example of the denaturation process. When vegetables are overcooked, they are less nutritious since many of the proteins are destroyed by denaturation; also the vitamins are destroyed by overcooking.

Classification of Proteins

Proteins can be classified according to their composition as simple or conjugated. *Simple proteins* give only amino acids upon hydrolysis. *Conjugated* proteins form other substances besides amino acids, such as carbohydrates, phosphoric acid, and nucleic acids (Section 27-5).

Proteins can also be classified according to their function. According to this classification, proteins are one of the following types:

1. Enzymes
2. Hormones
3. Antibodies
4. Structural proteins (found in skin, hair, and so on)
5. Contractile (found in the skeletal muscles)
6. Blood proteins

27-3 Lipids—Fats and Oils

Lipids are fatlike substances which are esters, acids, or alcohols and contain long chain organic groups. Thus, they are insoluble in water but soluble in nonpolar solvents, such as CCl_4. The most common examples of lipids, and the ones that we shall be mainly concerned with, are fats and oils.

Fatty acids are monocarboxylic acids, usually with long carbon chains (from about 10 to 20 carbon atoms). They can be either saturated or unsaturated. Some examples of fatty acids are palmitic acid ($C_{15}H_{33}COOH$) which is found in animal fats and oleic acid ($C_{17}H_{33}COOH$) which is found in olive oil.

Fats and oils are esters of glycerol (a trihydroxy alcohol) and fatty acids. They are formed by the following type of reaction:

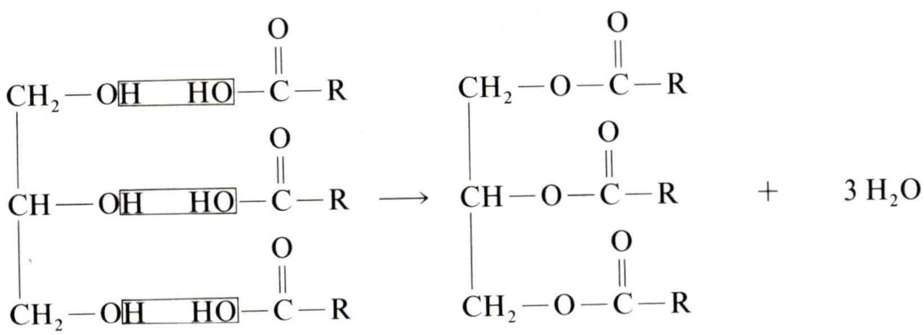

glycerol + fatty acids ⟶ fat or oil

The R groups can be the same or different. If they are all the same, the fat or oil is called a *simple glyceride;* if they are different, it is called a *mixed glyceride.* If the R groups are saturated (all single bonds), the esters are generally solids and are referred to as fats. If the R groups have multiple bonds in them (unsaturated), the melting point tends to be lower and the esters are more likely to be liquids and are thus called oils. For example, palmitic acid, which was mentioned above as being found in animal fats, has the formula $CH_3{-}(CH_2)_{14}{-}COOH$ and oleic acid is an oil with the formula $CH_3{-}(CH_2)_7{-}CH{=}CH{-}(CH_2)_7{-}COOH$.

However, there is no sharp dividing line between fats and oils. The oils can be converted to fats simply by hydrogenating them (reacting with hydrogen in the presence of a catalyst, such as platinum metal) to break the multiple bonds and to form a solid saturated compound (see Section 26-4). Solid shortening and oleomargarine are prepared in this manner from oils such as corn, soybean, peanut, and cottonseed oils. A considerable amount of research has been carried out on the relationship of fat consumption in humans and diseases of the arteries. Unsaturated glycerides (oils or fats obtained from partially hydrogenation of oils) appear to be helpful in lowering the incidence of certain heart diseases.

To find the number of double bonds that there are in a particular fat or oil, which tells the amount of unsaturation, the compound is reacted with iodine. The iodine adds to the double bonds (see Section 26-4) in the same way that hydrogen does in making a saturated fat from an unsaturated oil. By measuring the amount of iodine that will react with a given amount of fat or oil, the number of double bonds can be calculated. This gives what is called the *iodine number.* This number is the number of grams of iodine that will react with 100 grams of fat or oil.

Hydrolysis of fats and oils produces the fatty acids and glycerol that they are composed of. Hydrolysis in alkaline solution, which is called *saponification,* forms glycerol plus an alkali salt of the fatty acids, called a *soap.* To determine the molecular weight of a fat or oil, the number of milligrams of potassium hydroxide that will react with one gram of fat or oil is measured. This is called the *saponification number* of the fat.

The higher the molecular weight of the fat or oil, the fewer molecules that will be present in one gram and hence the less potassium hydroxide will be needed. Therefore, the lower the saponification number of the fat or oil, the higher is the molecular weight.

27-4 Enzymes

Enzymes are a type of protein; they act as catalysts for the many chemical reactions that take place in the body. Some enzymes are simple proteins and others are conjugated (contain nonprotein parts). In many cases these nonprotein parts of the enzyme are actually the active site in the catalysis process and often can be separated from the protein part. In these cases the nonprotein part is called a *coenzyme.*

Each chemical reaction that takes place in the body is catalyzed by a specific enzyme. Approximately 1000 enzymes have been identified. Without these enzymes the chemical reactions would not take place at a fast enough rate to maintain life. Thus, the absence of even one of these enzymes in the body can have serious consequences.

Since enzymes are proteins, high temperatures will cause them to become inactive. Thus, a temperature much above normal body temperature of 98.6°F (37°C) can be harmful to the enzymes in the body. Also, the activity of enzymes is dependent on pH. Each enzyme has a particular pH range at which it is most active. Generally these are around pH 6 to 8, but the range for each particular enzyme is quite small.

Classification of Enzymes

Enzymes are classified according to the type of reaction the enzyme catalyzes. For example, the following are some types of enzymes:

1. Oxidoreductases (catalyze oxidation-reduction reactions).
2. Hydrolases (catalyze hydrolysis reactions).
3. Isomerases (catalyze reactions involving the shift of groups within a molecule to form different structural isomers, such as *cis-* to *trans-* isomer conversions).
4. Synthetases (catalyze synthesis reactions).

27-5 Nucleic Acids

Nucleic acids are produced, along with proteins, upon hydrolysis of *nucleo-proteins,* which are the main constituents of genes and viruses. Hydrolysis of nucleic acids produces phosphoric acid, a nitrogen compound, and a pentose carbohydrate.

There are two types, or classes, of nucleic acids—*ribonucleic acids (RNA)* and *deoxyribonucleic acids (DNA).* Both of these classes of substances are composed of polymers of a phosphate pentose system (see Section 27-1). In the RNA's, the pentose is ribose; in the DNA's, it is 2-deoxyribose. The DNA's have molecular weights from 6 million to over 100 million. The RNA's are smaller, with molecular weights between 20,000 and 40,000.

DNA is composed of two strands, which form a double helix. It replicates itself by uncoiling and as it does this, each strand forms a similar complimentary strand through hydrogen bonding. Therefore, the key to the structure of this depends on the ability of the nitrogen bases within the molecules to form hydrogen bonds.

RNA is composed of single strands that fold back to form sections of double strands, again through hydrogen bonding. The soluble, or transfer, RNA's (*t*RNA) are responsible for selecting specific amino acids from the cell and transferring them to the messenger RNA. The messenger RNA (*m*RNA) transmits the genetic code from the DNA in protein synthesis.

In protein synthesis, the *t*RNA transports a specific amino acid to the *m*RNA; the amino acid on the *t*RNA then undergoes substitution with an amino acid on an adjacent RNA; and so on, forming a large protein molecule.

27-6 Metabolism of Carbohydrates

When food is taken into the body, it first undergoes **digestion,** in which the large molecules that make up the food are broken down into simple compounds. These are then absorbed by the blood and the various tissues in the body. In the tissues the food undergoes chemical changes, in the process called **metabolism.**

There are two types of metabolism. In *catabolism* the nutrients are broken down in order to produce energy. In the other type of metabolism, called *anabolism,* the products of digestion are converted to complex molecules that are either stored or used to produce tissue, enzymes, and other substances.

Carbohydrates are one of the major sources of energy and are usually consumed in the form of starches or sugars. Digestion of these starts with the saliva, is continued to a very small extent in the stomach, but takes place mainly in the small intestine. This converts the carbohydrate to glucose, fructose, and galactose; the fructose and galactose are converted to glucose in the liver. Glucose is the only sugar that circulates in the blood and is therefore known as *blood sugar.* The glucose is either passed directly into the circulatory system and transported to the cells or converted to liver glycogen and stored until it is needed.

In most animals the catabolism of glucose is an *anaerobic* reaction, meaning that no oxygen is involved. This reaction forms acids, which are converted to carbon dioxide and water through a series of reactions called the Krebs (or citric acid) cycle. This produces energy that is utilized by the biological system. In this process, about half the total energy available from the combustion of glucose is utilized by the body.

27-7 Metabolism of Proteins

In the stomach, proteins are denatured and hydrolyzed to polypeptides. This is followed by complete hydrolysis to amino acids in the small intestine. These amino acids are then either used for tissue building or broken down to ammonia, carbon dioxide and water, producing heat and energy in the process. This *deamination* as it is called takes place chiefly in the liver, usually by an oxidation process.

Since the body cannot store protein, which contains nitrogen, the amount of nitrogen ingested in food per day is generally equal to the amount of nitrogen excreted per day. When this is the case, the person is said to be in *nitrogen balance.* This only takes place in adults, however, since growing children have a greater need for amino acids in order to build the body's tissues. Therefore, the chief requirement of protein in the diet is to replace the nitrogen lost through excretion, as well as providing for growth or replacing tissues that have been destroyed.

Since ammonia is toxic, as it is produced in the deamination of the amino acids, it must be removed from the body. In the higher animals it is excreted as urea or as ammonium salts. The formation of urea takes place in the liver where the ammonia combines with carbon dioxide according to the following equation (although the reaction is actually much more complex than this, this is the basic chemical reaction that takes place):

$$2\,NH_3 + CO_2 \longrightarrow \underset{\text{(urea)}}{H_2N-\overset{\overset{\displaystyle O}{\|}}{C}-NH_2} + H_2O$$

27-8 Metabolism of Lipids

Like carbohydrates, fats and oils are also oxidized to produce energy in the body. They are an even richer source of energy than carbohydrates, producing about 9 kcal/g as compared to 4 kcal/g for carbohydrates.

The digestion of fats and oils does not occur until the fat reaches the small intestine. Here it reacts with the biosalts, is then hydrolyzed by pancreatic enzymes, and is absorbed through the wall of the intestine where recombination to triglycerides takes place. These triglycerides then undergo oxidative degradation that allows them to enter the Krebs cycle and to be converted to carbon dioxide and water. In the oxidation of fats, they are probably converted to complex lipids, which are then hydrolyzed to glycerol and fatty acids.

Fats can also be stored in the body. This acts not only as a source of reserve food but also as a support or cushion for the various organs in the body, as well as keeping the body warm.

27-9 Nutrition

The food that we eat consists of carbohydrates, fats, proteins, minerals, and *vitamins*. Vitamins are compounds that are essential to the normal metabolism of an organism but cannot be synthesized by the organism.

As food is digested, carbohydrates are hydrolyzed to monosaccharides, fats are hydrolyzed to glycerol and fatty acids, and proteins are hydrolyzed to amino acids, in each case with the necessary enzymes to catalyze the reactions.

Nutrition may be defined as the science of food and the process by which it is assimilated into the body and provides for growth, maintenance, and repair of living organisms.

There are two methods by which nutrients may be classified. The first of these, which we shall use in our considerations, is according to the function they perform in the body. These functions include

1. Provision of energy.
2. Provision of material for building or maintaining tissue (that is, body-building materials).
3. Provision of substances that act as regulators of body processes.

They could also be classified according to the categories of chemical types. These would be

1. Carbohydrates
2. Fats (lipids)
3. Proteins
4. Vitamins
5. Mineral elements
6. Water

Provision of Energy

All the vital processes of the body require energy to sustain them; this energy is provided by the food taken into the body. As a very rough general rule, it takes approximately 1 kcal*/kg of body weight per hour just

*The calorie referred to in nutrition is actually a kilocalorie. This is sometimes written as Calorie (Cal) and often called a *large calorie*. We shall continue to use kilocalorie (kcal) as we have done throughout this text. You should keep in mind, however, that the calories used in nutrition are actually kilocalories.

to carry on the vital life processes, although this varies with each individual. More energy is also needed, however, for any physical activity and just to provide for the various reactions by which food is converted to energy. Therefore, the total energy provided by food consumed by an individual each day must be sufficient to provide for all of these aspects. For the average adult this is roughly 3000 kcal per day.

These energy requirements are filled by the utilization of three main food types: carbohydrates, fats, and proteins. Table 27-2 compares the heats of combustion for each of these with the energy available to the body through metabolism. The difference in these two values is caused by the energy used up in digestion.

Body-Building Materials

Proteins, in addition to supplying energy, are necessary to supply the amino acids used to build new proteins (recall also that the essential amino acids must be part of the diet since they are not produced by the body). Children have a higher daily requirement of proteins than adults. A minimum requirement for adults is about 0.35 g of protein per kilogram of body weight, which is obtained in foods such as milk, eggs, meat, and fish, as well as certain plant foods such as soybeans.

Calcium constitutes about 2% of the total body weight. It and phosphorus are the two most important nonprotein body-building materials. About 99% of the Ca in the body is found in the bones, where it exists as the phosphate salt. The recommended minimum daily requirement varies from about 400 mg for adults to about 700 mg for children 10 to 15 years old. Milk and cheese are excellent sources of Ca, as well as dark green leafy vegetables and dried beans.

Phosphorus is the fifth most abundant element in the body (after C, H, O, and Ca), making up about 1% of the total weight. It occurs as the phosphate ion, mostly in the bones and teeth, but is also an essential part of every cell. Also, it is essential for the storage of energy. All foods that are rich in protein are also rich in phosphorus.

Regulators of Body Processes

Table 27-3 summarizes the properties of the elements that regulate body processes.

TABLE 27-2 *Heat of combustion versus energy available*

SUBSTANCE	HEAT OF COMBUSTION	ENERGY AVAILABLE FROM METABOLIC OXIDATION
Carbohydrate	4.1 kcal/g	4 kcal/g
Fats	9.4 kcal/g	9 kcal/g
Proteins	5.6 kcal/g	4 kcal/g

TABLE 27-3 *Elements that regulate body processes*

ELEMENT	FORM IT EXISTS IN	AMOUNT IN BODY	ROLE	APPROXIMATE DAILY REQUIREMENT	HOW OBTAINED	WHERE FOUND IN BODY
Na, K, and Cl	Na^+ and K^+ salts		Osmotic pressure, acid-base and water balance	2.5 g Na 2.5 g K 2.0 g Cl	NaCl (salt) K in most foods	K—in cells Na—fluids NaCl—in blood
Mg	Mg^{2+} salts	0.05% of body weight	Enzyme activator	350 mg	In many foods	Bones
S	Organic	0.25% of body weight	As amino acids			Amino acids
Fe	Fe^{2+} ion in compounds	0.004% of body weight	Oxygen carrying constituent of hemoglobin	10 mg	Meat, eggs, green vegetables	Red blood cells
Cu		Trace	Synthesis of hemoglobin	2 mg	Nuts, cereals, fruits, poultry	Blood
Co		Trace	Constituent of vitamin B_{12}			In vitamin B_{12}
I		Trace	Necessary for thyroid gland	0.14 mg	Foods and iodized salt	Thyroid gland
Zn		Trace	In enzymes and insulin	15 mg		With enzymes
Mo F		Trace		Minute 2 mg	Very widely distributed	Bones and teeth

Vitamins

Vitamins are compounds that occur in small quantities in foods. They do not provide energy, but are necessary for the use of other energy producing nutrients. Also, they are not part of the major body structure, but are necessary to regulate the building of these structures for proper growth.

There are two types of vitamins—fat soluble and water soluble ones. The fat soluble ones such as vitamin A, D, E, and K are stored in the body and therefore should not be taken in large amounts. The water soluble vitamins such as C, B_1, B_2, B_{12}, and niacin are not stored and if more is taken into the body than what is needed they just pass through.

Table 27-4 summarizes the properties of some of the more important vitamins.

TABLE 27-4 *Some important vitamins*

VITAMIN	DAILY REQUIREMENT	SOURCE	FUNCTION
C (ascorbic acid)	45 mg	Citrus fruits, green vegetables	Healthy gums, resistance to infection
B_1 (thiamine)	$\frac{0.5\text{ mg}}{1000\text{ cal}}$	Pork, liver, peanut butter whole grain cereals and breads	Digestion, appetite, and mental outlook
B_2 (riboflavin)	1.5 mg	Milk, meat, poultry, fish, legumes, whole grain breads	Skin and vision
B_{12}	3 mg	Meat, fish, poultry, milk, eggs	Formation of red blood cells
Niacin	$\frac{7\text{ mg}}{1000\text{ cal}}$	Whole grain cereal and bread, meat, poultry, fish	Digestion, skin, nervous system
A	5000 I.U.	Liver, milk, eggs	Bones, teeth, skin, night vision
D	400 I.U.	Milk, eggs, fish oils, sunlight	Bones and teeth
E	15 I.U.	Oils, whole grains, nuts	Red blood cells

27-10 Some Important Drugs and Their Action

We now want to look briefly at some chemicals that are used in the prevention, treatment, and cure of diseases and in the alleviation of pain. These chemicals are often referred to as *drugs.* We shall consider these under two classifications: first, pharmocodynamic agents, which are substances that alter normal body functions, either by stimulating or depressing them; second, chemotherapeutic agents, which destroy organisms that are harmful to the body. We shall consider a few of the more common examples of each of these types.

Pharmocodynamic Agents

There are two types of drugs in this category: those acting on the autonomic nervous system and those acting on the central nervous system.

Agents acting on the autonomic nervous system. This consists of the sympathetic system, which controls the mechanism for increased energy requirements of the body under stress, and the parasympathetic system, which controls the more passive processes of the body such as digestion, heartbeat, and salivary secretions. The drugs that affect these nervous systems can be classed as follows:

1. The *adrenergic agents,* which stimulate the sympathetic nervous system.
2. *Cholinergic agents,* which stimulate the parasympathetic system.
3. *Adrenergic blocking agents,* which depress the sympathetic system.
4. *Cholinergic blocking agents,* which depress the parasympathetic system.

Agents acting on the central nervous system. This consists of the brain and spinal cord and controls the sensory and motor systems of the body. Drugs that affect this fall into two classes: *depressants* and *stimulants.* These include the following.

1. *Analgesics* are drugs that affect the pain center in the brain. There are two types. The less potent, nonaddicting ones such as aspirin and the more potent addicting ones such as morphine. Aspirin, or acetylsalicylic acid, is probably the most commonly used drug (not including ethanol). It is both a mild analgesic (relieves pain) but is also an antipyretic (lowers fever). It is not so potent as morphine and is nonaddicting.
2. *Hypnotics and sedatives* are drugs that depress the central nervous system and either produce sleep (hypnotics) or simply promote relaxation (sedatives). The most common is the barbiturates, such as phenobarbital. These also, like morphine, are addicting with prolonged use. Another common one is the drug Dilantin, which is used in the treatment of epilepsy and depresses the central nervous system.
3. *Antihistamines* are drugs that counteract histamine action in the body; histamine causes the typical allergic reactions such as sneezing, coughing, and watering of the eyes.

Chemotherapeutic Agents

Chemotherapeutical agents are substances that can selectively kill or suppress the growth of harmful microorganisms. They are often referred to as *antibiotics.* Some common examples of these are penicillin, tetracyclines, streptomycin, and sulfa drugs.

SUMMARY

1. *Carbohydrates* are polyhydroxy aldehydes and ketones (called aldose and ketose) or compounds yielding these substances upon hydrolysis; they include sugars, starches, and cellulose.
2. Carbohydrates can be *monosaccharides* (containing one sugar unit) or *polysaccharides* (containing two or more units); those containing two units are often called *disaccharides.*
3. *Glucose* is one of the most common monosaccharides; it is the sugar found in the blood and hence is often referred to as blood sugar.
4. *Sucrose* (table sugar) is the most common disaccharide; it consists of a glucose unit and a fructose unit.
5. *Starch* (an important nutrient for man) is a polysaccharide made up of many glucose units.
6. Carbohydrates are formed from CO_2, H_2O, and chlorophyll, in the presence of sunlight, by the process of *photosynthesis.*
7. *Hydrolysis* of a polysaccharide yields the monosaccharides that it is made up of.
8. Carbohydrates undergo *fermentation* to produce alcohols.
9. *Reducing sugars* are identified by becoming oxidized by Cu^{2+} (which is reduced and hence the blue color disappears) in the Fehling's and Benedict's tests.
10. *Proteins* are the main constituents of living cells. They are made up of *amino acids* (acids that contain the amine group as well as the carboxyl group), linked through the nitrogen atom to form very large molecules with molecular weights as high as 50 million.
11. All the amino acids that have been isolated from proteins are *alpha amino acids* and hence have the general formula $\mathrm{R{-}\underset{}{\overset{NH_2}{\overset{|}{C}H}}{-}\overset{O}{\overset{\|}{C}}{-}OH}$; however, they can also have another $-NH_2$ or $-COOH$ as part of the R group. Also, the R group can contain an $-OH$ group or S atoms, and it can be aliphatic or aromatic.
12. The *essential amino acids* are those that are required by man but cannot be synthesized by the body.
13. The amino acids in a protein are joined together through the nitrogen atom on one amino acid and the carboxyl group on another, to form a *peptide linkage.* Hence proteins are sometimes called *polypeptides.*
14. The kind and order of amino acids in a protein determine the *primary structure,* and the type of chain and hydrogen bonding within the molecule determines the *secondary structure.* The *tertiary structure* has to do with the overall shape of the molecule, and the *quaternary structure* refers to interaction between protein molecules.
15. Since proteins contain both an amine group ($-NH_2$, which is basic) and the acid or carboxyl ($-COOH$) group, they are amphoteric and

act as buffers. They also form a type of double ion, called a *zwitterion.* The pH value where it is neutral is called the *isoelectric point.*

16. Proteins undergo *hydrolysis* to yield the amino acids that the protein is composed of; they can be *denatured* by heat, acid, base, or alcohol. This breaks the hydrogen bonds and thus changes the structure and reactivity of the protein.
17. Proteins can be *simple* (only contain amino acids) or *conjugated* (contain other groups such as phosphoric acid, carbohydrates, or nucleic acids). They can be classified according to their function as enzymes, hormones, antibodies, structural (in skin, hair, and so on), contractile (in skeletal muscles), or blood proteins.
18. *Lipids* are fatlike substances that are esters, acids, or alcohols. Fats, oils, and waxes are examples of lipids.
19. *Fatty acids* are long chain monocarboxylic acids that can be saturated or unsaturated.
20. *Fats and oils* are esters of glycerol and fatty acids; all R groups (from the acids) are the same in a *simple glyceride,* and they are different in a *mixed glyceride.* Fats have saturated R groups and in oils the R groups contain multiple bonds; an oil can be converted to a fat by hydrogenating it to form saturated R groups.
21. The number of double bonds in a fat or oil can be determined by the iodine number, which tells how many grams of iodine will react with 100 grams of fat or oil. The *saponification number* is used to determine the molecular weight of a fat, by measuring the number of milligrams of KOH that will react with one gram of fat, to form a *soap* (an alkali metal salt of a fatty acid).
22. *Waxes* are esters of high molecular weight monohydroxy alcohols.
23. *Simple lipids* are esters of fatty acids; *compound lipids* contain other groups (such as phosphoric acid, carbohydrates, and nitrogen compounds) along with the acid and alcohol; *derived lipids* are obtained by the hydrolysis of simple or compound lipids—for example, fatty acids and sterols (high molecular weight cyclic alcohols—such as cholesterol) are derived lipids.
24. *Enzymes* are proteins that act as catalysts for the biochemical reactions that take place in the body. There is a specific one for each reaction. Without these, the biochemical reactions would not take place fast enough to sustain life.
25. Enzymes are classified (and named) according to the type of reaction they catalyze, such as *oxidoreductases* (redox reactions), *hydrolases* (hydrolysis), *isomerases* (reactions that form different isomers of a compound), and *synthetases* (synthesis reactions).
26. The main constituent of genes and viruses in the body are *nucleoproteins,* which consist of proteins and *nucleic acids.* There are two classes of nucleic acids—ribonucleic acids (RNA) and deoxyribonucleic acids (DNA). These are responsible for the genetic code; trans-

fer RNA's (*t*RNA) select specific amino acids from a cell and transfer them to messenger RNA's (*m*RNA), which transmit the genetic code from the DNA in protein synthesis.

27. Food taken into the body first undergoes *digestion,* which breaks up the complex molecules into simple ones, and then *metabolism,* in which the digested food undergoes chemical changes in the tissues. There are two types of metabolism—*catabolism* produces energy from the nutrients and *anabolism* either stores the nutrients or produces proteins and other substances from them.
28. Carbohydrates, which are usually consumed as starches, are one of the main sources of energy in the body. By a series of processes they are converted to glucose and either used to produce energy (by forming CO_2 and water, plus energy, in the Krebs or citric acid cycle) or stored for later use.
29. In the metabolism of proteins, they are hydrolyzed to amino acids, which are then either used for tissue building or broken down to NH_3, CO_2 and H_2O, with the production of energy. The NH_3, which is toxic, is removed by converting it to urea in the liver. In adults there is a *nitrogen balance* between the amount taken in and the amount excreted; however in children this is not the case since some is used in tissue building.
30. Fats and oils are oxidized to produce energy in the body, being an even richer source than carbohydrates. Fats can also be stored in the body; these not only serve as a source of reserve food but also cushion the body organs.
31. The food we eat consists of carbohydrates, proteins, fats, minerals, water, and *vitamins* (compounds that are essential to an organism but cannot be synthesized by the organism).
32. *Nutrients* can be classified according to the following functions; (1) provision of energy, (2) body-building materials, and (3) regulators of the body processes. For the first of these, the three main food types are carbohydrates, proteins, and fats. For the second, proteins, calcium, and phophorus are used. And for the third, elements such as sodium, potassium, chlorine, magnesium, sulfur, and iron are necessary, along with traces of copper, cobalt, iodine, zinc, molybdenum, and fluorine.
33. *Drugs* fall into two categories: *pharmacodynamic agents,* which alter body functions by acting on the nervous system, and *chemotherapeutic agents,* which destroy organisms that are harmful to the body.

Upon completion of this chapter you should be able to do the following:

1. Given any of the following types of compounds, tell briefly what functional groups they contain and what the general structure is: carbohydrate, aldose, ketose, monosaccharide, polysaccharide, amino acid, alpha amino acid, protein, fatty acid, lipid, fat, oil, and wax.

2. Given any of the following types of compounds, tell what products would be produced upon hydrolysis: polysaccharide, protein, and fat.
3. Given any of the following types of substance, tell the function(s) of that substance: enzyme, nucleic acid, carbohydrate, protein, fat, vitamin, mineral element, and drug.
4. Given any of the terms in Exercise 1, give a definition or explanation in your own words.

EXERCISES

1. Define or explain the following terms:
   (a) carbohydrate
   (b) aldose
   (c) ketose
   (d) monosaccharide
   (e) disaccharide
   (f) polysaccharide
   (g) ketotriose
   (h) glucose
   (i) fructose
   (j) sucrose
   (k) starch
   (l) hydrolysis
   (m) photosynthesis
   (n) amino acid
   (o) alpha amino acid
   (p) protein
   (q) peptide linkage
   (r) essential amino acid
   (s) isoelectric point
   (t) zwitterion
   (u) lipid
   (v) simple lipid
   (w) triglyceride
   (x) iodine number
   (y) saponification number
   (z) unsaturated oil
   (aa) fatty acid
   (bb) fat
   (cc) oil
   (dd) wax
   (ee) soap
   (ff) enzyme
   (gg) coenzyme
   (hh) hydrolase
   (ii) DNA
   (jj) RNA
   (kk) *m*RNA
   (ll) *t*RNA
   (mm) metabolism
   (nn) catabolism
   (oo) anaerobic
   (pp) aerobic
   (qq) nitrogen balance
   (rr) anabolism
   (ss) nutrition
   (tt) vitamin
   (uu) drugs
   (vv) pharmacodynamic agents
   (ww) sympathetic system
   (xx) central nervous system
   (yy) antipyretic
   (zz) sedatives
   (aaa) antihistimines
   (bbb) chemotherapeutical agents (antibiotics)
   (ccc) autonomic nervous system
   (ddd) analgesics
   (eee) hypnotics
   (fff) barbiturates
2. Distinguish between each of the following:
   (a) simple and conjugated proteins
   (b) primary and secondary structure of proteins
   (c) simple glyceride and mixed glyceride
   (d) fat and oil
   (e) fat and wax
3. What is produced in the hydrolysis of a polysaccharide?

4. List the main products of each of the following reactions:
   (a) hydrolysis of sucrose
   (b) hydrolysis of starch
   (c) hydrolysis of cellulose
5. How do Benedict's and Fehling's solutions identify reducing sugars?
6. Why are proteins amphoteric?
7. What element is found in proteins that is not found in carbohydrates and lipids?
8. What are the common products of hydrolysis of proteins?
9. In the amino acid valine the R group is the isopropyl group. Write the formula for valine.
10. Tell what the products are for each of the following types of reactions:
    (a) hydrogenation of an oil
    (b) hydrolysis of a fat
11. Why are there so many different enzymes in the body?
12. In general, how are enzymes classified and named?
13. Where are nucleic acids found in the body?
14. What is the main function of DNA and RNA?
15. What is the function of *t*RNA and *m*RNA?
16. What happens to the acid in the citric acid cycle? What is the purpose of this reaction?
17. Are fats and oils a richer or poorer source of energy than carbohydrates? By how much?
18. In what part of the body does the major portion of the digestion of both carbohydrates and fats and oils take place?
19. Outline the steps in the catabolism of fats to CO_2 and H_2O.
20. Outline the steps in the digestion of proteins.
21. Why do children need to take in more nitrogen than they excrete, whereas in adults there is a nitrogen balance?
22. What are the two chief functions of proteins in the body?
23. What are the three final products (along with energy) in the catabolism of protein?
24. What are the main sources of energy in the diet of humans?
25. Nutrients may be classified according to the three types of functions they perform. List these.
26. List the six types of chemicals that nutrients contain.
27. What are the five most abundant elements in the body?
28. In what part of the body is most of the Ca found?
29. What four foods are richest in protein?
30. What foods are richest in calcium?
31. What are the three main body-building materials?
32. What three types of foods provide energy for the body?

REFERENCES

Cheldelin, Vernon H., and R. W. Newburgh, *The Chemistry of Some Life Processes,* New York: Reinhold Publishing Corporation, 1964.

Lappe, Frances Moore, "Protein from Plants," *Chemistry,* October 1973, **46,** p. 8.

Naves, Renee, and Barbara Strickland, "Barbiturates," *Chemistry,* March 1974, **46,** p. 15.

Routh, Joseph I., *Introduction to Biochemistry,* Philadelphia: W. B. Saunders Company, 1971.

Appendix I

Exponents, Scientific Notation, Significant Digits, and Logarithms

I-1 Exponents

An exponent is a whole number or symbol written as a superscript above another number or symbol, the base, denoting the number of times the base must be repeated as a factor. The number of times a base is repeated as a factor is called the *power of the base.* For example, in the symbol x^n, n is the exponent and x is the base. We read x^n as x raised to the nth power, which is equal to

$$\underbrace{x \cdot x \cdot x \cdot \cdot \cdot x}_{n \text{ factors}}$$

We read 10^6 as 10 raised to the sixth power, and it equals

$$\underbrace{10 \cdot 10 \cdot 10 \cdot 10 \cdot 10 \cdot 10}_{\text{six factors}} \quad \text{or} \quad 1{,}000{,}000$$

In *exponential notation,* numbers are written in two parts, a whole number or decimal and a power of 10. For example, 2×10^2, 2.5×10^8, 3.2×10^{-3}, and so forth. Recall that a positive exponent (that is, power of 10) means that the decimal point must be moved to the right and a negative exponent means that it is moved to the left.

Consider the following examples.

$$\begin{aligned} 100 &= 10^2 \\ 10 &= 10^1 \\ 1 &= 10^0 \\ \tfrac{1}{10} = 0.1 &= 10^{-1} \\ \tfrac{1}{100} = 0.01 &= 10^{-2} \end{aligned}$$

Multiplication and Division of Exponential Numbers

In multiplication, the numerical parts are simply multiplied together and the exponents are added. In division, the numbers are divided and the exponents are subtracted algebraically. Consider the following problem examples.

PROBLEM EXAMPLE I-1

Multiply the following numbers:
(a) $(2 \times 10^2)(2 \times 10^3)$
(b) $(3 \times 10^3)(2 \times 10^{-7})$
(c) $(0.02)(4000)$

SOLUTION:

(a) $(2 \times 10^2)(2 \times 10^3) = (2 \times 2)(10^2 \times 10^3) = 4 \times 10^5$
(b) $(3 \times 10^3)(2 \times 10^{-7}) = (3 \times 2)(10^3 \times 10^{-7}) = 6 \times 10^{-4}$
(c) $0.02 = 2 \times 10^{-2}$ $\quad 4000 = 4 \times 10^3$

$$(2 \times 10^{-2})(4 \times 10^3) = 8 \times 10^1 = \mathbf{80}$$

PROBLEM EXAMPLE I-2

Divide the following numbers:
(a) $\dfrac{4 \times 10^8}{2 \times 10^5}$

(b) $\dfrac{6 \times 10^{-5}}{3 \times 10^{-3}}$

(c) $\dfrac{0.008}{20{,}000}$

SOLUTION:

(a) $\dfrac{4 \times 10^8}{2 \times 10^5} = \dfrac{4}{2} \times \dfrac{10^8}{10^5} = \mathbf{2 \times 10^3}$

(b) $\dfrac{6 \times 10^{-5}}{3 \times 10^{-3}} = \dfrac{6}{3} \times \dfrac{10^{-5}}{10^{-3}} = \mathbf{2 \times 10^{-2}}$

(c) $\dfrac{0.008}{20{,}000} = \dfrac{8 \times 10^{-3}}{2 \times 10^4} = \mathbf{4 \times 10^{-7}}$

PROBLEM EXAMPLE I-3

Solve the following:
(a) $\dfrac{(4 \times 10^3)(3 \times 10^4)}{(6 \times 10^5)}$

(b) $\dfrac{(6 \times 10^5)(2 \times 10^{-8})}{(3 \times 10^{-2})(1 \times 10^{-4})}$

(c) $\dfrac{(600)(4000)(0.02)}{(0.002)(0.08)}$

SOLUTION:

(a) $\dfrac{(4 \times 10^3)(3 \times 10^4)}{(6 \times 10^5)} = \dfrac{12 \times 10^7}{6 \times 10^5} = \mathbf{2 \times 10^2}$

(b) $\dfrac{(6 \times 10^5)(2 \times 10^{-8})}{(3 \times 10^{-2})(1 \times 10^{-4})} = \dfrac{12 \times 10^{-3}}{3 \times 10^{-6}} = \mathbf{4 \times 10^3}$

(c) $\dfrac{(600)(4000)(0.02)}{(0.002)(0.08)} = \dfrac{(6 \times 10^2)(4 \times 10^3)(2 \times 10^{-2})}{(2 \times 10^{-3})(8 \times 10^{-2})} = \dfrac{48 \times 10^3}{16 \times 10^{-5}}$
$= \mathbf{3 \times 10^8}$

Addition and Subtraction of Exponential Numbers

In adding or subtracting numbers written in exponential form, the exponents must be the same power. The numbers are then added or subtracted and the exponent, or power of 10, is recorded. For example, to add 2.5×10^2 and 1.20×10^3, one of these must be changed, as

$$(2.5 \times 10^2) + (12.0 \times 10^2) = 14.5 \times 10^2 = 1.45 \times 10^3$$

or

$$(0.25 \times 10^3) + (1.2 \times 10^3) = 1.45 \times 10^3$$

Consider the following problem examples.

PROBLEM EXAMPLE I-4

Add the following numbers:

$$0.020 + (1.0 \times 10^{-3}) + (2.2 \times 10^{-2})$$

SOLUTION:

To add these, they must first all be written as exponents to the same power of 10, as

$$(2.0 \times 10^{-2}) + (0.10 \times 10^{-2}) + (2.2 \times 10^{-2})$$

Then the numbers are added, and the exponent is 10^{-2}, as

$$(2.0 + 0.10 + 2.2) \times 10^{-2} = \mathbf{4.3 \times 10^{-2}}$$

PROBLEM EXAMPLE I-5

Subtract the following numbers:

$$(4.5 \times 10^6) - (3.0 \times 10^5)$$

SOLUTION:

First we must change the 3.0×10^5 to 0.30×10^6 and then subtract, as

$$(4.5 - 0.3) \times 10^6 = \mathbf{4.2 \times 10^6}$$

Powers and Roots of Numbers Written in Exponential Form

To raise a number written in exponential form to a power, such as the second power (or squared) or the third power (or cubed), the exponent is multiplied by the power. Consider the following problem example:

PROBLEM EXAMPLE I-6

Raise each of the following numbers to the indicated power:
(a) $(1 \times 10^3)^2$
(b) $(2 \times 10^4)^3$
(c) $(1 \times 10^{-6})^2$

SOLUTION:
(a) $(1 \times 10^3)^2 = \mathbf{1 \times 10^6}$
(b) $(2 \times 10^4)^3 = (2)^3 \times (10^4)^3 = \mathbf{8 \times 10^{12}}$
(c) $(1 \times 10^{-6})^2 = \mathbf{1 \times 10^{-12}}$

To take the square root of an exponential number, the exponent must be divisible by 2. The square root of the number is found (either on a slide rule, from a table, on a calculator, or by using logs) and the exponent is divided by 2. For example, the square root of 1×10^6 is 1×10^3. Actually, if you will recall that the square root of a number can be written as the number raised to the $\frac{1}{2}$ power, it can readily be seen that we are doing the same thing here as done in the last problem example–multiplying the exponent times $\frac{1}{2}$ in order to raise the number to the $\frac{1}{2}$ power. Thus, in the example given,

$$\sqrt{1 \times 10^6} = (1 \times 10^6)^{1/2} = 1 \times 10^{6/2} = 1 \times 10^3$$

If the given exponent is not divisible by 2, the number must be written in such a way as to give an even exponent. For example, to take the square root of 1.6×10^5, the number is written as 16×10^4, and thus the square root is 4×10^2.

To take the cube root of a number, the exponent must be divisible by 3 (since you are actually raising the number to the $\frac{1}{3}$ power). As a general rule, we can thus say that to take any root of an exponential number the exponent must be divisible by that root; that is, to take the fourth root, it must be divisible by 4, and so forth.

Consider the following problem example.

PROBLEM EXAMPLE I-7

Take the indicated roots of the following numbers:
(a) $\sqrt{4 \times 10^{-8}}$
(b) $\sqrt{2.5 \times 10^{-7}}$
(c) $\sqrt[3]{8 \times 10^{6}}$
(d) $\sqrt[3]{2.7 \times 10^{-8}}$

SOLUTION:

(a) $\sqrt{4 \times 10^{-8}} = (4 \times 10^{-8})^{1/2} = \sqrt{4} \times (10^{-8})^{1/2} = \mathbf{2 \times 10^{-4}}$
(b) $\sqrt{2.5 \times 10^{-7}} = \sqrt{25 \times 10^{-8}} = \mathbf{5 \times 10^{-4}}$
(c) $\sqrt[3]{8 \times 10^{6}} = (8 \times 10^{6})^{1/3} = \sqrt[3]{8} \times (10^{6})^{1/3} = \mathbf{2 \times 10^{2}}$
(d) $\sqrt[3]{2.7 \times 10^{-8}} = \sqrt[3]{27 \times 10^{-9}} = \mathbf{3 \times 10^{-3}}$

I-2 Scientific Notation

In *scientific notation* the numbers are always written as a number greater than 1 and less than 10, with an exponent. Thus, there will always be one (and only one) digit to the left of the decimal.

Consider the following problem example:

PROBLEM EXAMPLE I-8

Express the following numbers in correct scientific notation:
(a) 250
(b) 0.00375
(c) 45×10^{8}
(d) 18.7×10^{-6}

SOLUTION:

Writing each as a number between 1 and 10 (that is, one number to the left of decimal point) with an exponent gives
(a) $250 = \mathbf{2.5 \times 10^{2}}$
(b) $0.00375 = \mathbf{3.75 \times 10^{-3}}$
(c) $45 \times 10^{8} = \mathbf{4.5 \times 10^{9}}$
(d) $18.7 \times 10^{-6} = \mathbf{1.87 \times 10^{-5}}$

I-3 Significant Digits

The significant digits in a number are the number of digits that give reasonably reliable information; that is, it is only the last digit that is in doubt. For example, if a number is expressed as 16.2 cm, it is only the 2 that is in doubt, and we can assume that this measurement was made accurately to within a few tenths. If this number had been expressed as 16.25 cm, however, then it would be only the 5 that is uncertain, and in this case we can assume that the measurement was made accurately to within a few hundredths. In the first case, we would, therefore, say that there are three significant digits; in the second case, there are four.

Significant digits are especially important any time experimental measurements are made and used in calculations. The proper number of significant digits must be made and used in calculations. The proper number of significant digits must be used to indicate how accurately the measurement was made, and in doing the calculations the final answer cannot contain any more significant digits than the least accurate measurement. For example, using a thermometer that is accurate to the nearest tenth (0.1) of a degree, the temperature of a substance appears to be exactly 75°. This should then be recorded as 75.0°, to indicate that it is the first decimal place that is in doubt. If it were simply recorded at 75°, it would appear as though the 5 were in doubt; if it were recorded as 75.00°, it would indicate that the thermometer could be read to hundredths of a degree.

To determine what digits are significant in a number, it is usually only necessary to count the number of digits expressed. The only exception to this is zeros before or after other numbers; if the zeros are between other numbers, they are always significant. *If the zero is only used to show where the decimal point is, it is not significant.* Therefore, if a zero before or after other numbers is not needed to show where the decimal point is, it must be significant; that is, it is there to show how accurately the measurement was made.

Terminal zeros, such as in 200, may be specified with a notation such as a bar above the last significant digit, such as $20\bar{0}$, which means the last one is significant. Therefore, in $20\bar{0}$ there are three significant digits. If they are not indicated, there is no way of knowing if the zeros are significant or not. Consider the following examples.

Determine the number of significant digits in the following numbers:

1. 25.27 = four significant digits
2. 200.5 = four significant digits
3. 0.0025 = two significant digits (the zeros before the 25 *only* show where the decimal point is)
4. 0.0250 = three significant digits (the 0 *after* the 25 is *not* needed to show where the decimal is; therefore, it is significant)
5. $600\bar{0}$ = four significant digits (since the last 0 is significant, as indicated by the bar, the others must be also)
6. 300—this is ambiguous; there is at least one (the 3) but could be two or three; there is no way of knowing from the way it is written.

Use of Significant Digits in Calculations

In *addition and subtraction,* the answer must not contain any more places (i.e., decimal, units, tens, and so on) than the smallest place common to all the numbers added or subtracted. The sum of 25.1 + 22.11 is 47.21, but the answer must be expressed to only the tenths decimal place since only the tenths decimal place is common to the two numbers; hence, the answer is 47.2. The reason becomes obvious if you note that the hundredths decimal place is not measured in the number 25.1 and, thus, could vary widely. Likewise, the difference of 4.732 − 3.62 is 1.112, but the answer must be expressed to only the hundredths decimal place because only the hundredths decimal place is common to the two numbers; hence, the answer is 1.11.

In *multiplication and division,* the answer must not contain any more significant digits than the least number of significant digits in the numbers used in the multiplication or division. The product of 22 × 2 is 44, but the answer must be expressed to only one significant digit since 2 has only one significant digit; hence, the answer is 40. The quotient of $\frac{22}{2}$ is 11, but the answer again must be expressed to only one significant digit since 2 has only one significant digit; hence, the answer is 10.

In using significant digits correctly, you must also keep in mind that some numbers are exact. For example, two atoms means exactly two, and thus this number has an infinite number of significant digits. Therefore in using a number of this type the two does not limit the number of significant digits. For example, if the mass of two atoms is 8.006 amu and we want to find the mass per atom, the answer is

$$\frac{8.006\text{ amu}}{2\text{ atoms}} = 4.003\frac{\text{amu}}{\text{atom}}$$

not 4 amu/atom.

Rounding Off

In rounding off the nonsignificant digits so as to arrive at the desired significant digits, the following rule may be applied:

If the nonsignificant digit is less than 5, it is dropped and the last significant digit remains the same. If it is greater than 5, the nonsignificant digits are dropped and the significant digit is increased by 1. If it is equal to 5, there are two rules that can be followed (you should choose one of these and use it consistently–this will tend to average out and not make any significant difference overall):

1. Simply increase the significant digit by 1 for a nonsignificant digit of 5 or greater.
2. If the last significant digit is even, leave it; if it is odd, increase it by 1 (to make it even).

Consider the following problem example.

PROBLEM EXAMPLE I-9

Solve the following and express the answer to the correct number of significant digits.
(a) 17.8 + 14.73 + 16

(b) $14.72 - 6.8$
(c) 13×752
(d) $\dfrac{180.8}{75}$

SOLUTION:

(a) The smallest place common to all three numbers is the units place; hence, the answer must be expressed to the units place.

$$\begin{array}{l} \downarrow\text{——(smallest common place)} \\ 17.8 \\ 14.73 \\ \underline{16} \\ 48.53 \end{array}$$

Rounded off to the units place, 48.53 is **49.**

(b) The smallest place common to the two numbers is the tenths decimal place; hence, the answer must be expressed to the tenths decimal place.

$$\begin{array}{r} \downarrow\text{——(smallest common place)} \\ 14.72 \\ \underline{-\ 6.8} \\ 7.92 \end{array}$$

Rounded off to the tenths decimal place, 7.92 is **7.9.**

(c) The number with the least number of significant digits is 13, which has only two; hence, the answer must be expressed to two significant digits.

$$752 \times 13 = 9776$$

Rounded off to two significant digits, 9776 is **9800.**

(d) The number with the least number of significant digits is 75, which has two; hence, the answer must be expressed to two significant digits. The division should be carried out to one more place than the required number of significant digits:

$$\frac{180.8}{75} = 2.41$$

Rounded off to two significant digits, 2.41 is **2.4.**

I-4 Logarithms

The common logarithm of a number is defined as the exponent to which 10 must be raised to give the number (N), as $N = 10^{\text{exponent}}$, where exponent $= \log N$. Thus, $\log 10^3 = 3$, $\log 10^{-4} = -4$, $\log 1 = 0$ (since $1 = 10^0$ or any nonzero number raised to the 0 power). A number that cannot be represented as 10 raised to an integer power will have a logarithm that is not an integer. Logarithms of such numbers must be obtained from Logarithms of Numbers, as found in Appendix VII. For example, the logarithm

4.40 is given in Appendix VII as 0.6435. Read down the N column to 4.4, and then go to the 0 column. Therefore, $4.40 = 10^{0.6435}$. Consider another example: The logarithm of 4.45 is 0.6484. Again, read down the N column to 4.4, and then go to the 5 column. Therefore, $4.45 = 10^{0.6484}$. Consider one more example: The logarithm of 8.62 is 0.9355. Again, read down the N column to 8.6 and then go to the 2 column. Therefore, $8.62 = 10^{0.9355}$.

For numbers that are not written in scientific notation (that is, one number before the decimal and a power of 10 or exponent), they must first be written in this form and then the log table is used. For example, to find the log of 0.001, it is first written as 1.0×10^{-3}. As another example, consider the number 0.0257. This is first written as 2.57×10^{-2}; then the log of each part of the number is found and they are added together (recall that to multiply exponential numbers the exponents are added). Thus the log of this number is $0.4099 - 2 = -1.5901$.

PROBLEMS

The answers to these are given (in parentheses) immediately following the problem. To use these, cover the answer, write your answer, and then check it.

Exponents and scientific notation

1. Express the following in scientific notation:
   (a) 7,020,000 (7.02×10^6)
   (b) 847 (8.47×10^2)
   (c) 7860 (7.86×10^3)
   (d) 72.4 (7.24×10^1)
   (e) 0.174 (1.74×10^{-1})
   (f) 0.000462 (4.62×10^{-4})
   (g) 0.00323 (3.23×10^{-3})
   (h) 0.0197 (1.97×10^{-2})
   (i) 0.0000000264 (2.64×10^{-8})
   (j) 0.00000184 (1.84×10^{-6})
2. Perform the following operations and express the answer in scientific notation:
   (a) $3.24 \times 10^3 + 2.40 \times 10^3 =$ (5.64×10^3)
   (b) $4.76 \times 10^2 + 8.7 \times 10^1 =$ (5.63×10^2)
   (c) $3.71 \times 10^{-2} + 2 \times 10^{-4} =$ (3.73×10^{-2})
   (d) $4.71 \times 10^{-3} - 2.61 \times 10^{-3} =$ (2.10×10^{-3})
   (e) $7.68 \times 10^{-5} - 9.7 \times 10^{-6} =$ (6.71×10^{-5})
   (f) $4.74 \times 10^6 - 3 \times 10^4 =$ (4.71×10^6)
   (g) $1.10 \times 10^4 \times 4.00 \times 10^3 =$ (4.40×10^7)
   (h) $6.00 \times 10^7 \times 3.00 \times 10^3 =$ (1.80×10^{11})
   (i) $8.40 \times 10^5 \times 2.00 \times 10^{-3} =$ (1.68×10^3)
   (j) $7.10 \times 10^{-3} \times 4.00 \times 10^2 =$ (2.84)
   (k) $\dfrac{3.00 \times 10^4}{2.00 \times 10^2} =$ (1.50×10^2)

(l) $\dfrac{2.40 \times 10^4}{6.00 \times 10^6} =$ (4.00×10^{-3})

(m) $\dfrac{6.40 \times 10^{-5}}{2.00 \times 10^2} =$ (3.20×10^{-7})

(n) $\dfrac{1.80 \times 10^5}{3.00 \times 10^{-3}} =$ (6.00×10^7)

(o) $\sqrt{9.00 \times 10^{-6}} =$ (3.00×10^{-3})
(p) $\sqrt{3.60 \times 10^4} =$ (1.90×10^2)
(q) $\sqrt{6.00 \times 10^5} =$ (7.75×10^2)
(r) $\sqrt{8.00 \times 10^{-3}} =$ (8.94×10^{-2})
(s) $\sqrt{50.0 \times 10^3} =$ (2.24×10^2)
(t) $\sqrt{30.0 \times 10^{-7}} =$ (1.73×10^{-3})

Significant digits

3. Tell how many significant digits there are in each of the following numbers:
(a) 102 (3)
(b) 2.000 (4)
(c) 0.015 (2)
(d) 45 (2)
(e) 45.0 (3)
(f) 0.003 (1)
(g) 0.0030 (2)
(h) 42.03 (4)
(i) 1.5×10^5 (2)
(j) 7.32×10^{-8} (3)

4. Express the answer to the correct number of significant digits:
(a) 45.2×6.00 (271)
(b) 0.005×3.25 (0.02)
(c) $\dfrac{3.2}{4.500}$ (0.71)
(d) $\dfrac{0.020}{4.325}$ (0.0046)
(e) 4.200×0.0026 (0.011)
(f) $\dfrac{1.2 \times 10^5}{4.37 \times 10^8}$ (2.7×10^{-4})
(g) $\dfrac{1.750 \times 10^{-8}}{8.50 \times 10^5}$ (2.06×10^{-14})
(h) $\dfrac{0.0022}{1.52}$ (0.0014)

Logarithms

5. Find the log of each of the following numbers:
(a) 2.42 (0.3838)
(b) 3.00 (0.4771)
(c) 6.40 (0.8062)
(d) 9.045 (0.9564)
(e) 4.25×10^5 (5.6284)
(f) 2.63×10^{-9} (−8.5800)
(g) 2500 (3.3979)
(h) 0.000725 (−3.1397)

Appendix II

Some Conversion Factors

Some Conversion Factors

1 angstrom (Å or A) = 10^{-8} cm = 10^{-10} m
1 amu = 1.66×10^{-24} g
1 atm = 760 torr (mm Hg) = 14.7 lb in.$^{-2}$ (psi)
1 erg = 2.39×10^{-8} cal
1 electron volt (ev) = 1.6×10^{-12} erg
1 joule = 1×10^{7} ergs
1 cal = 4.18 J = 4.18×10^{7} ergs
1 Btu = 252 cal

Appendix III

SI Units

III-1 Base Units

Length–meter (m)
Mass–kilogram (kg)
Time–second (s)
Temperature–kelvin (K*)
Amount–mole (mol)
Electric current–ampere (A)

III-2 Derived Units

Volume–m^3
(or liter, $\ell = 10^{-3}\ m^3$; therefore, $1\ m\ell = 1\ cm^3$)
Density–kg/m^3 (or $kg\ m^{-3}$)
Speed or velocity–m/s ($m\ s^{-1}$)
Concentration–mol/m^3 ($mol\ m^{-3}$)
Force–newton ($N = kg\ m/s^2$)
Pressure–pascal ($Pa = N/m^2$)
Energy–joule ($J = kg\ m^2/s^2$)
Power–watt ($W = J/s^2$)
Quantity of electricity–coulomb (coul = A s)
Electric potential (emf)–volt (V = W/A)

*For units that are named after a person, the abbreviation is capitalized but not the name.

Appendix IV

Thermodynamic Values-Enthalpy(ΔH), Free Energy(ΔG), and Entropy(S)

*Thermodynamic Values—
Enthalpy (ΔH),
Free Energy (ΔG),
and Entropy (S)*

SUBSTANCE	ΔH_f^0 (at 25°C, kcal/mol)	ΔG^0 (at 25°C, kcal/mol)	S^0 (at 25°C, cal/°K-mol)
Li	0.00	0.00	6.73
Na	0.00	0.00	12.24
K	0.00	0.00	15.34
Mg	0.00	0.00	7.81
Ca	0.00	0.00	9.95
B	0.00	0.00	1.40
Al	0.00	0.00	7.77
C (graphite)	0.00	0.00	1.36
C (diamond)	0.45	0.69	0.58
Si	0.00	0.00	4.50
Sn (white)	0.00	0.00	12.3
Pb	0.00	0.00	15.49
P (white)	0.00	0.00	9.82
S (rhombic)	0.00	0.00	7.60
Fe	0.00	0.00	6.49
Co	0.00	0.00	6.8
Ni	0.00	0.00	7.20
Cu	0.00	0.00	7.96
Zn	0.00	0.00	9.95
Ag	0.00	0.00	10.21
H_2 (gas)	0.00	0.00	31.21
O_2 (gas)	0.00	0.00	49.00
F_2 (gas)	0.00	0.00	48.44
Cl_2 (gas)	0.00	0.00	53.29
N_2 (gas)	0.00	0.00	6.96
H_2O (gas)	−57.80	−54.64	45.11
HCl (gas)	−22.06	−22.77	44.62
CO (gas)	−26.42	−32.81	47.30
CO_2 (gas)	−94.05	−94.26	51.06
NH_3 (gas)	−11.04	−3.976	46.01
SO_2 (gas)	−70.76	−71.79	59.40
SO_3 (gas)	−94.45	−88.52	61.24
NO (gas)	−21.60	−20.72	50.34
NO_2 (gas)	9.09	12.39	57.47
C_2H_2 (gas)	54.19	50.00	48.00
C_2H_4 (gas)	12.50	16.28	52.45
C_2H_6 (gas)	−20.24	−7.86	54.85
CH_4 (gas)	−17.89	−12.14	44.50

Appendix V

Vapor Pressure of Water at Various Temperatures

Vapor Pressure of Water at Various Temperatures

TEMPERATURE (°C)	PRESSURE (torr)	TEMPERATURE (°C)	PRESSURE (torr)
0	4.6	33	37.7
5	6.5	34	39.9
10	9.2	35	42.2
15	12.8	36	44.6
16	13.6	37	47.1
17	14.5	38	49.7
18	15.5	39	52.4
19	16.5	40	55.3
20	17.5	45	71.9
21	18.6	50	92.5
22	19.8	55	118.0
23	21.1	60	149.4
24	22.4	65	187.5
25	23.8	70	233.7
26	25.2	75	289.1
27	26.7	80	355.1
28	28.3	85	433.6
29	30.0	90	525.8
30	31.8	95	633.9
31	33.7	100	760.0
32	35.7		

Appendix VI

Equilibrium Constants (at 20°C)

VI-1
Solubility Product Constants

SUBSTANCE	K_{sp}
Acetates	
Silver acetate	4.0×10^{-3}
Bromates	
Silver bromate	6.0×10^{-5}
Bromides	
Lead(II) bromide	4.6×10^{-6}
Mercury(I) bromide	1.3×10^{-22}
Silver bromide	5.0×10^{-13}
Carbonates	
Barium carbonate	1.6×10^{-9}
Calcium carbonate	4.8×10^{-9}
Copper(II) carbonate	2.5×10^{-10}
Lead(II) carbonate	1.5×10^{-13}
Magnesium carbonate	4.0×10^{-5}
Mercury(I) carbonate	9.0×10^{-17}
Nickel carbonate	7.0×10^{-9}
Silver carbonate	8.2×10^{-12}
Strontium carbonate	7.0×10^{-10}
Zinc carbonate	6.0×10^{-11}
Chlorides	
Lead(II) chloride	1.6×10^{-5}
Mercury(I) chloride	1.1×10^{-18}
Silver chloride	1.8×10^{-10}
Chromates	
Barium chromate	1.2×10^{-10}
Calcium chromate	7.1×10^{-4}
Lead(II) chromate	2.0×10^{-14}
Mercury(I) chromate	2.0×10^{-9}
Silver chromate	4.0×10^{-12}
Strontium chromate	3.6×10^{-5}
Fluorides	
Barium fluoride	2.4×10^{-5}
Calcium fluoride	3.2×10^{-11}
Lead(II) fluoride	3.2×10^{-8}
Magnesium fluoride	6.5×10^{-9}
Strontium fluoride	7.9×10^{-10}
Hydroxides	
Aluminum hydroxide	5.0×10^{-33}
Calcium hydroxide	8.0×10^{-6}
Chromium(III) hydroxide	7.0×10^{-31}
Chromium(II) hydroxide	1.0×10^{-17}
Cobalt(II) hydroxide	2.0×10^{-16}
Copper(II) hydroxide	1.6×10^{-19}
Iron(III) hydroxide	4.0×10^{-38}
Iron(II) hydroxide	2.0×10^{-15}
Lead(II) hydroxide	4.0×10^{-15}

SUBSTANCE	K_{sp}
Hydroxides (cont.)	
Magnesium hydroxide	8.9×10^{-12}
Manganese(II) hydroxide	2.0×10^{-13}
Mercury(II) hydroxide	3.0×10^{-26}
Nickel hydroxide	1.6×10^{-16}
Silver hydroxide	2.0×10^{-8}
Tin(II) hydroxide	3.0×10^{-27}
Zinc hydroxide	5.0×10^{-17}
Iodides	
Lead(II) iodide	8.0×10^{-9}
Mercury(I) iodide	4.0×10^{-29}
Silver iodide	8.5×10^{-17}
Oxalates	
Barium oxalate	1.5×10^{-8}
Calcium oxalate	1.3×10^{-9}
Iron(II) oxalate	2.0×10^{-7}
Magnesium oxalate	8.6×10^{-5}
Manganese(II) oxalate	1.0×10^{-15}
Silver oxalate	1.0×10^{-11}
Strontium oxalate	5.6×10^{-8}
Zinc oxalate	1.5×10^{-9}
Phosphates	
Aluminum phosphate	6.0×10^{-19}
Barium phosphate	6.0×10^{-39}
Calcium phosphate	2.0×10^{-29}
Iron(III) phosphate	1.0×10^{-22}
Strontium phosphate	1.0×10^{-31}
Sulfates	
Barium sulfate	1.5×10^{-9}
Calcium sulfate	2.4×10^{-5}
Lead(II) sulfate	1.3×10^{-8}
Strontium sulfate	7.6×10^{-7}
Sulfides	
Bismuth sulfide	1.0×10^{-72}
Cadmium sulfide	6.0×10^{-27}
Cobalt(II) sulfide	5.0×10^{-22}
Copper(II) sulfide	4.0×10^{-38}
Copper(I) sulfide	2.5×10^{-50}
Iron(II) sulfide	1.0×10^{-19}
Lead(II) sulfide	4.0×10^{-26}
Manganese(II) sulfide	8.0×10^{-14}
Mercury(II) sulfide	3.0×10^{-53}
Mercury(I) sulfide	1.0×10^{-45}
Nickel(II) sulfide	1.0×10^{-22}
Silver sulfide	1.0×10^{-50}
Zinc sulfide	4.5×10^{-24}

VI-2 Ionization Constants for Weak Acids (for Ionization of One H^+)

ACID	IONIZATION CONSTANT
Acetic ($HC_2H_3O_2$)	1.8×10^{-5}
Benzoic (C_6H_5COOH)	6.6×10^{-5}
Boric (H_3BO_3)	6.0×10^{-10}
Carbonic (H_2CO_3)	4.2×10^{-7}
Hydrogen carbonate ion (HCO_3^-)	4.8×10^{-11}
Chlorous ($HClO_2$)	1.0×10^{-2}
Formic (HCOOH)	2.1×10^{-4}
Hydrocyanic (HCN)	4.0×10^{-10}
Hydrofluoric (HF)	6.9×10^{-4}
Hydrosulfuric (H_2S)	1.0×10^{-7}
Hydrogen sulfide ion (HS^-)	1.3×10^{-13}
Hypobromous (HBrO)	2.0×10^{-9}
Hypochlorous (HClO)	3.0×10^{-8}
Nitrous (HNO_2)	4.5×10^{-4}
Oxalic ($H_2C_2O_4$)	3.8×10^{-2}
Hydrogen oxalate ion ($HC_2O_4^-$)	5.0×10^{-5}
Phosphoric (H_3PO_4)	7.5×10^{-3}
Dihydrogen phosphate ion ($H_2PO_4^-$)	6.2×10^{-8}
Hydrogen phosphate ion (HPO_4^{2-})	1×10^{-12}
Hydrogen sulfate ion (HSO_4^-)	1.2×10^{-2}
Sulfurous (H_2SO_3)	1.3×10^{-2}
Hydrogen sulfite ion (HSO_3^-)	5.6×10^{-8}
Zinc hydroxide [$Zn(OH)_2$]	1.0×10^{-17}
Hydrogen zincate ion ($HZnO_2^-$)	2.0×10^{-13}

VI-3 Ionization Constants for Weak Bases

SUBSTANCE	IONIZATION CONSTANT
Ammonia [$NH_{3(aq)}$]	1.8×10^{-5}
Methylamine (CH_3NH_2)	4.4×10^{-4}
Phenylamine ($C_6H_6NH_2$)	4.5×10^{-10}
Zinc hydroxide [$Zn(OH)_2$]	5.0×10^{-13}

VI-4 Dissociation Constants for Complex Ions

ION	DISSOCIATION CONSTANT
AlF_6^{3-}	2×10^{-24}
$AgCl_2^-$	9×10^{-6}
$Ag(NH_3)_2^+$	6×10^{-8}
$Co(NH_3)_6^{3+}$	2×10^{-34}
$CuCl_2^-$	3×10^{-6}
$Cu(NH_3)_4^{2+}$	1×10^{-12}
$Fe(SCN)_6^{3-}$	2×10^{-4}
$Ni(NH_3)_4^{2+}$	1×10^{-8}
$Ni(NH_3)_6^{2+}$	6×10^{-9}
$Zn(NH_3)_4^{2+}$	2×10^{-9}
$Zn(OH)_4^{2-}$	4×10^{-16}

Appendix VII

Logarithms of Numbers

N	0	1	2	3	4	5	6	7	8	9
1.0	.0000	.0043	.0086	.0128	.0170	.0212	.0253	.0294	.0334	.0374
1.1	.0414	.0453	.0492	.0531	.0569	.0607	.0645	.0682	.0719	.0755
1.2	.0792	.0828	.0864	.0899	.0934	.0969	.1004	.1038	.1072	.1106
1.3	.1139	.1173	.1206	.1239	.1271	.1303	.1335	.1367	.1399	.1430
1.4	.1461	.1492	.1523	.1553	.1584	.1614	.1644	.1673	.1703	.1732
1.5	.1761	.1790	.1818	.1847	.1875	.1903	.1931	.1959	.1987	.2014
1.6	.2041	.2068	.2095	.2122	.2148	.2175	.2201	.2227	.2253	.2279
1.7	.2304	.2330	.2355	.2380	.2405	.2430	.2455	.2480	.2504	.2529
1.8	.2553	.2577	.2601	.2625	.2648	.2672	.2695	.2718	.2742	.2765
1.9	.2788	.2810	.2833	.2856	.2878	.2900	.2923	.2945	.2967	.2989
2.0	.3010	.3032	.3054	.3075	.3096	.3118	.3139	.3160	.3181	.3201
2.1	.3222	.3243	.3263	.3284	.3304	.3324	.3345	.3365	.3385	.3404
2.2	.3424	.3444	.3464	.3483	.3502	.3522	.3541	.3560	.3579	.3598
2.3	.3617	.3636	.3655	.3674	.3692	.3711	.3729	.3747	.3766	.3784
2.4	.3802	.3820	.3838	.3856	.3874	.3892	.3909	.3927	.3945	.3962
2.5	.3979	.3997	.4014	.4031	.4048	.4065	.4082	.4099	.4116	.4133
2.6	.4150	.4166	.4183	.4200	.4216	.4232	.4249	.4265	.4281	.4298
2.7	.4314	.4330	.4346	.4362	.4378	.4393	.4409	.4425	.4440	.4456
2.8	.4472	.4487	.4502	.4518	.4533	.4548	.4564	.4579	.4594	.4609
2.9	.4624	.4639	.4654	.4669	.4683	.4698	.4713	.4728	.4742	.4757
3.0	.4771	.4786	.4800	.4814	.4829	.4843	.4857	.4871	.4886	.4900
3.1	.4914	.4928	.4942	.4955	.4969	.4983	.4997	.5011	.5024	.5038
3.2	.5051	.5065	.5079	.5092	.5105	.5119	.5132	.5145	.5159	.5172
3.3	.5185	.5198	.5211	.5224	.5237	.5250	.5263	.5276	.5289	.5302
3.4	.5315	.5328	.5340	.5353	.5366	.5378	.5391	.5403	.5416	.5428
3.5	.5441	.5453	.5465	.5478	.5490	.5502	.5514	.5527	.5539	.5551
3.6	.5563	.5575	.5587	.5599	.5611	.5623	.5635	.5647	.5658	.5670
3.7	.5682	.5694	.5705	.5717	.5729	.5740	.5752	.5763	.5775	.5786
3.8	.5798	.5809	.5821	.5832	.5843	.5855	.5866	.5877	.5888	.5899
3.9	.5911	.5922	.5933	.5944	.5955	.5966	.5977	.5988	.5999	.6010
4.0	.6021	.6031	.6042	.6053	.6064	.6075	.6085	.6096	.6107	.6117
4.1	.6128	.6138	.6149	.6160	.6170	.6180	.6191	.6201	.6212	.6222
4.2	.6232	.6243	.6253	.6263	.6274	.6284	.6294	.6304	.6314	.6325
4.3	.6335	.6345	.6355	.6365	.6375	.6385	.6395	.6405	.6415	.6425
4.4	.6435	.6444	.6454	.6464	.6474	.6484	.6493	.6503	.6513	.6522
4.5	.6532	.6542	.6551	.6561	.6571	.6580	.6590	.6599	.6609	.6618
4.6	.6628	.6637	.6646	.6656	.6665	.6675	.6684	.6693	.6702	.6712
4.7	.6721	.6730	.6739	.6749	.6758	.6767	.6776	.6785	.6794	.6803
4.8	.6812	.6821	.6830	.6839	.6848	.6857	.6866	.6875	.6884	.6893
4.9	.6902	.6911	.6920	.6928	.6937	.6946	.6955	.6964	.6972	.6981
5.0	.6990	.6998	.7007	.7016	.7024	.7033	.7042	.7050	.7059	.7067
5.1	.7076	.7084	.7093	.7101	.7110	.7118	.7126	.7135	.7143	.7152
5.2	.7160	.7168	.7177	.7185	.7193	.7202	.7210	.7218	.7226	.7235
5.3	.7243	.7251	.7259	.7267	.7275	.7284	.7292	.7300	.7308	.7316
5.4	.7324	.7332	.7340	.7348	.7356	.7364	.7372	.7380	.7388	.7396

N	0	1	2	3	4	5	6	7	8	9
5.5	.7404	.7412	.7419	.7427	.7435	.7443	.7451	.7459	.7466	.7474
5.6	.7482	.7490	.7497	.7505	.7513	.7520	.7528	.7536	.7543	.7551
5.7	.7559	.7566	.7574	.7582	.7589	.7597	.7604	.7612	.7619	.7627
5.8	.7634	.7642	.7649	.7657	.7664	.7672	.7679	.7686	.7694	.7701
5.9	.7709	.7716	.7723	.7731	.7738	.7745	.7752	.7760	.7767	.7774
6.0	.7782	.7789	.7796	.7803	.7810	.7818	.7825	.7832	.7839	.7846
6.1	.7853	.7860	.7868	.7875	.7882	.7889	.7896	.7903	.7910	.7917
6.2	.7924	.7931	.7938	.7945	.7952	.7959	.7966	.7973	.7980	.7987
6.3	.7993	.8000	.8007	.8014	.8021	.8028	.8035	.8041	.8048	.8055
6.4	.8062	.8069	.8075	.8082	.8089	.8096	.8102	.8109	.8116	.8122
6.5	.8129	.8136	.8142	.8149	.8156	.8162	.8169	.8176	.8182	.8189
6.6	.8195	.8202	.8209	.8215	.8222	.8228	.8235	.8241	.8248	.8254
6.7	.8261	.8267	.8274	.8280	.8287	.8293	.8299	.8306	.8312	.8319
6.8	.8325	.8331	.8338	.8344	.8351	.8357	.8363	.8370	.8376	.8382
6.9	.8388	.8395	.8401	.8407	.8414	.8420	.8426	.8432	.8439	.8445
7.0	.8451	.8457	.8463	.8470	.8476	.8483	.8488	.8494	.8500	.8506
7.1	.8513	.8519	.8525	.8531	.8537	.8543	.8549	.8555	.8561	.8567
7.2	.8573	.8579	.8585	.8591	.8597	.8603	.8609	.8615	.8621	.8627
7.3	.8633	.8639	.8645	.8651	.8657	.8663	.8669	.8675	.8681	.8686
7.4	.8692	.8698	.8704	.8710	.8716	.8722	.8727	.8733	.8739	.8745
7.5	.8751	.8756	.8762	.8768	.8774	.8779	.8785	.8791	.8797	.8802
7.6	.8808	.8814	.8820	.8825	.8831	.8837	.8842	.8848	.8854	.8859
7.7	.8865	.8871	.8876	.8882	.8887	.8893	.8899	.8904	.8910	.8915
7.8	.8921	.8927	.8932	.8938	.8943	.8949	.8954	.8960	.8965	.8971
7.9	.8976	.8982	.8987	.8993	.8998	.9004	.9009	.9015	.9020	.9025
8.0	.9031	.9036	.9042	.9047	.9053	.9058	.9063	.9069	.9074	.9079
8.1	.9085	.9090	.9096	.9101	.9106	.9112	.9117	.9122	.9128	.9133
8.2	.9138	.9143	.9149	.9154	.9159	.9165	.9170	.9175	.9180	.9186
8.3	.9191	.9196	.9201	.9206	.9212	.9217	.9222	.9227	.9232	.9238
8.4	.9243	.9248	.9253	.9258	.9263	.9269	.9274	.9279	.9284	.9289
8.5	.9294	.9299	.9304	.9309	.9315	.9320	.9325	.9330	.9335	.9340
8.6	.9345	.9350	.9355	.9360	.9365	.9370	.9375	.9380	.9385	.9390
8.7	.9395	.9400	.9405	.9410	.9415	.9420	.9425	.9430	.9435	.9440
8.8	.9445	.9450	.9455	.9460	.9465	.9469	.9474	.9479	.9484	.9489
8.9	.9494	.9499	.9504	.9509	.9513	.9518	.9523	.9528	.9533	.9538
9.0	.9542	.9547	.9552	.9557	.9562	.9566	.9571	.9576	.9581	.9586
9.1	.9590	.9595	.9600	.9605	.9609	.9614	.9619	.9624	.9628	.9633
9.2	.9638	.9643	.9647	.9652	.9657	.9661	.9666	.9671	.9675	.9680
9.3	.9685	.9689	.9694	.9699	.9703	.9708	.9713	.9717	.9722	.9727
9.4	.9731	.9736	.9741	.9745	.9750	.9754	.9759	.9763	.9768	.9773
9.5	.9777	.9782	.9786	.9791	.9795	.9800	.9805	.9809	.9814	.9818
9.6	.9823	.9827	.9832	.9836	.9841	.9845	.9850	.9854	.9859	.9863
9.7	.9868	.9872	.9877	.9881	.9886	.9890	.9894	.9899	.9903	.9908
9.8	.9912	.9917	.9921	.9926	.9930	.9934	.9939	.9943	.9948	.9952
9.9	.9956	.9961	.9965	.9969	.9974	.9978	.9983	.9987	.9991	.9996

Glossary

(The number following the term refers to the section in the text.)

Acid (6-3, 8-4, 14-1, 26-10) A substance that produces H^+ ions or donates a proton; turns blue litmus red; and neutralizes a base (see also Arrhenius acid, Brønsted acid, and Lewis acid).

Acid anhydride (8-4) A nonmetal oxide that forms an acid when mixed with water.

Activation energy (15-3) The energy necessary for two colliding species to react.

Alcohol (26-8) An organic compound that contains the —OH group.

Alkali metal group (22-2) The Group IA elements in the periodic table.

Alkaline earth group (22-3) The Group IIA elements in the periodic table.

Alkane (26-3) A hydrocarbon that has all single bonds (a saturated hydrocarbon).

Alkene (26-4) A hydrocarbon that contains double bonds (a type of unsaturated hydrocarbon).

Alkyne (26-5) An unsaturated hydrocarbon that contains triple bonds.

Alpha particle (α) (25-1) A type of radiation that is identical to a helium nucleus (^4_2He).

Amine (26-12) An organic compound with the general formula $R{-}NH_2$; often referred to as the organic analog to ammonia.

Amino acid (27-2) An organic compound that contains both the amine group ($-NH_2$) and the acid or carboxyl group ($-COOH$).

Amorphous solid (12-3) A solid in which the particles are arranged in an irregular manner; thus it has no specific shape or structure.

Ampere (20-2) A rate at which electric current flows, equal to one coulomb per second ($A = coul/s$).

Amphiprotic substance (14-1) A substance that can either donate or accept a proton, such as HCO_3^-.

Amphoteric substance (14-1) A substance that can act either as an acid or a base.

Analytical chemistry (1-2) The field of chemistry pertaining to the examination or analysis of a substance, either qualitatively or quantitatively.

Angström (Å) (2-2) A unit of length equal to 10^{-8} cm; the nanometer (nm) is usually used in place of this unit.

Anhydride (8-4, 26-10) A substance that contains no water, such as an acid anhydride nonmetal oxide) or basic anhydride (metal oxide).

Anion (3-8, 7-1) A negatively charged atom or group of atoms (contains more electrons than protons).

Anode (20-1) The electrode at which oxidation takes place.

Aromatic compound (26-7) An organic compound that contains benzene (C_6H_6).

Arrhenius acid (14-1) A substance that produces H^+ ion in water solution.

Arrhenius base (14-1) A substance that produces OH^- ion in water solution.

Atmosphere (atm) (11-3) A unit of pressure equal to 760 torr (760 mm Hg), or 1.0×10^5 Pa.

Atom (3-1) The smallest unit of a particular element that undergoes chemical change; it is composed of a nucleus that contains protons and neutrons (and hence has a positive charge), surrounded by electrons in energy levels.

Atomic mass unit (amu) (3-4) The relative unit used to compare the masses of atoms, based on an atom of the isotope carbon-12 having a mass of exactly 12 amu.

Atomic number (3-6) The number of protons in an atom (which equals the number of electrons in a neutral atom).

Atomic weight scale (3-7) The average, relative mass of each element in the periodic table, compared to carbon-12 having an atomic weight of exactly 12.

Aufbau order (4-5) The order in which electrons occupy orbitals in an atom.

Avogadro's number (3-10) The number of units (atoms, ions, molecules, formula units, and so on) in 1 mole of a substance, equal to 6.02×10^{23}.

Balanced equation (8-1) A chemical equation that contains exactly the same number of each atom on each side of the equation.

Base (alkali, hydroxide) (6-4, 8-4, 14-1) A substance that produces OH^- ions or accepts a proton, turns red litmus blue, neutralizes an acid (see also Arrhenius base, Brønsted base, and Lewis base).

Basic anhydride (8-4) A metal oxide that forms a base when mixed with water, sometimes called a basic oxide.

Beta particle (β) (25-1) A type of radiation identical to an electron ($_{-1}^{0}e$).

Bidentate (24-1) A ligand, in a complex, that has two points of attachment to the central atom.

Binary acid (6-3, 21-2) An acid composed of hydrogen with only one other element.

Binary compound (6-1, 6-2) A compound composed of only two elements.

Biochemistry (1-2, chapter 27) The branch of chemistry concerned with the substances, and reactions of these substances, in living material.

Boiling point (12-2) The temperature at which the vapor pressure of a liquid equals atmospheric pressure.

Bond angle (7-6) The angle formed between three atoms in a molecule.

Bond length (7-2) The distance between the nuclei of two atoms joined together by a covalent bond.

Bond order (7-10) The number of bonds formed between two atoms in a compound, as found in the molecular orbital model of bonding, by taking the number of e^- pairs in BMO (bonding molecular orbitals) minus the number of e^- pairs in ABMO (antibonding molecular orbitals).

Bond strength (7-2) The force holding two atoms together in a covalent compound, measured by the amount of energy necessary to break the bond.

Boyle's law (11-5) The volume of a gas is inversely proportional to the pressure, at constant temperature.

Brønsted-Lowry acid (Brønsted acid) (14-1) A substance that can donate a proton (H^+) to another substance (proton donor).

Brønsted-Lowry base (Brønsted base) (14-1) A substance that can accept a proton (H^+) from another substance (proton acceptor).

Buffer (17-5) A mixture of a weak acid and a salt of the acid or a weak base and a salt of the base, which holds the pH of the solution relatively constant.

Calorie (cal) (2-5) The amount of heat necessary to raise the temperature of one gram of water one degree Celsius (1 cal = 4.18 J).

Catalyst (8-1, 15-4) A substance that alters the speed of a reaction without being consumed in the reaction; generally speeds up the reaction (see also inhibitor).

Cathode (20-1) The electrode at which reduction takes place.

Cation (3-8, 7-1) A positively charged atom or group of atoms; has more protons than electrons.

Charge density (21-1) The charge on an ion divided by the radius of the ion.

Charles' law (11-5) The volume of a gas is directly proportional to the kelvin (absolute) temperature, at constant pressure.

Chelate (24-1) A complex formed by a polydentate ligand, which forms a ring (cyclic) structure.

Chemical change (3-2) A change that alters a substance by forming a new substance.

Chemical equation (3-2, 8-1) A notation that shows what substances are reacting (reactants) and what new substances are being produced (products).

Chemical equilibrium (16-1) The point in a reversible reaction where the rate of the forward reaction is equal to the rate of the reverse reaction.

Chemical properties (3-2) Those properties of a substance that can be observed only when a chemical change takes place in the substance.

Chemistry (1-2) The study of matter and the changes that it undergoes.

Colloid (13-9) A mixture in which the dispersed particles are large enough to be seen but not large enough to settle out; intermediate between a true solution and a precipitate.

Combination reaction (8-4) A reaction in which two or more substances combine to form two or more new substances.

Compound (3-1) Two or more elements chemically combined in definite and constant proportions.

Condensation (12-2) Change from the gas to the liquid phase.

Conjugate pairs of acids and bases (14-1) The original acid (or base) and the new acid (or base) formed in a Brønsted-Lowry acid–base reaction.

Conversion factor (2-3) A factor that a number is multiplied by to change the unit but not the value; it is always equal to one, such as 1 m/100 cm or 100 cm/1 m.

Coordinate covalent bond (7-2) A covalent bond in which one atom furnished both electrons.

Coordination compound (complex) (24-1) A complex substance made up of two or more simpler substances, each of which can exist independently.

Coordination number (24-1) The number of bonds formed around the central atom in a coordination compound (complex).

Coulomb (coul) (20-2) A unit used to measure the amount of electrical energy that flows in one second by a current of one ampere (coul $= A \times s$).

Covalent bond (7-2) A bond formed by two electrons that are shared between two atoms; often called an electron pair bond.

Critical pressure (12-5) The pressure necessary to liquify a gas at the critical temperature.

Critical temperature (12-5) The temperature above which a gas cannot be liquified, no matter how much pressure is applied.

Curie (Ci) (25-8) The mass of a radioactive substance that will give 3.7×10^{10} disintegrations per second.

Dalton's Law of Partial Pressures (11-6) In a mixture of gases, each gas exerts the pressure it would exert if it were by itself; thus the total pressure is equal to the sum of the partial pressures of each gas in the mixture.

Decomposition reaction (8-5) A chemical reaction in which one substance breaks up into two or more new substances.

Delocalized π bond (7-5) A pi (π) bond that is spread out among three or more *p* orbitals rather than being localized between only two atoms.

Density (2-6) The mass of a substance ocupying a unit of volume: density = mass/volume.

Dibasic acid (18-1) An acid that contains two replaceable hydrogen ions.

Differentiating electron (5-5) The electron that makes an atom different from the one before it in the periodic table.

Digestion (27-6) The process whereby food is broken down into simpler substances in the body.

Dipole attraction (12-1) The forces of attraction between two polar molecules.

Distillation (12-2) Separation of substances with different boiling points by driving off the lower boiling portion and leaving everything else.

Electrolyte (3-12, 13-1) A substance that breaks up into ions when put into solution.

Electrolytic cell (electrolysis cell) (20-1) A cell in which electrical energy causes a chemical reaction to take place.

Electromotive series (activity series) (8-6, 21-3) Listing of metals (and hydrogen) in order of decreasing activity; anything higher on the series will displace something lower.

Electron (e^-) (3-5) A part of an atom with a -1 charge and negligible mass; found in energy levels around the nucleus.

Electron affinity (5-6) Energy released when an atom gains an electron to form a negative ion.

Electron capture (25-2) A type of nuclear reaction in which the nucleus captures an electron to form a neutron from a proton; sometimes called *K* capture (since the electron is generally from the first or *K* shell of electrons).

Electronegativity (5-6) The general tendency of an atom to attract electrons toward itself in a compound.

Element (3-1) A pure substance composed of only one kind of atom.

Empirical formula (9-3) The simplest formula of a compound; gives the smallest, whole number ratio of atoms that make up the compound.

Endothermic (10-1) A reaction that absorbs heat.

End point (14-1) The point where the indicator in an acid-base titration changes color.

Energy level (4-2) A region in space, of some general distance from the nucleus, where a group of electrons is most likely to be found.

Enthalpy of formation (ΔH_f) (10-2) The amount of heat transferred in a reaction for which 1 mole of a compound is formed from the elements in the standard state.

Enthalpy of reaction (ΔH_r) (10-2) The amount of heat transferred in a given reaction.

Entropy (ΔS) (10-6) A measurement of the amount of disorder of a system.

Enzyme (27-4) A substance that catalyzes a biological reaction in the body.

Equilibrium constant (K_{eq}) (16-2) The mathematical expression for the relationship between concentration of products and reactants in an equilibrium reaction.

Equivalent weight (20-2) The formula (molecular) weight of a substance divided by the number of electrons transferred in the reaction involved.

Equivalent weight of acid, base, or salt (14-2) The formula weight of an acid or base divided by the number of replaceable H^+ or OH^- ions present; for a salt it is the formula weight divided by the total positive or negative charge.

Essential amino acid (27-2) An amino acid that is needed by the body but cannot be synthesized by the body; thus it must be provided in the diet.

Ether (26-9) An organic compound with the general formula R—O—R.

Evaporation (12-2) The process whereby a liquid changes to a gas, due to molecules of the liquid escaping into the gas phase.

Exothermic reaction (10-1) A chemical reaction that gives off heat.

Family (5-2) A subgroup in the periodic table.

Faraday (F) (20-2) The amount of electrical energy necessary for the transfer of 1 mole of electrons in a reaction; equals 96,500 coulombs.

Fats and oils (27-3) Esters of glycerols and fatty acids. Fats have saturated carbon chains and oils have unsaturated carbon chains.

Fatty acids (27-3) Organic acids with relatively long carbon chains (approximately 10 to 20 carbons).

Filtrate (13-4) The substance that passes through the filter in a filtration process.

First Law of Thermodynamics (10-5) The Law of Conservation of Energy.

Formal concentration (*F*) (13-5) Equivalent to (*M*) concentration.

Formula (3-1) A notation that tells what atoms, in what proportions, are present in a compound.

Formula unit (3-1) The smallest unit that contains the number of atoms indicated by the formula for a substance (such as Na, H_2O, and NaCl).

Formula (molecular) weight (3-9) The relative mass of a substance, based on the atomic weight scale and the formula for the substance.

Free energy (ΔG) (10-6) The amount of energy available to do useful work in a chemical reaction; found from the enthalpy, entropy, and kelvin (absolute) temperature, using the equation $\Delta G = \Delta H - T\Delta S$

Freezing point (12-3) Same as melting point.

Gamma rays (25-1) Radiation that consists of rays of short wavelength, caused by the energy changes in nuclear reactions.

Gay-Lussac's Law (11-5) The pressure of a gas is directly proportional to the kelvin (K) temperature, at constant volume.

Graham's Law of Diffusion (11-10)

$$\frac{R_1}{R_2} = \sqrt{\frac{m_2}{m_1}} \qquad R = \text{rate of diffusion}, \quad m = \text{molecular weight}$$

Group (5-2) A vertical column in the periodic table.

Half-life ($t_{1/2}$) (25-4) The amount of time that it takes for one-half of a given amount of a radioactive species to decay.

Halogens (22-8) The Group VIIA elements in the periodic table.

Heat of fusion (12-3) The amount of heat necessary to convert a given amount of solid to liquid, at the melting point.

Heat of vaporization (12-2) The amount of heat necessary to convert a given amount of liquid to gas, at the boiling point.

Hess's Law (10-3) The ΔH of an overall reaction is equal to the sum of the ΔH values for each step in the reaction.

Heterogeneous equilibrium (16-2) A chemical equilibrium in which there are two phases present, such as gas and solid or solution and solid.

Hund's Rule (4-5) When electrons occupy orbitals of equal energy, one electron occupies each orbital before there is any pairing.

Hybrid orbital (7-7) A new type of orbital formed in the formation of a covalent bond such as sp, sp^2, sp^3, dsp^2, and d^2sp^3.

Hydrate (8-5) A salt that contains one or more molecules of water, as part of the crystal structure, for each formula unit, such as $BaCl_2 \cdot 2\,H_2O$.

Hydride (6-2, 22-1) An ionic compound formed from an active metal and hydrogen, in which the hydrogen is H^-.

Hydrocarbon (26-2) An organic compound composed of only carbon and hydrogen.

Hydrogen bond (12-6) The force of attraction between the hydrogen atom in one molecule and an electronegative atom in another molecule.

Hydrolysis constant (K_h) (18-2) The equilibrium constant for a hydrolysis reaction.

Hydrolysis of salts (18-2) The reaction of a salt with water to form a weak acid or weak base.

Hydronium ion (14-1) The H_3O^+ ion formed by H^+ in water solution; often written as $H^+_{(aq)}$ or just H^+ for simplicity.

Ideal gas (11-2) A gas that behaves according to the postulate of the kinetic theory of gases.

Immiscible (12-2) Substances that will not mix.

Indicator (14-1) A substance that changes color when the solution it is in changes from acid to base, or base to acid; used in titration of acids and bases.

Inhibitor (8-1, 15-4) A catalyst that slows down the speed of a reaction; sometimes called a negative catalyst.

Inner transition elements (5-3) The series of elements from 57 through 71, and 89 through 103.

Inorganic chemistry (1-2) That branch of chemistry having to do with the study of all elements and compounds other than carbon compounds.

Ion (3-8) An atom or group of atoms that has a charge due to losing or gaining one or more electrons (see also cation and anion).

Ionic bond (7-1) A chemical bond formed by the electrostatic attraction between ions in an ionic or electrovalent compound.

Ionic equation (8-7, 14-3) A chemical equation written in such a way that strong electrolytes in solution are written in the form of their ions.

Ionization constant (K_i, K_a or K_b) (17-2) The equilibrium constant for the equilibrium reaction representing the ionization of a weak acid (K_a) or weak base (K_b).

Ionization constant of water (K_w) (17-3) The equilibrium constant for the ionization of water ($H_2O \rightleftharpoons H^+ + OH^-$); $K_w = [H^+][OH^-] = 1.0 \times 10^{-14}$.

Ionization energy (5-6) The amount of energy necessary to remove an electron from an atom or ion.

Ion product (17-1) The product of the concentrations of the ions, raised to the proper powers, from the equation for the ionization of a slightly soluble substance.

Isotopes (3-7) Atoms of the same element with different masses, due to different numbers of neutrons.

Joule (J) (2-5) A unit of energy equal to 10^7 ergs; $J = kg\ m^2/s^2$ (4.18 J = 1 cal).

Ketone (26-11) An organic compound with the general formula $R-\overset{\overset{\displaystyle O}{\|}}{C}-R$.

Kinetic energy (3-3) Energy of motion.

Law of Conservation of Mass (Matter) (3-2) Matter (or matter-energy) cannot be created or destroyed but can change form.

Law of Definite Proportions (Constant Composition) (3-1) A chemical compound is always composed of the same elements in the same proportions.

Law of Mass Action (15-1) For the rate determining step in a reaction, the rate is proportional to the concentrations of the reactants, raised to the power of the coefficients in the balanced equation.

Le Châtelier's Principle (16-3) When a stress is placed on a system at equilibrium, the equilibrium will shift in such a way as to relieve the stress.

Lewis acid (14-1) A substance that accepts a pair of electrons in a reaction (electron pair acceptor).

Lewis base (14-1) A substance that furnishes a pair of electrons in a reaction (electron pair donor).

Lewis structure (7-3) An electron dot formula that shows the arrangement of the electron pairs in a compound.

Ligand (24-1) The species attached to the central atom in a coordination compound (complex).

Limiting reagent (9-4) The reagent that is all used up in a chemical reaction, and thus limits the amount of product that can be formed.

Lipids (27-3) Fats or oils.

Macromolecular substance (12-1) A substance in which covalent bonds must be broken to melt the substance, such as a diamond.

Magnetic quantum number (m_l) (4-8) The quantum number that specifies a particular orbital in a given set.

Mass number (3-6) The closest whole number to the relative mass of a particular isotope; equal to the number of protons and neutrons in the atom.

Matter (2-1) Anything that has mass and takes up space.

Mechanism of a reaction (15-5) Shows what steps the reaction goes through in forming the final product.

Melting point (12-3) The temperature at which the liquid and solid phases of a substance can exist in equilibrium.

Metabolism (27-6) The chemical changes in digested food that either produce energy (catabolism) or form complex molecules (anabolism) for storage or further production of other substances such as tissue.

Metathesis reaction (8-7) A double displacement reaction, in which two reactants form two products.

Millimeters of mercury (mm Hg) pressure (11-3) The pressure expressed in terms of a column of mercury that can be supported (1 atm = 760 mm Hg).

Miscible (12-2) Substances that will mix.

Mixture (3-1) Two or more substances, in variable proportions; can be homogeneous (as in a solution) or heterogeneous.

Molality (molal concentration—*m*) (13-5) Concentration of a solution expressed as moles of solute per kilogram of solvent (mol solute/kg solvent).

Molarity (molar concentration—*M*) (13-5) Concentration of solution expressed as moles of solute per liter of solution (mol/*l*).

Molar mass (3-9) The mass (in grams) of one mole of a substance.

Molar volume of a gas (11-7) The volume occupied by one mole of a gas, (22.4 *l* at STP).

Mole (mol) (3-10) The amount of a substance in grams, that contains Avogadro's number (6.02×10^{23}) of units (atoms, molecules, or formula units).

Molecular formula (9-3) The actual formula for one molecule of a compound (as opposed to the empirical formula).

Molecular orbital (7-2) The orbital formed by the overlap of atomic orbitals in the formation of a covalent bond; holds a pair of electrons.

Molecular (formula) weight (3-9) The sum of the atomic weights in the formula of a substance.

Molecule (3-8) The smallest particle of a gas; sometimes considered as the smallest particle of a substance that retains the properties of that substance.

Mole fraction (13-4) The moles of one component of a mixture (such as a solution) divided by the total number of moles in the mixture.

Monodentate (24-1) A ligand, in a complex, that has one point of attachment to the central atom.

Multiple bond (7-2) Two or three bonds formed between two atoms in a compound.

Net ionic equation (14-3) An ionic equation in which only the reacting species are shown; all spectator ions are eliminated.

Neutralization reaction (8-8) A reaction between an acid and a base; generally, forms water plus a salt.

Neutron (3-5) A particle with a relative mass of 1 amu and no charge, found in the nucleus of an atom.

Noble gas (5-3) A member of the last group of elements in the periodic table.

Noble metals (23-7) The least active metals; generally, the platinum metals (Ru, Rh, Pd, Os, Ir, Pt) and silver and gold.

Nodal plane (4-4) An area in space, within an atom or molecule, where there is zero probability of finding an electron (such as along the internuclear axis of a pi bond).

Nonelectrolyte (3-12, 13-1) A substance that does not break up into ions when put into solution.

Nonpolar molecule (7-8) A molecule in which the electron cloud is equally distributed over the molecule, and thus no one part of the molecule is more positive or more negative (see polar molecule).

Normal boiling point (12-2) The temperature at which the vapor pressure of a liquid is equal to 1 atm (760 torr).

Normality (*N*) (14-2) Concentration of solution expressed as equivalents of solute per liter of solution (equiv/l).

Nuclear fission (25-6) A nuclear reaction in which a nucleus splits into two or more parts of approximately equal mass and charge.

Nuclear fusion (25-7) A nuclear reaction in which two or more nuclei combine to form a new nucleus.

Nuclear reaction (25-2) A reaction in which there is a change in the nucleus of the reacting species.

Nucleus (3-6) The center of an atom; contains protons and neutrons.

Orbital (4-4) A region in space, within an atom or molecule, where there can be no more than two electrons.

Order of reaction (15-2) The exponents in a rate law, which tell how the rate changes as the concentration (or pressure) of a given reactant is changed.

Organic chemistry (1-2 and Chapter 26) The study of the compounds of carbon.

Osmosis (13-8) The process whereby a solvent flows through a porous membrane into a solution.

Osmotic pressure (13-8) The amount of pressure that must be applied to a solution to prevent osmosis.

Oxidation number (state) (5-8) The apparent charge on an atom in a compound; a way of accounting for where the electrons are most likely to be in a compound; used in nomenclature and balancing equations.

Oxidation potential (20-4) An electrode potential in which the reaction is written as an oxidation reaction (the reactant loses electrons).

Oxyacid (6-4, 21-2) An acid that contains oxygen, such as H_2SO_4.

Oxyanion (6-4) The anion of an oxyacid, such as $SO_4{}^{2-}$.

Pauli Exclusion Principle (4-8) States that no two electrons in a given atom can have exactly the same set of quantum numbers (which is why there can be no more than two electrons in any orbital, and these two must have opposite spins).

Percent concentration (by mass) (13-4)

$$\frac{\text{mass solute}}{\text{mass solution}} \times 100.$$

Percent yield (9-4) The actual yield in a reaction divided by the theoretical yield times 100:

$$\text{percent yield} = \frac{\text{actual yield}}{\text{theoretical yield}} \times 100.$$

Period (5-2) A horizontal row in the periodic table.

Periodic Law (5-1) The properties of the elements are periodic functions of their atomic numbers.

Periodic table (chart) (3-1, 5-1) A table of the elements, arranged in order of increasing atomic numbers.

pH (17-4) The negative log of the H^+ ion concentration; $pH = -\log [H^+]$.

Phase diagram (12-5) A graph of temperature versus pressure for a given substance, showing the relationship between the three states —gas, liquid, and solid.

Phenol (26-8) An organic compound containing the benzene ring and the —OH group, thus, an aromatic alcohol.

Physical change (3-2) A change that can be observed without changing the composition of a substance, such as a change of state.

Physical chemistry (1-2) The branch of chemistry having to do with the theoretical mathematical aspects of matter and its changes, such as in reaction rates, mechanisms, bonding and structure, and thermodynamics.

Physical properties (3-2) Those properties of a substance that can be observed without changing the composition of the substance, such as melting point, boiling point, and vapor pressure.

Pi (π) bond (7-5) A bond formed by the sideways overlap of *p* orbitals in a compound; formed after the sigma bond.

pOH (17-4) The negative log of the OH^- ion concentration; $pOH = -\log [OH^-]$.

Polar bond (7-2) A covalent bond in which the electron cloud is shifted more toward one atom, due to a difference in electronegativity of the atoms.

Polar molecule (7-8) A molecule in which the electron cloud is not equally distributed, and thus one part of the molecule is more positive and another part is more negative.

Polydentate (24-1) A ligand, in a complex, in which there is more than one point of attachment to the central atom.

Polyprotic acid (18-1) An acid that has more than one replaceable H^+ ion, such as in the dibasic acid H_2S.

Positron particle (β^+) (25-1) A type of radiation that is equivalent to a positive electron ($_{+1}^{0}e$).

Potential energy (3-3) Energy a particle contains due to its position in space.

Precipitate (17-1) A solid material that forms in a reaction and will eventually settle out of the solution.

Principal quantum number (*n*) (4-8) The quantum number that specifies the energy level.

Products (3-2) The substances formed in a chemical reaction; written on the right side of the arrow.

Protein (27-2) A complex organic molecule, made up of amino acid groups.

Proton (3-5) A particle with a mass of about 1 amu and a +1 charge, found in the nucleus of an atom.

Pure substance (3-1) An element or compound.

Radioactivity (25-1, 25-3) A property of certain unstable isotopes, which give off various types of radiation, such as alpha, beta, positron, and neutron.

Rare earth elements (23-2) The first inner transition series of elements; the lanthanide elements (atomic numbers 57 through 71).

Rate determining step (15-1) The slow step in a reaction that goes in more than one step; determines the rate at which final product can be formed.

Reactants (3-2) The substances that are reacting in a chemical reaction; written on the left side of the equation.

Reaction rate (15-1) The rate at which reactants are used up or products are formed in a chemical reaction.

Reduction potential (20-4) An electrode potential in which the reaction is written as a reduction (the reactant gains electrons).

Replacement reaction (8-6) A reaction in which an active metal displaces a less active metal, or hydrogen, from solution.

Representative elements (5-3) The A subgroups in the periodic table; valence electrons are in *s* and *p* orbitals; have only one incomplete energy level.

Resonance forms (7-4) Two equally correct Lewis structures that can be drawn for a molecule.

Reversible reaction (16-1) A reaction in which the products that are formed react to form the original reactants; written as an equilibrium reaction, such as $Ag^+ + Cl^- \rightleftarrows AgCl_{(s)}$.

Salt (8-8) The product formed when an acid reacts with a base; generally considered to be an ionic substance.

Saturated solution (13-3) A solution that contains as much of the solute as it can hold at the given temperature.

Secondary quantum number (l) (4-8) The quantum number that describes the sublevel, or type of orbital (s, p, d, f, and so on)

Sigma (σ) bond (5-7) The first bond formed in a covalent compound; formed by the direct, or head-on, overlap of orbitals.

Solubility product constant (K_{sp}) (17-1) The equilibrium constant for the equilibrium reaction between a solid and its ions in solution, such as for $AgCl \rightleftharpoons Ag^+ + Cl^-$

$$K_{sp} = [Ag^+][Cl^-]$$

Solute (3-12, 13-1) The substance being dissolved in a solution.

Solution (3-12, 13-1) A homogeneous mixture, in which the component parts (solute and solvent) can be varied.

Solvent (3-12,13-1) The dissolving medium in a solution.

Specific gravity (2-6) The ratio of the density of a substance to the density of a standard (usually water at 4°C); thus it is a pure number, equal to the density expressed in grams per milliliter.

Specific heat (heat capacity) (2-5, 12-4) The amount of heat necessary to raise the temperature of a given amount of substance by 1°; usually expressed as cal/g°C.

Spin quantum number (m_s) (4-8) The quantum number that specifies the spin of the electron in an orbital ($+\frac{1}{2}$ or $-\frac{1}{2}$).

Standard temperature and pressure (STP) (11-4) 0°C (273 K) and 1 atm (760 torr).

State function (10-2) A function that does not depend on the path by which the process takes place.

Stereoactive pairs of electrons (7-6) Pairs of electrons in a covalent compound that determine the shape of the molecule; sigma bonding and nonbonding (lone) pairs around the central atom.

Stoichiometry (Chapter 9) Chemically equivalent amounts, either from the formula for a compound or from a balanced equation.

Strong acid or base (14-1) An acid or base that is completely ionized.

Strong electrolyte (13-1) An ionic substance that is completely ionized in solution.

Sublevel (4 3) A subdivision of an energy level in an atom; made up of a set of orbitals, such as the s sublevel, p sublevel, and so on.

Sublimation (12-3) The process where a solid changes directly to a gas, without going through the liquid phase.

Supersaturated solution (13-3) A solution that contains more of the solute than it can normally hold at a given temperature.

Surface tension (12-2) The property of a liquid that draws the molecules at the surface of the liquid in toward the body of the liquid.

Symbol (3-1) A notation used to represent a particular element; the first letter of the name, and in some cases one other letter (some symbols come from the Latin name).

Theoretical yield (9-4) The calculated yield from a reaction, using the limiting reagent.

Thermochemical equation (10-1) A chemical equation that includes the amount of heat absorbed or given off for a particular reaction.

Titration (14-1) The process where the amount of solution necessary to complete a reaction is measured accurately into another solution, such as the amount of acid necessary to neutralize a given amount of base.

Torr (11-3) A unit of pressure equal to a millimeter of Hg (760 mm Hg = 760 torr).

Transition elements (5-3, chapter 23) The B subgroups in the periodic table; generally, the differentiating electron is in a d orbital; also, they generally have two incomplete energy levels of electrons.

Triple point (12-5) The point on a phase diagram where all three phases can exist in equilibrium.

Unsaturated solution (13-3) A solution that contains less solute than it can hold at a given temperature.

Valence electrons (5-4) The outermost electrons in an atom; the ones that are used in forming compounds.

Van der Waals forces (12-1) Weak forces of attraction between nonpolar molecules in a liquid or solid.

Vapor pressure (12-2) The pressure exerted by the vapor being given off by a liquid, at a given temperature.

Voltaic or galvanic cell (20-3) A system in which a chemical reaction produces electrical energy.

Weak acid or base (14-1) An acid or base that is only partially ionized.

Weak electrolyte (13-1) An electrolyte that is only partially ionized.

Answers to First Half of Problems

Chapter 2

1.

(a) 230 cm
(b) 120 mℓ
(c) 8500 g
(d) 1.85×10^6 μm
(e) 2.5×10^{-2} g

2.

(a) 1.80×10^6 mm
(b) 5.0×10^{-2} dm
(c) 3.2×10^4 μℓ
(d) 2.55×10^3 Å
(e) 2.8×10^3 cm^3

3.

(a) 20.8 cm
(b) 1.16×10^3 g
(c) 2.69 ℓ

4.

(a) 1.13×10^7 mg
(b) 82.0 ft
(c) 8.0×10^3 mℓ
(d) 4.92×10^3 yd
(e) 2.54×10^{12} μg

5.

(a) 1.02×10^2 ft/min
(b) 62.4 lb/ft^3
(c) 5.2×10^{-9} m^3
(d) 1.00×10^3 kg/m^3
(e) 2.2×10^{-19} ft^2

6.

(a) °F = 104°F; °K = 313°K
(b) °F = −112°F; °K = 193°K
(c) °F = 932°F; °K = 773°K

7.

(a) °C = 30.0°C; °K = 303°K
(b) °C = −28.9°C; °K = 244°K
(c) °C = −84.4°C; °K = 189°K

8.

(a) 7.0 kcal, 29.3 kJ

9.

(a) 0.369 cal/g°C
(b) 0.396 cal/g°C

10.

(a) 0.52 cal

11.

(a) 5.9 g/cm^3
(b) 2.0 g/cm^3
(c) 1.8 g/cm^3

12.

(a) 14.2 g ether

(b) 1.00×10^4 g CCl_4
(c) 1.8×10^3 g benzene

13.
(a) 50.0 mℓ
(b) 2.5×10^3 mℓ benzene
(c) 1.07×10^3 mℓ $CHCl_3$

14.
(a) 3.75
(b) 6.5
(c) 1.40

15.
(a) 44.2 lb/ft^3

16.
(a) 1.2×10^{-8} ℓ H_2O

17.
(a) 329 gal H_2SO_4

18.
(a) 4.72 g/cm^3

19. 17 min

21. 708 kg/m^3 ether

23. $\$5.29 \times 10^3$

25. 5.30×10^{-7}

27. 0.441 cal/g°C
1.84×10^3 J/kg°K

29. 453 mℓ alcohol

31. 37°C, 310°K

Chapter 3

1.
(a) 2 hydrogen, 1 oxygen
(b) 1 carbon, 4 hydrogen
(c) 1 nitrogen, 3 hydrogen
(d) 1 calcium, 2 oxygen, 2 hydrogen
(e) 1 magnesium, 2 nitrogen, 6 oxygen

2. (p, e^-, n)
(a) 6, 6, 6
(b) 35, 35, 45
(c) 1, 1, 1
(d) 83, 83, 131
(e) 82, 82, 124

3. (p, e^-)
(a) 11, 10
(b) 20, 18
(c) 35, 36
(d) 28, 25
(e) 8, 10

4.
(a) 69.72 amu

5.
(a) 111 amu
(b) 17 amu
(c) 171 amu
(d) 360 amu
(e) 138.6 amu

6.
(a) 156 g
(b) 136 g

7.
(a) 0.18 mol Al
(b) 25.5 mol NH_4Cl
(c) 3.32 mol Na
(d) 2.8×10^{-6} mol H_2O
(e) 8.2×10^{-8} mol Mg

8.
(a) 3.01×10^{23} atoms
(b) 4.2×10^{24} atoms
(c) 3.8×10^{21} atoms
(d) 4.9×10^{25} atoms
(e) 3.2×10^{25} atoms

9.
(a) 1.98×10^{23} molecules O_2
(b) 4.0×10^{25} molecules NH_3
(c) 1.9×10^{17} molecules SO_2
(d) 8.4×10^{25} molecules H_2O
(e) 4.8×10^{24} molecules CH_4

10.
(a) 79 g CO_2
(b) 0.731 g NaCl
(c) 0.673 g H_2O

(d) 2.33×10^{-23} g N
(e) 20 g SO_3

11.

(a) 15 cm^3
(b) 0.26 cm^3
(c) 1.4×10^3 cm^3
(d) 3.04×10^{-23} cm^3
(e) 5.5×10^2 cm^3

12.

(a) 43 g/mol
(b) 57 g/mol
(c) 3.0×10^2 g/mol
(d) 1.4×10^3 g/mol
(e) 1.2×10^3 g/mol

13.

(a) 16.4 mol O atoms
(b) 0.656 mol O atoms
(c) 7.20×10^{-2} mol O atoms
(d) 9.22×10^{-2} mol O atoms
(e) 11 mol O atoms

14.

(a) 9.05%
(b) 20.0%
(c) 17.6%
(d) 9.7%
(e) 17%

15.

(a) 4.66 *M*
(b) 0.40 *M*
(c) 0.787 *M*
(d) 2.3×10^{-2} *M*
(e) 7.8 *M*

16.

(a) 25 g NaOH
(b) 128 g $Ca(NO_3)_2$
(c) 8.4 g NaCl
(d) 29.2 g $CaCl_2$
(e) 108 g $NaNO_3$

17.

(a) 6.94×10^3 mmol H_2O
(b) 5.6×10^{-2} oz
(c) 4.8 lb CH_4
(d) 7.4×10^3 mmol Al
(e) 1.6×10^{25} molecules CCl_4
(f) 3.2×10^{-5} ℓ H_2O

18.

(a) 0.506 mol HCl
(b) 4.64 mol H_3PO_4
(c) 0.32 mol H_2SO_4

19.

(a) 38.1°F, 234°K
(b) 352°K; 173°F

20.

(a) 1.3×10^{-2}

21.

(a) 2.6×10^3 cal
(b) 2.5×10^2 cal

Chapter 4

1.

(a) $2(4^2) = 32$
(b) $2(7^2) = 98$

2.

(a) *s*
(b) *s*, *p*, *d*, *f*
(c) *s*, *p*, *d*, *f*, *g*, *h*

3.

(a) 5
(b) 1
(c) 9
(d) 13

4.

(a) 10
(b) 2
(c) 18
(d) 26

5. (a)

Principal Energy Level	Sublevel	No. of Orbitals in Sublevel	Max No. of e^- in Sublevel	Max No. of e^- in Energy Level
4	*s*	1	2	32
4	*p*	3	6	
4	*d*	5	10	
4	*f*	7	14	

6.

(a) P $1s^2$ $2s^2\,2p^6$ $3s^2\,3p^3$
(b) Fe $1s^2$ $2s^2\,2p^6$ $3s^2\,3p^6$ $4s^2\,3d^6$
(c) As $1s^2$ $2s^2\,2p^6$ $3s^2\,3p^6$ $4s^2\,3d^{10}\,4p^3$
(d) Sn $1s^2$ $2s^2\,2p^6$ $3s^2\,3p^6$ $4s^2\,3d^{10}\,4p^6$ $4d^{10}\,5s^2\,5p^2$
(e) Hg $1s^2$ $2s^2\,2p^6$ $3s^2\,3p^6$ $4s^2\,3d^{10}\,4p^6$ $5s^2\,4d^{10}\,5p^6\,6s^2\,4f^{14}\,5d^{10}$

7.

(a) Al 1
(b) Cl 1
(c) Pb 2
(d) Mn 5
(e) Ni 2

8.

(a) Cl $3s^2\,3p^5$
(b) S $3s^2\,3p^4$
(c) Ca $4s^2$
(d) Sn $5s^2\,5p^2$
(e) I $5s^2\,5p^5$

9.

(a) Ti $4s^2\,3d^2$
(b) Fe $4s^2\,3d^6$
(c) Co $4s^2\,3d^7$
(d) Hg $6s^2\,5d^{10}$
(e) Cu $4s^1\,3d^{10}$

10.

(a) Ba $6s^2$

(b) P $3s^2\,3p^3$

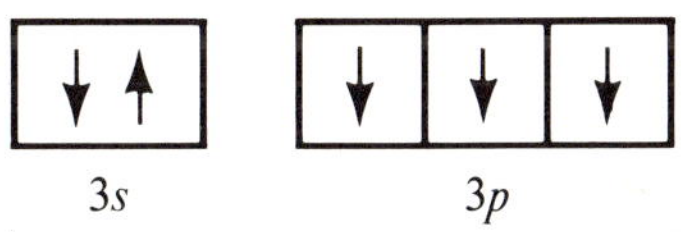

(c) Ni $3d^8\,4s^2$

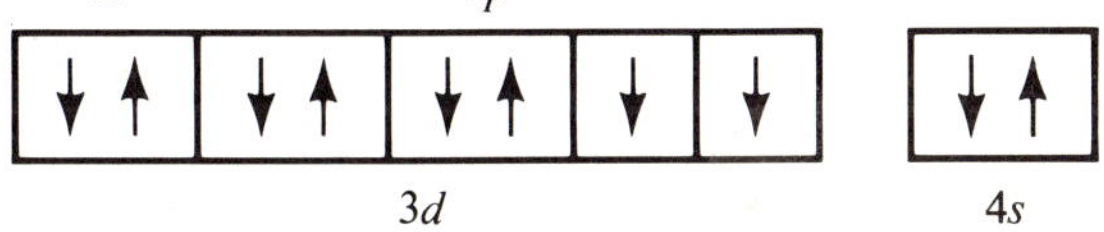

(d) Br $3d^{10}\,4s^2\,4p^5$

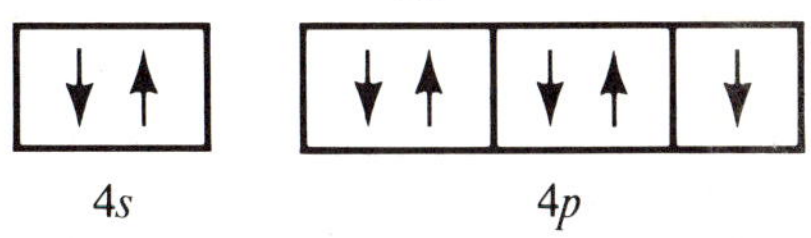

(e) Cr $3d^5\,4s^1$

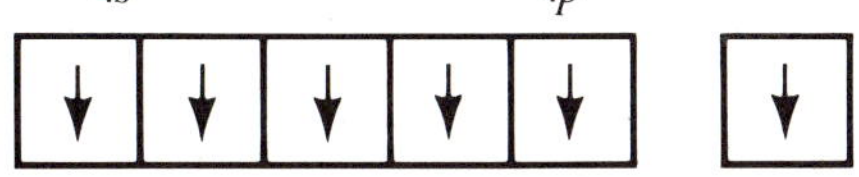

11.

(a) Sr
(b) S
(c) Fe
(d) Cu
(e) Xe

12.

(a) s, p sublevels; $l = 0$ or 1
(b) s, p, d, f, g; $l = 0, 1, 2, 3, 4$
(c) s, p, d, f, g, h, i, j; $l = 0, 1, 2, 3, 4, 5, 6, 7$

13.

(a) $l = 0$, $m_l = 0$; 1 orbital
(b) $l = 2$, $m_l = -2, -1, 0, 1, 2$; 5 orbitals
(c) $l = 4$, $m_l = -4, -3, -2, -1, 0, 1, 2, 3, 4$; 9 orbitals
(d) $l = 6$, $m_l = -6, -5, -4, -3, -2, -1, 0, 1, 2, 3, 4, 5, 6$; 13 orbitals

14.

(a) 1s
(b) 3d
(c) 6p
(d) 5f
(e) 7h

15. (a)

Value of n = Energy Level	Value of l and Type of Sublevel	Number of Values of l (and Allowed Values)	Maximum Number of Electrons in Sublevel	Maximum Number of Electrons in Energy Level
5	0 (s)	1(0)	2	
	1 (p)	3(-1, 0, 1)	6	
	2 (d)	5(-2, -1, 0, 1, 2)	10	50
	3 (f)	7(-3, -2, -1, 0, 1, 2, 3)	14	
	4 (g)	9(-4, -3, -2, -1, 0, 1, 2, 3, 4)	18	

16.

(a)
$$\begin{aligned} n = 3,\ l &= 0,\ m_l = 0 \text{ (1 orbital)} \times 2e^- = 2 \\ l &= 1,\ m_l = 1, 0, -1 \text{ (3 orbitals)} \times 2e^- = 6 \\ l &= 2,\ m_l = 2, 1, 0, -1, -2 \text{ (5 orbitals)} \times 2e^- = \underline{10} \\ & \qquad 18e^- \end{aligned}$$

(b) $n = 2$, $l = 1$, $m_l = -1, 0, 1 = 6e^-$
(c) $n = 4$, $l = 2$, $m_l = 0 = 2e^-$
(d) $n = 3$, $l = 2$, $m_l = 1$, $m_s = \frac{1}{2} = 1e^-$
(e)
$$\begin{aligned} n = 7,\ l = 0(s) &= 2e^- \\ 1(p) &= 6e^- \\ 2(d) &= 10e^- \\ 3(f) &= 14e^- \\ 4(g) &= 18e^- \\ 5(h) &= 22e^- \\ 6(i) &= \underline{26e^-} \\ & \ \ 98e^- \end{aligned}$$

17.

(a) When $n = 2$, l cannot be 2.
(b) When $l = 1$, m_l can only be -1, 0, 1 not 2.
(c) m_s cannot be 0.
(d) l cannot be $\frac{1}{2}$.
(e) n cannot be negative.

18.

(a) p orbitals do not have $8e^-$.
(b) 3s comes after 2p; there is no d in second energy level.
(c) Electrons occupy the lowest possible energy level so they would fill the 2p orbitals before the 3s.
(d) According to the Aufbau order, the 5s fills before the 4d.

19.

(a) 1, 0, 0, $+\frac{1}{2}$; 1, 0, 0, $-\frac{1}{2}$
(b) 1, 0, 0, $+\frac{1}{2}$; 1, 0, 0, $-\frac{1}{2}$; 2, 0, 0, $+\frac{1}{2}$; 2, 0, 0, $-\frac{1}{2}$

20.

(a) 27
(b) 9
(c) 3

21.

(a) Not possible; when $n = 2$, ℓ cannot be 2 (therefore, there is no 2d orbital).
(b) Possible; when $n = 4$, ℓ can be 3.
(c) Possible; when $n = 8$, ℓ can be 0.
(d) Possible; when $n = 7$, ℓ can be 4.
$n = 7$ $\ell = 0, 1, 2, 3, 4, 5, 6$
(s, p, d, f, g, h, i)
(e) Not possible; when $n = 5$, ℓ cannot be 5 (therefore, there is no 5h orbital).

22.

(a) When $n = 4$, ℓ can be 0, 1, 2, 3 (not 4).
(b) When $\ell = 2$, $m\ell$ can be $-2, -1, 0, 1, 2$.
(c) Would be i orbitals.
(d) $n = 5$, $\ell = 0, 1, 2, 3, 4$; therefore 5 sublevels.
(e) $\ell = 2$ is a d orbital.

23.

(a) 108 – $6d^6\ 7s^2$
(b) 116 – $7s^2\ 7p^4$
(c) 120 – $8s^2$

24.

(a) 4
(b) 2
(c) 0

25.

(a) 107
(b) 119
(c) 116

Chapter 5

1.

(a) IA
(b) VA
(c) IVA
(d) IIA
(e) VIIA

2.

(a) 2
(b) 6
(c) 5
(d) 4
(e) 6

3.

(a) representative
(b) transition
(c) noble gas
(d) inner transition
(e) transition

4.

(a) 7
(b) 5
(c) 3
(d) 6
(e) 4

5.

(a) $5s^2\ 5p^5$
(b) $2s^2\ 2p^4$
(c) $4s^2\ 4p^3$
(d) $4s^2$
(e) $5s^2\ 5p^2$

6.

(a) K·

(b) $\cdot\ddot{\text{S}}:$

(c) $\cdot\dot{\text{Si}}\cdot$

(d) $\cdot\ddot{\text{Cl}}:$

(e) $\text{H}\cdot$

7.

(a) $4s^2\ 3d^2$
(b) $4s^2\ 3d^6$
(c) $4s^2\ 3d^7$
(d) $6s^2\ 5d^{10}$
(e) $4s^1\ 3d^{10}$

8.

(a) $2p$
(b) $4s$
(c) $3p$
(d) $5p$
(e) $3d$

9.

(a) 1
(b) 3
(c) 2
(d) 1
(e) 6

10.

(a) Cl
(b) Li
(c) F

11.

(a) Zn is greater because it has a smaller radius and greater nuclear charge.
(b) Na^+ is greater because it has an electron configuration of a noble gas.
(c) Mg^+ is greater because of the + charge.

12.

(a) Mg, Na, K, Rb, Cs
(b) Ne, F, O, S, As

13.

(a) Rb, Na, Mg, N, O, F
(b) Sb, As, P, O, F

14.

(a) Se^{2-}, Kr, Rb^+, Sr^{2+}
(b) Cl^-, Ar, K^+, Ca^{2+}
(c) Te^{2-}, I^-, Cs^+, Ba^{2+}

15.

(a) +2
(b) +2, +4
(c) +3
(d) +2, +3
(e) +1, +2

16.

(a) +6, −2
(b) +5, −3
(c) +7, −1

17.

(a) +3
(b) +4
(c) +7
(d) +5
(e) +6

18.

(a) They each have the same number of electrons but K has one e^- in the s orbital, while Sc^{2+} has one e^- in the d orbital.

19.

(a) $4s^2\ 3d^{10}$ and $5s^2\ 4d^{10}$
(b) $2s^2$ and $5s^2$; $2s^2\ 2p^4$ and $4s^2\ 4p^4$

20.

(a) He, Ne, Zn, Cd, Te
(b) Be, O, F, Se, Sr

21.

(a) As
(b) Te
(c) Mg

22.

(a) Cs_2SO_4
(b) $Mg_3(AsO_4)_2$
(c) Na_2SeO_4
(d) Ga_2O_3
(e) $Ba_3(PO_4)_2$

23.

(a) Group VIB, $6d^4\ 7s^2$
(b) Group IVA, $7s^2\ 7p^2$
(c) Group IA, $8s^1$

24.

(a) +3, +5
(b) +2

Chapter 6

1.
(a) sulfur dioxide
(b) dinitrogen trioxide
(c) phosphorus pentabromide

2.
(a) PBr_3
(b) P_2O_5
(c) SO_3

3.
(a) magnesium chloride
(b) tin(IV) chloride or stannic chloride
(c) mercury(II) bromide or mercuric bromide
(d) lead(IV) sulfide or plumbic sulfide
(e) iron(II) oxide or ferrous oxide

4.
(a) $MgBr_2$
(b) Al_2O_3
(c) $FeBr_2$
(d) PbS
(e) SnO_2

5.
(a) ammonium chloride
(b) calcium cyanide

6.
(a) $Mg(CN)_2$
(b) $(NH_4)_2S$

7.
(a) hydrochloric acid
(b) hydrosulfuric acid

8.
(a) HBr
(b) HCN

9.
(a) magnesium hydroxide
(b) iron(III) hydroxide or ferric hydroxide
(c) zinc hydroxide
(d) tin(IV) hydroxide or stannic hydroxide

10.
(a) $Ba(OH)_2$
(b) $Al(OH)_3$
(c) $Hg_2(OH)_2$
(d) $Pb(OH)_2$

11.
(a) NH_4^+
(b) CO_3^{2-}
(c) ClO_2^-
(d) Sn^{2+}
(e) PO_3^{3-}
(f) BrO_3^-
(g) IO_4^-
(h) NO_3^-
(i) SO_3^{2-}
(j) ClO^-

12.
(a) sulfide
(b) sulfite
(c) phosphate
(d) lead(II) or plumbous
(e) perchlorate
(f) chloride
(g) chlorate
(h) hypobromite
(i) iodite
(j) chromate

13.
(a) sulfurous acid
(b) calcium phosphate
(c) phosphorous acid
(d) magnesium chlorate
(e) ammonium nitrite
(f) hypobromous acid
(g) lead(IV) sulfate or plumbic sulfate
(h) tin(II) iodate or stannous iodate
(i) nitrous acid
(j) aluminum carbonate

14.
(a) $Ca(NO_3)_2$
(b) $Fe_2(SO_4)_3$
(c) H_2SO_3
(d) $Hg_3(PO_3)_2$
(e) $Mg(ClO_3)_2$
(f) $Pb(CO_3)_2$
(g) $(NH_4)_3PO_4$
(h) $SnSO_3$
(i) $Fe(IO_2)_2$
(j) $HBrO_4$

15.

(a) aluminum hydrogen sulfate (or bisulfate—hydrogen is preferred)
(b) potassium dihydrogen phosphate
(c) copper(II) hydrogen carbonate or cupric hydrogen carbonate (or bicarbonate)
(d) calcium hydrogen phosphite
(e) ammonium hydrogen sulfite

16.

(a) $NaHSO_3$
(b) $Fe(HCO_3)_2$
(c) $Ca(H_2PO_4)_2$
(d) K_2HPO_4
(e) $(NH_4)_2HPO_3$

17.

(a) sodium calcium phosphate
(b) dipotassium barium sulfate

18.

(a) $NaAl(SO_4)_2$
(b) $Na_2Ca(NO_3)_4$

19.

(a) potassium selenate
(b) aluminum silicate

20.

(a) Ag_3AsO_4
(b) Cs_3P

21.

(a) nitrous acid
(b) calcium bromide
(c) dinitrogen tetroxide
(d) phosphoric acid
(e) chromium(III) hydroxide
(f) cadmium(II) nitrate
(g) iron(III) oxide or ferric oxide
(h) potassium hydroxide
(i) mercury(II) nitrite or mercuric nitrite
(j) perbromic acid
(k) potassium dihydrogen phosphate
(l) ammonium iodide
(m) magnesium cyanide
(n) iron(II) sulfate or ferrous sulfate
(o) tin(IV) sulfide or stannic sulfide
(p) silver phosphate
(q) magnesium nitride
(r) sodium potassium sulfate
(s) barium chromate
(t) sodium hydrogen sulfite
(u) lithium perbromate
(v) ammonium hydrogen carbonate (or bicarbonate)
(w) strontium silicate
(x) cobalt(II) arsenate
(y) potassium permanganate

22.

(a) Ag_3PO_4
(b) $Ca(NO_3)_2$
(c) SnI_4
(d) $Zn(HCO_3)_2$
(e) SO_2
(f) $KHSO_3$
(g) Li_2SO_3
(h) $Al_2(SO_4)_3$
(i) $(NH_4)_3N$
(j) PbS
(k) $HBrO_3$
(l) $KCaPO_4$
(m) $Fe(IO_3)_2$
(n) $Zn(CN)_2$
(o) PCl_3
(p) $Hg(OH)_2$
(q) $FeCl_2$
(r) $(NH_4)_2CO_3$
(s) $Pb(SiO_3)_2$
(t) $Cu(NO_2)_2$
(u) $SnCO_3$
(v) $FeBr_2$
(w) $Mg_3(AsO_4)_2$
(x) $Al(BrO)_3$
(y) $Ba(H_2PO_4)_2$

Chapter 7

1.

(a) $3s^2\ 3p^6$
(b) $3s^2\ 3p^6$
(c) $3s^2\ 3p^6$
(d) $3s^2\ 3p^6$
(e) $3d^{10}$

(f) $3d^{10}$
(g) $3d^5$
(h) $3d^3$
(i) $5s^2\ 4d^{10}$
(j) $3d^5$

2.

(a) Na^+
(b) Al^{3+}
(c) Br^-
(d) N^{3-}
(e) Ba^{2+}
(f) Ti^{2+}
(g) Fe^{2+}, Fe^{3+}
(h) Sn^{2+}

3.

(a) HF
(b) H_2O
(c) ICl
(d) PCl_3
(e) OF_2

4.

(a) N—N, N—O, N—F, Mg—O

5.

(a) 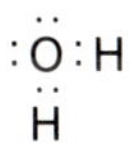

(b) 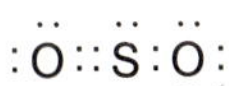

(c) 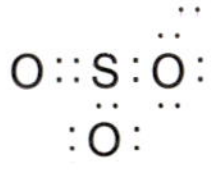

(d) 

(e) 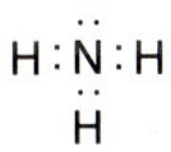

(f) 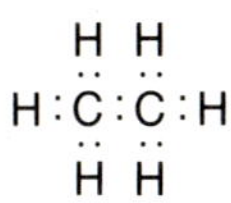

(g) 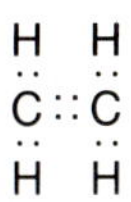

(h) H : C ::: C : H

(i) H : C ::: N :

(j) 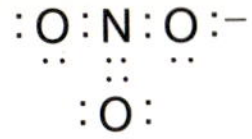

(k) 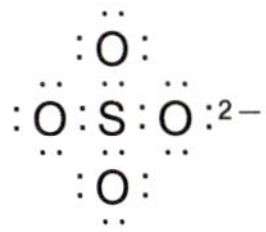

(l) 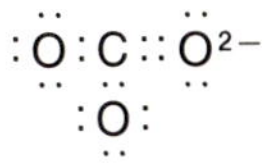

(m) 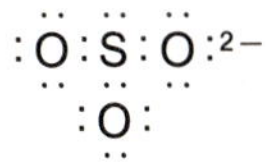

(n) 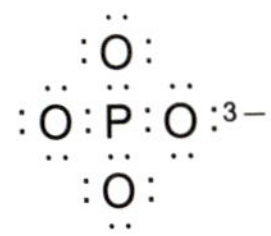

(o) 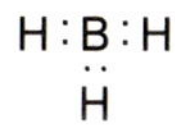

(p) 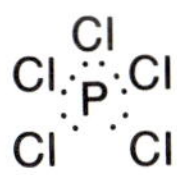

(q) 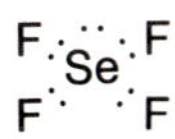

(r) 

(s) 

(t) 

6 through 10.

	Resonance Forms	Bond Angles (°)	Sigma (σ) Bonds	Pi (π) Bonds	Nonbonding Pairs	Hybrid Orbitals	Polar
(a)	No	< 109	2	0	2	sp^3	Yes
(b)	Yes—2	< 120	2	1	1	sp^2	Yes
(c)	Yes—3	120	3	1	0	sp^2	No
(d)	No	< 109	2	0	2	sp^3	Yes
(e)	No	< 109	3	0	1	sp^3	Yes
(f)	No	109	4 (each C)	0	0	sp^3	No
(g)	No	120	3 (each C)	1	0	sp^2	No
(h)	No	180	2 (each C)	2	0	sp	No
(i)	No	180	2	2	0	sp	Yes
(j)	Yes—3	120	3	1	0	sp^2	No
(k)	No	109	4	0	0	sp^3	No
(l)	Yes—3	120	3	1	0	sp^2	No
(m)	No	< 109	3	0	1	sp^3	Yes
(n)	No	109	4	0	0	sp^3	No
(o)	No	120	3	0	0	sp^2	No
(p)	No	90 and 120	5	0	0	dsp^3	No
(q)	No	< 90 and 120	4	0	1	dsp^3	Yes
(r)	No	90	6	0	0	d^2sp^3	No
(s)	No	< 90	5	0	1	d^2sp^3	Yes
(t)	No	90	4	0	2	d^2sp^3	No

11.

(a) CH_4 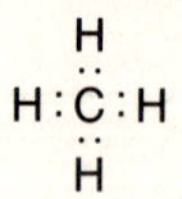
sp^3 orbitals, angles of 109°; 4 sigma bonds, 0 pi bonds; nonpolar

(b) SiO_2 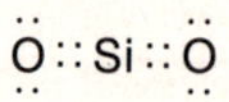
sp orbitals, angles of 180°; 2 sigma bonds, 2 pi bonds; nonpolar, three resonance forms

(c) SeO_2 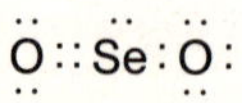
sp^2 orbitals, angles <120°; 2 sigma bonds, 1 pi bonds; polar, two resonance forms

(d) ClF_3 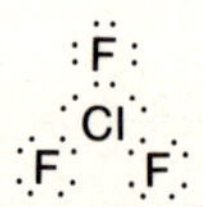
dsp^3 orbitals, angles <90°; 3 sigma bonds, 0 pi bonds; polar

(e) IF_5 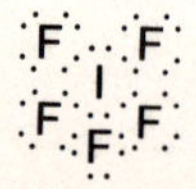
d^2sp^3 orbitals, angles <90°; 5 sigma bonds, 0 pi bonds; polar

12.

	Bond Order	Unpaired Electrons
(a)	1	0
(b)	0	0
(c)	3	0

	Bond Order	Unpaired Electrons
(d)	3	0
(e)	2.5	1

Chapter 8

1.

(a) $CaCl_{2(aq)} + (NH_4)_2CO_{3(aq)} \longrightarrow CaCO_{3(s)} + 2\ NH_4Cl_{(aq)}$

(b) $2\ KClO_{3(s)} \longrightarrow 2\ KCl_{(s)} + 3\ O_{2(g)}$

(c) $2\ Fe(OH)_{3(s)} + 3\ H_2SO_{4(aq)} \longrightarrow Fe_2(SO_4)_{3(aq)} + 6\ H_2O_{(l)}$

(d) $2\ K_{(s)} + 2\ H_2O_{(l)} \longrightarrow 2\ KOH_{(aq)} + H_{2(g)}$

(e) $3\ Mg_{(s)} + N_{2(g)} \longrightarrow Mg_3N_{2(s)}$

2.

(a) $CaCl_{2(aq)} + Pb(NO_3)_{2(aq)} \longrightarrow PbCl_{2(s)} + Ca(NO_3)_{2(aq)}$

(b) $Fe_2O_{3(s)} + 6\ HCl_{(aq)} \longrightarrow 2\ FeCl_{3(aq)} + 3\ H_2O_{(l)}$

(c) $3\ NaHCO_{3(aq)} + H_3PO_4 \longrightarrow Na_3PO_{4(aq)} + 3\ CO_{2(g)} + 3\ H_2O_{(l)}$

(d) $2\ Hg_{(l)} + O_{2(g)} \longrightarrow 2\ HgO_{(s)}$

(e) $Ba(NO_3)_{2(aq)} + H_2SO_{4(aq)} \longrightarrow BaSO_{4(s)} + 2\ HNO_{3(aq)}$

3.

(a) $2\ Ca_{(s)} + O_{2(g)} \longrightarrow 2\ CaO_{(s)}$

(b) $C_{(s)} + \text{excess } O_{2(g)} \longrightarrow CO_{2(g)}$

(c) $CaO_{(s)} + H_2O_{(l)} \longrightarrow Ca(OH)_{2(s)}$ [or $Ca(OH)_{2(aq)}$]

(d) $SO_{2(g)} + H_2O_{(l)} \longrightarrow H_2SO_{3(aq)}$

(e) $4\ Al_{(s)} + 3\ O_{2(g)} \longrightarrow 2\ Al_2O_{3(s)}$

(f) $2\ Al_{(s)} + N_{2(g)} \longrightarrow 2\ AlN_{(s)}$

(g) $Al_2O_{3(s)} + 3\ H_2O_{(l)} \longrightarrow 2\ Al(OH)_{3(s)}$

(h) $P_4O_{10(s)} + 6\ H_2O_{(l)} \longrightarrow 4\ H_3PO_{4(aq)}$

4.

(a) $2\ HgO_{(s)} \xrightarrow{\Delta} 2\ Hg_{(l)} + O_{2(g)}$

(b) $2\ KClO_3 \xrightarrow{\Delta} 2\ KCl_{(s)} + 3\ O_{2(g)}$

(c) $MgCO_{3(s)} \xrightarrow{\Delta} MgO_{(s)} + CO_{2(g)}$

(d) $Na_2CO_3 \cdot H_2O_{(s)} \xrightarrow{\Delta} Na_2CO_{3(s)} + H_2O_{(g)}$

(e) $Ca(HCO_3)_{2(s)} \xrightarrow{\Delta} CaO_{(aq)} + 2\ CO_{2(s)} + H_2O_{(l)}$

5.

(a) $Pb_{(s)} + 2\ HCl_{(aq)} \longrightarrow PbCl_{2(s)} + H_{2(g)}$

(b) $Ca_{(s)} + 2\ H_2O_{(l)} \longrightarrow Ca(OH)_{2(aq)} + H_{2(g)}$

(c) $Fe_{(s)} + CuCl_{2(aq)} \longrightarrow FeCl_{2(aq)} + Cu_{(s)}$

(d) $Cl_{2(g)} + 2\ NaBr_{(aq)} \longrightarrow 2\ NaCl_{(aq)} + Br_{2(l)}$

(e) $2\ Al_{(s)} + 6\ HCl_{(aq)} \longrightarrow 2\ AlCl_{3(aq)} + 3\ H_{2(g)}$

(f) $2\ Al_{(s)} + 3\ SnCl_{2(aq)} \longrightarrow AlCl_{3(aq)} + 3\ Sn_{(s)}$

(g) $Hg_{(l)} + SnCl_{2(aq)} \longrightarrow$ no reaction

6.

(a) $Pb(NO_3)_{2(aq)} + 2\ HCl_{(aq)} \longrightarrow PbCl_{2(s)} + 2\ HNO_{3(aq)}$

(b) $Pb(NO_3)_{2(aq)} + H_2S_{(aq)} \longrightarrow PbS_{(s)} + 2HNO_{3(aq)}$

(c) $Bi(NO_3)_{3(aq)} + 3\ NaOH_{(aq)} \longrightarrow 3\ NaNO_{3(aq)} + Bi(OH)_{3(s)}$

(d) $MnSO_{4(aq)} + (NH_4)_2S_{(aq)} \longrightarrow (NH_4)_2SO_{4(aq)} + MnS_{(s)}$

(e) $CaCO_{3(s)} + 2\ HCl_{(aq)} \longrightarrow CaCl_{2(aq)} + CO_{2(g)} + H_2O_{(l)}$

(f) $3\ ZnCO_{3(s)} + 2\ H_3PO_{4(aq)} \longrightarrow Zn_3(PO_4)_{2(s)} + 3\ CO_{2(g)} + 3\ H_2O_{(l)}$

7.

(a) $Zn(OH_{2(s)} + H_2SO_{4(aq)} \longrightarrow 2\ H_2O_{(l)} + ZnSO_{4(aq)}$

(b) $BaO_{(s)} + H_2SO_{4(aq)} \longrightarrow H_2O_{(l)} + BaSO_{4(s)}$

(c) $CO_{2(s)} + Ca(OH)_{2(aq)} \longrightarrow H_2O_{(l)} + CaCO_{3(s)}$

(d) $Al(OH)_{3(s)} + 3\ HCl_{(aq)} \longrightarrow 3\ H_2O_{(l)} + AlCl_{3(aq)}$

(e) $2\ KOH_{(aq)} + CO_{2(g)} \longrightarrow H_2O_{(l)} + K_2CO_{3(aq)}$

(f) $BaO_{(s)} + 2\ HCl_{(aq)} \longrightarrow H_2O_{(l)} + BaCl_{2(aq)}$

(g) $CaO_{(s)} + CO_{2(g)} \longrightarrow CaCO_{3(s)}$

(h) $NH_{3(g)} + HBr_{(g)} \longrightarrow NH_4Br_{(s)}$

8.

(a) $BaO_{(s)} + H_2O_{(l)} \longrightarrow Ba(OH)_{2(aq)}$ [or $Ba(OH)_{2(s)}$]

(b) $MgCO_{3(s)} \xrightarrow{\Delta} MgO_{(s)} + CO_{2(g)}$

(c) $Ca_{(s)} + ZnCl_{2(aq)} \longrightarrow CaCl_{2(aq)} + Zn_{(s)}$

(d) $NaCl_{(aq)} + Br_{2(g)} \longrightarrow$ no reaction

(e) $Mg_{(s)} + Br_{2(l)} \longrightarrow MgBr_{2(s)}$

(f) $P_2O_{5(s)} + 3\ H_2O_{(l)} \longrightarrow 2\ H_3PO_{4(aq)}$

(g) $Ca_{(s)} + 2\ HCl_{(aq)} \longrightarrow CaCl_{2(aq)} + H_{2(g)}$

(h) $ZnO_{(s)} + 2\ HClO_{3(aq)} \longrightarrow Zn(ClO_3)_{2(aq)} + H_2O_{(l)}$

(i) $H_2SO_{4(aq)} + BaCl_{2(aq)} \longrightarrow BaSO_{4(s)} + 2\ HCl_{(aq)}$

(j) $2\ KHCO_{3(s)} \xrightarrow{\Delta} K_2CO_{3(s)} + CO_{2(g)} + H_2O_{(g)}$

(k) $Sn_{(s)} + CaCl_{2(aq)} \longrightarrow$ no reaction

(l) $3\ CO_{2(g)} + Al_2O_{3(s)} \longrightarrow Al_2(CO_3)_{3(s)}$

(m) $BaCl_2 \cdot 2\ H_2O_{(s)} \xrightarrow{\Delta} BaCl_{2(s)} + 2\ H_2O_{(g)}$

(n) $3\ SO_{3(g)} + 2\ Fe(OH)_{3(aq)} \longrightarrow Fe(SO_4)_{3(aq)} + 3\ H_2O_{(l)}$

(o) $2\ HgO_{(s)} \xrightarrow{\Delta} 2\ Hg_{(l)} + O_{2(g)}$

(p) $Cu_{(s)} + AlCl_{3(aq)} \longrightarrow$ no reaction

(q) $2\ NaHCO_{3(s)} \xrightarrow{\Delta} CO_{2(g)} + Na_2CO_{3(aq)} + H_2O_{(g)}$

(r) $2\ Al_{(s)} + 3\ H_2SO_{4(aq)} \xrightarrow{\Delta} Al_2(SO_4)_{3(aq)} + 3\ H_{2(g)}$

(s) $Ca(HCO_3)_2 \xrightarrow{\Delta} 2\ CO_{2(aq)} + H_2O_{(g)} + CaO_{(s)}$

(t) $Na_2S_{(s)} + ZnCl_{2(aq)} \longrightarrow 2\ NaCl_{(aq)} + ZnS_{(s)}$

(u) $Ca(OH)_{2(s)} + 2\ HCl_{(aq)} \longrightarrow CaCl_{2(aq)} + 2\ H_2O_{(l)}$

(v) $2\ Al_{(s)} + 3\ FeSO_{4(aq)} \longrightarrow Al_2(SO_4)_{3(aq)} + 3\ Fe_{(s)}$

(w) $MgO_{(s)} + 2\ HNO_{2(aq)} \longrightarrow Mg(NO_2)_{2(aq)} + H_2O_{(l)}$

(x) $N_2O_{3(g)} + H_2O_{(l)} \longrightarrow 2\ HNO_{2(aq)}$

(y) $SO_{2(g)} + 2\ NaOH_{(aq)} \longrightarrow Na_2SO_{3(aq)} + H_2O_{(l)}$

Chapter 9

1.

(a) 57.5% Na
40.0% O
2.5% H

(b) 41.7% Mg
54.9% O
3.4% H

(c) 58.5% O
24.4% Ca
17.0% N

2.

(a) 65.2% O
(b) 37.0% O
(c) 43.0% O
(d) 39.6% O
(e) 43.1% O

3.

(a) 58.4%
(b) 71.8%
(c) 62.1%

4.

(a) 3.30 g Ca
(b) 135 g Ag
(c) 1.0 g $Zn(NO_3)_2$
(d) 159 mg Sn
(e) 0.703 g Fe

5.

(a) 2.5×10^2 g O
(b) 0.18 g O
(c) 426 g O
(d) 1.20 g O
(e) 1.5×10^{-8} g O

6.

(a) $FeBr_2$
(b) K_2SO_4
(c) $MnCl_2$
(d) Mg_3N_2
(e) CaS_2O_3

7.

(a) C_3H_3
(b) Hg_2Cl_2

8.

(a) $Na_2S_2O_2$
(b) N_2O_4
(c) C_4H_{10}

9.

(a) 3.60 mol Fe_2O_3
(b) 119 g SO_2
(c) 7.96×10^{-2} g O_2
(d) 0.231 mol SO_2
(e) 1.15×10^8 mg FeS

10.

(a) 2.10×10^{-2} mol C_2H_6
(b) 11.8 g O_2
(c) 2.1×10^5 mmol CO_2
(d) 5.36×10^9 mg CO_2
(e) 1.2×10^{-2} ℓ H_2O

11.

(a) 12.9 g $Ca_3(PO_4)_2$
(b) 335 g $Ca_3(PO_4)_2$
(c) 0.0546 g $Ca_3(PO_4)_2$

12.

(a) 9.13 g H_3PO_4 excess
(b) 263 g H_3PO_4 excess

(c) 0.284 g H_3PO_4 excess

13.

(a) 5.06 g $Al_2(SO_4)_3$
(b) 0.0444 mol H_2
(c) 1.45 g Al in excess

14.

(a) 86.5%
(b) 59.9%
(c) 78.1%

15.

(a) 76.1%
(b) 74.6%

16.

(a) 0.152 mol $Ca(NO_3)_2$
(b) 1.69×10^3 mg CO_2
(c) 60.8%
(d) 0.9 g HNO_3 excess
(e) 4.83 g $CaCO_3$

17.

(a) 5.67 g O_2
(b) 1.56 g O_2
(c) 25 g O_2
(d) 69.4 g O_2

18.

(a) Hg_2Br_2

Chapter 10

1.

(a) exothermic
(b) −347 kcal
(c) 138 kcal; 577 kJ
(d) 728 kcal; 3040 kJ
(e) 347 kcal
(f) −173.5 kcal

3.

(a) exothermic, −27 kcal
(b) exothermic, −683 kcal
(c) exothermic, −22.0 kcal
(d) exothermic, −305 kcal

4.

(a) 54 kcal

5.

(a) −189 kcal

6.

(a) $\Delta H = -22.0$
$\Delta S = -0.047$
$\Delta G = -8.0$
spontaneously
(b) $\Delta H = -135$
$\Delta S = -0.042$
$\Delta G = -122$
spontaneously
(c) $\Delta H = -600$
$\Delta S = -0.0468$
$\Delta G = -586$
spontaneously

7. 209 kcal

9.

(a) exothermic
(b) endothermic

Chapter 11

1.

(a) 0.999
(b) 522 K
(c) 597 torr
(d) 254 K = −19°C
(e) 0.758 atm

2.

(a) 0.949
(b) 0.648 l
(c) 214 K
(d) 0.868 l
(e) −225°C

3.

(a) 680 torr

4.

(a) 223 mℓ
(b) 538 mℓ

5.

(a) 1.96 g/ℓ
(b) 1.14 g/ℓ

6.

(a) 39.2 g/mol
(b) 75.3 g/mol
(c) 40 g/mol
(d) 85.2 g/mol

7.

(a) 4.36
(b) 3.12

8.

(a) 2.98 ℓ
(b) 4.23 ℓ

9.

(a) 10.8
(b) 2.12
(c) 52.2 ℓ
(d) 21.5 ℓ
(e) 0.386 ℓ

10.

(a) 0.0331
(b) 55.0
(c) 0.320

11.

(a) 3.8/1

12.

(a) 18 amu

13.

(a) 341 torr
(b) 1.7×10^3 torr
(c) 2.0×10^3 torr

14.

(a) -21.0
(b) 422

15.

(a) 2.3

16.

(a) 0.76
(b) 1.25

17.

(a) 5.31
(b) 6.54

18.

(a) 54
(b) 6.2

19.

(a) 4.28
(b) 366
(c) 167

20.

(a) 0.805
(b) 0.746

21.

(a) 0.0738

22.

(a) 138
(b) 21.8
(c) 1.7

Chapter 12

1. He, Br_2, NH_3, H_2S, H_2O, NaCl

2.

(a) 1.48
(b) 0.897
(c) 0.456
(d) 405
(e) 0.255
(f) 0.263
(g) 69.4
(h) 5.94

3. 2.17

5. 3.0×10^3 g

Chapter 13

1.
(a) 0.0385

2.
(a) 9.05%
(b) 20.0%

3.
(a) 61.8

4.
(a) 48.0

5.
(a) 571

6.
(a) 0.060
(b) 0.11

7.
(a) 5.85
(b) 3.57
(c) 16.4

8.
(a) 8.99

9.
(a) 167

10.
(a) 5.56
(b) 1.11
(c) 0.213
(d) 4.59
(e) 2.44

11.
(a) 2.0
(b) 0.56

12.
(a) 167
(b) 485
(c) 0.049

13.
(a) 37.6
(b) 115
(c) 76.9

14.
(a) 5.38
(b) 2.91
(c) 2.44

15.
(a) 23.8 torr

16.
(a) 78.1
(b) 210

17.
(a) $-0.37°C$; $100.1°C$
(b) $-5.0°C$; $101.4°C$
(c) $-1.4°C$; $100.38°C$

18.
(a) $-37°C$

19.
(a) 4.68%

20.
(a) 0.144 *m*

21.
(a) 66.7

22.
(a) 0.152 *M*

23.
(a) 2.13

24.
(a) 6.15 *M*, 7.57 *m*, 32.3%

25.
(a) 0.39

26.
(a) 1.2

27.
(a) 4.0 *M*

28.
(a) 1.1

29.
(a) 16

30.
(a) 69

31.
(a) 0.0982 *m*, 0.10 *M*, 0.00176

32.
(a) 31.4%, 5.68 *M*, 0.1138

33.
(a) 22.3%, 1.75 *m*, 0.0306

Chapter 14

1.
(a) V
(b) I, III, V
(c) I, III, IV, V, VI
(d) III, V
(e) II, IV, VI
(f) IV, VI

2.
(a) III
(b) I
(c) IV
(d) II
(e) I, II
(f) I, II

3.
(a) 0.303
(b) 2.37

4.
(a) 32.7 g
(b) 37.0 g
(c) 51.7 g
(d) 49.0 g

5.
(a) 0.694
(b) 1.55
(c) 0.320
(d) 8.0
(e) 3.66
(f) 6.9

6.
(a) 1.09
(b) 0.390
(c) 3.02
(d) 1.97
(e) 2.12

7.
(a) 82.4 mℓ
(b) 28.3 mℓ
(c) 50.3 mℓ

8.
(a) 92.6
(b) 60.4

9.
(a) $Ba^{2+} + 2\,Cl^- + 2\,Na^+ + CO_3^{2-} \longrightarrow \underline{BaCO_3} + 2\,Na^+ + 2\,Cl^-$
(net) $Ba^{2+} + CO_3^{2-} \longrightarrow \underline{BaCO_3}$
(b) $Ag^+ + NO_3^- + H^+ + Cl^- \longrightarrow \underline{AgCl} + H^+ + NO_3^-$
(net) $Ag^+ + Cl^- \longrightarrow \underline{AgCl}$
(c) $NH_4^+ + Cl^- + Ba^{2+} + 2C_2H_3O_2^- \longrightarrow$ no reaction
(d) $2\,H^+ + 2\,NO_3^- + CaCO_3 \longrightarrow CO_{2(g)} + H_2O + Ca^{2+} + 2\,NO_3^-$
(net) $2\,H^+ + CaCO_3 \longrightarrow CO_{2(g)} + H_2O$
(e) $H^+ + NO_3^- + NH_{3(aq)} \longrightarrow NH_4^+ + NO_3^- + H_2O$
(net) $H^+ + NH_3 \longrightarrow NH_4^+$
(f) $Na^+ + OH^- + NH_4^+ + Cl^- \longrightarrow NH_3 + H_2O + Na^+ + Cl^-$
(net) $NH_4^+ + OH^- \longrightarrow NH_3 + H_2O$

10.
(a) (i) 0.940
(ii) 3.36%
(b) (i) 1.04
(ii) 3.71%

11.
(a) 0.245

12.
(a) 143

13.

(a) 8.00%

14.

(a) 2.0
(b) 3.5

15. 38.8

17. 12.8

19. 2.7

21. 157

Chapter 15

1.

(a) rate $= k[N_2O_4]$
first order in N_2O_4; first order overall
(b) rate $= k[C_2H_4][H_2]$
first order in C_2H_4; first order in H_2; second order overall

2.

(a) $R = k[N_2O_5]$
There must be a slow, rate determining step, in which N_2O_5 forms an intermediate product which then reacts with more N_2O_5 to form the product.

3.

(a) four times as fast ($2^2 = 4$)
(b) eight times as fast ($2^2 \cdot 2^1 = 8$)
(c) $(3^2)(\frac{1}{2}) = (9)(\frac{1}{2}) = 4.5$ times as fast

Chapter 16

1.

(a) $K_{eq} = \dfrac{[HCl]^2}{[H_2][Cl_2]}$

(b) $K_{eq} = \dfrac{[SO_2]}{[O_2]}$

(c) $K_{eq} = \dfrac{[CO_2][H_2O]}{[CH_4][O_2]^2}$

(d) $K_{eq} = \dfrac{[SO_3]^2}{[SO_2]^2[O_2]}$

(e) $K_{eq} = \dfrac{[CO_2]^2}{[CO]^2}$

2.

(a) $K_{eq} = \dfrac{[C]^3[D]}{[A]^2[B]} = \dfrac{(0.20)^3(0.50)}{(0.30)^2(0.60)} = 0.074$

3.

(a) $K_{eq} = \dfrac{[NO]^2[O_2]}{[NO_2]^2}$
(b) right
(c) right
(d) right
(e) increase
(f) decrease
(g) right
(h) increase

5.

(a) 5.6

6.

(a) 0.395

7.

(a) 62.5
(b) 0.13

8.

(a) 5.6
(b) 0.42

Chapter 17

1.

(a) $K_{sp} = [Ag^+][Cl^-]$
(b) $K_{sp} = [Pb^{2+}][Cl^-]^2$
(c) $K_{sp} = [Bi^{3+}]^2[S^{2-}]^3$
(d) $K_{sp} = [Ba^{2+}]^3[PO_4^{3-}]^2$

2.

(a) 3.2×10^{-10}
(b) 3.9×10^{-5}
(c) 1.8×10^{-2}
(d) 3.7×10^{-5}

3.

(a) 8.5×10^{-15}
(b) 1.25×10^{-12}
(c) 1.0×10^{-7}

4.

(a) 1.6×10^{-9}
(b) 8.8×10^{-12}
(c) 1.01×10^{-50}
(d) 6.81×10^{-31}
(e) 5.16×10^{-13}

5.

(a) Precipitate will form.
(b) Precipitate will not form.
(c) Precipitate will not form.

6.

(a) 1.8×10^{-5}
(b) 1.8×10^{-5}
(c) 6.8×10^{-4}

7.

(a) 1.3×10^{-3}
(b) 4.5×10^{-6}

8.

(a) 4.2×10^{-4}

9.

(a) 3.6×10^{-5}
(b) 3.45×10^{-4}
(c) 1.6×10^{-10}

10.

(a) 1.3%
(b) 4.2%
(c) 6.5%

11.

(a) 0.25

12.

(a) 2.1

13.

(a) 7.2

14.

(a) 4.8

15.

(a) 0.11
(b) 6.4×10^{-3}

16.

(a) 1.0×10^{-10}
(b) 3.7×10^{-6}
(c) 1.0×10^{-5}
(d) 5.0×10^{-9}
(e) 1×10^{-6}
(f) 3.2×10^{-10}

17.

(a) 5
(b) 8.64
(c) 1.70
(d) 3.22
(e) 10
(f) 6.54
(g) 13.18
(h) 11.20

18.

(a) 10.42
(b) 9.10
(c) 8.93
(d) 13.36
(e) 7.0
(f) 4.92

19.

(a) 4.74
(b) 4.74
(c) 4.68
(d) 4.80

20.

(a) 0.25

21.

(a) 106

22.

(a) 6.9×10^{-5}

(b) 1.8×10^{-2}
(c) 2.6×10^{-4}
(d) 1.0×10^{-7}
(e) 7.7×10^{-10}

23.

(a) 0.46
(b) 14.40
(c) 1.66
(d) 10.98
(e) 4.74

24.

(a) 5.6×10^{-15}
(b) 7.18×10^{-2}
(c) 6.25×10^{-13}
(d) 6.3×10^{-9}
(e) 4.0×10^{-5}

25.

(a) 3.3×10^{-2}

26. 4.0×10^{-7}

Chapter 18

1.

(a) 1.3×10^{-13}
(b) 4.5×10^{-9}
(c) 1.3×10^{-17}
(d) 0.13
(e) 7.2×10^{-16}
(f) 2.2
(g) 1.0×10^{-12}
(h) 7.6×10^{-4}

2.

(a) 3.6×10^{-5}

3.

(a) 3.19

4.

(a) No precipitate forms.
(b) Precipitate forms.

5. 4.8×10^{-11}

6.

(a) 1.3×10^{-9} (= $[H^+]$)
(b) 7.5×10^{-6}
(c) 1.0×10^{-7}
(d) 1.0×10^{-7}

7.

(a) 9.08
(b) 5.48
(c) 7.00
(d) 7.00
(e) 8.81
(f) 5.00
(g) 9.25
(h) 4.46

8.

(a) 1.8×10^{-2}
(b) 4.5×10^{-2}

9.

(a) 3.8×10^{-7}

10.

(a) 0.16

11.

(a) 0.78

12.

(a) 5.8×10^{-20}
(b) 7.2×10^{-16}

13.

(a) 2.5×10^{-2}

14.

(a) 4.70
(b) 4.40
(c) 4.58
(d) 8.28
(e) 11.20
(f) 8.96
(g) 7.00
(h) 5.12

15. 8.8×10^{-11}
17. 10.62

Chapter 19

1. and 2.

	Oxidized (Reducing Agent)	Reduced (Oxidizing Agent)	Electrons Transferred
(a)	H_2	Cl_2	2
(b)	K	O_2	4
(c)	Zn	HCl	2
(d)	Al	O_2	12
(e)	Fe	S	6

3.

(a) 2, 1 $\longrightarrow$ 2, 1, 2
(b) 10, 1 $\longrightarrow$ 10, 2, 4
(c) 2, 3 $\longrightarrow$ 2, 3, 3
(d) 1, 5 $\longrightarrow$ 1, 5
(e) 4, 1 $\longrightarrow$ 1, 1, 2
(f) 2, 1 $\longrightarrow$ 1, 1 + 2 H_2O
(g) 2, 5, 16 $\longrightarrow$ 2, 5, 5 + 7 H_2O
(h) 1, 6, 14 $\longrightarrow$ 2, 6, 2 + 7 H_2O

4.

(a) 3, 1, 6 H^+ $\longrightarrow$ 3, 1 + 3 H_2O
(b) 5, 2, 6 H^+ $\longrightarrow$ 5, 2 + 3 H_2O
(c) 5, 2, 16 H^+ $\longrightarrow$ 10, 2 + 8 H_2O
(d) 2, 5, 14 H^+ $\longrightarrow$ 2, 5 + 7 H_2O
(e) 1, 1, 4 H^+ $\longrightarrow$ 1, 1 + 2 H_2O
(f) 3, 6 OH^- $\longrightarrow$ 1, 5 + 3 H_2O
(g) 2, 1, 4 OH^- $\longrightarrow$ 2 + 2 H_2O
(h) 4, 3, 2 H_2O $\longrightarrow$ 4, 3 + 4 OH^-

5.

(a) $8\,H^+ + MnO_4^- + 5e^- \longrightarrow Mn^{2+} + 4\,H_2O$
(b) $2\,OH^- + SnO_2^{2-} \longrightarrow SnO_3^{2-} + H_2O + 2e^-$
(c) $6e^- + 14\,H^+ + Cr_2O_7^{2-} \longrightarrow 2\,Cr^{3+} + 7\,H_2O$
(d) $2e^- + 2\,H^+ + ClO^- \longrightarrow Cl^- + H_2O$
(e) $8e^- + 7\,H_2O + NO_3^- \longrightarrow NH_4^+ + 10\,OH^-$

6.

(a) 6
(b) 3
(c) 6
(d) 8
(e) 4

7.

(a) 10, 2, 8 $\longrightarrow$ 5, 2, 1 + 8 H_2O
(b) 1, 2 $\longrightarrow$ 1, 1 + 2 H_2O
(c) 8, 5 $\longrightarrow$ 1, 4, 4 + 4 H_2O
(d) 1, 6, 8 $\longrightarrow$ 1, 3, 2 + 7 H_2O
(e) 2, 5, 3 $\longrightarrow$ 2, 5, 1 + 3 H_2O
(f) 2, 5, 16 $\longrightarrow$ 2, 5, 5 + 7 H_2O
(g) 3, 1, 7 $\longrightarrow$ 3, 1, 1 + 7 H_2O
(h) 6, 4 $\longrightarrow$ 1, 2 + 3 H_2O
(i) 2, 5, 2H_2O $\longrightarrow$ 1, 2, 2
(j) 4 $\longrightarrow$ 1, 3

8.

(a) 1, 1, 4 H^+ $\longrightarrow$ 1, 1 + 2 H_2O
(b) 1, 1, 2 H^+ $\longrightarrow$ 2, 1, 1
(c) 3, 1, 14 H^+ $\longrightarrow$ 3, 2 + 7 H_2O
(d) 1, 6, 14 H^+ $\longrightarrow$ 2, 6 + 7 H_2O
(e) 8, 3, 5 OH^- + 2 H_2O $\longrightarrow$ 3, 8
(f) 2, 3, 2OH^- $\longrightarrow$ 2, 3 + H_2O

9.

	Oxidized (Reducing Agent)	Reduced (Oxidizing Agent)
(a)	Cu	SO_4^{2-}
(b)	CdS (S^{2-})	I_2
(c)	Zn	$Cr_2O_7^{2-}$
(d)	Fe^{2+}	$Cr_2O_7^{2-}$
(e)	Al	NO_3^-
(f)	CrO_2^-	ClO^-

Chapter 20

1.

(a) 32.7
(b) 1.0
(c) 9.0
(d) 59.4
(e) 127

2.

(a) 0.466 g Al
(b) 5.6 g Al

3.

(a) 1.09×10^4 s
(b) 7.85×10^3 s
(c) 1.29×10^3 s
(d) 53.3 s

4.

(a) 9.0
(b) 154
(c) 11.2

5.

(a) 1.03×10^5 coul
(b) 398 coul

6.

(a) 0.32
(b) 0.59
(c) 1.22

7.

(a) 0.51; yes
(b) -0.63; no
(c) 1.26; yes

8.

(a) 0.25
(b) -0.22

9.

(a) 1.13
(b) 2.30
(c) -1.43

10.

(a) 2.8×10^6
(b) 4.6×10^{-1}

11.

(a) anode: $2\,Cl^- \longrightarrow Cl_2 + 2e^-$
cathode: $2\,Na^+ + 2e^- \longrightarrow 2\,Na$

12.

(a) -0.45; no

13. 5.6 ℓ Cl_2

15. 156

17. 5.31×10^5 coul

19. 1.2

21. -1.95

Chapter 21

1.

(a) F
(b) F
(c) K

2.

(a) 5
(b) 2
(c) 7

3.

(a) 3*p*
(b) 3*d*
(c) 5*p*

4.

(a) +2
(b) +3
(c) +2, +4

5.

	Low	*High*
(a)	K^+	Al^{3+}
(b)	Cs^+	Sn^{2+}
(c)	Cs^+	Pb^{2+}
(d)	K^+	Mg^{2+}

6.

	Weak	*Strong*
(a)	NH_3	HF
(b)	HF	HBr
(c)	H_2S	HBr
(d)	HBrO	$HBrO_4$

(e) HIO_2 $HClO_2$
(f) sulfurous acid sulfuric acid

7.

(a) Basic
(b) Basic
(c) Acidic
(d) Basic
(e) Acidic

8.

(a) Most: K_2O, Least: SO_2
(b) Most: CaO, Least: SnO_2

9.

(a) $2\,HNO_3 + Ca(OH)_2 \longrightarrow 2\,H_2O + Ca(NO_3)_2$
(b) $Na_2O + 2\,HCl \longrightarrow H_2O + 2\,NaCl$
(c) $MgO + SO_2 \longrightarrow MgSO_3$
(d) $2\,HCl + FeO \longrightarrow H_2O + FeCl_2$
(e) $2\,Al(OH)_3 + 3\,N_2O_3 \longrightarrow 2\,Al(NO_2)_3 + 3\,H_2O$

10.

(a) $2\,CuS + 3\,O_2 \xrightarrow{\Delta} 2\,CuO + 2\,SO_2$
(b) $CuO + C \xrightarrow{\Delta} Cu + CO$
(c) $CuCl_2 \longrightarrow Cu + Cl_2$

11.

(a) $2\,Al + 3\,FeCl_2 \longrightarrow 3\,Fe + 2\,AlCl_3$
(b) $SnO_2 + 2\,H_2 \xrightarrow{\Delta} Sn + 2\,H_2O$
(c) $Ca + H_2 \xrightarrow{\Delta} CaH_2 + \text{heat}$
(d) $Mg + 2\,HCl \longrightarrow H_2 + MgCl_2$
(e) $2\,Al + 6\,H_2O \xrightarrow{\Delta} 3\,H_2 + 2\,Al(OH)_3$

12.

(a) nuclear charge—same energy level
(b) size
(c) size
(d) Na_2O is more ionic.
(e) more oxygen
(f) more oxygen in bromic
(g) HNO_2 is a weak acid; HNO_3 is strong.
(h) HClO is a weaker acid, due to less oxygen.

13.

(a) $2\,Al + 3\,ZnCl_2 \longrightarrow 3\,Zn + 2\,AlCl_3$
(b) $Ca + 2\,H_2O \longrightarrow H_2 + Ca(OH)_2$
(c) $Ba + H_2 \xrightarrow{\Delta} BaH_2$
(d) $Al_2O_3 + 6\,HNO_3 \longrightarrow 3\,H_2O + 2\,Al(NO_3)_3$
(e) $Al_2O_3 + 2\,NaOH \longrightarrow H_2O + 2\,NaAlO_2$
(f) $N_2O_3 + Ca(OH)_2 \longrightarrow H_2O + Ca(NO_2)_2$
(g) $HClO_3 + KOH \longrightarrow H_2O + KClO_3$
(h) $2\,HClO_2 + Ca \longrightarrow H_2 + Ca(ClO_2)_2$
(i) $Zn(OH)_2 + 2\,NaOH \longrightarrow Na_2ZnO_2 + 2\,H_2O$

Chapter 22

1.

(a) $2\,H_2 + O_2 \longrightarrow 2\,H_2O$
(b) $3\,H_2 + N_2 \longrightarrow 2\,NH_3$
(c) $H_2 + 2\,Na \longrightarrow 2\,NaH$
(d) $NaH + H_2O \longrightarrow H_2 + NaOH$
(e) $Zn + 2\,HCl \longrightarrow H_2 + ZnCl_2$

2.

(a) $2\,K + 2\,H_2O \longrightarrow H_2 + 2\,KOH$
(b) $2\,Cs + H_2 \longrightarrow 2\,CsH$
(c) $2\,Na + O_2 \longrightarrow Na_2O_2$
(d) $4\,Li + O_2 \longrightarrow 2\,Li_2O$
(e) $Na_2O + 2\,HNO_3 \longrightarrow H_2O + 2\,NaNO_3$

3.

(a) $2\,Ca + O_2 \longrightarrow 2\,CaO$
(b) $3\,Mg + N_2 \longrightarrow Mg_3N_2$
(c) $Ca + H_2 \longrightarrow CaH_2$
(d) $Ba + 2\,H_2O \longrightarrow H_2 + Ba(OH)_2$
(e) $Be + H_2 \longrightarrow$ no reaction

4.

(a) $4\,Al + 3\,O_2 \longrightarrow 2\,Al_2O_3$
(b) $2\,Al + 6\,HNO_3 \longrightarrow 3\,H_2 + 2\,Al(NO_3)_3$
(c) $2\,B + 3\,Cl_2 \longrightarrow 2\,BCl_3$
(d) $Al(OH)_3 + NaOH \longrightarrow NaAlO_2 + 2\,H_2O$
(e) $Ga + H_2 \longrightarrow$ no reaction

5.

(a) $C + O_2 \longrightarrow CO_2$
(b) $CO_2 + 2\,NaOH \longrightarrow Na_2CO_3 + H_2O$
(c) $SnO + 2\,KOH \longrightarrow K_2SnO_2 + H_2O$
(d) $Pb(OH)_2 + 2\,NaOH \longrightarrow Na_2PbO_2 + 2\,H_2O$
(e) $Na_2CO_3 + 2\,HNO_3 \longrightarrow H_2O + CO_2 + 2\,NaNO_3$

6.

(a) $N_2 + 3\,Mg \longrightarrow Mg_3N_2$
(b) $N_2O_3 + H_2O \longrightarrow 2\,HNO_2$
(c) $N_2 + 3\,H_2 \longrightarrow 2\,NH_3$
(d) $P_2O_3 + 6\,NaOH \longrightarrow 2\,Na_3PO_3 + 3\,H_2O$
(e) $Bi_2O_3 + 6\,HNO_3 \longrightarrow 2\,Bi(NO_3)_3 + 3\,H_2O$

7.

(a) $S + O_2 \longrightarrow SO_2$
(b) $SO_2 + 2\,KOH \longrightarrow H_2O + K_2SO_3$
(c) $SeO_3 + H_2O \longrightarrow H_2SeO_4$
(d) $Na_2S + Zn(NO_3)_2 \longrightarrow \underline{ZnS} + 2\,NaNO_3$
(e) $K_2S + 2\,HCl \longrightarrow H_2S + 2\,KCl$

8.

(a) $3\,Cl_2 + 2\,Fe \longrightarrow 2\,FeCl_3$
(b) $2\,NaBr + Cl_2 \longrightarrow 2\,NaCl + Br_2$
(c) no reaction
(d) $Cl_2O_3 + 2\,KOH \longrightarrow 2\,KClO_2 + H_2O$
(e) $AgNO_3 + NH_4Cl \longrightarrow \underline{AgCl} + NH_4NO_3$

9.

(a) too active
(b) small size of Li atom (and hence high ionization energy, etc.)
(c) doesn't form any common precipitate
(d) ionization energy (due to size of Li atom)
(e) $Al(OH)_3$ is a weak base.
(f) nonmetal oxide $\longrightarrow$ metal oxide
(g) SO_2 is a nonmetal oxide; the oxidation state of S in SO_2 is $+4$ and hence H_2SO_3 is formed (in which S also has a $+4$ oxidation state).
(h) F is the most electronegative element.
(i) HCl is a strong acid, but HF is weak.
(j) Limestone is $CaCO_3$; when a carbonate is heated, it forms an oxide and CO_2; lime is CaO.
$CaCO_{3(s)} \xrightarrow{\Delta} CaO_{(s)} + CO_{2(g)}$
(k) $HCO_3^- + H^+ \longrightarrow (H_2CO_3) \longrightarrow H_2O + CO_2$
$HCO_3^- + OH^- \longrightarrow H_2O + CO_3^{2-}$
(l) $Cl_2O_3 + H_2O \longrightarrow 2\,HClO_2$
(nonmetal oxide)
(m) $H_2S \rightleftharpoons 2\,H^+ + S^{2-}$
The greater the $[H^+]$, the lower the $[S^{2-}]$.

10.

(a) $2\,Al_{(s)} + 6\,HCl_{(aq)} \longrightarrow 3\,H_{2(g)} + 2\,AlCl_{3(aq)}$
(b) $Zn_{(s)} + 2\,NaOH_{(aq)} \longrightarrow H_{2(g)} + Na_2ZnO_{2(aq)}$
(c) $Mg_{(s)} + H_{2(g)} \longrightarrow MgH_{2(s)}$
(d) $2\,Na_{(s)} + H_{2(g)} \longrightarrow 2\,NaH_{(s)}$
(e) $NaHCO_{3(s)} + HCl_{(aq)} \longrightarrow H_2O_{(l)} + CO_{2(g)} + NaCl_{(aq)}$
(f) $2\,K_{(s)} + 2\,H_2O_{(l)} \longrightarrow H_{2(g)} + 2\,KOH_{(aq)}$
(g) $Na_2CO_{3(aq)} + CaCl_{2(aq)} \longrightarrow CaCO_{3(s)} + 2\,NaCl_{(aq)}$
(h) $FeS_{(s)} + 2\,HCl_{(aq)} \longrightarrow H_2S_{(g)} + FeCl_{2(aq)}$

11.

(a) $2\,Bi^{3+} + 3\,H_2S_{(g)} \longrightarrow Bi_2S_{3(s)} + 6\,H^+$
(b) $NH_{3(aq)} + HC_2H_3O_{2(aq)} \longrightarrow NH_4^+ + C_2H_3O_2^-$
(c) $ZnS_{(s)} + 2\,H^+ \longrightarrow H_2S_{(g)} + Zn^{2+}$
(d) $NH_4^+ + OH^- \longrightarrow NH_{3(aq)} + H_2O_{(l)}$
(e) $CaO_{(s)} + 2\,HC_2H_3O_{2(aq)} \longrightarrow Ca^{2+} + 2\,C_2H_3O_2^- + H_2O_{(l)}$
(f) $3\,Zn^{2+} + 2\,Al \longrightarrow 3\,Zn + 2\,Al^{3+}$
(g) $2\,H^+ + Na_2O_{(s)} \longrightarrow H_2O_{(l)} + 2\,Na^+$
(h) $HCO_3^- + H^+ \longrightarrow H_2O_{(l)} + CO_{2(g)}$
(i) $Ca^{2+} + CO_3^{2-} \longrightarrow CaCO_{3(s)}$
(j) $PbO_{(s)} + 2\,OH^- \longrightarrow PbO_2^{2-} + H_2O_{(l)}$
(k) $Ga^+ + Cl_{2(g)} \longrightarrow Ga^{3+} + 2\,Cl^-$
(l) $SO_{2(g)} + 2\,OH^- \longrightarrow H_2O_{(l)} + SO_3^{2-}$
(m) $FeS_{(s)} + 2\,H^+ \longrightarrow H_2S_{(g)} + Fe^{2+}$

Chapter 24

1.

(a) 4; Cl^-; square planar or tetrahedral; no; potassium tetrachlorozincate(II)
(b) 6; NH_3 and F^-; octahedral; yes; difluorotetraammineplatinum(II)
(c) 6; NH_3; octahedral; no; hexaamminechromium(III) chloride

(d) 6; NH_3 and H_2O; octahedral; no; pentaammineaquocobalt(III) iodide
(e) 6; H_2O and NH_3; octahedral; yes; tetraaquodiamminecobalt(III) chloride
(f) 4; H_2O; square planar or tetrahedral; no; tetraaquoplatinum(II) chloride
(g) 6; H_2O and Cl^-; octahedral; yes; dichlorotetraaquoplatinum(II)
(h) 4; ethylenediamine and F^-; square planar or tetrahedral; yes; difluoroethylenediaminecobalt(II)
(i) 6; NH_3 and Cl^-; octahedral; yes; calcium tetrachlorodiamminenickelate(II)
(j) 4; NH_3 and H_2O; square planar or tetrahedral; yes; diamminediaquochromium(III) perchlorate

2. Answers given in order of formula, coordination number, ligands, possible shape, if stereoisomers are possible.

(a) $Na_2[PtBr_6]$; 6; Br^-; octahedral; no
(b) $K_2[ZnCl_4]$; 4; Cl^-; square planar or tetrahedral; no
(c) $[Cr(H_2O)_4](ClO_3)_3$; 4; H_2O; square planar or tetrahedral; no
(d) $[Cr(NH_3)_6]Cl_3$; 6; NH_3; octahedral; no
(e) $[Co(NH_3)_5(H_2O)]I_3$; 6; NH_3 and H_2O; octahedral; no
(f) $K[Ni(NH_3)_3Cl_3]$; 6; NH_3 and Cl^-; octahedral; yes
(g) $[Co(en)Cl_2]$; 4; ethylenediamine and Cl^-; square planar or tetrahedral; yes
(h) $Na_2[PtCl_4]$; 4; Cl^-; square planar or tetrahedral; no
(i) $[Pt(H_2O)_4Cl_2]$; 6; H_2O and Cl^-; octahedral; yes
(j) $K_2[Ni(NH_3)_2F_4]$; 6; NH_3 and F^-; octahedral; yes

Chapter 25

1.

(a) ${}^{87}_{36}Kr \longrightarrow {}^{0}_{-1}e + {}^{87}_{37}Rb$
(b) ${}^{232}_{92}U \longrightarrow {}^{4}_{2}He + {}^{228}_{90}Th$
(c) ${}^{140}_{59}Pr \longrightarrow {}^{0}_{1}e + {}^{140}_{58}Ce$
(d) ${}^{16}_{8}O + {}^{1}_{0}n \longrightarrow {}^{4}_{2}He + {}^{13}_{6}C$
(e) ${}^{9}_{4}Be + {}^{1}_{1}H \longrightarrow {}^{4}_{2}He + {}^{6}_{3}Li$
(f) ${}^{176}_{71}Lu + {}^{1}_{0}n \longrightarrow {}^{177}_{71}Lu + \gamma$

2.

(a) ${}^{55}_{27}Co$

3.

(a) 1.00 mg
(b) 6.00 years

4.

(a) 20 minutes

5.

$$1.40 \times 10^{9}\ \frac{\text{kcal}}{\text{mol}}$$

6.

(a) ${}^{24}_{11}Na \longrightarrow {}^{0}_{-1}e + {}^{24}_{12}Mg$
(b) 3.12×10^{-8} g

8.

$1.00 \times 10^{-6}\%$

10.

(a) ${}^{32}_{14}Si \longrightarrow {}^{0}_{-1}e + {}^{32}_{15}P$
(b) ${}^{240}_{96}Cm \longrightarrow {}^{4}_{2}He + {}^{236}_{94}Pu$
(c) ${}^{18}_{10}Ne \longrightarrow {}^{0}_{+1}e + {}^{18}_{9}F$

11.

(a) ${}^{7}_{4}Be + {}^{0}_{-1}e \longrightarrow {}^{7}_{3}Li$
(b) ${}^{208}_{84}Po \longrightarrow {}^{204}_{82}Pb + {}^{4}_{2}He$
(c) ${}^{66}_{29}Cu \longrightarrow {}^{66}_{30}Zn + {}^{0}_{-1}e$
(d) ${}^{16}_{8}O + {}^{1}_{0}n \longrightarrow {}^{13}_{6}C + {}^{4}_{2}He$
(e) ${}^{9}_{4}Be + {}^{1}_{1}H \longrightarrow {}^{6}_{3}Li + {}^{4}_{2}He$
(f) ${}^{10}_{5}B + {}^{1}_{0}n \longrightarrow {}^{3}_{1}H + 2\,{}^{4}_{2}He$
(g) ${}^{6}_{3}Li + {}^{1}_{0}n \longrightarrow {}^{3}_{1}H + {}^{4}_{2}He$
(h) ${}^{60}_{27}Co \longrightarrow {}^{60}_{28}Ni + {}^{0}_{-1}e$

Chapter 26

1.

(a) All sp^3 hybrid orbitals—sigma bonds around each carbon.

(b) sp^2 hybrid on second and third carbon, with one *p* orbital left to form pi bond.

2.

(a) $CH_3-CH_2-CH_2-CH_3$ 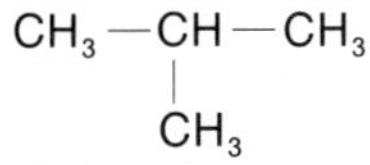

(b) $CH_3-CH_2-CH_2-CH_2-CH_2-CH_3$

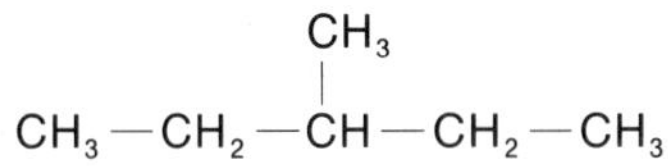

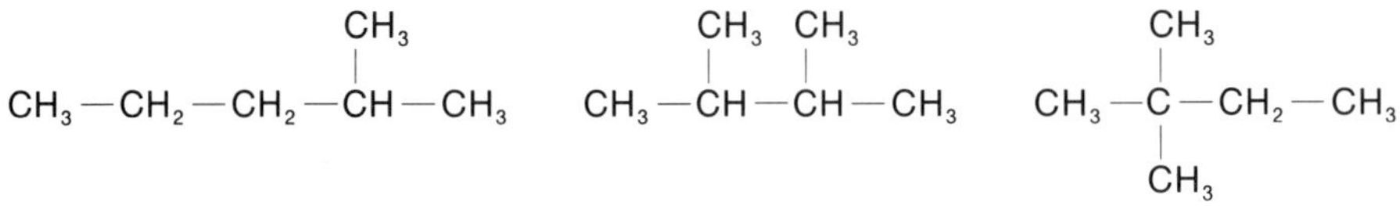

3.

(a) alkene
(b) aromatic
(c) ether
(d) alkyne
(e) acid
(f) amine
(g) aldehyde
(h) alcohol
(i) alkane
(j) ketone

4.

(a) 1-butene
(b) 2-pentyne
(c) 2-methylpentane
(d) 2-bromobutane (sec-butyl bromide)
(e) 2,3-dimethylbutane
(f) methylcyclobutane
(g) 2-chloropropane (isopropyl chloride or *sec*-propyl chloride)
(h) 2-methyl-2-iodopropane
(i) 3-bromo-2-nitropentane
(j) 2-methyl-2-pentene
(k) 4-chloro-6-nitro-2-heptene
(l) 2,5-dimethyl-3-heptyne

5.

(a) 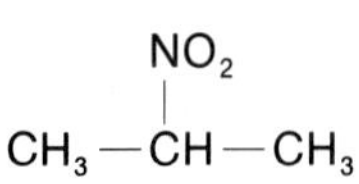

(b) 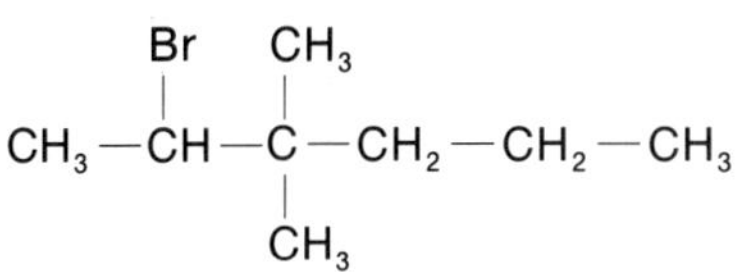

(c) 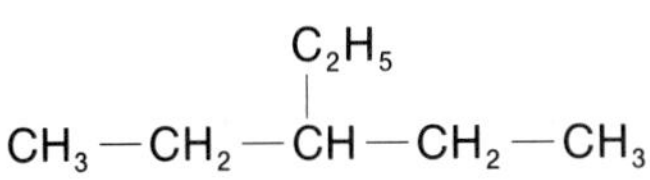

(d) 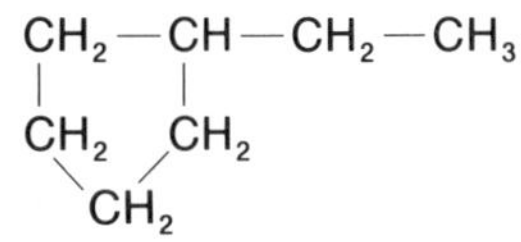

(e) 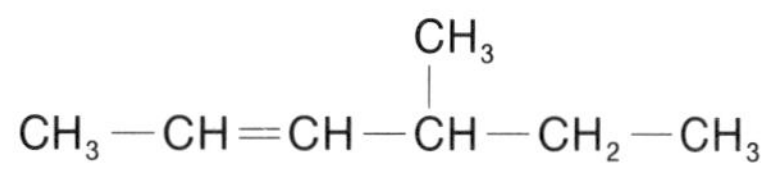

(f) 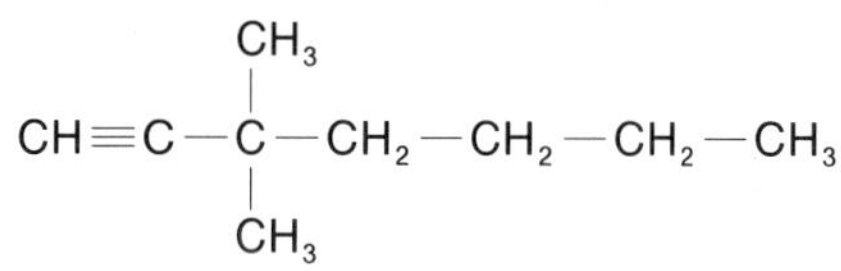

6.

(a) $CH_3-CH=CH_2 + H_2 \longrightarrow CH_3-CH_2-CH_3$

(b) $CH\equiv C-CH_2-CH_3 + 2\,Br_2 \longrightarrow$

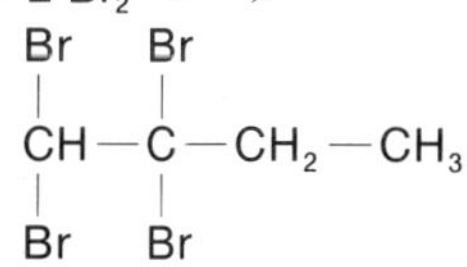

(c) $CH_2=CH_2 + HCl \longrightarrow CH_3-CH_2-Cl$

(d) $CH_3-CH_2-CH_2-CH_2-CH_3 + 8\,O_2 \xrightarrow{\Delta} 5\,CO_2 + 6\,H_2O$

(e) 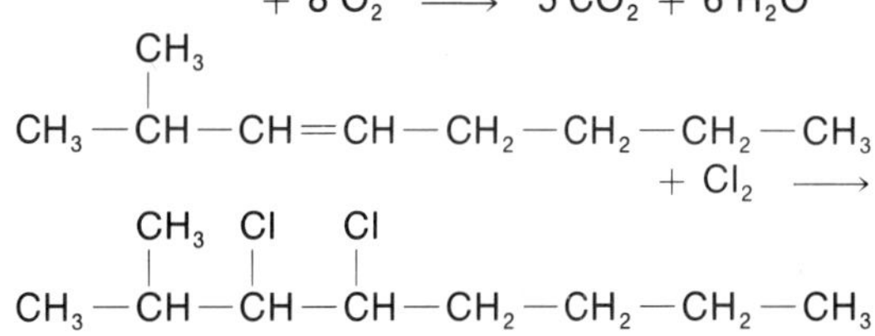

7.

(a) 2-butyne
(b) 2-methylpropane
(c) chloroethane
(d) 1-bromo-3-methylbutane

(e) tetrachloromethane
(f) 2-bromo-2-methylpropane
(g) 6-chloro-3-octene
(h) 2-chloropropane

8.

(a) bromobenzene
(b) 1,2-dibromobenzene (or *o*-dibromobenzene)
(c) 2-bromonitrobenzene
(or 2-nitrobromobenzene,
or *m*-bromonitrobenzene)

9.

(a)

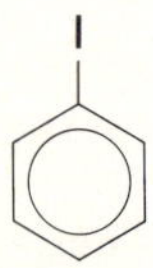

(b)

(c)

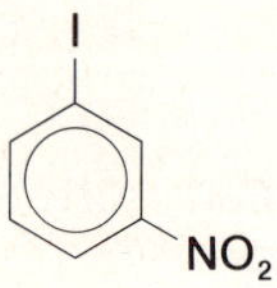

(d)

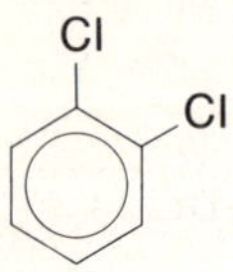

(e)

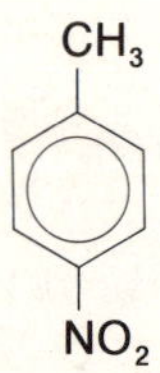

10.

(a) 1-butanol (1°)
(b) 3,3-dimethyl-2-pentanol (2°)
(c) 2-methyl-2-butanol (3°)
(d) 2-methylphenol
(e) 3-phenyl-2-hexanol (2°)

11.

(a) $CH_3-\underset{|}{\overset{OH}{CH}}-CH_2-CH_2-CH_2-CH_3$

(b) $CH_3-\overset{OH}{\overset{|}{CH}}-\overset{CH_3}{\overset{|}{CH}}-CH_2-CH_3$

(c) $CH_3-CH_2-CH_2-OH$

(d) $CH_3-\underset{OH}{\overset{CH_3}{C}}-CH_3$

(e) $CH_3-\underset{OH}{CH}-CH_3$

(f)

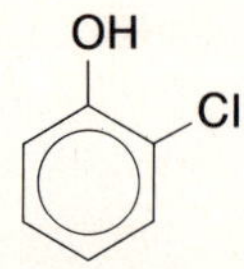

(g)

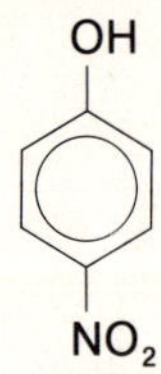

12.

(a) 2-methyl-2-propanol
(b) 2-methyl-1-propanol
(c) 2-propanol

13.

(a) methyl ethyl ether
(b) ethyl phenyl ether
(c) *tert*-butyl isopropyl ether

14.

(a) CH_3-O-CH_3

(b) $CH_3-CH_2-\overset{CH_3}{CH}-O-CH_2-\overset{CH_3}{CH}-CH_3$

(c) 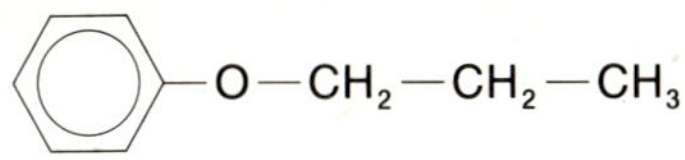

15.

(a) butanoic acid
(b) benzoic acid
(c) 3-methylbutanoic acid
(d) 5,5-dichloro-3-methylhexanoic acid
(e) 3,4-dimethylbenzoic acid

16.

(a) 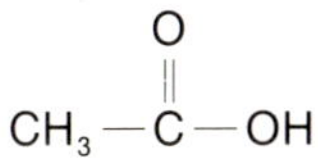

(b) 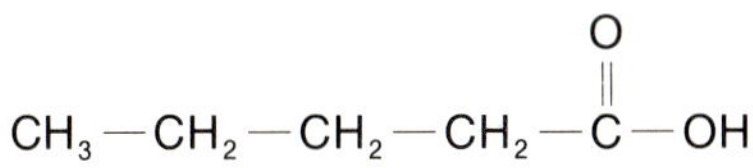

(c) 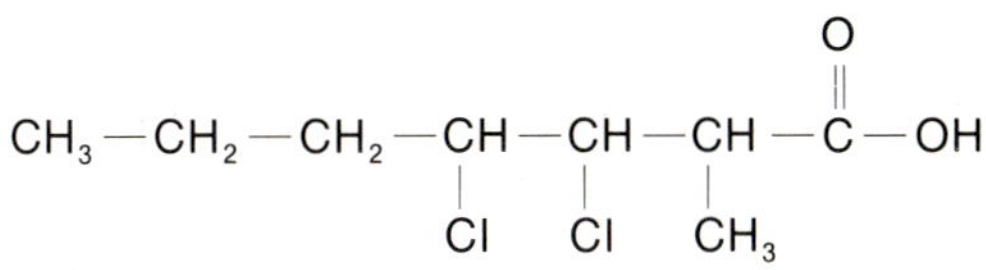

(d) 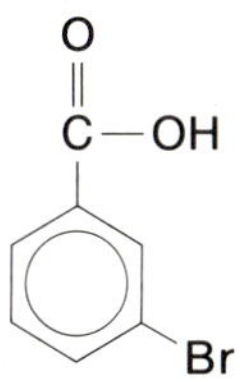

(e) 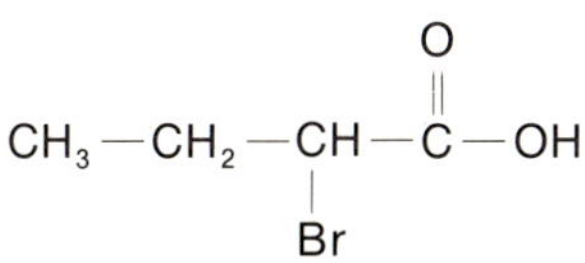

17.

(a) ethyl butanoate
(b) ethyl benzoate

18.

(a) 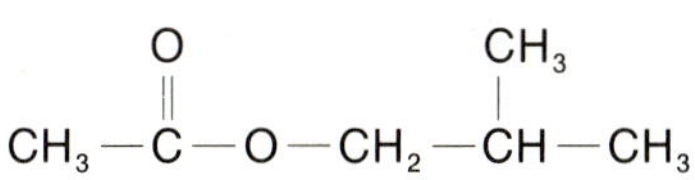

(b) 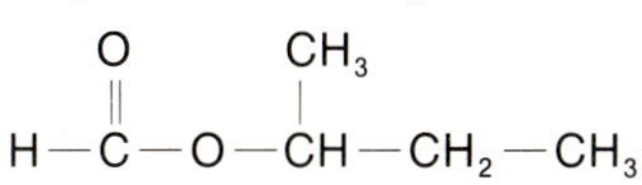

(c) 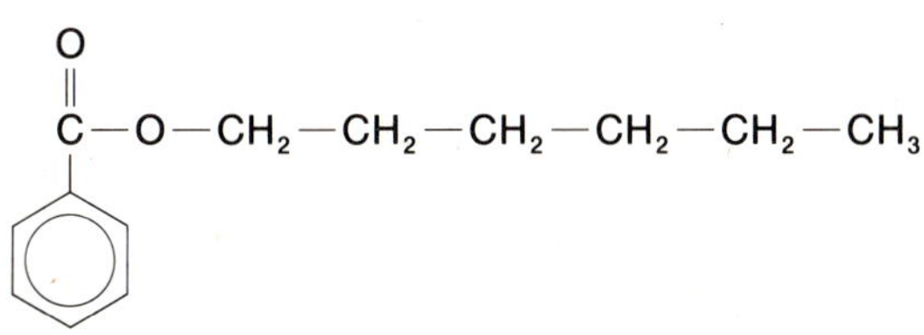

19.

(a) ethanoic anhydride (acetic anhydride)
(b) butanoic methanoic anhydride
(c) benzoic propanoic anhydride

20.

(a) 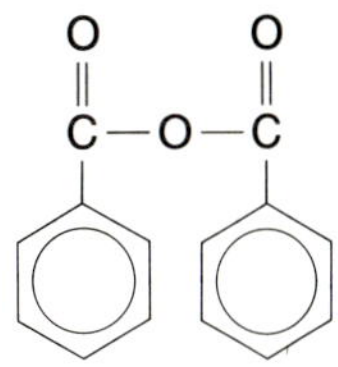

(b) 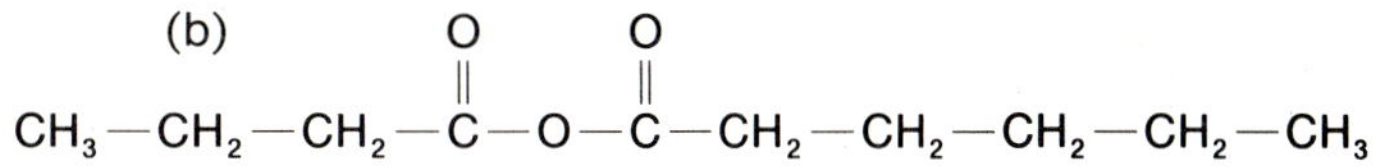

(c) 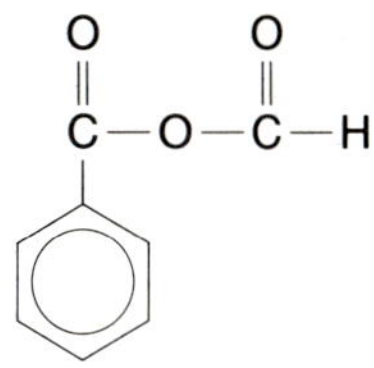

21.

(a) propanamide
(b) methanamide
(c) 3,4-dichloropentanamide

22.

(a) 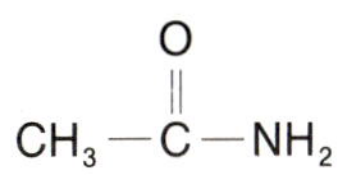

(b) 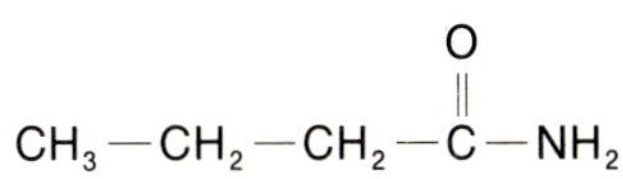

(c) 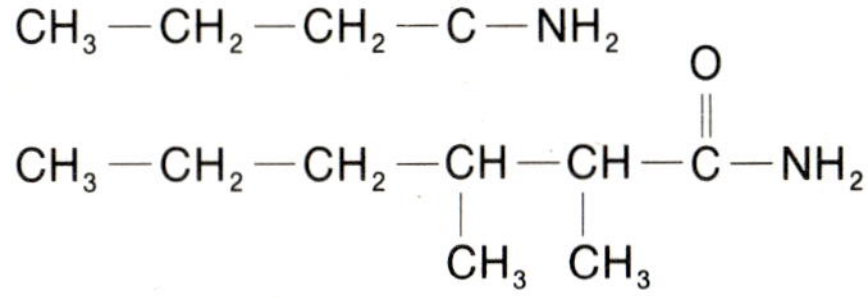

23.

(a) propanal
(b) benzaldehyde
(c) 2-butanone
(d) 3-pentanone

24.

(a)
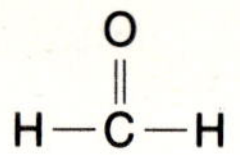

(b)
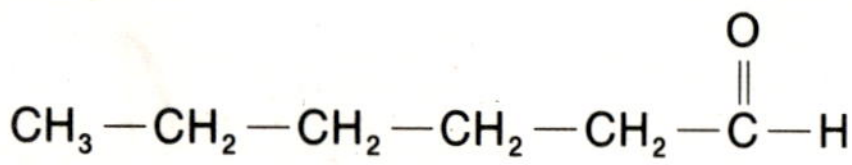

(c)
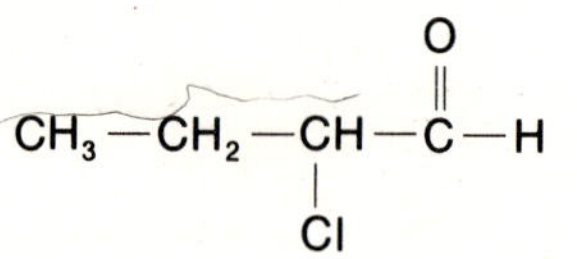

(d)
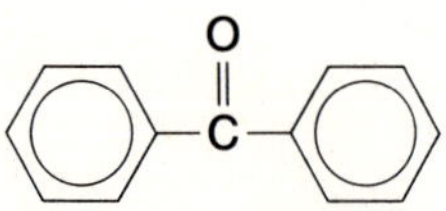

(e)
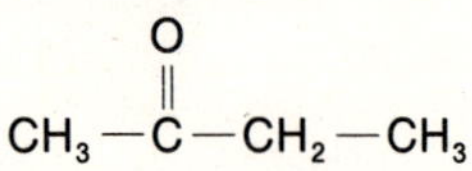

(f)
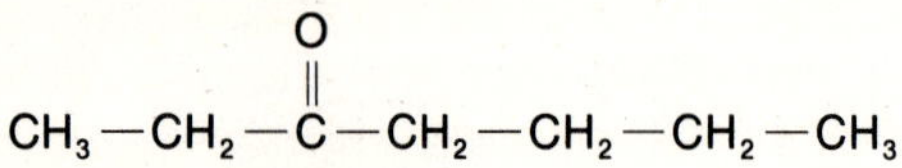

25.

(a) methylamine (1°)
(b) dimethylamine (2°)
(c) isopropylamine (1°)
(d) methylethylpropylamine (3°)
(e) 2,4-dichloroaniline (1°)

26.

(a)
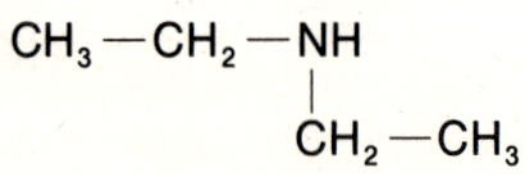

(b)
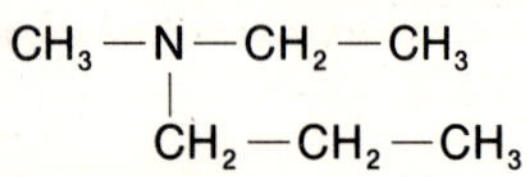

(c)
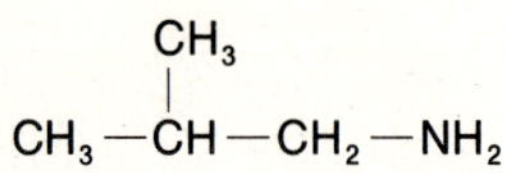

(d)
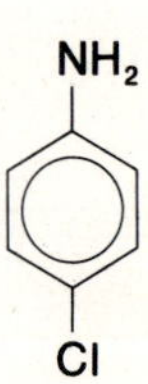

(e)
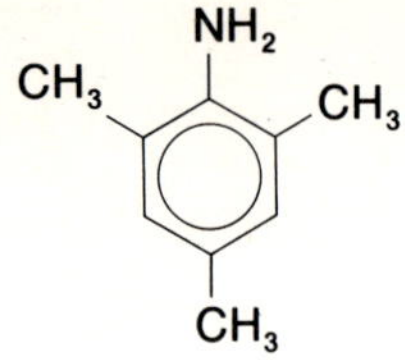

27.

(a) alkene
(b) alkyne
(c) alkane

28.

(a)
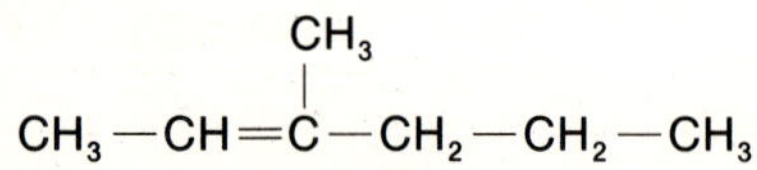
(alkene)

(b) $CH_3-C\equiv C-CH_3$ (alkyne)

(c)
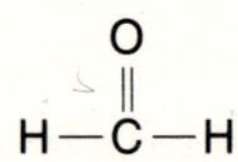
(aldehyde)

(d)
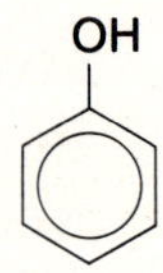
(aromatic alcohol)

(e)
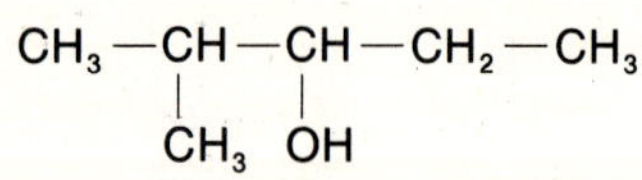
(alcohol)

(f)
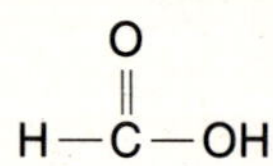
(acid)

(g) $CH_2=CH-CH=CH-CH_2-CH_2-CH_2-CH_3$ (alkene)

(h)
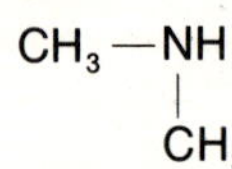
(amine—2°)

(i)
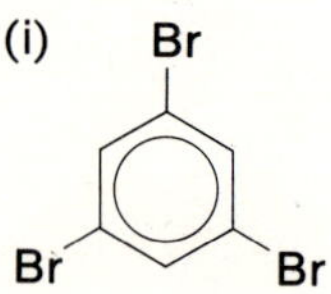
(aromatic)

(j)
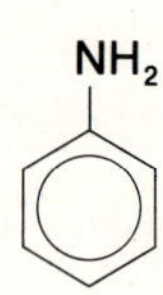
(aromatic amine)

(k)

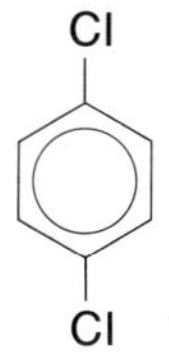

(aromatic)

(l)

$$CH_3-\overset{\overset{\displaystyle O}{\|}}{C}-CH_2-CH_2-CH_2-CH_3 \text{ (ketone)}$$

29.

(a) 2-propanol
(b) 1,3,4-trimethylbenzene
(c) 2-butanone
(d) 4,5-dimethyl-2-hexene
(e) ethanal
(f) 2-butanol
(g) 2-methyl-2-propanol
(h) ethyl methanoate

Index

A

Absolute (kelvin) temperature scale, 21
Absolute zero, 23, 245
Accelerator, 545
Acetaldehyde (ethanal), 579
Acetamide, 578
Acetic acid (ethanoic acid), 576
Acetic anhydride, 577
Acetone, 580
Acetylene, 567
Acetylenes, 567 (*see also* Alkynes)
Acid anhydride, 188, 578
Acid carbonate, 134
Acids:
  amphoteric, 329
  Arrhenius, 327–28
  binary, 131–32, 471–72
  Brønsted-Lowry, 329
  conjugate acids, 330
  definition, 131
  equivalent weight, 333–36
  general properties, 326–27
  hydrogen compounds, 471–72
  ionization constants (*table*), 642
  Lewis, 331
  oxides, 472–74
  oxyacids, 472
  weak versus strong, 328
Acid–base titration, 331–32
Acid oxide, 188
Acid salts, 134
Acid strength:
  hydrogen compounds, 471–72
  oxide, 472–73
  oxyacids, 471–72
Actinides (actinium series), 515
Activation energy, 353–54
Activity series of metals, 190, 453
Actual yield, 210
Aerosol, 313
Alanine, 603
Alchemy, 7
Alcohols, 563, 574–75
Aldehydes, 563, 578–79
Aldohexoses, 590
Aldoses, 598
Aliphatic compounds, 564
Alkalies, 132 (*see also* Bases; Hydroxides, nomenclature)
Alkali metals, 485–88
  activity, 485–86
  hydroxides, 486

Alkali metals (cont.)
occurrence, 486
properties (*table*), 487
reactions, 487–88
Alkaline earth metals, 488–91
activity, 488–89
occurrence, 488
properties (*table*), 491
uses, 489
Alkanes, 563–66
Alkenes, 563, 566–67
Alkyl groups, 568
Alkynes, 563, 567
Allotropic forms (allotropes), 279
Alpha amino acids, 601
Alpha carbon, 576
Alpha particles, 539
Aluminum, 490–93
hydrolysis of salts, 492
occurrence, 490
reaction with acid, 491
reaction with base, 491
reaction with oxygen, 491
Aluminum hydroxide, 492
Aluminum ions, 492
Amalgams, 520
Amides, 577
Amines, 563, 580–81
Amino acids, 601–02
alpha, 601
essential, 602
Ammonia, 580
Lewis structure, 152
preparation, 362–63
reaction with acid, 193, 497
reaction with water, 497
Ammonium ion, 497
Ammonium salt, 193
Amorphous, 279
Ampere, 450
Amphiprotic, 329
Amphoteric, 329, 473
Amu, 45
Amyl acetate, 577
Amyl (pentyl) group, 569
Anabolism, 607
Anaerobic, 608
Analgesics, 613
Analytical chemistry, 6
Angstrom, 15, 632
Anhydride:
acid, 188
basic, 187
Anion, 49
Anode, 447
Anode reactions, 448
Antibodies, 604
Antibiotics, 613
Anti-bonding molecular orbitals, 171
Antihistamines, 613
Antimony, 496–99
occurrence, 496
oxidation states, 496
reactions, 497
Aqueous solution, 57
Aromatic compounds, 563, 571–74
Arrhenius acid, 327–28
Arrhenius base, 327–28
Arrhenius, Svante August, 327
Arsenic, 496–99
occurrence, 496
oxidation states, 496
reactions, 497
Arsine, 497
Artificial radioactivity, 541–43
Ascorbic acid (vitamin C), 612
Aspartic acid, 603
Atmosphere, pressure, 241
Atom, 38
basic structure, 45–46
electronic structure, 83–87
Atomic mass unit (amu), 45
Atomic number, 46
Atomic size, 109
Atomic structure, theories of, 70–73
Atomic weight scale, 47–9
Attractive forces in liquids and solids, 270–72
Aufbau order, 80

Auto oxidation–reduction, 436
Avogadro, Amedeo, 53
Avogadro's hypotheses, 251–55
Avogadro's number, 53

B

Balance, 12–13
Balanced equations, 43
  calculations from, 206–11
Balancing chemical equations, 183–85, 433–37
Barium, 488–91
  hydroxide, 489
  ion, 490
  oxide, 489
  reaction with nitrogen, 489
  reaction with oxygen, 489
  reaction with water, 489
Barometer, 241
Base(s):
  Arrhenius, 327–28
  Brønsted-Lowry, 329
  conjugate pairs, 330
  equivalent weight, 333–36
  general properties, 326–27
  ionization constants (*table*), 642
  Lewis, 331
  nomenclature, 132
  weak versus strong, 328
Basic anhydride, 187
Basic oxide, 187
Becher, Johann, 7
Becquerel, 70
Benedict's test, 600
Benzene, 571
Benzene ring, 564, 571
Benzoic acid, 576
Beryllium, 488–91
Beryllium oxide, 489
Beta carbon, 576
Beta particles, 539–40
Bicarbonates, 134 (*see also* Hydrogen carbonates)
  decomposition of, 189
Bidentate ligand, 527
Binary acids, 131–32
Binary compounds, 128–32
Binding energy, 550
Biochemistry, 6
Bismuth, 496–99
  occurrence, 496
  oxidation states, 496
  reactions, 497
Black, Joseph, 7
Blood proteins, 604
Blood sugar, 607
Body-building materials, 610
Bohr, Neils, 72
Boiling point, 275–76
  water, 22–23
Boiling point elevation, 310–11
Bond:
  covalent, 149
    in macromolecular substances, 272
  double, 149
  electrovalent, 144
  energy, 149, 226–67
  hydrogen, 285
  ionic, 144
  length, 149
  metallic, 271–72
  multiple, 149
  order, 171
  pi, 156
  sigma, 156
  triple, 149
Bonding molecular orbitals, 171
Born-Haber cycle, 148–49
Boron, 490–93
Boyle, Robert, 7, 243
Boyle's law, 243–44
Bromide ion, 505
Bromine, 503–04
  occurrence, 503
  preparation, 503

Bromine (cont.)
reactions, 504
2-Bromobutane, 329
Brønsted, Johannes, 329
Brønsted-Lowry acid, 329
Brønsted-Lowry base, 329
Brown, Robert, 315
Brownian movement, 315
Buffer solutions, 399–401
Buret, 14, 331–32
Butane, 565
Butanoic acid (butyric acid), 576
1-Butanol (*n*-butyl alcohol), 574
2-Butanol (butyl alcohol), 574
1-Butene, 566
2-Butene, 566
n-Butyl alcohol (1-butanol), 574
sec-Butyl alcohol (2-butanol), 574
tert-Butyl alcohol (2-methyl-2-butanol), 574
n-Butylbromide, 569
sec-Butyl chloride, 570
Butyl group, 569
Butyric acid (butanoic acid), 576

C

Cadmium, 520–21
Calcium, 488–91, 610
carbonate (limestone), 489
hydroxide, 489
ion, 490
oxide, 489
reaction with hydrogen, 489
reaction with nitrogen, 489
reaction with oxygen, 489
reaction with water, 489
Calorie, 25
Calorimeter, 227
Calorimetry, 227
Carbohydrates, 598–601
disaccharide, 600
fermentation, 600
hydrolysis, 598
metabolism, 607–08
monosaccharides, 599
polysaccharides, 600
reactions, 600–01
Carbon, 493–95
occurrence, 493
oxides, 494
uses, 493
Carbonates, decomposition of, 189
Carbonate ion, 494
Carbon-14 dating, 551
Carbon tetrachloride (tetrachloromethane), 565
Carboxylic acids (organic acids), 563, 575–76
Carnotite, 516
Catabolism, 607
Catalyst, 355–56
in chemical equations, 183
negative, 183 (*see also* Inhibitor)
Cathode:
definition, 447
reactions, 448–49
Cathode-ray tube, 70
Cation, 49
Cavendish, Henry, 7
Cellulose, 600
Celsius, Anders, 22
Celsius (centigrade) temperature scale, 21
Centigrade (celsius) temperature scale, 21
Central atom, 526
Cesium, 485–88
Chain reaction, 547–48
Charge density, 270, 471
Charles, J., 245
Charles' Law, 245–46
Chelate, 528
Chelating agent, 528
Chemical change, 42

Chemical equations, 43 (*see also* Equations)
- acid–base reactions, 192–93
- balancing, 183–85
- calculations from, 206–11
- decomposition reactions, 188–89
- displacement reactions, 190
- double displacement reactions, 191–92
- ionic, 191, 339–42
- metathesis reactions, 191–92
- neutralization reactions, 192–93
- replacement reactions, 190
- thermochemical, 220–22
- types, 186

Chemical equilibrium, 362 (*see also* Equilibrium)
Chemical kinetics, 350–57
Chemical property, 42
Chemical properties of metals, 476–77
Chemical reaction:
- definition, 43
- energy changes, 222–23

Chemistry:
- analytical, 6
- biochemistry, 6
- definition, 5
- inorganic, 6
- organic, 6
- physical, 6

Chemotherapeutic agents, 613
Chlorate, decomposition of, 189
Chlorate ion, 133
Chloric acid, 133
Chloride ion, 505
Chlorine, 503–04
- occurrence, 503
- preparation, 449, 503
- reactions, 504

Chlorite ion, 133
Chloroform (trichloromethane), 565
Chloromethane (methyl chloride), 565
o-Chlorophenol (1-chlorophenol), 574
p-Cholorophenol (3-chlorophenol), 574
Chlorophyll, 600
1-Chloropropane (*n*-propyl chloride), 570
Chlorous acid, 133
Chromium, 517
Chromium(III) hydroxide, 517
Chromium(III) ion, 517
Cis- isomer, 529
Citric acid cycle, 608
Classification of proteins, 604
Cobalt, 518–19
Cobalt(II) ion, 519
Coenzyme, 606
Colligative properties of solutions, 309–12
Collision theory, 353–54
Colloids, 313–15
- Brownian movement, 315
- definition, 313
- dispersed particles, 313
- dispersing medium, 313
- electrical charge effect, 315
- motion effect, 314–15
- optical effect, 314
- Tyndal effect, 314
- types, 313–14

Colorimeter, 357
Combination reactions, 186–88 (*see also* Chemical equations)
Combined gas laws, 248–50
Common ion effect, 385–86
Complex, 526 (*see also* Coordination compounds)
Complex ion, 526 (*see also* Coordination compounds)
- dissociation constants (*table*), 643

Compound, 40
- comparison to mixture, 41

Concentrated solution, 300
Concentration of solutions:
- definition, 57
- molal, 305
- molar, 58, 306–08

Concentration of solutions (cont.)
mole fraction, 304
normal, 333–39
percent by mass, 57, 301–03
percent by volume, 303
Conjugated proteins, 604
Conjugate pairs of acids and bases, 330
Conservation of energy, law of, 43
Conservation of mass (matter), law of, 43
Constant composition, law of, 40
Contractile proteins, 604
Conversion factors, 17
common relationships, 632
use, 18
Coordination compounds, 526–33
bidentate ligand, 527
central atom, 526
chelate, 528
coordination number, 527
determination of formulas, 530
geometry, 528–29
isomers, 528
ligand, 526
monodentate ligand, 527
nomenclature, 531–33
polydentate ligand, 527
theories of bonding, 530–31
Coordination number, 527
Copper, 519–20
purification, 449–50
Copper(II) ion, 520
Copper, oxides, 519–20
Cottrell precipitator, 315
Coulomb, 450
Covalent bond, 149
Critical mass, 547–48
Critical pressure, 284
Critical temperature, 284
Crookes tube, 70
Cryolite, 503
Crystal field theory, 531
Crystal structure, 279–80
Crystallization, 299–300
Curie (Ci), 549
Cyclo- compounds, 566
Cyclotron, 545
Cysteine, 603

D

Dalton, John, 8, 38
Dalton's Law of Partial Pressures, 250–51
Deamination, 608
Decane, 565
Decomposition reactions, 188–89
Decyl group, 569
Definite proportions, law of, 40
Delocalized pi bond, 158, 571
Denaturation of proteins, 604
Density, 27
calculations, 28
table, 29
Deoxyribonucleic acids, 607
Depressants, 613
Detection of radiation, 548–49
o-Dichlorobenzene (1,2-dichlorobenzene), 573
1, 2-Dichlorobenzene (*o*-dichlorobenzene), 573
p-Dichlorobenzene (1,4-dichlorobenzene), 573
1,4-Dichlorobenzene (*p*-dichlorobenzene), 573
Dichloromethane (methylene chloride), 565
Differentiating electron, 108, 470
Diffusion of gases, 259
Digestion, 607
Dihydrogen phosphate, 134
Dihydroxy acetone, 599
Dilute solution, 300
Dimethylacetylene, 570–71
Dimethyl ether (methoxymethane), 575
Dipole, 168
Dipole attraction, 271

Disaccharides, 600
Displacement reactions, 190
Disproportionation, 436
Dissociation constants of complex ions (*table*), 643
Dissolving precipitates, 386–87
Distillation, 277–78
DNA, 607
Double bond, 149
Double displacement reactions, 191–92
Drugs, 612–13
Dry cell, 460–61
dsp^3 hybrid orbitals, 166–67
d^2sp^3 hybrid orbitals, 167

E

E, 222
ΔE, 222
Einstein, Albert, 44
Einstein equation, 44
Electrolysis:
- cells, 447–50
- definition, 447
- preparation of metals, 476

Electrolyte, 57, 295
Electrolytic cells, 447–50
Electrolytic conduction, 446
Electromotive series of metals, 190, 453
Electron(s):
- arrangement, 83–87
- charge, 45
- definition, 45
- mass, 45
- quantum numbers, 88–93
- spin, 91
- unpaired, 78
- valence, 107–08, 471

Electron affinity, 113–14
Electron cloud, 73
Electron configuration of atoms, 83–87
Electron pair bond, 149
Electron dot:
- formula, 151–56
- notation of elements, 107

Electronegativity, 115
Electron pair acceptor, 331
Electron pair bond, 149
Electron pair donor, 331
Electron structure of atoms:
- arrangement in energy levels, 74
- arrangement in sublevels, 75

Electrovalent bond, 144
Element, 39, 105, 546
Elements:
- inner transition, 107
- periodic table, 39, 56
- representative, 105
- transition, 106

Elements that regulate body processes (*table*), 611
Emf, 634
Empirical (simplest) formulas, 203
- determination of, 203–04

Emulsion, 313
Endothermic reaction, 220
End point, 331
Energy:
- activation, 353–54
- binding, 550
- free, 229
- ionization
- kinetic, 43
- potential, 43

Energy changes:
- in chemical reaction, 222–23
- in formation of ionic compounds, 148
- in nuclear reactions, 550

Energy levels:
- arrangement of electrons, 74
- definition, 73–74
- maximum number of electrons, 75

Enthalpy, 222–25, 636 (*table*)

Entropy, 228–31, 636 (*table*)
Enzymes, 606
Equation of state, 256–58
Equations, 182–93 (*see also* Chemical equations)
  nuclear, 541–42
  oxidation-reduction, 432–33
  products, 43
  reactants, 43, 182
Equilibrium:
  buffer solutions, 399–401
  common ion effect, 385–86
  constants, 363–67
  definition, 362
  dissolving precipitates, 386–87
  heterogeneous, 367
  homogeneous, 367
  hydrolysis of salts, 415–24
  ionization constants, 388–93
  ionization of hydrogen sulfide, 412–15
  ionization of water, 393–94
  ion-product, 382
  Le Châtelier's principle, 368
  pH, 94–98
  predicting precipitation, 387–88
  redox equations, 457–60
  shift, 368–71
  slightly soluble compounds, 380
  solubility product, 380–88
  weak acids and bases, 388–93
Equilibrium constants:
  dissociation of complex ions (*table*), 643
  ionization of acids (*table*), 642
  ionization of bases (*table*), 642
  solubility product (*table*), 640–41
Equivalence point, 331
Equivalent, 334
Equivalent weight:
  acids and bases, 333–36
  electrolytic reactions, 450
  salts, 335
Erg, 550
Essential amino acids, 602
Esters, 576–77
Esters of glycerol, 605
Ethanal (acetaldehyde), 579
Ethane, 565
Ethanoic acid (acetic acid), 576
Ethers, 563, 574–75
Ethyl acetate, 577
Ethylene, 566
Ethylenediamine, 528
Ethyl group, 569
Ethyl phenyl ether, 575
Evaporation, 273–74
Excess reagent, 209–11
Exclusion principal, Pauli, 91
Exothermic reaction, 220
Exponential notation, 621
Exponents, 620–25
  addition and subtraction, 623–24
  division, 622–23
  multiplication, 622–23
  powers and roots, 624–25
  scientific notation, 625

F

Factor-unit method, 18
Fahrenheit, Gabriel Daniel, 21
Fahrenheit temperature scale, 21
Family, 105
Faraday, 450
Fats, 605–06
Fatty acids, 605
Fehling test, 600
Ferric hydroxide, 518
Ferrous hydroxide, 518
First Law of Thermodynamics, 228
Fission (nuclear), 547–48
Flame test:
  barium, 490
  calcium, 490
  potassium, 487
  sodium, 486
  strontium, 490
Flask, volumetric, 14

Fluorine:
occurrence, 503
preparation, 503
reactions, 504
Fluorite, 503
Foam, 313
Fool's gold, 518
Formal concentration, 306
Formaldehyde, 576
Formic acid (methanoic acid), 576
Formula:
definition, 40
empirical, 203–04
molecular, 203–05
Formula unit, 40, 50, 145
Formula weight, 50–51 (*see also* Molecular weight)
Francium, 485–88
Free energy, 229, 636 (*table*)
Freezing point, 280
Freezing point depression, 310–12
Fructose, 599
Fuel cell, 461–62
Functional groups, 563
Fusion (nuclear), 548

G

Galvanic cell, 452–53
Gamma rays, 540
Gases:
Avogadro's hypothesis, 251–55
Boyle's law, 243–44
Charles' law, 245–46
collection over water, 250–51
combined gas laws, 248–50
comparison to liquids and solids, 238–40
Dalton's law of partial pressures, 250–51
Gay Lussac's law, 247–48
Graham's law of diffusion, 259
ideal, 240
Gases (cont.)
ideal gas law, 257–58
kinetic theory, 240
molar volume, 251–55
noble, 56, 105–06
pressure, 241–42
properties of, 238–40
relationships involving moles, 255–57
standard temperature and pressure, 242
Gas law relationships involving moles, 255–57
Gay-Lussac's law, 247–48
Geiger-Mueller counter, 548–49
Genetic code, 607
Geometry of molecules, 159–63
Germanium, 493–95
Glucose, 599
Glyceraldehyde, 589
Glycerol, 605
Glycine, 601
Gold, 519–20
Goldstein, Eugene, 70
Graduated cylinder, 14
Graham's law of diffusion, 259
Gram, 14
Greek prefixes, 129
Group, 105
Group IA metals, 485–88 (*see also* Alkali metals)
Group IIA metals, 488–91 (*see also* Alkaline earth metals)
Group IIIA elements, 490–93
occurrence, 490
properties (*table*), 491
Group IVA elements, 493–95
oxidation states, 493
oxides, 494–95
properties (*table*), 495
Group VA elements, 496–99
occurrence, 496
oxidation states, 496
properties (*table*), 499
reactions, 497
Group VIA elements, 499–502
properties (*table*), 502

Group VIIA elements, 503–06
(*see also* Halogens)
Group IB elements (Cu subgroup), 519–20
Group IIB elements (Zn subgroup), 520–21
Group IIIB elements, 515–16
Group IVB elements, 516
Group VB elements, 516–17
Group VIB elements, 517
Group VIIB elements, 517–18
Group VIII elements, 518–19
Guldberg, Cato, 351

H

H, 222
ΔH, 222
ΔH_f, 223
ΔH_r, 223
Haber, Fritz, 363
Haber process, 362–63
Hafnium, 516
Hahn, Otto, 546
Hahnium, 546
Half-life, 544–45
Half-reactions balancing, 437–39
Halogens (Group VIIA elements), 116, 503–06
  occurrence, 503
  physical properties (*table*), 504
  preparation, 503
  properties (*table*), 505
  reactions, 504
Hard water, 489
Heat capacity (specific heat), 25
Heat flow, 220
  in thermodynamic system, 220–21
Heat of combustion of nutrients, 610
Heat of formation, 223–25
Heat of fusion, 281
Heat of reaction, 223–25
Heat of vaporization, 276–77
Heat transformations in changes of state, 281–83
Hematite, 518
Henry, William, 297
Henry's law, 297
2,5-Heptadiene, 566
Heptane, 565
Heptyl group, 569
Hess's law, 226–27
Heterogeneous equilibrium, 367
Hexane, 565
2-Hexene, 566
Hexyl group, 569
Homogeneous equilibrium, 367
Hormones, 604
Hund's rule, 80
Hybridization, 164–68
Hybrid orbitals, 164–68
Hydrate, decomposition of, 189
Hydrated, 294
Hydrocarbons, 563–64
Hydrogen, 484–85
  bond, 285
    in water, 285–86
  occurrence, 484
  preparation by electrolysis, 449
  properties (*table*), 485
Hydrogen carbonates (*see* Bicarbonates)
Hydrogen halides, 504
Hydrogen sulfide, 578
  ionization, 412–15
Hydrolases, 606
Hydrolysis:
  calculations, 419–24
  carbohydrates, 598
  constant, 417
  definition, 415
  metal ions, 424–25
  proteins, 604
  salts, 415–24
  salt of weak acid and strong base, 416–17
  salt of weak acid and weak base, 418–19
  salt of weak base and strong acid, 417–18

Hydronium ion, 327
Hydroxides, nomenclature, 132
Hydroxyl group, 574
Hypertonic, 313
Hypnotics, 613
Hypochlorite ion, 133
Hypochlorous acid, 133
Hypothesis, 5
Hypotonic, 313

I

Ideal gas, 240
Ideal gas law, 257–58
Immiscible, 273
Indicator, 331
Inhibitor, 183, 355–56
Inner complex, 531
Inner transition elements, 107
Inorganic chemistry, 6
Iodide ion, 505
Iodine:
  occurrence, 503
  preparation, 503
  reactions, 504
Iodine number, 606
Ion:
  definition, 49
  formation, 146
  monatomic, 116, 146
Ion product, 382
Ionic bond, 144
Ionic equations, 191, 339–42
  balancing, 434–37
Ionic potential, 471
Ionic size, 110–11
Ionization:
  constant for water, 393–94
  constants, 388–93
  hydrogen sulfide, 412–15
  percent, 389–90
  polyprotic acids, 412–15
  water, 393–94
  weak acids and bases, 388–93
Ionization constants of weak acids (*table*), 642
Ionization constants of weak bases (*table*), 642
Ionization energy, 111–12
Iridium, 519
Iron, 518–19
Iron(II) hydroxide, 518
Iron(III) hydroxide, 518
Iron(III) ion, 519
Iron pyrite, 518
Iron triad, 518–19
Isobutane (2-methylpropane), 570
Isobutyl alcohol (2-methyl-l-propanol), 574
Isobutylbromide, 569
Iso- compounds, 569
Isoelectric point, 604
Isoelectronic species, 116–17
Isoleucine, 602
Isomerases, 606
Isomers, 528, 561
Isopentyl bromide (1-bromo-3-methylbutane), 570
Isopropyl ether, 575
Isotonic, 313
Isotopes, 47

J

Joule, 25

K

K_a, 388
K_b, 388
K_{eq}, 363–67
K_h, 417–24
K_i, 388–93
K_{redox}, 457

K_{sp}, 380–88
K_w, 393
Kelvin (absolute) temperature scale, 21, 245
Kelvin, Lord, 23
Kernel notation, 107
Ketones, 563, 578–79
Ketoses, 598
Kinetic energy, 43
Kinetic theory of gases, 240
Kinetics, 350–57
Krebs cycle, 608

L

Le Châtelier's principle, 368
Lanthanide contraction, 515
Lanthanides (Lanthanum series), 515
Lanthanum series (rare earth elements), 515
Lattice energy, 148
Lavoisier, Antoine Laurant, 7
Law:
  Boyle's, 243–44
  Charles', 245–46
  conservation of energy, 43
  conservation of mass (matter), 43
  constant composition, 40
  Dalton's (partial pressures), 250–51
  definite proportions, 40
  Gay-Lussac's, 247–48
  Graham's (of diffusion), 259
  Henry's 297
  Hess's, 226–27
  Ideal gas, 257–58
  mass action, 351
  periodic, 105
  Raoult's, 309–10
  of thermodynamics, first, 228
  of thermodynamics, second, 228–31
Lawrencium, 546
Lead, 493–95
  occurrence, 493
  oxide, 494
Lead(IV) hydroxide, reaction with base, 494
Lead(II) ion, 494
Lead storage battery, 460–62
Lewis acid, 331
Lewis base, 331
Lewis structure, 151–56
Ligand, 526
Ligand field theory, 531
Lime (calcium oxide), 489
Limestone (calcium carbonate), 489
Limewater, 494
Limiting reagent, 208–11
Linear accelerator, 546
Lipids, 605–06 (*see also* Fats and Oils)
Lipids, metabolism, 608–09
Liquid(s):
  boiling point, 275–76
  comparison to gases and solids, 238–40
  distillation, 277–78
  evaporation, 273–74
  heat of vaporization, 276–77
  mixability, 273
  normal boiling point, 275–76
  properties of, 272–73
  surface tension, 278
  vapor pressure, 274–75
Liquid state, 272–78
Lithium, 485–88
  reaction with oxygen, 486
Lithium hydroxide, 486
Lithium oxide, reaction with water, 488
Litmus, 327
Logarithms, 394–95
  table, 645–46
  use, 628–29
London dispersion forces, 271
Lone pairs of electrons, 149
Lowry, Thomas, 329
Lysine, 603

M

Macromolecular substances, 272
Magic numbers, 542
Magnesium, 488–91
  reaction with nitrogen, 489
  reaction with oxygen, 489
  reaction with water, 489
Magnesium oxide, 489
Magnetic quantum number, 89
Maltose, 600
Manganese, 517–18
Manganese(II) ion, 518
Markownikoff's rule, 568
Mass, 12
Mass–energy relations, 43–44
Mass number, 46
Matter:
  classification, 42
  definition, 5, 12
Mechanism of reaction, 356–57
Melting point, 280
Mercury, 520–21
Mercury(I) ion, 521
Mercury(II) ion, 521
Messenger RNA, 607
Meta compounds, 572
Metabolism:
  carbohydrates, 607–08
  lipids, 608–09
  proteins, 608
Metal oxide:
  decomposition of HgO, 189
  reaction with nonmetal oxide, 188
  reaction with water, 187
Metallic bond, 271–72
Metallic conduction, 446
Metallic properties, 109
Metalloids, 471
Metallurgy, 475–76
Metals:
  activity series, 477
  chemical properties, 476–77
  definition, 56
Metals (cont.)
  metallurgy, 475–76
  occurrence, 475
  physical properties, 474–75
  transition, 106, 514–22
Metathesis reactions, 191–92
Methane, 565
Methanoic acid (formic acid), 576
Methanol (methyl alcohol), 574
Methyl alcohol (methanol), 574
Methyl amine, 580
Methylbenzene (toluene), 572
Methyl chloride
    (chloromethane), 565
Methyl group, 569
3-Methylhexane, 569
Methyl orange, 331
Methylphenyl amine, 580
2-Methylpropane (isobutane), 570
Methyl propyl ether, 575
Methylene chloride
    (dichloromethane), 565
Metric system:
  definition, 14
  metric-English equivalents, 17
  prefixes, 15
Milliken, Robert, 71
Millimole (mmol), 53
Miscible, 273
Mixability of liquids, 273
Mixed ether, 574
Mixed glyceride, 605
Mixed salts, 135
Mixture:
  comparison to compound, 41
  definition, 41
  heterogeneous, 41
  homogeneous, 41
Molal boiling point elevation
    constant, 310
Molal concentration, 305
Molal freezing point depression
    constant, 310
Molar concentration, 57–59, 306–08
Molar mass, 53
Molar volume of a gas, 251–55

Mole:
calculations, 53–56
definition, 52
number of particles, 53
Mole fraction, 304
Molecular (covalent) compound, 49
Molecular formula, 203–05
Molecular orbital, 149
Molecular orbital theory, 170–73, 531
Molecular weight, 50 (*see also* Formula weight)
Molecule(s), 39
bond angles, 159–63
geometry, 159–63
Lewis structure 151–56
polar, 168
shape, 159–63
Molybdenum, 517
Monatomic ions, 116, 146
Monodentate ligand, 527
Monosaccharides, 599
Multiple bonds, 149

N

Natural radioactivity, 542
Negative catalyst, 355–56
Nernst equation, 455–57
Net ionic equation, 340
Neutralization reactions, 192–93, 331–33
Neutron, 45
Neutrons in nuclear reactions, 540
Newton, 634
Niacin, 612
Nickel, 518–19
Nickel dimethylgloxime, 519
Nickel(II) ion, 519
Nielsborium, 546
Niobium, 516
Nitrate ion:
identification, 497
Lewis structure, 153
Nitric acid, 497
Nitrogen, 496–99
Lewis structure, 153
molecular orbital theory of bonding, 172
occurrence, 496
oxidation states, 496
reactions, 497
Nitrogen balance, 608
Noble gas, 56, 105–06
Noble metals, 518
Nomenclature, inorganic:
binary compounds, 128–32
oxyacids and salts, 132–34
Roman numeral system, 130
stock system, 130
use of periodic table, 135
Nonane, 565
Nonbonding molecular orbitals, 171
Nonbonding pairs of electrons, 149
Noncritical mass, 548
Nonelectrolyte, 57, 295
Nonessential amino acids, 602
Nonmetal, 56
Nonmetal oxide:
reaction with metal oxide, 188
reaction with water, 188
Nonpolar molecule, 168
Nonyl group, 569
Normal boiling point, 275–76
Normal concentration, 333–39
Normality, 333–39
Nuclear chemistry:
applications, 551
energy changes, 550
half-life, 544
nuclear stability, 542–43
reactions, 541–42
transmutation of elements, 545
Nuclear equations, 541–42
Nuclear fission, 547–48
Nuclear fusion, 548
Nuclear radiations:
alpha, 539
beta, 539–40
gamma, 540
positron, 540

Nuclear reactions, 541
Nuclear stability, 542–43
Nucleic acids, 607
Nucleoproteins, 607
Nucleus, 45
Nutrients, 609
Nutrition, 609–12

O

Octane, 565
Octet, 151
Octyl group, 569
Oil drop experiment, 71
Oils, 605–06
Orbital:
  definition, 73–76
  molecular, 149
  shapes, 82
Order of reaction, 352–53
Organic acids, 563, 575–76 (*see also* Carboxylic acids)
Organic chemistry, 6, 560–83
Organic compounds, 560–83
  aliphatic compounds, 564
  alkyl groups, 568
  aromatic compounds, 563, 571–74
  benzene, 564
  cyclo compounds, 566
  functional groups, 563
  hydroxyl group, 574
  iso- compounds, 569
  isomers, 561
  meta compounds, 572
  normal compounds, 569
  ortho compounds, 572
  para compounds, 572
  phenyl group, 572
  properties, 561
  radicals, 568
  resonance in benzene, 571
  saturated hydrocarbons, 564
  structural formulas, 561
  unsaturated hydrocarbons, 564
Ortho compounds, 572
Osmium, 519
Osmosis, 312–13
Osmotic pressure, 313
Outer complex, 531
Oxidation:
  definition, 432–33
  electrolytic cells, 447–49
  sugars, 600
Oxidation number, 17–20 (*see also* Oxidation state)
  definition, 117–18
  periodic variations, 119–20
  rules for determining, 118–19
Oxidation potentials, 453
Oxidation state, 117–20 (*see also* Oxidation number)
Oxide:
  acid, 188
  basic, 187
Oxidizing agent, 432–33
Oxidoreductases, 606
Oxyacids, 472
  nomenclature, 132–34
  salts of, 132–34
Oxyanion, 132–33
Oxygen:
  abundance, 499
  discovery, 7
  molecular orbital theory of bonding, 172
  occurrence, 499
  physical properties, 500
  preparation, 500

P

Paired electrons, 78
Palladium, 519
Para- compounds, 572
Pascal, B., 242
Pauli exclusion principle, 91
Penicillin, 613
Pentane, 565

Pentyl (amyl) group, 569
2-Pentyne, 567
Peptide linkage, 603
Percent:
  composition of compounds, 200
  concentration, 57
  concentration (mass), 301–03
  concentration (volume), 303
  ionization, 389–90
  yield, 207–11
Period, 83, 105
Periodic law, 105
Periodic table:
  definition, 39, 56
  development, 104
  families, 105
  groups, 105
  periodic variations, 109–15
  periods, 83, 105
  relationship to electron structure of atoms, 83–87
  subgroups, 105
  types of elements, 105
  use in predicting properties, 116–17
Periodic variations, 109–15
  acid–base strength, 471–74
  atomic size, 109–10, 470
  differentiating electron, 108
  electron affinity, 113–14
  electronegativity, 115, 470
  ionic size, 109–110
  ionization energy, 111–12, 470
  metallic properties, 109
  names and formulas, 117
  oxidation states (numbers), 119–20
Permanent hardness, 489
Permanganate ion, 517–18
pH, 394–98
Pharmocodynamic agents, 613
Phase diagram, 283–84
Phenolphthalein, 331
Phenols, 563, 574–75
Phenyl group, 572
Phenylalanine, 601
Phlogiston theory, 7
Phosphate ion, 498
Phosphine, 497
Phosphite, 497
Phosphorus, 496–99, 610
  occurrence, 496
  oxidation states, 496
  reactions, 497
Photosynthesis, 600
Physical change, 41
Physical chemistry, 6
Physical properties, 41
Pi bond, 156
Pilot plant, 207
Pipet, 14
Platinum, 519
Platinum metals, 518–19
Plumbate ion, 494
Plumbite ion, 494
pOH, 394
Polar covalent bond, 150
Polar molecule, 168
Polarity of covalent bonds, 150
Polarity of molecules, 168
Polarizability, 471
Polydentate ligand, 527
Polypeptides, 603
Polyprotic acids, 412
Polysaccharides, 600
Positrons, 540
Potassium, 485–88
  reaction with hydrogen, 486–88
  reaction with oxygen, 486
Potassium ion, 486
Potassium oxide, reaction with water, 488
Potential energy, 43
Practical cells, 460–62
Precipitation, 380
Precipitation of sulfide, 413–15
Pressure, critical, 284
Pressure of gases, 241–42
  atmosphere (atm), 240
  cm Hg, 241–42
  mm Hg, 241–42
  pascal (Pa), 242

Pressure of gases (cont.)
pounds per square inch, 241–42
torr, 242
Priestly, Joseph, 7
Primary amine, 580
Primary carbon, 570
Primary structure of proteins, 603
Principal quantum number, 88
Products in a chemical equation, 43, 182
Propane, 565
n-Propyl chloride (1-chloropropane), 570
Propyl group, 569
Propylene, 566
Propyne, 567
Proteins, 601–04
classification, 604
hydrolysis, 604
metabolism, 608
peptide linkage, 603
properties, 603–04
structure, 602–03
synthesis, 607
Proton:
charge, 45
definition, 45
mass, 45
Proton (H^+) acceptor, 329
Proton (H^+) donor, 329
Provision of energy, 609–10
Pure substance, 39
Purification by crystallization, 299
Pyrosulfuric acid, 501

Q

Quantization of energy, 72–74
Quantum mechanics, 73
Quantum numbers:
definition, 88
magnetic, 89
principal, 88
problems, 93
secondary, 89
spin, 91
Quaternary structure of proteins, 603

R

Radicals, 568
Radioactive isotopes, 538, 542
Radioactivity:
artificial, 541–43
definition, 538
natural, 542
Radium, 488–91
Raoult's law, 309–10
Rare earth elements (Lanthanum series), 515
Rate constant, 350–51
Rate determining step, 351
Rate law, 350–51
Reactants, 43
Reactants in chemical equation, 182
Reactions:
acid-base, 192
actual yield, 210
combination, 186–88
complete, factors of, 380
decomposition, 188–89
displacement, 190
double displacement, 191–92
endothermic, 220
excess reagent, 209–11
exothermic, 220
heat flow, 220
hydrocarbons, 564
limiting reagent, 208–11
mechanism, 356–57
metathesis, 191–92
neutralization, 192–93
nuclear, 541–42
order, 352–53
percent yield, 207–11

Reactions (cont.)
rate of, factors, 354–56
replacement, 190
reversible, 362
theoretical yield, 206
Reaction rate, 350–52
Redox equations, 432
Reducing agent, 432–33
Reducing sugars, 600–01
Reduction:
definition, 432–33
electrolytic cells, 447–49
Reduction potentials, 453
Regulators of body processes, 610
Relationships from the formula of a compound, 202–03
Replacement reactions, 190
Representative elements:
definition, 105
oxidation states, 119–20
Resonance, 115–56
Resonance forms, 571
Reversible reactions, 362
Rhenium, 517
Rhodium, 519
Riboflavin (vitamin B_2), 612
Ribonucleic acids, 607
RNA, 607
Roasting, 475
Roentgen, 70
Roman numeral system of nomenclature, 130
Rounding off, 627
Rubidium, 485–88
Ruthenium, 519
Rutherford, Ernest, 72

S

Salt, 132, 192
Salt bridge, 452–53
Saponification, 606
Saponification number, 606
Saturated hydrocarbons, 564
Saturated solution, 298–99
Scandium, 516
Scheele, Carl, 7
Schrödingen wave equation, 88
Science, 5
Scientific law, 5
Scientific method, 5
Scientific notation, 625
Second law of thermodynamics, 228–31
Secondary amine, 580
Secondary carbon, 570
Secondary quantum number, 89
Secondary structure of proteins, 603
Sedatives, 613
Semimetals, 471
Selenic acid, 501
Selenous acid, 501
Serine, 603
Shift in equilibrium, 368–71
SI units, 16, 634 (*table*)
Sigma bond, 156
Significant digits (significant figures), 625–27
Silicon, 493–95
occurrence, 493
oxide, 494
Silver, 519–20
Silver–ammonia complex ion, 520
Silver ion, 520
Simple ether, 574
Simple glyceride, 605
Simple proteins, 604
Slightly soluble compounds (*table*), 640–41
Smoke, 313
Soap, 606
Sodium, 485–88
reaction with oxygen, 486
reaction with water, 486–88
Sodium hydride, 486
Sodium hydride, reaction with water, 486
Sodium ion, 486
Sodium hydroxide, 486
preparation by electrolysis, 449

Sodium oxide, reaction with water, 488
Sodium peroxide, reaction with water, 488
Sol, 313
Solids:
 comparison to gases and liquids, 238–40
 crystalline, 279–80
 heat of fusion, 280–81
 melting point, 280
 sublimation, 281
Solubility, 296–70
Solubility product constant, 381, 640–41 (*table*)
Solubility product equilibria, 380–88
Solubility rates, 191
Solute, 57, 294
Solution(s):
 aqueous, 57
 colligative properties, 309–12
 concentrated versus dilute, 300
 definition, 41, 57, 294
 electrolyte versus nonelectrolyte, 295
 osmosis, 312–13
 rate of, factors, 298
 saturated, 298–99
 solubility factors, 296–97
 stoichiometry, 308–09
 supersaturated, 299–300
 types, 295
 unsaturated, 299
 weak and strong electrolytes, 295
Solvent, 57, 294
sp hybrid orbitals, 165–67
sp^2 hybrid orbitals, 165–67
sp^3 hybrid orbitals, 164–67
sp^3d^2 hybrid orbitals, 167
sp^3d hybrid orbitals, 166–67
Specific gravity:
 calculations, 29–30
 definition, 29
 table, 29
Specific heat:
 calculations, 26–27
Specific heat (cont.)
 definitions, 25
 table, 26
Spectator ions, 340
Spin quantum number, 91
Square planar compound, 529
Stahl, George Ernst, 7
Standard electrode potentials, 453
Standard state, 223, 455
Standard temperature and pressure (STP), 242
Stannate ion, 494
Stannite ion, 494
Starch, 600
State function, 222–23
Stereoactive pairs of electrons, 159–63
Stimulants, 613
Stock system of nomenclature, 130
Stoichiometry:
 calculations from balanced equations, 206–11
 definition, 200
 relationships from the formula of a compound, 202–03
 involving gases, 254–55
 mass-volume, and volume-volume, 254–55
 solution, 308–09
Streptomycin, 613
Strong electrolyte, 295
Strontium, 488–91
 reaction with hydrogen, 489
 reaction with nitrogen, 489
 reaction with oxygen, 489
 reaction with water, 489
Strontium hydroxide, 489
Strontium ion, 490
Strontium oxide, 489
Structure of atoms, 70–95 (*see also* Atomic Structure)
Structure of proteins, 602–03
Structural formulas, 561
Structural isomers, 561
Structural proteins, 604
Subatomic particles, 45
Subcritical mass, 548

Subgroups, 105
Sublevel(s):
 definition, 75
 designation, 75
 maximum number of electrons in, 76
 number in each energy level, 75
 order of occurrence, 78–80
Sublimation, 281
Substance, 39 (*see also* Pure substance)
Sucrose, 600
Sulfa drugs, 613
Sulfate ion, 501
Sulfide ion, 501
Sulfur:
 occurrence, 500
 oxides, 501
 preparation, 500
Sulfur dioxide, 153, 501
Sulfur trioxide, 501
Sulfuric acid, 501
Superoxide, 486, 500
Supersaturated solution, 299–300
Surface tension, 278
Symbols:
 common elements, 39
 definition, 39
 Latin names, 40
Synthetases, 606

T

Tantalum, 516
Technetium, 517
Temperature, critical, 284
Temperature scales:
 absolute, 21
 Celsius, 21
 conversion, 22–25
 Fahrenheit, 21
 kelvin, 21, 245
Temporary hardness, 489
Terminal elements, 520
Ternary acids, 132–33 (*see also* Oxyacids)
Tertiary amine, 580
Tertiary carbon, 570
Tertiary structure of proteins, 603
Tetrachloromethane (carbon tetrachloride), 565
Tetracyclines, 613
Tetrahedral compound, 529
Theoretical yield, 206
Theory, 5
 molecular orbital, 170–73
 valence shell electron pair repulsion, 159–60
Thermochemical equations, 220–22
Thermodynamic values (*table*), 636
Thermodynamics:
 first law, 228
 second law, 228–31
Thiamine (vitamin B_1), 612
Thioacetamide, 578
Thompson, J. J., 71
Thompson, William (*see* Lord Kelvin)
Tin, 493–95
 occurrence, 493
 oxides, 494
Tin(II) hydroxide, reaction with base, 494
Tin(II) ion, 494
Titanium, 516
Titration, 331–32
Tollens test, 601
Toluene (methylbenzene), 572
Torr, 242
Torricelli, E., 241
Trans- isomer, 529
Transfer RNA, 607
Transition elements, 106, 514–22
 general properties, 514–15
 oxidation states, 120, 515
Transmutation of elements, 545–47
Trichloromethane (chloroform), 565
2,2,3-Trichloropentane, 569
Trimethylamine, 580

Triple bond, 149
Triple point, 283
Tungsten, 517
Tyndall effect, 314
Tyndall, John, 314
Types of chemical reactions, 186

U

Unpaired electrons, 78
Unsaturated hydrocarbons, 564
Unsaturated solution, 299

V

Valence bond theory, 530
Valence electrons, 107–08, 471
Valence shell electron pair repulsion theory (VSEPR), 159–60
Van der Waals force, 271
Vanadium, 516
Vapor pressure, 274–75
Vapor pressure of water (*table*), 638
Vitamins, 612
- A, 612
- B_1 (thiaminc), 612
- B_2 (riboflavin), 612
- B_{12}, 612
- C (ascorbic acid), 612
- D, 612
- E, 612
- Niacin, 612

Voltaic cell, 452–53
Volume, 14–15
Volume percent concentration, 303
Volumetric flask, 14

W

Waage, Peter, 351
Water:
- decomposition of, 189
- density at various temperatures (*graph*), 286
- hydrogen bonding, 285–86
- Lewis structure, 152
- physical properties, 284–85
- vapor pressure (*table*), 638

Wave mechanics, 73
Weak acids and bases, 388–93
Weak electrolyte, 295
Weight, 12

Z

Zinc, 520–21
Zinc hydroxide, 521
Zinc ion, 521
Zinc subgroup, 520–21
Zirconium, 516
Zwitterion, 604